FOOD MICROSCOPY

FOOD SCIENCE AND TECHNOLOGY

A SERIES OF MONOGRAPHS

Maynard A. Amerine, Rose Marie Pangborn, and Edward B. Roessler, PRINCIPLES OF SENSORY EVALUATION OF FOOD. 1965.

C. R. Stumbo, THERMOBACTERIOLOGY IN FOOD PROCESSING, second edition. 1973.

Gerald Reed (ed.), ENZYMES IN FOOD PROCESSING, second edition. 1975.

S. M. Herschdoerfer, QUALITY CONTROL IN THE FOOD INDUSTRY. Volume I – 1967. Volume II – 1968. Volume III – 1972.

Hans Riemann, FOOD-BORNE INFECTIONS AND INTOXICATIONS. 1969.

Irvin E. Liener, TOXIC CONSTITUENTS OF PLANT FOODSTUFFS. 1969.

Martin Glicksman, GUM TECHNOLOGY IN THE FOOD INDUSTRY. 1970.

L. A. Goldblatt, AFLATOXIN. 1970.

Maynard A. Joslyn, METHODS IN FOOD ANALYSIS, second edition. 1970.

A. C. Hulme (ed.), THE BIOCHEMISTRY OF FRUITS AND THEIR PRODUCTS. Volume 1 – 1970. Volume 2 – 1971.

G. Ohloff and A. F. Thomas, GUSTATION AND OLFACTION. 1971.

George F. Stewart and Maynard A. Amerine, INTRODUCTION TO FOOD SCIENCE AND TECHNOLOGY. 1973.

Irvin E. Liener (ed.), TOXIC CONSTITUENTS OF ANIMAL FOODSTUFFS. 1974.

Aaron M. Altschul (ed.), NEW PROTEIN FOODS: Volume 1, TECHNOLOGY, PART A – 1974. Volume 2, TECHNOLOGY, PART B — 1976. Volume 3, ANIMAL PROTEIN SUPPLIES, PART A — 1978.

S. A. Goldblith, L. Rey, and W. W. Rothmayr, FREEZE DRYING AND ADVANCED FOOD TECHNOLOGY. 1975.

R. B. Duckworth (ed.), WATER RELATIONS OF FOOD. 1975.

A. G. Ward and A. Courts (eds.), THE SCIENCE AND TECHNOLOGY OF GELATIN. 1976.

John A. Troller and J. H. B. Christian, WATER ACTIVITY AND FOOD. 1978.

A. E. Bender, FOOD PROCESSING AND NUTRITION. 1978.

D. R. Osborne and P. Voogt, THE ANALYSIS OF NUTRIENTS IN FOODS. 1978.

Marcel Loncin and R. L. Merson, FOOD ENGINEERING: PRINCIPLES AND SELECTED APPLICATIONS. 1979.

Hans Riemann and Frank L. Bryan (eds.), FOOD-BORNE INFECTIONS AND INTOXICATIONS, Second Edition. 1979.

N. A. Michael Eskin, PLANT PIGMENTS, FLAVORS AND TEXTURES: THE CHEMISTRY AND BIOCHEMISTRY OF SELECTED COMPOUNDS. 1979.

J. G. Vaughan (ed.), FOOD MICROSCOPY. 1979.

In preparation

J. R. A. Pollock (ed.), BREWING SCIENCE. Volume 1.

Food Microscopy

Edited by J. G. VAUGHAN

Biology Department,
Queen Elizabeth College,
London, U.K.

1979

ACADEMIC PRESS
London · New York · San Francisco

A Subsidiary of Harcourt Brace Jovanovich, Publishers

ACADEMIC PRESS INC. (LONDON) LTD
24–28 Oval Road,
London NW1

U.S. Edition published by
ACADEMIC PRESS INC.
111 Fifth Avenue,
New York, New York 10003

Library of Congress Catalog Card Number: 78 18033
ISBN: 0 12 715350 0

Printed in Great Britain by
T. & A. CONSTABLE LTD, EDINBURGH

Contributors

R. E. ANGOLD, *The Lord Rank Research Centre, Lincoln Road, High Wycombe, Bucks., U.K.*

K. G. BERGER*, *Lyons Central Laboratories, 149 Hammersmith Road, West Kensington, London W14, U.K.*

B. E. BROOKER, *National Institute for Research in Dairying, Shinfield, Reading, U.K.*

A. D. EVERS, *Flour Milling and Baking Research Association, Chorleywood, Rickmansworth, Herts., U.K.*

F. O. FLINT, *Procter Department of Food Science, University of Leeds, Leeds, U.K.*

P. HOWGATE, *Torry Research Station, 135 Abbey Road, Aberdeen, U.K.*

G. G. JEWELL‡, *The Leatherhead Food Research Association, Randalls Road, Leatherhead, Surrey, U.K.*

GEORGINA H. JOLLIFFE, *Department of Pharmacy, Chelsea College, University of London, Manresa Road, London SW3, U.K.*

G. O. JOLLIFFE, *Department of Pharmacy, Chelsea College, University of London, Manresa Road, London SW3, U.K.*

D. F. LEWIS, *The Leatherhead Food Research Association, Randalls Road, Leatherhead, Surrey, U.K.*

R. J. M. POLLITT§, *Lyons Central Laboratories, 149 Hammersmith Road, West Kensington, London W14, U.K.*

H. E. C. POWERS, *34 Wordsworth Avenue, South Woodford, London E18, U.K.*

* Present address: *Palm Oil Research Institute of Malaysia, 18th Floor, Angkasa Raya, Jalan Ampang, Kuala Lumpur, Malaysia.*

‡ Present address: *Research and Development Department, Confectionery Division, Cadbury Limited, Bournville, Birmingham, U.K.*

§ Present address: *24 Rothesay Avenue, Richmond, Surrey, U.K.*

D. S. ROBINSON, *Procter Department of Food Science, University of Leeds, Leeds, U.K.*

C. G. SMITH, *Unilever Research Laboratory, Colworth House, Sharnbrook, Bedford, U.K.*

JANE A. STUBBS, *Royal Botanic Gardens, Kew, Richmond, Surrey, U.K.*

J. G. VAUGHAN, *Biology Department, Queen Elizabeth College, University of London, Campden Hill Road, London W8, U.K.*

H. VOEHRINGER, *Landw. Untersuchungs- und Forschungsanstalt, Weberstr. 59–61, 5300 Bonn 1, West Germany.*

C. A. VOYLE, *Agricultural Research Council, Meat Research Institute, Langford, Bristol, U.K.*

G. W. WHITE, *Lyons Central Laboratories, 149 Hammersmith Road, West Kensington, London W14, U.K.*

Preface

Microscopy has always played an important part in the practice of food science. As soon as the light microscope became available, those engaged in the quality control of food found it most useful for the detection of adulteration or admixture. Indeed, the earlier works on food microscopy were produced to enable the scientist to carry out identification. A number of books were written on the subject but special mention should be made of the publications of Andrew L. Winton and Kate Barber Winton in the 1930s.

In the intervening years, useful modifications of the light microscope have appeared and, of course, the electron microscope (both transmission and scanning) has been developed. All these methods of microscopy are now being used by food scientists. However, as far as is known, no comprehensive account of food microscopy has appeared in recent years. The object of the present book is an attempt to fill this gap.

So much work has been carried out in food microscopy that it would be difficult for one person to survey the whole field. Consequently the various chapters of the book have been produced by a number of experts. As has been stated, the earlier books dealt mainly with identification. The present volume shows how microscopy may be used to investigate problems of food processing. Nevertheless, the chapters on animal feed contain information pertaining to contemporary problems of identification in international commerce.

The literature on food microscopy is now enormous. For this reason, some authors have only been able to refer to a restricted number of publications. Articles on food microscopy appear in diverse journals. A valiant attempt to help the food microscopists as regards the background literature has been carried out under the direction of G. W. White (Annotated Bibliography of Food Microscopy: *J. Assoc. Publ. Analysts* **8,** 74; **12,** 27, 77; **13,** 34, 135; **14,** 113; **15,** 33).

The authors of the chapters have been drawn from the food industry, research stations and the universities. Special mention might be made of H. E. C. Powers for the chapter on sugar, based on his fifty years' experience in the industry.

Acknowledgements are given at the end of the various chapters but, in the general production of the book, the assistance of J. Pacey of Queen Elizabeth College in the production of a number of figures and Emily Wilkinson and Tim Lincoln of Academic Press in editorial matters is acknowledged.

J. G. VAUGHAN

January, 1979

Contents

3. CEREALS AND BAKERY PRODUCTS

R. E. Angold

4. CEREAL STARCHES AND PROTEINS

A. D. Evers

5. MEAT
C. A. Voyle

6. MEAT PRODUCTS
D. F. Lewis

7. MILK AND ITS PRODUCTS
B. E. Brooker

8. THE DOMESTIC HEN'S EGG

D. S. Robinson

9. FISH

P. Howgate

10. ANIMAL FEEDS—PLANT CONSTITUENTS

J. G. Vaughan and Jane A. Stubbs

11. ANIMAL FEEDS—ANIMAL CONSTITUENTS

H. Voehringer

12. OILS AND FATS

K. G. Berger, G. G. Jewell and R. J. M. Pollitt

13. ICE CREAM

K. G. Berger and G. W. White

14. NOVEL PROTEIN FOODS

F. O. Flint

15. SUGAR

H. E. C. Powers

16. COMPUTER-AIDED IDENTIFICATION OF FOOD MATERIALS

Georgina H. Jolliffe and G. O. Jolliffe

1. Fruits and Vegetables

G. G. JEWELL

The Leatherhead Food Research Association, Leatherhead, U.K.

I. INTRODUCTION

When one considers that fruits and vegetables can be consumed at virtually every meal as major or minor constituents, either raw, cooked or processed, it is perhaps surprising that so little work has been undertaken on the structure of these food materials, particularly the ultrastructure using electron microscopy techniques.

Historically one of the most comprehensive attempts to discuss the structure of food plants must be the work of Winton and Winton (1935) which followed upon the useful but not as wide ranging work of Greenish (1923). We then have a series of papers published over a period of about 40 years until Reeve (1970) produced an excellent review article on histological structure and texture in fresh and processed fruits and vegetables.

The aim of this chapter is to provide an insight into the types of cells found in plant tissues and their role in texture, and then consider the changes which may be brought about by a variety of food processing conditions.

II. STRUCTURE OF PLANTS

A. General Aspects

A schematic representation of a plant cell is shown in Fig. 1 and illustrates the major structure observed in plant cells, i.e. the occurrence of a thick cell wall surrounding the living part of the cell the protoplast, and the presence of vacuoles within the cytoplasm. Such a structure can be readily studied using conventional histological procedures and employing light microscopy. The advent of electron microscopy techniques led to the identification of greater structural detail both within the protoplast and within the cell wall.

The cell wall has received a great deal of attention from numerous workers. Using light microscopy and staining techniques, it was obvious that the wall contained cellulose (stained with iodine and potassium iodide in sulphuric acid), pectin (stained with ruthenium red) and sometimes lignin (stained with phloroglucinol). The appearance of birefringence when viewed under crossed polars indicated that the arrangement within the wall gave rise to an anisotropic structure, but it was only through the use of electron microscopy techniques that the detailed and intricate architecture of the cell wall was revealed. Thus, Roelofsen (1965) elegantly demonstrated the alternating layered orientation of the cellulose fibres, whilst Heyn (1966, 1969) studied the detailed structural arrangements of the cellulose fibres. It is possible to stain the pectins histochemically using iron hydroxylamine (Albersheim and Killias, 1963), and Jewell and Saxton (1970) introduced a method for staining the hemicellulose components at the electron microscope level. Zones of secondary thickening may be identified, as can deposits of callose. Traversing the walls are the plasmodesmata and their detailed ultrastructure has been studied by Northcote (1968) and Robards (1971). Thus, we now realize that the cell wall is a highly ordered structure with alternating layers of cellulose microfibrils embedded in a matrix of hemicelluloses; the central region of the wall, the middle lamella, which acts as a cementing layer, is rich in pectins. The whole wall may be further strengthened by layers of secondary thickening which may or may not be lignified. As will become apparent later, it is the nature of the cell walls which to a large extent controls the final texture of fruits and vegetables that have been subjected to different processing conditions.

The protoplasts of all plant cells are bounded by a thin semi-permeable membrane, the plasmalemma. Within the protoplast is located the cytoplasm which contains a variety of organelles to undertake the metabolic activity of the cell; these include the nucleus, mitochondria, golgi bodies, endoplasmic reticulum and ribosomes, and, depending on the cell type, chloroplasts, chromoplasts, starch granules, protein bodies, crystalline inclusions and lipid droplets may also be present. It is beyond the scope of this chapter

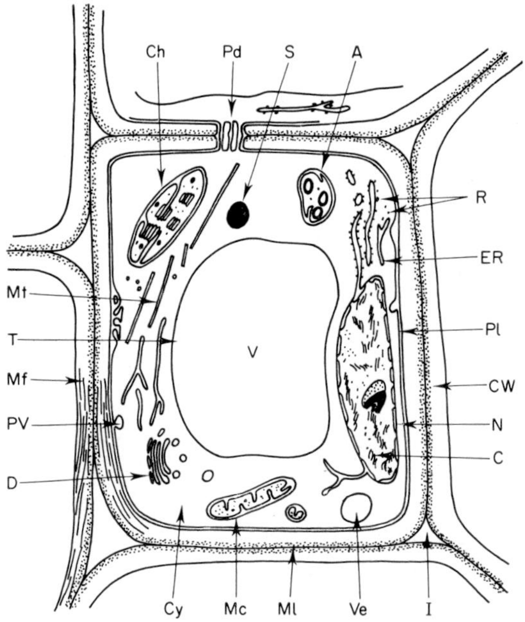

FIG. 1. A schematic representation of a plant cell. (In collaboration with Dr E. G. Jordan, Queen Elizabeth College.) A (amyloplast), starch-storing plastid; C (chromatin), deoxyribonucleoprotein fibrils, the chromosomes at interphase; Ch (chloroplast), double-membraned organelle with photosynthetic lamellae; CW (cell wall), rigid extracellular layer composed of microfibrils in a matrix; Cy (cytoplasm), the fluid matrix and associated molecules between vacuole and wall; D (dictyosome), membrane system of a few disc-shaped cisternae with secretory vesicles. The golgi apparatus is the cell's whole complement of dictyosomes; ER (endoplasmic reticulum), flattened cisternae and tubules of membrane sometimes with attached ribosomes (R); I (intercellular space), the space formed at the junction of three or more adjacent cell walls; Mc (mitochondrion), a cell organelle responsible for cellular respiration; Mf (microfibril), the cellulose strands of the cell wall, having a diameter of 10–30 nm; ML (middle lamella), a more darkly staining zone between adjacent cells, rich in pectin; Mt (microtubules), 20 nm wide tubules which may be involved in movement in the cytoplasm; N (nucleus), a double-membraned compartment enclosing the chromosomes, nucleolus and nucleoplasm. The envelope has nuclear pores; Pd (plasmodesmata), intercellular cytoplasmic connections; Pl (plasmalemma), cell membrane; PV (pinocytotic vesicle), an 'infolding' of the cell membrane; S (spherosomes), single-membraned vesicles with stainable contents; T (tonoplast), vacuole membrane; V (vacuole), aqueous, membrane-bound area within the cytoplasm, Ve (vesicle), membrane-bound cytoplasmic inclusion.

to give detailed consideration to the structure and functions of these organelles and inclusions; for this the reader is referred to standard botanical texts such as Esau (1965) and Frey-Wyssling and Mühlethaler (1965) or an atlas of ultrastructure such as Dodge (1968). Some plant cells have extensive central vacuoles which contain so-called cell sap; this is usually a complex mixture of sugars, polyphenols and other materials.

In the living plant it is the osmotic pressure generated within the cell sap which gives rise to the so-called turgor pressure which maintains the protoplast firmly pressed against the cell wall and hence keeps the tissue firm or turgid. In living plants which are subjected to stress from water loss or deficiency the tissues frequently become flaccid and this is related to loss of turgor pressure within the cell, brought about by a process known as plasmolysis. As will be seen later, the degree of turgor pressure plays an important role in determining the texture of the tissue, and plasmolysis is frequently observed during the processing of fruits and vegetables.

B. Cell Types

The various types of organs that are commonly consumed are illustrated in Fig. 2, and here one will recognize familiar names such as leaf, e.g. cabbage or lettuce, root, e.g. carrot, stem, e.g. asparagus, modified stem, e.g. potato, and so on.

All these different structures are built up from a number of common cell types and in general it is the number, distribution and organization of these different cell types which dictate the type of tissue formed.

1. Parenchyma

This is one of the commonest tissues found in edible plants and makes up the cortex and pith of stems and the mesophyll of leaves. The appearance of the cells in transverse section is usually round or oval and their walls are thin. The shape and contact between adjacent cells leads to intercellular spaces and in some cases these may be extensive, giving rise to highly aerated tissues such as the albedo region of citrus fruit skins (i.e. the white, spongy zone found beneath the coloured outer layer). The parenchyma cells are normally concerned with nutrition by either synthesis or storage of food materials. Although they are thin-walled cells with limited inherent mechanical strength, parenchyma cells can impart a considerable degree of strength and support when turgid.

2. Collenchyma

These cells are frequently found beneath the epidermis of many stems, leaf-stalks (petioles) and midribs, and a good example of collenchyma cells

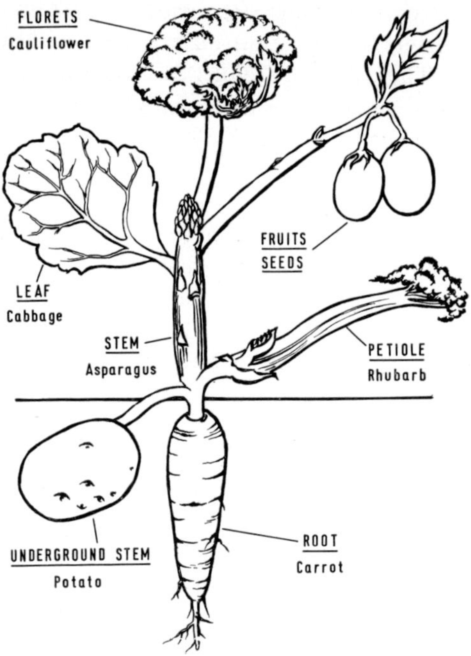

FIG. 2. A schematic representation of the various parts of plants consumed as food.

can be found in celery. The cells are usually elongated, being as much as five-to-ten times long as wide, and develop marked wall thickening in the angles where adjacent cells join; collenchyma thus imparts mechanical strength. According to Reeve (1970), the wall thickenings are high in pectin substances which, following processing involving dehydration, greatly influence the toughness of the reconstituted product.

3. Sclerenchyma

Two types of cells are found in sclerenchyma, namely fibres and sclereids (stone cells). The fibres are usually developed in bundles or layers, but can occur singly. The individual fibres are elongated cells with pointed ends which tend to interlock with the ends of other fibres in the same bundle. The walls of the cells are thick and composed of cellulose and are usually lignified. The walls frequently have marked pits and the presence of so-called cross-pits is considered characteristic of sclerenchyma fibres. It has been found that extended sclerenchyma fibres are elastic and return to their original lengths when the load is removed, whereas stretched collenchyma remains permanently extended (Esau, 1936).

The sclereids or stone cells are frequently found in many foodstuffs, e.g. as the gritty clusters of pear and quince, and within the seed coat of beans and peas, etc. The cells are frequently isodiametric, but they may be slightly elongated. The cell wall is thick, stratified and strongly encrusted with either lignin or hemicellulose materials.

4. Xylem

These cells make up part of the vascular tissue of the plant. Their function is primarily concerned with the transport of water, and the cell walls are thickened and usually lignified. Various characteristic wall-thickening patterns are observed, e.g. annular, spiral, reticulate and pitted, and these features are extremely useful diagnostic aids in the microscopic identification of vascular tissue.

5. Phloem

This type of tissue also contributes to the vascular system but is concerned primarily with the transport of organic food materials. The two principal types of cell recognized are sieve tubes and companion cells. The sieve tubes are elongated cells placed end-to-end. The side walls are relatively thin, whereas the end walls are thickened and modified to form sieve plates, these plates being characteristic of the sieve tubes. At certain times the sieve plate is covered in a layer of callose and the presence of excessive quantities of callose may cause problems during processing. For example, in the production of pickled beetroot the leaves should always be twisted off, rather than cut off. A cut will not only lead to extensive 'bleeding' and loss of colour, but also usually results in the formation of a callose cap in the wound area which is very difficult to remove in conventional peeling machines.

6. Laticiferous tissues (milk tubes)

These are specialized types of cells found in certain groups of plants, e.g. onions, which are capable of producing a milky exudate called latex.

7. Glandular cells

Again, these are specialized cells, in this case capable of producing a variety of secretory materials, e.g. gum, resin, oil, tannin, etc. An example relevant to food is the oil glands found in the skins of citrus fruits.

8. Surface cells

The cells found on the external surfaces of plants are frequently flattened and normally possess an external coat of a wax-like material (the cuticle). Their function is to protect the inner tissues from the outside environment and they serve as a defence against both mechanical damage and attack by pathogenic organisms. The external layers are frequently built up into multi-layers and may also contain cork deposits, e.g. in the periderm of a mature potato as opposed to the thin epidermal layer of a new potato.

III. PROCESSING

A. General Aspects

With the increasing trend towards convenience prepacked foods, a wide range of fruit and vegetables are processed in some form or another. The types of processing can be divided into three broad categories, namely thermal, chemical and mechanical. The most widely practised of these is thermal processing which includes cooking, blanching, pasteurizing and canning on the heating side, and on the cooling side chilling, freezing and freeze-drying. Chemical methods are usually associated with some form of preservation, i.e. brining, syruping, and pickling in vinegar. The mechanical aspects of processing are very diverse and cover a variety of treatments from peeling and slicing to juicing or comminuting. It is obvious that all three forms of processing may have been used to produce a single finished product. For example, canned carrots are subjected to peeling (mechanical or steam) and canning (thermal), and the canning liquor may contain salt (chemical).

Every consumer is aware that one of the major effects of the processing of fruits and vegetables is a change in the texture of the product; this may be of a major or minor nature which is regarded as desirable or sometimes undesirable, but some degree of textural alteration is, of necessity, unavoidable. Hence, one could argue that cooking most root crops produces a major and desirable change in texture, whereas the change in texture between fresh and frozen or canned strawberries, equally unavoidable, is also major but much less desirable.

With high manufacturing and marketing costs, the processor is striving to achieve a consistent level of quality in his products and this requires a knowledge of the optimum processing conditions. In turn, this necessitates an understanding of the effect of processing on all aspects of the product, including structure which, as has already been indicated, controls texture. It is the desire for knowledge concerning optimum processing that has prompted most of the investigations cited in the remainder of this section; these fall into two main areas. Firstly, work which has been aimed at an understanding of the changes manifested during processing, since a precise knowledge of the mechanisms involved must lead to better control of the process. Secondly, studies which have attempted to establish why defects occurred in products which usually process correctly. The second approach can frequently be resolved by recourse to conventional light-microscope histology, whereas the elucidation of detailed changes in structure has necessitated the application of electron microscopy usually coupled with some other objective evaluation such as chemical or texture analysis.

B. Specific Problems

1. Thermal processing

The two most widely studied areas in this field are cooking and freezing. Very extensive literature can be found on the cooking of potatoes, and Warren and Woodman (1974) have published a review on this subject. During the cooking process a series of structural changes take place. As the temperature approaches 60°C, some of the starch granules will swell and lose birefringence (i.e. gelatinization is occurring). By 65°C most of the starch has gelatinized, although at temperatures as high as 70°C some of the smaller granules may still be intact (Reeve, 1967a, b). Following closely upon starch gelatinization, degradation and breakdown of the pectin in the middle lamella of some cells will occur. This cell wall breakdown has been found to occur earlier in varieties that cook to the so-called mealy texture (Reeve, 1943, 1954c, d, 1963, 1967a, b; Whittenberger and Nutting, 1950; Sterling, 1955, 1963, 1966). The cell wall breakdown leads to marked cell separation, which, according to Reeve (1954c), starts in the cortical tissue. In contrast, little or no cell separation is observed in varieties which remain firm in texture. Two factors are thought to be responsible for this cellular separation. Firstly, the cooking temperature may break hydrogen bonds in pectins and other cell wall polysaccharides (Sterling, 1963); secondly, the gelatinized and swollen starch tends to distend the cell, thus subjecting the weakened cell wall to strain and split (Reeve, 1954c, d; Sterling, 1955). The degree of starch swelling and subsequent cell wall rupture has been found to be related to stickiness in mashed and processed potato products (Reeve, 1954c, d, e). Various other processes have been applied to potato products and they all influence the starch content. Longree (1950) found that freezing and thawing of cooked potatoes result in a very mealy texture described as 'dry cottony'. The effect was reduced by a further heating step. French-fried chips have been studied by Spiruta and Mackey (1961), Reeve *et al.* (1968) and Mohr (1972). Mohr found that the soggy region in the centre of certain chips was related to the inclusion of the pith region which is deficient in starch. Further aspects of freeze–thaw regimes in starch-containing products is related to the process of starch retrogradation which, in turn, influences the water-holding capacity of the starch gel. Reeve (1954c, d, 1967a, b) found that a process involving heat treatment which allowed a certain degree of retrogradation during the cooling stage yielded potato pieces with minimal cell separation. This treatment has also been applied to the production of dehydrated potato flake, with good rehydration performance (Cording *et al.*, 1955, 1959).

The behaviour of the potato during processing is intimately associated with the effects on the starch content, and similar observations have been made on other starch-containing materials such as sweet potatoes, yams, Jerusalem artichokes, bread fruit, rice and popped cereals (Reeve, 1970). However, many processed fruits and vegetables contain little or no starch, and thus other factors are likely to control the cellular response to thermal treatment.

Brown (1967) studied the effect of freezing rates on the texture of green beans and concluded that rapid freezing in liquid nitrogen produced least structural and textural deformation, whilst slower rates of freezing resulted in poor texture associated with various degrees of cell separation. Similar conclusions were reached by Rahman *et al.* (1971) in a study on the freezing of carrots.

The ultrastructural changes associated with freeze–thaw and blanching have been studied by Mohr and by Bassi and Crivelli in several types of tissue: cauliflower (Bassi and Crivelli, 1968), haricot bean (Bassi and Crivelli, 1969), pepper (Crivelli and Bassi, 1969), rhubarb, mung bean and ripe black fig (Mohr, 1971) and pea pod, spinach leaf and green bean (Mohr, 1974). In general terms all these workers noted a gradual breakdown in the organization of the protoplasmic structure and, in most cases, the rupture of the plasmalemma with subsequent loss of turgor pressure; changes occurring in spinach leaf are shown in Figs 3, 4, 5 and 6. Some degradation and separation of cell walls was also noted. A similar response was also observed in the blanching of carrots (Mirza and Jewell, 1976). At the light microscopy level, water-blanched carrots can be seen to have suffered some cell collapse and wall rupture when compared with the fresh tissue (Figs 7 and 8). At the ultrastructure level, the water blanching has produced a loss of fine structural detail and a pronounced separation of cell walls in the region of the middle lamella (Fig. 9). In contrast, steam blanching produced protoplasmic degradation and the formation of numerous vesicles, but cell wall degradation was minimized (Fig. 10). These differences in structure resulted in differences in texture as assessed by texture measurement using an Instron Universal Tester and a taste panel. Carrots with the least structural deformation are regarded as the most acceptable.

Dehydration processes were studied by Lee *et al.* (1967) using apple as a test system and they showed that freeze-dehydration produced minimal changes in structure and texture when compared with hot air drying or dehydro-freezing. The effect of incorporating sodium chloride into freeze-dried carrots was studied by Curry *et al.* (1976) who found that the incorporation

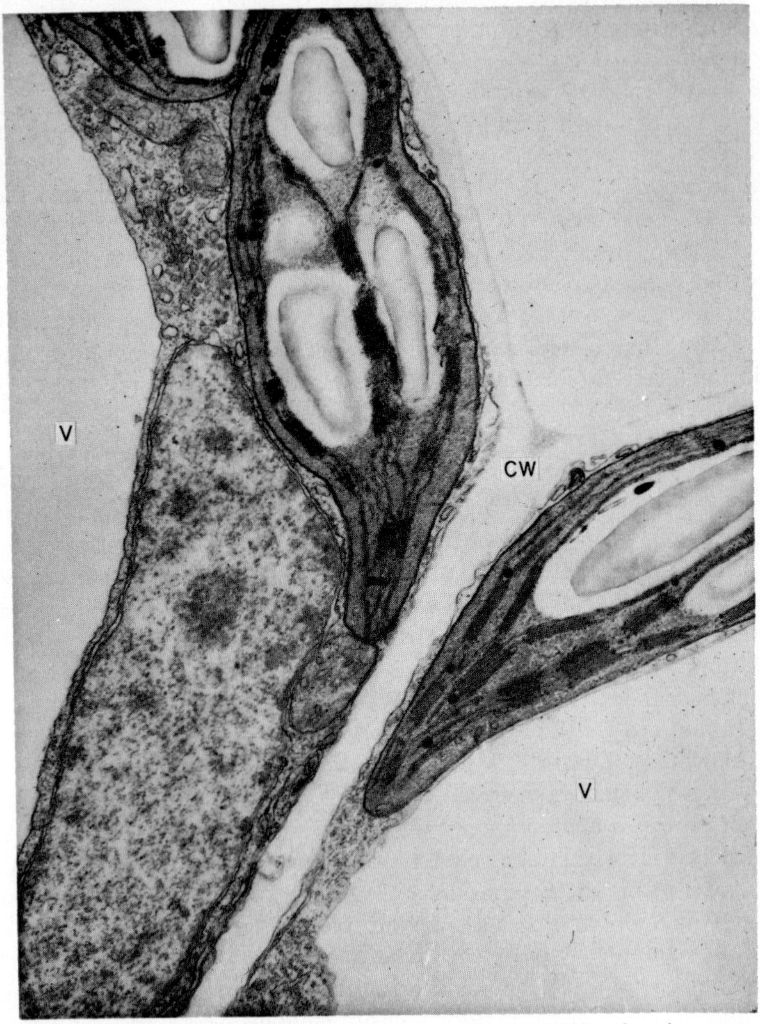

FIG. 3. Fresh spinach leaf parenchyma, showing cell wall (CW), plasmic structure and extensive vacuoles (V), ×14 000. (By courtesy of Dr W. P. Mohr and the *Journal of Texture Studies*)

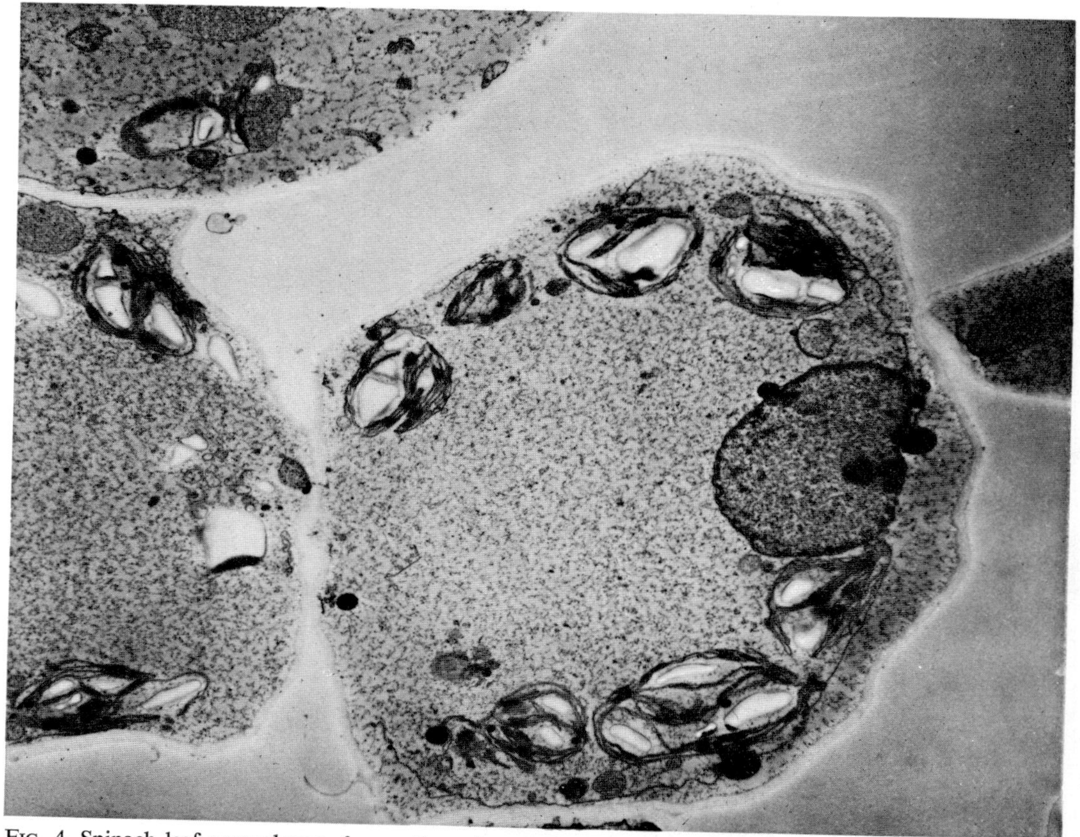

FIG. 4. Spinach leaf parenchyma, frozen–thawed, × 5000. The freezing stage was carried out rapidly (2–3s). The protoplasmic components are distorted and disrupted and the tonoplast is ruptured. (By courtesy of Dr W. P. Mohr and the *Journal of Texture Studies*.)

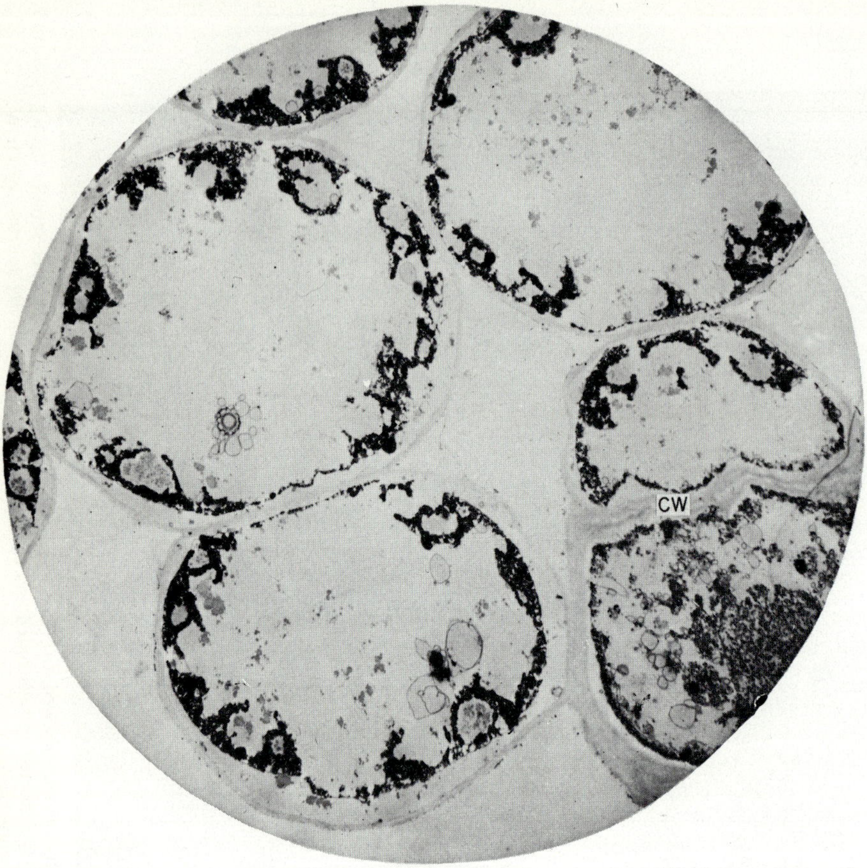

FIG. 5. Spinach leaf parenchyma, blanched, ×3100. The fine structure is almost unrecognizable, although the protoplasmic layer remains peripherally situated. (By courtesy of Dr W. P. Mohr and the *Journal of Texture Studies*.)

of 0·2M NaCl gave the highest rehydration ratios. They concluded from results obtained using a scanning electron microscope that some of the NaCl recrystallized in certain cell types during dehydration, and that it was the effect of water on crystalline salt which gave rise to the improvement in rehydration. Similarly, Shipman *et al.* (1972) showed that inclusion of glycerol into dehydrated celery improved its rehydration properties.

2. Chemical processing

The two most widely practised chemical methods of processing are the preservation of fruits or vegetables in brine (for storage or pretreatment

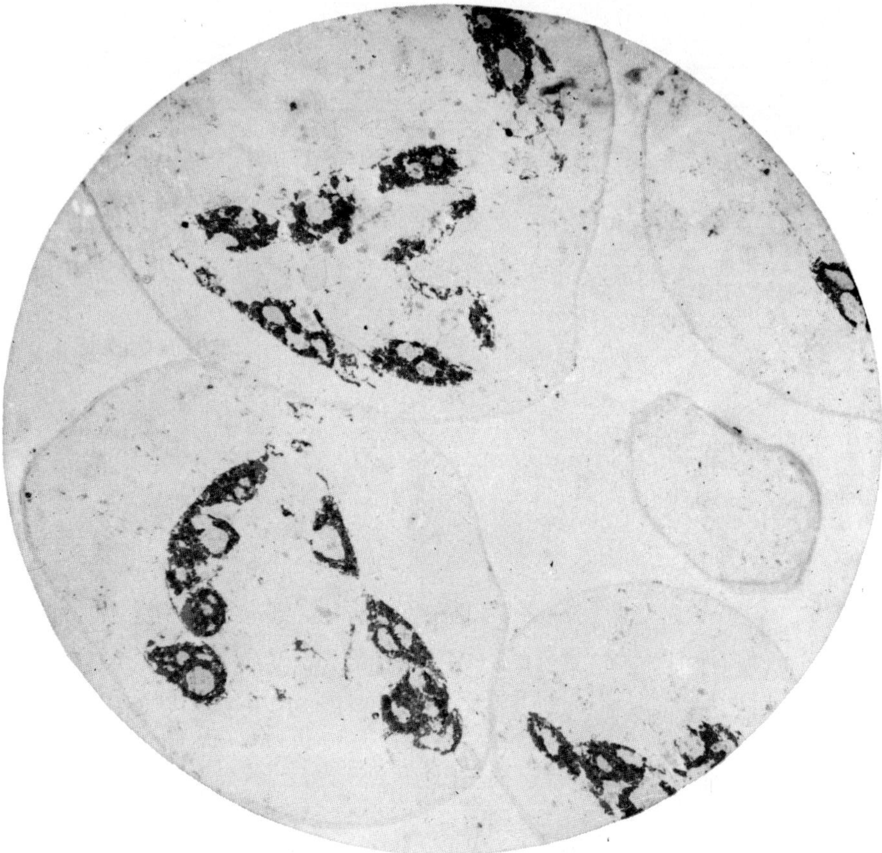

FIG. 6. Spinach leaf parenchyma, blanched–frozen–thawed, ×3100. The extensively damaged protoplasm appears similar to that just blanched (Fig. 5) except for its more central location within the cells. (By courtesy of Dr W. P. Mohr and the *Journal of Texture Studies*.)

prior to pickling) and their treatment with syrup (to produce candied fruits or canning liquors). A common effect of high concentrations of either salt or sugar is severe plasmolysis of the cells. In most commercial processes this results in the rupture of the plasma membrane and hence loss of turgor pressure (Jewell, 1972).

In studies on the changes produced during the pickling of cauliflower (Saxton and Jewell, 1969) and onions (Jewell, 1972) it was seen that brining (in 16% NaCl) produced a gradual degradation of the fine structural organization within the cells which eventually left just a network of cell walls (Figs 11, 12 and 13). After placing in vinegar, some modification of cell wall

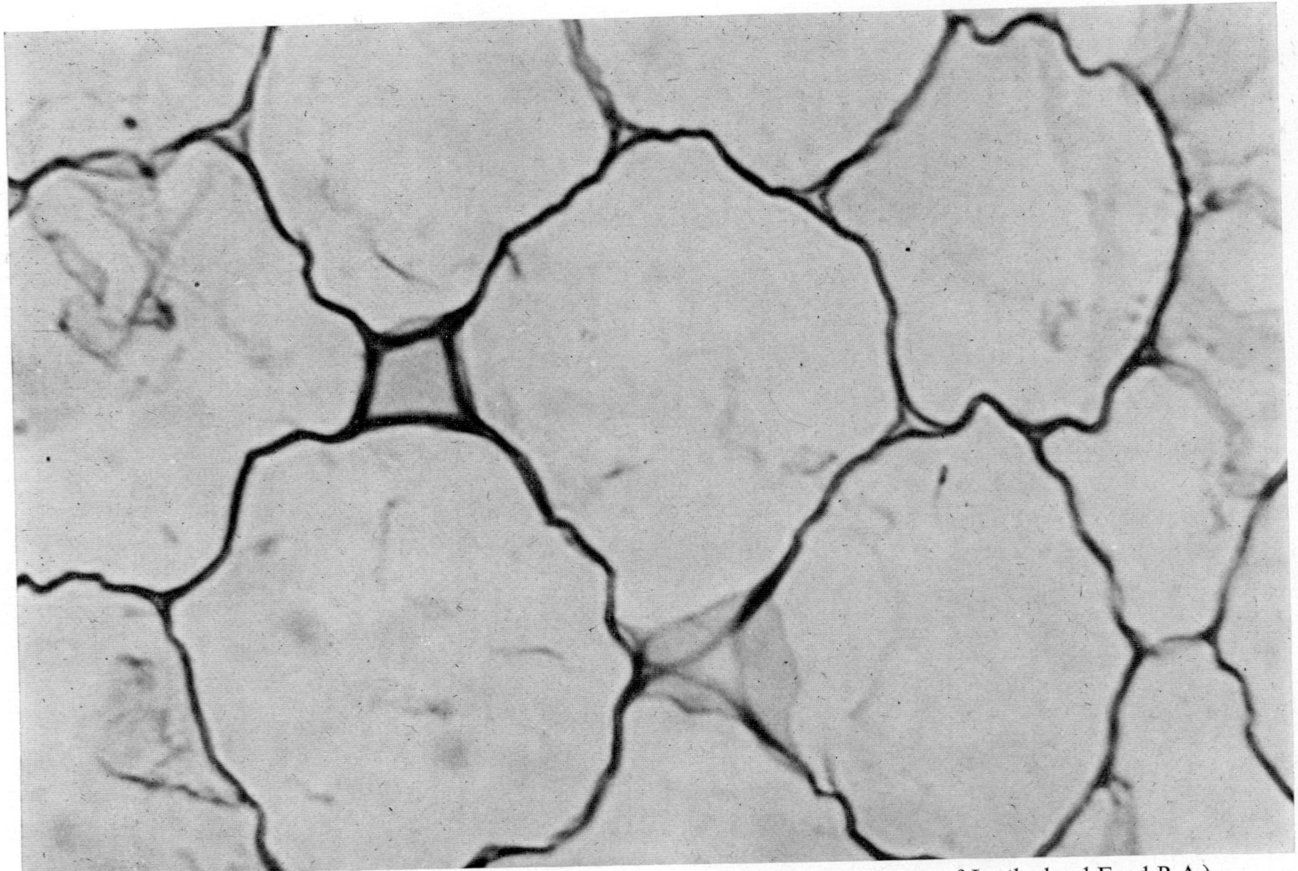

FIG. 7. Fresh carrot showing intact, well-organized cell structure, $\times 600$. (By courtesy of Leatherhead Food R.A.)

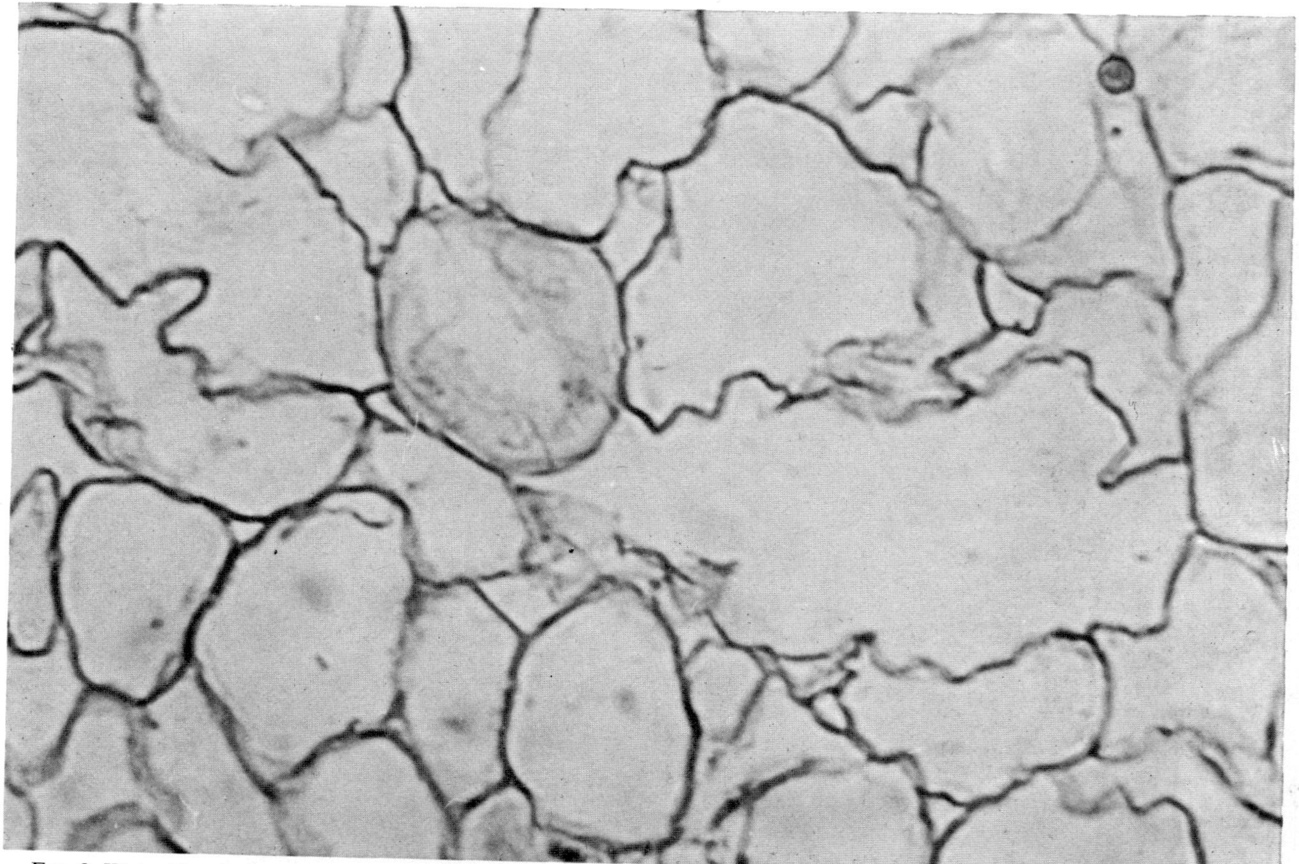

FIG. 8. Water-blanched carrot, showing some cellular collapse and disruption, $\times 600$. (By courtesy of Leatherhead Food R.A.)

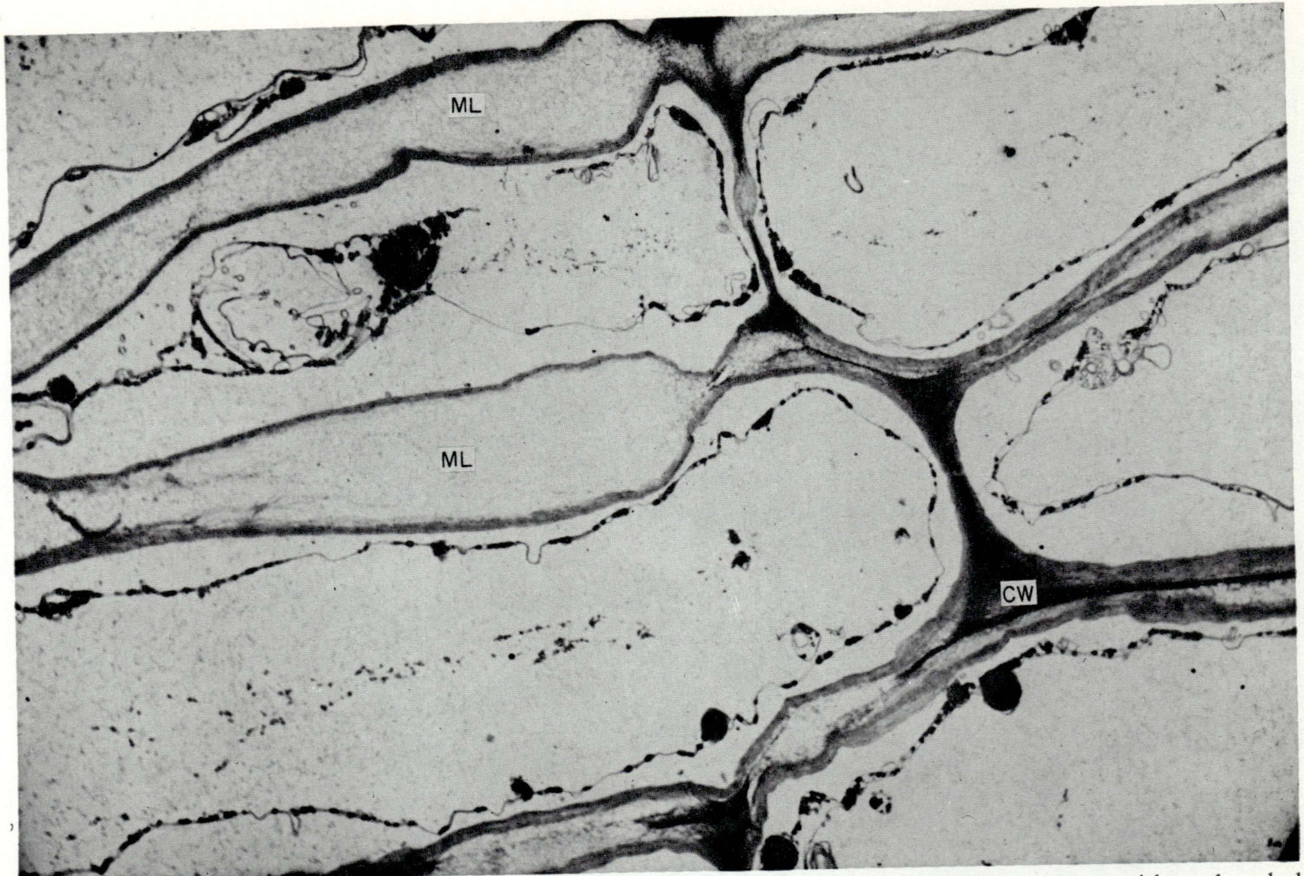

FIG. 9. Water-blanched carrot, showing some plasmolysis, rupture of the plasmalemma, formation of numerous vesicles and marked separation of the middle lamella (ML) region of the cell wall (CW), ×5500. (By courtesy of Leatherhead Food R.A.)

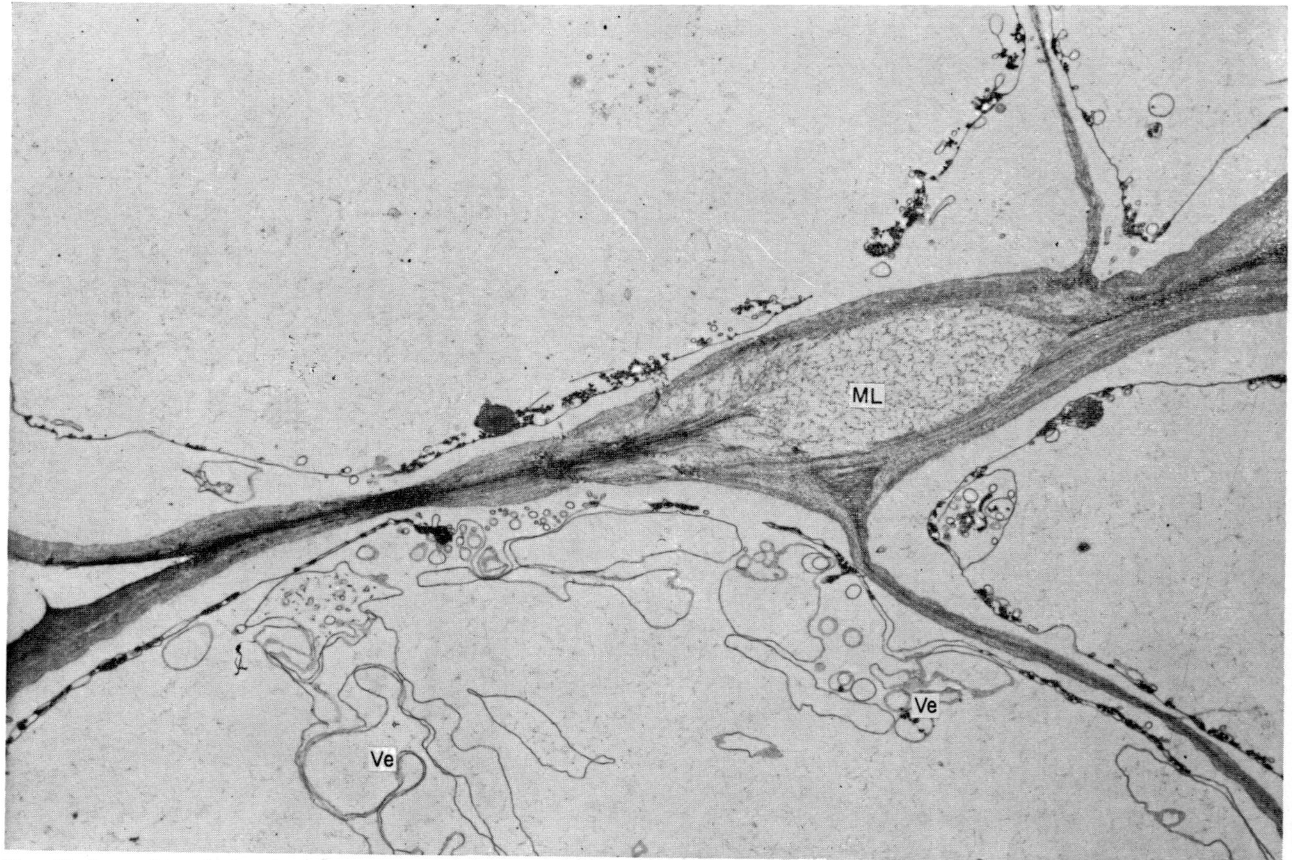

FIG. 10. Steam-blanched carrot, showing protoplasmic degradation and formation of vesicles (Ve) but less separation of the middle lamella (ML), ×6200. (By courtesy of Leatherhead Food R.A.)

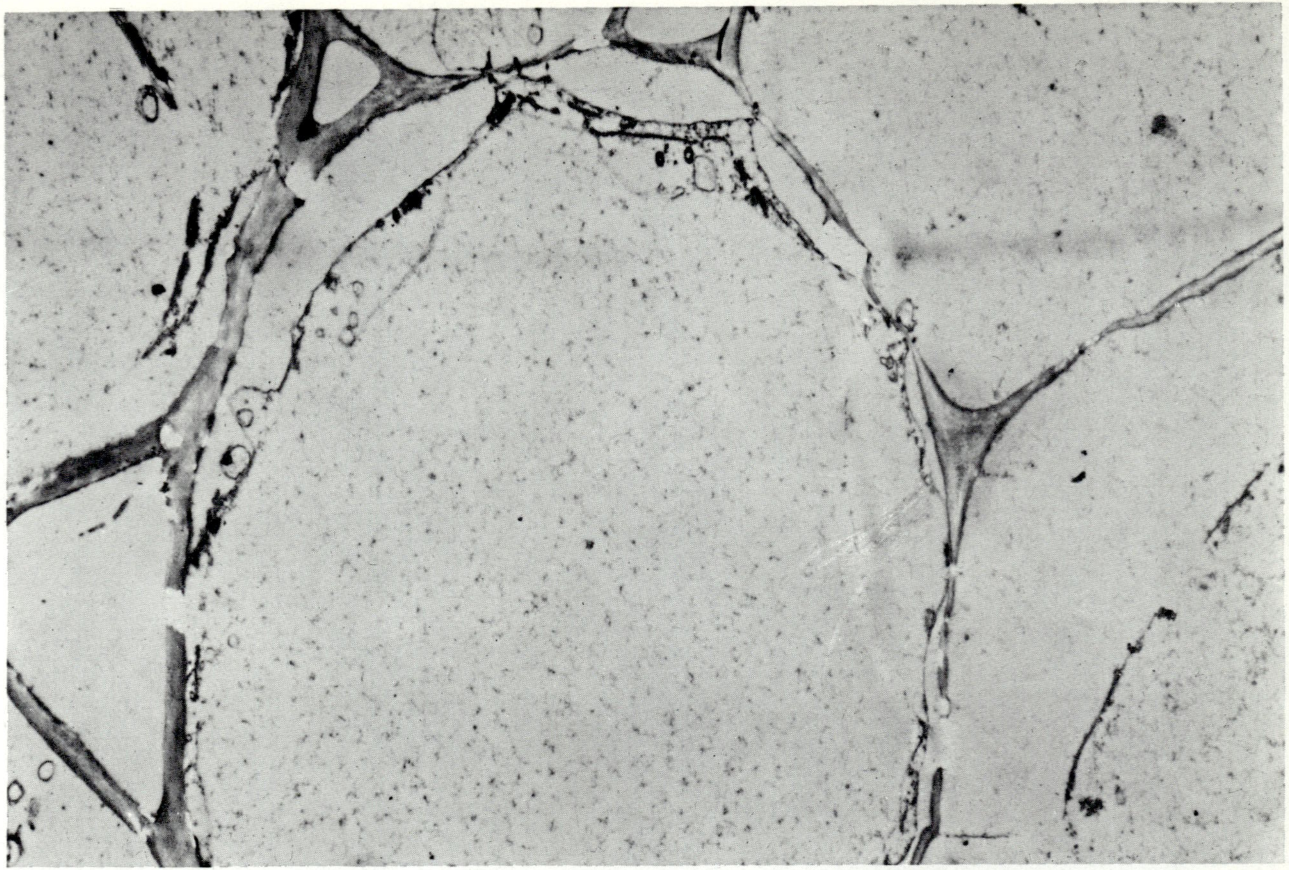

FIG. 11. Parenchyma of brined (8 h) cauliflower stem, $\times 7500$. Various degrees of plasmolysis are evident, as is loss of fine structural detail. (By courtesy of Leatherhead Food R.A.)

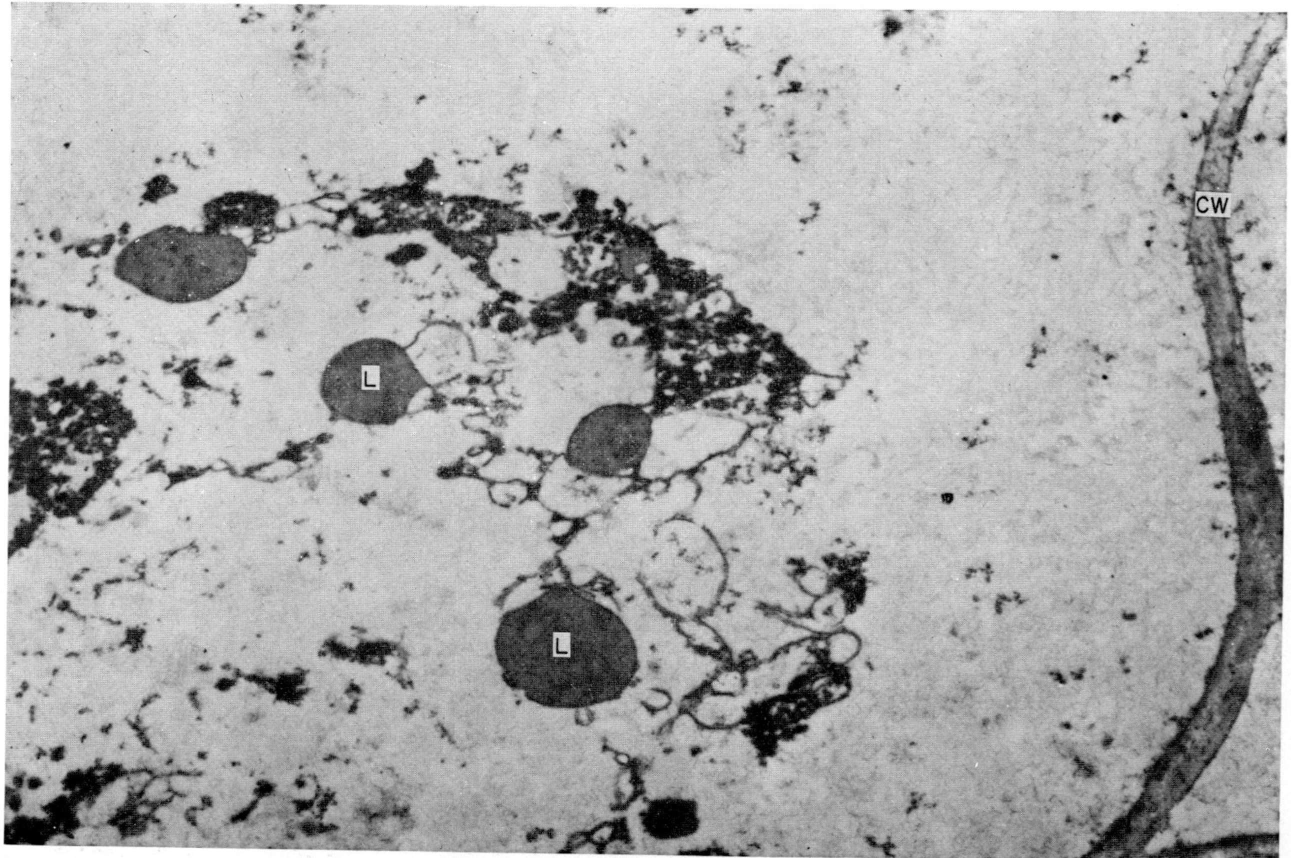

FIG. 12. Parenchyma of brined (7 d) cauliflower stem, ×16 800. Note the absence of the plasmalemma and degraded organelles; only the darkly staining lipid droplets (L) appear to remain intact. (By courtesy of Leatherhead Food R.A.)

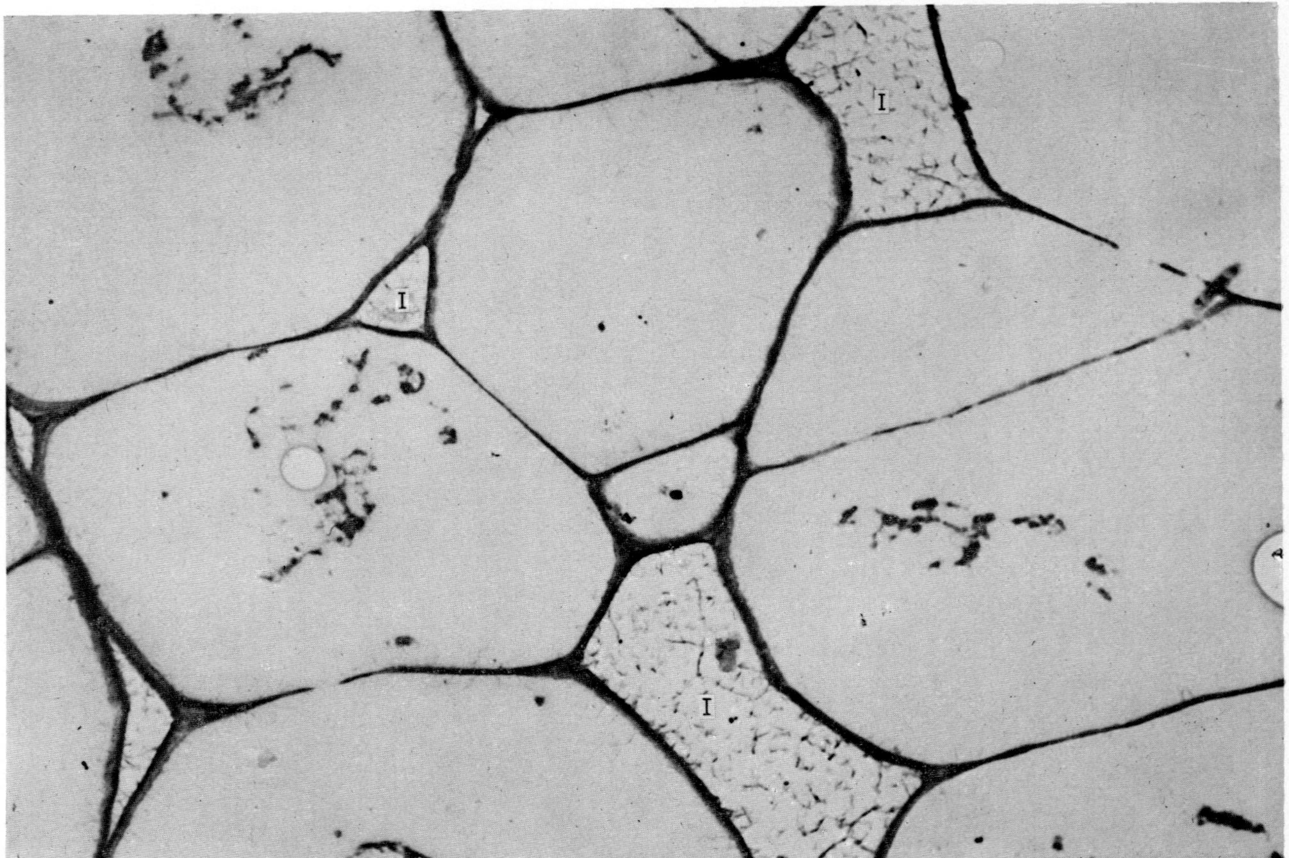

FIG. 13. Parenchyma of brined (3 week) cauliflower stem, $\times 6700$. Note the virtual absence of protoplasmic material and the enlarged intercellular spaces (I). (By courtesy of Leatherhead Food R.A.)

materials was noted which involved the pectin hemicellulose and cellulose fractions. The changes in texture were studied concurrently with the structural changes. It was found that the most marked change in texture (Fig. 14) was associated with the loss of turgor pressure (a reduction of 75% of original firmness). Changes in cell wall materials during pickling produced only small effects when compared with the loss of turgor pressure (a further reduction of 15%). But the cell wall effects become more significant, since

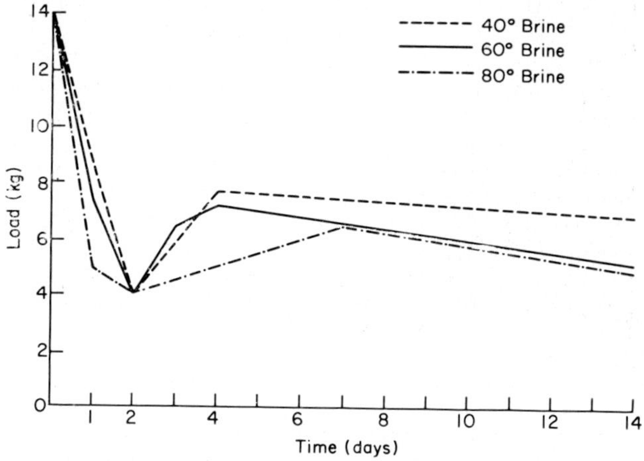

FIG. 14. A force deformation curve showing the change in hardness (load) of cauliflower stem during brining for various times in three different brine strengths, 40° (11% NaCl), 60° (16% NaCl) and 80° (22% NaCl).

during prolonged storage in brine or in the finished pickle there will be a continued degradation of the cell wall resulting in a softer product.

3. *Mechanical processing*

The particle size in many comminuted products, e.g. apple sauce or tomato paste, will influence the texture of the product, very grainy sauces being regarded as unacceptable. Mohr and Adair (1970) showed that the variety of apple influenced the graininess of the product. Following this work, Mohr (1973) applied scanning microscopy techniques to show that the difference in graininess was related to the particle size distribution, different size particles being formed from clusters of cells (Figs 15, 16, 17 and 18). The role of the cell wall materials (pectins and cellulose) in tomato paste has been studied by Whittenberger and Nutting (1957, 1958). They found that provided the pectin substances have not been too degraded during processing, the product viscosity is related to the particle size distribution of the cellulose materials. In general, thicker consistency results from finely divided

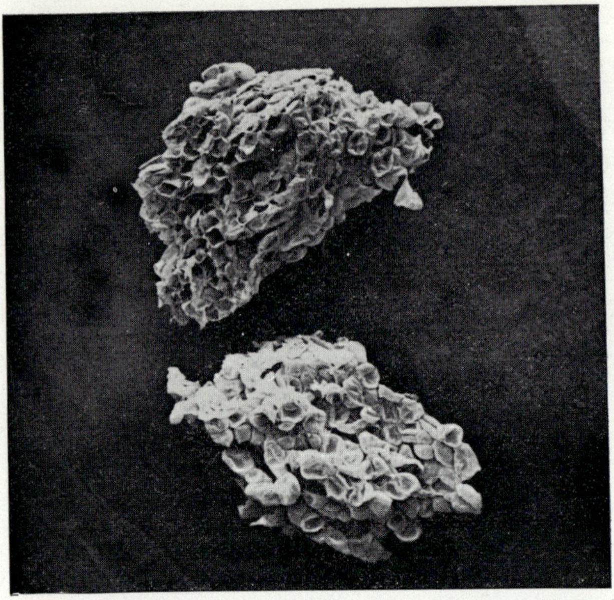

FIG. 15. Particles from apple sauce sieved to yield large particles, ×30. Each particle is a cluster of many cells. (By courtesy of Dr W. P. Mohr and the *Journal of Texture Studies.*)

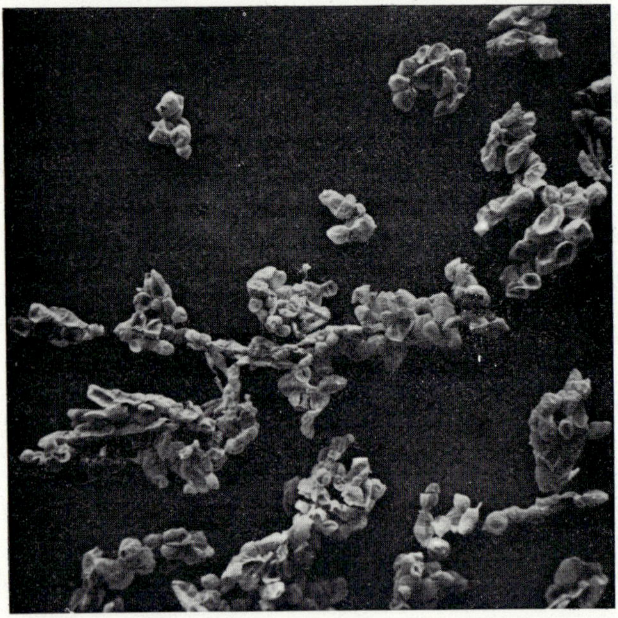

FIG. 16. Particles from apple sauce sieved to yield medium particles, ×30. Each particle is a cluster of cells. (By courtesy of Dr W. P. Mohr and the *Journal of Texture Studies.*)

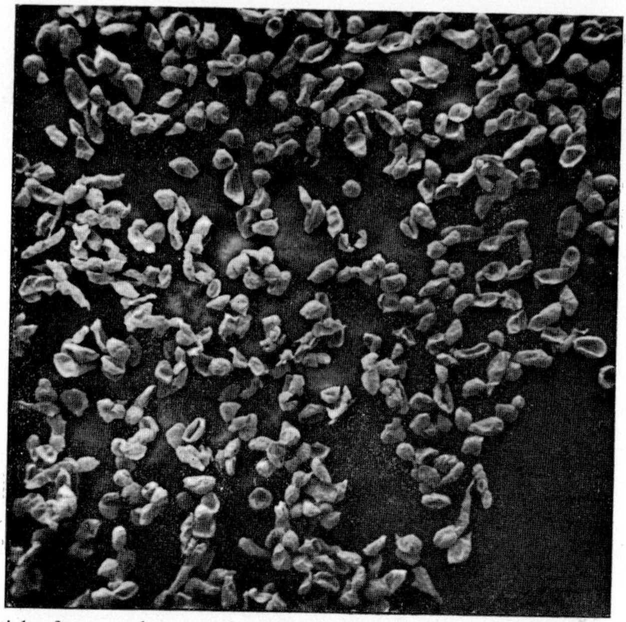

FIG. 17. Particles from apple sauce sieved to yield small particles, ×30. Note each particle is either an individual cell or a cluster of only a few cells. (By courtesy of Dr W. P. Mohr and the *Journal of Texture Studies*.)

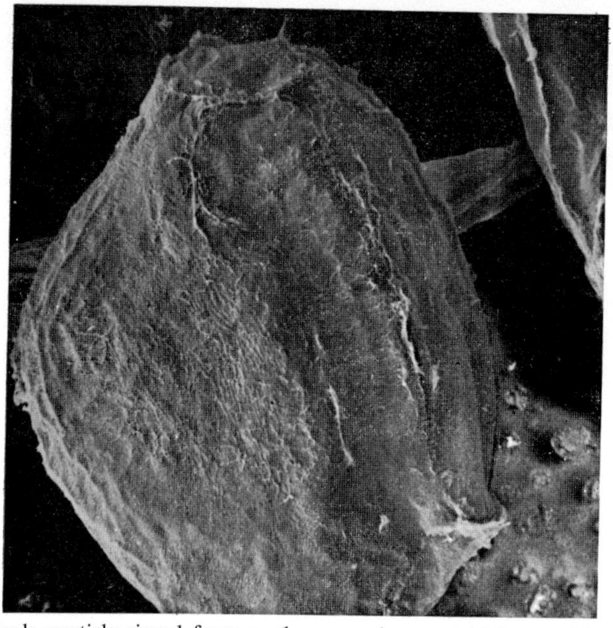

FIG. 18. A single particle sieved from apple sauce shown at high magnification, ×280. (By courtesy of Dr W. P. Mohr and the *Journal of Texture Studies*.)

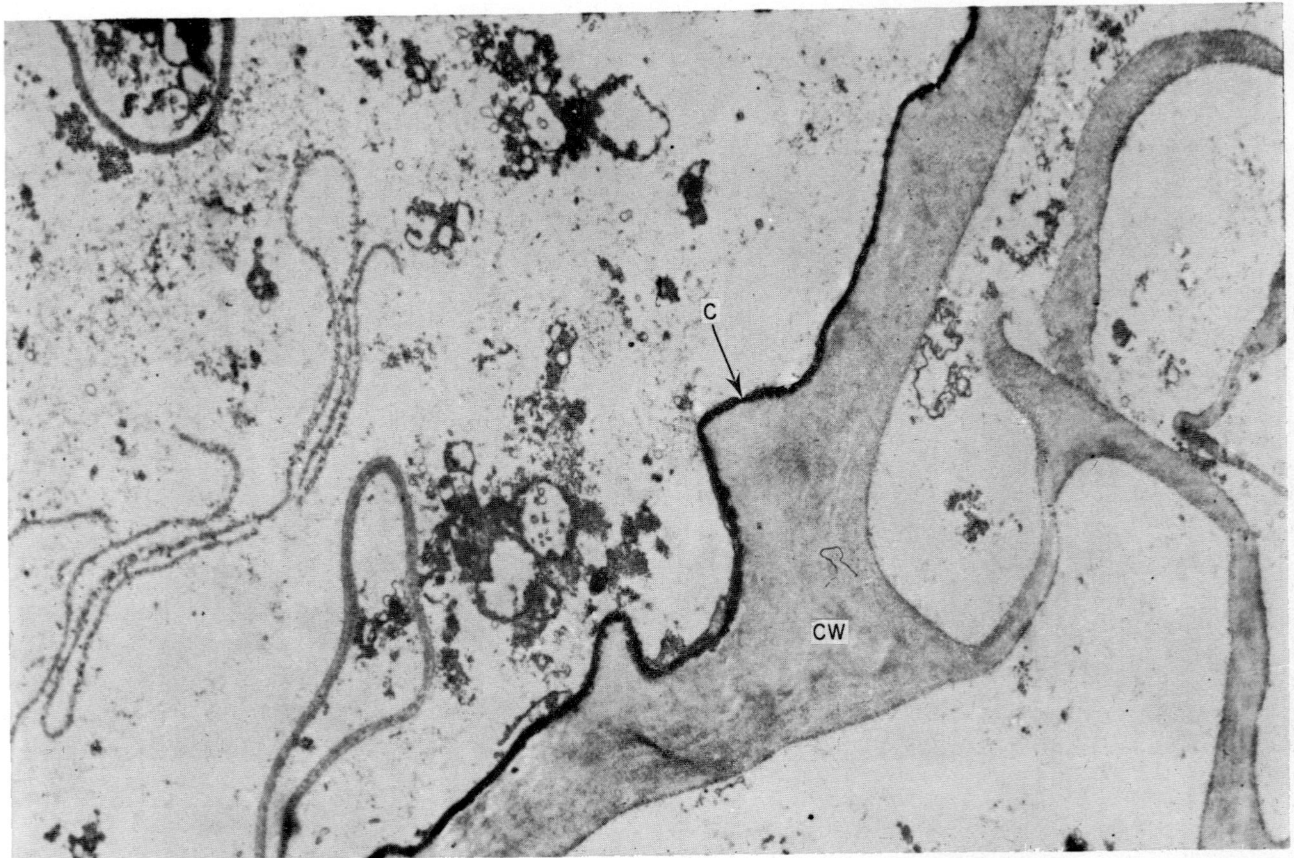

FIG. 19. A thin section of orange juice, ×8600. Note the range of cell wall sizes (CW), the darkly staining cuticle (C) on the outer wall of the juice sac and the disordered and clumped protoplasmic remnants. (By courtesy of Leatherhead Food R.A.)

FIG. 20. A freeze-etched sample of orange juice, $\times 63\ 000$. Note the globules and regular form of the crystalline inclusions. (By courtesy of Leatherhead Food R.A.)

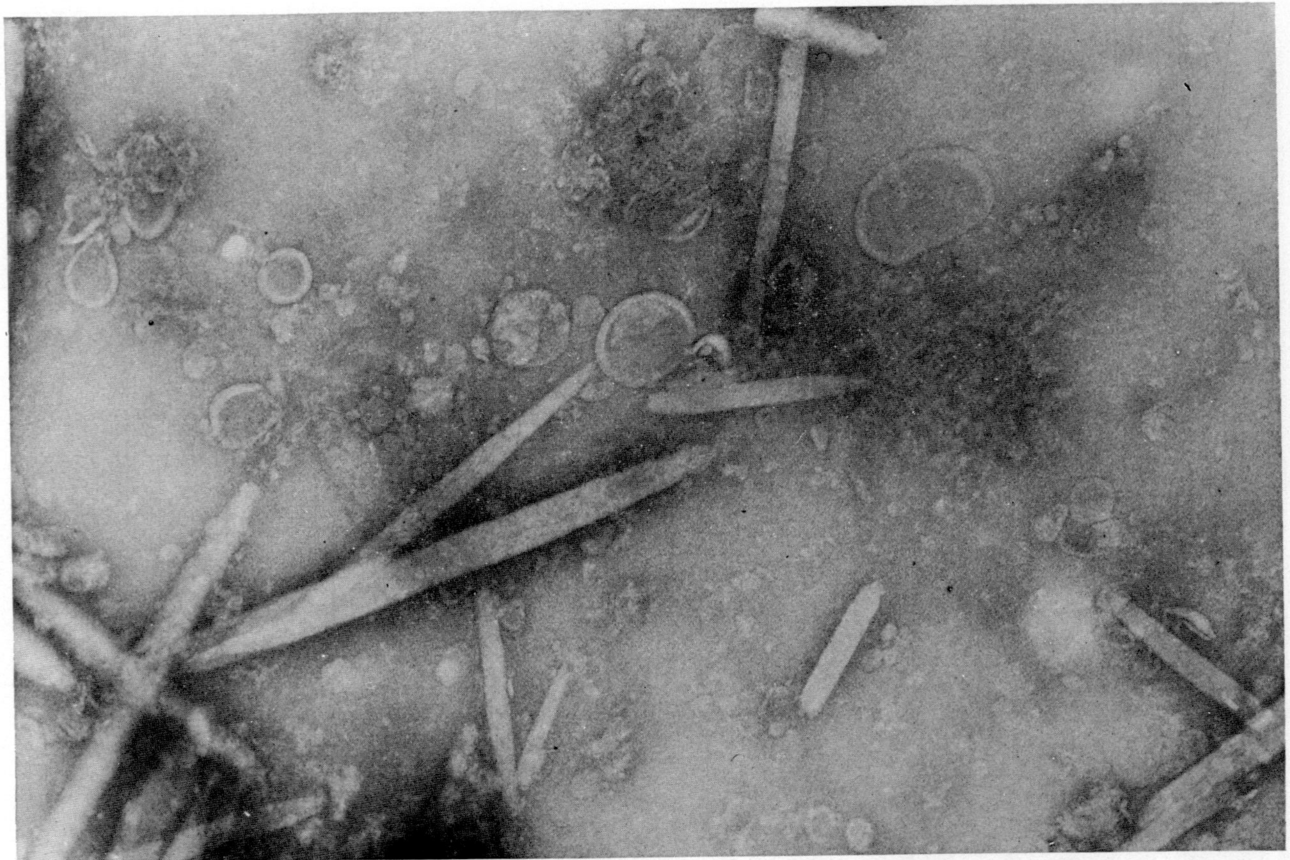

FIG. 21. A negatively stained preparation of orange juice, ×65 000. Note the numerous vesicles and crystalline bodies. (By courtesy of the Leatherhead Food R.A.)

cellulose materials. In contrast, Reeve (1956) found that a high frequency of intact cells was desirable for optimum consistency in certain fruit *purées* such as apricot.

Another area of mechanical processing which has recently received some attention is the comminution of citrus fruits to yield juices. The annual value of citrus juices is about £250 million and as such accounts for rather more than half of the total world business in fruits of all kinds. In view of the value of these commodities, it is rather surprising that so little microscopic study has been carried out upon them. Resch and Schara (1970) published a general guide, based upon light microscopy, to the anatomy of the orange and the types of particles found in its juice. Mizrahi and Berk (1970) published one electron micrograph showing some of the particles found in orange juice. Jewell (1975) utilized a wide range of light and electron microscopy techniques to study the tissue of the orange and orange juice. Lewin and Jewell (1976) studied the particles in lemon and grapefruit juices by electron microscopy, whilst Venolia *et al.* (1974) undertook a study of the particle sizes in lemon juice. Jewell (1975) noted that a wide range of particle types and sizes could be found in orange juice and that no single electron microscopy preparation method revealed all the particle types. Thus, it was necessary to use thin sectioning, freeze-etching and negative staining (Figs 19, 20 and 21). The comminution process results in the rupture of cells and cell walls, and the fine structural detail of the cytoplasmic organelles is generally lost Numerous vesicles are formed, and these and fragments of cell walls are the most readily identified structures. It was of interest to note that many of the particles were $<0{\cdot}5\ \mu m$ in diameter and thus could not readily be detected using light microscopy. One of the characteristic features of the outer cell wall of the citrus juice sac is the cuticle with its adherent wax (Fig. 19). This cuticular layer has also been studied by Fahn *et al.* (1974) uing both transmission and scanning electron microscopy (Figs 22 and 23). These workers concluded that the epicuticular wax serves as an adhesive which holds the juice vesicles in each fruit segment together and thus influences their retention of integrity during processing. Further work on the structure (Figs 24 and 25) and composition of cuticular waxes has been given by Hamilton and Hamilton (1972). When blueberries are canned and then subsequently cooled without agitation or rolling, the wax on the cuticle of the fruit fuses and the berries stick together tightly (Powers *et al.*, 1958). The fused berries cannot be readily separated without tearing of the skins, but simple rolling of the cans during cooling completely eliminates the problem.

Canned and frozen raspberries sometimes suffer from a texture defect, they become crumbly. The crumbliness is caused by the abortion of successive

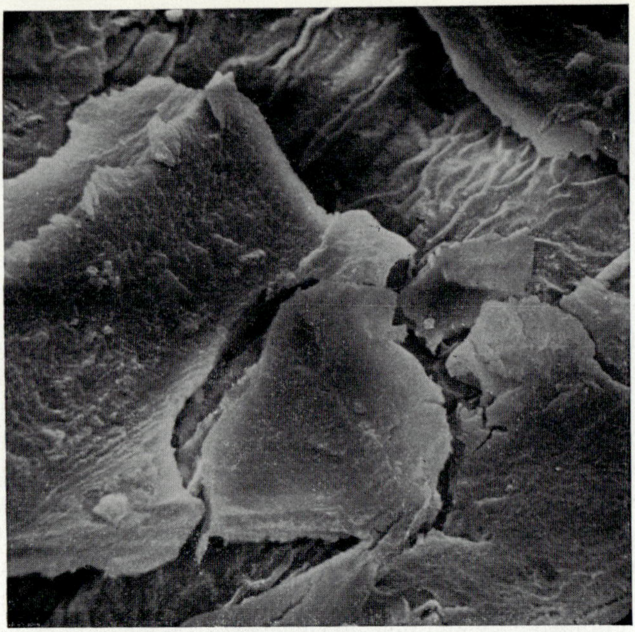

FIG. 22. Surface of the juice vesicle from a Shamouti orange, ×2000. (By courtesy of Professor A. Fahn and *Annals of Botany*.)

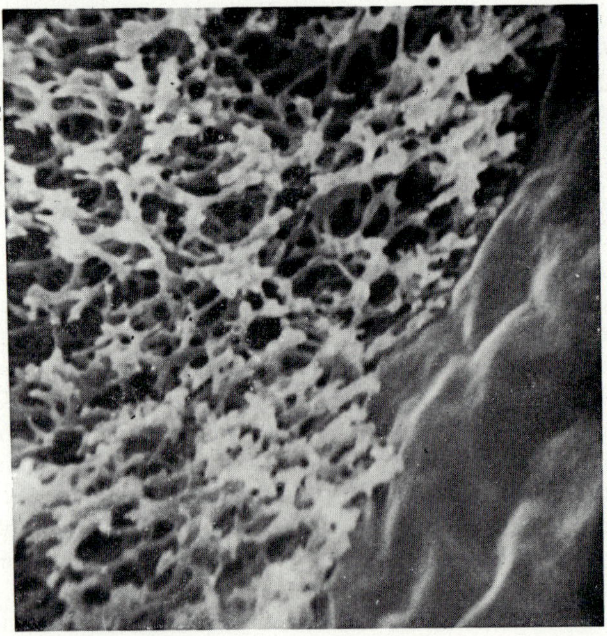

FIG. 23. Surface of the juice vesicle from a Marsh seedless grapefruit, ×4400. (By courtesy Professor A. Fahn and *Annals of Botany*.)

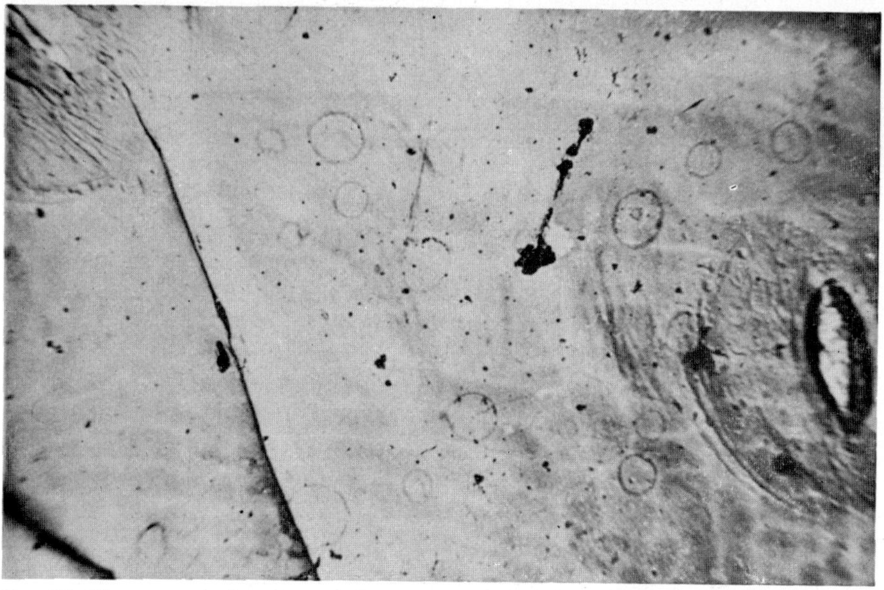

FIG. 24. The upper leaf surface of the common beet (*Beta vulgaris*), ×16 750. Note the wax layer is relatively smooth. (By courtesy of Drs S. and R. J. Hamilton and the publishers of 'Topics in Lipid Chemistry', Vol. 3, Paul Elek (Scientific Books).)

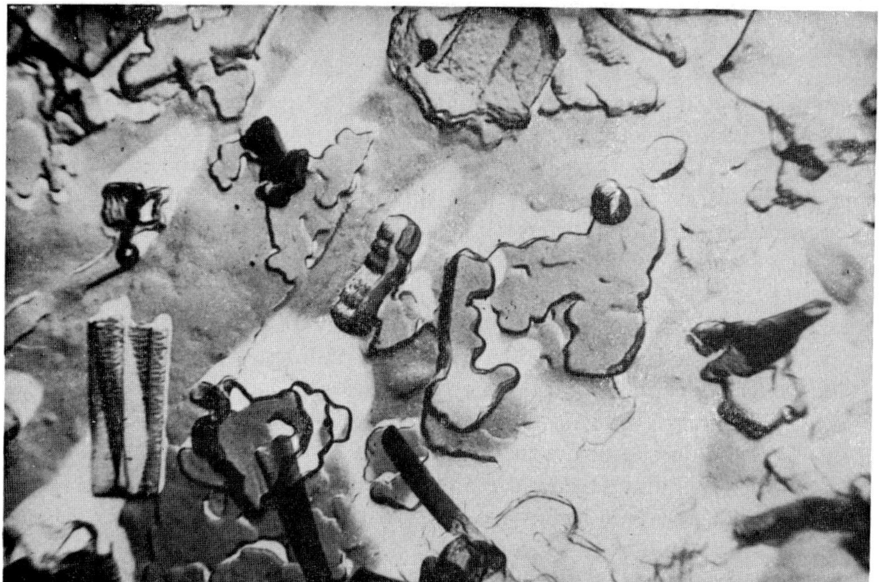

FIG. 25. The upper leaf of cabbage (*Brassica oleracea*) var. January king, ×16 750. Note the tube-like growths of wax. (By courtesy of Drs S. and R. J. Hamilton and the publishers of 'Topics in Lipid Chemistry', Vol. 3, Paul Elek (Scientific Books).)

numbers of drupelets, which remain small and dry (Reeve 1954a, b; Reeve *et al.*, 1965). These aborted drupelets lead to regions of structural weakness and hence, during the mechanical stages involved in preparing the fruit for canning or freezing, some disintegration and crumbling of the berries occurs.

As mentioned earlier, strawberries undergo considerable loss of texture during processing. Szczesniak and Smith (1969) have published some histological data on strawberries, and Jewell *et al.* (1973) related the structure of strawberries to the type of breakdown observed in strawberry jam.

Thus, one can conclude that in many of the processing operations the cytoplasmic structure will be destroyed, plasmolysis will occur and turgor pressure will be lost and not normally recovered. The cell walls will be weakened and some degree of cell wall separation may occur. The loss of turgor pressure and the changes in the cell wall usually have profound effects on the texture of the product.

IV. CHANGES DURING GROWTH

The changes in structure which occur during development, growth and ripening have been studied in a number of fruits and vegetables, e.g. in oranges (Rogers, 1971; Albrigo, 1972), lemons (Soni and Ronhawa, 1969), apples (Nelmes and Preston, 1968; Simons, 1968), tomatoes (Mohr and Stein, 1969) and pears (Bain and Mercer, 1964). In general terms, these authors have observed changes within the protoplast and cell wall. As fruit approaches maturity there is a general degeneration of the protoplasmic components and numerous vesicles are formed. The cell walls tend to separate along the middle lamella and in some fruits adjacent cells may become completely separated. These alterations in cell wall organization are related to changes in cell wall composition. Nelmes and Preston (1968) studied both the composition and the structure of apple cell walls during growth and noted changes in cell wall morphology, an increase in hemicelluloses and decreases in pectin content prior to harvest. During storage, however, they observed a decrease in hemicellulose and an increase in pectins.

In a review on the ultrastructure of plant senescence, Butler and Simon (1971) concluded that degeneration and death of the cell, whatever the cause, follows a characteristic and consistent pattern. In their study they did not include cell death due to food processing, but they suggested the following generalized sequence of events which is clearly in agreement with the observations on the changes that are brought about by processing, whether it be thermal, mechanical or chemical. The first detectable changes are usually a decrease in the number of ribosomes and the start of chloroplast breakdown. The free ribosomes are the first to disappear, followed by those attached to the endoplasmic reticulum. Next, the chloroplast stroma disappears, the thylakoids swell and disintegrate, and there is an increase in the number of osmophilic globules. Some degeneration of the mitochondria is noted. The endoplasmic reticulum swells, vesiculates and disappears, as do the dictyosomes. The tonoplast breaks down long before the organelles have completely degenerated, but the plasmalemma is usually one of the last recognizable components. The nucleus remains relatively stable, but eventually breaks down and the chromatin disappears. Death of the cell is normally preceded by loss of the plasmalemma. In thermal and chemical food processing, changes in the cell wall structure, such as splitting along the middle lamella due to extraction or modification of the pectins and hemicellulose, are normally also observed.

In many of the food processes, particularly thermal processing, the degradation may proceed too quickly to permit the identification of all the steps listed above. The gradual breakdown of organelle structure is frequently replaced by the rapid production of heat-coagulated material.

V. FUTURE PROSPECTS

The continuing pressures from increases in raw commodity prices and the desire to achieve consistent quality with minimum waste must be balanced with the inevitably variable nature of the raw materials; hence, the food processor is faced with a considerable problem. There would appear to be two methods of minimizing the problem. Firstly, as stated earlier, a greater knowledge of the changes occurring during processing should enable a better control of the process. Secondly, we have already seen the introduction of reformed fruits, where in essence fruits are disintegrated, mixed with a suitable gelling agent and then reformed to a reproducible product and a gel formed. This process permits the utilization of material of variable quality and also cuts down on waste. However, the technology is still relatively in its infancy and some products are fairly crude attempts to mimic the subtle textures of natural fruits and vegetables. Thus, with both these approaches more detailed information concerning the relationship between plant structure, particularly cell wall architecture and texture, is required, and this, coupled with the obvious application of the microscopical identification of contaminants, should ensure a heavy work load for the food microscopist in the future.

ACKNOWLEDGEMENTS

I am indebted to the director of the Leatherhead Food R.A., Dr A. W. Holmes, for permission to prepare this contribution and include some previously unpublished photographs. I am particularly grateful to those authors and journals or publishers who have given me permission to reproduce their work, namely Dr W. P. Mohr, Professor A. Fahn and Drs S. and R. J. Hamilton. I am also pleased to acknowledge my former colleagues in the microscopy unit at the Leatherhead Food R.A. for their dedication and work over the years.

REFERENCES

Albersheim, P. and Killias, U. (1963). *Am. J. Bot.* **50** (7), 732.
Albrigo, L. G. (1972). *J. Am. Soc. Hort. Sci.* **97**, 761.
Bain, J. M. and Mercer, F. V. (1964). *Aust. J. Biol. Sci.* **17**, 78.
Bassi, M. and Crivelli, M. G. (1968). *Revue Gén. Froid*, **59**, 1211.
Bassi, M. and Crivelli, M. G. (1969). *Revue Gén. Froid*, **60**, 1289.
Brown, M. S. (1967). *J. Sci. Fd Agric.* **18**, 77.
Butler, R. D. and Simon, E. W. (1971). *In* 'Advances in Gerontological Research' (B. L. Strehler, ed.), Vol. 3, p. 73. Academic Press, London and New York.
Cording, J., Jr, Willard, M. J., Jr, Eskew, R. K., Edwards, P. W. and Sullivan, J. F. (1955). ARS 73–9. U.S. Department of Agriculture.
Cording, J., Jr, Sullivan, J. F. and Eskew, R. K. (1959). ARS 73–25, U.S. Department of Agriculture.
Crivelli, M. G. and Bassi, M. (1969). *Revue Gén. Froid*, **60**, 421.
Curry, J. C., Burns, E. E. and Heidelbaugh, N. D. (1976). *J. Fd Sci.* **41**, 176.
Dodge, J. D. (1968). 'An Atlas of Biological Ultrastructure'. Edward Arnold, London.
Esau, K. (1936). *Hilgardia*, **10**, 431.
Esau, K. (1965). 'Plant Anatomy', 2nd Edition. John Wiley and Sons, New York.
Fahn, A., Shomer, I. and Ben-Gera, I. (1974). *Ann. Bot.* **38**, 869.
Frey-Wyssling, A. and Mühlethaler, K. (1965). 'Ultrastructural Plant Cytology'. Elsevier, Amsterdam.
Greenish, H. G. (1923). 'Food and Drugs'. J. and A. Churchill, London.
Hamilton, S. and Hamilton, R. J. (1972). *In* 'Topics in Lipid Chemistry' (F. D. Gunstone, ed.), Vol. 3, p. 199. Paul Elek (Scientific Books), London.
Heyn, A. N. J. (1966). *J. Cell Biol.* **29** (2), 181.
Heyn, A. N. J. (1969). *J. Ultrastruct. Res.* **26**, 52.
Jewell, G. G. (1972). *J. Fd Technol.* **7**, 387.
Jewell, G. G. (1975). *Leatherhead Food R.A. Res. Rep.* **230**.
Jewell, G. G. and Saxton, C. A. (1970). *Histochem. J.* **2**, 17.
Jewell, G. G., Rantsios, A. and Scholey, J. (1973). *J. Text. Stud.* **4**, 363.
Lee, C. Y., Salunkhe, D. K. and Nury, F. S. (1967). *J. Sci. Fd Agric.* **18**, 89.
Lewin, Y. A. and Jewell, G. G. (1976). *Leatherhead Food R.A. Res. Rep.* **251**.

Longree, K. (1950). *Fd Technol., Champaign*, **4**, 98.
Mirza, S. and Jewell, G. G. (1976). *Leatherhead Food R.A. Res. Rep.* **253**.
Mizrahi, S. and Berk, Z. (1970). *J. Sci. Fd Agric.* **21**, 250.
Mohr, W. P. (1971). *J. Text. Stud.* **2** (3), 316.
Mohr, W. P. (1972). *Can. Inst. Fd Sci. Technol. J.* **5** (4), 179.
Mohr, W. P. (1973). *J. Text. Stud.* **4**, 263.
Mohr, W. P. (1974). *J. Text. Stud.* **5**, 13.
Mohr, W. P. and Adair, R. G. (1970). *Can. Dept. Agric. Publ.* **1424**.
Mohr, W. P. and Stein, M. (1969). *Can. J. Pl. Sci.* **49**, 549.
Nelmes, B. J. and Preston, R. D. (1968). *J. Exp. Bot.* **19** (60), 496.
Northcote, D. H. (1968). *In* 'Plant Cell Organelles' (J. B. Pridham, ed.), p. 179. Academic Press, New York and London.
Powers, M. J., Reeve, R. M., Leinbach, L. R., Talburt, W. F. and Brekke, J. E. (1958). *Fd Technol., Champaign*, **12**, 99.
Rahman, A. R., Henning, W. L. and Westcott, D. E. (1971). *J. Fd Sci.* **36**, 500.
Reeve, R. M. (1943). *Fd Res.* **8**, 128.
Reeve, R. M. (1954a). *Am. J. Bot.* **41**, 152.
Reeve, R. M. (1954b). *Am. J. Bot.* **41**, 173.
Reeve, R. M. (1954c). *Fd Res.* **19**, 323.
Reeve, R. M. (1954d). *Fd Res.* **19**, 333.
Reeve, R. M. (1954e). *Fd Res.* **19**, 340.
Reeve, R. M. (1956). *Fd Res.* **21**, 329.
Reeve, R. M. (1963). *J. Fd Sci.* **28**, 198.
Reeve, R. M. (1967a). *Econ. Bot.* **21**, 294.
Reeve, R. M. (1967b). Proc. 16th Ann. Potato Util. Conf. (F. E. Collins, ed.). ARS 74-40, U.S. Department of Agriculture.
Reeve, R. M. (1970). *J. Text. Stud.* **1**, 247.
Reeve, R. M., Wolford, E. and Nimo, C. C. (1965). *Fd Technol., Champaign*, **19**, 78
Reeve, R. M., Feinberg, B., Boyle, F. P. and Notter, G. K. (1968). *Fd Technol., Champaign*, 22, 94.
Resch, A. and Schara, A. (1970). *Confructa*, **15** (3), 166.
Robards, A. W. (1971). *Protoplasma*, **72**, 315.
Roelofsen, P. A. (1965). *In* 'Advances in Botanical Research' (R. D. Preston, ed.), Vol. 2, p. 69. Academic Press, New York and London.
Rogers, B. J. (1971). *Planta*, **97** (4), 358.
Saxton, C. A. and Jewell, G. G. (1969). *J. Fd Technol.* **4**, 363.
Shipman, J. W., Rahman, A. R., Segars, R. A., Kapsalis, J. G. and Westcott, D. E. (1972). *J. Fd Sci.* **37** (4), 568.
Simons, R. K. (1968). *Proc. Am. Hort. Sci.* **93**, 775.
Soni, S. L. and Ronhawa, G. S. (1969). *Indian J. Agric. Sci.* **39** (8), 813.
Spiruta, S. L. and Mackey, A. (1961). *J. Fd Sci.* **26**, 656.
Sterling, C. (1955). *Fd Res.* **20**, 474.
Sterling, C. (1963). *In* 'Recent Advances in Food Science' (J. M. Leitch and D. N. Rhodes, eds), Vol. 3, p. 259. Butterworth, London.

Sterling, C. (1966). Proc. Plant Sci. Symp. Campbell Inst. Agr. Res., Campden N.J., p. 11.
Szczesniak, A. S. and Smith, B. J. (1969). *J. Text. Stud.* **1**, 65.
Venolia, W., Peak, S. and Payne, F. (1974). *J. Agric. Fd Chem.* **22** (1), 133.
Warren, D. S. and Woodman, J. S. (1974). *J. Sci. Fd Agric.* **25**, 129.
Whittenberger, R. T. and Nutting, G. C. (1950). *Fd Res.* **15**, 331.
Whittenberger, R. T. and Nutting, G. C. (1957). *Fd Technol., Champaign*, **11**, 19.
Whittenberger, R. T. and Nutting, G. C. (1958). *Fd Technol., Champaign*, **12**, 420.
Winton, A. L. and Winton, K. B. (1935). 'The Structure and Composition of Foods', Vol. 2. John Wiley and Sons, New York.

2. Oil Seeds

C. G. SMITH

Unilever Research Laboratory,
Bedford, U.K.

I. INTRODUCTION

The microstructure of oil-bearing seeds is in many respects similar to many other seeds, but by virtue of the fact that some species of oil-bearing seeds contain oil in quantities large enough to be extracted economically, they are a very important commercial crop. Superimposed upon this fact, and of equal commercial importance, is their ease of culture and harvest, adaptability to climate and the characteristics of the oil yielded.

In 1975 commercial oil seeds yielded 30 million tons of vegetable oil (Langstraat, 1976), with nearly 90% of the total world production of these seeds being provided by five main crops: soybean, groundnut, cotton, rape and sunflower (Roberts, 1976). As well as for the valuable oil they provide, many oil-bearing seeds have now become equally, or more, important as a source of vegetable proteins. Oil seeds, nuts and pulses now provide 13% of the world's total supply of proteins, being of equal importance to meat and poultry. The products derived from commercial oil-bearing seeds have wide application, the edible oils and protein forming the basis of many foodstuffs and animal feeds, whilst the non-edible oils have important industrial applications, including their use as raw material in paint and varnish manufacture.

Published works of reference on the microstructure of oil-bearing seeds are sparse. Two of the more notable published books are the early work of Winton and Winton (1932) and the later publication of Vaughan (1970); the latter is a detailed and comprehensive reference work, including the morphology, histology and utilization of most, if not all, of the oil-bearing seeds. Rather than reiterate the contents of these books, this chapter will endeavour to review the more recent ultrastructural studies of oil-bearing seeds and, in particular, the structure and development of oil bodies and protein bodies, together with examples of the application of microscopy to seed processing and the products derived from oil seeds.

II. MICROSCOPY OF SEEDS

Published studies on the ultrastructural aspects of oil seed structure and their products are relatively few, being devoted only to selected species. This probably reflects the commercial importance of the seeds studied.

Whereas the preparation of oil-bearing seeds for optical microscopy appears to be amenable to the standard methods of fixation and embedding, they present many problems in the context of transmission electron microscopy. In a comprehensive study on the preparation of seed material for electron microscopy (Mollenhauer and Totten, 1971), it was stated that 'seed tissues are unusually difficult to process for electron microscopy', a situation which appears to be made doubly difficult in the case of oil-bearing seeds because of the oil present. To a certain extent, some of these problems have been alleviated by the introduction of new low-viscosity resins (Spurr, 1969).

As indicated by Mollenhauer and Totten, no single fixation method is suitable for every type of seed tissue, and the optimum method of fixation, staining and embedding must be formulated for a particular tissue and for the information required from that tissue. Fixation of tissue can often give rise to misleading conclusions concerning the ultrastructure of a particular organelle, as shown in an ultrastructural study on protein bodies (Lott and

Vollmer, 1973). Many other published papers also draw attention to this fact.

Probably the use of two methods of fixation (or an alternative method of preparation) would help in the interpretation of ultrastructure. One such alternative technique to the conventional preparative methods for transmission electron microscopy is the freeze-etch technique (Steere, 1957; Moor *et al.*, 1961). It has previously been suggested (Mollenhauer and Totten, 1971) that the freeze-etch technique may well be suited to the ultrastructural study of seeds but, judging from the paucity of published results using this technique on oil-bearing seed tissues, it would seem that only a few workers have heeded this suggestion or that the technique is as difficult as the conventional methods.

Scanning electron microscopy now offers a further alternative method and has proved very valuable in the study of oil-bearing seeds and, in particular, the products derived from them (Wolf, 1970; Wolf and Baker, 1972, 1973; Stanley *et al.*, 1976). This instrument offers many advantages including ease of specimen preparation, high resolution, the facility for examination of large areas of the specimen and a three-dimensional image which allows the spatial arrangement of structures to be studied.

III. SEED STRUCTURE AND DEVELOPMENT

During the development of oil-bearing seeds two major intracellular food reserves are laid down, which when the seed germinates provide nourishment for the developing seedling. In most species of oil-bearing seeds the oil and protein are found in the cells of the embryo, e.g. rapes and cotton, whilst in a few species they are found only in the cells of the endosperm, e.g. oil palm and coconut, or in both the embryo and endosperm, e.g. flax.

At the early stages of seed development the cells of the embryo or endosperm appear as typical meristematic plant cells (Fig. 1) with all the usual associated organelles, but as the seed develops oil and protein are laid down simultaneously until the cells become almost entirely filled with oil bodies (Fig. 2). Although the basic microstructure of mature oil seed cells has been known for over a century, it is only much more recently that some insight has been gained into the structural transformation of meristematic cells into the mature cell structure of oil-bearing seeds.

IV. OIL BODIES AND SPHEROSOMES

The structure and development of oil bodies in oil-bearing seeds has been the subject of many of the published papers, but certain aspects of their

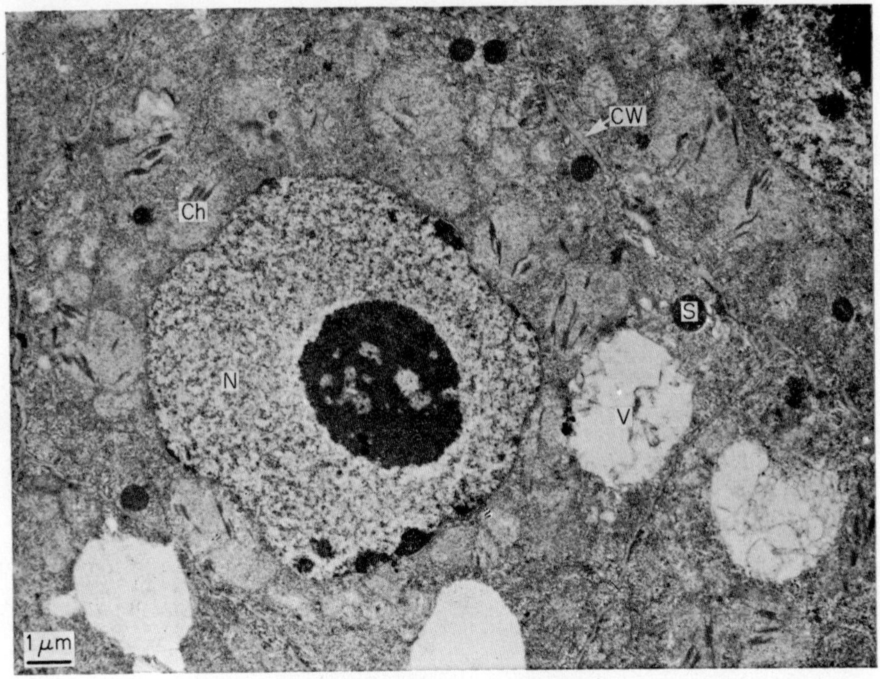

FIG. 1 (*Caption on opposite page*)

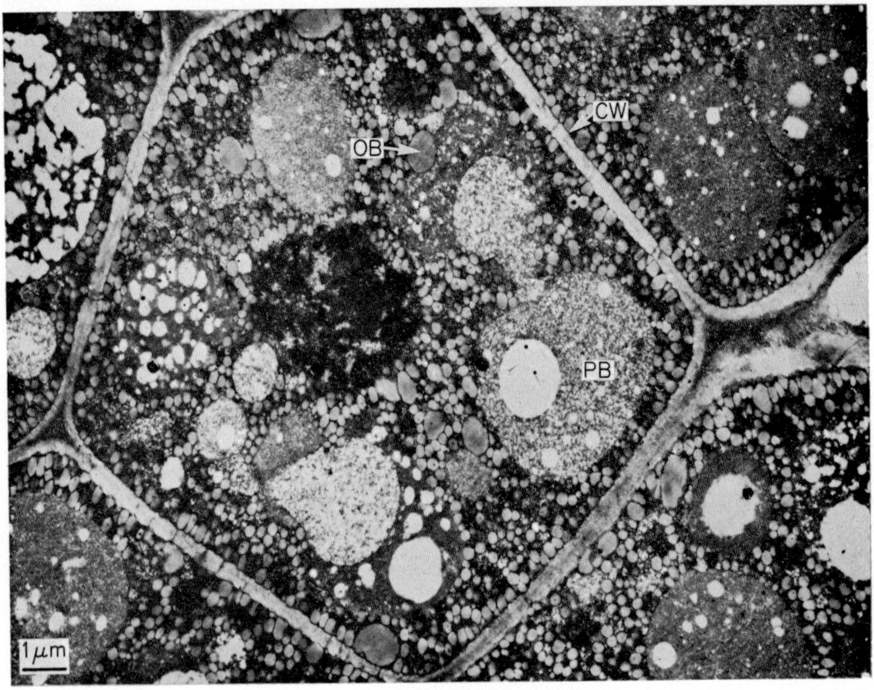

FIG. 2 (*Caption on opposite page*)

origin and development remain uncertain. The subject has been further complicated by the apparent confusion concerning the spherosome and oil body development and the interchange of terminology used to describe them. Much of this confusion has probably arisen because of the similarity in the ultrastructural appearance of the two bodies. Various authors have presented evidence in support of claims that the spherosome is the precursor of the oil body, is synonymous with the oil body or is a separate entity with entirely different functions. Given this background, it is impossible to describe the structure and development of oil bodies without involving a discussion of the spherosome.

A. The Spherosome

Spherosomes were first described by Hanstein in 1880 as being small (0·5–1·0 μm) highly refractile bodies in the cytoplasm of plant cells. They appear as brightly shining granules when viewed under dark field illumination, and as black granules exhibiting Brownian movement when viewed under phase contrast. Hanstein originally referred to spherosomes as microsomes, but this name was relinquished after Claude (1943) used the same term to describe small osmophilic particles found in liver homogenates. Perner (1953) eventually adopted the name spherosome, a term originally used by Dangeard (1921), to describe the structure first described by Hanstein.

Because the spherosome had been shown to stain in a similar manner to oil bodies with Sudan III, Sudan black and Rhodamine and Nile blue, and also gives a positive Nadi reaction, Perner (1952, 1953) concluded that the spherosome was not simply an ergastic particle but was rather an enzyme-active organelle capable of oxidation–reduction reactions. Later, Drawert (1953) was able to disprove these results when he demonstrated that the oxidation–reduction reaction was taking place in the mitochondria, and that the end product of the Nadi reaction, being lipophilic, accumulated in the spherosomes. This led to the conclusion (Drawert and Mix, 1962) that

---

FIG. 1. Embryo and cotyledonary cell of *Crambe abyssinica* 5 d after petal fall. The cells illustrate many of the ultrastructural features associated with meristematic plant cells, including, a large nucleus (N), thin cell walls (CW), vacuoles (V) and spherosomes (S). Chloroplasts (Ch) are also present since the cotyledons of *C. abyssinica* are the first photosynthetic organs.

FIG. 2. Mature cotyledonary cell of *Crambe abyssinica* showing the typical structure of mature cells of oil-bearing seeds. The cells are bounded by a thickened cell wall (CW), and are almost entirely filled with protein bodies (PB) and oil bodies (OB).

spherosomes are lipid bodies with no appreciable amount of protein, and are comparable to ergastic oil bodies.

Realizing the confusion surrounding the nature and function of spherosomes, Frey-Wyssling *et al.* (1963) studied various tissues, including *Brassica napus* and *Sinapis alba*, using the electron microscope, to firstly determine the origin of spherosomes in plant cells and secondly establish whether or not spherosomes are organelles with special structure and function, or simply inert bodies of ergastic material. Their results indicated that spherosomes are formed in a terminal vesicle of the endoplasmic reticulum, are bound by a unit membrane, and contain a granular matrix when potassium permanganate or osmium tetroxide is used as the fixative. In the developing cells of the oil-bearing seeds studied, they found that oil bodies pass through the same development as the spherosomes, and therefore concluded that oil bodies must be formed from spherosomes. As the oil body develops the granulation of the spherosome gradually disappears from the centre outwards until a homogenous oil body is formed. Because of the staining by heavy metals, the authors argued that the granular material of the spherosome is not lipid, and that the striking refractivity of the spherosome cannot be caused by lipid. They tentatively suggested that spherosomes may contain lipoprotein and concluded that spherosomes are the precursors of oil bodies, both having a unit membrane.

Ultrastructural work on the developing endosperm of *Ricinus communis* by Schwarzenbach (1971) again indicated that spherosomes are formed from the endoplasmic reticulum and that they are the precursors of oil bodies. However, Schwarzenbach changed the terminology, referring to spherosomes as prospherosomes and oil bodies as spherosomes. This author further suggested that during seed maturation the prospherosome (spherosome) unit membrane separates into an inner and outer layer, with lipid synthesis, it was claimed, taking place between the two layers of the membrane. As the spherosome (oil body) develops, the inner half of the unit membrane becomes invisible, leaving the outer half of the unit membrane to form the limiting membrane of the spherosome (oil body).

In a series of papers, Walek-Czernecka (1962, 1965), Sorokin and Sorokin (1966), and Sorokin (1967), employing histochemical and selective staining methods at the light microscope level, shed further light on the structure and function of the spherosomes, and their association, if any, with the oil bodies. Walek-Czernecka was able to demonstrate the presence of several hydrolytic enzymes, acid phosphatase and non-specific esterases in the spherosomes of onion scale epidermis cells, and concluded that spherosomes are involved in intracellular hydrolysis, and not neutral lipid synthesis. Examining tissues of various oil-bearing seeds (*Helianthus annuus*, *Brassica napus* and *Cocos nucifera*), Sorokin (1967) found that spherosomes do not stain with the usual fat stains but do give a positive reaction with phospholipid stains. Again, the author concluded that spherosomes could not be the direct precursors of oil bodies. She further drew attention to the fact that whereas spherosomes are a common feature of most vegetative cells of higher plants, even in cells not producing oil, oil bodies are restricted to oil-producing cells. Additionally, she noted that spherosomes have a limited size range within a given species, whereas the size of oil bodies can vary greatly within a single species.

Yatsu, *et al.* (1971) isolated spherosomes from *Brassica capitata* and *Allium cepa* and compared them with the oil bodies of *Gossypium hirsutum*. Although the authors claim that spherosomes and oil bodies are morphologically, ultrastructurally and chemically very similar and should be renamed 'oleosomes', the results presented indicate a large percentage difference in the non-lipid residues and protein content of the cabbage spherosomes, as compared with the cotton seed oil bodies. Acid phosphatase activity was not found in either the spherosomes or oil bodies, but, as pointed out by the authors, this negative result is inconclusive since enzyme activity appears to be dependent upon physiological status. Similar results had been noted previously (Sorokin and Sorokin, 1968), when it was found that acid phosphatase activity fluctuates with the changes in lighting conditions to which the plant is subjected.

In an ultrastructural study on the developing embryo of *Crambe abyssinica*, Smith (1974) observed that spherosomes were present in the cells of the youngest stages of embryo development he examined (Fig. 1), i.e. before the onset of oil body formation when the cells of the embryo resembled typical meristematic plant cells. The author concurred with the findings of Frey-Wyssling *et al.* (1963) and Schwarzenbach (1971), since he found that spherosomes develop from terminal vesicles of the endoplasmic reticulum (Fig. 3); however, he also noted that spherosomes are not the precursors of oil bodies. Acid phosphatase activity was demonstrated in the spherosomes, which allowed them to be differentiated from oil bodies when the two occurred together in the same cell (Fig. 4). Using the acid phosphatase marker it was also possible to demonstrate one function, if not the only function, of spherosomes; during the development of the embryo it was found that some spherosomes became associated with the vacuole in which the aleurone grain develops and, on association with the vacuole (Fig. 5), the spherosome itself appeared to vacuolate, forming a thin layer of spherosomal material which was closely associated with the vacuolar membrane. Earlier, Poux (1963, 1965), observing the association of acid phosphatase activity with the aleurone grains of *Triticum vulgare* Vill., suggested that the

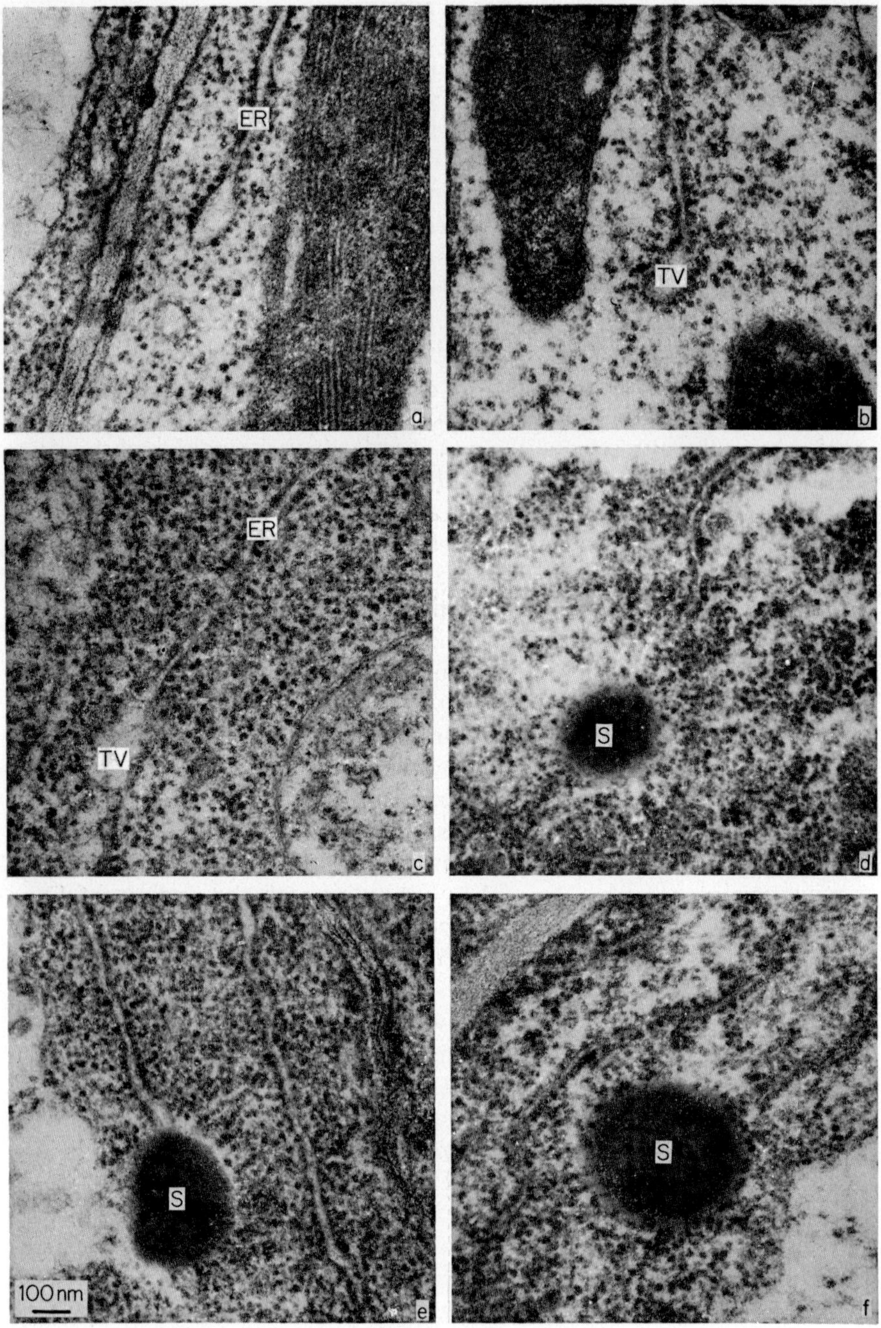

FIGS 3a–f. *Crambe abyssinica*. Development of the spherosome (S) from the terminal vesicle (TV) of the endoplasmic reticulum (ER).

aleurone grain may acquire acid phosphatase activity by incorporation of the spherosome. Yatsu and Jacks (1968) confirmed the results of Poux and were able to demonstrate the presence of alcohol dehydrogenase and acid proteinase, as well as acid phosphatase, in the aleurone grains of ungerminated seeds of *Gossypium hirsutum*.

Because of the association of hydrolytic enzymes with the spherosome, many authors have drawn an analogy between spherosomes and the lysosomes of animal cells. Berjak (1972) proposed that it may be more appropriate

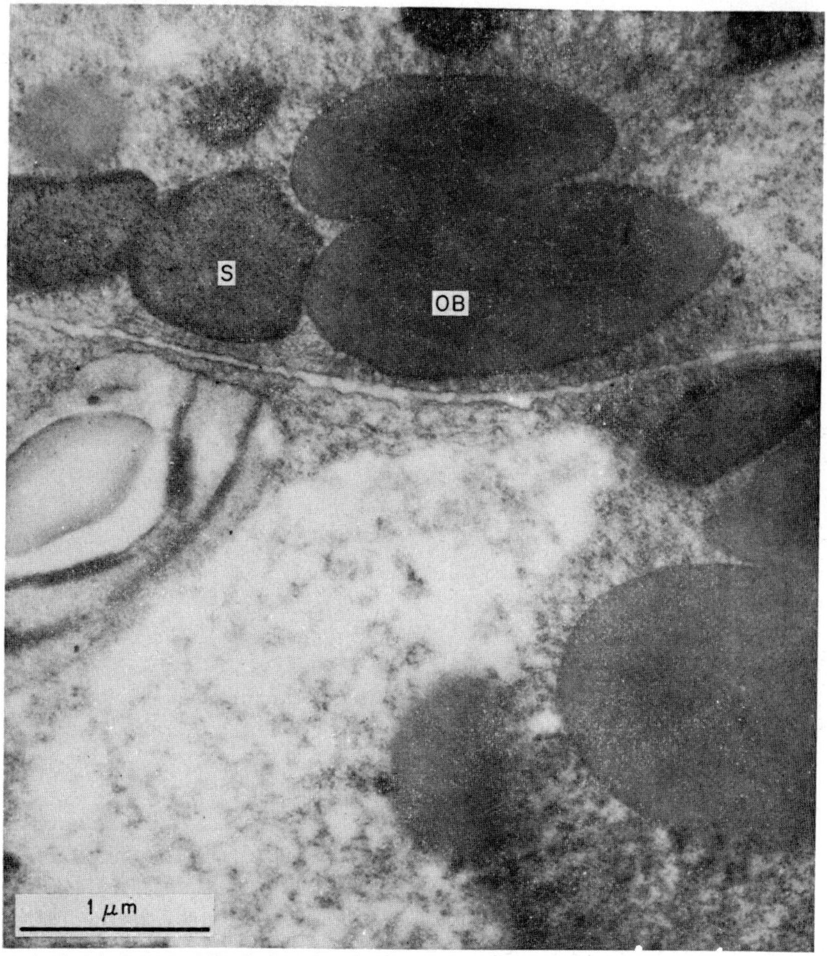

FIG. 4. Cotyledonary cell of *Crambe abyssinica* containing spherosomes (S) showing acid phosphatase reaction product, together with oil bodies (OB) with no acid phosphatase reaction product.

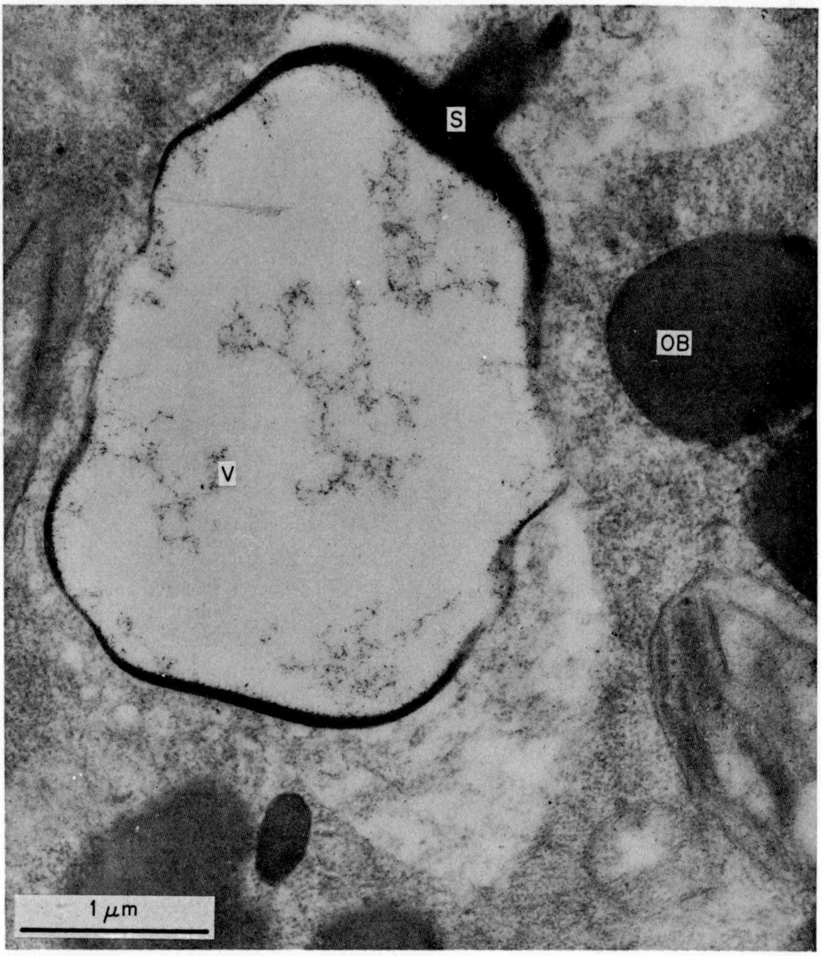

FIG. 5. *Crambe abyssinica.* Spherosomes (S), with acid phosphatase reaction product, becoming associated with a vacuole (V). Note that adjacent oil bodies (OB) have no reaction product.

to refer to a lysosomal system in the plant cell, rather than to define a particular subcellular component as a lysosome, since hydrolytic enzymes are found at sites other than strictly definable lysosomes, e.g. vacuoles. In the context of a lysosomal system, Matile and co-workers had previously described spherosomes (Matile and Spichinger, 1968), vacuoles (Matile and Moor, 1968) and aleurone vacuoles (Matile, 1968; Yatsu and Jacks, 1968) as constituents of the lysosomal system.

B. Oil Bodies

More recent ultrastructural studies on the development of oil-bearing seeds (e.g. on soybean (Düvel *et al.*, 1969), *Ricinus communis* (Harwood *et al.*, 1971) and *Crambe abyssinica* (Smith, 1974)) have demonstrated, unlike previous findings, that oil bodies are separate entities and develop independently of the spherosomes. Düvel *et al.*, studying the development of oil bodies in the cotyledons of soybean, found that the oil bodies develop in the cytoplasm without connection to the endoplasmic reticulum or spherosomes. The author described how the oil-body membranes are formed by an arrangement of subunits consisting of two types of ribosomes and suggested that one type of ribosome contains the enzymes required for neutral lipid synthesis, whilst the second type has no enzymatic activity and forms the basic structure of the oil-body membrane. The substrate for neutral lipid synthesis passes through the membrane structure containing the enzymes, and the end product, neutral lipid, is formed within the membrane. As the oil body increases in volume, additional subunits become associated with the membrane. A similar mechanism of oil-body formation was found in the developing cotyledons of *Crambe abyssinica* (Smith, 1974). The oil bodies were found to arise in the cytoplasm from small, particulate electron-dense masses which, in some examples, were found to be closely associated with the endoplasmic reticulum (Fig. 6). The oil body itself forms in the centre of the particulate mass, which gradually decreases as the oil body develops. This arrangement of a central developing oil body surrounded by a particulate mass suggested to the author that the particulate mass is the site of enzymes required for oil synthesis. The nature of the particulate mass was not identified but it may represent specialized ribosomal material. Also, the close proximity of the endoplasmic reticulum to the particulate mass may indicate that the endoplasmic reticulum is involved in oil-body formation. Association of the endoplasmic reticulum with oil-body formation in *Brassica napus*, *Brassica campestris* and *Crambe abyssinica* has been noted by Ström (1975). She drew attention to the fact that when potassium permanganate is used as a fixative the oil body develops in vesicles connected to the endoplasmic reticulum, whereas when osmium tetroxide is used the developing oil body appears in vesicles unconnected to the endoplasmic reticulum. Unfortunately, no experimental data or micrographs are available to support these results, which remain unclear since it is known that potassium permanganate fixation does not preserve the ribosomes of the endoplasmic reticulum or the oil of oil bodies. This therefore makes interpretation of ultrastructure very difficult.

The work by Harwood *et al.* (1971) on the ultrastructural development of

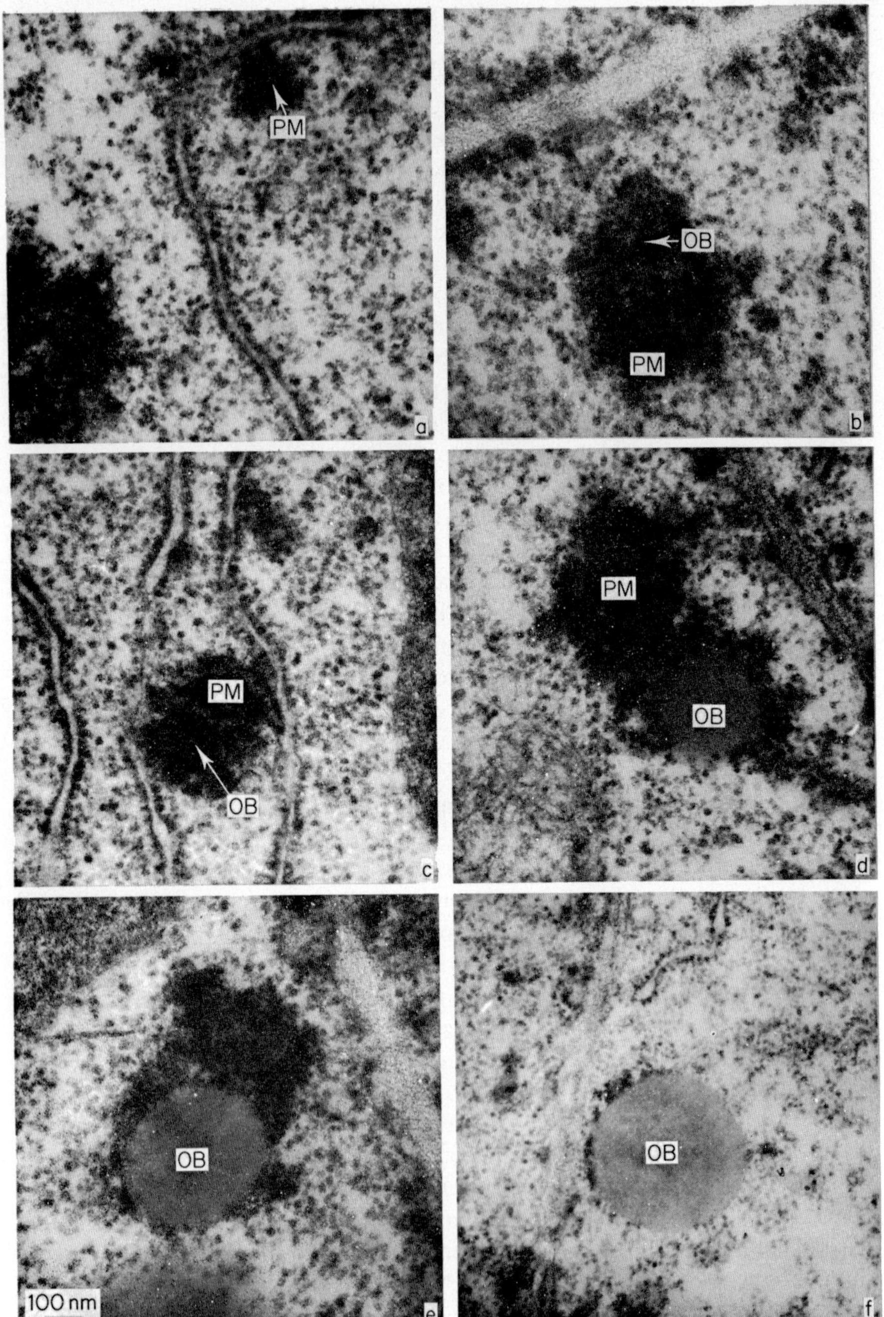

FIGS 6a–f. Oil-body formation in *Crambe abyssinica*. Oil bodies (OB) develop from an electron-dense particulate mass (PM) in the cytoplasm. Note the close association of the endoplasmic reticulum with the particulate mass in some examples.

oil bodies in *Ricinus communis* indicates a different method of oil-body formation from that described by Schwarzenbach (1971), although both workers used the same species. Harwood *et al.* found that oil bodies, which they state 'cannot be equated with spherosomes', develop around a cluster of enzymes. Electron micrographs of both isolated and *in vivo* oil bodies showed small vacuole-like inclusions, containing small particulates, the site of enzymes. The proposed method of oil-body formation requires that the substrate must pass through the end product to reach the enzymes. This arrangement appears at first sight to be rather unsatisfactory, but the authors argue that since the oil bodies are in a state of constant flux and agitation the inclusions (containing enzymes) may contact the peripheral interface of the droplet, therefore allowing both substrate and enzymes to become available due to their proximity to the oil-body interface.

The presence or absence of a limiting membrane surrounding the developing oil body still remains to be established beyond doubt, despite the many studies made on the subject. Investigations which indicate that oil bodies originate from spherosomes consequently imply that the oil bodies retain the unit membrane of the spherosome derived originally from the endoplasmic reticulum. The method of spherosome formation and oil-body formation proposed by Schwarzenbach (1971) differs slightly from this, since he suggests that the oil body is surrounded by a half unit membrane. Düvel *et al.* indicated that the oil bodies of soybean are membrane bound, but that the membrane differs in structure from that of the spherosome membrane. It was found that the oil-body membrane was narrower and less electron dense than the usual tripartite membrane, and that there appeared to be an interlocking of the membranes between adjacent oil bodies. A similar, unique, half unit membrane has also been proposed by Yatsu and Jacks (1968) for the spherosome (oil body). Again, it was noted that when adjacent oil bodies touched the two half unit membrane structures (surrounding each oil body) formed the normal tripartite membrane structure. Morphologically, the authors interpreted the unique membrane structure as having polar surfaces adjacent to the cytoplasm, and lipoidal non-polar surfaces in contact with the storage lipid.

Harwood *et al.* (1971) and Smith (1974) were unable to demonstrate a membrane although they noted a sharply defined interface. Likewise, Rest and Vaughan (1972) could not actually resolve a membrane but suggested that a membrane is present, probably in the form of a monomolecular layer of phospholipids (Fig. 7). Circumstantial evidence suggests that a limiting structure of some nature must be present surrounding the oil body, since oil bodies do not normally coalesce even when subjected to pressures within the cell; this is often seen in electron micrographs when the normally circular or ovoid bodies become squared or rectangular in cross-section. Also, when potassium permanganate is used as the fixative, and the oil of the oil body is lost, a 'honeycomb' pattern of material remains indicating the outline of the oil body.

There has been little reported work on the use of the freeze-etch technique to investigate the ultrastructure of oil bodies in oil-bearing seeds. In one study, Buttrose and Swift (1973) examined the oil bodies of various seeds

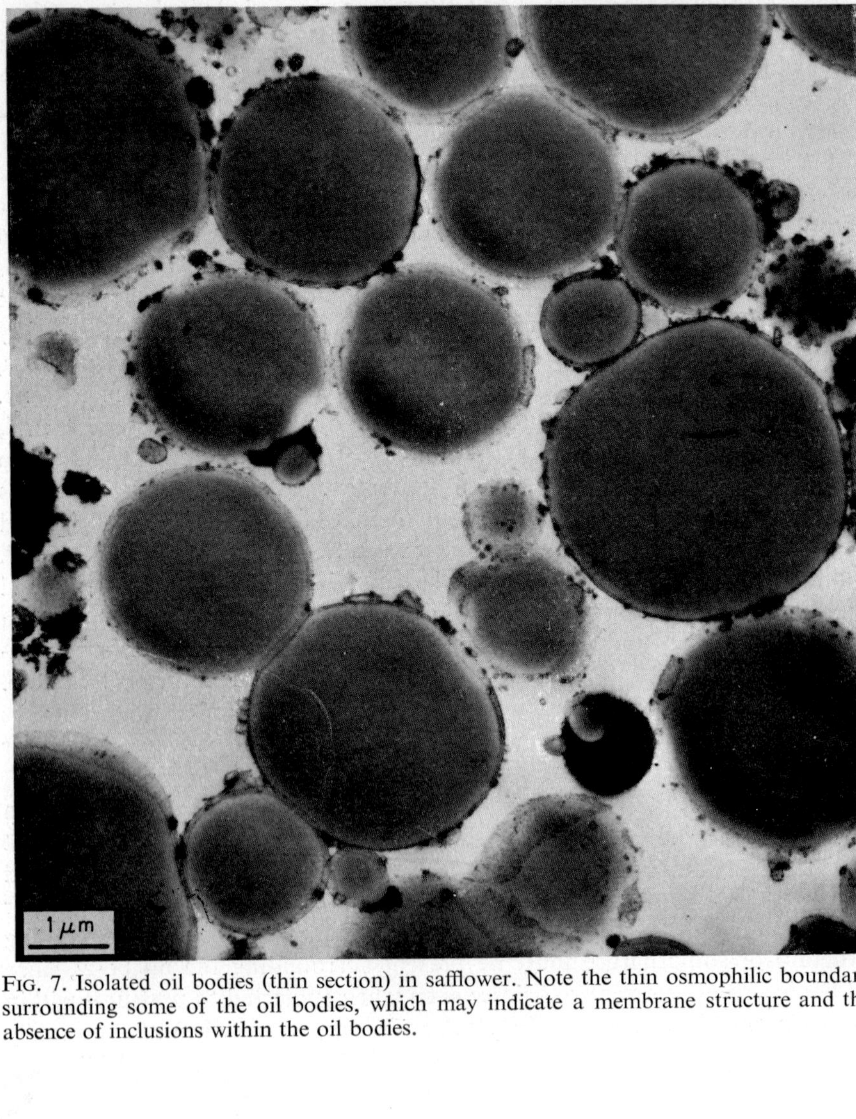

FIG. 7. Isolated oil bodies (thin section) in safflower. Note the thin osmophilic boundary surrounding some of the oil bodies, which may indicate a membrane structure and the absence of inclusions within the oil bodies.

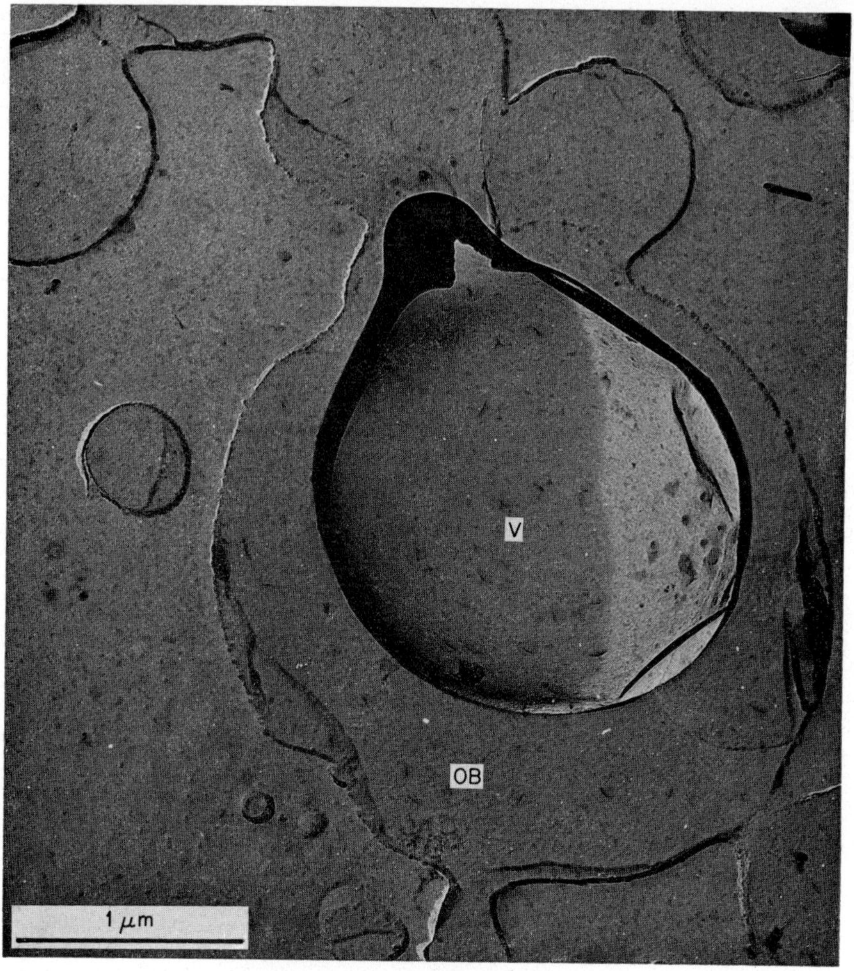

FIG. 8. Isolated oil bodies (freeze-etched) in safflower, illustrating the central inclusion or vacuole (V), found with the oil bodies (OB) after freeze-etching.

and found that each oil body contained a depression (always concave) which, it was thought, was caused by sublimation of material from the specimen. However, comparative studies with other organisms and model systems indicated that the depressions represented gaseous inclusions. Similar results were obtained for the isolated oil bodies of safflower (J. M. Stubbs, unpublished results) (Fig. 8). In both cases no explanation or functional role could be accredited to the gaseous inclusion, although Buttrose and Swift did note that the inclusions appeared during the final stages of seed maturation and disappeared during the first hours of germination.

Clearly, there are certain aspects of oil-body ultrastructure and development which remain unresolved, and authors will probably continue to 'beg

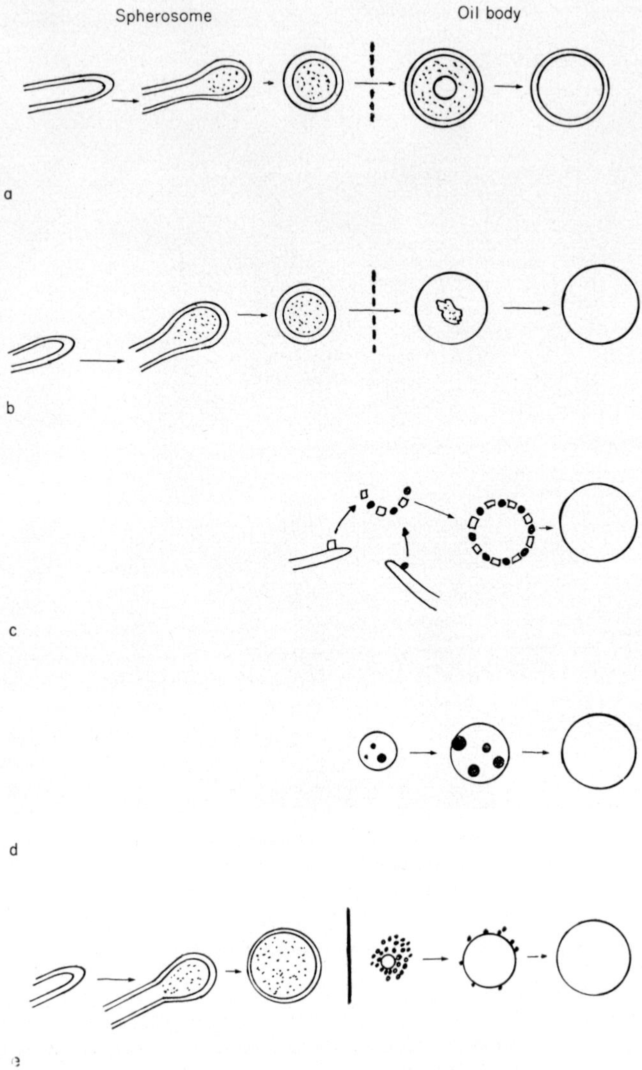

FIG. 9. Schematic diagram of spherosome and oil body formation, and their interrelationship, if any, according to various authors. (a) Frey-Wyssling *et al.* (1963); (b) Schwarzenbach (1971); (c) Düvel *et al.* (1969); (d) Harwood *et al.* (1971); (e) Smith (1974).

to differ' (as reflected in the interchange of terminology which still persists) regarding the interrelationship, if any, of the spherosome and the oil body. The more recent ultrastructural studies clearly favour the concept that spherosomes and oil bodies are separate entities, the spherosome, lysosomal-like in function, being a common feature of most cells of higher plants, whilst the oil body is restricted to oil-producing cells and is simply a site of oil storage. A diagrammatic representation of the theories concerning the formation of spherosomes and oil bodies is shown in Fig. 9.

V. PROTEIN BODIES

A. Introduction

Protein bodies, which are considered to be ergastic particles, are found in the embryos and endosperm of seeds (Esau, 1960) and are a prominent structural feature of mature seeds of many species (Pfeffer, 1872; Rost, 1972), including oil-bearing seeds. Protein bodies are, in fact, the most obvious subcellular structure of mature oil seeds (with notable exceptions).

As early as 1855, Hartig successfully isolated 'proteinaceous granules' from various oil seeds by a non-aqueous technique, and demonstrated that they contain protein. Later, Hartig (1856) used the term 'aleurone grain' to describe the protein bodies because the isolated material resembled flour in appearance. The protein grains, which are now generally recognized as the site of storage of the major seed proteins, are bound by a unit membrane of size range 0·5–2·0 μm, and contain a matrix of protein. The protein matrix may or may not contain crystalline inclusion 'bodies'.

The most common of these inclusions is the globoid body, first observed by Pfeffer (1872), which is not membrane bound although the shape of an interface line is often visible (Yatsu, 1965). Protein bodies with globoid inclusion bodies have been described for many oil-bearing seeds, including: *Ricinus communis* (Sobolev, 1966; Prokofev *et al.*, 1967); *Gossypium hirsutum* (Yatsu, 1965; Englemann, 1966; Lui and Altschul, 1967); *Cucurbita maxima* (Lott *et al.*, 1971); and sunflower (Agaev and Agaev, 1976). No globoid bodies were found in the protein bodies of: *Glycine max* (Bils and Howell, 1963; Tombs, 1967); *Arachis hypogaea* (Altschul *et al.*, 1964); and *Zea mays* (Khoo and Wolf, 1970). Dieckert *et al.* (1962) reported that they were able to demonstrate protein bodies both with and without globoids in *Arachis hypogaea*, whilst in an ultrastructural study of the same species Bagley *et al.*

(1963) failed to differentiate two types of protein body. Further examples of oil-bearing seeds containing two types of protein bodies (i.e. with and without globoids) have been demonstrated by ultrastructural studies in *Sinapis alba* (Rest and Vaughan, 1972), *Crambe abyssinica* (Smith, 1974) (Fig. 10), and *Brassica napus* and *Brassica campestris* (Hofsten 1970, 1974;

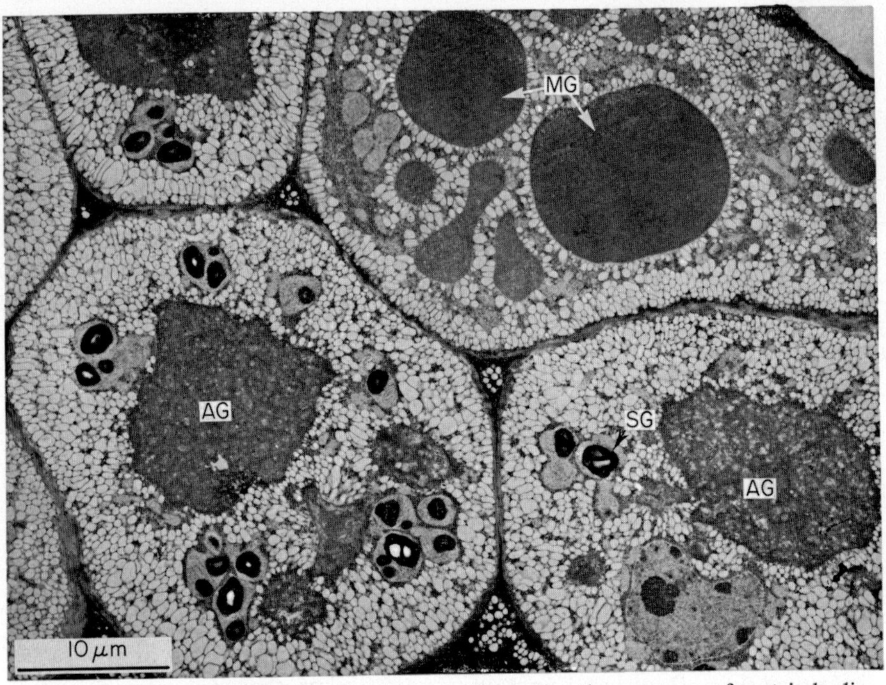

FIG. 10. Cotyledonary cell of *Crambe abyssinica* illustrating the two types of protein bodies, aleurone grains (AG) and myrosin grains (MG). Note the absence of plastids and starch grains (SG) in cells containing myrosin grains.

Bengtsson *et al.*, 1972; Ström, 1975). In order to differentiate between the two types of protein body in the context of microscopy, Rest and Vaughan suggested that the term aleurone grain should be used to describe protein bodies with globoid inclusions, whilst the protein bodies without inclusions should be termed myrosin grains. Cells containing myrosin were first noted by Heinricher (1884) because of the greater intensity of staining obtained with Millon's reagent compared to that of cells containing aleurone grains. This type of cell was later equated with the myrosin tubules found in cruciferous plants (Guignard, 1890), and they thus became known as myrosin cells.

Many protein bodies are also known to contain other crystalline inclusions, one notable example being the protein bodies of hemp seed which contain distinct crystalloid-type inclusions (St Angelo *et al.*, 1968). Isolation and purification of these crystals showed them to be essentially pure edestin, the major storage protein of hemp seed.

B. Development of Protein Bodies

Although the presence of intracellular protein-containing bodies, in many species of seeds, has been recognized for some time, their origin remained somewhat speculative until recently. Some early workers indicated a plastid origin for these protein bodies (Mottier, 1921); others (Rendle, 1888; Wieler, 1943) considered their origin to lie elsewhere in the cytoplasm, whilst Guilliermond (1921a, b; 1941) suggested a vacuolar origin.

Whilst many authors have described the ultrastructure of the mature protein body in oil-bearing seeds (Yatsu, 1965; Tombs, 1967; Mitsuda and Kishida, 1967; Hofsten, 1974), Englemann (1966) was probably the first author to study their ultrastructural development. Through investigation of three stages of the seed development of *Gossypium hirsutum*, it was demonstrated that protein bodies are formed in membrane-bound vesicles, which themselves are derived from the rough endoplasmic reticulum. At the earliest stages of seed development studied, Englemann noted that the ultrastructure of the cells closely resembled that of meristematic plant cells, i.e. they contained all the usual organelles but no vacuoles or protein bodies. As the seed matures, numerous vacuoles develop in the cytoplasm and these may remain as discrete vacuoles, as in *Gossypium hirsutum*, or coalesce to form fewer vacuoles which themselves subsequently divide, e.g. *Sinapis alba* (Rest and Vaughan, 1972), *Crambe abyssinica* (Smith, 1974) and *Ricinus communis* (Dangeard, 1921), each subdivision becoming filled with protein.

The role of rough endoplasmic reticulum in protein production is now virtually undisputed (Birbeck and Mercer, 1961), and many authors have noted that with the onset of protein body formation there is a rapid proliferation of rough endoplasmic reticulum around the vacuoles. However, the method by which the protein is transported to, or produced and accumulated in, the protein body remains unexplained. Englemann (1966) observed that when potassium permanganate was used as the fixative there was a striking resemblance between the staining of the contents of the rough endoplasmic reticulum cisternae and the vacuoles still attached. A tentative suggestion made as a result of this observation was that protein may be transported directly from the rough endoplasmic reticulum (site of production) to the vacuole. Unfortunately, because ribosomes are not preserved by permanganate fixation, the problem as to whether protein is synthesized on the membrane leading to the vacuole, or whether synthesis takes place elsewhere in the cell, remained unanswered.

Rest and Vaughan (1972) were also unable to determine the site of protein

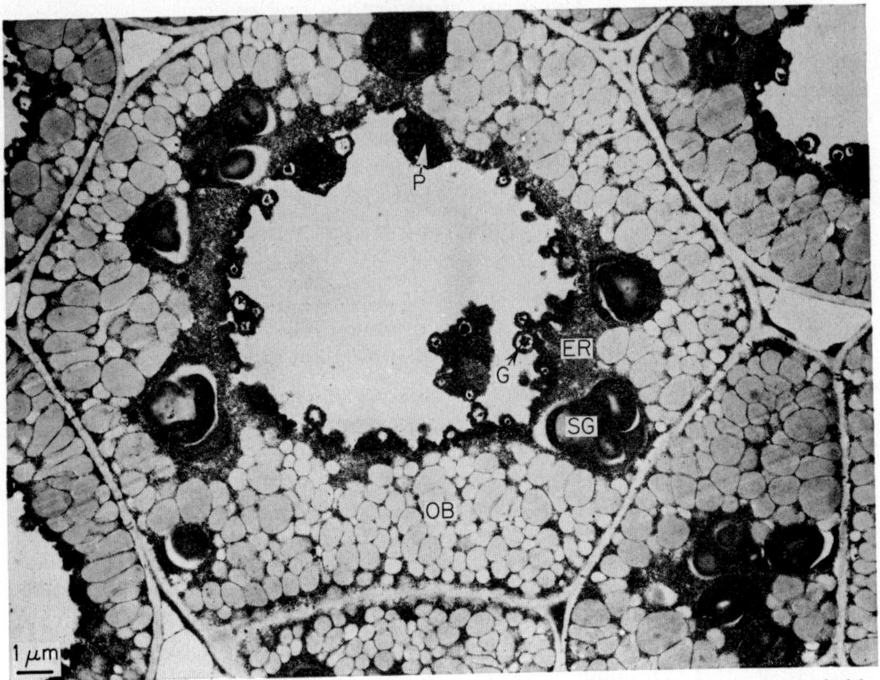

FIG. 11. Developing aleurone grain in *Crambe abyssinica*. Note the vacuole surrounded by endoplasmic reticulum (ER), and protein (P) formation within the vacuole. Globoid bodies (G) are present in the protein bodies.

production in cells of *Sinapis alba* containing aleurone grains. Although they noted a proliferation of the rough endoplasmic reticulum, reserve protein was first observed within the tonoplast as lumps of electron-dense material (Fig. 11), whilst the surrounding rough endoplasmic reticulum only stained lightly, probably indicating a change in the nature of the protein within the vacuole. In cells of *Sinapis alba* containing myrosin grains, Rest and Vaughan, like Englemann, noted a similarity in the staining of the cisternae and vacuolar contents of the rough endoplasmic reticulum. This was taken as evidence that the protein of the myrosin-type body accumulates in the rough endoplasmic reticulum before being transferred to the vacuole. The mode of transport of the protein from the cisternae to the vacuole remained unknown, however, since no connection was found between the two.

A series of recent papers by Dieckert (1969, 1971) and Dieckert and Dieckert (1972, 1975, 1976) has made a major contribution to the understanding of how proteins, and subsequently protein bodies, are formed in seeds. Dieckert drew attention to the fact that in animal cells protein bodies are also known to form in vacuole-like structures (Dieckert, 1971), the formation of zymogen granules by acinar cells of the pancreas being a particularly well-understood example (Palade, 1959, 1975). From ultrastructural studies, Dieckert and Dieckert (1972, 1976) proposed that a similar method to that found in animal cells is present in developing seeds, including oil-bearing seeds, for the development of protein bodies. This hypothesis suggests that proteins are synthesized by ribosomes on the cytoplasmic side of the rough endoplasmic reticulum. The protein is then transported by way of cavities to the Golgi apparatus (dictyosomes), where it is concentrated and possibly modified. Following this, membrane-bound protein droplets are formed at the periphery of the Golgi bodies, eventually becoming 'pinched off' and forming protein bodies in the cytoplasm. These 'free' protein droplets then fuse with the vacuole in which the protein body subsequently develops. For this hypothesis to be validated certain cytomorphological features must be present in cells producing protein bodies, namely, a highly developed rough endoplasmic reticulum, elements and vacuoles of the dictyosomes containing protein, and vacuoles limited by a single membrane containing protein.

Certain of these cytomorphological features associated with the onset of protein body development had previously been noted by other authors. An ultrastructural study of the development of protein bodies in peanut (Dieckert and Dieckert, 1972) supported many of the facts proposed in their own hypothesis. They found, in common with earlier observations, that in the early stages of peanut seeds the expected apparatus for protein production was missing and no protein deposits were present. However, in the later stages of seed development, when protein bodies were being formed, there was an abundance of rough endoplasmic reticulum, dictyosomes with protein droplets and 'free-floating', small, membrane-bound, protein-containing sacs, which fused with the large vacuoles already containing protein. In order to demonstrate beyond doubt that the electron-dense material observed in the various organelles and vesicles was protein, the authors subjected alternate sections of material prepared for transmission electron microscopy to a mixture of proteolytic enzymes (pronase). This treatment removed proteins from the sections and, from comparison with untreated sections, clearly showed the presence of protein in the various structures. By use of this technique the authors were also able to demonstrate similar patterns of aleurone grain development in cotton and coconut (Dieckert and Dieckert, 1975), where the aleurins form as crystals. This work also indicated that the classical aleurone grain is formed upon desiccation of the mature seed, and in seeds which do not become desiccated at maturity, e.g. the coconut, the proteins remain in a watery vacuole.

An interesting, but as yet unexplained, feature of the protein body is the effect of the type of fixative used (Figs 12 and 13) on the ultrastructural appearance of the body. This phenomenon was first noted in oil-bearing seeds by Englemann (1966), and subsequently by Lott *et al.* (1971), Rest and Vaughan (1972) and Smith (1974). When potassium permanganate is used as the fixative, a uniformity in the electron density of the protein body is noted, whereas glutaraldehyde fixation gives rise to 'lumping' of the protein in the aleurone grain but not in myrosin grains. There is therefore a correlation between 'lumping' and the presence of globoid bodies in the protein body, and this may reflect a difference in chemical composition.

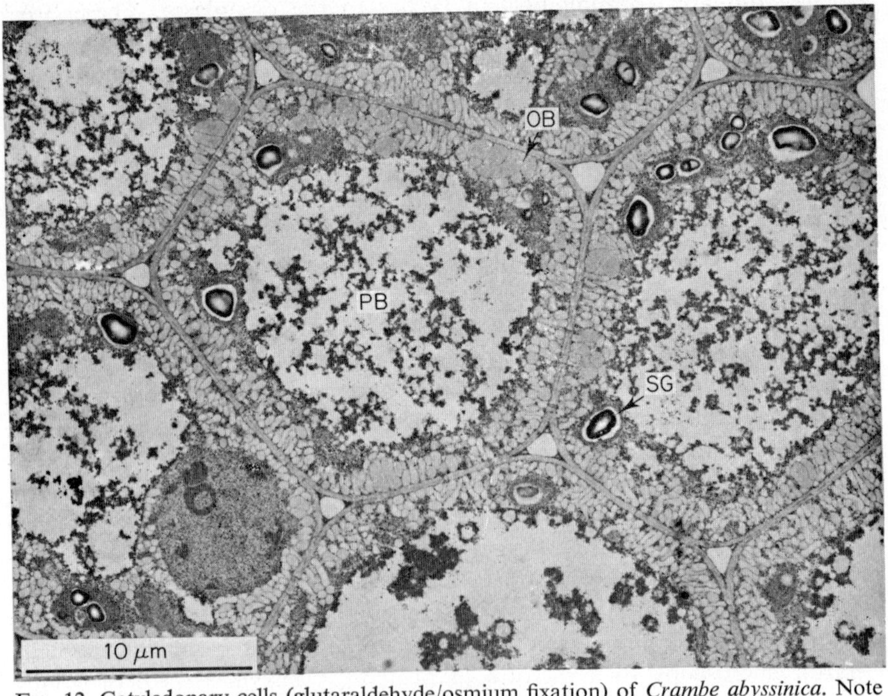

FIG. 12. Cotyledonary cells (glutaraldehyde/osmium fixation) of *Crambe abyssinica.* Note the granular nature of the protein bodies (PB) and the appearance of the starch grains (SG) and oil bodies (OB) (cf. Fig. 13).

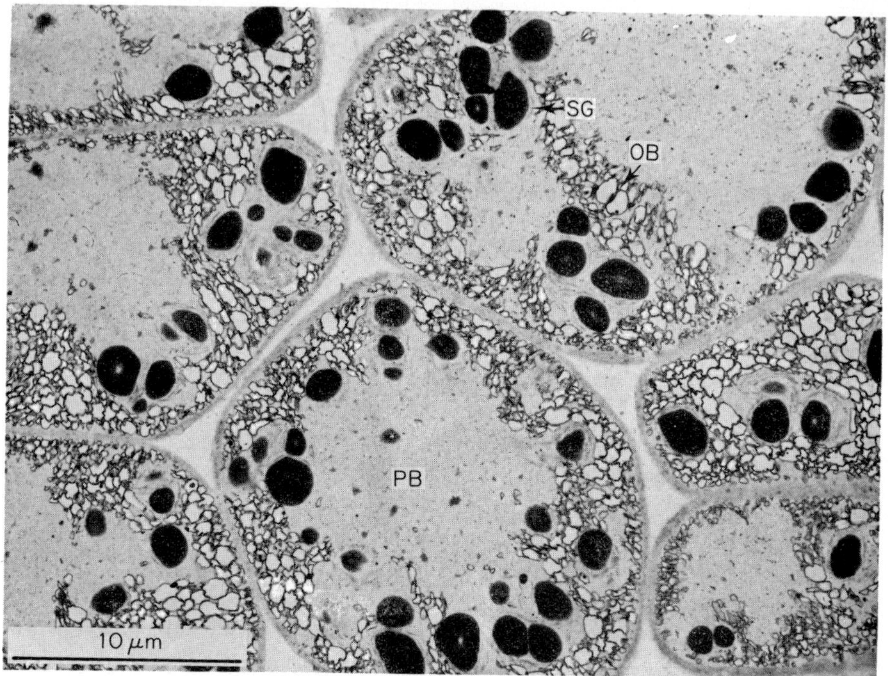

FIG. 13. Cotyledonary cell (potassium permanganate fixation) of *Crambe abyssinica*. The protein bodies (PB) have an amorphous appearance; the oil bodies (OB) have lost their contents and have an irregular shape. The starch grains (SG) are electron dense (cf. Fig. 12).

C. The Globoid Body and Enzymes

As indicated previously, the globoid inclusion body is the most commonly occurring inclusion found within the protein body or vacuole. Despite the fact that globoid bodies have been known for over a century (Pfeffer, 1872), there is virtually no information available regarding the development of globoids and their function.

Rest and Vaughan (1972) found no evidence from their study of *Sinapis alba* to suggest the mechanism of globoid body formation. Globoid bodies first appeared in the aleurone grains as spherical bodies within clear angular areas of the protein matrix (Fig. 12). As the seed matured, the globoids themselves became angular and increased in number. Additionally, there was no evidence to suggest that globoids are formed before the protein (as was indicated by Frey-Wyssling and Mühlethaler (1965)), but rather Rest and Vaughan found that the two are laid down simultaneously.

In a comprehensive ultrastructural study of protein bodies in *Cucurbita maxima* (Lott *et al.*, 1971), evidence was presented indicating that the globoid body is composed of two parts, the central globoid crystal (containing phytin) and the surrounding soft globoid of unknown composition. This structure of the globoid was demonstrated using fixation methods (glutaraldehyde/osmium) and freeze-etching (Lott and Vollmer, 1973).

More evidence is available concerning the composition of the globoid. Pfeffer was able to demonstrate the presence of magnesium and phosphorous compounds in the globoid body. More recently, Dieckert *et al.* (1962) and Sharma and Dieckert (1974) confirmed and extended the earlier results when they demonstrated that globoids from peanuts contain phytic acid and a large fraction of the cellular potassium, magnesium, manganese and copper. Similar results were obtained by Lui and Altschul (1967) for cotton and by Lott (1975) for *Cucurbita maxima*, the latter using the relatively new technique of micro X-ray analysis. Using X-ray analysis, the elements sulphur, magnesium and calcium have also been detected in the globoid bodies of *Crambe abyssinica* (Hofsten, 1973). Studying globoids isolated from the aleurone grains of castor, Suvorov *et al.* (1970) found that, in addition to their phytin content, globoids from mature seeds also contain certain amounts of protein, whilst globoids from immature seeds also contain phospholipid and RNA localized at their surface. The work further indicates that the globoid is membrane bound. In aleurone grains where globoids are not found, e.g. soybean (Tombs, 1967), it was suggested that the phytate is combined with the protein rather than being compartmentalized.

Together with the protein matrix, globoid inclusion bodies and various crystalline inclusions, enzymes have also been found in aleurone grains (see Ory (1972) for a detailed review). Yatsu and Jacks (1968) were able to demonstrate the presence of acid phosphatase and acid proteinase in the aleurone grains of ungerminated seeds of *Gossypium hirsutum*. Isolation and analysis of these subcellular organelles indicated that they contain 77% of the acid phosphatase activity and 100% of the acid proteinase activity of the cell. Smith (1974) (see Section IV), using ultrastructural cytochemical techniques, also demonstrated that the aleurone grains of *Crambe abyssinica* contain acid phosphatase which is acquired from the spherosome. The reason for this association of acid phosphatase activity with the aleurone grain is unclear, but may be explained if the vacuole, and consequently the aleurone grain which develops within it, are thought of as compartments of the cell lysosomal system, as proposed by Berjak (1972). In considering the aleurone vacuole as a lysosome, Matile (1968) attributes several functions to the aleurone vacuole, one of which is its transformation, on germination of the seed, to a lysosome in which protein reserves are mobilized within the vacuolar membrane. These mobilized reserves are then released into the cellular cytoplasm as and when required during germination of the seed.

In his summary on enzymes in protein bodies, Ory (1972) suggests that the enzymes present have the following roles when seed germination takes place; acid phosphatases are involved in the release of the inorganic phosphate from phytic acid (which is required for the phosphorylation of sugars), whereas acid proteases release amino acids from the reserve proteins for new protein synthesis in the developing seedling.

D. Protein Body Degradation

Several studies have been carried out on the ultrastructural changes which take place within the protein bodies of oil-bearing seeds upon germination of the seed. Results indicate that the ordered sequence of events leading to the degradation of protein bodies is very similar ultrastructurally, although in reverse, to that of their formation. Examination of protein bodies in the cotyledons of germinating *Arachis hypogaea* seeds (Bagley *et al.*, 1963) showed that the first change to take place was a large increase in the size of the protein grain, caused by the imbibition of liquid. This, in turn, led to the formation of cavities within the protein body, which by 10–12 days after germination had become a mass of small fragments. By 18–24 days after germination it was found that most of the protein body fragments had disappeared, leaving only a few scattered fragments at the cell periphery. A similar sequence of events has been described for the degradation of aleurone grains in the germinating seeds of *Sinapis alba* (Werker and Vaughan, 1974), although the results differ from those of Bagley *et al.* with respect to the presence or absence of a membrane surrounding the protein body. Werker and Vaughan, in common with other authors, indicate that the aleurone grain is membrane bound and that protein body degradation takes place within the membrane. Following the breakdown of the aleurone grains, vacuoles are formed which subsequently coalesce to form one large vacuole. Bagley *et al.* noted that in *Arachis hypogaea* the protein bodies were continuous with the cytoplasm, and no evidence of a protein body membrane was found. The work of Werker and Vaughan also describes the degradation of myrosin grains in the cotyledons of *Sinapis alba*. After imbibition of liquid each myrosin grain appears as a separate vacuole filled with myrosin. The membrane surrounding each individual myrosin grain then breaks down, leading to the formation of one large myrosin-filled vacuole.

In order to study the degradation of protein bodies in the germinating seeds of *Cucurbita maxima*, Lott and Vollmer (1973) employed the freeze-etch technique, together with conventional fixation methods. Earlier work (Lott *et al.*, 1971) had shown that protein bodies from *Cucurbita maxima* are composed of four major components: the proteinase matrix, the protein crystal, the soft globoid and the globoid crystal. Upon germination of the seed it was found that the first structural change was the appearance of small areas within the protein matrix, similar in appearance to the soft globoid (site of phytic acid). These areas rapidly became distributed throughout the protein matrix, indicating that the protein matrix is aqueous. After two days of germination, it was noted that breakdown of the globoid crystal had started, whilst after 3–4 days the protein body had become a largely aqueous vacuole containing a few irregular shaped pieces of protein material. As germination continued, the protein material disappeared completely, leaving a series of vacuoles which then fused to form one large vacuole.

VI. APPLICATION OF MICROSCOPY TO THE STUDY OF OIL SEED PROCESSING AND THE PRODUCTS DERIVED FROM OIL SEEDS

Optical microscope, and more recently electron microscope, studies, together with biochemical techniques, have provided much information regarding the structural development of oil-bearing seeds. A logical extension of this type of work would appear to be the application of this knowledge and the available microscopy techniques to oil seed processing and the structure of food materials derived from products of oil-bearing seeds. Unfortunately, judging from the paucity of published papers, only relatively few workers have as yet applied microscopy to these aspects of oil-bearing seeds, a fact also pointed out by Wolf (1972) with respect to the processing and products of soybean.

In the broadest terms, the processing of oil-bearing seeds involves dehulling (removal of the seed coat) followed by the application of mechanical pressures and then solvents to extract the oil from the broken cells of the seed. The residue, or cake, is then further processed to give other products. In practice, the extraction procedure and subsequent processing is obviously more complex than this because of the great variability of seed structure. Many of the different types of extraction processes have been described in detail, e.g. for palm fruit (Ramecourt, 1976), soybean and cotton seed (Hutchins, 1976), sunflower, groundnuts, palm kernel, rape seed and copra (Stein and Glaser, 1976), peanuts and cotton seed (Garcia, 1976); Ward (1976) has discussed extraction processes in general. One of the few published examples of the application of microscopy to the changes in oil-seed structure during processing (Stanley *et al.*, 1976) describes the effect of mechanical dehulling on the structure of rape seed (*Brassica campestris*).

As well as being an important source of vegetable oil, rape seeds have more recently become an important source of vegetable protein (Altschul, 1970). One problem associated with the flour prepared from rape seed is the presence of toxins which make it unsuitable for human consumption. This problem has now been overcome (Ohlson, 1972) and detoxified protein products can now be prepared, thus allowing a valuable new source of vegetable protein to be exploited. Stanley *et al.* (1976) also drew attention to a further problem associated with the utilization of rape seed, namely the relatively high fibre content of the defatted meal or flour, a consequence of the presence of hull material. The monitoring, by microscopy, of a process to mechanically dehull rape seed using pneumatic attrition, clearly indicated that loss of cellular material occurred during the operation. This cellular damage may promote myrosinase activity, thus leading to the formation of toxins and thereby apparently making the process unsuitable for the removal of hull material. Although the work only details one application of microscopy to seed processing, the authors conclude that much more research of a similar nature is required to equate seed ultrastructure to behaviour during processing.

Some preliminary work investigating the changes in the ultrastructure of rape seed (*Brassica campestris*) associated with seed processing has been undertaken (A. S. Rodrigues-Pereira, personal communication).

Examination of seed material after the first processing stage (heat conditioning in the presence of moisture) showed that the oil bodies had coalesced into larger bodies and that the protein bodies had become 'lumped' to form very electron-dense bodies. Also, transverse fractures of the cell wall were noted which had probably been caused by the previous flaking process. Passage of this material through the 'first expeller' caused further breakdown and compaction of the cell wall, but the oil still remained within the cell walls in the form of large 'oil pools'. A second heat conditioning of the material caused no further major change in ultrastructure, but a second expeller treatment resulted in the typical structure of pressed cake before extraction, with highly folded and broken cell walls containing pockets of oil. After extraction of the pressed cake, the results clearly showed the loss of oil, leaving the highly folded and contorted cell walls enclosing residual material which was probably derived from protein bodies and cytoplasm (Figs 14a–d).

The soybean is now well established as a commercially important source of vegetable oil and protein and, consequently, has received relatively more attention from research workers. A detailed scanning electron microscopy study of soybean (Wolf and Baker, 1972) indicated several external and internal structural features of importance in seed processing. Examination of the

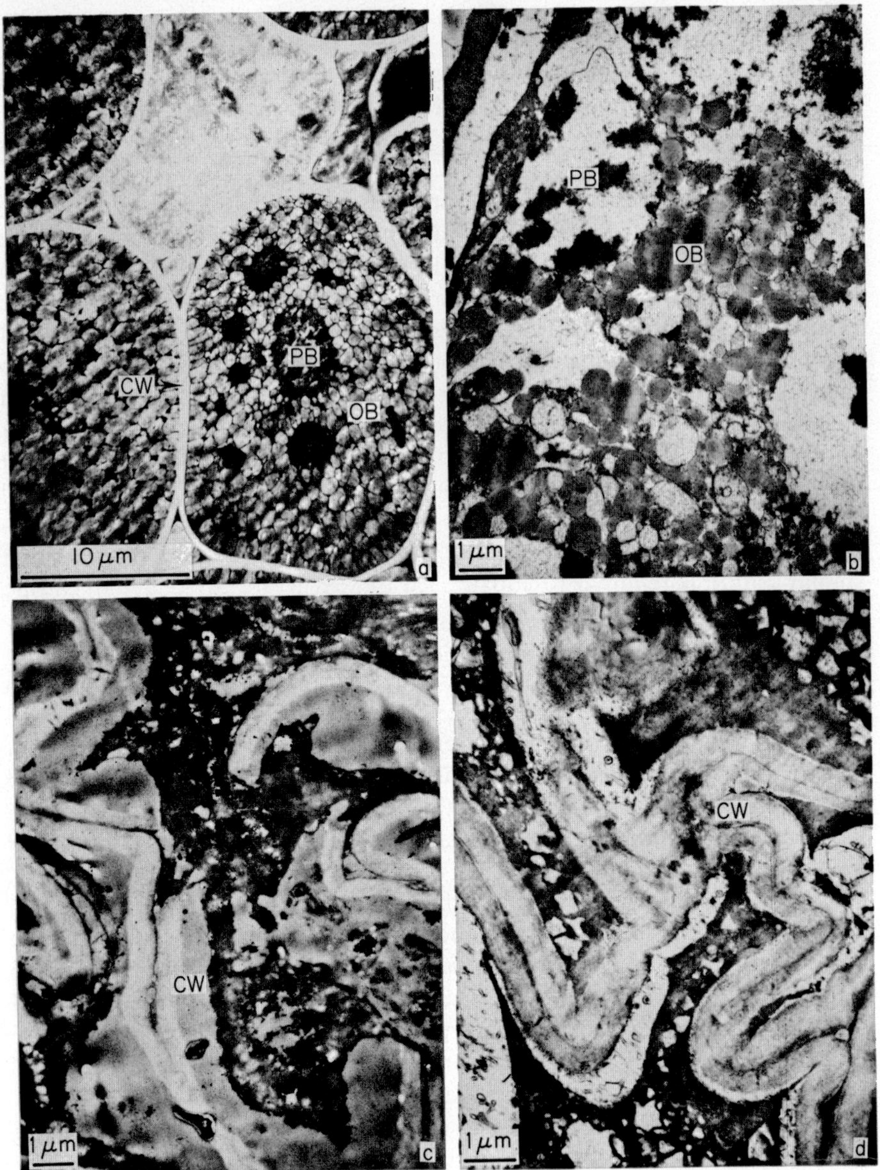

FIG. 14. *Brassica campestris.* (a) Untreated mature seeds, showing typical cell structure of oil-bearing seeds, with oil bodies (OB) and protein bodies (PB). (b) Sample taken from lower part of a stack cooker (steam treatment). The membranes of the oil bodies are broken and their contents, in most cases, have merged into larger droplets. (c) Sample taken at the exit of the expeller. Cell walls (CW) are broken and folded. The oil still contained within the 'press cake' forms one large pool. (d) Cellular fragment in desolventized meal, i.e. 'press cake', after extraction. Folded cell wall fragments (CW) are present together with residual material, probably derived from protein bodies and cytoplasm.

external surface of the seed coat revealed numerous pits or pore-like structures which appeared to end in elongated slot-like structures, features not previously observed by other methods of microscopy. Hofsten (1974) and Stanley *et al.* (1976) have more recently demonstrated similar elaborate structures in the seed coats of *Brassica napus* and *Brassica campestris*. Also, several of the samples examined revealed defective seed coats; these could be potential sites of attack by fungi and insects and allow the rapid absorption of water during seed storage (Figs 15a, b and 16a, b). Internally, the fracture faces of broken cotyledons, defatted with hexane, showed that the protein bodies are of two types: individual particles 5 μm in diameter or larger, and other particles, up to 5 μm in diameter, made up of smaller particles 1–2 μm in diameter. Evidence from this study and from the earlier work of Wolf and Baker (1972) indicated that when the two types of protein body are subjected to isolation techniques (Tombs, 1967), the larger bodies are unstable, giving rise to amorphous material, whilst the smaller protein bodies remain stable. In a later but more comprehensive study (Wolf and Baker, 1973), the scanning electron microscope was again used to examine the structure of the soybean and commercially produced soy flour, protein concentrate and protein isolate. Whereas in the earlier study Wolf and Baker (1972) reported the structure of soybean cotyledons as observed on mechanical fracture surfaces, in this later work a freezing method to produce fracture surfaces for examination was employed. Better preservation of cotyledon ultrastructure was obtained using the freezing method, thus allowing more detailed observations to be made of cellular structure (Figs 15c and 16c). The results clearly revealed protein bodies covered by oil bodies and a network of cytoplasm, as was the inner surface of the cell wall. Treatment of the fractured surface with hexane (in order to remove the oil bodies) permitted closer examination of the protein bodies; a crater-like surface, probably caused by the residue of cytoplasm, and depressions left by the oil bodies were evident. Examination of fractured, undefatted cotyledon surfaces, after treatment with water, showed that the entire cell contents are removed. The ease with which the cell contents are washed out (following cell wall damage) had previously not been known, although, as the authors indicate, this fact may have been unknowingly exploited in the traditional oriental preparation of soymilk. Results obtained from the examination of two soy products, full-fat flour and defatted flour, indicated that some of the structural features of soybean cotyledon cells survive the processing. Full-fat soy flour was found to contain both forms of protein body (Figs 18a, b). Three types of defatted soy flour (subjected to various degrees of heating during preparation) were examined in this study, and it was found that the structural appearance of the protein bodies in the flour was directly related

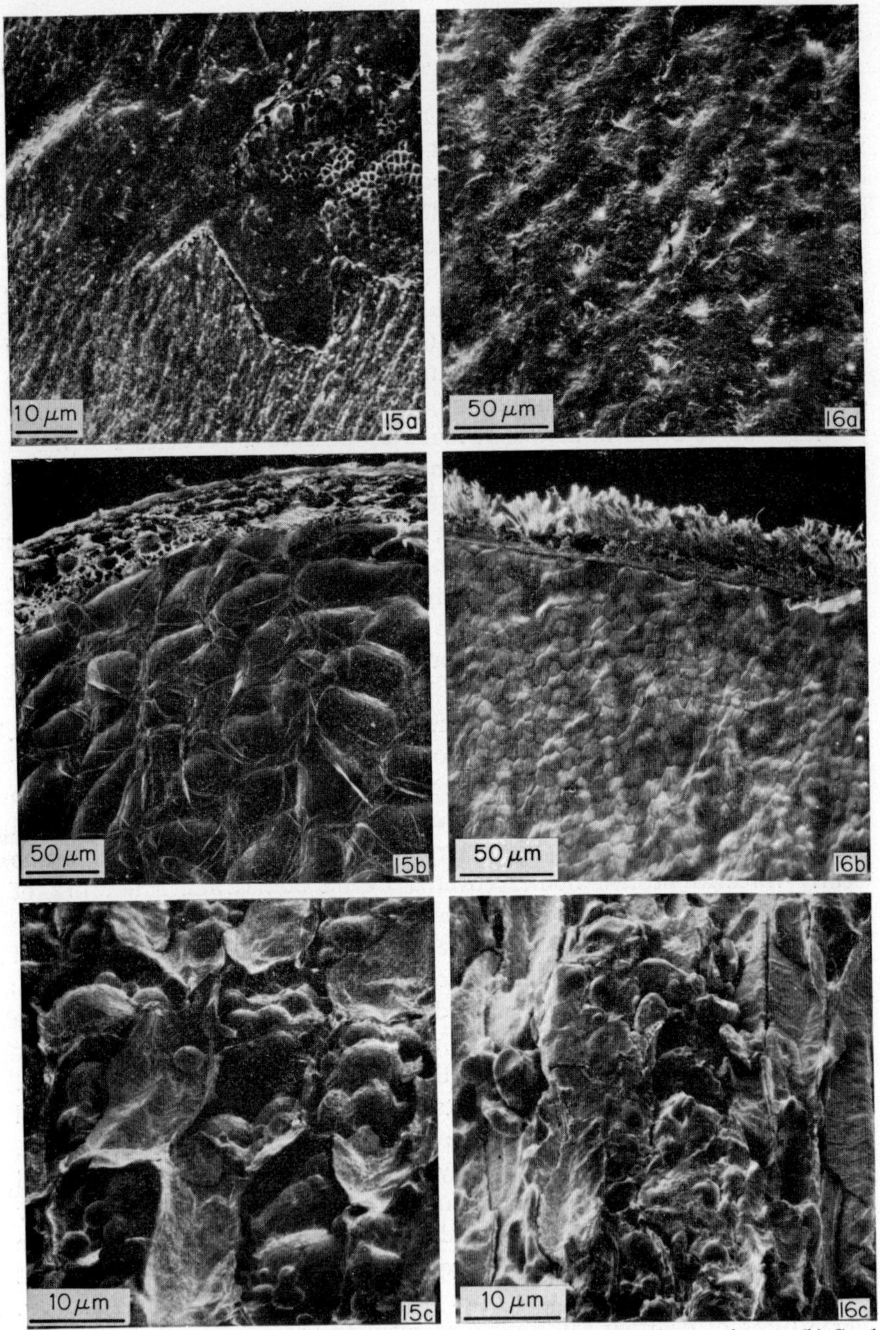

FIG. 15. *Crambe abyssinica.* (a) Seed testa outer surface showing damaged area. (b) Seed testa, inner surface and cross-section. (c) Fracture surface, defatted cotyledon, showing cellular structure and protein bodies.

FIG. 16. Soybean. (a) Seed testa outer surface. (b) Seed testa, inner surface and cross-section. (c) Fracture surface, defatted cotyledon. Note cellular structure and protein bodies.

Figures 15 and 16 are original SEM photographs by C. G. Smith. The situation in *Crambe* is similar to that in *Brassica* spp.

FIGS 17a, b. Soy protein isolate, spray-dried, composed of differing sizes, often collapsed.
FIGS 18a, b. Soy, defatted flour illustrating broken cellular structure and protein bodies.
FIGS 19a, b. Soy protein concentrate, spray-dried, illustrating the typical structure of spray-dried material, i.e. spheres of differing sizes.

to the degree of heating the flour had undergone, i.e. the greater the degree of heating, the greater the structural disruption of the protein body.

In contrast to the results obtained for the two types of flour, the observed structure of the protein concentrate (Fig 19a, b) and protein isolate (Figs 17a, b) showed that the new structural elements are formed during the processing, with none of the cotyledonary cell structural features being retained. The protein concentrate was found to be made up of partially collapsed spheres, plus other contorted particles covered by a continuous film of material. Also, the protein isolate (depending upon the method of preparation) was found to contain spheres 2–40 μm in diameter, some partially collapsed, and with both rough and smooth surfaces. Examples of other soy products are shown in Figs 20a–c, 21a, b and 22.

In a review on the ultrastructures of cotton seed and peanut and their relationship to processing (Vix *et al.*, 1972), examples are given of four potentially important products from oil-bearing seeds, made possible because of a clear understanding of seed microstructure. Three of the products described, cotton seed flour, hexane-extracted cotton seed flour and hexane-extracted peanut flour, demonstrate the application of optical microscopy, whereas the fourth example, partially defatted peanuts (Vix *et al.*, 1967), illustrates the use of transmission electron microscopy to elucidate what is being accomplished during processing. Electron micrographs of peanut cells before and after treatment (i.e. pressing followed by expansion in hot water) clearly indicated that the cell is compressed, oil is lost (it is thought through the plasmodesmata) and the aleurone grains are compressed. After treatment with hot water, it was shown that the cell swells, the cell wall increases in thickness, starch grains swell and the aleurone grains remain coalesced. The authors cautiously suggest that the starch grains may have gelled and this, together with the coalescence of the protein body, may contribute to the crunchiness and the crispness of the product.

The transmission electron microscope has also been employed to investigate changes which occur during the processing of soybeans into traditional Japanese food (Saio and Watanabe, 1968). It was found that overnight soaking of soybeans in water caused no change in the seed ultrastructure, whilst steaming of the beans caused rupturing of the protein bodies, thus releasing the oil. Incubation of the steamed beans with *Bacillus natto* to produce ‘natto’, partially destroyed the cell walls and reduced some of the oil bodies to a meshwork. Further examples of other foods prepared from soybean are described by Saio and Watanabe, indicating how the application of microscopy can lead to a better understanding of the process.

The production of foodstuffs, and particularly meat analogues, based upon textured vegetable protein has now become an important aspect of food

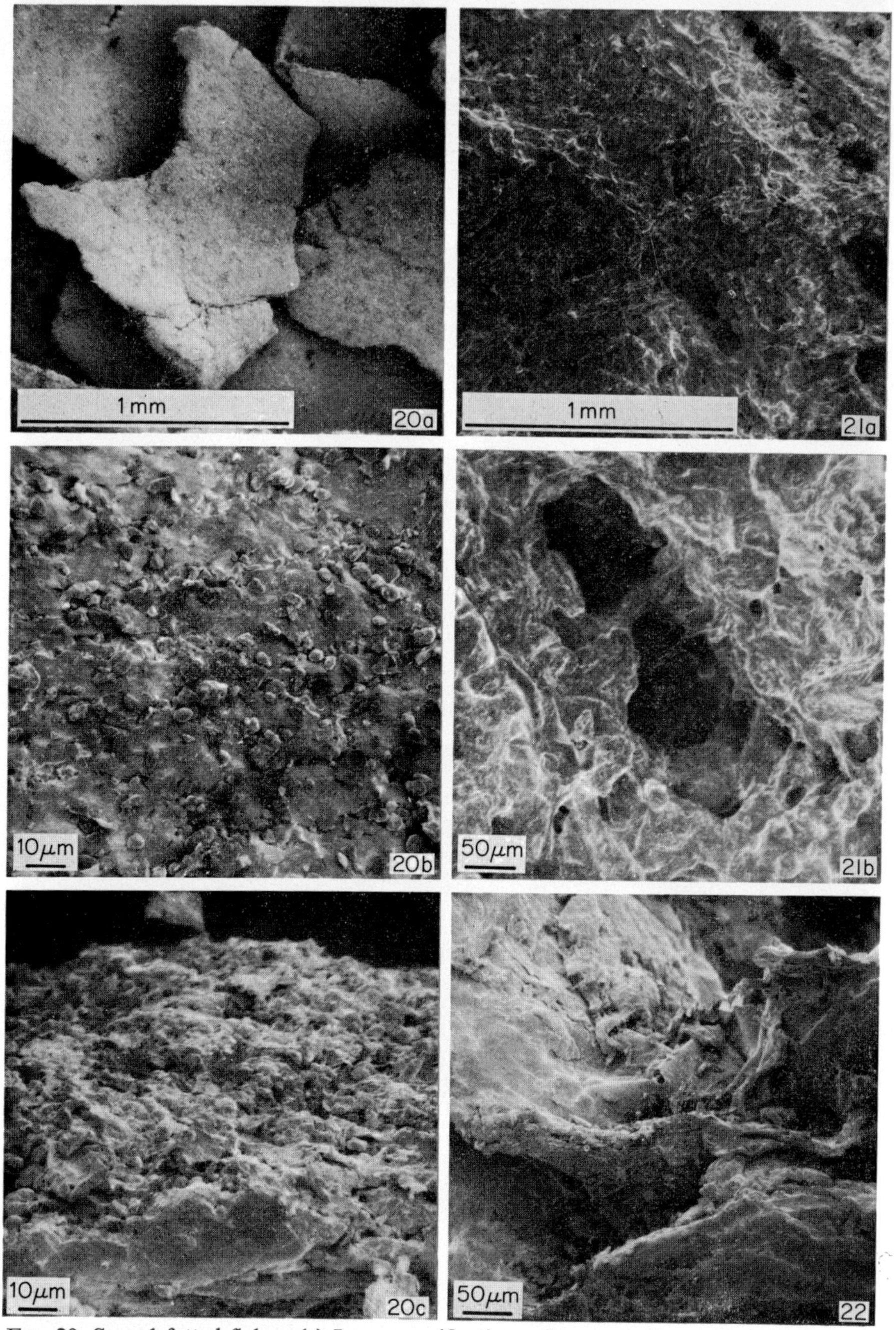

FIG. 20. Soy, defatted flakes. (a) Low magnification micrograph illustrating the size and shape of the flakes; (b) surface of a defatted flake showing the outline of cells and protein bodies; (c) cross-section of a flake.

FIGS 21a, b. Soy, textured concentrate.

FIG. 22. Soy, textured flour.

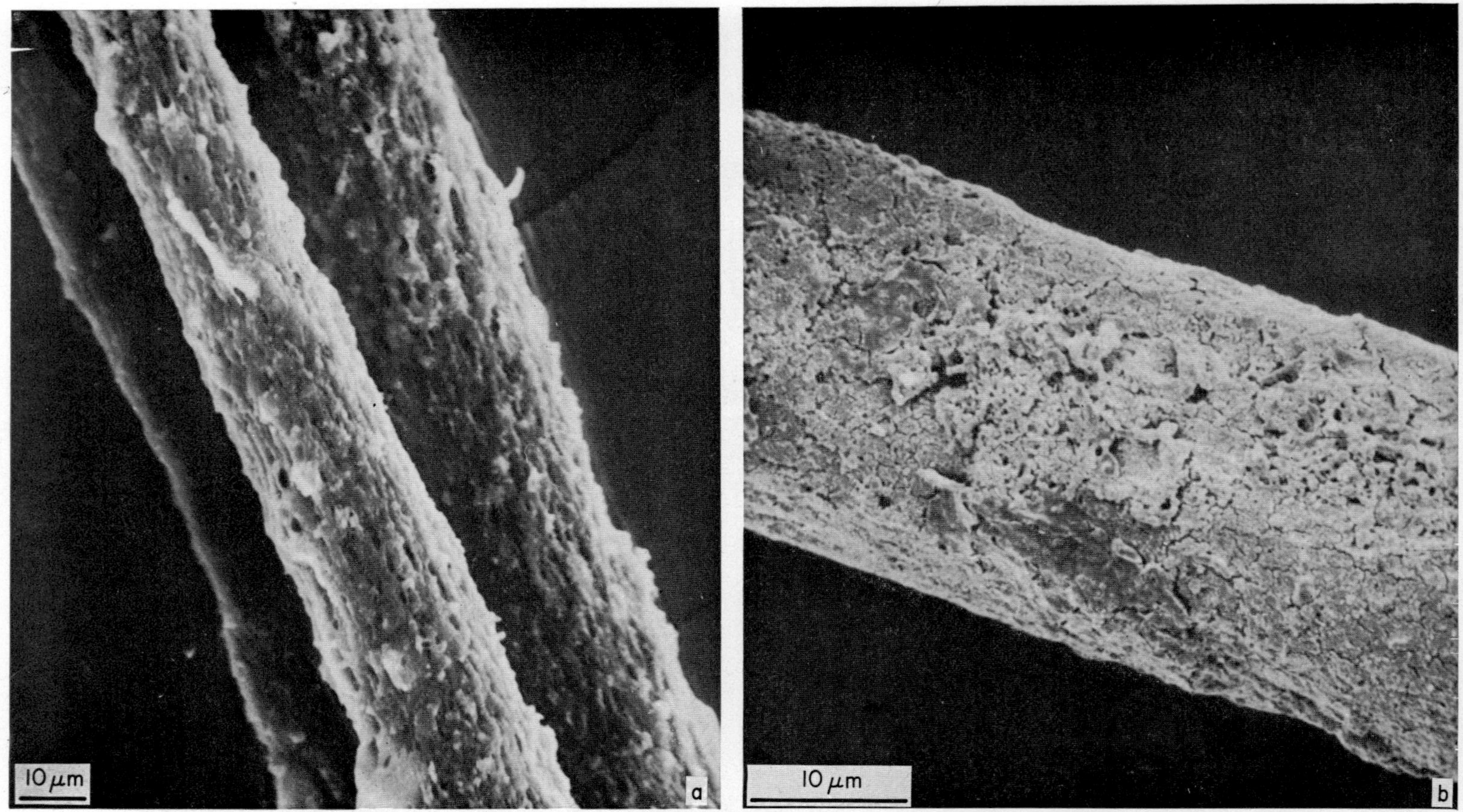

FIGS 23a, b. Scanning electron micrographs of extruded soy protein fibres, illustrating differences in surface texture of the fibres.

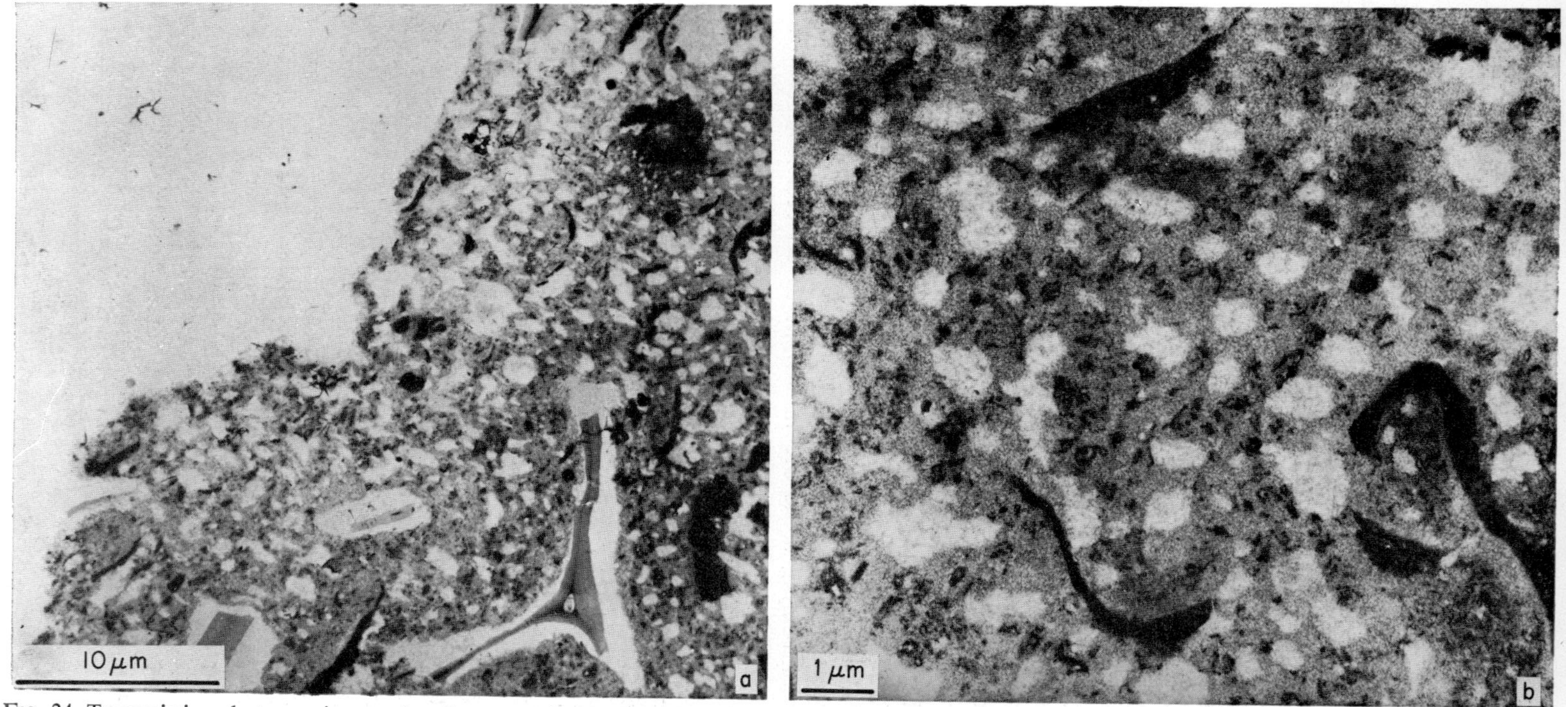

FIG. 24. Transmission electron micrographs of soy protein fibres. (a) Illustrating surface profile. (b) Section illustrating structure at the core of the fibre; note the fragments of cell wall material.

technology. These products are based upon defatted flours, grits or protein concentrates derived predominantly from soybeans which have been subjected to various physical, chemical and mechanical processes to obtain the desired end product. During such processes, structural changes take place (which can be monitored by microscopy) in the raw material, giving rise to a structurally different end product. One such study (Aguilera *et al.*, 1976) followed the ultrastructural changes occurring as soybean grit is subjected to thermoplastic extrusion or extrusion cooking under predetermined conditions. The results indicated that although the structure of starting material was morphologically similar to intact cotyledons, this structure was completely destroyed by extrusion. By detailed examination of the material at various stages of extrusion (using scanning and transmission electron microscopy) the authors were able to determine when and where the various structure changes take place, thus giving some insight into the mechanics of the process. Similar work has been reported for an extruded soy product (Cummings *et al.*, 1972) and rehydrated spun soybean fibres (Stanley *et al.*, 1972). Examples of extruded fibres are illustrated in Figs 23a, b and 24a, b.

A slightly different but excellent example of the application of microscopy to the processing of seeds has been reported by Wood and Vaughan (1976a, b). Many oil-bearing seeds are important as an economic crop for the production of dessert nuts. Preparation of nuts for such use involves the removal of the pericarp and testa to give a clean kernel surface, i.e. the testa must be completely removed. One problem which arises during the processing of cashew nuts, whether by hand or mechanically, is that whereas most of the testa can be removed easily from the cotyledon surface, there are certain areas where the testa cannot be easily separated. This makes the product commercially unacceptable. Wood and Vaughan (1976a) studied this problem using transmission electron microscopy in order to determine the ultrastructure of unprocessed seeds, processed seeds and, in particular, the interface of the cotyledon and the testa. Their results showed that a cuticle is present on the surface of the cotyledons, the ultrastructure of which, whilst retaining its basic structure, differed in response to a wide range of temperature and humidity. Processing was found to drastically affect the ultrastructure of the testa, cuticle, epidermis and subepidermis of the cotyledon. Further work (Wood and Vaughan, 1976b) revealed that ultrastructural variations occur at the testa–cotyledon interface of the nut; these are reflected in the difficulty of removing the entire testa from the cotyledon. Their results also indicated that where a cuticle is present on the surface of the cotyledon, the testa can be removed easily. If, on the other hand, the cuticle has failed to develop in certain areas because of abnormalities in seed development, the testa cannot be easily separated from the cotyledon.

Although very limited, the examples quoted above illustrate the wide and varied application of microscopy to the area of oil-bearing seed processing and the products derived from these seeds.

ACKNOWLEDGEMENTS

The author wishes to thank Dr Rodrigues-Pereira for the use of unpublished results and Figs 14a–d, the editors of *Planta* for permission to publish Figs 1, 3 and 6, Mrs V. Crutcher for typing the manuscript and Mrs L. Luddington for preparation of the illustrations.

REFERENCES

Agaev, M. and Agaev, A. A. (1976). *Tsitologiya*, **18** (6), 739.

Aguilera, J. M., Kosikowski, F. V. and Hood, L. F. (1976). *J. Fd Sci.* **41**, 1209.

Altschul, A. M. (1970). *In* 'Evaluation of Novel Protein Products' (A. E. Bender, R. Kihlberg, B. Lofqvist and L. Munck, eds), p. 41. Pergamon Press, Oxford.

Altschul, A. M., Neucere, N. J., Woodham, A. A. and Dechary, J. M. (1964). *Nature, Lond.* **4944**, 501.

Bagley, B. W., Cherry, T. H., Rollins, M. L. and Altschul, A. M. (1963). *Am. J. Bot.* **50** (6), 523.

Bengtsson, L., Hofsten, A. v. and Loof, B. (1972). *In* 'Rapeseed' (L. A. Appelqvist and R. Ohlson, eds), p. 34. Elsevier, Amsterdam.

Berjak, P. (1972). *Ann. Bot.* **36**, 73.

Bils, R. F. and Howell, R. W. (1963). *Crop Sci.* **3**, 304.

Birbeck, M. S. C. and Mercer, E. H. (1961). *Nature, Lond.* **189**, 558.

Buttrose, M. S. and Swift, J. G. (1973). *Protoplasma*, **78**, 471.

Claude, A. (1943). *Science*, **97**, 451.

Cummings, D. B., Stanley, D. W. and de Mann, J. M. (1972). *Can. Inst. Fd Sci. Technol. J.* **5** (3), 124.

Dangeard, P. A. (1921). *C. R. Hebd. Séanc. Acad. Sci.*, Paris, **173**, 857.

Dieckert, J. W. (1969). Abstr. XI International Botanical Congress, Seattle, Washington.

Dieckert, J. W. (1971). *In* 'Chemistry of the Cell Interface' (H. D. Brown, ed.), p. 33. Academic Press, New York and London.

Dieckert, J. W. and Dieckert, M. C. (1972). *In* 'Seed Proteins' (G. E. Inglett, ed.), p. 52. Avi. Publ. Co., Westport, Connecticut.

Dieckert, J. W. and Dieckert, M. C. (1975). Proc. Workshop, Gen. Improv. Sd Prot. National Academy of Sciences.

Dieckert, J. W. and Dieckert, M. C. (1976). *J. Fd Sci.* **41**, 475.

Dieckert, J. W., Snowden, J. E., Jr, Moore, A. T., Heinzelman, D. C. and Altschul, A. M. (1962). *J. Fd Sci.* **27**, 321.

Drawert, H. (1953). *Ber. Dt. Bot. Ges.* **66**, 134.
Drawert, H. and Mix, M. (1962). *Ber. Dt. Bot. Ges.* **75**, 128.
Düvel, D. Von, Jacob, J. and Grimmer, G. (1969). *Beitr. Biol. Pfl.* **46**, 209.
Englemann, E. M. (1966). *Am. J. Bot.* **53**, 231.
Esau, K. (1960). 'The Anatomy of Seed Plants'. John Wiley and Sons, New York.
Frey-Wyssling, A. and Mühlethaler, K. (1965). 'Ultrastructural Plant Cytology'. Elsevier, London.
Frey-Wyssling, A., Grieshaber, E. and Mühlethaler, K. (1963). *J. Ultrastruct. Res.* **8**, 506.
Garcia, E. J. (1976). *J. Am. Oil Chem. Soc.* **53** (6), 259.
Guignard, L. (1890). *C.R. Hebd. Séanc. Acad. Sci., Paris,* **111**, 249.
Guilliermond, A. (1921a). *C. R. Séanc. Soc. Biol.* **85**, 1033.
Guilliermond, A. (1921b). *C.R. Ass. Fr. Avanc. Sci.* **45**, 578.
Guilliermond, A. (1941). 'The Cytoplasm of Plant Cells'. Chronica Botanica Co., Waltham, Massachusetts.
Hanstein, J. V. (1880). *Botan. Abhandl. Morphol. Physiol., Bonn,* **4**, 1.
Hartig, T. (1856). *Bot. Ztg,* **14**, 257.
Harwood, J. L., Sodja, A., Stumpf, P. K. and Spurr, A. R. (1971). *Lipids,* **6**, 851.
Heinricher, E. (1884). *Ber. Dt. Bot. Ges.* **2**, 463.
Hofsten, A. v. (1970). *In* 'Proc. Int. Conf. Sci. Technol.: Marketing of Rapeseed and Rapeseed Products', Ste. Adele Quebec, p. 70. Rapeseed Association of Canada, Winnipeg.
Hofsten, A. v. (1973). *Physiologia Pl.* **29** (1), 76.
Hofsten, A. v. (1974). *Svensk Bot. Tidskr.* **68**, 153.
Hutchins, R. P. (1976). *J. Am. Oil Chem. Soc.* **53** (6), 279.
Khoo, U. and Wolf, W. J. (1970). *Am. J. Bot.* **57**, 1042.
Langstraat, A. (1976). *J. Am. Oil Chem. Soc.* **53** (6), 241.
Lott, J. N. A. (1975). *Pl. Physiol.* **55**, 913.
Lott, J. N. A. and Vollmer, C. M. (1973). *Protoplasma,* **78**, 255.
Lott, J. N. A., Larsen, P. L. and Darley, J. J. (1971). *Can. J. Bot.* **49**, 1777.
Lui, N. S. T. and Altschul, A. M. (1967). *Archs Biochem. Biophys.* **121**, 678.
Matile, Ph. (1968). *Z. PflPhysiol.* **58**, 365.
atile, Ph. and Moor, H. (1968). *Planta,* **80**, 159.
tile, Ph. and Spichinger, J. (1968). *Z. PflPhysiol.* **58**, 277.
Mitsuda, H. and Kishida, H. (1967). *Kyota Univ. Coll. Agr. Mem.* **92**, 17.
Mollenhauer, H. H. and Totten, C. (1971). *J. Cell Biol.* **48**, 387.
Moor, H., Mühlethaler, K., Waldner, H. and Frey-Wyssling, A. J. (1961). *J. Biophys. Biochem. Cytol.* **10**, 1.
Mottier, D. M. (1921). *Ann. Bot.* **35**, 349.
Ohlson, R. (1972). *J. Am. Oil Chem. Soc.* **49**, 522.
Ory, R. L. (1972). *In* 'Seed Proteins' (G. E. Inglett, ed.), p. 86. Avi. Publ. Co., Westport, Connecticut.
Palade, G. (1959). *In* 'Subcellular Particles' (T. Hayashi, ed.), p. 64. Ronald Press Co., New York.
Palade, G. (1975). *Science,* **189**, 347.

Perner, E. S. (1952). *Bot. Zentr.* **71**, 43.
Perner, E. S. (1953). *Protoplasma*, **42**, 457.
Pfeffer, W. (1872). *Jb. Wiss. Bot.* **8**, 429.
Poux, N. (1963). *J. Microscopie*, **2**, 557.
Poux, N. (1965). *J. Microscopie*, **4**, 771.
Prokofev, A. A., Sveshnikova, I. N. and Sobolev, A. M. (1967). *Fiziologiya Rast.* **14**, 889.
Ramecourt, B. de (1976). *J. Am. Oil Chem. Soc.* **53** (6), 256.
Rendle, A. B. (1888). *Am. J. Bot.* **2**, 161.
Rest, J. A. and Vaughan, J. G. (1972). *Planta*, **105**, 245.
Roberts, R. (1976). *J. Am. Oil Chem. Soc.* **53** (6), 302.
Rost, T. L. (1972). *Am. J. Bot.* **59** (6), 607.
Saio, K. and Watanabe, T. (1968). *J. Fd Sci. Technol., Tokyo*, **15**, 290.
Schwarzenbach, A. M. (1971). *Cytobiologie*, **4**, 145.
Sharma, C. B. and Dieckert, J. W. (1974). *Physiologia Pl.* **30**, 178.
Smith, C. G. (1974). *Planta*, **119**, 125.
Sobolev, A. M. (1966). *Fiziologiya Rast.* **13**, 193.
Sorokin, H. P. (1967). *Am. J. Bot.* **54** (8), 1008.
Sorokin, H. P. and Sorokin, S. (1966). *Protoplasma*, **62**, 216.
Sorokin, H. P. and Sorokin, S. (1968). *J. Histochem. Cytochem.* **16** (12), 791.
Spurr, A. R. (1969). *J. Ultrastruct. Res.* **26**, 31.
St Angelo, A. J., Yatsu, L. Y. and Altschul, A. M. (1968). *Archs Biochem. Biophys.* **124**, 199.
Stanley, D. W., Cummings, D. B. and de Mann, J. M. (1972). *Can. Inst. Fd Sci. Technol. J.* **5** (3), 118.
Stanley, D. W., Gill, T. A., de Mann, J. M. and Tung, M. A. (1976). *Can. Inst. Fd Sci. Technol. J.* **9** (2), 56.
Steere, R. L. (1957). *J. Biophys. Biochem. Cytol.* **3**, 44.
Stein, W. and Glaser, F. W. (1976). *J. Am. Oil Chem. Soc.* **53** (6), 283.
Ström, B. (1975). *Abstr. J. Ultrastruct. Res.* **54**, 493.
Suvorov, V. I., Buzulukova, N. P., Sobolev, A. M. and Sveshnikova, I. N. (1970). *Sov. Pl. Physiol.* **17**, 1020.
Tombs, M. P. (1967). *Pl. Physiol.* **42**, 797.
Vaughan, J. G. (1970). 'The Structure and Utilization of Oil Seeds'. Chapman and Hall, London.
Vix, H. L. E., Pominski, J., Pearce, H. M., Jr and Spadaro, J. J. (1967). *Peanut J. Nut Wld*, **46** (3), 10; (4), 10, 18; (6), 10.
Vix, H. L. E., Gardner, H. K. Jr, Lambou, M. G. and Rollins, M. L. (1972). *In* 'Seed Proteins' (G. E. Inglett, ed.), p. 212. Avi. Publ. Co., Westport, Connecticut.
Walek-Czernecka, A. (1962). *Acta Soc. Bot. Pol.* **31**, 539.
Walek-Czernecka, A. (1965). *Acta Soc. Bot. Pol.* **34**, 573.
Ward, J. A. (1976). *J. Am. Oil Chem. Soc.* **53** (6), 261.
Werker, E. and Vaughan, J. G. (1974). *Planta*, **116**, 243.
Wieler, A. (1943). *Protoplasma*, **38**, 21.

Winton, A. L. and Winton, K. B. (1932). 'The Structure and Composition of Foods', Vol. 1. John Wiley and Sons, New York.
Wolf, W. J. (1970). *J. Am. Oil Chem. Soc.* **47**, 107.
Wolf, W. J. (1972). *In* 'Seed Proteins' (G. E. Inglett, ed.), p. 231. Avi. Publ. Co., Westport, Connecticut.
Wolf, W. J. and Baker, F. L. (1972). *Cereal Sci. Today*, **17** (5), 125.
Wolf, W. J. and Baker, F. L. (1973). *Cereal Chem.* **52**, 387.
Wood, B. and Vaughan, J. G. (1976a). *Ann. Bot.* **40**, 213.
Wood, B. and Vaughan, J. G. (1976b). *Ann. Bot.* **40**, 1111.
Yatsu, L. Y. (1965). *J. Cell Biol.* **25**, 193.
Yatsu, L. Y. and Jacks, T. J. (1968). *Archs Biochem. Biophys.* **124**, 466.
Yatsu, L. Y., Jacks, T. J. and Hensarling, T. P. (1971). *Pl. Physiol.* **48**, 675.

3. Cereals and Bakery Products

R. E. ANGOLD

The Lord Rank Research Centre,
High Wycombe, U.K.

I. INTRODUCTION

The cereals form the staple diet of mankind. Cereals, and their products, are so fundamental to the human economy that in several religions they carry symbolism suggesting a spiritual analogy to the needs of the physical nutrition of man.

The cereals are grasses and are grown in all regions of the world. Seventy per cent of the world's cultivated acreage is devoted to cereal production.

The eight principal cereals are listed in Table 1, in order of quantity produced. Although a greater tonnage of wheat than rice is grown, a greater part of the world's population subsists on rice. Triticale, a man-made cereal derived from a cross between wheat and rye, has generated interest over the last ten years, and is now being grown and evaluated in several countries (Tsen, 1974).

The portion of the plant most valued as food by both man and animals is the fruit. As grasses, the cereals (members of the most successful family of the Monocotyledonae) produce a dry, indehiscent fruit in which the pericarp and testa are fused. Botanists refer to this fruit as a caryopsis. The moisture content of the mature caryopsis is low, usually between 10% and 20%. It is this characteristic, coupled with a relatively high protein content, which gives the cereals their particular value to mankind. Storage of cereals in the mature, dry state is a relatively easy process.

TABLE 1. Principal cereals (source, F.A.O.)

Cereal	Systematic name	1976 Harvest (millions of tonnes)	Principal areas grown
Wheat	*Triticum aestivum* *Triticum durum*	417	U.S.S.R., N. America, Europe, India, China
Rice	*Oryza sativa*	343	India, China, Japan, Bangladesh, Brazil
Maize	*Zea mays*	335	N. America, U.S.S.R., Europe, S. America
Barley	*Hordeum vulgare*	190	W. Europe, Asia, U.S.S.R.
Millet	*Panicum miliaceum* *Setaria italica* *Panicum typhoideum* *Eleusine coracana*	57[a]	Africa, India, U.S.A., China
Sorghum	*Sorghum vulgare*	51[a]	U.S.A., India, Africa
Oats	*Avena sativa*	50[a]	China, N. America, W. Europe, U.S.S.R.
Rye	*Secale cereale*	27[a]	U.S.S.R., Europe

[a] F.A.O. estimate.

The major component of the fruit is the starchy endosperm. This is a triploid tissue derived from the fusion of one of the two male gametes with the two polar nuclei of the egg. The starchy endosperm contains, as its name implies, starch, which is stored as a reserve carbohydrate to be used by the embryo as a carbon source at germination. In addition to the starch, storage protein is laid down in the endosperm cells which, at maturity, are densely packed with these two components and little else. Figure 1 shows the

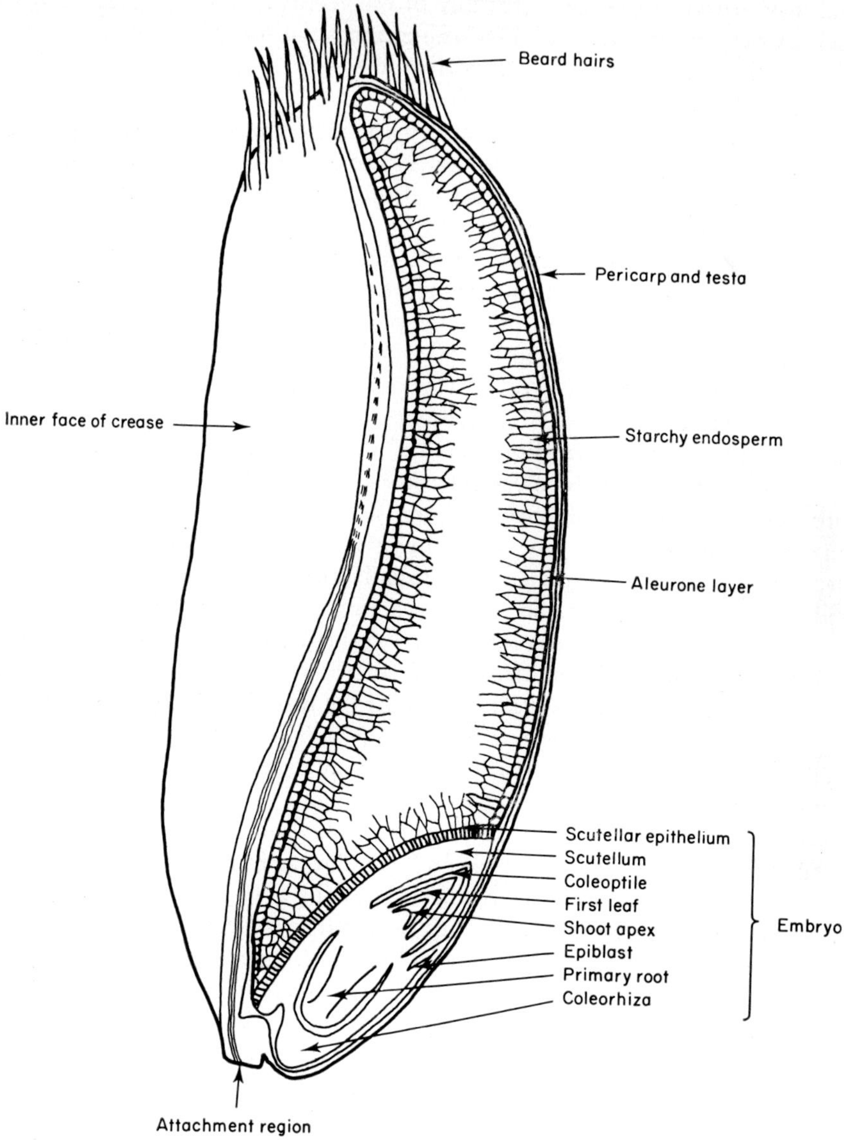

FIG. 1. Wheat caryopsis. Median longitudinal section.

structure of the wheat grain, but this diagram, with only changes of shape and proportion, would serve for any of the cereals. Although some cereals, such as rice and maize, are eaten whole, the main use of cereals for human consumption is in the preparation of products based on grinding the caryopsis into flour. This process may be elaborated to separate the components of the fruit. Thus, white wheat flour consists almost entirely of starchy endosperm (up to 80% of the total weight of the caryopsis). The aleurone layer, the testa and the pericarp are separated off as the 'bran'. The embryo, with the scutellum, forms the 'germ'.

Microscopic investigation of wheat, maize and rice has been extensive because of their economic importance as food. Barley, too, has received attention because of its use in the preparation of beers, and the financial strength of the brewing industry is sufficient for extensive research. Less attention, however, has been paid to the other cereals.

Techniques used in cereal microscopy have included light microscopy, and scanning and transmission electron microscopy. Transmission electron microscopy has been used most extensively in biologically oriented research into cereals; that is, treating the grain as a biological system which will germinate and develop. The work involving the use of the transmission electron microscope has, therefore, usually been of a fundamental nature and has not emphasized the grain as a raw material for the processes of milling and baking. The advent of the scanning electron microscope in the 1960s has shifted this emphasis, and there are now very many papers describing the structure of cereals as seen by the scanning electron microscope. The type of image produced is more easily interpreted than that from the transmission instrument. The difficulties of specimen preparation for transmission electron microscopy have inhibited the use of the latter, but the cereals are easy to study by scanning electron microscopy, being dry enough to put into the instrument and examine without any treatment other than the application of an electrically conducting layer. The use of a beam of electrons to image the specimen has a secondary effect, the production of X-rays from the specimen. The X-ray wavelength is dependent on the element emitting it and this overcomes one of the inherent disadvantages of the scanning electron microscope. In light microscopy coloured stains may be used, and for transmission electron microscopy electron-dense stains are available which can indicate the nature of cellular components; however, neither of these techniques can be employed for the scanning instruments. This phenomenon of X-ray fluorescence, nevertheless, does provide the potential for the elemental 'staining' of some cereal components for the scanning electron microscope. The benefits of this techinque for the examination of foodstuffs are limited at present by the limitations of the technique.

Carbon and nitrogen have too small an atomic weight to be detected, and oxygen is detected only with difficulty and then with low efficiency. Thus, the principal elements of living matter are beyond the range of the equipment.

II. MICROSCOPY TECHNIQUES

The preparation of cereal grains, cereal products and baked goods for light and electron microscopy requires a wide range of techniques if the maximum amount of information with a minimum of artefacts is to be achieved.

A. Light Microscopy

1. Cereal grains

The preparation of samples from whole grains must involve sectioning unless only the surface is to be examined. For sectioning a rigid matrix throughout the grain is required. This matrix can be introduced, after chemical fixation to stabilize cellular structure. Embedding in molten wax which, when cool, solidifies, is the conventional approach used by histologists (Clayden, 1971). The infiltration of the tissues with a plastic monomer, which can be polymerized *in situ*, is an increasingly used technique; this method was pioneered by electron microscopists but is also useful to the light microscopist (Feder and O'Brien, 1968). The freezing of a fixed or unfixed specimen avoids the introduction of an infiltration medium by solidifying the water present in the specimen and using that as the rigid support. The frozen specimen is then sectioned on a freezing microtome or on a cryostat. The latter, although more expensive, does have the advantage of precise specimen and knife temperature control, which may be essential when a caryopsis with a low moisture content is to be cut. The specimen may be frozen in a deep freeze, in the cryostat itself, by the use of liquid CO_2 (Flint and Moss, 1970) or by immersion in liquid nitrogen.

2. Flours

These can be examined directly, by incident light, or mounted in a liquid and viewed by transmitted light. The appropriate stains for the components of interest may be added to the mountant. (See Chapter 4 for a thorough description of the microscopy of the principal flour components.)

3. Doughs and batters

The fixation and stabilization of doughs and batters for light microscopy poses many problems. It is not possible, for instance, to put a batter into an aqueous fixative without immediately introducing artefacts due to the presence of water. Batters can, however, be spread on the slide and viewed directly, although distortion of gas bubbles present will be produced by the presence of the slide and coverslip. Doughs can also be spread onto the slide, but this does result in changes in the gluten structure since the act of spreading the film of dough puts a high shear into the system. It is this visco-elastic gluten system which makes dough difficult to fix, as well as examine directly, since very often the microscopist is seeking knowledge of the structure of the dough during the mixing process.

The removal of the dough from the mixer, the immersion of a sample in fixative and the penetration of the fixative into the specimen all takes time, and during this time the structure of the gluten network will change. Freezing of the specimen of batter or dough overcomes these difficulties, but introduces problems of its own. Fast freezing minimizes structural changes but, in our experience, the freezing of any dough or batter sample of more than 5mm thickness in liquid nitrogen sets up internal stresses in the specimen which make subsequent cryostat section cutting difficult. Liquid CO_2 (Flint and Moss, 1970) is slower, and is effective for specimen preparation for the cryostat. Any slower freezing than this permits the growth of ice crystals in the specimen to a size sufficient to introduce structural artefacts. The specimen may then be cut frozen, or placed in a fixative which is at a low enough temperature to prevent the thawing of the ice in the specimen. As the fixative solution containing an antifreeze dissolves the ice in the specimen, the structure is immediately chemically stabilized by the fixative; ethylene glycol with glutaraldehyde was found to be useful for this purpose by Moss (1974). Simple thawing of the specimen in a fixative is not adequate (see Figs 36 and 37, pp. 127 and 128). After fixation has been completed, the fixative and specimen are brought to room temperature, and dehydration and embedding for sectioning can take place. Specimen collection from doughs must be made with great caution whatever method of fixation is to be employed, since the handling of the material and the use of implements to extract a sample will introduce shear, and therefore structural change, into the dough.

4. *Baked goods*

Raised products and products containing gelatinized starch are difficult to prepare for sectioning. Any aqueous fixative used will distort the structure of a product containing gelatinized starch since the latter swells rapidly on contact with water. Freezing the specimen, and using the cryostat or freezing microtome, is not adequate, since the gas cells in the system prevent sectioning. Freeze-fixation as described for doughs and batters, followed by rinsing in water and embedding in a supporting medium such as 'Tissue-tek' (Lab-Tek Products, Miles Laboratories Inc., Naperville, Illinois 60540) will give good sections. Alternatively, a freeze-drying process can be employed, and the dry material may then be emdedded directly into wax by a gentle vacuum infiltration. After they have been dewaxed, sections prepared in this way must undergo fixation. Gelatinized starch will still swell, but the disruption of the foam structure of the product will be minimal. Flint and Moss (1970) 'humidified' baked goods over a saturated aqueous solution of Na_2HPO_4 at 8°C for 2 days before immersion for 10 days in phosphate-buffered 4% glutaraldehyde. After rinsing in water the specimen was embedded in 'Tissue-Tek' and cryostat-sectioned. The prehydration over Na_2HPO_4 solution allows the starch to swell slowly, so that rapid swelling and distortion does not occur on immersion in the fixative; nevertheless, the whole process does produce substantial swelling.

B. Electron Microscopy

1. Transmission electron microscopy

Specimens for the transmission electron microscope must be sufficiently thin for the electron beam to pass through them. The constraints which apply to specimen preparation for the light microscope also apply to transmission electron microscopy, and to these are added the problems of ultrathin sectioning.

Fixatives used in the preparation of cereal specimens for transmission electron microscopy include the oxidizing fixatives, such as osmium tetroxide and potassium permanganate, and the cross-linking fixatives such as the aldehydes, formaldehyde and glutaraldehyde. In addition to acting as fixatives, osmium tetroxide and potassium permanganate will also provide staining. Both of these compounds react with unsaturated lipids, giving contrast to sections cut from the fixed material. Osmium tetroxide is the better fixative, since it also reacts with protein and thus preserves more cell components, but potassium permanganate does emphasize membrane systems well (largely because other cellular components are not preserved at all). The aldehyde fixatives are more effective in preserving cellular components but, containing no heavy elements, they have no staining function. The most effective compromise is a double fixation procedure as described by Sabatini *et al.* (1963) and modified by later users to provide a completely isotonic fixation procedure with a common buffer throughout (Angold, 1967).

The fixation of cereal grains for transmission electron microscopy is a problem, for the low water content of the dry cereal is one of the factors enabling its survival in an inert state for long periods. On hydration, rapid induction of hormones and enzymes will occur, and it is necessary to fix the seed before too many biochemical and physiological changes can take place. Immersion of the dry seed in glutaraldehyde or formol glutaraldehyde for periods of not less than 24 hours has given satisfactory results, the fixation occurring as hydration proceeds without any evidence of enzyme induction occurring. Once the tissue is fixed, the next problem is the infiltration process of the embedding procedure. The thick walls and dense contents of mature aleurone cells make resin embedding difficult, and the use of low-viscosity resins (Spurr, 1969) does not entirely overcome this problem. A method which has given fairly satisfactory results in the author's laboratory is the immersion of the specimen in a 25% solution of the Spurr resin in 1:2 epoxy propane. The uncovered vessel is left overnight at 30°C, at the end of which most of the solvent has evaporated, leaving the material in concentrated resin ready for polymerization.

Once embedded, sectioning the material can be a problem. The endosperm tissues are rich in starch; in the transmission electron microscope starch is a difficult and disappointing subject. The density of the granule is such that the penetration of fixatives and resin does not occur, so that when sections have been obtained the starch granules in them hydrate whilst floating on the surface of the water in the boat. The resultant swelling gives rise to radial folds which were thought by some to be details of granule structure until Gallant and Guilbot (1971) pointed out the process by which this artefact occurred. Cutting sections onto butanol instead of water (Angold, 1975) overcomes the problem but is a tedious process.

Flour is difficult to handle for transmission microscopy and this has been investigated by Crozet (1977). He concluded that solution of protein occurs during fixation in aqueous fixatives and during dehydration in ethanol. In our own laboratory, flours for transmission electron microscopy are fixed in 1% osmium tetroxide (buffered to pH 6·8) and rinsed in water by centrifugation. Following this, the pellet is resuspended in an equal volume of 4% agar which has been melted and cooled to around 45°C. The mix is then spread on a glass slide to a depth of about 1 mm and allowed to solidify. The resultant gel is cut into 1 mm cubes which are then relatively easy to handle during the dehydration and embedding process. This avoids the loss of some of the smaller components which would occur during a series of solution changes by centrifugation.

As it does for the light microscope, freezing can provide a method of stabilizing the specimen without the problems of aqueous fixation. Freeze-drying is little used for transmission electron microscopy, but freeze-etching (Moor *et al.*, 1961) is of value. Swift and Buttrose (1972) have used this technique to avoid artefacts due to the hydration of cereal specimens.

2. Scanning electron microscopy

Flours are very simply prepared for this instrument, usually by being dusted dry onto a stub which has been coated with adhesive, and since Aranyi and Hawrylewicz (1968) there have been numerous papers on flour structure as seen in the scanning electron microscope. The mature cereal caryopsis, with its low water content, can be fractured dry and examined directly. Baked products are often simply air-dried, or vacuum-dried, and mounted directly onto stubs. Doughs and batters are more difficult, being hydrated. The use of a freezing stage in the scanning microscope is of little value, since the water in the specimen is not transparent to electrons; thus, the water must be removed. Varriano-Marston (1977) has explored many techniques for dough specimen preparation for the scanning electron microscope, and concluded that freeze-drying or vacuum-drying gives the best result. Critical point drying (Boyde and Wood, 1969), as used by Varriano-Marston (1977), did not give a satisfactory result. This may have been due to an aqueous fixation before drying. Where specimens are to be examined dry it is unwise to hydrate the specimen at any stage during the preparation process. In our own laboratory, the most effective method of dough or batter specimen preparation is freezing as rapidly as possible, in isopentane cooled in liquid nitrogen, or in melting nitrogen slush (Sleytr, 1970; Umrath, 1974) which does not boil round the specimen and thus insulate it from contact with the cryogen. Having frozen the specimen rapidly, minimizing ice crystal formation, the material is then freeze-dried, ideally at $-100°C$ and preferably not above $-75°C$, to reduce the rate of growth of ice crystals in the material during the drying stage.

III. THE STRUCTURE OF THE CEREAL CARYOPSIS

One of the older, and yet still useful, tracts upon the microscopic structure of the cereals is Winton and Winton (1932). Although an early work, this is a very thorough and comprehensive text in relating the microstructure of cereals to their food use. The microscopic characters of cereals used in feeds are considered elsewhere in this book (Chapter 10) and there is little point in reiterating here what has already been throughly described. This section, therefore, will concentrate on those aspects of cereal structure which have been revealed by the use of electron microscopes.

A. Wheat (*Triticum aestivum*, bread wheat, and *Triticum durum*, macaroni wheat)

The morphology of this grain is illustrated in Fig. 1. Bread wheat and durum wheat are the most widely grown species, but there are, in all, about 20 species of wheat grown throughout the world. Bennett *et al.* (1975) have followed the microscopic structure of wheat from fertilization onwards. Their work, together with that of Evers (1970) on the development of the endosperm, provides a clear account of the main structural changes taking place during the development of the wheat grain. At maturity, the grain has a pericarp consisting of five layers of cells, which are: (i) the epicarp, (ii) the hypoderm, (iii) a layer which is not continuous, the scattered groups of intermediate cells, (iv) the cross cells and (v) the tube cells. The cells have charac-

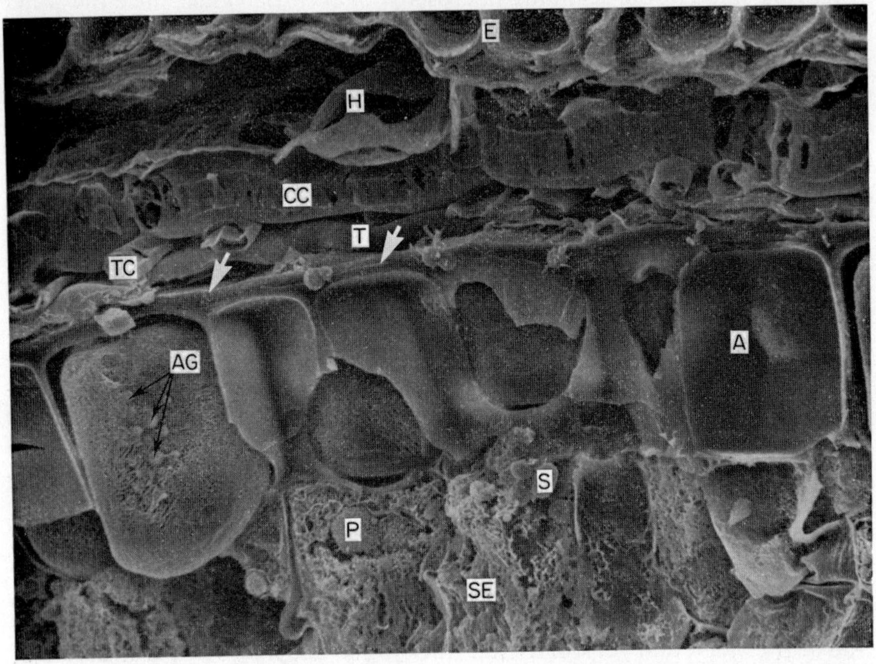

FIG. 2. Wheat (*Triticum aestivum*), SEM ×850. Imbibed, freeze-fractured, freeze-dried. The outer region of an imbibed wheat grain after freezing in nitrogen slush, followed by freeze-drying. The empty cells of the epicarp (E) cover the slightly collapsed hypodermal cells (H). The cross cells (CC) lie parallel to the plane of fracture, whilst the tube cells (TC) are seen transversely fractured. Below the tube cells (the innermost pericarp layer), is the testa (T) consisting of collapsed cells in two layers. The nucellar envelope (arrowed) is almost indistinguishable from the outer wall of the aleurone cells (A). Within the aleurone cells, the cytoplasm appears lacy. This is a result of ice crystal formation during the freezing process. Subsequent freeze-drying removes the ice, leaving needle-shaped voids in the cytoplasm which appear as round holes in cross-section or, where fractured longitudinally, as parallel voids. The aleurone grains (AG), having a low water content, remain intact after freeze-drying. Beneath the aleurone cells, the starchy endosperm cells (SE) also show freezing damage. Starch granules (S) which do not imbibe much water in the cold, remain smooth and intact. Protein bodies (P) which become hydrated appear fenestrated after freeze-drying.

teristic shapes which enable the identification of pericarp fragments in cereal (and other) products (Vaughan, 1970). Within the epicarp is a residual testa, and below this the perisperm is composed of a single, collapsed nucellar epidermis. The endosperm in wheat is composed of two very distinct cell types. The outermost layer of cells is the aleurone layer which is composed of rectangular cells with thick walls and densely packed contents, rich in mineral elements, protein and lipid (Figs 2, 3, 4 and 5). The remainder of the endosperm is the starchy endosperm, thin-walled, prism-shaped cells, packed with starch and protein. The protein content of endosperm varies from below 8% to above 15%, being higher in the durum varieties. The protein content of the starchy endosperm cells also varies within each wheat grain, being highest near the surface of the grain. The aleurone cells of the mature wheat

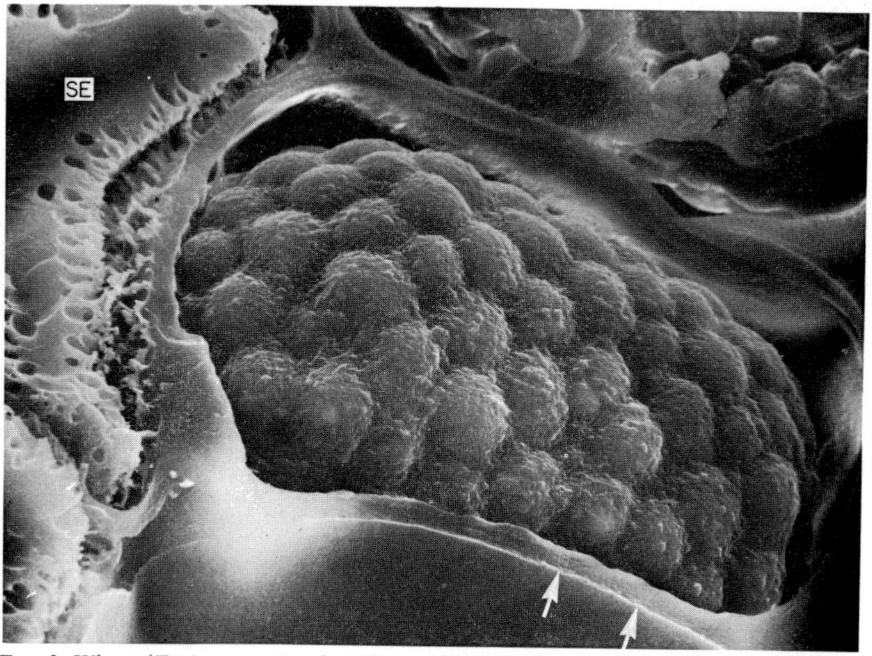

FIG. 3. Wheat (*Triticum aestivum*), SEM ×3500. Imbibed, freeze-fractured, freeze-dried. Parts of two aleurone cells can be seen, with the cell wall broken away to reveal the cell contents. In the lower cell, the cell membrane lies over the cytoplasmic contents of the cell. Aleurone grains show as bumps in the cell membrane. These bumps appear rough because of the lipid droplets which surround the grains. In the upper cell, the cytoplasm has fractured revealing the aleurone grains as rounded, smooth structures. The starchy endosperm (SE) cell to the left was fully hydrated before freeze-drying, and the cytoplasm has shrunk away from the cell wall during the drying process, leaving fragments of cytoplasm still adhering to the wall. Plasmodesmata can just be seen in the fractured radial walls of the aleurone cells (arrowed).

grain are illustrated in Figs 2, 3, 4 and 5. The aleurone cell walls are thick and during development have numerous plasmodesmata on the radial walls, but these are less apparent at maturity. The wall is two-layered, the inner layer staining more intensely by the periodic acid–Schiff technique. The walls are autofluorescent due to the presence of a ferulic acid–carbohydrate complex (Fulcher *et al.*, 1972).

The cytoplasm of the cells is densely packed and the main component is the aleurone grains. These are surrounded by lipid droplets, which also line the cell boundaries. The remaining cytoplasm is occupied by the nucleus and a few mitochondria. Short strands of endoplasmic reticulum can be found, but in the quiescent seed the ribosomes are dispersed through the cytoplasm and line the lipid droplets. The aleurone grains themselves consist of two components: the ground material, regarded by Morrison *et al.* (1975) to be largely

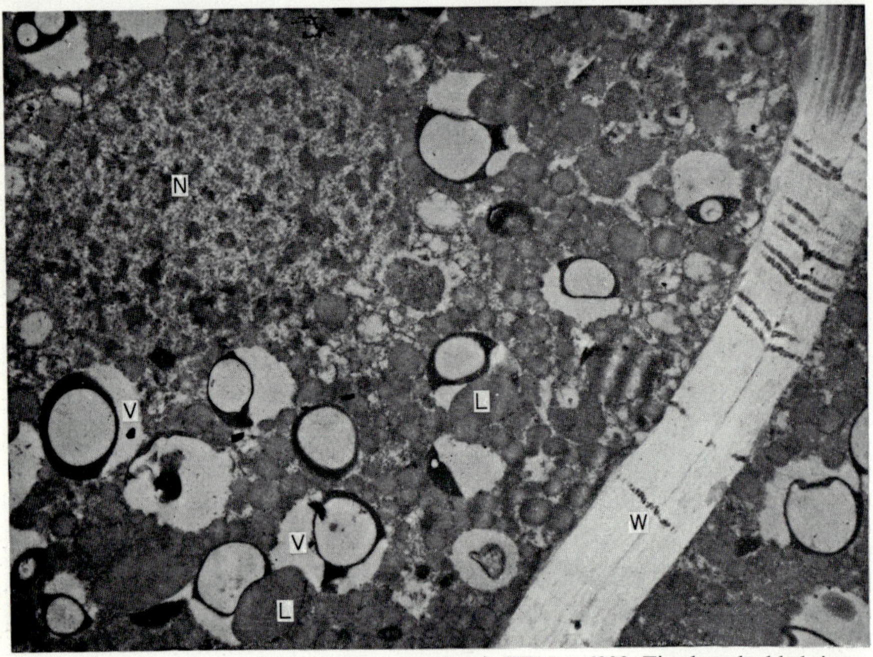

FIG. 4. Aleurone cells of wheat (*Triticum aestivum*), TEM ×6000. Fixed, embedded tissue. Glutaraldehyde and osmium fixation, followed by embedding in Spurr (1969) resin and staining with lead citrate (Reynolds, 1963). The dividing radial wall (W) between the cells shows numerous plasmodesmata. The cytoplasm within the cells is heavily populated with lipid droplets of various sizes (L). The nucleus (N) of one of the cells is clearly visible. Developing aleurone grains can be seen within the numerous vacuoles (V). This stage of development was reached 20 d after fertilization. In this particular experiment, the aleurone cells appeared fully mature by 35 d, however the grain was not ready for harvest until 45 d after fertilization.

composed of protein (stains with bromophenol blue, amido black and ponceau 2R), and the inclusions within the proteinaceous ground substance of the aleurone grains, said to be composed of inositol phosphate (the 'type 1' inclusion of Morrison *et al.* (1975)), which are crystalline; it is this latter feature which renders the grain so difficult to cut on an ultramicrotome.

The aleurone layer is modified in the region of the base of the crease. It is through this part of the aleurone layer, which lies between the vascular strand and the bulk of the starchy endosperm, that all the substrates for the storage reserves of the grain must pass. At maturity, the cells of the aleurone layer in this region (Fig. 5) contain few aleurone grains and those that are present are quite different in structure to normal grains, being enveloped in vacuoles and not having the type 1 inclusions. The remaining cytoplasm is relatively free of lipid, those droplets present, however, lining the vacuoles surrounding the aleurone grains and the cell boundary. Within the cytoplasm are numerous profiles of rough endoplasmic reticulum and large numbers of

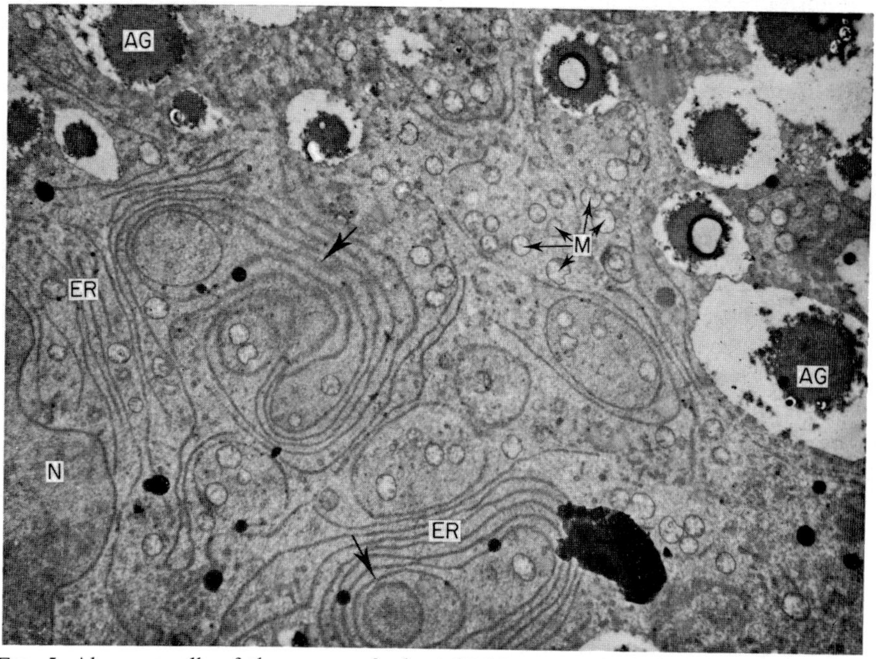

FIG. 5. Aleurone cells of the crease of wheat (*Triticum aestivum*), TEM ×4700. Fixed, embedded tissue. In aleurone cells at the base of the crease of the mature caryopsis there are relatively few aleurone grains (AG) which differ from those elsewhere in the aleurone tissue. Mitochondria (M) are numerous but small. Between the nucleus (N) and the aleurone grains are numerous profiles of endoplasmic reticulum (ER), often arranged in concentric layers (arrowed).

small mitochondria. These mitochondria exhibit circular profiles in all section orientations, indicating that they are spherical (Fig. 5).

The starchy endosperm is the tissue which contains the protein and carbohydrate reserves of the seed. It is the starchy endosperm which is milled to give white flour, and it comprises about 80% of the total weight of the caryopsis. Since the pioneering work of Moeller (1886), the structure and development of the endosperm have been described by many workers. The early work, based on light microscopy, is summarized by McMasters *et al.* (1971). The ultrastructure of the endosperm was first investigated in detail by Buttrose (1963). Simmonds (1972) extended this work to relate the microscopic appearance of the endosperm to the microstructure of the flour resulting from the milling of that endosperm. (See Chapter 4 for an account of the microstructure of starch and protein.) As well as starch and protein, wheat endosperm, and thus the resultant flour, contains about 1% lipid. This lipid is present in the protein matrix where its content may be as high as 10% Bound lipid (comprising about 40% of the total lipid present) was shown by Seckinger and Wolf (1967) to be present within the protein matrix as small inclusions. They suggested that these are the remains of cytoplasmic organelles. The remainder of the lipid present, the free lipid, appeared to be uniformly distributed throughout the protein.

The microstructure of the endosperm appears to affect the rate of penetration of water into the seed. In an autoradiographic study of the uptake of tritiated water, Stenvert and Kingswood (1977) showed that the density of the packing of the endosperm cells appeared to be correlated with a resistance to water penetration. Although this work was a study on wheat 'tempering' (control of water content for milling), the conclusions drawn also relate to imbibition prior to germination. The nature of the testa and its effect on the uptake of water has been studied by Hinton (1955). The embryo is attached to the endosperm by the scutellum. This structure, the name of which derives from its shield-like shape as seen from the endosperm after excision, is, morphologically, the cotyledon of the embryo (Fig. 6). In the Gramineae it serves as a 'placenta' between the developing seedling and the storage reserves of the starchy endosperm. Swift and Buttrose (1972) used freeze-etching in a study of the scutellum since the water content in air-dried mature grains is very low (about 8%), and earlier experience with this material had indicated that fixing in aqueous fixatives could give rise to hydration artefacts (Swift and O'Brien, 1972). Protein bodies present in the scutellum are rounded in ultrathin section and resemble aleurone grains. Freeze-etching without hydration reveals an irregular outline to the protein body. Swift and Buttrose suggest that any ultrathin section of tissue fixed in aqueous media is an image of the hydrated tissue, the hydration occurring

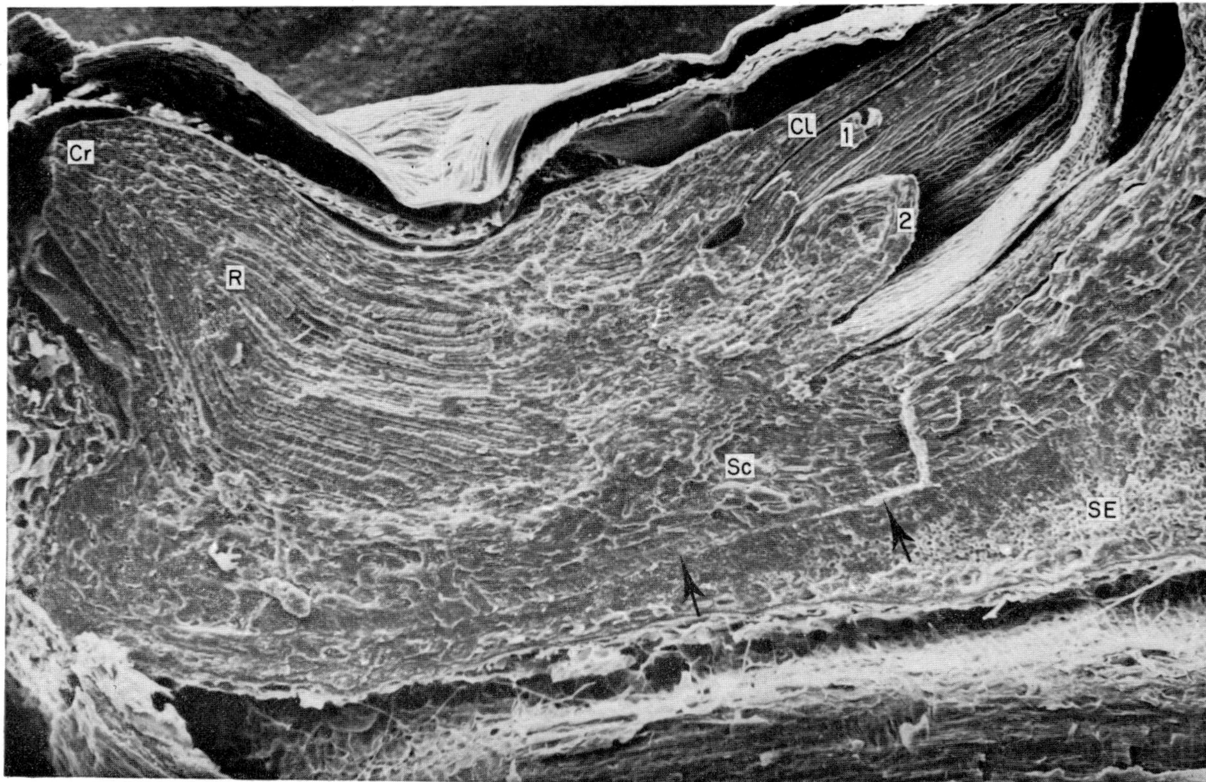

FIG. 6. Wheat (*Triticum aestivum*), SEM ×64. Dry-fractured. Longitudinal section through the embryo end of the caryopsis. This scanning electron micrograph shows the coleorhiza (Cr) covering the primary root (R). The coleoptile (Cl) and first (1) and second (2) leaves of the shoot are visible. The embryonic axis is connected to the starchy endosperm (SE) by the scutellum (Sc). The line of cells making up the scutellar epithelium can be seen (arrowed).

during the time interval between the imbibition of water by the cell and the diffusion of the fixative in sufficient concentration to stabilize the structures present. Fractures through dry wheat caryopses in preparation for scanning electron microscopy reveal that the aleurone grains within the aleurone layer are rounded whether the cells are hydrated or not.

Much of the microscopy undertaken on the embryo has been in support of studies on the physiology and biochemistry of germination. Marinos and Fife (1972) concentrated on the ultrastructure of the shoot apex. They found that in the mature, dry caryopsis the apical cells differ from the typical meristematic cell in that, together with large nuclei and dense cytoplasm, there is a population of lipid droplets lining the cell boundary, a pattern also observed in the aleurone cells (Fig. 4). The mitochondria and plastids have little internal structure. (See also Fig. 5, the aleurone cells of the crease.) On hydration of the tissue, the lipid droplets disperse and the mitochondria and plastids assume a 'normal' appearance.

B. Rice (*Oryza sativa*)

The members of the Oryzae are the main (and often the only) cereal used by nations of the Far East. Rice has been cultivated in China for 5000 years. *Oryza sativa* is the commonest species of rice. Other species grown are *O. rufipogon* (red rice), *O. mutica* (dry rice or mountain rice), *O. praecox* (early rice) and *O. glutinosa* (glutinous rice), but these latter species form only a small proportion of rice production. *Oryza sativa* falls into two groups, the 'indica', which are long grained (over 6 mm), and the 'japonica', the short grained (under 6 mm).

The rice inflorescence is a panicle. The branches of the panicle terminate in a single fertile flower. After fertilization, the flower develops into a single grain with a brown husk. The threshing of the inflorescence produces the brown grains termed 'paddy'. This rough rice, still bearing the lemma and palea of the floret, is not suitable as a foodstuff, the lemma and palea being siliceous. The rice is hulled to remove these floral parts and the result is brown rice. Decortication of the brown rice removes the pericarp, embryo and much of the aleurone layer of the endosperm, the resultant rice being the familiar white grains of polished rice. The consumption of brown rice is increasing since, much of the protein and most of the vitamin B located in the aleurone layers are lost during the polishing process. Beriberi, resulting from vitamin B_1 deficiency amongst those relying upon a diet largely consisting of polished rice, is often treated by the use of 'tikitiki extract'; this is a preparation made from the bran which was removed from rice during the polishing process.

The microstructure of the rice caryopsis has been extensively covered. Bechtel and Pomeranz (1977) have started a series of studies on this grain with an examination of the caryopsis coat down to the aleurone layer, i.e. those layers lost in the polishing process. They found, contrary to the observation of Winton and Winton (1932), that there is a layer of crushed nucellar cells in the caryopsis coat. The multicellular aleurone (Fig. 7) is reduced to a single modified layer, where it extends around the embryo. Over the endosperm the cells are cuboid and separated by walls about 0·5 μm thick which contain numerous plasmodesmata. Within the aleurone cytoplasm, aleurone grains, similar to those of wheat, are present, and these and the cell boundary are lined with lipid droplets. The aleurone grains and lipid droplets are the dominant cell components. There is a central nucleus and, in some parts of the cell, small areas of cytoplasm which contain plastids, mitochondria and a few profiles of endoplasmic reticulum. Dictyosomes were not observed. There are numerous ribosomes, not attached to endoplasmic reticulum, but lying free in the cytoplasm and also forming a lining between adjacent lipid droplets (as in wheat, Fig. 4). Over the embryo and scutellum the aleurone cells lack aleurone grains, have less densely packed cytoplasm and contain fewer and smaller lipid droplets. Mitochondria and plastids are scattered throughout the cytoplasm. This work was undertaken on the dry caryopsis, and the plastids, seen in the transmission electron microscope, appeared 'fenestrated', with regions of the ground cytoplasm appearing as membrane-bound areas within the plastid profile. Opik (1972) observed similar plastid structures in the ungerminated coleoptile of rice. Using the scanning electron microscope, Watson and Dikeman (1977) have examined the endosperm, aleurone, germ and hull of paddy and of parboiled rice. They found that the ultrastructure of long-, medium- and short-grained rice did not differ. They noted that the aleurone layer could be up to seven cells deep on the dorsal side of the grain, although none of their micrographs showed this many cell layers. Microscopy of parboiled grains revealed that the structure of the cell contents in the aleurone layer was completely obliterated. The method of dehydration for the examination of parboiled caryopsis was not given however, and it is possible that surface tension forces generated during an air-drying process would result in a loss of cellular detail. Tanaka *et al.* (1976) used X-ray microanalysis coupled with the scanning electron microscope to demonstrate that the distribution of P, Mg and K in the scutellar epithelium resembled that of the aleurone layer. Subsequently, using the transmission electron microscope with X-ray analysis, the same group (Tanaka *et al.*, 1977) was able to show that the electron-dense bodies within the cells were the sites of the elements revealed by the scanning electron microscope.

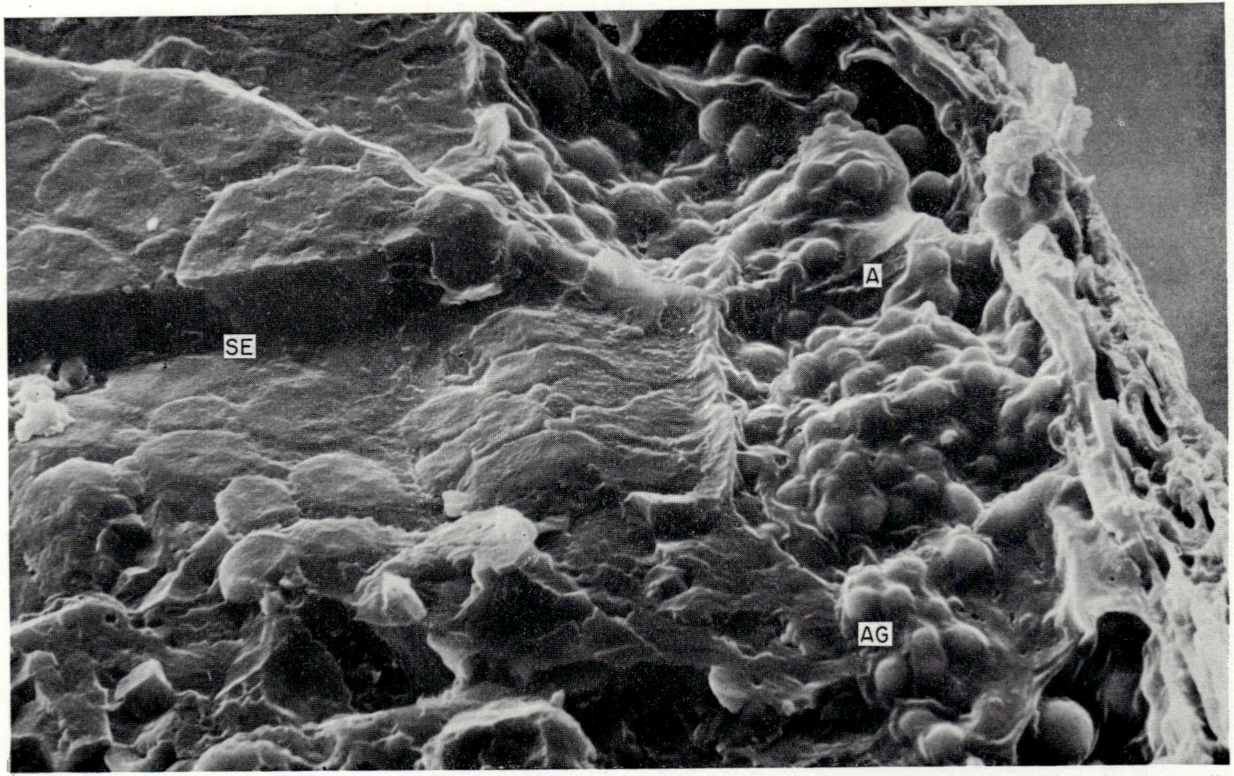

FIG. 7. Rice (*Oryza sativa*), SEM ×1800. Dry-fractured. Aleurone grains (AG) show clearly in the cytoplasm of the aleurone cells (A). The fracture plane passes between the outermost cells of the starchy endosperm (SE). The outlines of compound starch granules show through the relatively thin walls of the starchy endosperm.

The endosperm cells of rice are prism-shaped and elongated radially (Fig. 8). The cell contents are tightly packed throughout the starchy endosperm, not breaking down into a central mealy area to the same extent as some other members of the cereals. Starch granules within these cells are compound (Fig. 9) and there are few protein bodies.

C. Maize (*Zea mays*)

Maize is a member of the Tribe Maydeae. It has been cultivated in the Americas for 4000 years. Maize is a monoecious annual with flowers arranged in separate inflorescences. The female inflorescence is a spike and the axis of the spike is broadened to form the fusiform cob (or spadix) with extremely short internodes. The female florets are borne on spikelets. Each spikelet consists of two florets, of which only one is fertile. Corn may be milled wet, to give starch, corn oil and animal feeding products. Cornflour, used in the U.K. as a base for custard powders and as a general thickener, is corn starch prepared by the wet milling process. Dry milling of corn gives a flour which, in the U.S.A., is known as cornflour. It consists of finely ground endosperm with most of the germ and husk removed. American cornflour, therefore, contains most of the protein and some of the fat from the whole caryopsis. A coarse meal similar to semolina is also produced by dry milling. Hominy grits are very coarsely milled corn. However, 80% of the maize produced in the U.S.A. is used as animal feed.

The cultivars of maize fall into several groups. The main type grown in the Corn Belt of the U.S.A. is dent maize (*Z. mays* var. *americana*) which has a depression on top of the grain caused by contraction of the endosperm. Flint maize, (*Z. mays* var. *praecox*) has a more rounded form of caryopsis and the indentation is missing; this cultivar is the maize normally grown in Europe. *Zea mays* var. *saccharata* is the sweet corn, normally grown for human consumption as corn-on-the-cob. *Zea mays* var. *everta* is used for the production of popcorn. This variety has small grains with a hard and glossy outer endosperm. Corn for human consumption is also prepared by rolling and roasting grits from the dry milling process to produce cornflakes.The starch of most maize cultivars contains about 28% amylose, but amylopectin only is present in waxy maize. *Zea mays* var. *amylacea* is soft maize and is not grown at all extensively save by American Indians for their own use. The starch of soft maize is unusual amongst other members of the species in that it is rounded rather than angular, and may exceed 30 μm in diameter (Winton and Winton, 1932). Robutti *et al.* (1974) used the scanning electron microscope in a study of normal and opaque 2 mutant maize caryopses. (Opaque 2 is a high-lysine mutant in which the endosperm is soft.) In normal

FIG. 8. Rice (*Oryza sativa*), SEM ×225. Dry-fractured. This fracture passes across the central long axis of the caryopsis. The cells of the starchy endosperm in this area are arranged in a very regular radial fashion. The density of the starchy endosperm is such that relatively few cells have fractured open.

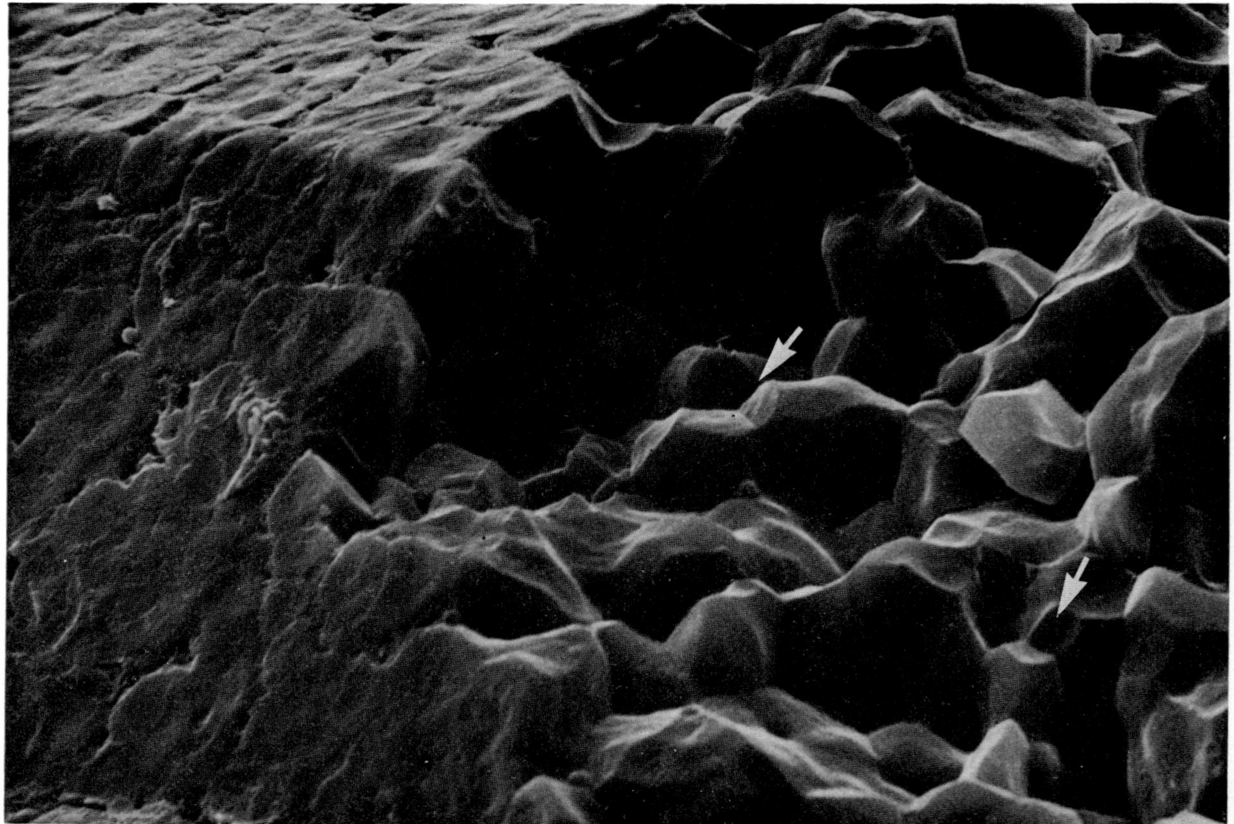

FIG. 9. Rice (*Oryza sativa*), SEM $\times 3250$. Dry-fractured. One of the starchy endosperm cells has broken open. The cell contents consist of densely packed, angular starch granules. These granules form compound groups. Individual granules have been broken out of some groups to leave depressions (arrowed).

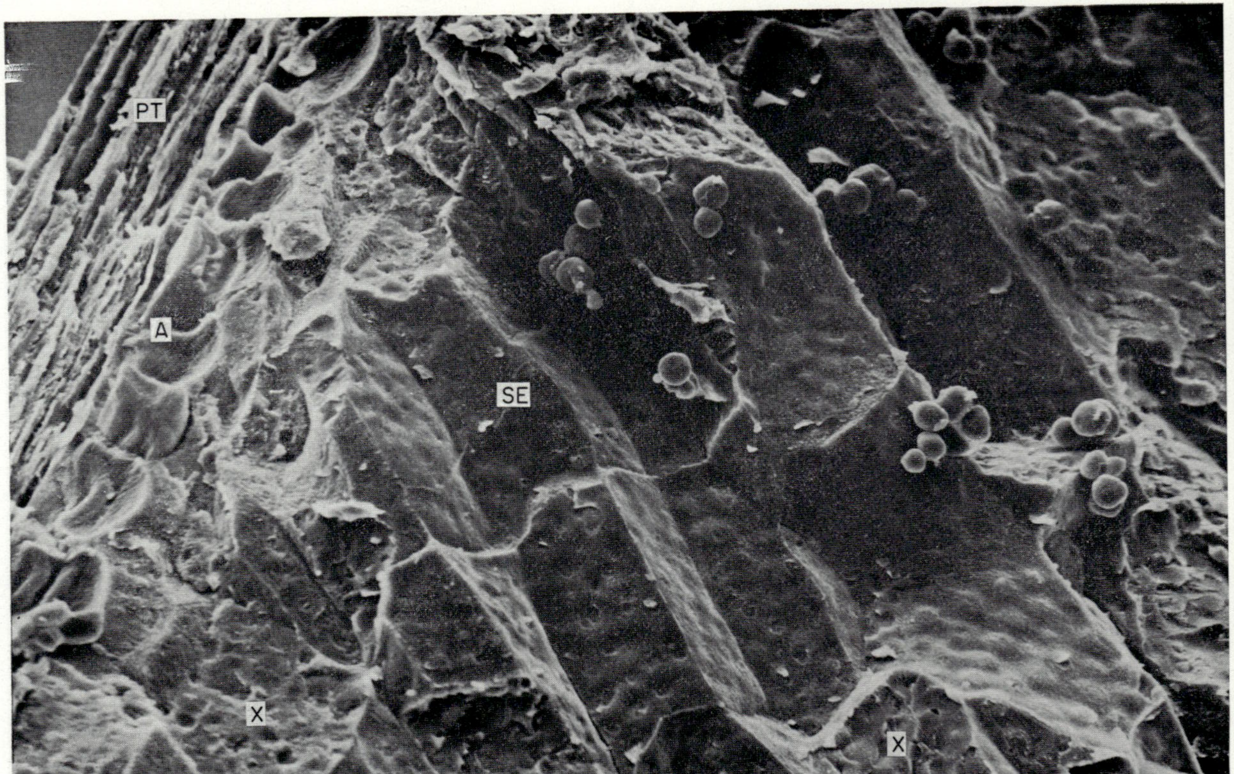

FIG. 10. Maize (*Zea mays*), SEM ×425. Dry-fractured. Beneath the pericarp and testa (PT) is the aleurone layer (A) of the endosperm which, in maize, is a single layer of brick-shaped cells. The starchy endosperm (SE) consists of prism-shaped cells, radially orientated. The fracture through the grain tends to run round the cells, but some cells (X) have fractured open. These cells are densely packed with protein and starch; there are no air spaces within the cells.

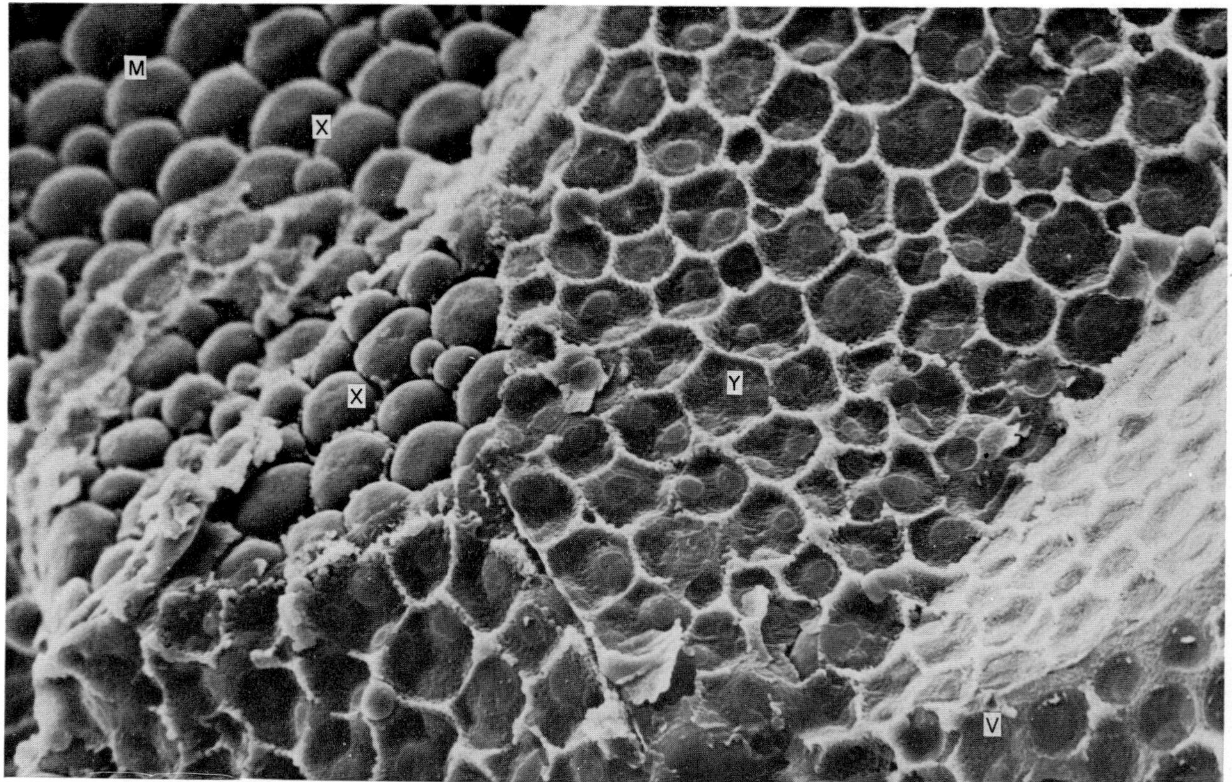

FIG. 11. Maize (*Zea mays*), SEM ×850. Dry-fractured. Boundary between the vitreous (V) and mealy (M) region of the grain. Areas where the cell wall has been stripped away (X) are interspersed with areas where the cell has been lifted off, leaving the inner face of the wall exposed (Y). In these areas the honeycomb pattern of the protein bodies can be seen.

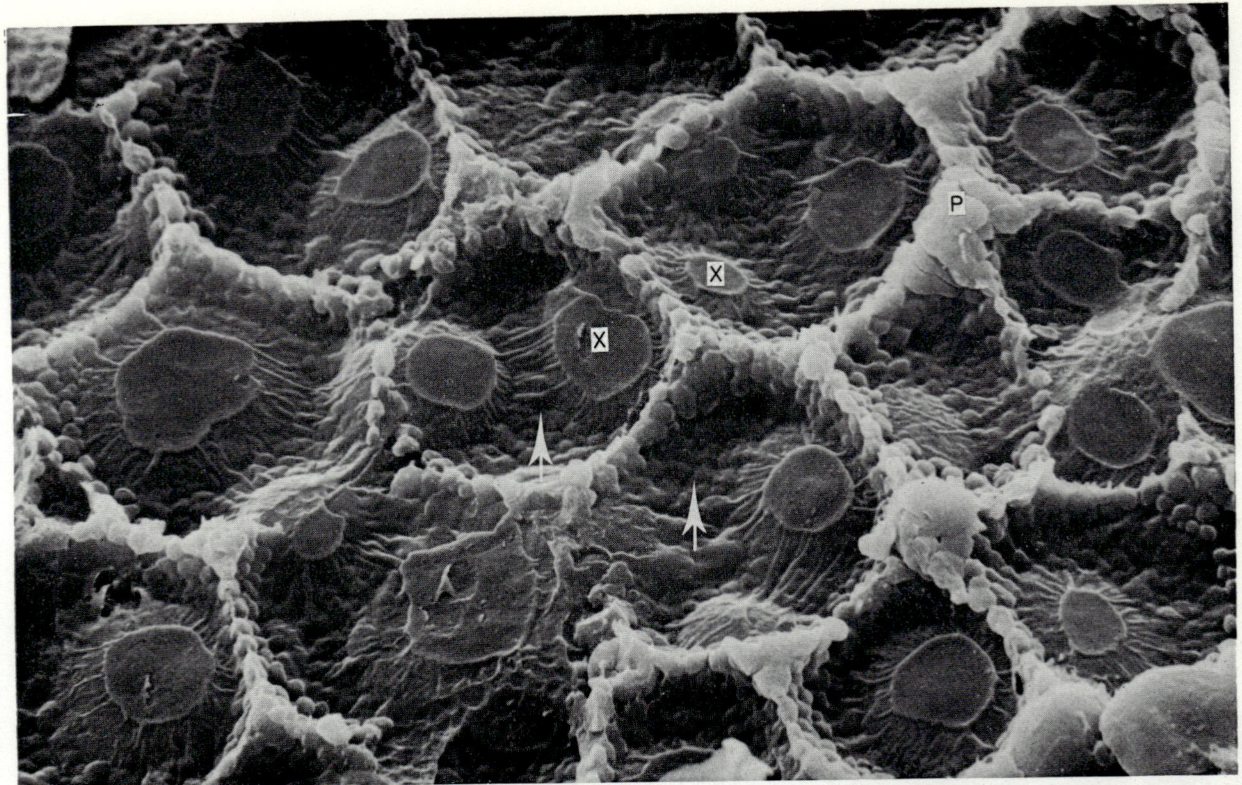

FIG. 12. Maize (*Zea mays*), SEM ×5000. Dry-fractured. The inner face of the cell wall between two endosperm cells. Protein bodies ('Zein granules') (P) are arranged in a honeycomb pattern. These granules were lining the spaces between the starch granules (removed during the fracturing process). The flattened areas (X) are contact faces between the starch granule and the cell boundary. Outlines of the starch granules in the cell on the far side of the wall can be seen, with, between them, outlines of the protein bodies (arrowed). The flattened contact areas mark the places where the starch granules are closely appressed to both sides of the wall.

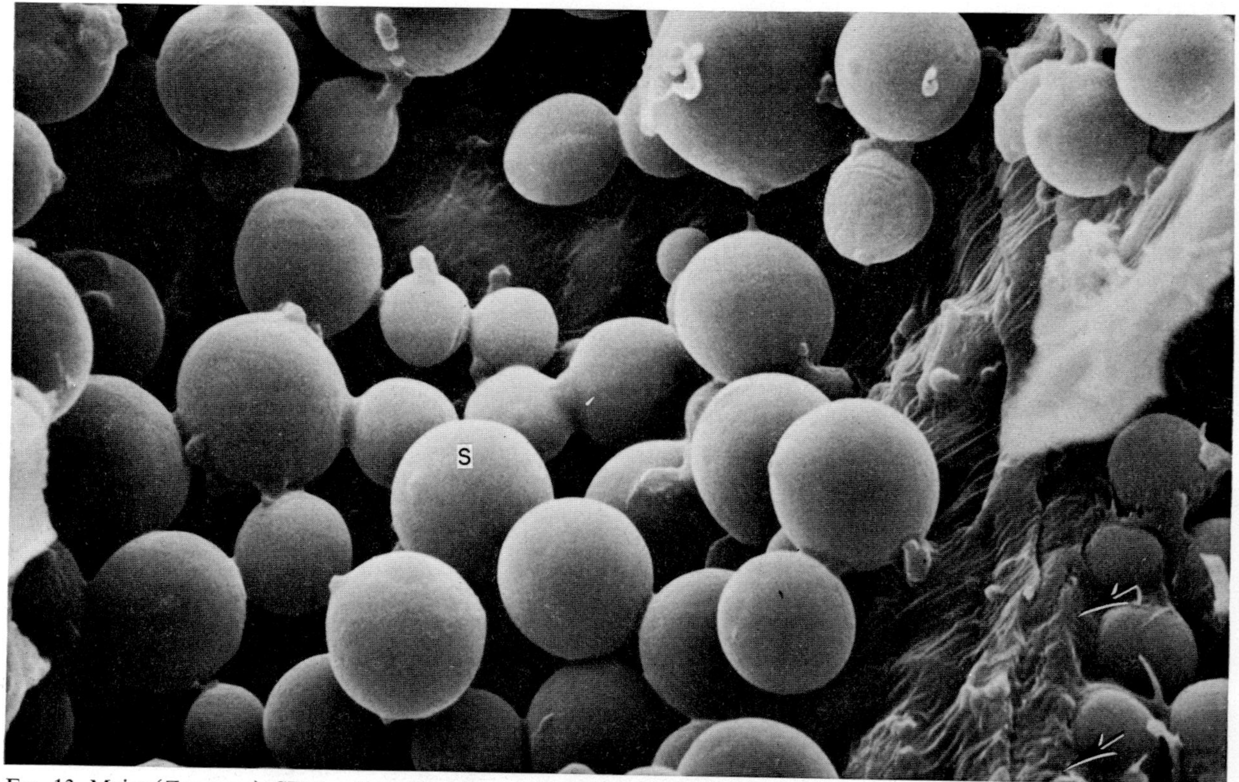

FIG. 13. Maize (*Zea mays*), SEM $\times 3250$. Dry-fractured. Fractured cell from the mealy area of the caryopsis. Here, the starch granules (S) are rounded, and protein is present as strands and as a continuous matrix (arrowed). These cells appear white when the caryopsis is fractured because there are numerous air spaces within them. The space between the rounded starch grains is air-filled when the grain is mature.

Figs 14 and 15 (*Captions on opposite page*)

corn the starch granules were shown to be tightly packed polygonal structures, with no air spaces. Opaque 2 endosperm, however, showed nearly round starch granules existing in a continuous protein matrix. Numerous air spaces (giving many refracting surfaces) account for the opacity of the endosperm. The aleurone layer of corn is a single layer of brick-shaped cells (Fig. 10); beneath this the endosperm consists of radially elongated cells, vitreous at the periphery of the starchy endosperm and mealy in the central area, with rounded starch granules loosely packed in the cells (Fig. 13). The small, angular protein bodies are largely confined to the vitreous areas of the endosperm (Figs 11 and 12).

D. Barley (*Hordeum vulgare*)

Barley is a member of the Triticeae. Although the cultivars include a two- and a six-rowed form (*Hordeum distichum* and *Hordeum hexastichum*, respectively), the two forms are the same species in that they produce fertile hybrids when intercrossed and exhibit only single-gene differences. The inflorescence consists of one-flowered spikelets arranged in fan-like groups of three inserted alternately on either side of a zig-zag, jointed rachis. In the two-rowed form only the middle spikelet of each three is fertile (Fig. 14; cf. Fig. 15). The caryopsis is usually invested with the lemma and palea at maturity, but naked forms in which the pericarp is exposed are also grown. The furrow, after removal of the chaff (where present), is shallow. The naked forms tend to have a thicker epicarp and hypoderm.

Pomeranz and Sachs (1972a) studied most of the tissues making up the barley caryopsis using a scanning electron microscope. They observed that the aleurone grains, at 1·5 μm, are smaller in preparations for the SEM than

---

FIG. 14. Flowering apex of barley (*Hordeum vulgare*), SEM at -100°C, $\times 100$. Fresh, frozen tissue. This micrograph (left) and Fig. 15 were prepared using fresh, frozen tissue mounted on a freezing stage in the scanning electron microscope. The specimen was held at -100°C and the uncoated material was examined at a low accelerating voltage (800 V) in a field emission scanning electron microscope. In barley, three one-flowered spikelets are alternately inserted at each node along the rachis. On the extreme left of the spike can be seen a row of initials which will form fertile florets. Already, anthers are beginning to differentiate on the lowermost initials (arrowed). In this two-rowed cultivar (Maris otter) the lateral rows of initials are suppressed during development. At the base of the spike, where the first-formed florets are inserted, the difference between the fertile and non-fertile florets is becoming apparent.

FIG. 15. Flowering apex of wheat (*Triticum aestivum*), SEM at -100°C, $\times 90$. Fresh, frozen tissue. This flowering tiller had reached this stage of development in early April in the southern U.K. Twenty lateral spikelets and the apical spikelet can be seen. Each spikelet bears five potential florets, giving the ear a potential of 105 grains. The basal florets are beginning to differentiate glumes, but the anthers are not yet visible.

they are in thin section in the TEM (2·4 μm) (Jones, 1969). It was concluded that this is probably a result of hydration of the cells during the fixation of the material for the transmission electron microscope. Pomeranz and Sachs also discovered modified stoma-like structures. Figure 16 shows the outer tissues of the caryopsis after cross-fracture. The lemma is fused to the pericarp. Beneath the warty epidermis of the lemma is a fibre layer which has annular and scalariform lignified thickening in a series of cells running parallel to the long axis of the caryopsis. The aleurone layer is packed, as in wheat, with the aleurone grains which are themselves surrounded by lipid droplets. The aleurone cells may be up to four layers deep. Beneath them, the sub-aleurone layer of the starchy endosperm has little starch, but is densely packed with protein and some lipid. The bulk of the endosperm is composed of radially elongated cells filled with the characteristic dual-size population

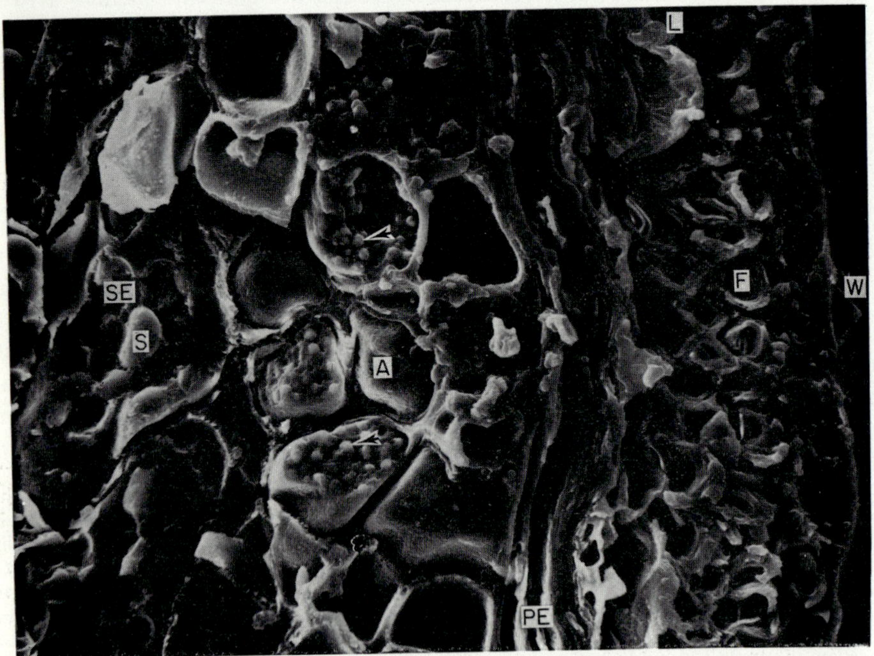

FIG. 16. Barley (*Hordeum vulgare*), SEM ×750. Dry-fractured. The lemma (L) is still adhering, and at its surface shows one of the cuticular 'warts' (W). The fibre layer (F) shows scalariform thickenings in the walls, visible down the empty lumena of the cells. Beneath the compressed pericarp (PE), the aleurone layer (A) is several cells thick. The cell walls of the aleurone layer are relatively thick, and within these cells numerous aleurone grains can be seen (arrowed) where the cytoplasm has cross-fractured. The starchy endosperm (SE) immediately below the aleurone layer appears as thin-walled cells in which there is relatively little starch. Starch granules (S) are more abundant further into the starchy endosperm.

of starch granules in a proteinaceous matrix (Fig. 17). The cell walls are relatively thin, and a dry fracture through the caryopsis of barley passes through, rather than round, the endosperm cells.

E. Millet

Millet is the common name for a number of genera of cereals grown in several parts of the world. *Eleusine coracana* is finger millet, or African millet. It is the only member of the millets which belongs to the Chloridae, all the others being members of the Panicae. It is also unusual amongst the cereals in that the fruit is an utricle and not a caryopsis. The pericarp does not envelop the fruit but is confined to the attachment end. The seed is roughly globular to oval and 1–1·5 mm in diameter (Figs 18, 19, 20 and 21).

Setaria italica is foxtail millet, and, like all millets other than coracan, belongs to the Panicae. It is also known as Italian, German, Hungarian or Siberian millet, depending on its country of origin. Winton and Winton (1932) are of the opinion that it is a cultivated variety of green foxtail,

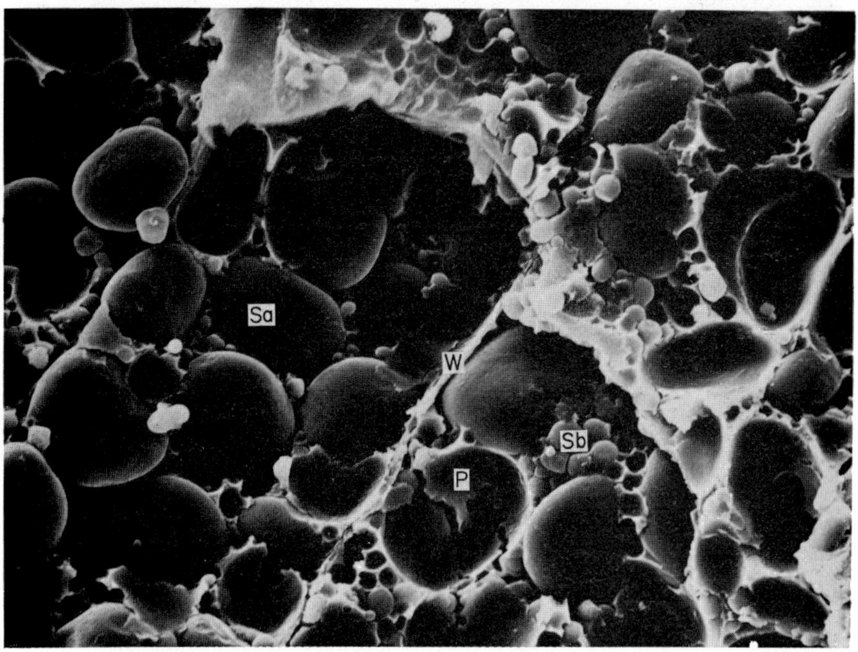

FIG. 17. Barley (*Hordeum vulgare*), SEM ×1000. Dry-fractured. Starchy endosperm of barley. Within these radially elongated cells can be seen the large (Sa) and small (Sb) starch granules, and the protein matrix (P) into which the starch granules are set in the mature dry caryopsis. Part of the cell wall (W) of one of the cells can also be seen.

FIG. 18. Coracan (*Eleusine coracana*), SEM ×280. Dry-fractured. The centre of the starchy endosperm is at the right of the micrograph. In this area the endosperm is mealy (M). In the vitreous zone (V) of the starchy endosperm, the fracture has run around the prism-shaped endosperm cells.

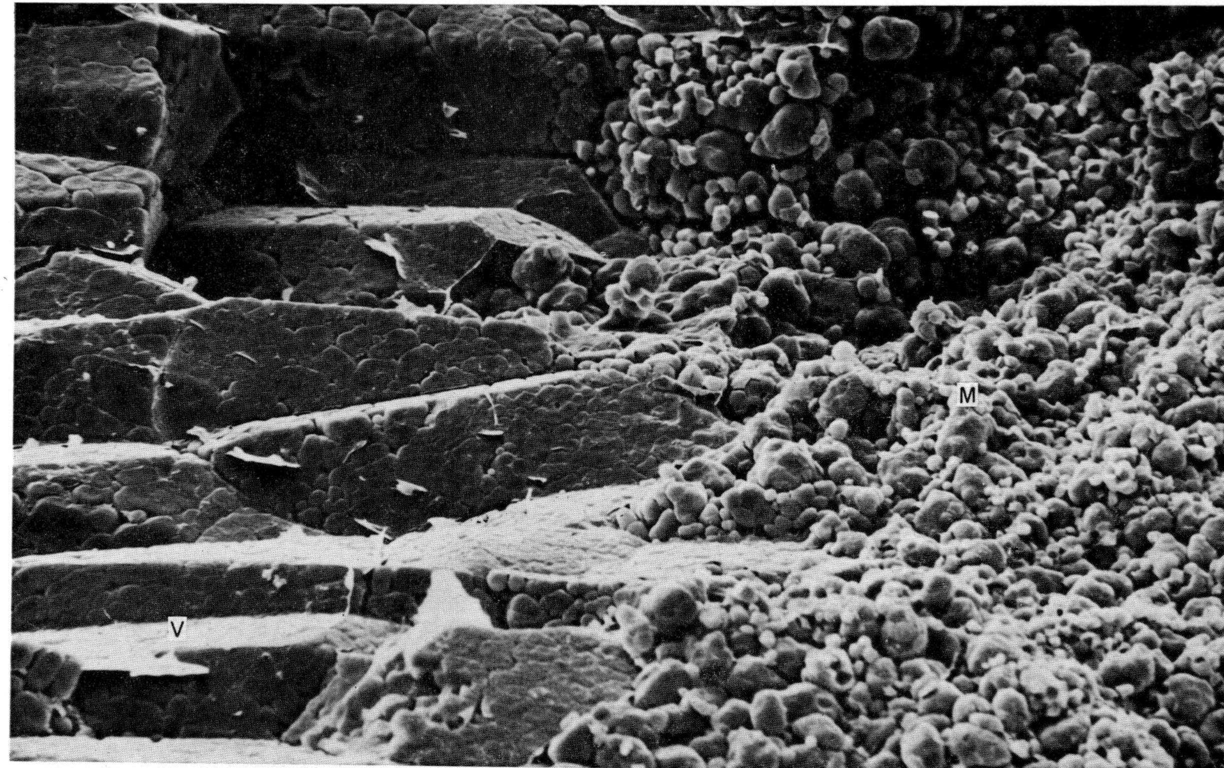

FIG. 19. Coracan (*Eleusine coracana*), SEM ×500. Dry-fractured. The boundary between the vitreous (V) and mealy (M) portions of the endosperm. The prismatic cells of the vitreous endosperm have sufficiently dense contents to fracture around their periphery. The more loosely packed cell contents of the mealy area allow the fracture to pass through the cells, and the cell outlines are lost.

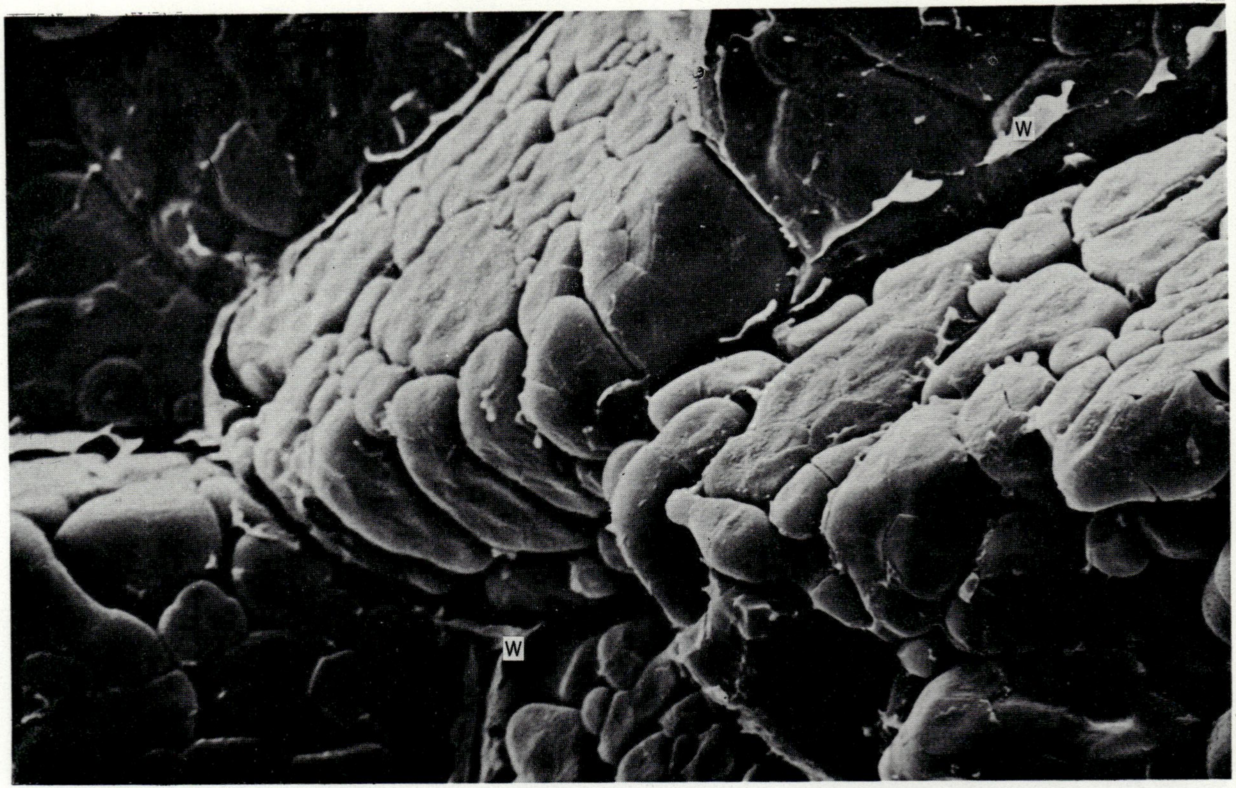

FIG. 20. Coracan (*Eleusine coracana*), SEM $\times 1400$. Dry-fractured. Part of the vitreous region of the endosperm seen at high magnification. The starch granules present are compound, each mass of starch being flattened at the cell surface to fill most of the available space within the cell. Fragments of cell wall (W) can be seen. Little protein is visible as protein bodies.

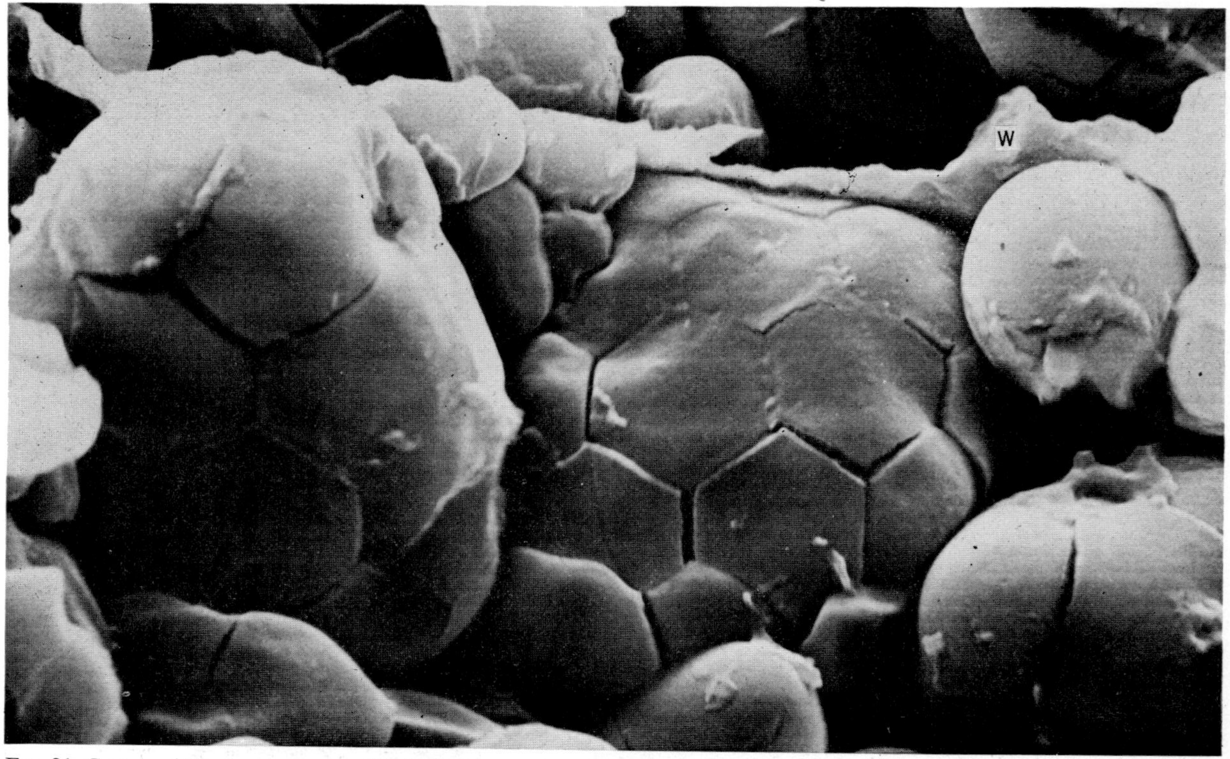

FIG. 21. Coracan (*Eleusine coracana*), SEM × 5000. Dry-fractured. Part of a cell from the mealy endosperm. The starch granules are rounded and compound. The subunits of the starch granules are tightly packed, revealing hexagonal faces at the surface. The granules are not flattened against the cell walls (W) and there are air spaces between the compound granules.

Setaria viridis. The caryopsis, which is about 2 mm in length, can be white, red, yellow, brown or black (Figs 22 and 23). *Panicum miliaceum* is the common millet and is the true millet of the Roman era, its Latin name being 'milium'. It is known in the U.S.A. under its Russian name of proso. The caryopsis is about 2 mm long and 2 mm broad, and is covered by a smooth, hard and shiny lemma and palea. *Pennisetum typhoideum*, also known as *P. glaucum*, is pearl millet or bulrush millet. Its inflorescence is a spike-like panicle, explaining one of its common names. The caryopsis, which may be up to 5 mm long, is broad at the apex, tapering to the base to give a tear-drop shape. The fruit is of a pearl-grey colour, giving the other common name. The coloration is darker toward the base, with a dark spot on the side opposite the embryo near to the base.

Badi *et al.* (1976) used the scanning electron microscope in a study of pearl millet. They observed the differing cell content of the hard and soft regions of the endosperm. The outer part of the endosperm is hard, and between the angular starch granules are the protein bodies. Few protein bodies were found in the soft area of the endosperm, and the starch granules are rounded, with air spaces between them. Figures 24 and 25 show the structure of the cross-fractured caryopsis. As in maize, the outer endosperm cells contain protein bodies which are particularly concentrated at the periphery of the cell.

F. Sorghum (*Sorghum vulgare*)

Sorghum belongs to the Andropogonae. It is often, and erroneously, called millet. A tropical plant, it is a native of Africa and Asia where a number of varieties are grown. *Sorghum vulgare* var. *durra* is durha, with a caryopsis 5–6 mm long and 5 mm broad. Kaffir corn, *S. vulgare* var. *caffrorum*, has a caryopsis some 4 mm long and nearly globular; it is cultivated in Africa. *Sorghum vulgare* var. *rexburgii* is shallu, the Indian sorghum, and *S. vulgare* var. *nervosum* is the Chinese cultivar, kaoliang. In China, India and Africa sorghum is an important human food, being used to make porridge, and bread, in which it may be blended with wheat flour.

Work on the ultrastructure of sorghum has been undertaken by Seckinger and Wolf (1973) who employed transmission and scanning electron microscopy in a study of the protein granules. They found that the peripheral, vitreous area of the kernel is rich in protein, and that most of the protein present is in the form of granules some 2–3 μm in diameter. The subaleurone cells were found to contain mostly protein with little or no starch, and the remaining vitreous cells to contain twice as much protein as the floury cells. However, this peripheral, vitreous endosperm is nutritionally inferior since it has a high prolamine and low lysine content (Shoup *et al.*, 1969). Sullins and Rooney (1974) used the microscope in a study of sorghum lines differing in endosperm characteristics. After acrolein fixation and Araldite–Epon embedding, sections were treated with α-amylase, pronase (a proteolytic enzyme of bacterial origin) and a buffered rumen fluid to assess the digestibility of the various cellular components. They found that the waxy-grain forms had less peripheral endosperm than other grains, and postulated that this was the reason for the improved value of waxy sorghum as a feedstuff.

The ultrastructure of *S. vulgare*, as seen by the SEM is shown in Figs 26 and 27. The outer endosperm cells have many small protein bodies. The angular starch granules have depressions along the edge between the flattened contact faces and it is within these depressions that the protein bodies lie. The protein bodies within the vitreous region of the endosperm are particularly concentrated at the periphery of the cells, and those starch granules lining the cell boundary exhibit more depressions. In the mealy part of the endosperm, the protein that is present is seen, in the dry caryopsis, as strands laid over the surface of the starch granules. Starch from the mealy region is rounded and the granules are larger.

G. Oats (*Avena sativa*)

Oats are generally cultivated in the cooler areas of the Northern Hemisphere which are marginal for wheat and barley cultivation. It is grown mainly for feeding stock, especially horses, but is also used to make oatmeal and rolled oats. Porridge, associated with the Scots, and oat pudding (a Russian food) are similar dishes made by boiling oatmeal in water (or milk, although a Scot might frown on this practice).

The caryopsis of oats is longer and narrower than that of wheat and is usually thrashed out with the lemma and palea still attached. Like barley, the aleurone layer may be two or more cells deep. Pomeranz and Sachs (1972b) studied the oat caryopsis with a scanning electron microscope. This study revealed little not already documented by Winton and Winton (1932) but it was observed that the interior surface of the lemma is totally different from the interior surface of the palea. Where the lemma covers the crease, the lemma abaxial surface is covered with fine wrinkles ('hairs', according to Pomeranz and Sachs). The microstructure of the oat grain is illustrated in Figs 28 and 29.

H. Rye (*Secale cereale*)

Rye is a member of the Triticeae. It is the tallest of the temperate cereals, growing to 2 m in height, and produces a spike made up of two-flowered

FIG. 22. Foxtail millet (*Setaria italica*), SEM ×100. Dry-fractured. This fracture through the whole grain (which is a utricle) passes through both the starchy endosperm and the germ. Beneath the testa (T) (the pericarp is confined to the base of the fruit) is the starchy endosperm. The prism-shaped cells of foxtail millet are less elongated in the radial direction than are those of coracan (see Fig. 19) and there is not a clear distinction between the vitreous and mealy regions of the endosperm. The embryonic axis has been removed, and the depression in the scutellum (Sc) occupied by the shoot can be seen. Between the bulk of the scutellum and the starchy endosperm there is a line of block-shaped cells forming the scutellar epithelium (arrowed).

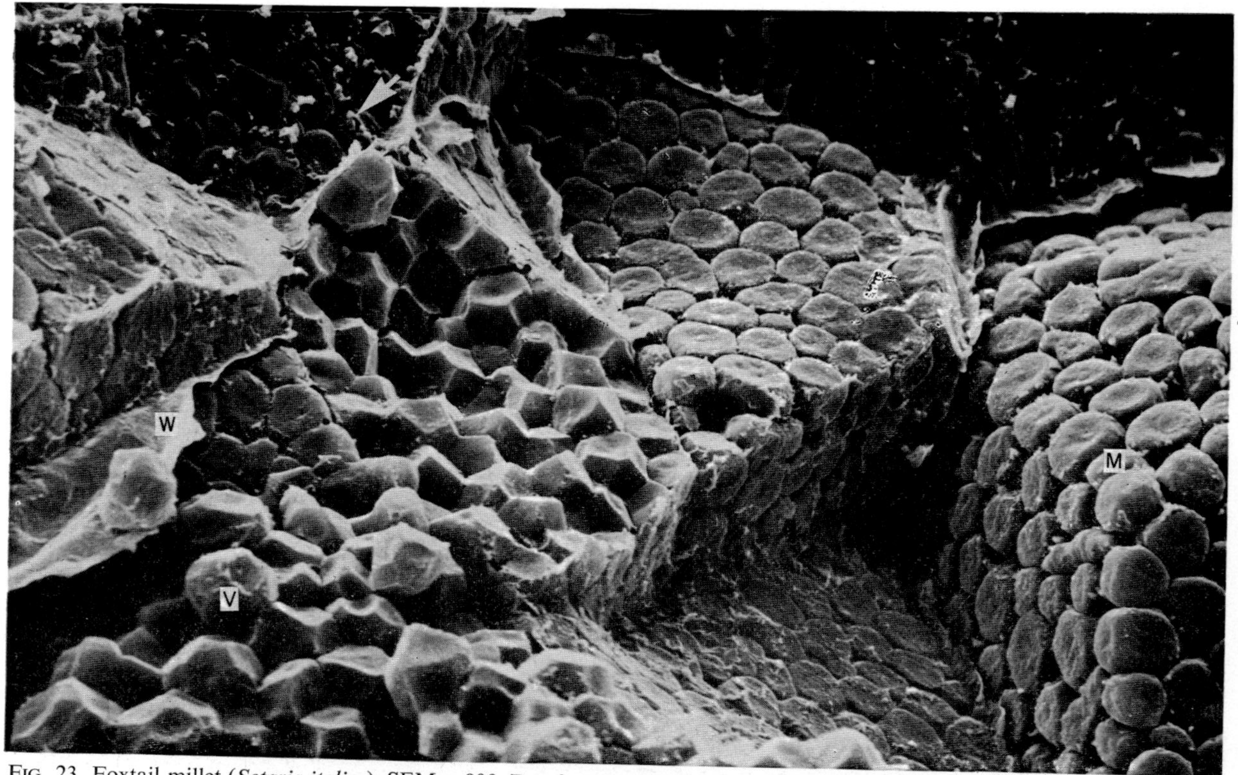

FIG. 23. Foxtail millet (*Setaria italica*), SEM ×990. Dry-fractured. The boundary between the vitreous (V) and the mealy (M) regions of the endosperm. Fragments of cell wall (W) are visible. Protein bodies (arrowed) are not present in such high numbers as in *Pennisetum* or sorghum.

FIG. 24. Pearl millet (*Pennisetum typhoideum*), SEM $\times 210$. Dry-fractured. Cross-fracture through the dry, mature caryopsis. The relatively thin pericarp and testa (PT) overlying the aleurone layer (A) can be seen on the left. The starchy endosperm (SE) is largely vitreous, although a small area of mealy tissue (M) can be seen on the right of the micrograph, which is at the centre of the grain.

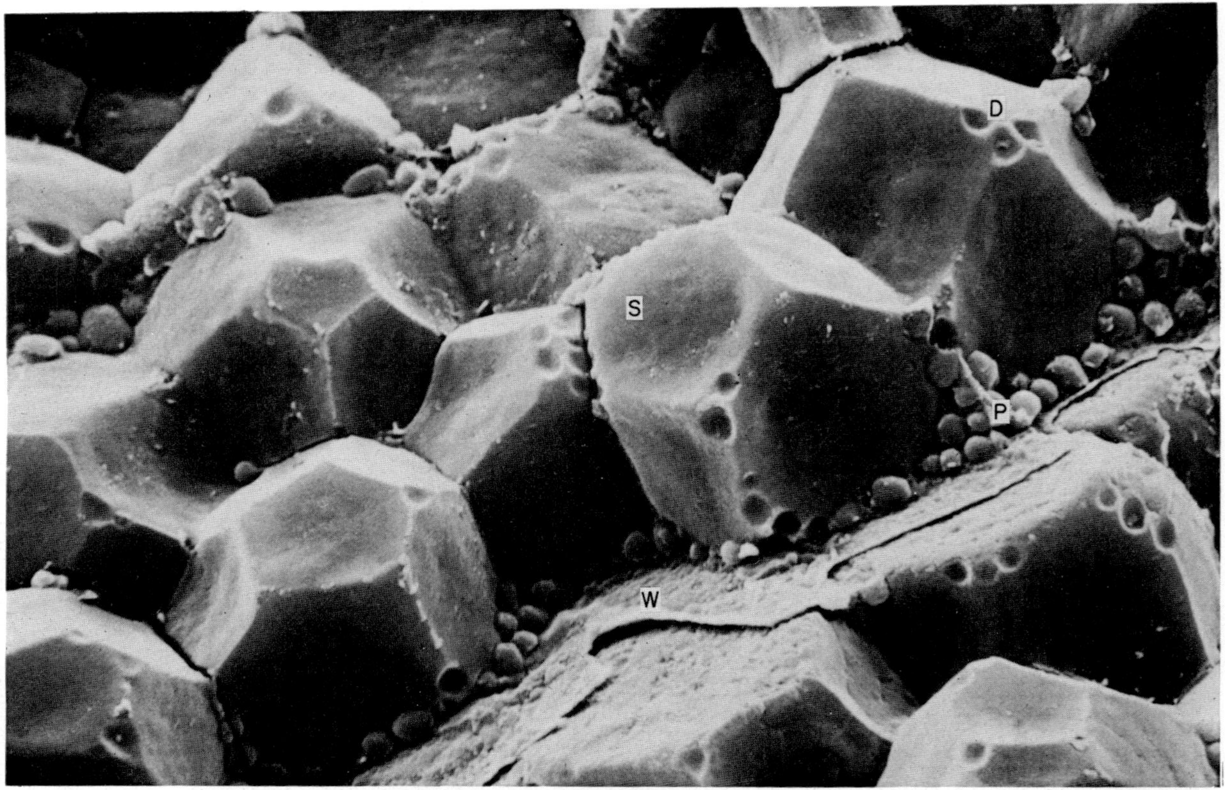

FIG. 25. Pearl millet (*Pennisetum typhoideum*), SEM ×4250. Dry-fractured. Part of two cells from the vitreous region of the endosperm. The angular starch granules (S) are densely packed. Protein bodies (P) lie along the edges of the starch granules, particularly near the cell wall (W). Pits in the starch granules (D) are left where the protein bodies have been removed during fracturing.

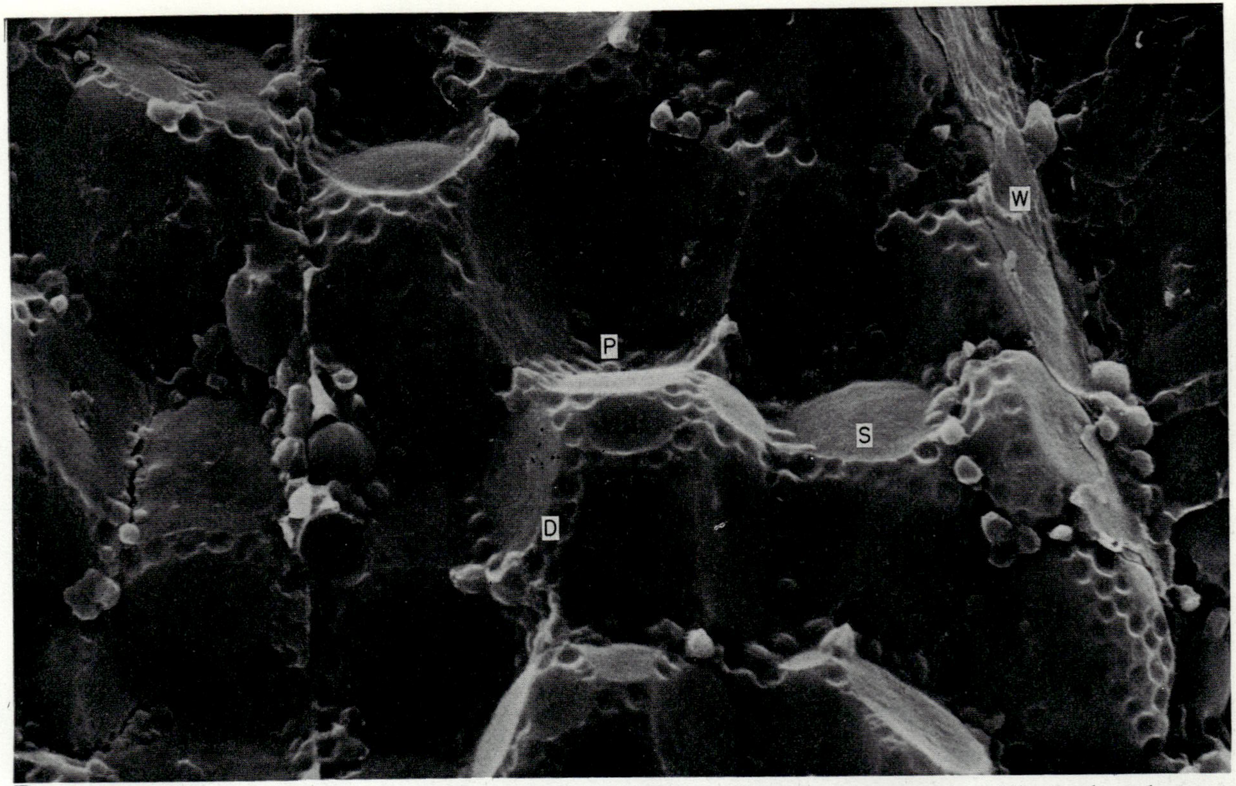

FIG. 26. Sorghum (*Sorghum vulgare*), SEM × 2400. Dry-fractured. Cells from the outer, vitreous region of the starchy endosperm. Angular starch granules (S) have protein bodies (P) packed along their edges. Depressions (D) along these edges and on parts of the starch granules abutting the cell wall (W) can be seen; these indicate where protein bodies have been removed during specimen preparation.

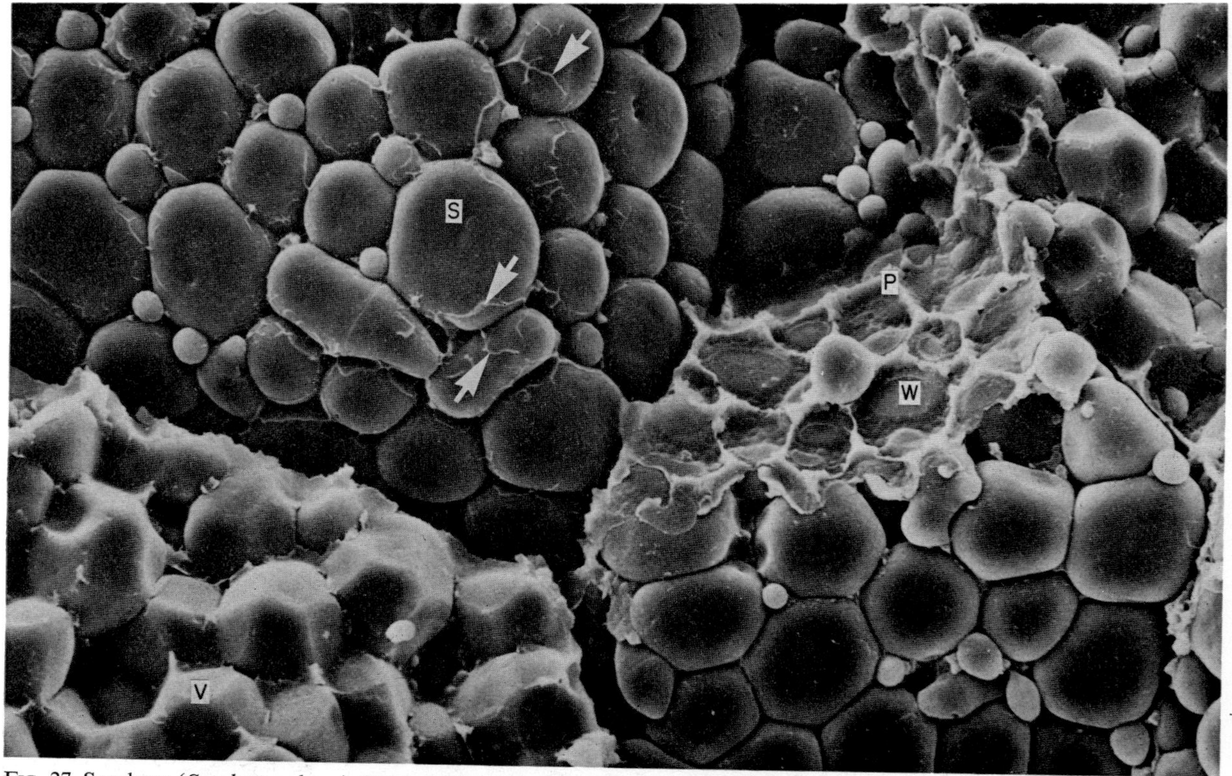

FIG. 27. Sorghum (*Sorghum vulgare*), SEM $\times 1060$. Dry-fractured. The boundary between the vitreous and mealy portion of the endosperm. A portion of a vitreous cell (V), fractured open, shows at bottom left. Within the other cells, the starch granules (S) have a more rounded appearance. Protein bodies (P) are present at vitreous cell boundaries still attached to the portion of cell wall (W) centre right, but much of the protein present is as a reticulum of fine strands over the surface of the starch granules (arrowed).

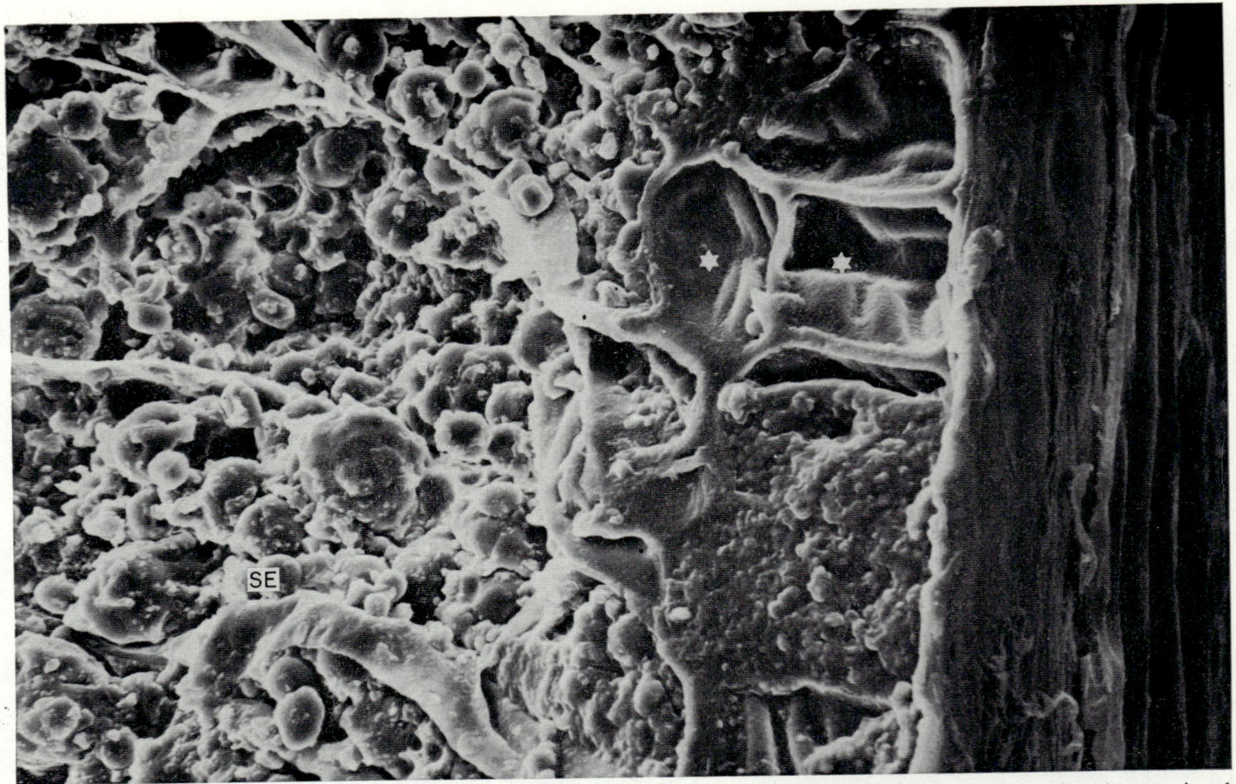

FIG. 28. Oats (*Avena sativa*), SEM × 500. Dry-fractured. The aleurone layer is largely a single layer of cells, but has the occasional tangential division to give two cells in depth (starred). The starchy endosperm (SE) is rich in lipid (about 8%), and the cells show the presence of this fat, which tends to obscure the clear outlines of starch granules present.

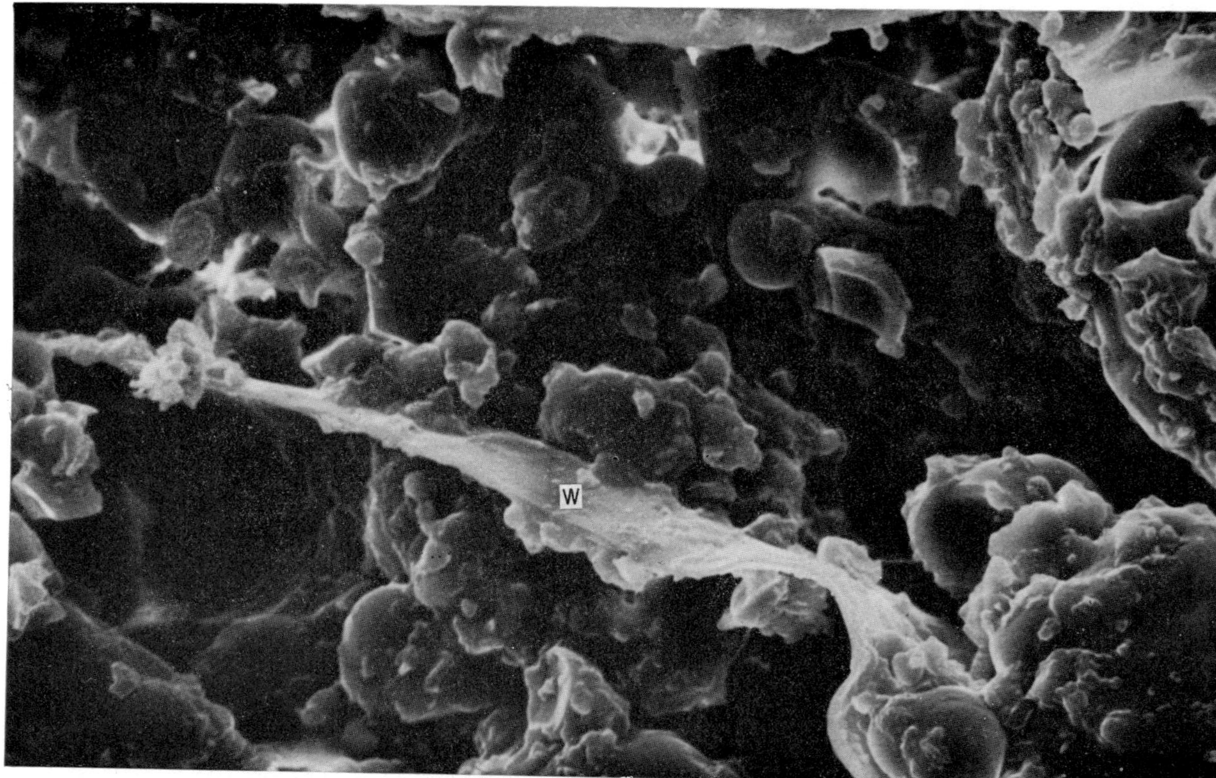

FIG. 29. Oats (*Avena sativa*), SEM ×1700. Dry-fractured. Within the endosperm, near the centre of the kernel, part of two cells which have fractured open. Oats tend to fracture through, rather than round, the endosperm cells. The cell components, starch and protein, are not so clearly displayed as by the other members of the Triticeae, lipid masking their outlines. W = cell wall.

spikelets with long-awned lemmas. There are few cultivars, and some 90% of the world's production of this cereal comes from Europe. Rye is made into bread in Germany, Austria, Czechoslovakia, Poland and Russia. The caryopsis is of a similar size and shape to that of wheat, although more tapered at the ends, but the presence of anthocyanin in the aleurone layer results in a grey to bluish-green coloration. Rye is one parent of triticale (wheat is the other).

The caryopsis has the same number of pericarp layers as wheat. The beard hairs, however, differ from those of wheat in having a broad lumen right to the tip (Winton and Winton, 1932) and this character is claimed to be useful in differentiating between wheat and rye in baked products. The microstructure of rye is shown in Figs 30 and 31.

IV. FOODSTUFFS PREPARED FROM CEREALS

Some cereals are consumed whole. Corn-on-the-cob is the whole spike. Brown and polished rice have already been discussed. Wheat, dehulled but whole, is boiled with milk, sugar and cinnamon to make frumenty.Pearl barley, from which the husk, pericarp, testa and aleurone have been removed, is added to soups. Most cereal products consumed by man, however, are based on ground caryopses. Simple grinding may suffice, as in the production of oatmeal which forms the basis of porridge. Normally, however, the grinding of cereals in developed countries involves the separation of the components of the caryopsis. Thus, white flour from wheat consists almost entirely of ground starchy endosperm. The pericarp, testa and aleurone layer are separated as the bran, or wheatfeed, whilst the embryo and scutellum form the 'germ' fraction.

A. Flour

The light microscope is of great use in the study of flours. The size and shape of the starch granules may indicate the species of cereal used in the preparation of the flour. Numerous atlases of starch granules seen by the light microscope exist, e.g. Seidemann (1966), and these are of value in identifying raw flours. Other cereal components may be present; bran fragments and hairs may enable the microscopist to identify the cereal used even after the flour-based product has been cooked and the starch granules have been gelatinized beyond recognition (see Chapter 10).

The proliferation of publications on the scanning electron microscopy of flours from various cereals, starting with Aranyi and Hawrylewicz (1968), is a good indication of the simplicity of preparation of flour for the scanning instrument. Selective staining, however, cannot be undertaken in the SEM, so that all interpretation of structure is based on morphology. Some components are easy to identify, e.g. aleurone cells, with or without their contents, and starch granules. Endosperm protein fragments of wheat flour are less reliably identified, but are usually assumed, somewhat dangerously, to be that which is not starch or any other identifiable component. Cell wall fragments may be mistaken for protein fragments, and few lipids are ever indicated in SEM images of flour. Thus the light microscope must be used, together with the appropriate stains, to identify the constituents of flours. The combination of this information, together with the three-dimensional image given by the SEM, can be a powerful method in the analysis of the behaviour of cereals in a mill. Figures 32, 33 and 34 show the effects of milling hard and soft wheat endosperm.

B. Doughs and Batters

A dough is a cereal flour which has been mixed with water and worked. Wheat flour protein has the greatest ability of all to form a developed gluten network, and the majority of reports on the microstructure of dough are based on wheat flour doughs. The typical dough has a gluten network giving it its integrity and elasticity; the input of work to produce this is an essential step in the breadmaking process.

The inclusion of additional components (fats, egg, milk, etc.) results in the production of a batter which forms the basis for flour confectionery such as cakes, buns and pastries. The simplest way of looking at a batter with the light microscope is to take a few drops of it and put it on a glass slide. Starch granules present may be identified using polarized light, and air or gas bubbles are clearly visible due to their high refractivity. Other components require staining to reveal their identity, but much information can be gleaned from this simple and direct examination.

The application of heat to the slide during the observation of the batter has been used in an effort to understand more about the baking process. Bell *et al.* (1975) used this technique and, by staining the fat with oil red O before making the batter, coupled with cine and video recording, watched the process of baking in sponge cake, Madeira cake and high-ratio cake batters. Time lapse photography was used to follow the development of structure within the model system. The result confirmed that the air cells in cakes form in, and in the baked cake are lined by, the lipid component of the batter. The authors point out, however, that the restriction of the system between two planes of glass is an artificial situation, and they conceded that the gas cells in their 'bakes' are only one-sixth of the size of those found in the full size cake. As well as prestaining the components to be used in the batter or dough,

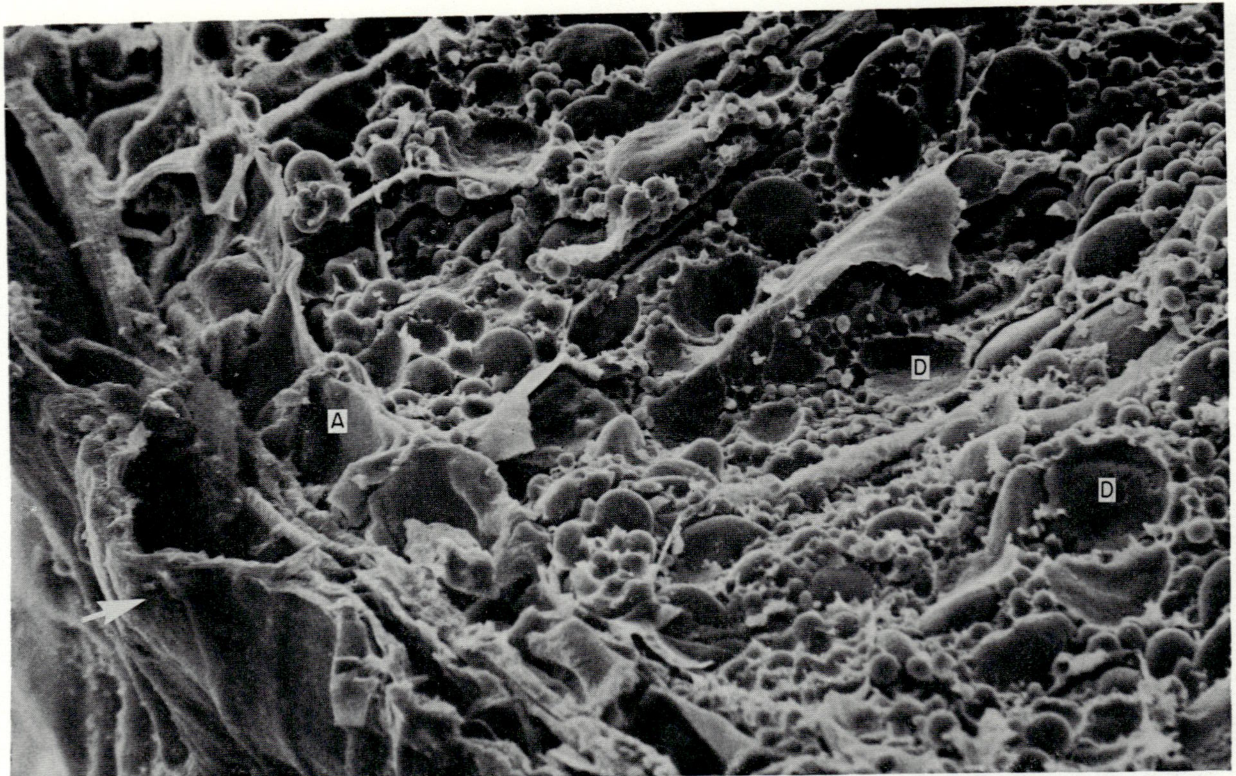

FIG. 30. Rye (*Secale cereale*), SEM ×425. Dry-fractured. The pericarp tissues (arrowed) in their dehydrated state form the outer boundary of the grain. The single line of aleurone cells (A) are brick-shaped. Within the aleurone layer, the starchy endosperm cells are linear, with their long axes orientated radially. Within the endosperm cells are the large and small starch granules, and, in places, depressions (D) on the protein matrix of the cells where starch granules have been torn out during the fracturing process.

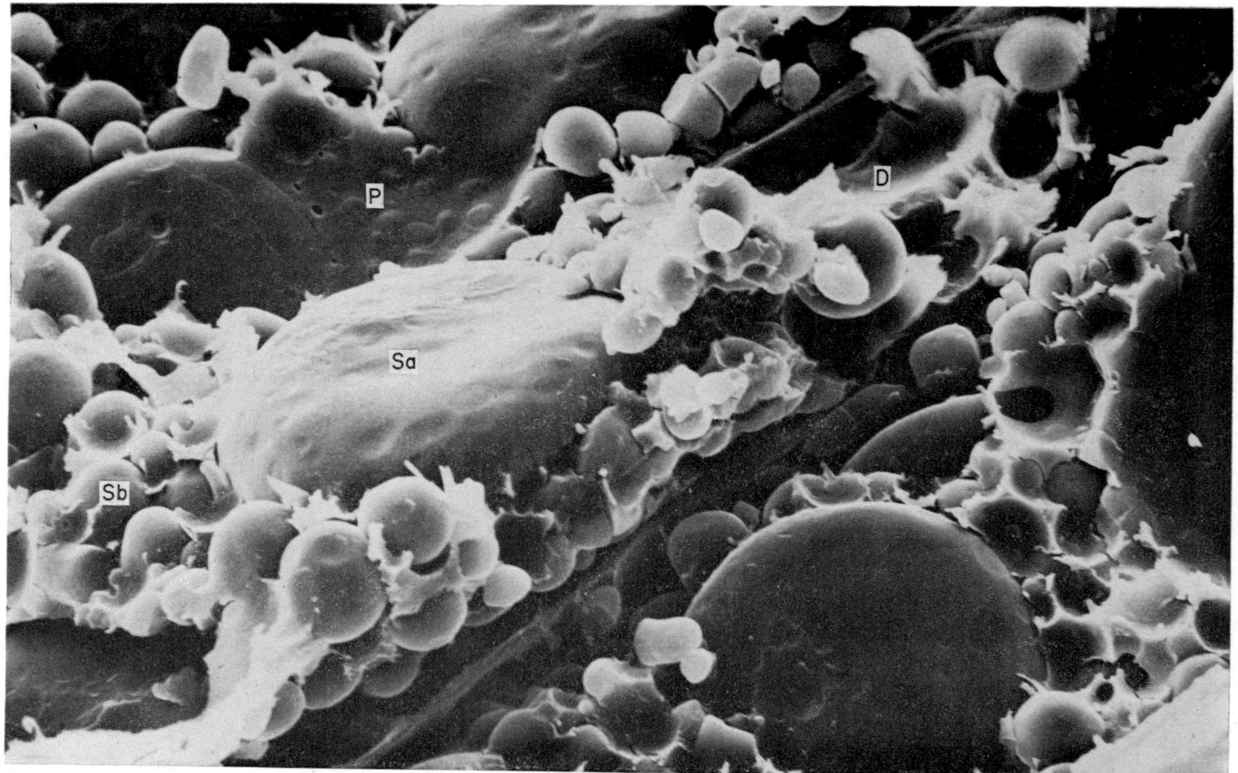

FIG. 31. Rye (*Secale cereale*), SEM ×1625. Dry-fractured. In this view of the cross-fractured caryopsis the contents of parts of three of the endosperm cells can be seen. The starch granules are of two very distinct types, the large, lenticular granules (Sa), and the smaller, isodiametric granules (Sb). The protein matrix (P) shows depressions (D) where granules have broken out during the fracturing process.

a number of workers have developed techniques for staining the components after some form of processing for light microscopy (Flint and Moss, 1970). Figures 35 and 36 illustrate the structure of a dough after staining with light green; these micrographs also show the effects of varying the preparation technique. The structure of a bread dough as seen by the scanning electron microscope is illustrated in Fig. 37. Work on dough structure is beginning to provide insight into its behaviour. All forms of microscopy are in use in this field, and three-dimensional structural relationships are being elucidated by the use of the scanning electron microscope. Caution, however, must be exercised in the interpretation of the micrographs for the reasons already stated in Section II. Cumming and Tung (1975) investigated the microstructure of gluten in wheat flour doughs and concluded that the presence of the starch in a dough is a necessary adjunct to the formation of a gluten network; without the starch the gluten forms a structureless mass as seen through the scanning electron microscope. When viewed in the transmission electron microscope. however, orientation of fibrils within the gluten mass was observed. Moss (1972, 1974) has used the light microscope in a series of studies on the microstructure of bread doughs, and has concluded that the dough mixing process starts with a diffuse continuum of protein and is complete when the majority of starch granules are enclosed in a reticulum of gluten fibrils. Evans *et al.* (1977) used the scanning electron microscope in a study of dough structures in the presence of emulsifiers and single cell protein. These authors used critical point drying to prepare glutaraldehyde-fixed specimens. A fine sheeting of protein, covering the starch granules, was shown, but not all of the granules were entirely enveloped. Fine fibrils of gluten joined those granules which were not veiled with sheets of gluten (Fig. 37).

In the manufacture of pasta, the durum semolina is mixed with water to a moisture level of about 35% and is then extruded through a die by a pressure of about 70 bars (1000 pounds per square inch). Insufficient work is done to develop a gluten network, and so the structure of pasta is of starch granules embedded in a protein continuum (Fig. 38). The surface of the pasta is affected by the type of die used for the extrusion; a Teflon-faced die giving a smooth surface, with starch granules embedded in the protein, whilst the traditional bronze die gives a much rougher surface, upon which lie starch granules linked by strands of protein (Figs 39 and 40).

C. Baked Products

The light microscopy of bread and dough was pioneered by Sheffer (1916). Butterworth and Colbeck (1938) produced some of the earliest micrographs.

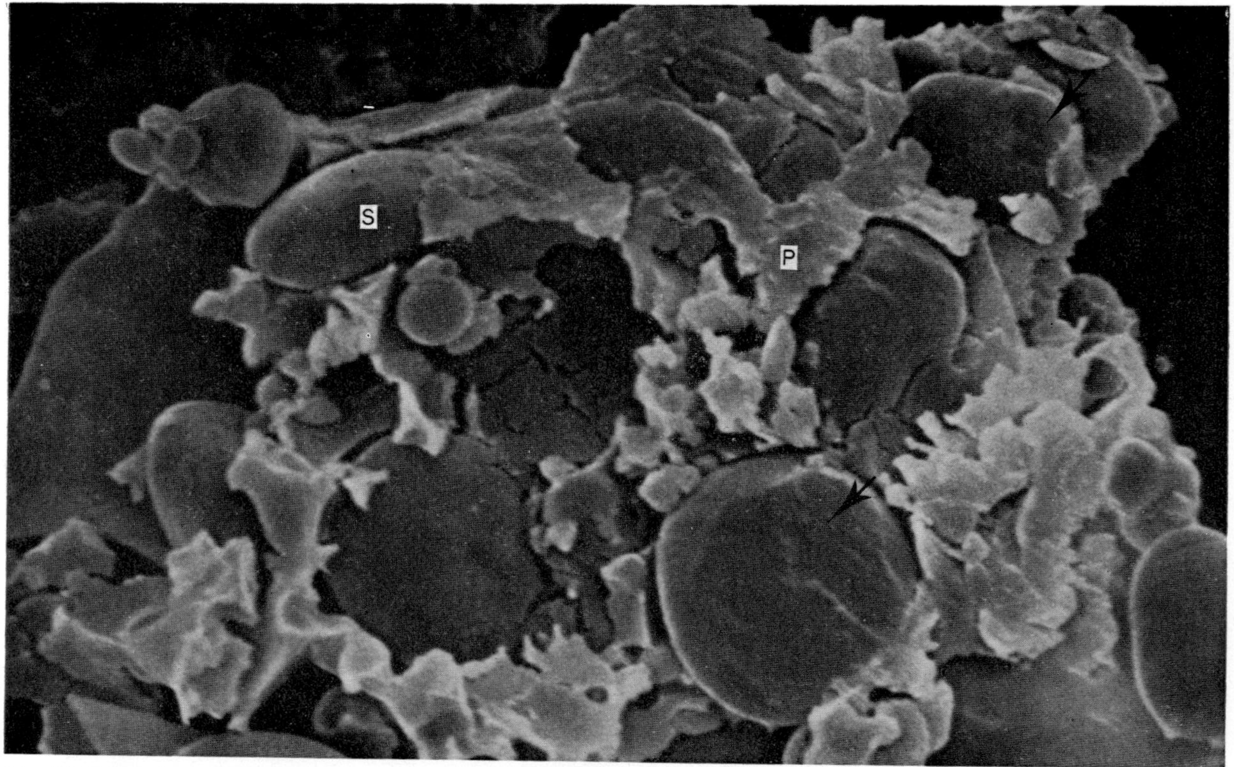

FIG. 32. Wheat flour, SEM ×2250. Dry preparation. Flour particle from second reduction rolls of the Buhler mill. The grist was Cardinal, a hard wheat. This particle of flour shows starch granules (S) embedded in the protein matrix (P). Some of the granules have been held firmly in the protein whilst being abraded by the mill (arrowed). This is one of the ways in which hard wheat mills to give a high starch damage

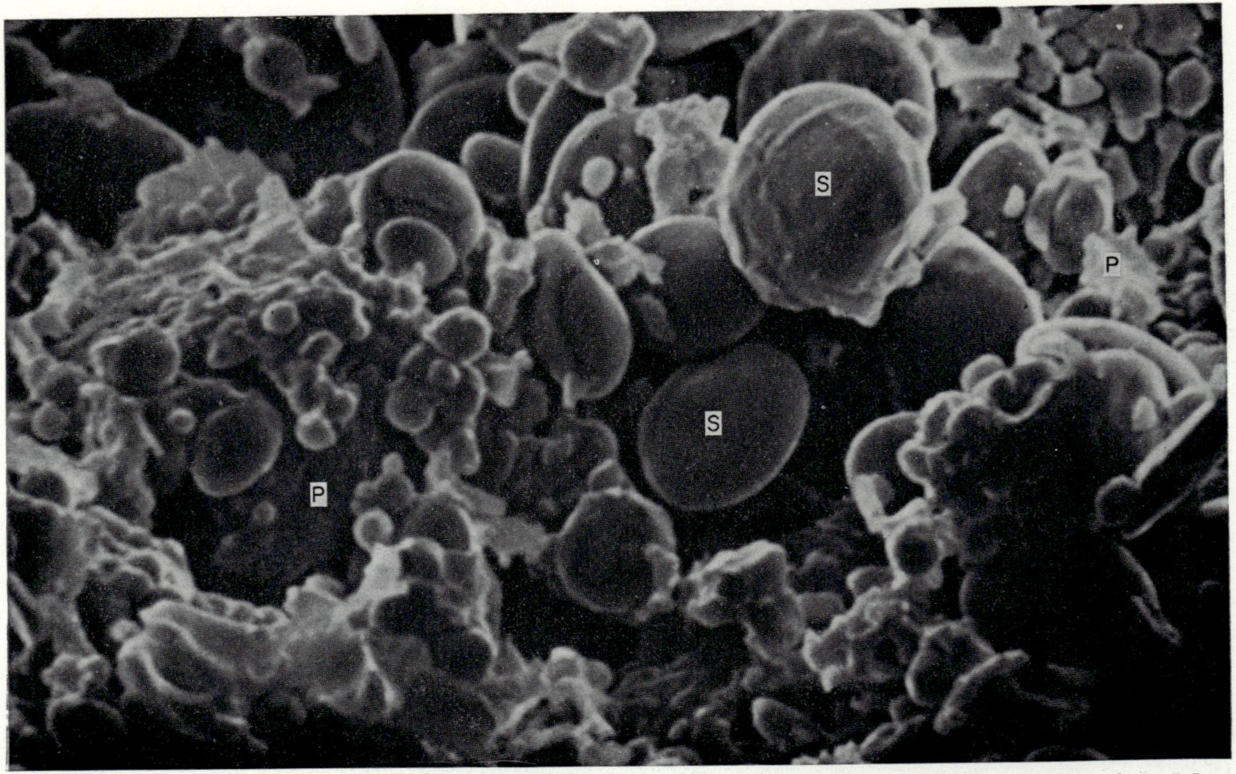

FIG. 33. Wheat flour, SEM ×1770. Dry preparation. Flour from second reduction rolls of the Buhler mill. The grist was Joss Cambier, a soft wheat. Here, the endosperm has broken down into a loose assembly of starch granules (S) and protein (P). This wheat produces a flour with low starch damage.

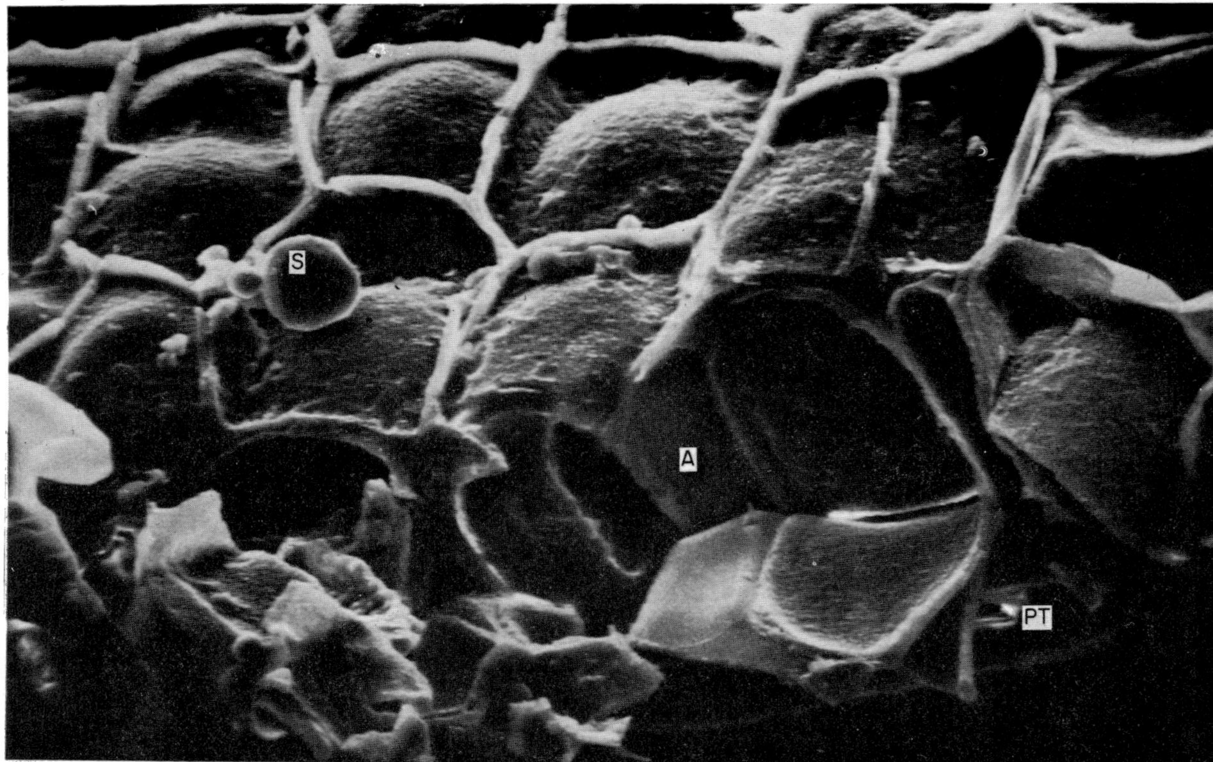

FIG. 34. Wheat bran, SEM ×1625. Dry preparation. This particle of bran is from Maris widgeon grown with high added nitrogen to give a high protein content. The wheat was Buhler milled. Bran consists of aleurone cells (A), usually still attached to the pericarp and testa (PT). Some of the aleurone cells in this fragment have been broken open to reveal the still intact contents. The endosperm of the wheat was vitreous, and the aleurone has sheared off at the outer edge of the subaleurone cells. Only the outlines of the cell walls are present, none of the starchy endosperm save a few starch granules (S) remaining with the bran. This wheat gave a high extraction rate in the mill.

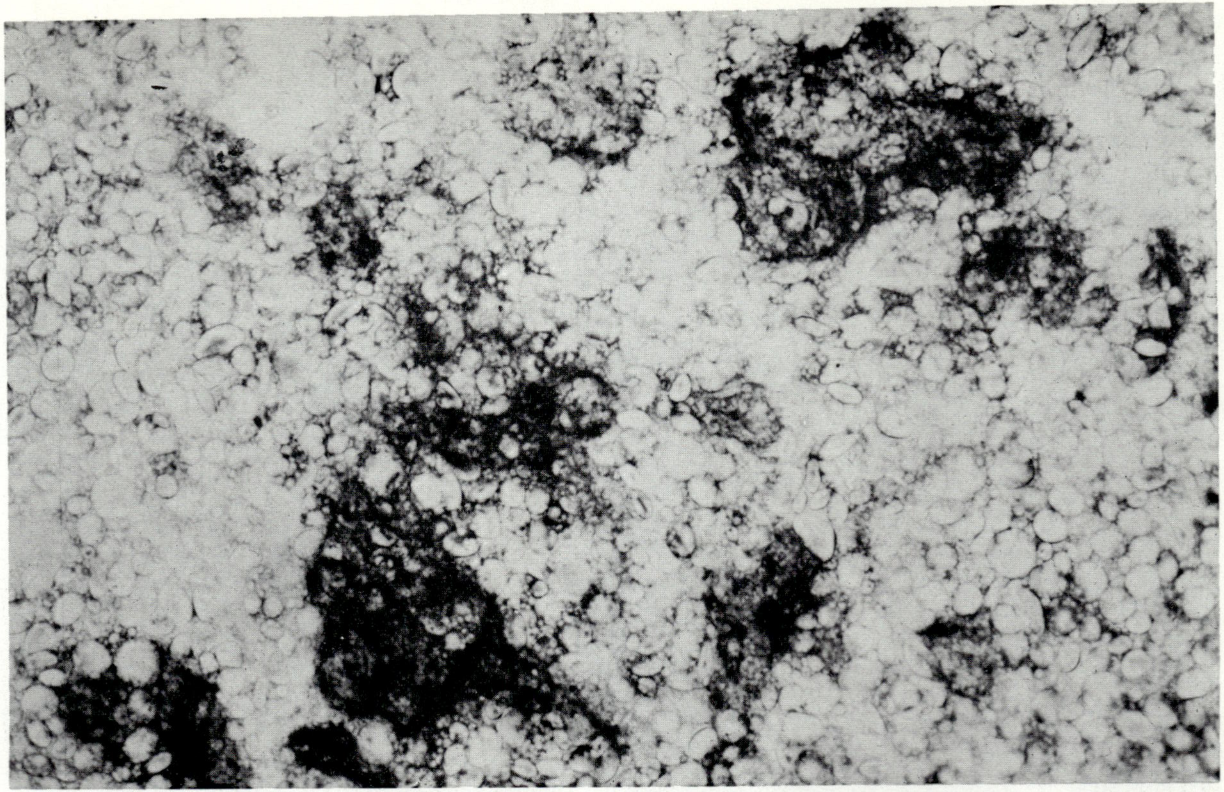

FIG. 35. Dough from wheat flour, LM $\times 240$. Liquid nitrogen frozen, cryostat-sectioned. Dough from a gluten washing process, extracted from the mixer before complete development of the gluten network. After freezing in liquid nitrogen slush and sectioning at 15 μm on a cryostat at $-30°C$, the sections were stained with light green and photographed through a red filter. The gluten, stained dark in this micrograph, appears as diffuse patches with starch enmeshed in the gluten strands.

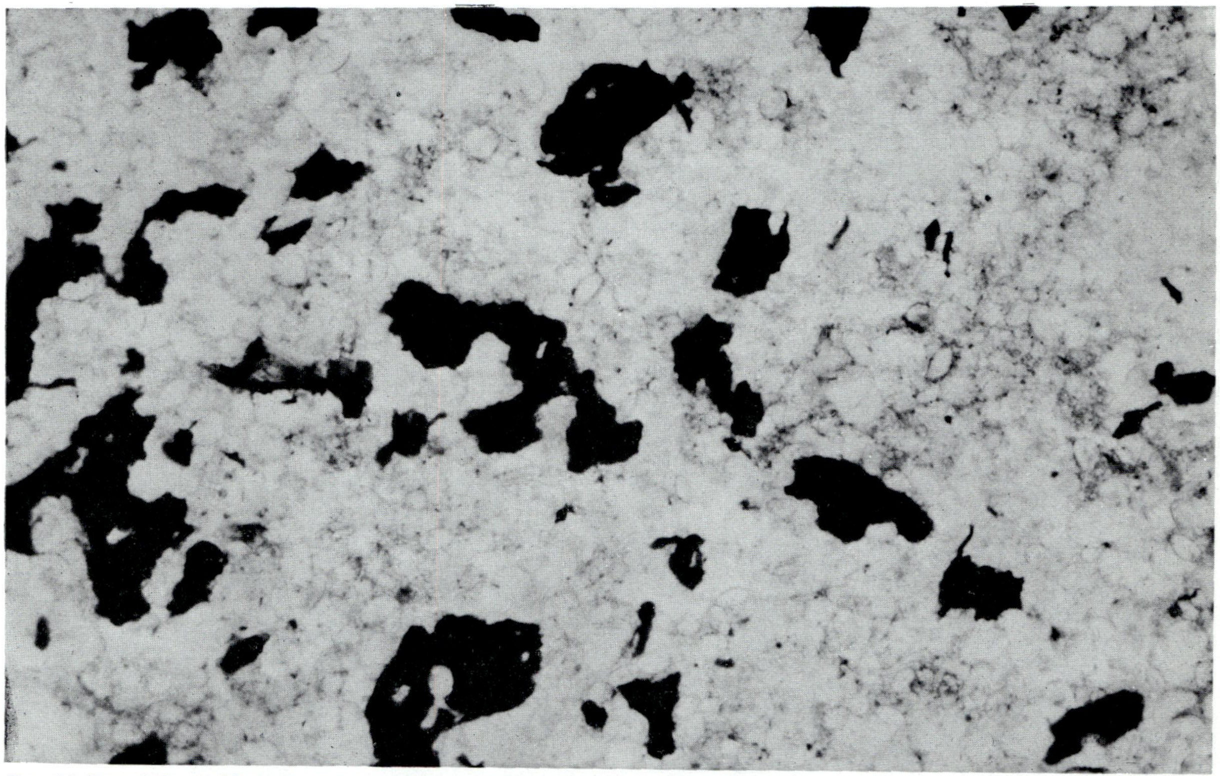

FIG. 36. Dough from wheat flour, LM ×240. Fixed in glutaraldehyde, cryostat-sectioned. As Fig. 35, but fixed in 2·5% glutaraldehyde. The gluten has relaxed into compact lumps and little starch is enmeshed in the gluten.

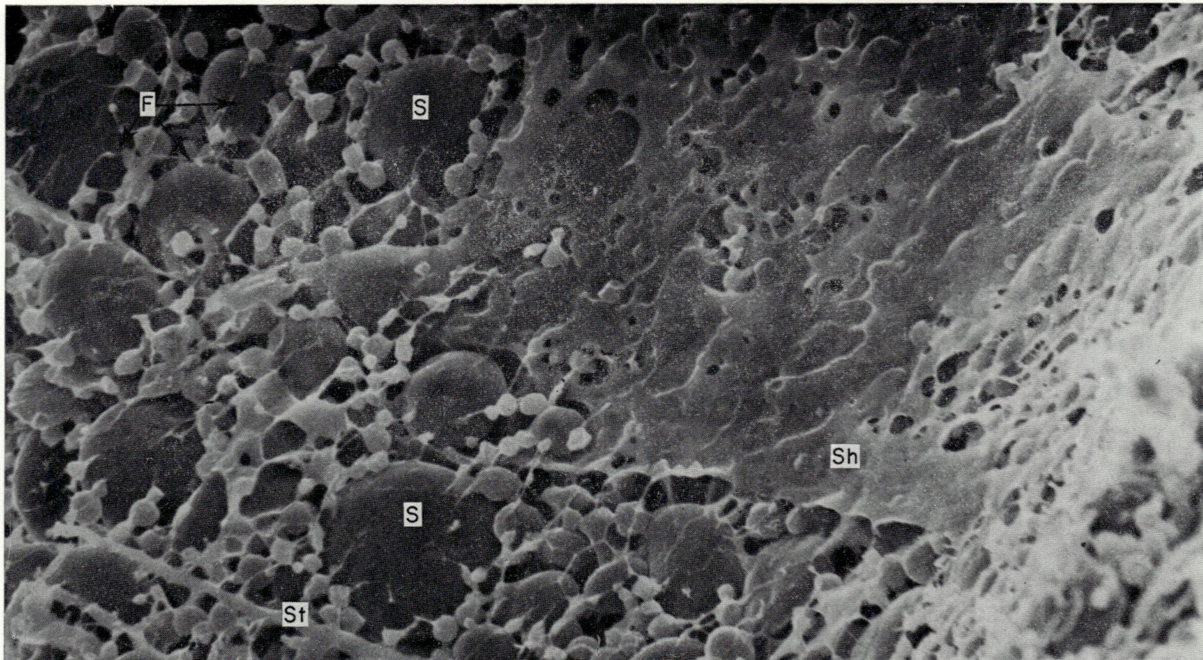

FIG. 37. Dough from wheat flour, SEM ×990. Freeze-dried. Bread dough was mixed in a Tweedy mixer. After proving, a sample of the dough was frozen in nitrogen slush and freeze-dried. The area shown is the inner surface of a gas bubble. Water in the dough having been removed by the freeze-drying process, the relationship between the starch and gluten can be seen. Starch granules (S) are held together by gluten which is in three forms. There are sheets of gluten (Sh) lining the gas bubble (top right), together with strands of gluten (St). All the starch granules visible are held together by fine filaments of gluten (F) which appear to be adhering to the surface of the starch granules.

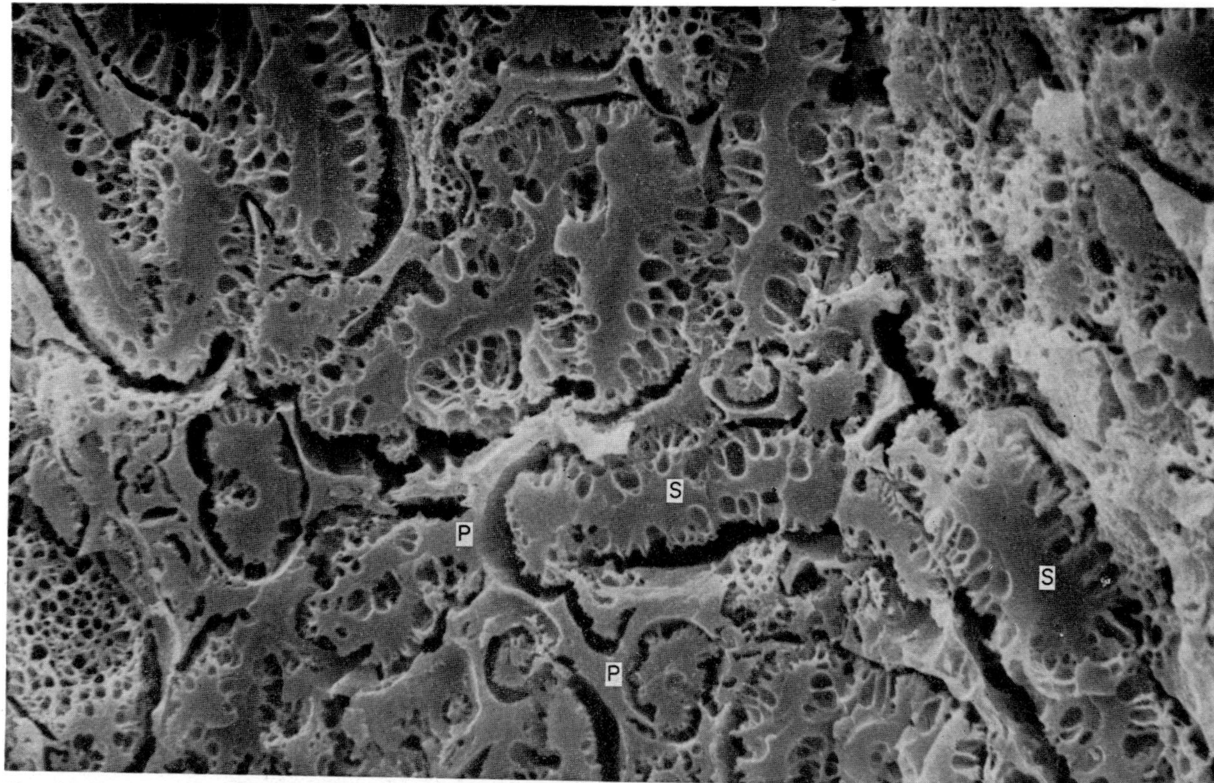

FIG. 38. Pasta made from durum wheat, SEM $\times 3100$. Freeze-dried. This micrograph is of spaghetti cooked for 9 min. After cooking, the sample was frozen in nitrogen slush and freeze-dried. The dried pasta was cross-fractured to reveal the internal structure. The protein (P) forms the continuous phase of the product, surrounding the starch granules (S). The starch appears honeycombed after freeze-drying and there are spaces between the starch granules and the protein. These voids mark those parts of the cooked product where most of the water was present before freeze-drying. (See Section II.A.3 for a discussion on freezing artefacts in specimen preparation.)

FIG. 39. Pasta made from durum wheat, SEM ×850. Dry specimen. Surface of pasta extruded through a Teflon-coated die. Starch granules (S) appear as outlines largely included in a relatively smooth product surface. The continuous phase surrounding the starch granules is the protein from the durum semolina.

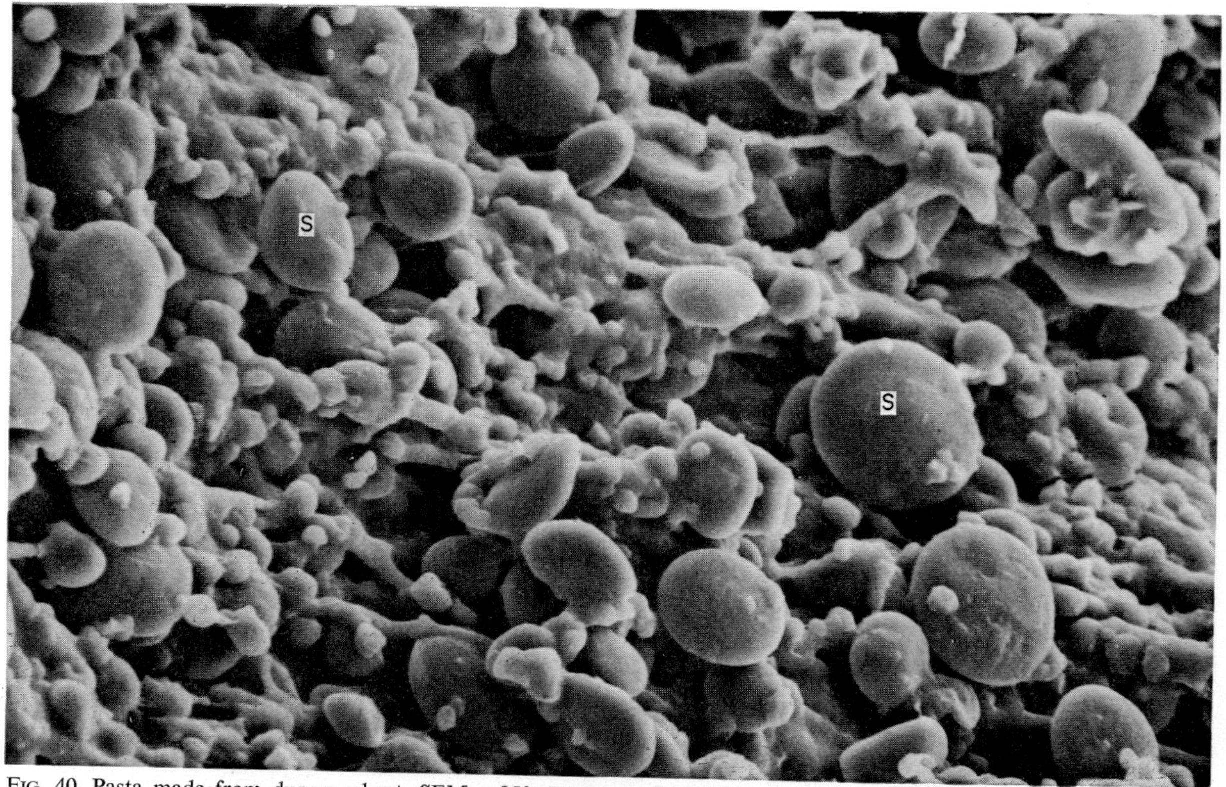

FIG. 40. Pasta made from durum wheat, SEM ×850. Dry specimen. This pasta surface results from extrusion through a bronze-faced die. The roughness of the die surface has resulted in a much rougher pasta surface. The starch granules (S) project above the general surface, and strands of protein show the direction of extrusion (from left to right).

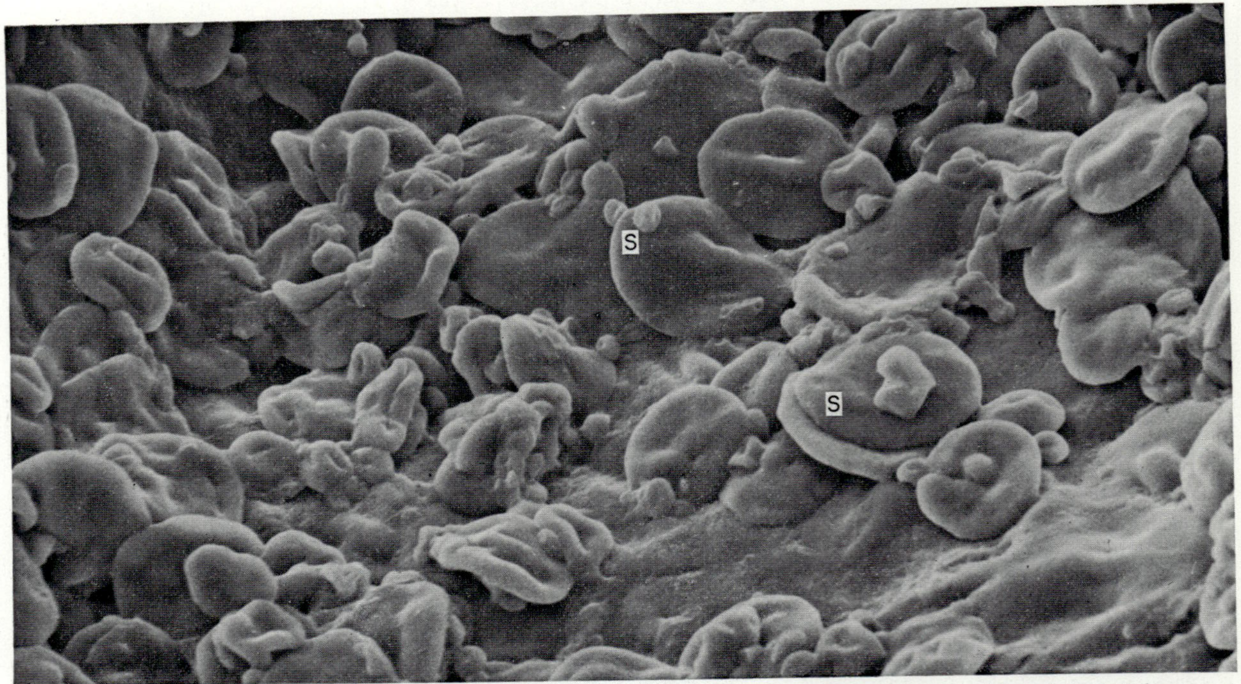

FIG. 41. Bread made from wheat flour, SEM ×990. Air-dried. Bread baked from a grossly overmixed dough (8 watt/hours per pound, or 63 kilojoules per kilogram, at 800 revolutions per minute in a Tweedy mixer). Many of the starch granules (S) in the crumb lie upon the gluten film and are not enmeshed within it. This crumb lacked elasticity and the loaf was of poor volume (cf. Fig. 42).

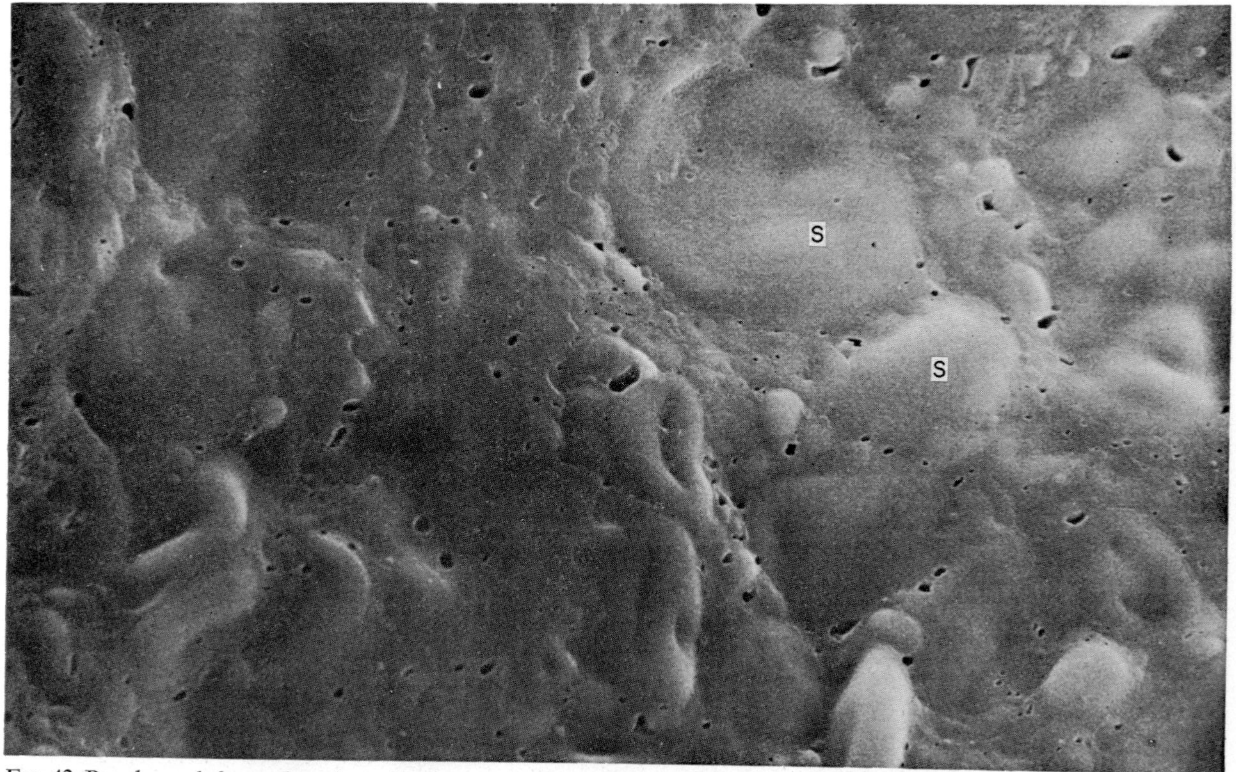

FIG. 42. Bread crumb from wheat flour, SEM $\times 1700$. Air-dried. The region shown is part of the wall separating the component gas cells. The starch granules (S), although gelatinized and expanded, are still recognizable. The continous phase is the gluten, and it appears to cover the starch granules as a thin film. There are numerous small perforations in the gluten film through which pressure equilibration takes place during and after baking.

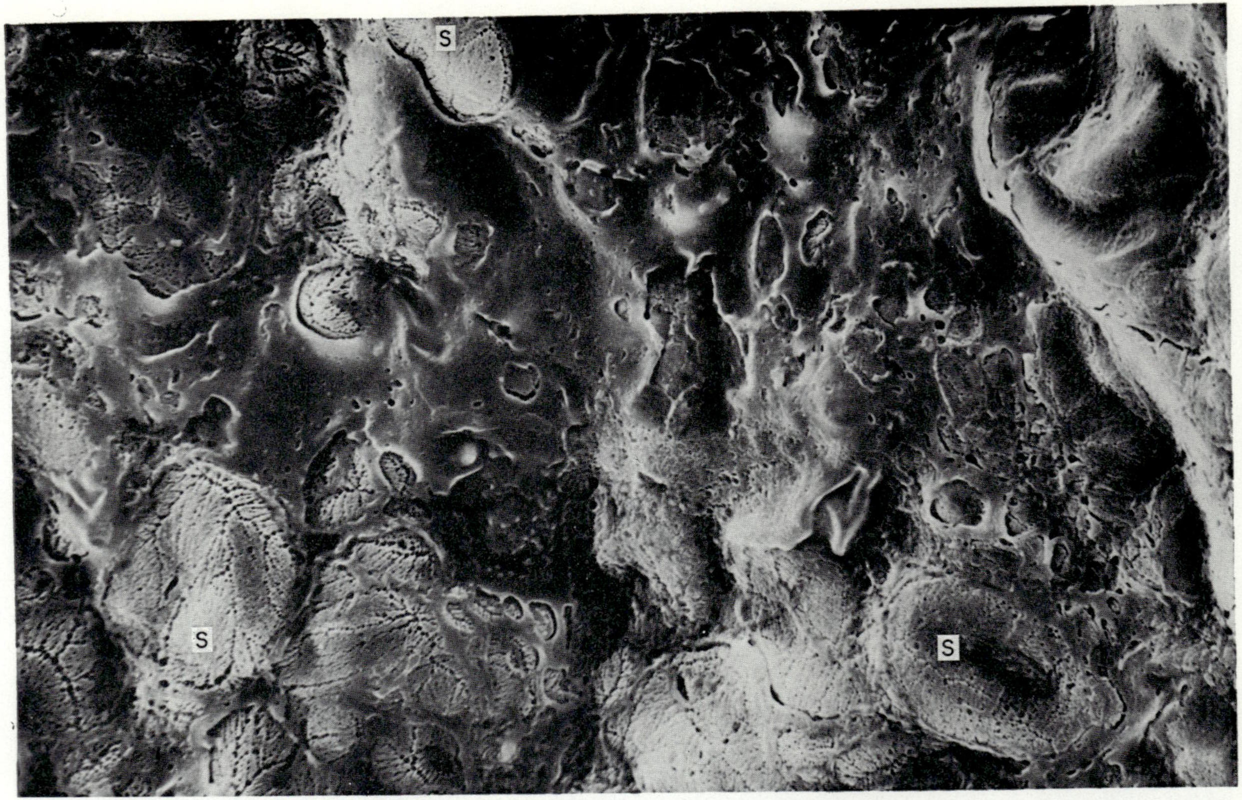

FIG. 43. Bread crumb from wheat flour, SEM $\times 1700$. Freeze-dried. A piece of bread crumb approximately 10 mm^3 was plunged into nitrogen slush, and freeze-dried. The gelatinized starch granules (S) contained sufficient water to freeze out as ice crystals. The drying process has resulted in the appearance of patterns of holes in the starch granules marking the position occupied by the ice. The gluten film shows a similar structure to that of the air-dried material, but does not cover all of the surfaces of the starch granules (cf. Fig. 42).

Burhans and Clapp (1942) undertook extensive studies into the process of mixing and that of gas cell formation during proving, and also investigated the crumb structure of the baked product. They observed that 'whiteness' in baked bread is enhanced by the presence of large numbers of small gas cells in the finished loaf.

Sandstedt *et al.* (1954) avoided problems of specimen hydration in bakery products by plastic embedding followed by exposure of the specimen by grinding. Reflected light was used for specimen examination (as with geological material). The authors found that cotton blue, iodine, acid and basic fuchsin in 80% glycerol could be used to stain the exposed surface of the block with little specimen distortion. They found that starch granules retain their integrity in the bread crumb, but that they are deformed and flattened so that their long axes lie parallel to the surface of the gas cells within the bread. Francis and Groves (1962) used the light microscope in a study of bread, cake and biscuits. These authors used gelatin embedding, and were mindful of the risks of artefact generation by the hydration of baked goods. They concluded that in all baked goods, from biscuits to cakes, a gluten network exists; exceptions to this are 'short sweet' biscuits, where the structure is made up by fat and sugar bridges, and crispbread, where large fragments are present and dough development has not occurred. In 'high-ratio' cakes (those with a greater weight of sugar than flour in the mix) which contain flour which has been chlorinated, Francis and Groves found it difficult to establish whether starch or protein was the principal structure-forming material. Stevens (1976) used the scanning electron microscope to show that, in wafer biscuits, the continuous phase, and structure-forming component, is the starch.

Moss (1975) used light microscopy in a study of the effects of cysteine and oxidizing agents on bread structure. It was found that the gas cell walls were thicker when cysteine was used. During this work it was noted that the hydration necessary for embedding in an aqueous medium ('Tissue-Tek') caused disruption of the protein network due to the swelling of gelatinized starch. Chlorazol violet was used to assess the degree of starch gelatinization.

The use of the scanning electron microscope to study the structure of bread crumb reveals that the gas cell walls have a continuous protein phase, with starch granules more or less embedded in it (Figs 41 and 42). Most work on the microstructure of bread crumb has been undertaken on air-dried or vacuum-dried bread. When freeze-drying is employed, the shrinkage of starch granules does not occur (Fig. 42). The starch granules show patterns of voids, which indicate where ice crystals formed in the gelatinized, and still partially hydrated, starch granules during freezing. Wassermann and Dorfner (1974) examined freeze-dried bread in the scanning electron microscope. Their specimens included crumb derived from wheat and rye flour, and they observed that the protein film lining the gas cells in rye bread is spongy, with numerous pores occupying half the area of the gas cell surfaces. Belderok (1975) also used scanning electron microscopy in a study on the effects of bread improvers, and concluded that the microstructure of the crumb of baked wheaten bread was affected by the use of 'Emulthin'. Flours with a low protein content could be used, and these still gave the same fine structure as bread made from a high-protein ('strong') flour. Khoo *et al.* (1975), in work covering both doughs and baked bread crumb, used scanning and transmission electron microscopy and observed that enzymic digestion of the starch granules occurs during dough proving, and that the large starch granules are distorted more than small granules during the baking process.

The scanning electron microscope will be used more and more in the exploration of the relationship between texture, as perceived by the consumer, and microstructure. At all times during studies of this type it is absolutely essential to have an understanding of the effects of specimen collection and processing upon the material, and to be certain that sufficient material is examined for the interpretation of the micrographs obtained to be based on a representative sample. The use of the light microscope, with appropriate stains for the identification of the components of the product, is a necessary adjunct to any electron microscopic investigation.

ACKNOWLEDGEMENT

I wish to thank Dr J. Edelman (Director of Research, R.H.M. Research Ltd.) for permission to write this chapter.

REFERENCES

Angold, R. E. (1967). *Rev. Palaeobot. Palynol.* **3**, 205.

Angold, R. E. (1975). *In* 'Bread' (A. Spicer, ed.), p. 141. Applied Science Publishers, London.

Aranyi, C. and Hawrylewicz, E. I. (1968). *Cereal Chem.* **45**, 500.

Badi, S. M., Hoseney, R. C. and Casady, A. I. (1976). *Cereal Chem.* **53**, 478.

Bechtel, D. B. and Pomeranz, Y. (1977). *Am. J. Bot.* **64**, 966.

Belderok, B. (1975). *Mühle Mischfuttertechnik*, **19**, 251.

Bell, A. V., Berger, K. G., Russo, J. V., White, G. N. and Weathers, T. L. (1975). *J. Fd Technol.* **10**, 147.

Bennett, M. D., Smith, J. B. and Barclay, I. (1975). *Phil. Trans. R. Soc. Ser. B*, **272**, 13.

Boyde, A. and Wood, C. (1969). *J. Microscopie*, **90**, 221.

Burhans, M. E. and Clapp, J. (1942). *Cereal Chem.* **19**, 196.

Butterworth, S. W. and Colbeck, W. J. (1938). *Cereal Chem.* **15**, 475.

Buttrose, M. S. (1963). *Aust. J. Biol. Sci.* **16**, 305.

Clayden, E. C. (1971). 'Practical Section Cutting and Staining'. Churchill Livingstone, London.

Crozet, N. (1977). *Cereal Chem.* **54**, 1108.

Cumming, D. B. and Tung, M. A. (1975). *Can. Inst. Fd Sci. Technol. J.* **8**, 67.

Evans, L. G., Volpe, T. and Zabik, M. E. (1977). *J. Fd Sci.* **42**, 70.

Evers, A. D. (1970). *Ann. Bot.* **34**, 547.

Feder, N. and O'Brien, T. P. (1968). *Am. J. Bot.* **55**, 123.

Flint, F. O. and Moss, R. (1970). *Stain Technol.* **45**, 75.

Francis, B. and Groves, C. H. (1962). *Jl R. Microsc. Soc.* **81**, 53.

Fulcher, R. G., O'Brien, T. P. and Lee, J. W. (1972). *Aust. J. Biol. Sci.* **25**, 23.

Gallant, D. and Guilbot, B. (1971). *Stärke,* **23**, 244.

Hinton, J. J. C. (1955). *Cereal Chem.* **32**, 296.

Jones, R. L. (1969). *Planta,* **87**, 119.

Khoo, U., Christianson, D. D. and Inglett, G. E. (1975). *Bakers' Dig.* **49**, 24.

Marinos, N. G. and Fife, D. N. (1972). *Protoplasma,* **74**, 381.

McMasters, M. M., Hinton, J. J. C. and Bradbury, D. (1971). *In* 'Wheat, Chemistry and Technology' (Y. J. Pomeranz, ed.), p. 51. A.A.C.C., St Paul, Minnesota.

Moeller, J. (1886). 'Mikroscopie der Nahrungs und Genusmittel aus dem Pfkenzenreiche'. Julius Springer, Berlin.

Moor, H., Mühlethaler, K., Waldner, H. and Frey-Wyssling, A. (1961). *J. Biophys. Biochem. Cytol.* **10**, 1.

Morrison, I. N., Kuo, J. and O'Brien, T. P. (1975). *Planta,* **123**, 105.

Moss, R. (1972). *C.S.I.R.O. Fd Res. Q.* **32**, 50.

Moss, R. (1974). *Cereal Sci. Today* **19**, 557.

Moss, R. (1975). *Cereal Fds Wld,* **20**, 289.

Opik, H. (1972). *Planta,* **102**, 61.

Pomeranz, Y. and Sachs, I. B. (1972a). *Cereal Chem.* **49**, 1, 7.

Pomeranz, Y. and Sachs, I. B. (1972b). *Cereal Chem.* **49**, 11, 20.

Reynolds, E. S. (1963). *J. Cell Biol.* **17**, 208.

Robutti, J. L., Hoseney, R. C. and Wassom, C. E. (1974). *Cereal Chem.* **51**, 173.

Sabatini, D. D., Miller, F. and Barnett, R. J. (1963). *J. Histochem. Cytochem.* **12**, 57.

Sandstedt, R. M., Schaumburg, L. and Fleming, J. (1954). *Cereal Chem.* **31**, 43.

Seckinger, H. L. and Wolf, M. J. (1967). *Cereal Chem.* **44**, 669.

Seckinger, H. L. and Wolf, M. J. (1973). *Cereal Chem.* **50**, 455.

Seidemann, J. (1966). 'Starch Atlas: Principles of Starch Microscopy and Descriptions of Important Varieties of Starch'. Parey, Berlin.

Sheffer, W. (1916). *Z. Ges. Getreidew.* **8**, 6.

Shoup, F. K., Dyre, C. W., Campbell, J. and Parrish, D. B. (1969). *Cereal Chem.* **46**, 164.

Simmonds, D. H. (1972). *Cereal Chem.* **49**, 212.

Sleytr, U. B. (1970). *Protoplasma,* **71**, 295.

Spurr, A. R. (1969). *J. Ultrastruct. Res.* **26**, 31.

Stenvert, N. L. and Kingswood, R. (1977). *Cereal Chem.* **54**, 627.

Stevens, D. J. (1976). *Stärke*, **28**, 5.

Sullins, R. D. and Rooney, L. W. (1974). *Cereal Chem.* **52**, 361.

Swift, J. G. and Buttrose, M. S. (1972). *J. Ultrastruct. Res.* **40**, 378.

Swift, J. G. and O'Brien, T. P. (1972). *Aust. J. Biol. Sci.* **25**, 9.

Tanaka, K., Ogawa, M. and Kasai, Z. (1976). *Cereal Chem.* **53**, 643.

Tanaka, K., Ogawa, M. and Kasai, Z. (1977). *Cereal Chem.* **54**, 684.

Tsen, C. C. (1974). 'Triticale: The First Man-made Cereal'. A.A.C.C., St Paul, Minnesota.

Umrath, W. (1974). *J. Microscopie*, **101**, 103.

Varriano-Marston, E. (1977). *Fd Technol., Champaign*, **31**, 32.

Vaughan, J. G. (1970). 'The Structure and Utilization of Oil Seeds'. Chapman and Hall, London.

Wassermann, L. and Dorfner, H. H. (1974). *Getreide Mehl Brot*, **28**, 324.

Watson, C. A. and Dikeman, E. (1977). *Cereal Chem.* **54**, 120.

Winton, A. L. and Winton, K. B. (1932). 'The Structure and Composition of Foods', Vol. 1. John Wiley and Sons, New York.

4. Cereal Starches and Proteins

A. D. EVERS

Flour Milling and Baking Research Association,
Rickmansworth, U.K.

Starch is better preserved in alcohol than in formalin or acidic solutions which might hydrolyse it. Usually however, it is preferable to be rid of it.

(J. Brontë-Gatenby and T. S. Painter in 'The Microtomist's Vade-mecum'.)

I. INTRODUCTION

A. Literature Coverage

Light microscopy (LM), transmission electron microscopy (TEM) and scanning electron microscopy (SEM) have all been applied widely and effectively to the study of physical aspects of the starch and protein of cereals. These two major components of cereal endosperm have been studied either *in situ* by examination of thin sections of kernels, or after their removal and possible modification, intentional or otherwise. A brief guide to the literature concerning the microscopy of starch and cereal proteins has recently been published (White and Shenton, 1975) and much information on microscopical aspects of these cereal components can be found in general texts on starch (Whistler and Paschall, 1965, 1967; Seidemann, 1966; Radley, 1968, 1976; Badenhuizen, 1969; Banks and Greenwood, 1975) and seed proteins (Inglett, 1972; Norton, 1978). Publications dealing with anatomical aspects are listed by Rost and Lersten (1973), and fundamental and technological aspects of individual cereals are reviewed in specialized monographs published by the American Association of Cereal Chemists (Pomeranz, 1971; Houston, 1972; Bushuk, 1976). The bulk of the available information concerns bread wheat (*Triticum aestivum*), but in many cases data gained through the study of this cereal is considered to be applicable to other members of the Tribe Triticeae (= Hordeae) which includes other wheat species such as *T. durum*, *T. sphaerococcum*, etc. (see Chapter 3 and Peterson, 1965), barley (*Hordeum vulgare*), rye (*Secale cereale*) and the intergeneric hybrid triticale. Corn or maize (*Zea mays*) is well covered, there being much information on both the normal type and the mutant types; the latter have been bred for specific requirements and are characterized by starch with higher or lower amylose: amylopectin ratios than normal or proteins with higher lysine content than normal. Sorghum (*Sorghum vulgare*), with its corresponding mutants, are less well covered, but as their importance as ingredients of processed foods has increased, both this cereal and the millets (*Pennisetum* spp. *Setaria* spp., etc.) have recently attracted more attention from microscopists. Considerable information on the structure of rice (*Oryza sativa*) kernels has been available for some time but the advent of electron microscopes of both types has been particularly helpful to the study of the starch of rice because of the small size of its granules. As with maize and sorghum, mutants with abnormal amylose : amylopectin ratios are covered. Undoubtedly the most neglected cereal, certainly as far as microscopy is concerned, is the oat (*Avena sativa*), information at both the light and electron microscope levels being extremely sparse.

B. Microscopical Techniques

1. Light microscopy

a. Modes of observation. Techniques employed in the study of cereal starches and proteins range from the simplest to the highly sophisticated. With the simplest technique, individual components removed from ground kernels or baked products may be examined in aqueous or other mounts, with or without stains, in order to reveal their characteristic shape and size. Examination of similarly mounted sections of endosperm or baked products can reveal, in addition to the characteristics of the individual components, the relationship borne by one to the other. Similarly, the nature of the observations is variable, reasons for microscopical examination may be merely to gain a mental image of the component, for which the simplest apparatus is adequate. Alternatively, a precise quantitative assessment of one or more measurable features may require the computer-aided analysis of data which has been manually or automatically collected. The quantitative and qualitative aspects of microscopy require different approaches on the part of the microscopist and are hence considered separately in the present review.

b. Fixation and embedding. Fixation is carried out in order to preserve material against decay and undesirable changes which might otherwise result from contact with reagents used in preparing specimens for observation. Of the two components considered here, fixation is applicable, for the most part, only to protein, since starch neither requires nor responds to fixatives, owing to its natural impermeability. For the permanent mounts used in light microscopy, fixation, followed by infiltration, embedding, thin sectioning on a microtome or ultramicrotome, staining and mounting, is necessary. In general, the techniques listed in histology manuals (e.g. Jensen, 1962) are applicable; however in the case of maize kernels, because of their large size, hardness and the crumbly nature of some regions, special methods are required (Sass, 1958). These techniques can be applied to other kernels if required. Methods of preparation of cereal seeds for microscopy were reviewed by Hanssen (1955). A simple method for preparation of relatively thick, temporary mounts of wheat kernel sections was used by Kent (1966), paraffin wax being used to support damped kernels which did not require fixation or infiltration. General methods of fixation were discussed by Feder and O'Brien (1968). Their recommendation of aldehyde fixatives for light microscopy has since been followed widely for cereal kernels. These authors also introduced an embedding technique to light microscopy in which fixed specimens were infiltrated and embedded in the resin glycol methacrylate. Glycol methacrylate infiltrates sufficiently rapidly to be used with specimens of suitable size for light microscopy. At the same time, it allows cutting of sections thin enough for histochemical purposes. Starch granules are firmly held in the resin matrix and their displacement, which occurs frequently in softer embedding media, is avoided. A further advantage of glycol methacrylate embedding is the water permeability of the medium. Thus sections of tissues and surrounding resin can be mounted on slides and brought into contact with suitable stain solutions. Purified resin does not itself become stained.

c. Staining. Stains suitable for use with starch and protein are too numerous to be covered in detail here; however, a selection of frequently used preparations is listed below. For temporary mounts an extremely useful dye is a 0·3% solution of iodine in 6% potassium iodide, in this form or as a component of chlor-zinc iodide, iodine stains normal starch blue-violet and protein a straw colour; starches consisting of a higher than normal amylopectin content can be identified by their red coloration and regions of different composition within the same granule can be distinguished by this differential staining effect (Meyer, 1886). A very weak solution of iodine (0·03%) in potassium iodide can be used for distinguishing granules which have become physically damaged or gelatinized as a result of being heated in the presence of water (Jones, 1940).

The chemistry of the iodine reaction with starches has been discussed by Hollo (1968) and Banks and Greenwood (1975), and the red-staining starches by Hixon and Brimhall (1968). Iodine is extremely volatile and is hence not suitable for use in the TEM or normally in permanent mounts for light microscopy. A method for use of iodine vapour in permanent mounts has however been described (Little, 1957). With hydrogen iodide it is possible to distinguish potato starch from wheat starch (Czaja, 1961). It is more difficult to distinguish between the starches of cereal species by staining. Rye and wheat starch granules are particularly difficult to differentiate microscopically because of their similarity in size and shape, but a procedure described by Unna (1918) is said to achieve this distinction. Details of this and other staining methods have been given by Moss (1976).

The chlorazol series of dyes has been widely used in staining both protein and starch. Only granules which have become damaged or gelatinized take up these stains, which include Congo red. For identifying damaged starch a solution of 2–3% in water is suitable, and cereal proteins stained with Congo red become more orange than damaged starch granules. In cases where a greater distinction between the two components is required the reactive amino groups on proteins may be selectively blocked by pretreatment with a dye-resist agent such as Tanninol ADR (Flint and Moss, 1970). More sophisticated methods for distinguishing damaged granules include the use of fluorescent dyes such as acridine orange and primulin (Czaja, 1961). Results, though striking and aesthetically pleasing, barely justify, in routine work, the greater complexity of procedure and equipment required, since as much useful information is obtained by simpler methods (see however further application of fluorescent dyes in Section V). Dyes such as Congo red can be readily removed from damaged starch granules by washing with water, thus, by alternate irrigation with dye solution and pure water, staining and destaining can be achieved (Jones, 1940). This is because the chromophore is only attached to hydroxyl groups in the starch by weak hydrogen bonds. In another type of dye, described by Stute (1973) as reactive dyes, the chromophore becomes covalently bonded to the hydroxyl groups. These include the Levafix E and R series and the Procion M series. Unfortunately, covalent bonds are also formed with the amino groups in proteins and hence suffer from the same disadvantages as the dyes of the chlorazol series. A permanent method of staining starch granules which does not distinguish between damaged and undamaged granules is the histochemical periodic acid–Schiff (PAS) reaction (Humason, 1967). In this reaction cleavage of α-glycols by periodic acid oxidation produces aldehyde groups; these react under acid conditions and in the presence of sulphur dioxide with the dye basic fuchsin to form a chromaphore *in situ.* This method is frequently used in preparations made with glycol methacrylate as an embedding medium. Good results are achieved with PAS used with the polychromatic toluidine blue or the similar azur A (Himes and Moriber, 1956; Feder and O'Brien, 1968). In addition to their polychromatic reaction with nucleic acids and cell wall components, these dyes stain proteins blue. Fast green may also be used with the PAS technique. Details of these methods are given by Feder and O'Brien (1968). Besides those already described, stains often used for identification of protein include haematoxylin (both Heidenhein's and Delafield's) and orange G. Pomeranz and Shellenberger (1961) tested 29 of the basic and acid dyes cited for protein staining in the literature. In their opinion the most satisfactory one was ponceau 2R. Bromophenol blue, used with mercuric chloride, is another frequent choice for proteins which are not water soluble.

2. Transmission electron microscopy

The penetration of those fixatives and stains used in transmission electron microscopy into starch granules is extremely limited; hence in the present context the discussion is limited to the treatment of proteins. Fixatives in common use with the TEM can be grouped as osmium fixatives, aldehyde fixatives and permanganate fixatives. In the first group the essential ingredient is osmium tetroxide, but variations in composition abound as a result of the use of different buffers and concentrations. Osmium fixatives are fast acting but penetrate relatively slowly, and in addition to fixing they provide some electron staining through the selective deposition of osmium ions in unsaturated lipids, phospholipid membranes and proteins. Permanganate fixatives also have a dual role, both preserving and staining lipoprotein membranes; other components of tissues are not well preserved however. Aldehyde fixatives include those based on formaldehyde, glutaraldehyde and acrolein. These are purely fixatives and since they contain no heavy atoms they do not selectively affect electron scattering power. Double fixation, using, for example, glutaraldehyde followed by osmium, can be used to produce better results than either fixative alone.

The function of osmium and potassium permanganate as stains has already been mentioned. Other stains useful for protein examination include phosphotungstic acid, uranyl acetate (protein stain), lead citrate and lead hydroxide (membranes, etc.).

3. Scanning electron microscopy

The image obtained with the SEM is of the specimen surface. Information can be obtained by the use of the simplest methods of preparation such as dusting starch powder onto the surface of a specimen stub. Preparations of whole kernels, or kernel sections or fragments can also be examined. Sections prepared for light microscopy and the TEM can be used in the SEM, although in the case of embedded sections it is necessary to remove the embedding medium before examination. Removal of naturally occurring components by chemical or enzymic means (Briarty, 1974) may also be desirable in order to study a specific component in detail. As with biological material in general, preparations of cereal products, for best results with the SEM, require coating by evaporation or sputtering of a conducting layer of metal which also serves as a source of secondary electrons (Hearle *et al.*, 1972). Natural or artificial colours, such as those imparted by dyes, do not help to identify individual components since only the shape of specimens can be detected.

The three microscopical methods are clearly complementary, LM offering the unique facility for examining wet specimens, TEM providing the greatest degree of resolution and SEM giving three-dimensional information on natural or artificially created surfaces. A method for use of LM and SEM on the same specimen has been described by Hall and Sayre (1973).

II. OCCURRENCE AND MATURE STRUCTURE

A. Introduction

Starch and protein occur in many tissues and organs of the cereals and other plants. Starch is present at certain stages of development or germination in most parts of the kernel, including the scutellum, the pericarp and the aleurone layer or layers. Similarly, proteins as enzymes or as major constituents of the cellular cytoplasm are universally distributed throughout the plant. For human food usage however, it is the proteins and starches of the starchy endosperm which are most important, and it is their development, mature structure and modification which are discussed here.

Within a cereal endosperm the relative proportions of starch and protein vary, and complex concentration gradients exist in some cases (Boyer *et al.*, 1976a). In all cereal endosperms, however, a gradient exists from high starch and low protein content in the centre of the grain, to low starch and high protein content at the periphery (Cobb, 1902; Hinton, 1953). The inverse relationship between the two major constituents results from the dilution of protein and other cell constituents by starch. This develops to a greater extent in the older cells near the centre than in the younger cells at the outside (Evers, 1970).

B. Starch

Endosperm starch occurs as microscopic granules, the shape and size of which are characteristic of their species. Granules are colourless translucent bodies, the surfaces of which are generally rounded. Angularity and concave depressions sometimes arise as a result of plastic deformation due to pressure of other granules or other cellular components (see Chapter 3, Figs 25, 26 and 31). All cereal endosperms contain starch granules which are near-spherical or, due to compaction, angular with a variable number of flattened sides. In cereals in which horny and floury regions can be distinguished, both spherical and angular granules abound, the angular granules occurring in the more compact horny regions. In addition to spherical and angular granules (which can be together regarded as comprising a single population), some species contain simple or compound granules of entirely different shapes; thus while sorghum, maize and the millets have only one type of granule, oats and rice also contain large, compound, ellipsoid granules. Members of the Tribe Triticeae, viz. wheat, barley, rye and triticale, contain an additional granule population of lenticular form. Intraspecific variation occurs particularly in species where mutants with abnormal amylose : amylopectin ratios occur. In general, high amylopectin content is associated with greater angularity and high amylose with greater variability in shape.

In their natural state cereal starches do not exhibit the concentric banded appearance of some other types, although a marked ring structure can be revealed by a number of treatments in normally grown granules. In plants grown under continuous illumination, no ring structure can be shown (Bakhuyzen, 1926) (Figs 1 and 2). In some types a hilum can be seen, while in others the hilum is revealed only when the granules are extremely dehydrated. The hilum is a small void occurring at or near the centre of the

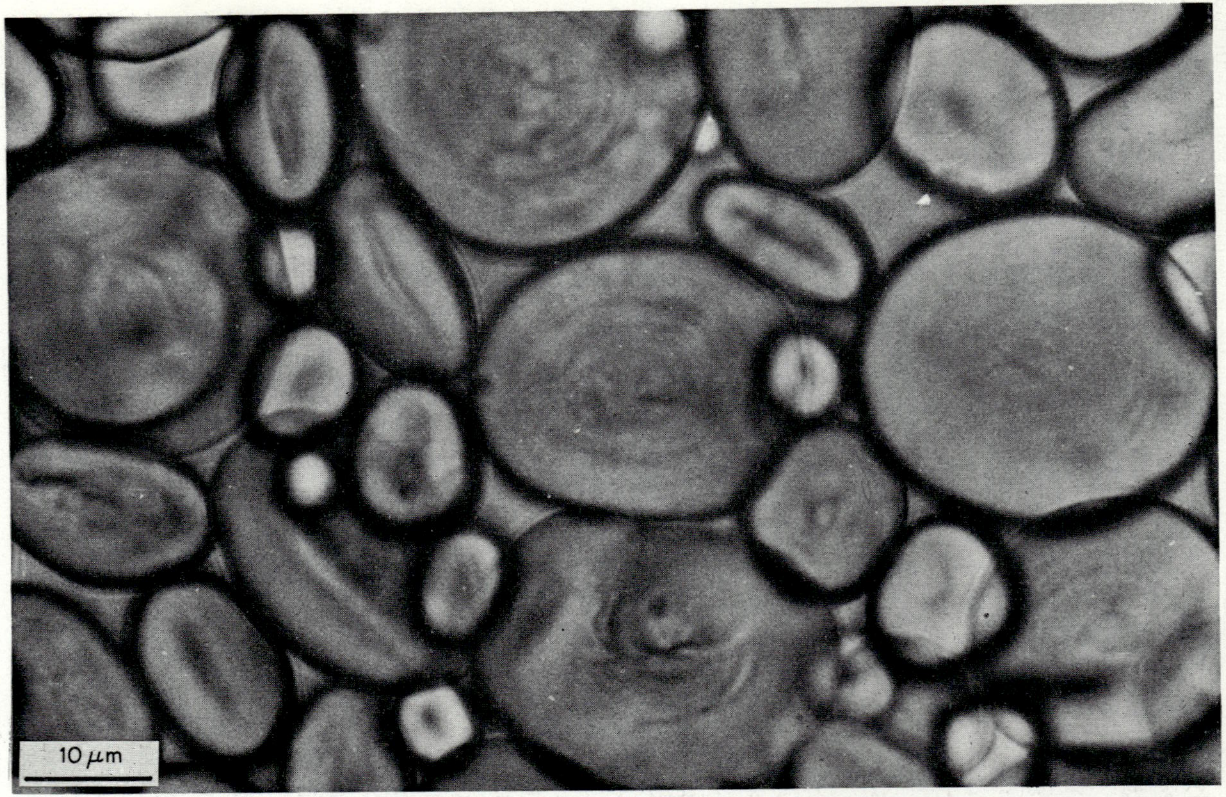

FIG. 1. Starch granules from the endosperm of wheat grown in the field. The concentric ring structure is not normally visible in cereal starches but it has been revealed here in an aqueous suspension which has been warmed slightly. Some swelling has also resulted. Light micrograph, unstained.

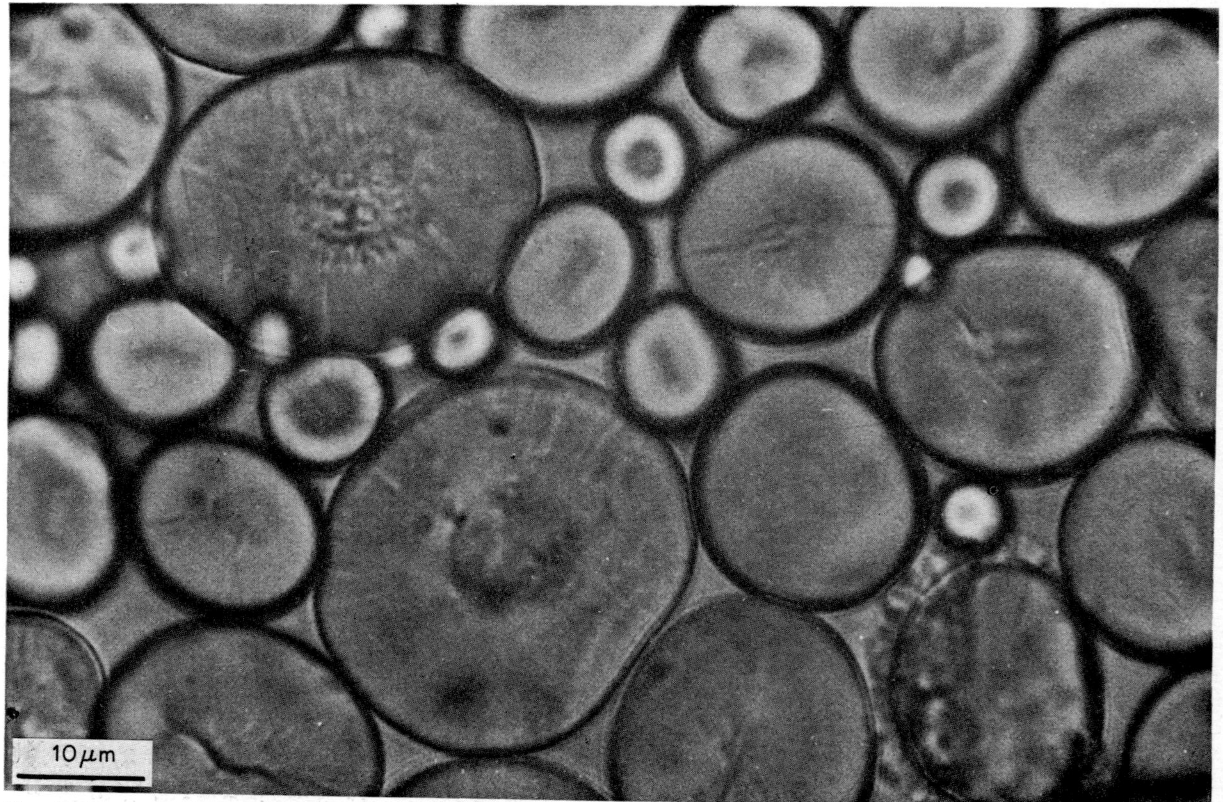

FIG. 2. Starch granules from the endosperm of a wheat plant grown, from anthesis to maturity, under conditions of continuous illumination. These granules have been treated as those in Fig. 1 but no ring structure has been revealed. Light micrograph, unstained.

granule. It is considered to correspond to the growth centre of the granule around which accumulation occurs.

All cereal starch granules except those with a very high amylose content exhibit marked birefringence under crossed polars. This phenomenon is a manifestation of a high degree of molecular orientation (though not necessarily crystallinity; see Banks and Greenwood (1975)). The centre of the characteristic 'Maltese cross' seen on starch granules viewed between crossed polars corresponds to the position of the hilum. Illustration of starches under normal light and crossed polars are to be found in many works of reference (Reichert, 1913; Kerr, 1950; Seidemann, 1966; Whistler and Paschall, 1967; Radley, 1968; Banks and Greenwood, 1975). Aesthetic improvement to the black and white images normally seen under polarizing conditions can be obtained by the interposition, between the polarizer and specimen, of a wafer of gypsum, cellophane or other material which suitably retards the incident beam. The use of a material retarding first-order red light has been recommended by Czaja (1972). As a result, colours are introduced by which the sign of the birefringence can be determined by reference to a similar material in which the orientation is known. Such methods have been used in demonstrating that, in large starch granules from wheat, molecules were perpendicular to the surface at all points (French, 1972). The sharpness of the birefringence in starch granules varies according to the area of the granule considered. In the case of the large granules from the Triticeae, the sharpest birefringence occurs at the granule centre (Czaja, 1972). Birefringence can be detected in granules which are lightly stained with iodine, and the combination of properties in mixtures of waxy and other starches can be of use in some industrial applications (Schoch and Maywald, 1967).

C. Protein

Under the light microscope cereal protein examined in thin section is colourless, translucent material with a somewhat granular appearance. In some cereals (maize, sorghum, millets) identifiable protein bodies are present (Hartz, 1885; Mottier, 1921; Duvick, 1961; Rooney and Sullins, 1977). These are spherical and about one-tenth of the size of starch granules found in these species (see Chapter 3, Figs 25 and 26). In other cereals proteins are present as an apparently uniform reticulum which lies among the starch granules (see Chapter 3, Figs 17, 30 and 31). In the outermost layers of starchy endosperm uninterrupted cores of protein are frequent (Kent, 1966). Most of the protein which constitutes the structures described above is of the type known as storage protein, the name deriving from the assumption that it is laid down as a food reserve for the requirements of the germinating embryo. However, other protein types are present as enzymes and cytoplasmic constituents and it is customary to distinguish between the types by their solubility in a sequence of extractants (Osborne, 1907). Albumins and globulins are those proteins which dissolve in water and sodium chloride solution respectively, prolamines dissolve in 70% alcohol, and glutelins in dilute alkali. Of the storage proteins, it is the prolamines which form the protein bodies. The relationship between the constituents of cells of maize endosperm is well described by Duvick's analogy (1961) of marbles (starch granules) and buckshot (zein, the prolamine of maize) set in a transparent glue (clear viscous cytoplasm).

At a TEM level, examination of maize protein bodies revealed the presence of a bounding membrane and, as with starch granules, compaction led to distortion into non-spherical shapes (Khoo and Wolf, 1970). Staining reactions were found to vary among protein bodies according to their location; in glutaraldehyde–osmium fixed tissue stained with lead compounds, non-uniform staining occurred in bodies located in the subaleurone layer, but in those present in the central cells only specks in the middle became stained. Adams *et al.* (1976) found that small bodies stained uniformly, while larger ones were vacuolate with densely staining materials present in the vacuoles.

The solubility of isolated protein bodies in alcohol (Christianson, *et al.* 1969) confirmed the identification of protein bodies as sites of zein storage, a conclusion based previously on their greater readiness to stain in alcoholic than aqueous dye solutions at a light microscope level (Duvick, 1961). Protein of the matrix stained more readily than the zein, even with aqueous dyes, although isolated matrix protein had the solubility characteristics of glutelin (Christianson *et al.*, 1969). In mutants of maize in which less than normal amounts of zein were present, both the size and number of protein bodies were reduced (Christianson, 1970; Robutti *et al.*, 1974).

In millet (*Eleusine coracana*) protein bodies were seen by TEM as discrete capsules possibly surrounded by a membrane. Some were homogenously stained, while others showed a darker staining central area. In prosso millet a central dark core and two or more dark bands were noted in protein bodies examined in endosperm sections (Jones *et al.*, 1970). Protein bodies of *Setaria lutescens* were membrane bound and roughly spherical, although some were lobed. Alternating light and dark bands were observed in sections stained by lead citrate (Rost, 1971). The significance of the layered appearance of these or other protein bodies (e.g. rice, maize, prosso millet) remains obscure. In *Pennisetum typhoidium* isolated protein bodies had discrete edges, but membranes were not definitely identified and concentric rings of darker staining dots suggested a layered internal structure. The bodies were frequently lobed, possibly as a result of their composite nature due to fusion of individual bodies (Adams *et al.*, 1976). Protein bodies of pearl millet (*Pennisetum americanum*) were shown by SEM to be soluble in 70% ethanol, but not in tertiary butyl alcohol which did however remove matrix protein (Badi *et al.*, 1976). Conversely, protein bodies of sorghum failed to dissolve in 60% ethanol but were readily removed with 60% tertiary butyl alcohol (Seckinger and Wolf, 1973). The proteolytic enzyme 'pronase' had little effect on sorghum protein bodies although it digested matrix protein within two hours. After prolonged incubation partial degradation of bodies occurred, in some cases the differential susceptibility of parts of the granules suggested a laminar structure.

Isolated protein bodies from sorghum were mainly spherical, with a definite boundary (probably membranous). Concentric rings were present (Adams *et al.*, 1976) and the general appearance was similar to bodies examined *in situ* (Seckinger and Wolf, 1973).

Protein bodies of two types were isolated from rice. When fixed in glutaraldehyde and osmium and stained with uranyl acetate some bodies had a lamellate appearance, while others were uniform (Mitsuda and Murakami, 1969). The banding was considered by Hoshikawa (1970) to be a feature which developed with increasing degree of maturation. Bounding membranes were present on about half of the isolated bodies (Mitsuda and Murakami, 1969).

The appearance of protein bodies from oats at a TEM level has not been recorded. However numerous bodies which gave a positive reaction for arginine with two histochemical methods were identified with the LM (Fulcher *et al.*, 1972). The same histochemical techniques applied to wheat endosperm gave evidence of only a few discrete spherical bodies, mostly in the subaleurone layer. Thin strands resembling endoplasmic reticulum were also stained. The methods of isolating protein bodies from endosperm (Adams *et al.*, 1976) of other cereals have not been successfully applied to mature wheat; nor have discrete protein bodies been identified in mature endosperm tissue by TEM (e.g. Hughes, 1976). Seckinger and Wolf (1967) described a single, undifferentiated, amorphous protein with randomly dispersed inclusions extending up to the surface of starch granules. Adams *et al.* (1976) concluded that the protein bodies seen during development are replaced by a matrix-type protein.

Protein bodies isolated from barley (Adams *et al.*, 1976) resembled those described by earlier authors as having come from aleurone cells (Ory and Henningsen, 1969). Vesicles containing dense osmiophilic bodies were observed in barley grown under low light intensity, but these may have been lipid droplets (Buttrose, 1960). In mature triticale and rye the situation as regards protein bodies is considered to be most like that in wheat (Simmonds, 1974). In wheat and its close relatives changes clearly occur whereby discrete bodies, easily identifiable during development, become closely integrated into a continuum with other cellular components including membrane materials and structural proteins (Seckinger and Wolf, 1967; Simmonds, 1972a). Wheat proteins can be separated, however, into those which adhere closely to starch granule surfaces and those which can be isolated as discrete chunks. Thus, Hess (1954) described 'adherent' and 'wedge' proteins, the latter term describing the shape imposed on protein particles due to their occurring in the interstices among large starch granules. Surface replicas examined under the TEM showed that adherent protein was fibrous, while wedge protein showed no such structure (Hess *et al.*, 1955; Hess, 1960). Material adhering to starch surfaces has been equated with residues of membranes (principally from amyloplasts) (Buttrose, 1963; Kent, 1970).

Some water-soluble proteins, possibly enzymes associated with starch synthesis, were concentrated at the starch–protein interface (Barlow *et al.*, 1973). The above findings endorse the concept of the starch–protein–lipid relationships envisaged by Buttrose (1963), the latter described discrete protein deposits (wedge protein) partly surrounded by membrane residues; similar residues, derived from desiccated plastid stroma and cytoplasmic membranes and ground substance (adherent protein), being present on starch granule surfaces. The degree of adhesion of membrane residues to starch granules was greater in hard than in soft wheats (Simmonds, 1972a).

III. MODIFICATION

A. Starch

1. During extraction

Strictly, most of the starch ever examined has been modified, since even the process of extraction from the endosperm cannot be achieved without altering some of the properties of the granules. Thus, when starches and their reactions are compared it is important that due consideration be given to the effects of different methods of preparation. Extraction of 'pure' starch invariably involves the use of water, and the manner in which the starch is dried can be an important determinant of the degree of permeability and the reactivity with oxidizing agents, both factors being greater in lyophilized than in air-dried starches (Whistler *et al.*, 1958, 1959). Technological properties have also been shown to depend upon the isolation procedure, contact with solvent being a particularly important factor (Cauvain *et al.*, 1977).

2. Thermal modification

At normal temperatures the changes which occur in starch suspended in water are not easily observed, but at and above the gelatinization temperature they are dramatic. The gelatinization temperature (which is specific to each starch type but generally around 60°C) is that at which swelling becomes irreversible, birefringence is lost and permeability to water and dissolved substances, including dyes, increases greatly; susceptibility to amylolysis and oxidation are similarly increased. In a large excess of water further swelling occurs above the gelatinization temperature, and ultimately (at about 100°C) granules lose their identity. The observed changes in microscopic appearance are manifestations of changes in crystallinity (Collison, 1968), as shown by the different X-ray patterns given by granules in the various conditions described (Katz and Muschter, 1933). Birefringence was found to return in granules from most cereals and other sources when iodine solution was added to a suspension, even after heating to 10–15°C above the gelatinization temperature (MacMasters, 1953; Gough and Pybus, 1973). In limited water and in the presence of solutes in the aqueous medium in which starch granules are suspended, gelatinization temperatures and mode of gelatinization are different from those in excess pure water. Qualitative variables include the degree to which ring structure is revealed and the area of the granules at which disorganization begins (Gough and Pybus, 1973) (Fig. 3). In baked products, the degree of gelatinization depends upon the amount of water present and the nature of the dissolved substances (Derby *et al.*, 1975). The amount of water available to granules is not uniform throughout the product since evaporation is greater from the superficial layers. Hence the greatest swelling occurs in the innermost region (Flint *et al.*, 1970). In many baked products partial gelatinization occurs (Figs 4 and 5). In stale bread surface 'decoration' has been observed (Angold, 1975).

The viscosity of starch pastes increases above gelatinization temperatures due to two factors: the greater friction as swollen granules are pushed past each other (Schoch, 1965), and the release into solution of an exudate which can be recognized by both LM and SEM (Miller *et al.*, 1973). The greater importance of the latter is demonstrated by the fact that the largest increase in viscosity occurs after swelling ceases.

Amylomaize starch has anomalous gelatinization properties in that the three-dimensional appearance is retained up to 10°C above the temperature at which birefringence is lost. In some granules no change is apparent after a further 10°C rise (Hill and Dronzek, 1973).

Extreme effects of heat and moisture are found in wafers in which granule structure is totally disrupted (Stevens, 1976) (Fig. 6). In puffed products the appearance of maize, millet and rice starches have been examined by SEM (Babichenko, 1973). As swelling occurred separate layers peeled off the outside of granules, leading to their acquiring a bud-like appearance. In popped barley products granules were uniformly gelatinized, while in popped maize

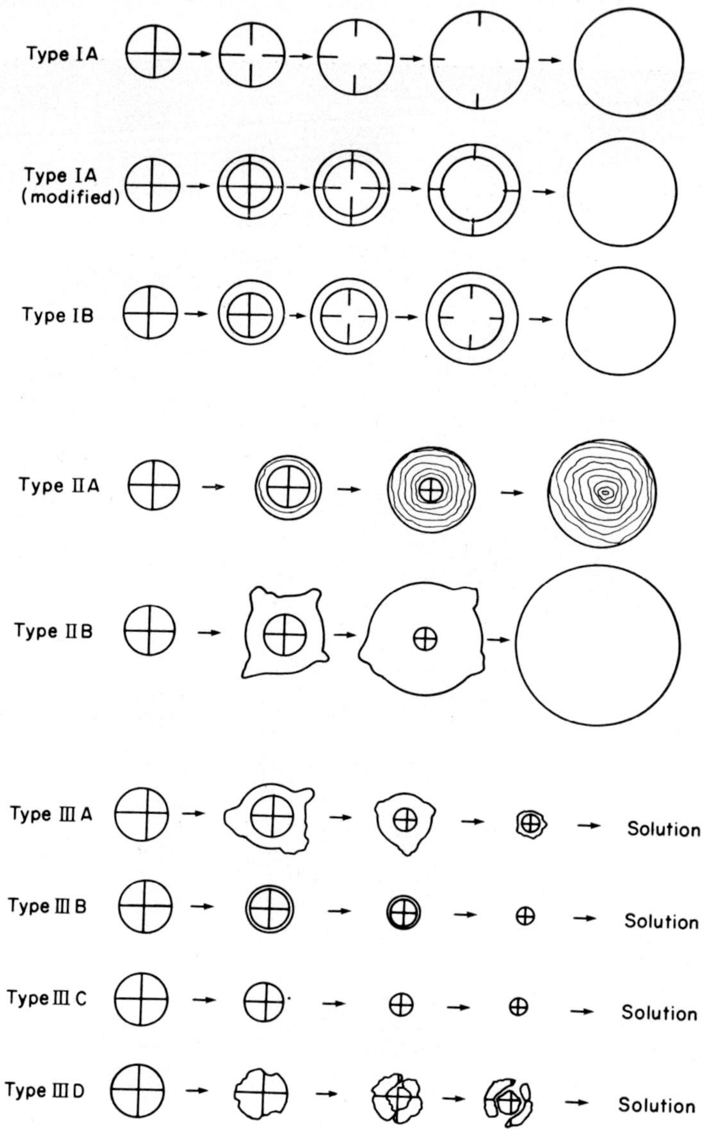

FIG. 3. Mechanisms by which wheat starch granules have been observed to gelatinize. The diagrams represent starch granules, viewed between crossed polars, undergoing the various types of gelatinization.

Type I, tangential swelling with simultaneous loss of birefringence, progressing from the hilum to the periphery.

Type II, swelling and loss of birefringence, both proceeding from the periphery to the hilum.

Type III, apparent solubilization of the granule to give a starch solution. (Gough and Pybus, 1973.)

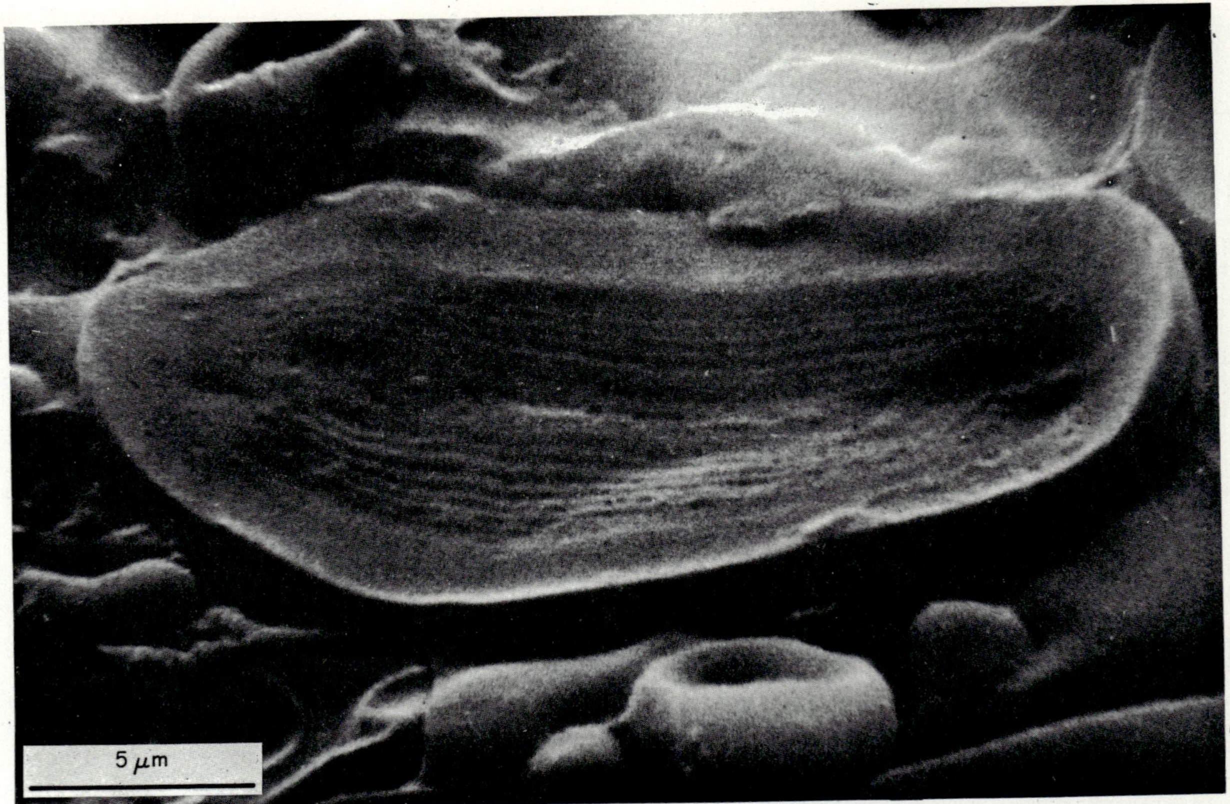

Fig. 4. Section of a starch granule in cooked pastry. Partial gelatinization has led to the revelation of ring structure within the granule. Scanning electron micrograph, formaldehyde vapour fixation. (Courtesy K. K. Barlow.)

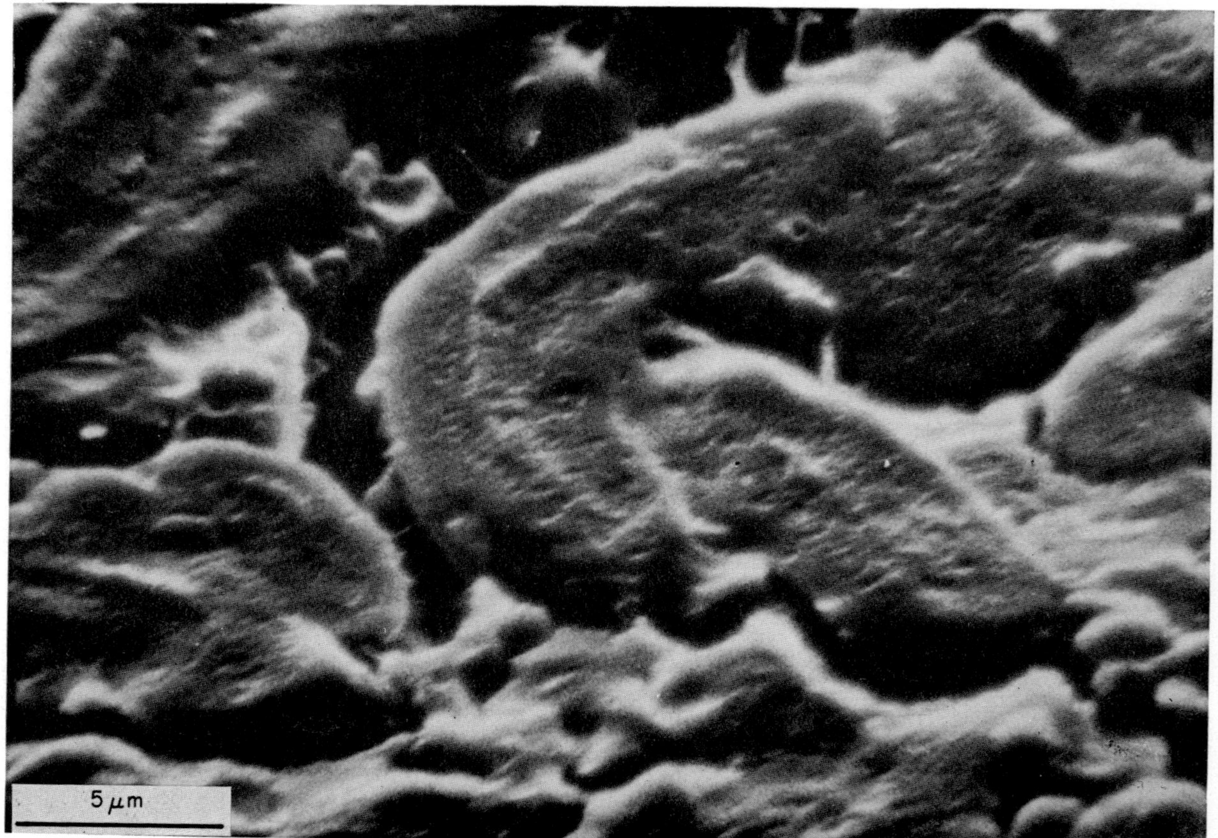

FIG. 5. Cut face of a frozen block of breadcrumb. Gelatinization has proceeded further than shown in Fig. 4. No ring structure is apparent but a septum is present along the longitudinal axis of each cut face. Scanning electron micrograph.

FIG. 6. Broken edge of a wafer made from a recipe in which flour was replaced by starch alone. During cooking extreme heat and pressure are applied and as a result starch granules completely lose their individual identity. Scanning electron micrograph (Stevens, 1976).

products a gradient existed with greatest gelatinization occurring in the central endosperm where the 'popping' had actually occurred. Starch granules were only slightly deformed in flaked products (Farber and Gallant, 1976). In flaked and micronized flours, granules of sorghum starch were found to be much distended (McNeil *et al.*, 1975).

3. Mechanical modification

Granules subjected to pressure and stress during their passage through grinding rolls may become modified in a number of ways which may be recognized under the light microscope. Williams (1968) described four types found in wheat flour: radial cracking, chipping and splitting, abrasions, and a further type which has been described as 'ghost' formation (Jones, 1940). It is generally accepted that it is this last type which most changes the properties of granules. The name ghosts was coined to describe the indistinct outline and the flimsy non-refractive microscopical appearance of such granules. While ghosts can be produced in most starches, it is in wheat and rye that they are most important since the proportion of granules so damaged helps to determine the amount of water absorbed by a flour in forming a dough. Like gelatinized granules, ghosts are not birefringent in aqueous medium when viewed under crossed polars, and they absorb water, soluble dyes and enzyme solutions far more readily than intact or otherwise mechanically damaged granules (Alsberg and Griffing, 1925). Under the SEM they appear flattened (Fig. 7).

The organized structure of granules is not totally disrupted until contact is made with water, as shown by the limited birefringence of heavily damaged starches mounted in liquid paraffin (Lampitt *et al.*, 1941). The nature of the changes occurring in ghost formation is little understood since they cannot be produced by cutting granules with, for instance, a microtome. This produces damage similar to that described as chipping and splitting, in which only areas adjacent to the cut are affected rather than the whole granule changing as in ghost formation (Sandstedt and Schroeder, 1960). Granules cut into thin slices have been shown to retain their organization to an extent sufficient to exhibit dichroism under crossed polars (French, 1972). Also, the cut faces of granules, observed under the SEM both before and after exposure to water, have concave profiles, suggesting that tension existed throughout the structure (Fig. 8). On the other hand, ultrathin sections of granules expand and fold on contact with water, causing difficulties and artefactual structures in TEM specimens (Gallant and Guilbot, 1971). The concave nature of cut faces of undamaged granules is contrasted with the level cut surfaces of both damaged and gelatinized granules, an example of the latter type being found

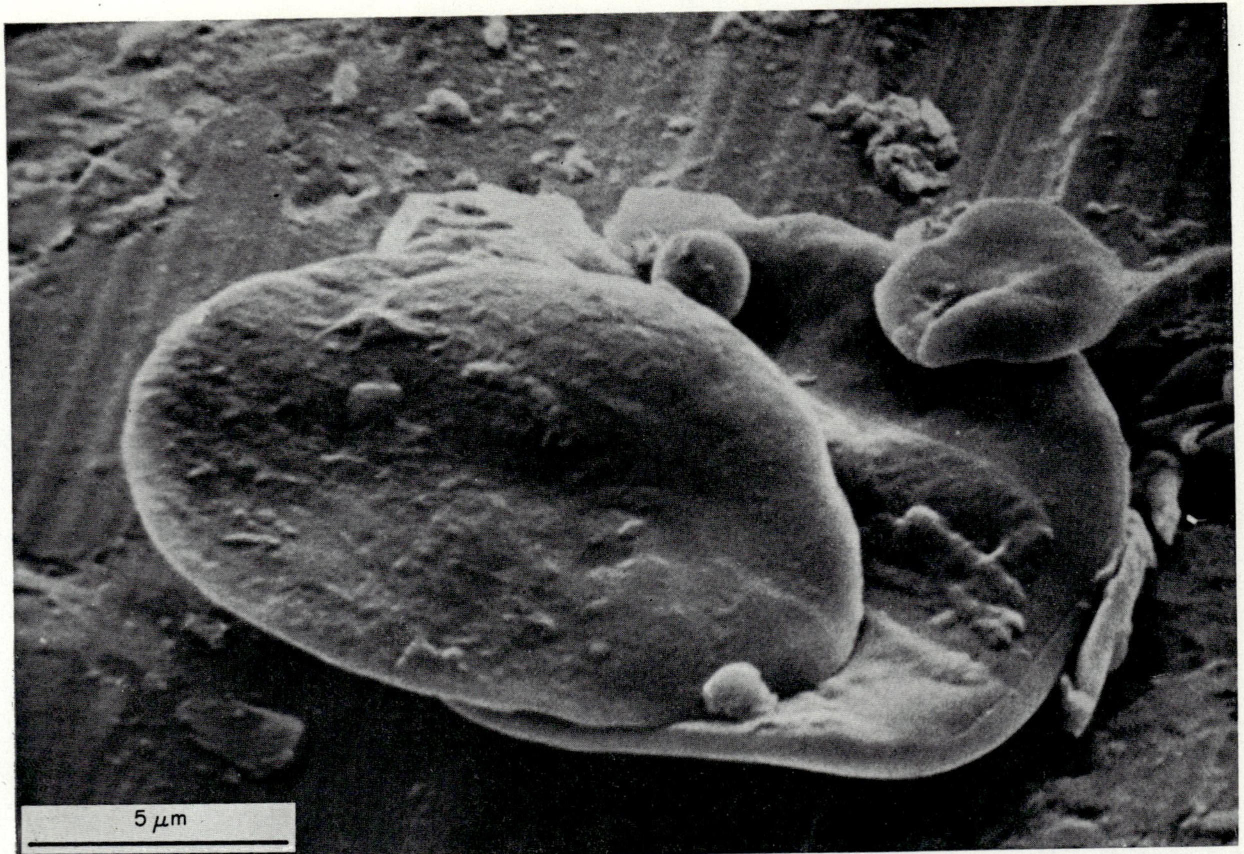

FIG. 7. Mechanically damaged starch granules. These granules have been flattened as a result of pressure and shear. Granules so damaged do not exhibit birefringence after coming into contact with water. Scanning electron micrograph.

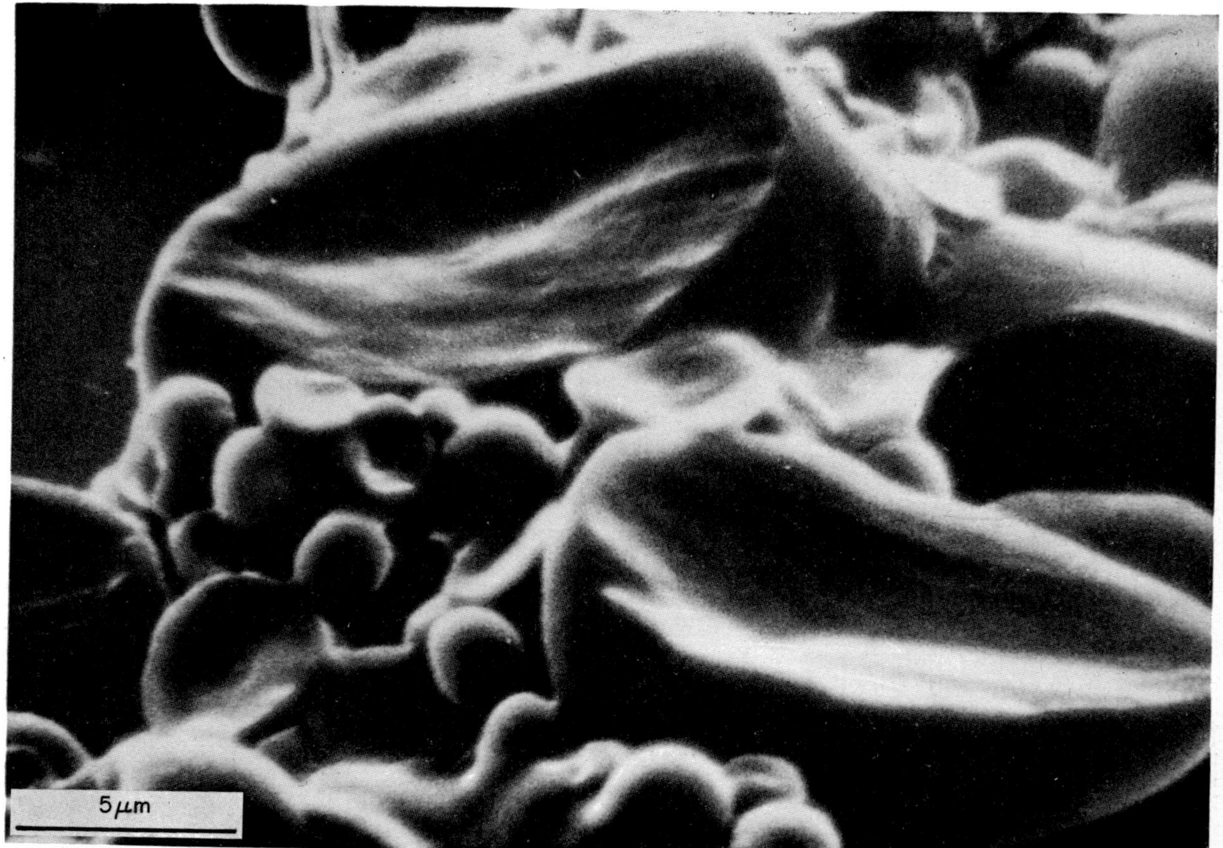

FIG. 8. Cut face of a frozen block of dough. The cut faces of the starch granules are concave owing to the effect of internal tension. No evidence of such tension can be seen in granules in Figs 4 and 5, suggesting that the tension is relieved in granules in which gelatinization has begun. Scanning electron micrograph.

in bread crumb (Fig. 5). A cut face of each of the above types is typically divided by a septum in the mid-line of the face.

4. Chemical modification

The potential for chemical modification of starch is clearly enormous and the following does not pretend to be an exhaustive review but merely gives an indication of the types of modification which have been achieved in cereal starches. A brief review of modified starches used in the food industry has been given by Selby (1977).

One of the best known and most widely studied chemical modifications to cereal starches is hydrolysis by mineral acid. 'Lintnerization' involves prolonged treatment in a weak solution of hydrochloric acid. Solubility is increased as a result, and under the LM and TEM the lamellate structure of granules is clarified owing to the differential susceptibility of alternate layers of the granule to hydrolysis (Sandstedt, 1955; Buttrose, 1960; Mussulman and Wagoner, 1968; Gallant and Guilbot, 1973). Concentric shells of starch were revealed by TEM in a series of rice starches but penetration of the solution into the granule occurred without apparent damage to the surface since no changes in surface appearance were noted between untreated and lintnerized granules under the SEM (Evers and Juliano, 1976). The causes of the differing susceptibilities of alternate layers are not fully explained. Amylopectin has been said to be the more readily hydrolysed component (Arbuckle and Greenwood, 1958), while some believe the susceptible layers to be the more amorphous (Innamorati, 1966). The former view has been revised (Banks and Greenwood, 1975), and the latter is not compatible with the observation that many of the amorphous amylomaize granules resist hydrolysis (Mussulman and Wagoner, 1968). The small granules of barley do not normally display marked lamellation, but those from phosphorus-deficient plants did (Buttrose, 1960). Residual rings were easily hydrolysed by α-amylase (as are granules from normal plants) and β-amylase (unlike those from normal plants).

Chlorination of flour has a profound effect upon the behaviour of starch (Gough *et al.*, 1978), but the changes which result are not recognizable microscopically (Scallet and Sowell, 1967; Hall *et al.*, 1971) unless the granules are heated in water, when, on swelling, chlorinated granules do not exude amylose in the same way as untreated granules (Miller *et al.*, 1973). Sodium stearyl fumarate and sodium stearyl-2-lactylate are also said to prevent exudation, although it is not suggested that the mode of action of chlorine and the surfactants is similar. Greater swelling and amylose retention were recorded in starches cross-linked with phosphorus oxychloride and epichlorhydrin than in unmodified corn and waxy maize starches. Starches derivatized with propylene oxide or acetic anhydride formed large bubbles on swelling.

Collapse of these led to a wrinkled appearance under the SEM (Chabot *et al.*, 1976) (Figs 9, 10, 11 and 12). Polyacrylonitrile graft co-polymerization of wheat starch granules did not alter the appearance of the granules. Removal of starch from these showed that low add-on graft co-polymers were surface grafted, while in cases of higher add-on grafting occurred throughout the granule interior (Fanta *et al.*, 1973).

5. Enzymic digestion

Consideration has already been given to some aspects of the construction of the starch granule in a series of concentric rings. Occasionally these rings can be seen in apparently unmodified cereal starches, their clarity being enhanced in granules which are heated in water to a temperature below that at which gelatinization occurs. Treatment with dilute hydrochloric acid reveals that alternate layers are respectively more and less susceptible to oxidation, but it is the partial digestion of intact granules by enzymes which produces the most dramatic manifestations of the concentric shell structure (Fig. 13). Many amylolytic enzymes exist, and while the ultimate effect of all of them, working alone or together, is total digestion, the route taken to this objective and the speed with which it is achieved varies with the class of enzyme or enzymes involved, the enzyme's source and the botanical source, the chemical composition and the condition of the granular substrate.

In addition to variation among different starches, variation occurs among individual granules from the endosperm of the same species (Leach and Schoch, 1961). Thus, while large granules from the same cell were attacked by native α-amylase according to the same pattern and at the same rate, those from adjacent cells were dissimilar in both of these respects (Sandstedt, 1955). In view of the enormous potential of permutations afforded by all these variables, it is not surprising that no systematic catalogue exists from which it is possible to predict details of any given combination. However, a number of combinations have been examined and as a result hypotheses on the mature structure and development of granules and the mode of action of enzymes have been advanced.

The groups of amylolytic enzymes and their mode of cleavage of the molecular chains have been widely described. As far as modification of whole starch granules is concerned, two classes of enzyme are important, the α-amylases (Robyt and Whelan, 1968) and glucoamylases or amyloglucodidases (Fleming, 1968). Beta-amylase does not attack intact granules (Walker and Hope, 1963). The physical effects of the first group have been the more widely studied. In general, the enzyme solution digests channels from selected points on the surface towards the centre of the granule; on reaching the centre attack proceeds outwards over a broader front. During both

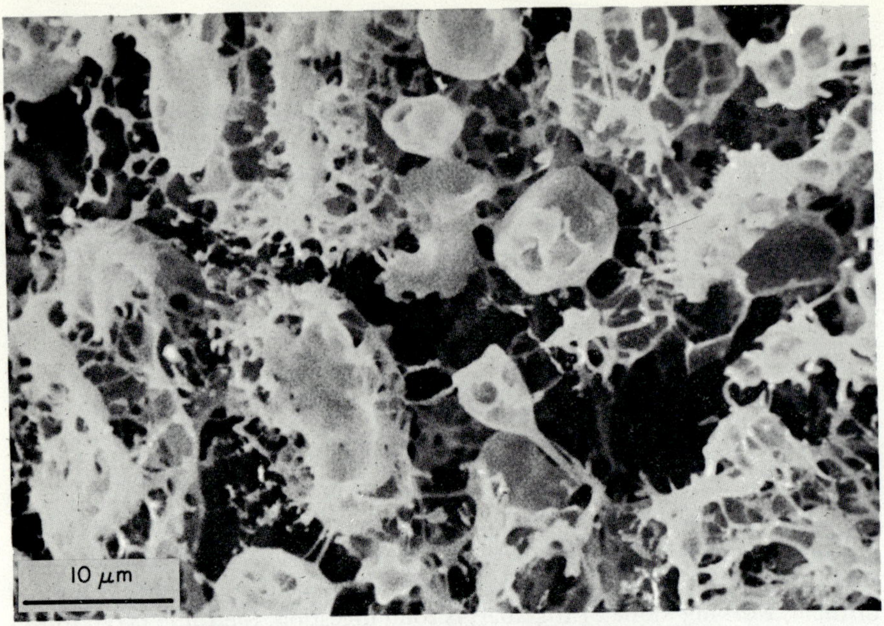

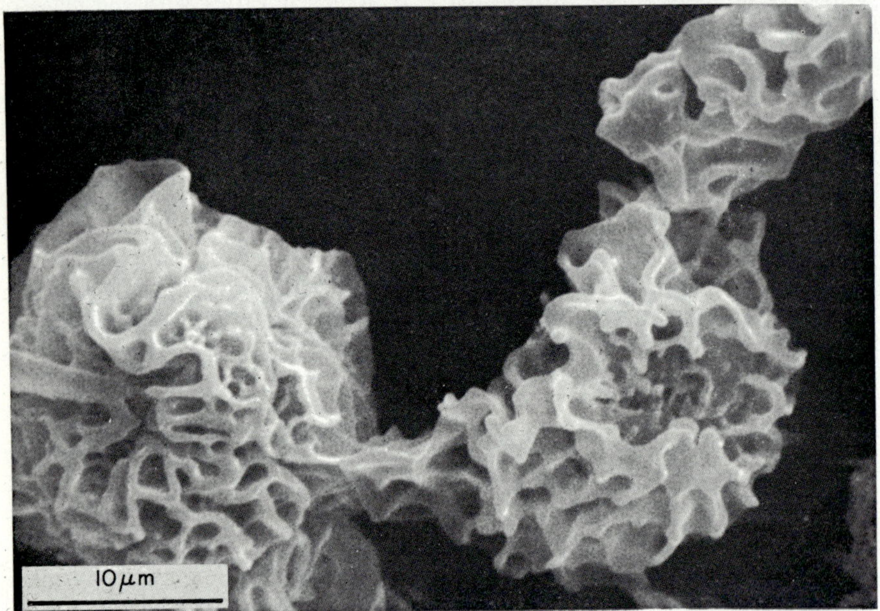

FIGS 9–12. Cooked starches derived from maize and waxy maize. In Fig. 9 (upper left) unmodified waxy maize starch cooked to 95°C is shown. Granules have disintegrated and some coarse fibrils are formed. The granules of hydroxypropyl distarch glycerol waxy maize shown in Fig. 10 (lower left) have retained their granular form and the result of swelling and subsequent collapse is seen as a much convoluted surface. Cross-linked maize

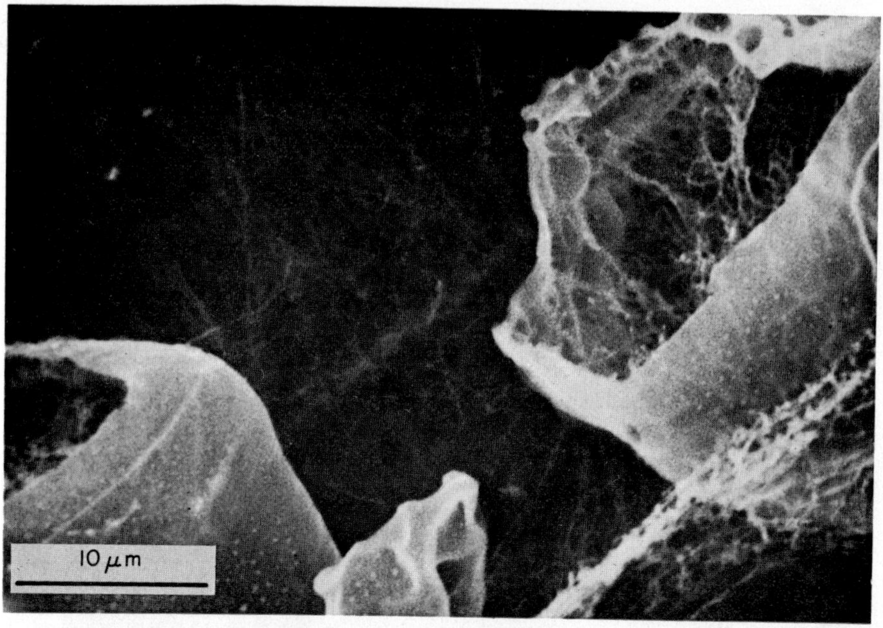

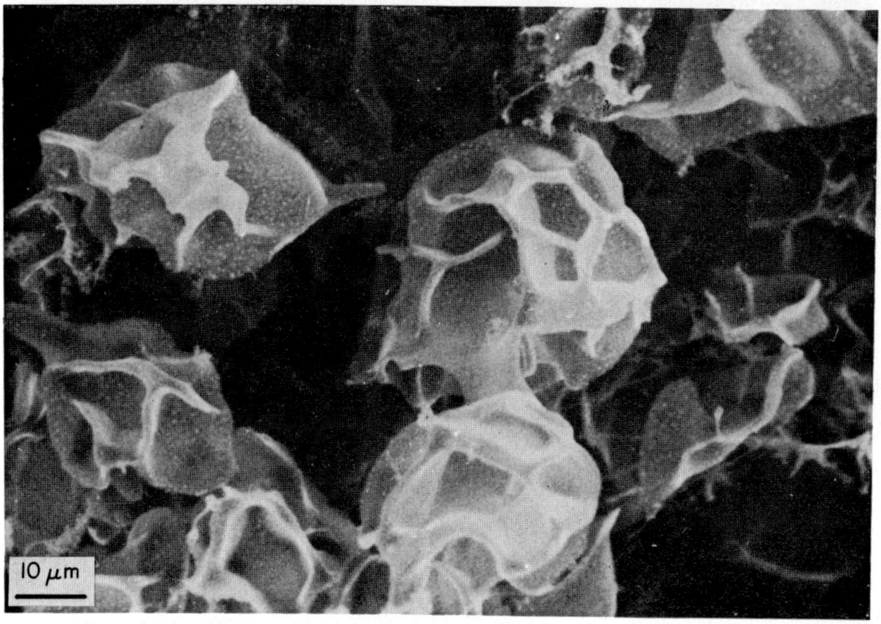

starch (distarch phosphate corn) cooked to 90°C shows many fibrils both inside and outside of the ripped open granular envelopes (Fig. 11, upper right). In Fig. 12 (lower right) cross-linked and substituted maize starch (hydroxypropyl distarch phosphate corn) is shown. The granules formed folded balloons and only a few extragranular fibrils can be identified. Scanning electron micrographs (Chabot *et al.*, 1976).

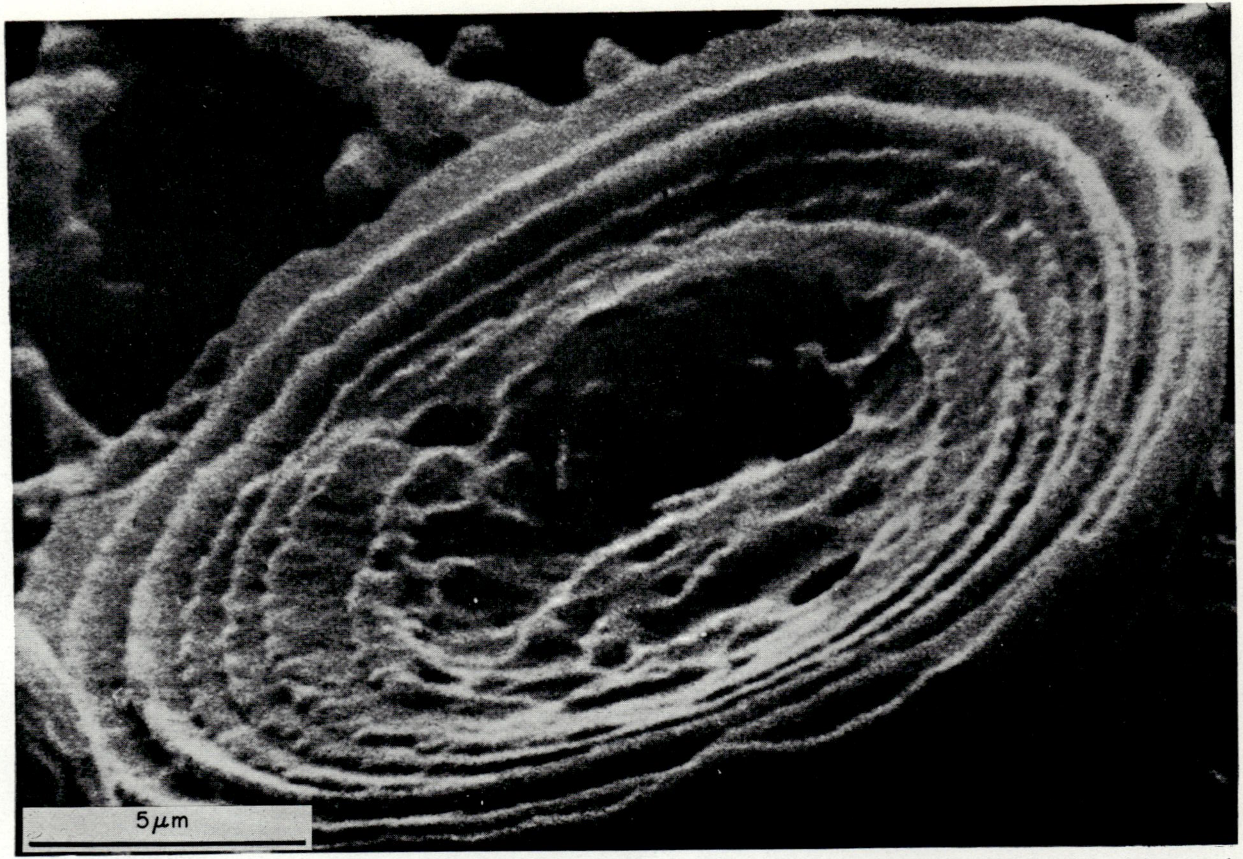

FIG. 13. Part of a large granule of wheat starch which has been attacked by hog pancreatic α-amylase. Separation of the granule into two halves has occurred at the plane of weakness in its centre. Alternate layers of greater and lesser susceptibility to hydrolysis are clearly revealed. Scanning electron micrograph.

phases of enzyme attack it is clear that some regions are more readily digested than others. One result of this is the clear revelation of the ring structure, since surrounding the completely digested areas are regions in which only alternate layers remain. This produces a saw-tooth profile in transverse section (Nikuni and Whistler, 1957; Buttrose, 1962). Alternate layers of greater and lesser susceptibility are also a feature of attack by glucoamylose, but here attack has been commonly found to occur centripetally across a broad front. Thus, under the light microscope granules attacked by glucoamylose have a solid appearance with a surface covered with many shallow pits, while those attacked by α-amylase have a hollow centre linked to the surface by a few radial channels (Sandstedt, 1955; Evers *et al.*, 1971).

Within a cereal species the greatest source of variability in granule susceptibility is the amylose : amylopectin ratio. The most readily attacked starches are those from waxy cereals and the least easily digested are the high-amylose varieties. So resistant are granules with extremely high amylose content that little superficial attack, no ring structure and only rare radial channels have been observed (Gallant *et al.*, 1972, 1973).

In general, the patterns of attack by native α-amylase are similar in *in vitro* studies to those found in granules from germinating kernels. However, the relative rates of attack may vary according to the conditions existing in local areas of the endosperm. In germinating grains enzymes are present which digest protein and other cell components, thus making granule surfaces available to the α-amylase (e.g. MacLeod and Palmer, 1966). Since the enzymes are produced at the periphery of the endosperm, it is the granules in this region which are digested earliest (Palmer, 1971). The reverse is true when only amylolytic enzymes are applied to cut endosperm surfaces; in this case the peripheral granules are least susceptible since they are most nearly completely surrounded by protein which prevents access of the enzyme (Rooney and Sullins, 1977). The presence of more protein on the surfaces of small starch granules than on large ones has also been invoked to explain the apparently greater resistance of small granules to α-amylolysis in germinating wheat (Palmer, 1972; Lineback and Ponpipom, 1977). It is paradoxical that it is the small granules which are said to be the more susceptible type in germinating barley (Palmer, 1972), and some doubt must remain as to the validity of conclusions based on SEM observations to which no truly quantitative techniques were applied. A hazard which should also be borne in mind when interspecific comparisons are made is that cultivars vary in their levels of α-amylase activity (Stewart, 1964; Gale and Marshall, 1973) and in the particle size distribution of their starch granules (Evers, 1973; Goering *et al.*, 1973).

Few studies have been performed on oat starches. There is however a measure of agreement that these are among the most resistant to attack by α-amylase, both *in vitro* (Buttrose, 1960) and *in vivo* (Lineback and Ponpipom, 1977). Even after 14 days germination, and in spite of dissolution of the protein matrix and separations of the granuli, only mild corrosion was noted on their surfaces. Pearl millet is also considered to be attacked relatively slowly by both its native α-amylase (Lineback and Ponpipom, 1977), which incidentally has a particularly high specific activity, and that of the hog pancreas (Sullins and Rooney, 1977).

In sorghum the degree of starch digestibility with hog pancreatic α-amylase depends upon the number of waxy alleles in the endosperm genome. In a homozygous series of grains with an increasing contribution by waxy alleles, the digestibility *in vitro* increased progressively. In starches where the recessive waxy genes were present there were fewer, but more uniformly thick, concentric rings revealed (Rooney and Sullins, 1977). This feature was also found to be characteristic of waxy maize starch attacked by bacterial α-amylase (Gallant *et al.*, 1972).

Five patterns of attack were identified in rice starch granules attacked by hog pancreatic α-amylase: granule surfaces became defaced by a single large hole, many medium-sized holes, 'pepper-potting', sponge-like erosion or uniform surface erosion (Evers and Juliano, 1976) (Figs 14, 15 and 16). As in corn and sorghum, waxy granules were the most readily digested *in vitro* but, unlike the other cereals, two high-amylose types showed greater digestibility than those of intermediate amylose contents. The high resolution facilities of TEM and SEM have been particularly helpful in the study of rice starch, since, as Juliano (1972) wrote, 'few histological studies have been made on the rice starch granule because it is very small compared with corn and wheat'. Hence, in wheat, barley and rye much knowledge of the physical aspects of amylolysis had already been accumulated before the advent of electron microscopes, and the types of pitting resulting from both α-amylase and the enzyme systems of *Aspergillus oryzae* (probably glucoamylose) were well established (Sandstedt, 1955; Leach and Schoch, 1961). The benefits of SEM and TEM have been the revelation of greater detail and, in the former case, the addition of the third dimension to the light microscopical image. It is in the attack of large A-type granules that this has been of most benefit in that the important role played by the equatorial groove in allowing α-amylase into the granule interior has been emphasized (Evers and McDermott, 1970; Hall and Sayre, 1970; Dronzek *et al.*, 1972; Jones and Bean, 1972; Gallant *et al.*, 1973; Smith and Lineback, 1976). While there is some variation of opinion as to whether or not the groove exists before amylolysis commences (Evers, 1971; Smith and Lineback, 1976), there is general consensus regarding its preferential susceptibility. Such unanimity is based on the radial channels

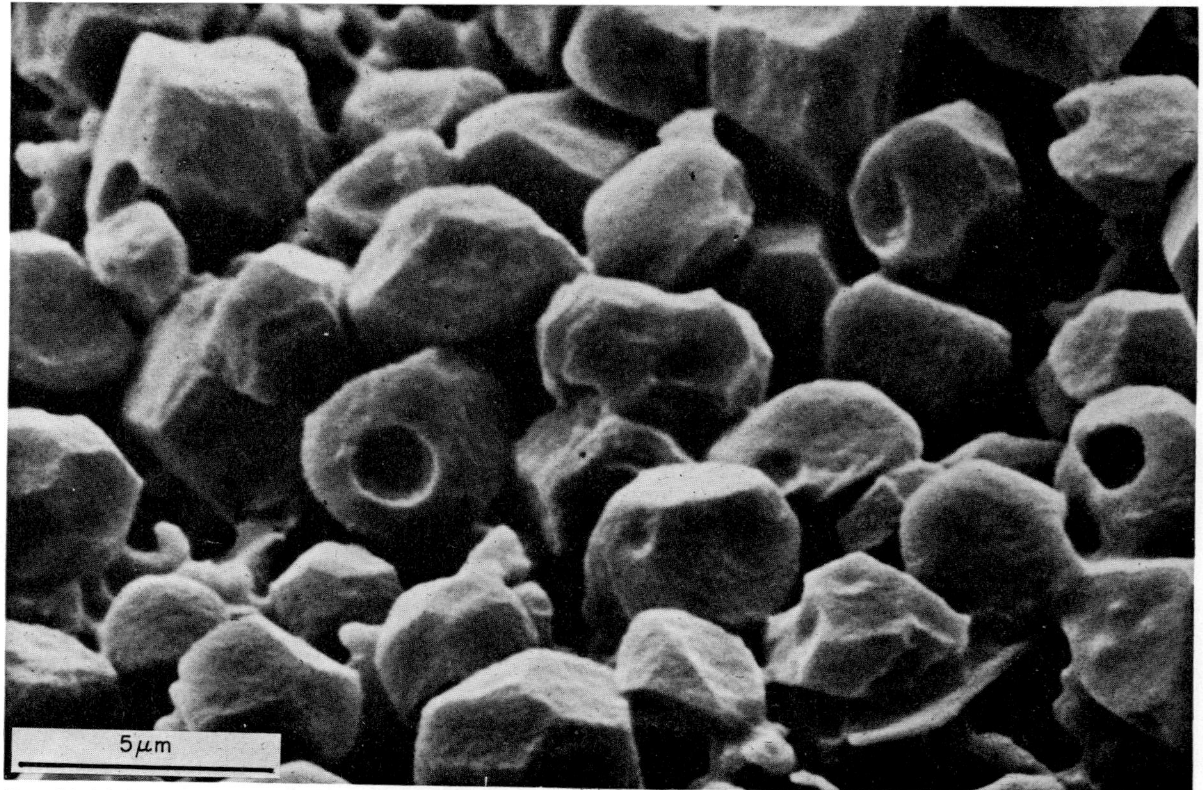

FIGS 14–16. Starches from different types of rice, subjected to partial hydrolysis by hog epancreatic α-amylase. Each type of starch has been attacked in a different way. Scanning electron micrographs (Evers and Juliano, 1976).

FIG. 14. Non-waxy rice, only one large hole per granule.

FIG. 15. Waxy rice, many holes in each granule leaving a sponge-like residue which shows layering.

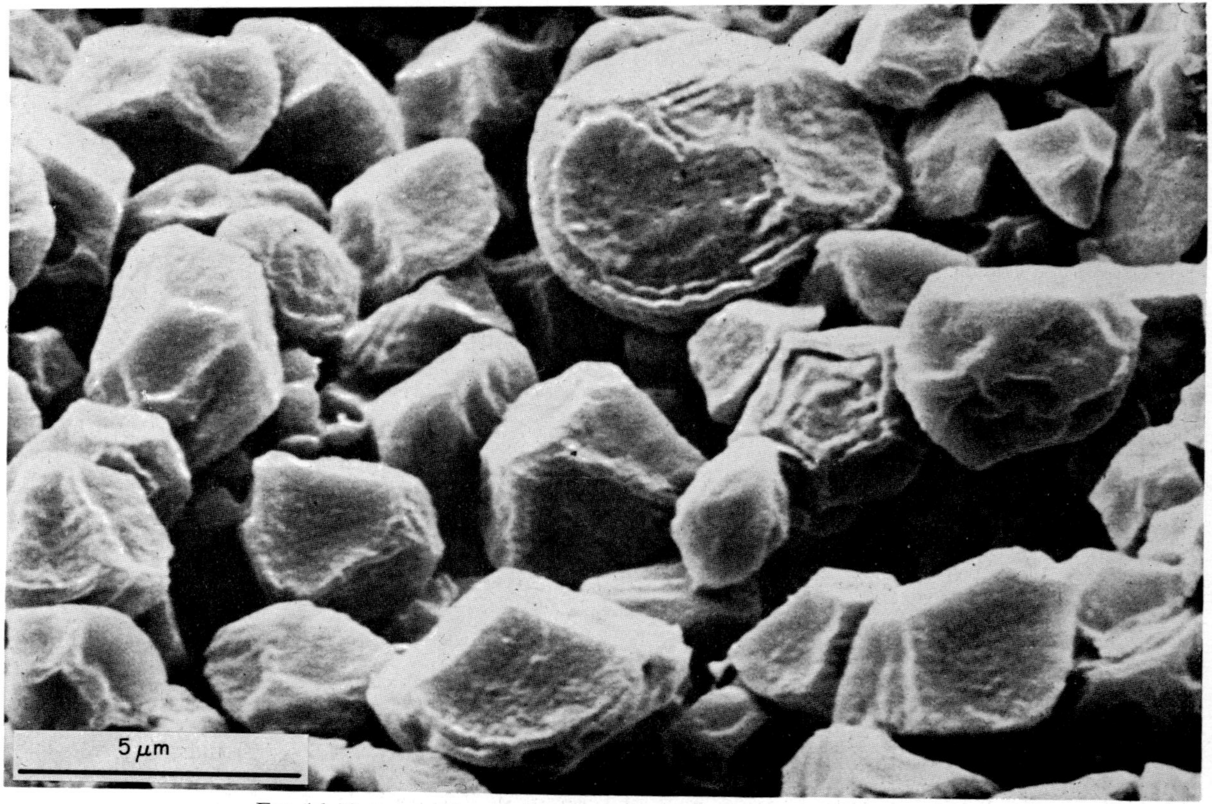

FIG. 16. Non-waxy rice, surface corrosion revealing some ring structure.

seen under the light microscope, the preferential surface attack and the readiness of granules to separate into two halves in the plane of the groove, as shown by SEM (Fig. 17) (Evers and McDermott, 1970), and the weakness of the structure revealed by fractures perpendicular to that plane by SEM (Fig. 18) and TEM (Evers and McDermott, 1970; Buttrose, 1960). The structure observed in profiles of this nature reflects the ontogeny of the granule, with both the thickness of each successively apposed layer and the depth of its equatorial groove declining (Evers and McDermott, 1970). Differences between the patterns of attack by α-amylases and glucoamylases have been seen as being compatible with a radial arrangement of molecules within the granule, with non-reducing ends located at the surface (Evers *et al.*, 1971). Further evidence possibly supporting the fibrillar concept of granule structure was obtained by experiments with purified glucoamylase iso-enzymes (Smith and Lineback, 1976). Clear interpretation of the results of attack by both α-amylases and glucoamylases is prevented by the incompleteness of our understanding of the behaviour of enzymes at a molecular level, in particular the differences between iso-enzymes in this respect. While there is much accord among the results of similar experiments, there remain areas where observations are apparently contradictory. Thus Sandstedt (1955) found that glucoamylase attack on rye starch proceeded as decribed above, but Brudka and Jakubczyk (1976) detected a pattern more like that described for α-amylase. Also, Smith and Lineback (1976), using two glucoamylases prepared from *Aspergillus niger*, found that in one case the centres of granules collapsed under reduced pressure, indicating an α-amylase-like attack on the interior of the granule, and a similar observation was reported by Rasper *et al.* (1974). Clearly, a systematic investigation on a broader scale than any attempted so far is required before generalized conclusions can be drawn regarding patterns of attack and their relationship to granule structure.

B. Protein

The most important modification of cereal protein is the formation of gluten which results from the mixture of water with wheat protein and, to a lesser extent, the proteins of rye and triticale. Much microscopical research has been directed to gain an understanding of the special characteristics of these proteins and to explain them in terms of their physical features. Early observations have been reviewed by Hanssen (1955). More recently, using the LM, dynamic studies have been performed in which protein present in thin sections of endosperm were brought into contact with an encroaching front of water (Bernardin and Kasarda, 1973a). On contact, fibrils streamed into the aqueous phase, explosively at first but more slowly later, and became linked into a network. While this sequence of events occurred with wheat, rye and triticale, it did not do so with endosperm sections of corn, rice or barley. More fibrils were produced with wheat than by the other gluten-forming cereals. Freeze-dried fibrils examined under the TEM and SEM were finer and more profuse than those seen under the light microscope. They appeared as strands drawn out from sheets, sometimes with torn sheet-fragments along their length. The transformation from sheet to fibrillar form was examined by TEM and SEM by Bernardin and Kasarda (1973b). The fibres observed were not considered to correspond to those which imparted a fibrillar nature to adherent protein (Hess, 1960).

Seckinger and Wolf (1970) dispersed purified gliadins (the prolamines of wheat) onto the surface of water. The films thus formed were dried and examined under the TEM and were found to consist of small particles 20–80Å in diameter. Glutenin preparations took the form of strands and had no such particles. Dispersions of unclassified proteins from different wheat types had particles of different sizes, those of soft wheats being larger than those of hard and durum wheats. Only hard wheat particles were increased in size by mild treatment with sodium metabisulphite. In sections of kernels deposited on potassium permanganate solution, spreading of protein occurred less in hard than in soft wheats.

In ultrathin section, glutenin from soft wheats was heterogeneous with zones of granular structure and fibrillar felting; the fibrils were 100–200Å thick. Gliadin was homogeneous, smooth and compact, but the albumins and globulins more closely resembled glutenin. Total gluten consisted of a smooth compact matrix with fibrous and granular zones containing lipid inclusions (Crozet *et al.*, 1966, 1974). In sections of flour particles fibrillar felting was located at the edges of the particles and adjacent to starch granules. The globular form of lipids in the fibrous areas was considered to result from osmium fixation. Striations observed in the protein matrix have been variously interpreted as linked fibrils (Shkvarkina *et al.*, 1965) or artefacts resulting from sectioning.

It is generally considered that the presence of lipids, and particularly phospholipid, is essential for the formation of gluten with desirable visco-elastic properties, and many attempts to elucidate the relationship between the protein and lipid components have been made. Thus Grosskreutz (1961) examined dough before and after fat extraction with *n*-butanol. Extraction did not destroy the basic protein platelet structure of normal dough but it affected the ability of platelets to bond in coherent sheets capable of extensive plastic deformation. The hypothesis that lipoprotein slip-planes exist among the protein platelets is an attractive one, but neither Grosskreutz nor subsequent investigators have found conclusive supporting evidence through use of the TEM.

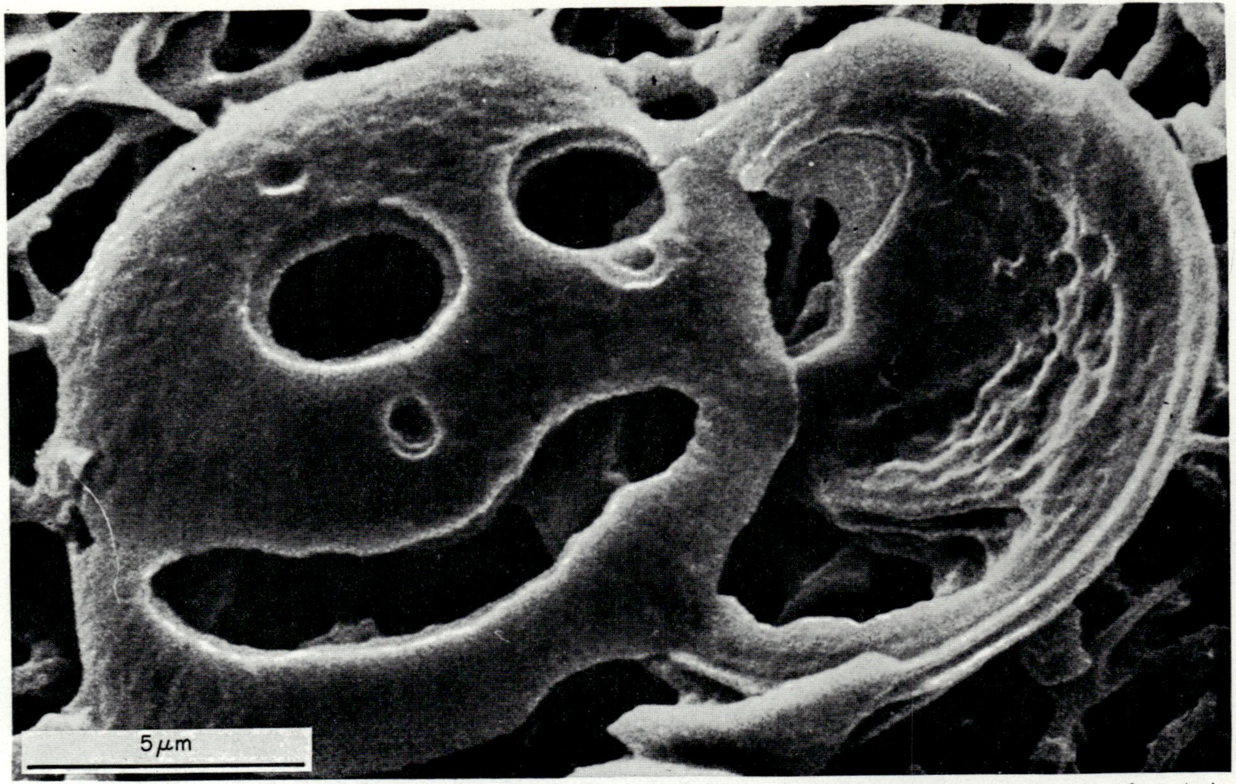

FIG. 17. Wheat starch granule attacked by hog pancreatic α-amylase. Corrosion channels are present on the upper face. As in Fig. 13, separation of the two halves has occurred at the equatorial plane of weakness. The preferential attack at the centre of the lower half can be seen; also, a ring structure is evident at the edges of the channels and at the periphery of the lower half-granule. Scanning electron micrograph.

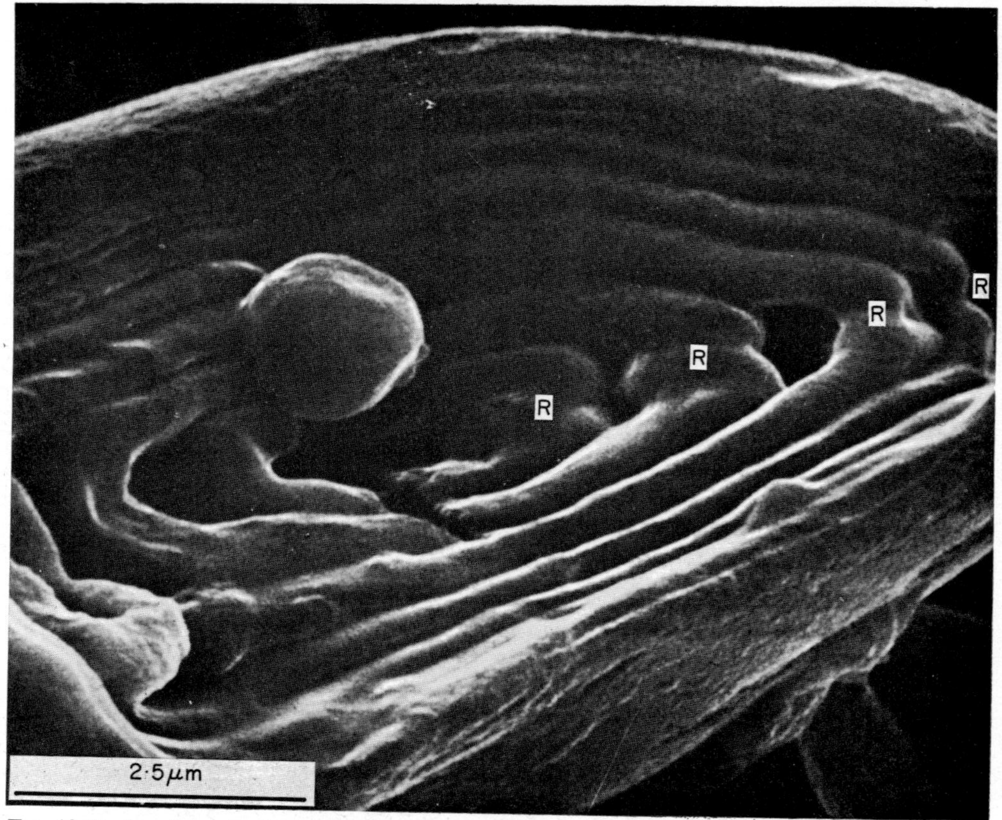

FIG. 18. Broken face of a large wheat starch granule, split perpendicularly at the equatorial plane of weakness. Attack by wheat α-amylase has removed the more susceptible layers, revealing the nature of the junctions between the two saucer-shaped halves of the granule. The depth of the re-entrant regions (R) declines towards the outside of the granule. Scanning electron micrograph (Evers and McDermott, 1970).

The effect of changes in relationships between the constituents of endosperm on dough formation have been studied by SEM (Aranyi and Hawrylewicz, 1968) and TEM (Simmonds, 1972b). Extraction of dough with petroleum ether removed globular lipid inclusions (type II) but not remnants of membrane from the endoplasmic reticulum (type I inclusions). The endoplasmic reticulum was considered to be a major source of both types of inclusion, although contamination of endosperm by lipids from the germ, and aleurone layer, known to be present in dough (Stevens, 1959) probably contributed to the type II inclusions. Examination of what were essentially glutenin preparations revealed a concentration of type I inclusions, structural preservation of which was better in the better breadmaking flours (Simmonds, 1972b).

Glutens in relaxed and stressed doughs were examined by SEM and TEM (Cumming and Tung, 1975). Sheets and fibres were present in the stressed condition. Adherence of gluten to starch was increased when doughs were made from defatted flours. Comparisons between doughs and vital gluten preparations suggested that the fibrous network present in doughs owes its existence to the presence of starch granules interrupting the gluten matrix Nevertheless, inclusions present in the stressed vital gluten showed considerable orientation (Cumming and Tung, 1975; Crozet *et al.*, 1974).

Glutenins isolated from a series of wheat and rye species were studied by Orth *et al.* (1973) using the SEM. Characteristic features of bread wheats were thick (10 μm) and thin (1 μm) fibrous strands, of circular cross-section, intertwined and unidirectionally aligned. Some filmy material also occurred, especially at points of contact between strands. Durum glutenin gave ribbon-like strands with more filmy material. The tetraploid parent of a synthetic hexaploid bread wheat was similar to the durum examined, and the fibrous nature of the other parent (*Aegilops squarrosa*) (which contributed the D genome) suggested that it is this species which is responsible for the fibrous nature of bread wheats. Addition of the D genome also resulted in improved baking quality. Rye glutenin showed the greatest variation from bread wheat gluten; short rod-like structures, exhibiting no entanglement, were typical. Gluten of a hexaploid triticale showed structures typical of both its durum wheat and rye parents (Dronzek *et al.*, 1974).

IV. DEVELOPMENT

A. Starch

In common with those from storage organs of other higher plants, cereal starch granules originate and grow in amyloplasts (Badenhuizen, 1969).

Whereas in the cotyledons of some legumes the same plastids are capable of functioning both as amyloplasts and chloroplasts (Whatley, 1974), in cereal endosperms amyloplasts are committed to the unique function of starch accumulation. Compared with most organelles, cereal endosperm amyloplasts are extremely simple. In corn, sorghum and the millets they appear to consist merely of a double membrane enclosing a ground substance, termed the stroma, in which starch granules develop. It is generally accepted that the increase in size of starch granules, which occurs during endosperm maturation, occurs by apposition; i.e. it results from the deposition of new starch material on the surface of pre-existing granules. Coacervate material, taken to be an intermediate between soluble carbohydrate and accreted starch material, has been reported in some amyloplasts. Membrane figures described as pockets have frequently been observed, and granules may form in these; however they are apparently not essential for granule formation. Amyloplasts are capable of division into units in each of which a single granule develops. Alternatively, more than one granule can occur in each daughter amyloplast, in which case 'compound' or 'polydelphous' granules (Sterling, 1968) are formed, the components of which are termed granuli. Some compound granules probably occur in all cereal types but they are particularly characteristic of oats and rice (see Chapter 3, Fig. 21). In oats, the amyloplasts which produce compound granules have, among the granuli, a complex of tubules (Buttrose, 1963). These tubules, consisting of membrane material and, in some cases at least, connected to the inner amyloplast membrane, are homologous with the membranes which form the pockets in the maize type of amyloplasts. They are considered, albeit from circumstantial evidence only, to be the sites of greater concentrations of starch-synthesizing enzymes than elsewhere in the stroma (Badenhuizen, 1969) since their location coincides with the areas of the granule at which most growth occurs. It is in members of the Triticeae that the nature and function of the tubuli are best understood. The presence of tubuli in the equatorial plane of A-type granules suggests that their lenticular shape arises from preferential growth in this plane (Buttrose, 1963; Jenkins *et al.*, 1974; Hughes, 1976). When their function with regard to A-type granules is finished, tubules play a further role in the Triticeae: they evert through the amyloplast wall and small, spherical (B-type) granules develop within them (Buttrose, 1963; Gallant and Guilbot, 1973). Structures resembling further tubuli or pockets have been reported inside the everted original tubuli by Hughes (1976), but these are generally considered to be simple structures.

Microscopical techniques, and indeed all other techniques, have so far failed to directly expose the location or the mechanisms of starch synthesis. Evidence from the formation of crystallites by synthetic polymers suggests that under favourable circumstances the formation of crystallites can occur; the near-spherical shape to which most granules tend may be the result of such physico-chemical influences (Banks and Greenwood, 1975). Additionally though, amyloplast shape appears to be capable of influencing granule shape. In single amyloplasts the relative strength of the two influences appear to vary according to the amylose : amylopectin ratio. The higher the amylose proportion, the less the influence of physico-chemical conditions and the greater the dependence of granule shape on that of the amyloplast. In the endosperm of high amylose maize the first-formed starch granules are spherical but those initiated later show a wide variety of forms (Boyer *et al.*, 1976a).

In the A-type granules of the Triticeae, dramatic changes in shape occur. In the earliest stages of endosperm development (Figs 19a–c) only tiny spherical granules or those with semi-circular outline are found (Teller, 1938 Bice *et al.*, 1945; Sandstedt, 1946; Evers, 1971; Dronzek *et al.*, 1974). With increasing maturation, the shape changes as a result of preferential deposition in the equatorial plane to the familiar lenticular form (Figs 19d–f). An equatorial groove, prominent in early development, becomes progressively shallower as the granule increases in size, and ultimately it becomes little more than a superficial score (Fig. 19g). An even earlier change in shape has been observed in proplastids in the meristematic layer from which the endosperm cells originate (Hughes, 1976).

A further influence upon granule shape becomes increasingly important in the later stages of endosperm development: this is the space limitation which exists as a result of close packing of storage products. Thus, in millet, corn, rice and sorghum, indentations due to pressure by protein bodies occur. They are not present in high-lysine mutants such as opaque 2 (Robutti *et al.*, 1974). In the Triticeae, indentations due to pressure from adjacent B-type granules can be seen on A-type granules (see Chapter 3, Fig. 31), and the presence of wedge-shaped B-type granules suggests that the entire form of those which develop last is determined by the space available. In cereals where regions of different packing density exist, granules in the tightly packed horny region are polyhedral, while those in the less dense floury region are spherical (see Chapter 3, Fig. 27). However, in rice it has been reported that some granuli are polyhedral before space limitations influence granule form (Hoshikawa, 1961).

B. Protein

Although not identifiable as individual bodies at maturity, protein bodies were observed with TEM in developing wheat endosperm (Graham *et al.*, 1962). It was deduced that the protein contained in these was synthesized by

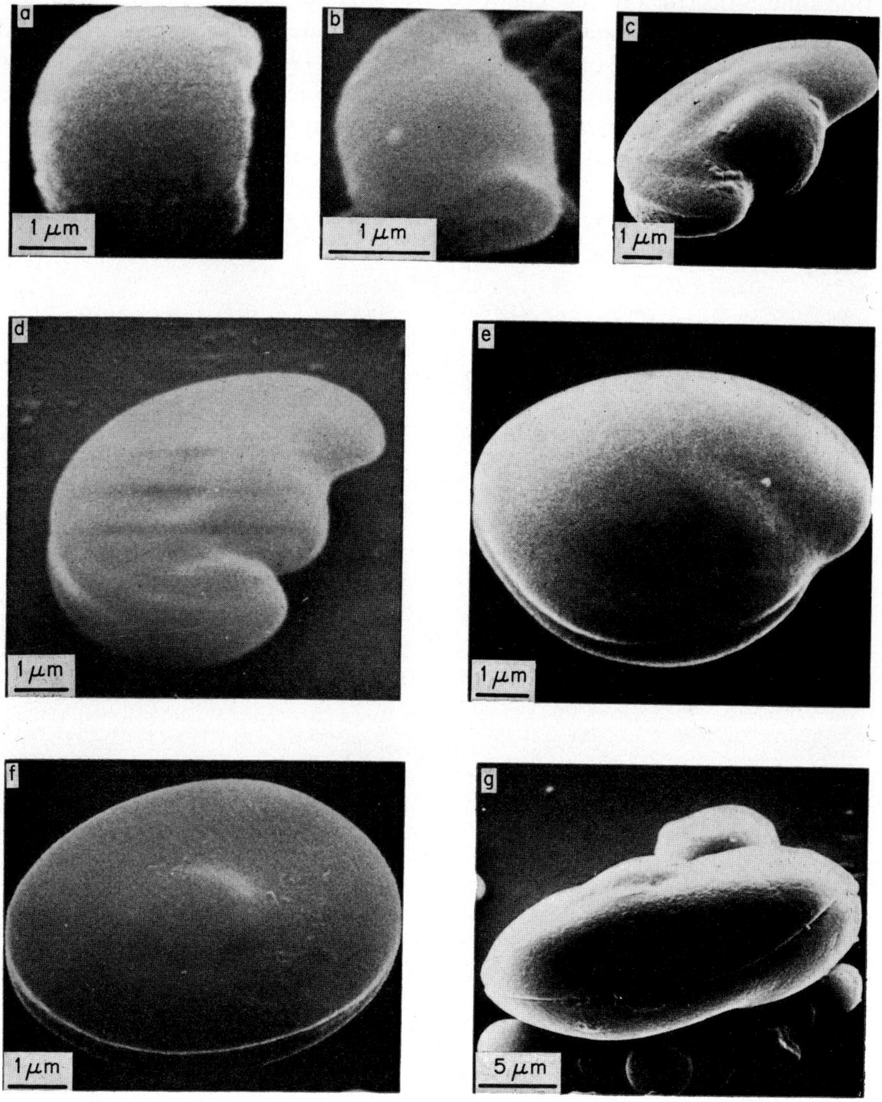

FIGS 19a–g. Proposed growth stages in the development of large lenticular starch granules of the Triticeae. For description see text. Scanning electron micrographs (Evers, 1971).

ribosomes located on the endoplasmic reticulum. The bodies were not always bounded by membranes (a feature shared with those present in mature endosperms of other cereals) (Buttrose, 1963). Association with the Golgi body (dictyosome) suggested that this organelle participated in protein body formation (Buttrose, 1963) (Fig. 20). Many aspects of protein bodies were studied in detail and extraction of protein from isolated bodies suggested

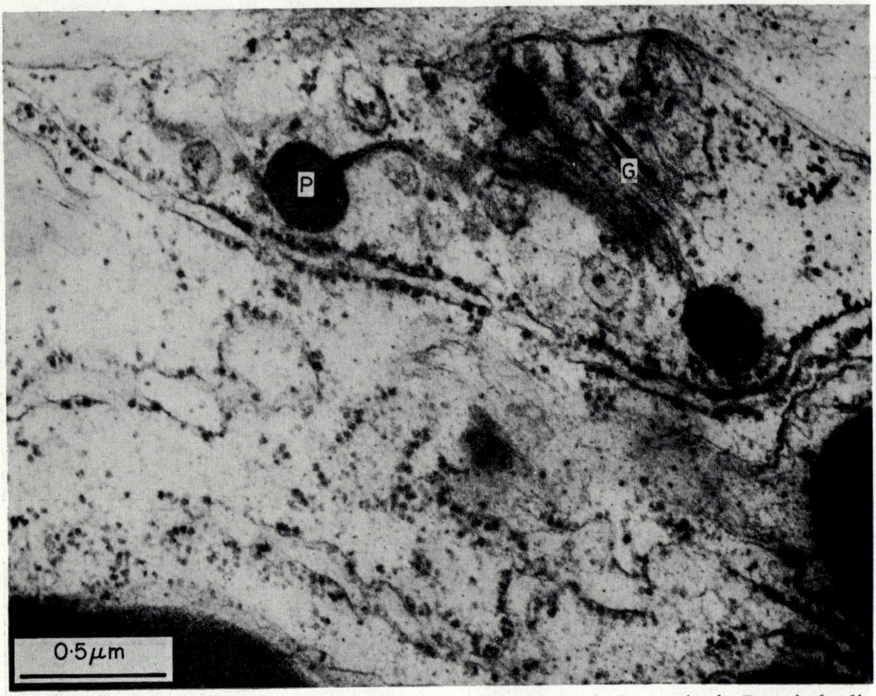

FIG. 20. Ultrathin section of wheat endosperm cell about 12 d after anthesis. Protein bodies (P) are closely associated with cisternae of the Golgi body (G). Transmission electron micrograph (Hughes, 1976).

both glutenin and gliadin were present (Graham *et al.*, 1962). Greater proportions of gliadin were present in a large body fraction than in a small body fraction. An alternative means of protein accumulation to that described above was by both synthesis and storage occurring within specialized organelles, described as proteoplasts (Morton *et al.*, 1964). No further evidence of the existence of these structures in wheat has been obtained however, and the more conventional view of synthesis by ribosomes on the rough endoplasmic reticulum has been endorsed in two recent studies (Barlow *et al.*, 1974; Hughes, 1976). Coalescence of protein bodies occurred (Fig. 21) and

Simmonds (1972a) suggested that the bodies were transported via concentric channels of endoplasmic reticulum. Possible modes of origin and development were summarized by Hughes (1976) (Fig. 22). Briarty (1978) drew attention to the problem of explaining transport when some bodies occurred in vacuoles and others in cisternae of the rough endoplasmic reticulum. Hoshikawa

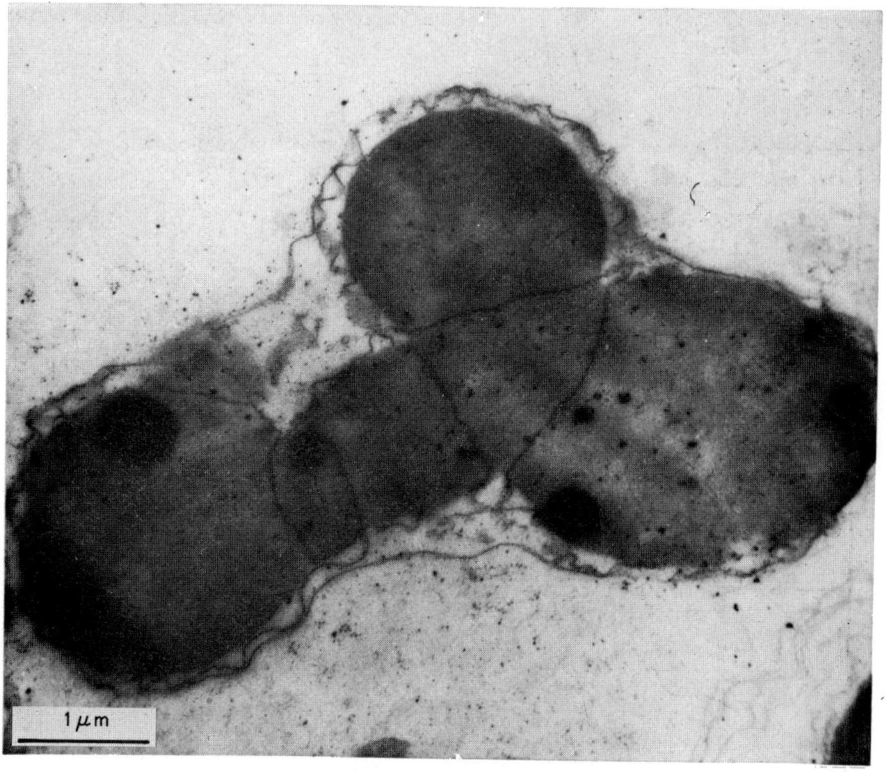

FIG. 21. Protein deposit from wheat endosperm some 22 d after anthesis. Six protein bodies have apparently fused. The entire structure is bounded by a membrane but it is not clear whether each body is bounded by a further membrane. Transmission electron micrograph (Hughes, 1976).

(1970) described proteoplasts found in developing rice endosperm, the amoeba-like constricted forms of some proteoplasts suggesting active division and hence self-propagation. On the other hand, the evidence of Khoo and Wolf (1970) favoured the non-plastid accumulation in corn endosperm, vesicles formed from ends of cisternae of the endoplasmic reticulum forming vesicles in addition to those originating with dictyosomes.

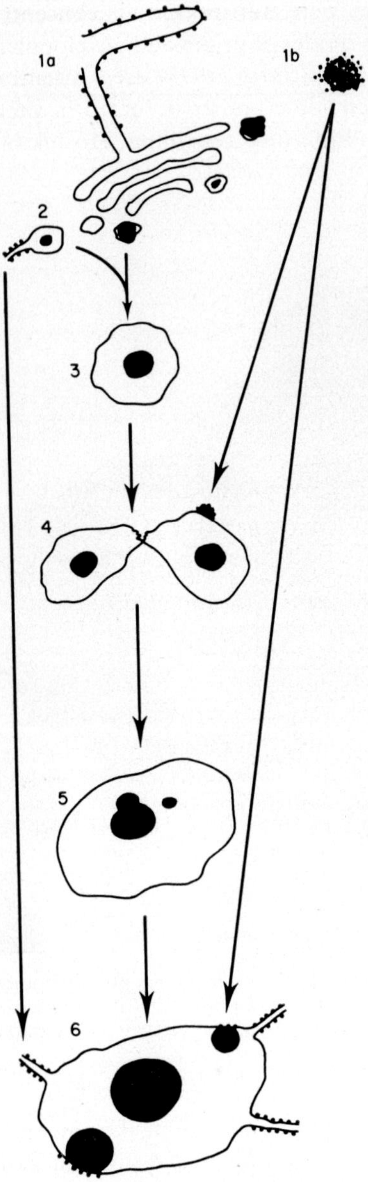

1a. Protein synthesis on ribosomes of the RER. Synthesized protein is released into the ER cisternae.
1b. Ribosomes associated with structures believed to be protein bodies.

2. Packaging of protein deposits by Golgi bodies. Production of membrane-bounded protein bodies.

3. Increase in size of protein-containing vacuole/vesicle and protein body.

4. Coalescence of vacuoles/vesicles. Possible entry of apparently non-vacuolate-contained protein bodies into protein-containing vacuoles or vesicles.

5. Coalescence of protein bodies and a further increase in the size of protein-containing vesicle or vacuoles.

6. Possible connections of RER with protein-containing vacuoles or vesicles, which may well have been present throughout development. Some protein bodies, when in direct contact with the vacuole/vesicle membrane, appear to be in close association with ribosomes.

FIG. 22. Diagram showing proposed mode of development of protein bodies in a wheat endosperm cell (Hughes, 1976).

V. QUANTITATIVE ASPECTS

A. Starch

1. Size distribution

Qualitative aspects of starch microscopy are concerned with the detailed appearance of representative individual granules and with the changes which occur therein during development and as a result of modification. Quantitative aspects, on the other hand, frequently concern attributes of populations of granules. Even so, in collecting quantitative data it is not possible to make measurements or observations on all granules in the population, and hence a representative sample is examined and data obtained from the sample are taken to apply to the whole population. For microscopical and other purposes a representative sample is best obtained from a well-mixed suspension in a suitable liquid. When a cover slip is applied to the drop of suspension on a slide the smaller granules tend to move to the outside edges and thus an uneven distribution arises; additionally, it is unlikely that a single microscope field contains a representative selection of granules. Therefore it is essential that a number of fields are considered in different areas of the preparation. The number of fields considered, the density of suspension and the total number of granules measured depend on the amount of detail required. If only the size range is required, as it might be for identification purposes, it is only necessary to measure the largest and smallest granules of frequent occurrence. For quality control purposes, where maximum size is a criterion of acceptance, it has been shown statistically that 137 is an adequate number of readings for assessing corn starches (Castelino and Youngken, 1972). In some cases, samples cannot be distinguished merely on the basis of their size range and it may be necessary to determine the mean diameter (Schoch and Maywald, 1967). For corn and sorghum starches, reproducible curves, from which 'number average diameters' could be calculated, were obtained by measuring the diameters of 600–1000 granules. This number corresponds with the recommendations of the British Standards Institution (B.S. 3406, Part 4, 1963) for complete particle size analysis of powders in general. This number is suitable for powders in which a single population is present, but it is inadequate for assessing distributions in the starches in which two distinct populations exist. For example, in wheat, where 97% of the granules are under 10 μm in size (Evers and Lindley, 1977), a count of 600 granules may include as few as 20 granules above 10 μm. Since these range between about 15 μm and 30 μm in diameter this would not be a sound basis on which to base a size distribution analysis, especially if the count is performed to provide a means of calculating a weight or volume distribution. On the other hand, measurement of a population including 600 of the larger granules would involve the measurement of 19 000 smaller granules, so some compromise has to be achieved. Counting of large numbers of particles has been facilitated by the availability of automated methods of particle measurement such as the Quantimet Image-analysing Computer (Cambridge Instruments Ltd., Melbourn, Royston, Herts.). As yet, however, the application of this or similar instruments to the measurement of starch populations has not been wide (Evers and Lindley, 1977). Recommendations for sampling, measuring and analysing are given in specialized works on particle size measurement such as that of Allen (1974). In this details are given of devices such as field finders, comparators and standard grids which facilitate measurements made manually. However, Nathan *et al.* (1971–2) found determinations made by inexperienced microscopists with a linear graticule to be less subject to reproducibility errors than those in which image-splitting devices or globe and circle comparators were used.

Results of microscopical methods of particle sizing rarely coincide with those from methods such as sedimentation or Coulter counting (Coulter Electronics Ltd., Harpenden, Herts.) in spite of the fact that microscopical sizing is the basis of calibration of the Coulter method. This lack of agreement is to be expected in view of the difference in criteria assessed by the various methods. For example, microscopical assessment depends upon linear measurements (or possibly area in automated methods) while the Coulter Counter depends upon direct measurements of volume. Conversion of one to the other generally involves the assumption that granules are spherical and this can be misleading in the cases of large granules from the Triticeae and amylomaize granules.

All methods of particle size analysis are subject to errors, the total error being the sum of the counting statistics error, the human error and the error of the method (Seldon, 1977). The counting statistics error can be quantified, and in a hypothetical case where ten intervals are chosen to contain about 10% each of the counts counting and sizing of 2000 particles will result in a counting statistics error of about $\pm 13{\cdot}4\%$ per size interval. Subjective errors causing differences among microscopists have been investigated; thus, Bucsky (1975) found a reproduction error of less than 2% in microscopical assessments.

The field of particle size analysis of starch populations is a confused one; while some values in the literature have been obtained by well-controlled methods which are statistically acceptable, there are many others which are little more than casual impressions.

Preparation of starches for particle size analysis in cereals which do not form a gluten can simply involve washing and grinding of the endosperm.

In the cases of wheat and rye, proteolysis may be necessary (Evers, 1973; Hanssen *et al.*, 1953). Alternatively, protein can be removed by shaking the starch suspension with toluene (Adkins and Greenwood, 1966). Deterioration due to fungal and bacterial contamination can be avoided by inclusion of mercuric chloride, thiomersal, etc. Granules which have become swollen thermally or mechanically can be eliminated by treatment of the entire suspension with α-amylase (Schoch and Maywald, 1967). Brownian movement can be prevented by the use of a 1 : 2 glycerol–water mix containing a trace of sodium chloride (Rasper, 1971).

Electron microscopy has not been applied widely and offers few advantages over the light microscope in sizing free starch granules, although an SEM technique has been used for confirmation of results obtained by differential interference contrast microscopy (Bluhm and Haller, 1976). Scanning electron microscopy may be useful for sizing such starches as rice, the granules of which are too small to be accurately measured with the light microscope.

The technique of stereology, only recently applied to biological materials (Briarty, 1975), has been used to determine the particle size distribution of wheat starch granules in sections of endosperms (Hughes and Briarty, 1976). A contribution of 33% by B-type granules to total starch volume was found. Detailed changes occurring during endosperm development were quantified with the technique, and a good log relationship between amyloplast tubule volume and starch volume between 6 and 22 days after anthesis provided evidence for the involvement of amyloplast tubuli in starch synthesis (Hughes and Briarty, 1975).

To some extent granule dimensions depend upon environment. They are usually measured in aqueous suspension and in this condition they are larger than in a vacuum-dry condition. Thus, corn starch expanded linearly by 9% and waxy maize starch by 22·7% when transferred from a dry to a water-saturated environment (Hellman *et al.*, 1952). Lenticular wheat starch granules expanded by 30% (Hanssen *et al.*, 1953), but granules extracted from immature barley endosperm did not expand even after several hours in water (May and Buttrose, 1959). Measurements made in 5% sucrose solution gave identical results to those obtained in water. In a brief investigation, Cornford (1961) found that chlorination of flour caused no increase in the size of the starch granules present until the flour had been wetted. Other workers found no change in size distribution of starch treated with chlorine, even after the starch had been wetted (Anon., 1977). Mechanically damaged granules are particularly susceptible to swelling in water (Jones, 1940). Granules expanded by one-half of their original diameter when compressed under a coverslip in an aqueous mount.

2. Gelatinization

Another quantitative application of the light microscope involves the determination of gelatinization temperature. In one method (MacMasters, 1964), a suspension of starch in distilled water (0·5% w : v) is heated at a controlled rate in a water bath. A drop of suspension is removed and examined at each 0·5°C rise in temperature. Gelatinization temperature is recognized either by loss of birefringence or by uptake of a dye such as Congo red. Alternatively a drop of a 1–2% w : v suspension, encapsulated between a slide and coverslip and surrounded by oil, is heated on a microscopical hot-stage (Watson, 1964). Gelatinization does not occur uniformly, so the temperature is recorded when 2, 10, 25, 50, 75, 90 and 98% of granules no longer exhibit birefringence. The 98% point is taken as the birefringence end-point temperature (BEPT). Fields of 100–200 granules are ideal, and determinations in duplicate or triplicate are desirable. Gelatinization temperature provides a useful means of helping to identify starches of different plant sources and types. It also provides a means of defining the suitability of samples for food and industrial applications (Webb and Stermer, 1972). A device designed to eliminate the subjective element of BEPT determinations incorporated a photo-cell to detect the disappearance of birefringence (Berry and White, 1966). Gelatinization temperatures could be measured more accurately with the device which was sufficiently sensitive to detect a decline in light transmitted at temperatures as low as 20–30°C. The effects on gelatinization temperature of the addition to the starch suspension of various constituents of cakes were studied. All caused an increase, but two different mechanisms were involved: oils merely reduced the amount of water available by physically replacing it, while the other substances, including sugars, proteins and egg solids, also bound water which caused an enhanced effect.

Another method for achieving objectivity in the determination of gelatinization temperature involved photography of a field on a slide being heated on a hot-stage. Measurements were made on prints of photographs taken at intervals to determine the degree of swelling, which was then plotted on a graph against temperature. The effects of sodium chloride concentration upon gelatinization temperature were studied by this technique (Collison, 1961). Salts in general can have a marked elevating or depressing effect upon gelatinization temperature; thus in a boiling sodium sulphate solution starch will not gelatinize, but it will do so at room temperature in sodium iodide (Banks and Greenwood, 1975). The effects of a range of concentrations of various chlorides on wheat starch gelatinization have been studied in some detail (Gough and Pybus, 1973). With most salts, increasing concentration resulted in gelatinization temperatures first being raised, then depressed, and then raised again.

There are also marked effects upon gelatinization temperature resulting from the prolonged maintenance of granules even in distilled water at a temperature just below that at which gelatinization normally begins. After 72 hours at 50°C the temperature range of wheat starch gelatinization was much contracted from 9°C to 0·5°C. There was an overall elevation (Gough and Pybus, 1971).

3. *Surface pore size*

A subtle interpretation of the results of staining with fluorescent dyes led to the conclusion that pores of size 0·7–20Å existed on starch granule surfaces (Sterling, 1964). The speculation was based on the ability of dimers of primulin to enter the granule. Such pores have not been detected by high-resolution TEM, but this may be due to the failure of the replica-making process to resolve the pores rather than to their absence.

B. Protein

Few applications of systematic quantitative techniques to the microscopy of cereal proteins have been recorded. The application of stereology, at a TEM level by Hughes and Briarty (1976), to protein body development in wheat endosperm has thus provided a welcome precedent, which, it is to be hoped will be followed by wider exploitation. Hughes and Briarty related the increase in storage protein volumes per cell to that in rough endoplasmic reticulum (RER). Their results showed a parallel increase in these two factors, thus providing circumstantial evidence for the involvement of RER in storage protein synthesis. The surface-to-volume ratio of RER was also determined; this rose rapidly before the onset of protein synthesis, suggesting that the RER was becoming flattened, and then dropped due to swelling as storage protein accumulated.

Microscopists in general are turning increasingly to quantitative techniques as a means of testing the validity of qualitative impressions or extracting information which is not obtainable by purely qualitative techniques. In the sphere of cereal microscopy, quantitative techniques are not widely used, and indeed not all are readily applicable to the starch and protein components. However, many opportunities exist for the further exploitation of well-tried and reliable quantitative methods, and if these opportunities are taken there is much benefit to be gained.

ACKNOWLEDGEMENTS

I wish to thank Professor B. Spencer, Director General of the Flour Milling and Baking Research Association, for permission to contribute this chapter. I am also grateful to colleagues both within and outside the Research Association for their helpful advice and co-operation.

REFERENCES

Adams, C. A., Novellie, L. and Liebenberg, N. V. D. W. (1976). *Cereal Chem.* **53**, 1.

Adkins, G. K. and Greenwood, C. T. (1966). *Stärke*, **18**, 213.

Allen, T. (1974). 'Particle Size Measurement.' Chapman and Hall, London.

Alsberg, C. L. and Griffing, E. P. (1925). *Cereal Chem.* **2**, 325.

Angold, R. (1975). *In* 'Bread' (A. Spicer, ed.), p. 141. Applied Science Publishers, London.

Anon. (1977). Flour Milling and Baking Research Association Annual Report, Chorleywood.

Aranyi, C. and Hawrylewicz, E. J. (1968). *Cereal Chem.* **45**, 500.

Arbuckle, A. W. and Greenwood, C. T. (1958). *J. Chem. Soc.* 2629.

Babichenko, L. V. (1973). *Izv. Vyssh. Ucheb. Zaved. Pishch. Teknol.* **6**, 36.

Badenhuizen, N. P. (1969). 'The Biogenesis of Starch Granules in Higher Plants.' Appleton-Century-Crofts, New York.

Badi, S. M., Hoseney, R. C. and Cassidy, A. J. (1976). *Cereal Chem.* **53**, 478.

Bakhuyzen, H. L., van de sande (1926). *Proc. Soc. Exp. Biol. Med.* **23**, 302.

Banks, W. and Greenwood, C. T. (1975). 'Starch and its Components.' Edinburgh University Press, Edinburgh.

Barlow, K. K., Buttrose, M. S. and Simmonds, D. H. (1973). *Cereal Chem.* **50**, 443.

Barlow, K. K., Lee, J. W. and Vesk, M. (1974). *In* 'Mechanisms of Regulation of Plant Growth' (R. L. Bideski, A. R. Fergusson and M. M. Cresswell, eds), p. 799. Bull. 12, Royal Society of New Zealand.

Bernardin, J. E. and Kasarda, D. D. (1973a). *Cereal Chem.* **50**, 529.

Bernardin, J. E. and Kasarda, D. D. (1973b). *Cereal Chem.* **50**, 735.

Berry, G. K. and White, G. W. (1966). *J. Fd Technol.* **1**, 249.

Bice, C. W., MacMasters, M. M. and Hilbert, G. E. (1945). *Cereal Chem.* **22**, 463.

Bluhm, J. W. and Haller, M. N. (1976). *Int. Lab.* Sept.–Oct., 23.

Boyer, C. D., Daniels, R. R. and Shannon, J. C. (1976a). *Crop Sci.* **16**, 298.

Boyer, C. D., Shannon, J. C., Garwood, D. L. and Creech, R. G. (1976b). *Cereal Chem.* **53**, 327.

Briarty, L. G. (1974). *In* 'Principles and Techniques of Scanning Electron Microscopy' (M. A. Hayat, ed.), Vol. 1, p. 206. Van Nostrand Reinhold, New York.

Briarty, L. G. (1975). *Sci. Prog., Oxf.* **62**, 1.

Briarty, L. G. (1978). *In* 'Plant Proteins' (G. Norton, ed.), p. 81. Butterworth, London.

Brudka, J. and Jakubczyk, T. (1976). *Acta Aliment. Pol.* **11** (26), 287.

Bucsky, G. (1975). *Process Eng*. July, 52.
Bushuk, W. (1976). 'Rye'. A.A.C.C., St Paul, Minnesota.
Buttrose, M. S. (1960). *J. Ultrastruct. Res.* **4**, 231.
Buttrose, M. S. (1962). *J. Cell. Biol.* **14**, 159.
Buttrose, M. S. (1963). *Aust. J. Biol. Sci.* **16**, 305.
Castelino, N. J. and Youngken, H. W. (1972). *Stärke*, **24**, 412.
Cauvain, S. P., Gough, B. M. and Whitehouse, M. E. (1977). *Stärke*, **29**, 91.
Chabot, J. F., Hood, L. F. and Allen, J. E. (1976). *Cereal Chem.* **53**, 85.
Christianson, D. D. (1970). Proc. 3rd Ann. SEM Symp., I.I.T. Res. Inst., Chicago, p. 161.
Christianson, D. D., Nielsen, H. C., Khoo, U., Wolf, M. J. and Wall, J. S. (1969). *Cereal Chem.* **46**, 372.
Cobb, N. A. (1902). *Agric. Gaz. N.S.W.* **13**, 74.
Collison R. (1961). *Stärke*, **13**, 9.
Collison, R. (1968). *In* 'Starch and its Derivatives' (J. A. Radley, ed.), p. 168. Chapman and Hall, London.
Cornford, S. J. (1961). *J. Sci. Fd Agric.* **12**, 693.
Crozet, N., Gallant, D., Godon, B., Guilbot, A. and Petit, L. (1966). *Inds. Aliment. Agric.* **83**, 559.
Crozet, N., Godon, B., Petit, L. and Guilbot, A. (1974). *Cereal Chem.* **51**, 288.
Cumming, D. B. and Tung, M. A. (1975). *J. Can. Sci. Technol. Aliment.* **8**, 67.
Czaja, A. Th. (1961). *Stärke*, **13**, 357.
Czaja, A. Th. (1972). *Stärke*, **24**, 77.
Derby, R. I., Miller, B. S., Miller, B. F. and Trimbo, H. B. (1975). *Cereal Chem.* **52**, 702.
Dronzek, B. L., Hwang, P. and Bushuk, W. (1972). *Cereal Chem.* **49**, 232.
Dronzek, B. L., Orth, R. A. and Bushuk, W. (1974). *In* 'Triticale: First Man-made Cereal' (C. C. Tsen, ed.), p. 91. A.A.C.C., St Paul, Minnesota.
Duvick, D. N. (1961). *Cereal Chem.* **38**, 374.
Evers, A. D. (1970). *Ann. Bot.* **34**, 547.
Evers, A. D. (1971). *Stärke*, **23**, 157.
Evers, A. D. (1973). *Stärke*, **25**, 303.
Evers, A. D. and Juliano, B. O. (1976). *Stärke*, **28**, 160.
Evers, A. D. and Lindley, J. (1977). *J. Sci. Fd Agric.* **28**, 98.
Evers, A. D. and McDermott, E. E. (1970). *Stärke*, **22**, 23.
Evers, A. D., Gough, B. M. and Pybus, J. N. (1971). *Stärke*, **23**, 16.
Fanta, G. F., Baker, F. L., Burr, R. C., Doane, W. M. and Russell, C. R. (1973). *Stärke*, **25**, 157.
Farber, B. and Gallant, D. (1976). *Annls Zootech.* **25**, 13.
Feder, N. and O'Brien, T. P. (1968). *Am. J. Bot.* **55**, 123.
Fleming, I. D. (1968). *In* 'Starch and its Derivatives' (J. A. Radley, ed.), p. 498. Chapman and Hall, London.
Flint, F. O. and Moss, R. (1970). *Stain Technol.* **45**, 75.
Flint, F. O., Moss, R. and Wade, P. (1970). *Fd Trade Rev.* **40**, 32.
French, D. (1972). *J. Jap. Soc. Starch Sci.* **19**, 8.

Fulcher, R. G., O'Brien, T. P. and Simmonds, D. H. (1972). *Aust. J. Biol. Sci.* **25**, 487.
Gale, M. D. and Marshall, G. A. (1973). Proc. 4th Int. Wheat Genetics Symp., Missouri Agr. Exp. Sta., Columbia.
Gallant, D. and Guilbot, A. (1971). *Stärke*, **23**, 244.
Gallant, D. and Guilbot, A. (1973). *Stärke*, **25**, 335.
Gallant, D., Mercier, C. and Guilbot, A. (1972). *Cereal Chem.* **49**, 354.
Gallant, D., Derrien, A., Aumaitre, A. and Guilbot, A. (1973). *Stärke*, **25**, 56.
Goering, K. J., Fritts, D. H. and Eslick, R. F. (1973). *Stärke*, **25**, 297.
Gough, B. M. and Pybus, J. N. (1971). *Stärke*, **23**, 210.
Gough, B. M. and Pybus, J. N. (1973). *Stärke*, **25**, 123.
Gough, B. M., Greenwood, C. T. and Whitehouse, M. (1978). 'Critical Reviews in Food Science and Nutrition.' Chemical Rubber Company, Cleveland, Ohio. (In press.)
Graham, J. S. D., Jennings, A. C., Morton, R. K. and Raison, J. K. (1962). *Nature, Lond.* **196**, 967.
Grosskreutz, J. C. (1961). *Cereal Chem.* **38**, 336.
Hall, D. M. and Sayre, J. G. (1970). *Text. Res. J.* **40**, 256.
Hall, D. M. and Sayre, J. G. (1973). *Stärke*, **25**, 119.
Hall, D. M., Van Patten, E., Brown, J. L., Harnon, G. R. and Nix, G. H. (1971). *Ind. Engng Chem., Proc. Res. Develop.* **10**, 191.
Hanssen, E. G. (1955). *Mikroskopie*, **10**, 155.
Hanssen, E. G., Dodt, E. and Niemann, E. G. (1953). *Kolloidzeitschrift*, **130**, 19.
Hartz, C. O. (1885). 'Landwirtschaftliche Samenkunde.' Paul Parey, Berlin.
Hearle, J. W. S., Sparrow, J. T. and Cross, P. M. (1972). 'The Use of the Scanning Electron Microscope.' Pergamon Press, Oxford.
Hellman, N. N., Boesch, T. F. and Melvin, E. H. (1952). *J. Am. Chem. Soc.* **74**, 348.
Hess, K. (1954). *Kolloidzeitschrift*, **136**, 84.
Hess, K. (1960). *Qualitas Pl. Mater. Veg.* **6**, 312.
Hess, K., Mahl, H., Gutter, E. and Dodt, E. (1955). *Mikroskopie*, **10**, 6.
Hill, R. D. and Dronzek, B. L. (1973). *Stärke*, **25**, 367.
Himes, M. B. and Moriber, L. (1956). *Stain Technol.* **31**, 67.
Hinton, J. J. C. (1953). *Cereal Chem.* **30**, 441.
Hixon, R. M. and Brimhall, B. (1968). *In* 'Starch and its Derivatives' (J. A. Radley, ed.), p. 247. Chapman and Hall, London.
Hollo, J. (1968). *In* 'Starch and its Derivatives' (J. A. Radley, ed.), p. 203. Chapman and Hall, London.
Hoshikawa, K. (1961). *Proc. Crop. Sci. Soc., Japan*, **29**, 415.
Hoshikawa, K. (1970). *Proc. Crop Sci. Soc., Japan*, **39**, 295.
Houston, D. F., ed. (1972). 'Rice Chemistry and Technology.' A.A.C.C., St Paul, Minnesota.
Hughes, C. E. (1976). 'The Developing Endosperm of *Triticum aestivum* (L). An Ultrastructural and Morphometric Study.' Ph.D. Thesis, University of Nottingham.

Hughes, C. E. and Briarty, L. G. (1975). Fourth Int. Cong. for Stereology, Gaithersberg, Maryland, p. 385.

Hughes, C. E. and Briarty, L. G. (1976). *Stärke*, **28**, 337.

Humason, G. (1967). 'Animal Tissue Techniques.' W. H. Freeman, San Francisco.

Inglett, G. E. (1972). 'Seed Proteins.' Avi. Publ. Co., Westport, Connecticut.

Innamorati, M. (1966). *Caryologia*, **19**, 343.

Jenkins, L. D., Loney, D. P., Meredith, P. and Finneran, B. A. (1974). *Cereal Chem.* **51**, 717.

Jensen, W. A. (1962). 'Botanical Histochemistry.' W. H. Freeman, San Francisco.

Jones, C. R. (1940). *Cereal Chem.* **17**, 133.

Jones, F. T. and Bean, M. M. (1972). *Microscope*, **20**, 333.

Jones, R. W., Beckwith, A. C., Khoo, U. and Inglett, G. E. (1970). *J. Agric. Fd Chem.* **18**, 37.

Juliano, B. O. (1972). *In* 'Rice Chemistry and Technology' (D. F. Houston, ed.), p. 16. A.A.C.C., St Paul, Minnesota.

Katz, J. R. and Muschter, F. J. R. (1933). *Biochem. Z.* **252**, 385.

Keller, A. (1959). *J. Polym. Sci.* **39**, 151.

Kent, N. L. (1966). *Cereal Chem.* **43**, 585.

Kent, N. L. (1970). *In* 'Proteins as Human Food' (R. A. Lawrie, ed.), p. 280. Butterworth, London.

Kerr, R. W. (1950). 'Chemistry and Industry of Starch.' Academic Press, New York and London.

Khoo, U. and Wolf, M. J. (1970). *Am. J. Bot.* **57**, 1042.

Lampitt, L. H., Fuller, C. H. and Goldenberg, N. (1941). *J. Soc. Chem. Ind., Lond.* **60**, 1.

Leach, H. W. and Schoch, T. J. (1961). *Cereal Chem.* **38**, 34.

Lineback, D. R. and Ponpipom, S. (1977). *Stärke*, **29**, 52.

Little, R. R. (1957). *Stain Technol.* **32**, 7.

MacLeod, A. M. and Palmer, G. H. (1966). *J. Inst. Brew.* **72**, 580.

MacMasters, M. M. (1953). *Cereal Chem.* **30**, 63.

MacMasters, M. M. (1964). *In* 'Methods in Carbohydrate Chemistry' (R. L. Whistler, ed.), Vol. 4, p. 233. Academic Press, New York and London.

McNeil, J. W., Potter, G. D., Riggs, J. K. and Rooney, L. W. (1975). *J. Anim. Sci.* **40**, 335.

May, L. H. and Buttrose, M. S. (1959). *Aust. J. Biol. Sci.* **12**, 147.

Meyer, A. (1886). *Ber. Dt. Bot. Ges.* **4**, 337.

Miller, B. S., Derby, R. I. and Trimbo, H. B. (1973). *Cereal Chem.* **50**, 271.

Mitsuda, H. and Murakami, K. (1969). *Physiol. Pl., Japan*, **8**, 1.

Morton, R. K., Palk, B. A. and Raison, J. K. (1964). *Biochem. J.* **91**, 522.

Moss, G. E. (1976). *In* 'Examination and Analysis of Starch and Starch Products' (J. A. Radley, ed.), p. 1. Applied Science Publishers, London.

Mottier, D. M. (1921). *Ann. Bot.* **35**, 349.

Mussulman, W. C. and Wagoner, J. A. (1968). *Cereal Chem.* **45**, 162.

Nathan, I. F., Barnett, M. I. and Turner, T. D. (1971–2). *Powder Tech.* **5**, 105.

Nikuni, Z. and Whistler, R. L. (1957). *J. Biochem., Tokyo*, **44**, 227.

Norton, G. (1978). 'Plant Proteins.' Butterworth, London.
Orth, R. A., Dronzek, B. L. and Bushuk, W. (1973). *Cereal Chem.* **50**, 688.
Ory, R. L. and Henningsen, K. W. (1969). *Pl. Physiol., Lancaster*, **44**, 1488.
Osborne, T. B. (1907). 'The Proteins of the Wheat Kernel.' Publ. No. 84, Carnegie Institute of Washington, Washington D.C.
Palmer, G. H. (1971). Proc. 13th Cong. of European Brewery Convention, Estoril, p. 59.
Palmer, G. H. (1972). *J. Inst. Brew.* **78**, 326.
Peterson, R. F. (1965). 'Wheat, Botany, Cultivation and Utilization.' Interscience Publishers, New York.
Pomeranz, Y. (1971). 'Wheat, Chemistry and Technology.' A.A.C.C., St Paul, Minnesota.
Pomeranz, Y. and Shellenberger, J. A. (1961). *Cereal Chem.* **38**, 109.
Radley, J. A., ed. (1968). 'Starch and its Derivatives.' Chapman and Hall, London.
Radley, J. A. (1976). 'Examination and Analysis of Starch and Starch Products.' Applied Science Publishers, London.
Rasper, V. (1971). *J. Sci. Fd Agric.* **22**, 572.
Rasper, V., Perry, G. and Duitschaever, G. L. (1974). *Can. Inst. Fd Sci. Technol. J.* **7**, 166.
Reichert, E. T. (1913). 'The Differentiation and Specificity of Starches in Relation to Genera, Species, etc.' Publ. No. 173 (1), (2), Carnegie Institute of Washington, Washington D.C.
Robutti, J. L., Hoseney, R. C. and Wassom, C. E. (1974). *Cereal Chem.* **51**, 173.
Robyt, J. F. and Whelan, W. J. (1968). *In* 'Starch and its Derivatives' (J. A. Radley, ed.), p. 430. Chapman and Hall, London.
Rooney, L. W. and Sullins, R. D. (1977). *In* 'Proceedings: Symposium on Sorghum and Millets' (D. A. V. Dendy, ed.), p. 91. Tropical Products Institute.
Rost, T. L. (1971). *Protoplasma*, **73**, 475.
Rost, T. L. and Lersten, N. R. (1973). *Iowa St. J. Res.* **48**, 47.
Sandstedt, R. M. (1946). *Cereal Chem.* **23**, 337.
Sandstedt, R. M. (1955). *Brewers' Dig.* **30**, 49.
Sandstedt, R. M. and Schroeder, H. (1960). *Fd Technol., Champaign*, **14**, 257.
Sass, J. E. (1958). 'Botanical Microtechnique.' Iowa State University Press, Ames, Iowa.
Scallet, B. L. and Sowell, E. A. (1967). *In* 'Starch Chemistry and Technology' (R. L. Whistler and E. F. Paschall, eds), Vol. 2, p. 237. Acacemic Press, New York and London.
Schoch, T. J. (1965). *Bakers' Dig.* **39**, 48.
Schoch, T. J. and Maywald, E. (1967). *In* 'Starch Chemistry and Technology' (R. L. Whistler and E. F. Paschall, eds), Vol. 2, p. 637. Academic Press, New York and London.
Seckinger, H. L. and Wolf, M. J. (1967). *Cereal Chem.* **44**, 669.
Seckinger, H. L. and Wolf, M. J. (1970). *Cereal Chem.* **47**, 236.
Seckinger, H. L. and Wolf, M. J. (1973). *Cereal Chem.* **50**, 455.
Seidemann, J. (1966). 'Principles of Starch Microscopy and Description of Important Varieties of Starch.' Paul Parey, Berlin. (In German.)

Selby, K. (1977). *Chemy Ind.* 494.
Seldon, M. G. (1977). *Microscope*, **26**, 127.
Shkvarkina, T. I., Ivanova, E. A. and Lunev, V. A. (1974). *Selekts. Semanov.* **5**, 46.
Simmonds, D. H. (1972a). *Cereal Chem.* **49**, 212.
Simmonds, D. H. (1972b). *Cereal Chem.* **49**, 324.
Simmonds, D. H. (1974). *In* 'Triticale: First Man-made Cereal' (C. C. Tsen, ed.), p. 105. A.A.C.C., St Paul, Minnesota.
Smith, J. S. and Lineback, D. R. (1976). *Stärke*, **28**, 243.
Sterling, C. (1964). *Protoplasma*, **59**, 180.
Sterling, C. (1968). *In* 'Starch and its Derivatives' (J. A. Radley, ed.), p. 139. Chapman and Hall, London.
Stevens, D. J. (1959). *Cereal Chem.* **36**, 452.
Stevens, D. J. (1976). *Stärke*, **28**, 25.
Stewart, B. A. (1964). *Nature, Lond.* **204**, 1088.
Stute, R. (1973). *Stärke*, **25**, 409.
Sullins, R. D. and Rooney, L. W. (1977). *In* 'Proceedings: Symposium on Sorghum and Millets' (D. A. V. Dendy, ed.), p. 79. Tropical Products Institute.
Teller, G. L. (1938). *Pl. Physiol., Lancaster*, **13**, 227.
Unna, E. (1918). *Z. Nahr. Genussm.* **36**, 49.
Walker, G. J. and Hope, P. M. (1963). *Biochem. J.* **86**, 452.
Watson, S. A. (1964). *In* 'Methods in Carbohydrate Chemistry' (R. L. Whistler, ed.), Vol. 4, p. 240. Academic Press, New York and London.
Webb, B. D. and Stermer, R. A. (1972). *In* 'Rice Chemistry and Technology' (D. F. Houston, ed.), p. 102. A.A.C.C., St Paul, Minnesota.
Whatley, J. M. (1974). *New Phytol.* **73**, 1097.
Whistler, R. L. and Paschall, E. F., eds (1965). 'Starch Chemistry and Technology', Vol. 1. Academic Press, New York and London.
Whistler, R. L. and Paschall, E. F., eds (1967). 'Starch Chemistry and Technology', Vol. 2. Academic Press, New York and London.
Whistler, R. L., Spencer, W. W., Goatley, J. L. and Nikuni, Z. (1958). *Cereal Chem.* **35**, 331.
Whistler, R. L., Goatley, J. L. and Spencer, W. W. (1959). *Cereal Chem.* **36**, 84.
White G. W. and Shenton, A. J. (1975). *J. Ass. Public. Anal.* **13**, 34.
Williams, P. C. (1968). *Bull. Ass. Operative Millers*, July, 3048.

5. Meat

C. A. VOYLE

Agricultural Research Council, Meat Research Institute,
Langford, Bristol, U.K.

I. INTRODUCTION

Meat is generally defined as the flesh of animals used for food. More specifically, it is those parts of the musculature of certain mammals and birds which are found to be acceptable to the consumer in terms of palatability, nutrition and economics. Additionally, other tissues and organs of animal origin may be regarded as suitable items of diet, e.g. liver, kidneys, tongue, etc., and these are generally classified as 'edible offal'.

This chapter is devoted to a consideration of the structure of muscle as meat, and will describe macroscopic and microscopic structure in 'normal' tissue. The effect on structure of such treatments as freezing, freeze-drying and cooking will also be discussed.

Methods of preparation of tissue for microscopic examination are not described. This information can be obtained from a textbook of histological technique, e.g. Disbrey and Rack (1970).

In more primitive times man would obtain the meat component of his diet by hunting, as dictated by his personal needs or those of the community in which he lived. With the development of his skills at husbandry and his involvement in other activities arising from the increasing complexities of life, methods of preservation and storage of meat became a necessity, so that a temporary surplus could be retained against seasonal shortages. Such methods of preservation as drying, freezing and curing have an effect on the structure of meat, which in turn will affect its eating quality.

Whereas it is exceptional to eat meat in the raw state, the use of heat in the preparation of meat for eating is both commonplace and important to palatability. The heat may be in the form of fire, hot air or a heated liquid medium. Inevitably, exposure to this sort of treatment will markedly alter the structure of the tissue. Some of these aspects will be examined in this chapter.

Most of the meat produced for human consumption is derived from cattle, sheep and pigs, with a significant contribution from poultry. The Annual Report of the National Food Survey Committee for 1976 shows that the average weight of carcass meat purchased per person per week in the United Kingdom was 6·3% of the total weight of food purchased. Poultry accounted for a further 2·5% of purchases. About 17% of the average weekly expenditure on food was spent on these items. The consumption of meat from other sources, such as rabbit, is comparatively small but there is an increasing interest in the farming of other meat producing species, for example red deer in Scotland.

II. STRUCTURE OF MUSCLE

A. General Considerations

The structure of the skeletal muscles of mammals and birds is basically similar in all species. The muscles within a species vary in shape and in their detailed micro-anatomy, but the structural hierarchy and the macromolecular units of structure are common to all. The primary function of muscles is to effect locomotion. They must, therefore, be securely attached to those components of the body, mostly bones, where movement is required. About three hundred anatomically distinct muscles are recognizable in the main meat-producing species.

Muscles function by means of contraction, involving a complex series of biochemical reactions initiated by a stimulus from the nervous system. For a detailed description of the contractile mechanism a textbook of animal physiology should be consulted. Reviews of the subject have been written by Bendall (1969) and Huxley (1972). A summary of the main factors involved is included here for the sake of completeness.

As living tissue, the muscle is supplied with chemical energy which it converts into mechanical energy. The chemical energy is derived from hydrolytic reactions involving adenosine triphosphate (ATP), phosphocreatine (PC) and glycogen. In the presence of calcium and magnesium ions ATP is split to form adenosine diphosphate (ADP) and inorganic phosphate (P_i). The chemical energy thus released is utilized in the contractile process as mechanical energy. The supply of ATP is regenerated by the transphosphorylation of ADP, the additional phosphate required being provided by PC. This reaction is mediated by the enzyme creatine phosphotransferase. The reserves of PC are restored by the breakdown of glycogen derived from food, a process known as glycolysis. In living muscle an end-product of glycolysis is pyruvic acid. The removal of hydrogen from this substance and from the fatty acids which are present in the tissue is effected by the mechanism known as the Krebs or tricarboxylic acid cycle. This process contributes to the restoration of the PC supply, as does the activity of the cytochrome chain of iron-containing proteins in mitochondria.

After death, the capacity for contraction and relaxation diminishes and the muscle sets in the condition known as rigor mortis. This condition is achieved when ATP is no longer available to provide energy for the contractile reaction. The absence of a supply of oxygen, following death, results in a limited synthesis of ATP from the products of glycolysis compared with the amount produced under aerobic conditions. The muscle tissue becomes acidified due to the formation of lactic acid. Resting muscle in the live animal has a pH of about 7·1, whereas the pH of a muscle in full rigor is about 5·5.

Mammalian muscle structure has been fully described by Bloom and Fawcett (1975) and by other authors. Sections II.B and II.C contain a detailed summary of this information.

B. Gross Structure

Muscle tissue is largely made up of bundles of cells having a very specialized structure and known generally as myofibres. The cell bundles are separated from each other, and collectively held together by sheaths or membranes of connective tissue comprised of collagen and elastin. Strong attachments between muscles and the bony framework of the body are formed by tendons, which also consist of connective tissue. Figure 1 is a diagrammatic representation of muscle showing a cut surface through several bundles of myofibres. Each myofibre is surrounded by a fine connective tissue membrane (the endomysium) which forms a delicate supporting network throughout the entire muscle. The bundles of myofibres are separated from one another by coarser connective tissue membranes (the perimysium), Included in these membranes are small arterioles, venules and capillaries of the blood supply system, together with small bundles of nerve fibres. Figure 2, which is a micrograph at low magnification, shows these structures in a stained section. A sheath of white connective tissue (the epimysium) surrounds the entire

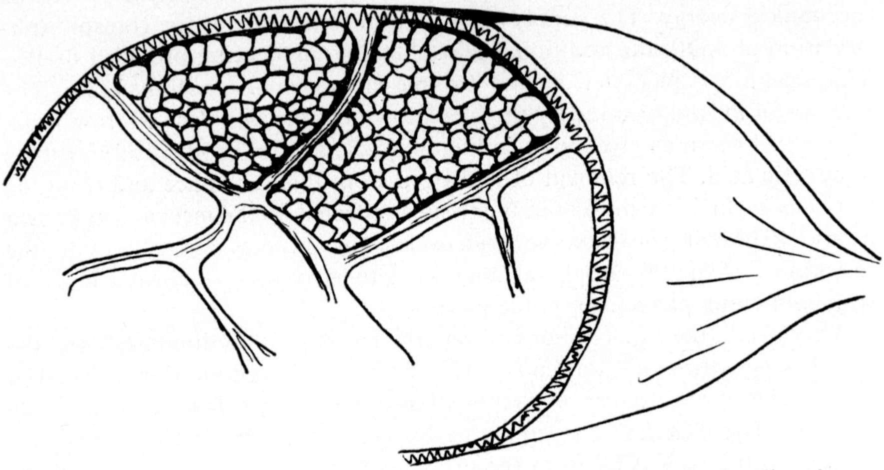

FIG. 1. Diagrammatic representation of a cut surface of muscle showing bundles of myofibres (not to scale).

muscle. The terminal processes of this sheath form the tendons. It is via these attachments that muscle action is translated to the bony framework of the animal.

Just as muscles vary in size and shape according to their function, so the arrangement of the fibres which comprise a muscle is related to the nature of the movement required. In some muscles the myofibres run parallel to the long axis, individual fibres extending from the origin (proximal attachment) to the insertion (distal attachment). In other muscles the fibres run from a centrally positioned septum of connective tissue, which is continuous with one tendon, to the surface of the muscle. Here the ends of the myofibres are attached to the epimysium which is continuous with the other tendon. Some muscles present a very complex arrangement with several tendinous attachments to the skeleton. A more detailed description of the arrangement of

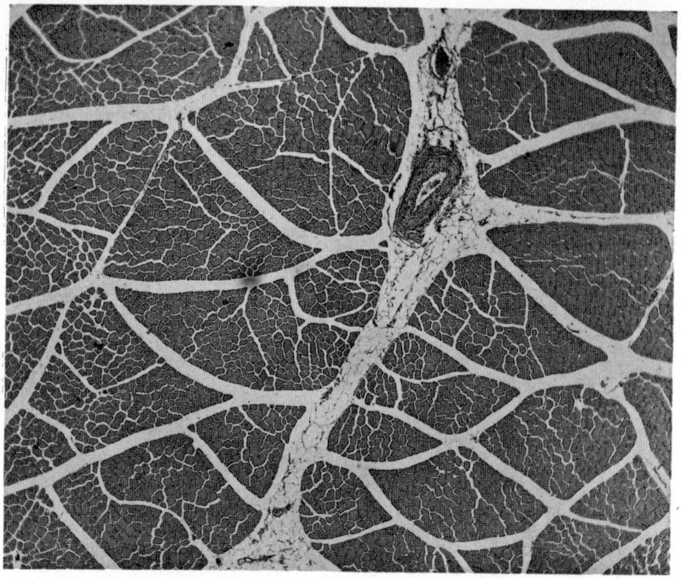

FIG. 2. Muscle bundles in transverse section, ×30. The thick, light spaces contain perimysium.

myofibres within muscles may be found in textbooks of veterinary anatomy, e.g. Getty (1975).

The proteins collagen and elastin, which are the major components of the supporting connective tissue, are laid down in bundles of fibres which may be easily observed with the naked eye. Frequently, layers of these bundles are assembled together to form sheets which give considerable strength and support to the entire organ. The proportion of collagen to elastin in muscles is variable, depending on the nature of the work which the muscle is required to perform. The variable connective tissue content, particularly with respect to the collagen component, is reflected in the texture of meat (Bendall, 1967; Cross *et al.*, 1973). Figure 3 shows a section of *Musculus semitendinosus* from a beef animal. This muscle has a substantial amount of elastin, seen in the micrograph as darkly stained fibres, as well as collagen which is less darkly stained. A section of *M. longissimus dorsi* from the same source is shown in Fig. 4 for comparison. It will be observed that the connective tissue content of this muscle is much less than that of *M. semitendinosus.*

C. Microscopic Structure

The myofibres comprising the fibre bundles referred to above are characteristically long, thin, multi-nucleated cells. They may be over 100 mm in length

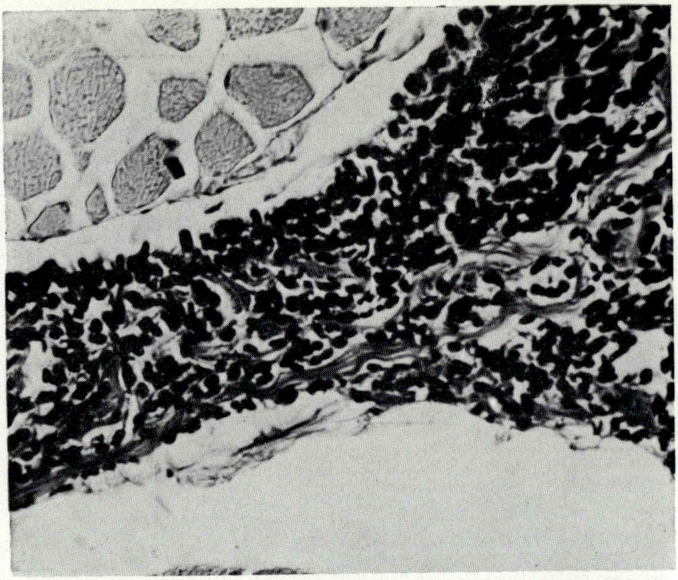

FIG. 3. Perimysial connective tissue in *M. semitendinosus* of ox, ×500. Elastic fibres are stained black. Collagen fibres are finer and less densely stained.

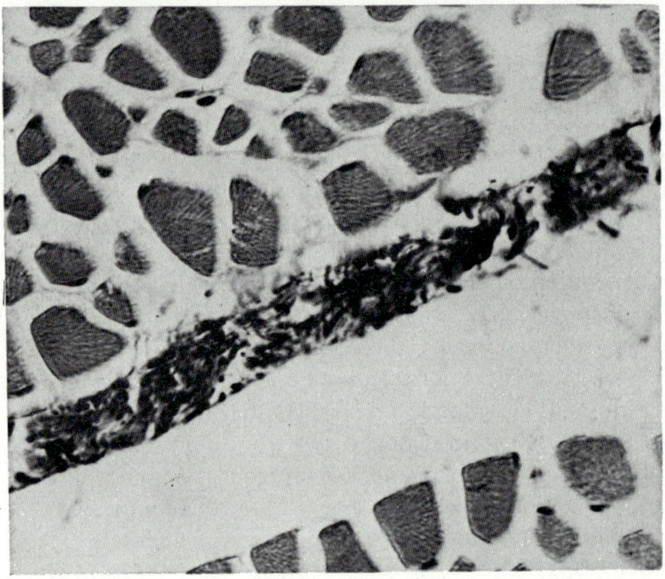

FIG. 4. Perimysial connective tissue in *M. longissimus dorsi*, ×500. There is very little elastic tissue compared with Fig. 3.

but only 0·1 mm in diameter. The fibres are circular or elliptical in cross-section, and in very young animals the diameter may be as little as 5 μm. As well as varying with age, the size of myofibres is found to be related to species, breed, sex and nutritional status (Hammond, 1932; Joubert, 1956). Furthermore, it differs between muscles within an animal and even within a muscle. The latter variation is associated with biochemical differences between myofibres (Gauthier, 1969).

The myofibre is distinct from all other cells. Instead of a single nucleus within the cytoplasm, there are many nuclei which, in a mature cell, are situated just beneath the cell wall. In addition there is a complex longitudinal array of contractile microfilaments bathed in the so-called sarcoplasmic fluid. In times when the study of cell structure was limited to what could be observed with the light microscope, the cell wall was seen as a homogenous, apparently structureless membrane. Bowman (1840) called it the 'sarcolemma'. By the use of the electron microscope, this apparently structureless membrane has been found in more recent times to consist of two parts. In modern terminology the inner layer is the plasmalemma, equivalent to the cell membrane of other cells. It has a thickness of about 7·5 nm and shows as a dense, linear structure in electron micrographs prepared from very thin, stained sections. This is typical of the appearance of a lipid–protein membrane, which with high-resolution electron microscopy can be seen as a three-layered structure. External to the plasmalemma is a thicker layer which has a fuzzy, amorphous appearance and is composed of non-fibrous collagen embedded in mucopolysaccharide. This layer is about 100 nm wide and forms the so-called basement membrane or basal layer. Associated with this structure is an array of extra-fine collagen fibres, generally known as reticulin, which can be observed with the light microscope after staining with a silver–amine complex. Preliminary treatment of the tissue section with an oxidizing agent such as periodic acid is necessary. Similar treatment, followed by exposure of the section to Schiff's reagent, stains the mucopolysaccharide-containing basement membrane. The chemical differences between these fibres are currently being investigated. Figure 5 is an electron micrograph showing the composite structure of the cell wall of a myofibre. The micrograph shows the plasmalemma and basement membrane as well as fine connective tissue fibres of reticulin. Figure 6 illustrates, by light microscopy, the distribution of reticulin around each myofibre. It also shows the difference in coarseness of this network between *M. psoas* and *M. sternomandibularis* of the ox.

Contained within each myofibre are the fibrous proteins of the contractile system, assembled together to form a highly organised array, the myofibril. Each myofibril is 1–2 μm in diameter. By suitable staining, such as

Heidenhain's iron haematoxylin, the myofibrils can be shown to have a repeating striated structure in which light and dark bands alternate regularly. This appearance can also be observed using phase contrast microscopy. There is a reversal in the optical density of the striations using polarized light. Figure 7 shows the microscopic appearance of skeletal muscle using each of these methods of presentation. Thus it is easy to see why skeletal muscle is described

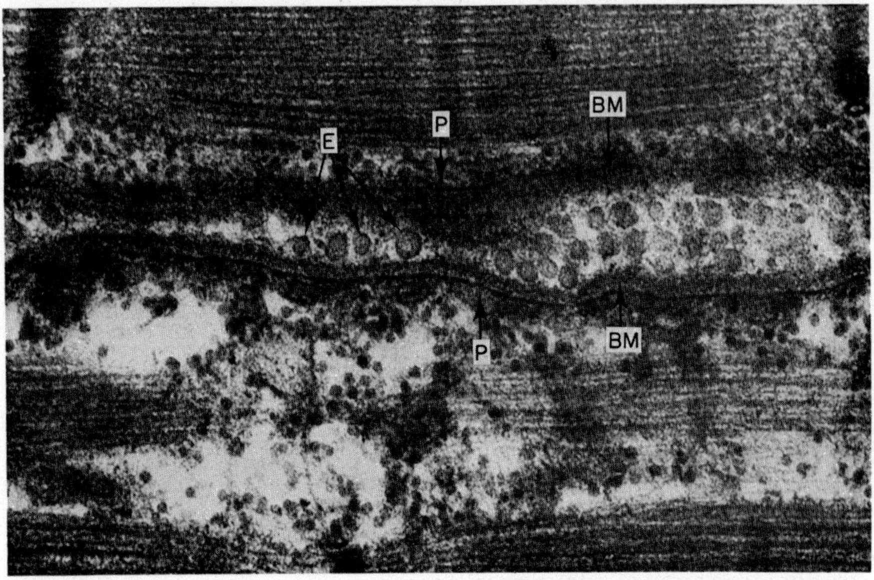

FIG. 5. Electron micrograph showing the composite structure of the cell wall of a myofibre, × 60 000. Note the thin, dense plasmalemma (P), the broader, less-dense basement membrane (BM) and the endomysial connective tissue fibres (E) shown in transverse section.

as 'striated' muscle, differentiating it from the 'smooth' muscle found in such tissues as the gut wall, urinary bladder and the vessels of the blood supply systems.

The reason for the striated appearance of skeletal muscle becomes clear at the higher magnification and resolution provided by the electron microscope. The repeating units of the myofibrillar structure are known as sarcomeres, and the structural components of each unit have a distinct morphology. It is helpful to describe these individually before attempting to relate them to the total structure, (Fig. 8). A diagrammatic representation of this electron micrograph is shown in Fig. 9, indicating the structural features of the sarcomere referred to in the text.

1. Light band (I-band)

Thin filaments consisting mainly of the protein actin can be observed by electron microscopy in the light band. It is frequently described as the I-band because it is isotropic in polarized light, showing only slight

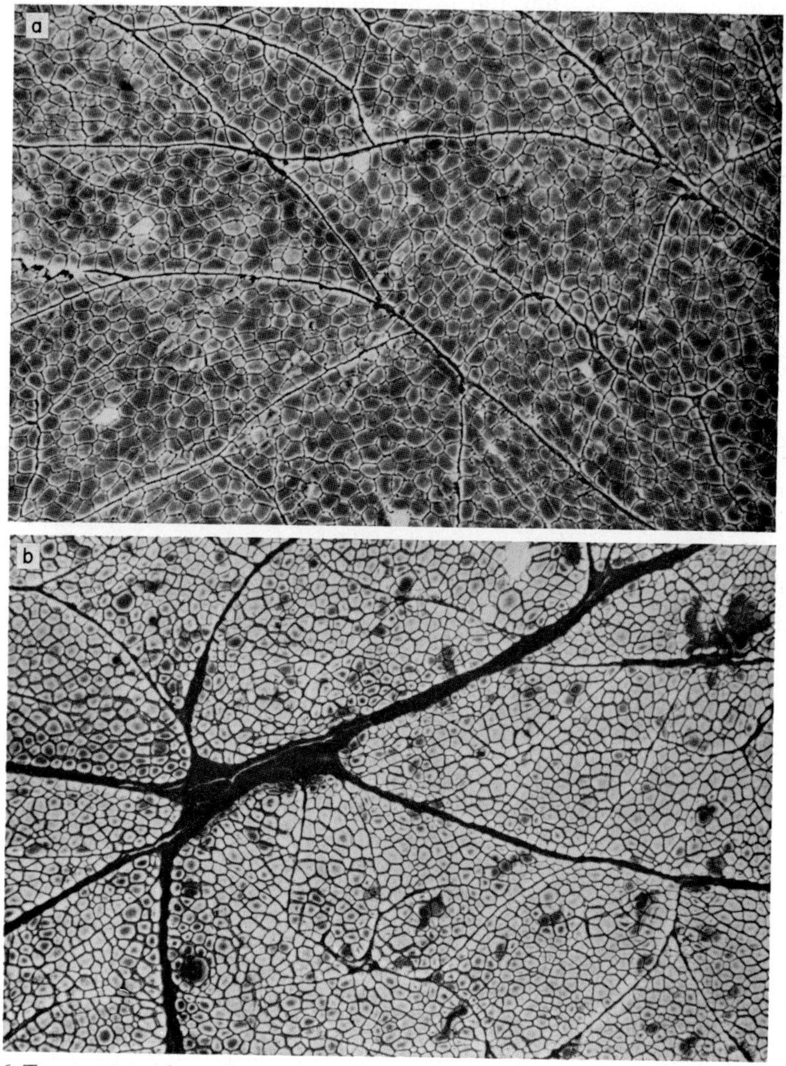

FIG. 6. Transverse sections of muscle stained with silver to show reticulin in the perimysium and endomysium, ×38. (a) Ox *M. psoas*. (b) Ox *M. sternomandibularis*. Note the difference in the recticulin content of the two muscles.

birefringence (Fig. 7c). The thin filaments are attached to either side of a thin, dense structure which lies transversely across the myofibril. This structure, known as the Z-band or disc, bisects the I-band and defines the limits of each sarcomere. The Z-band derives its name from the German word 'zwischenscheibe' meaning 'intermediate disc'. Actin-containing filaments have a diameter of about 5–7 nm and high-resolution microscopy shows that the filaments consist of a double-stranded helix of globular units of G-actin

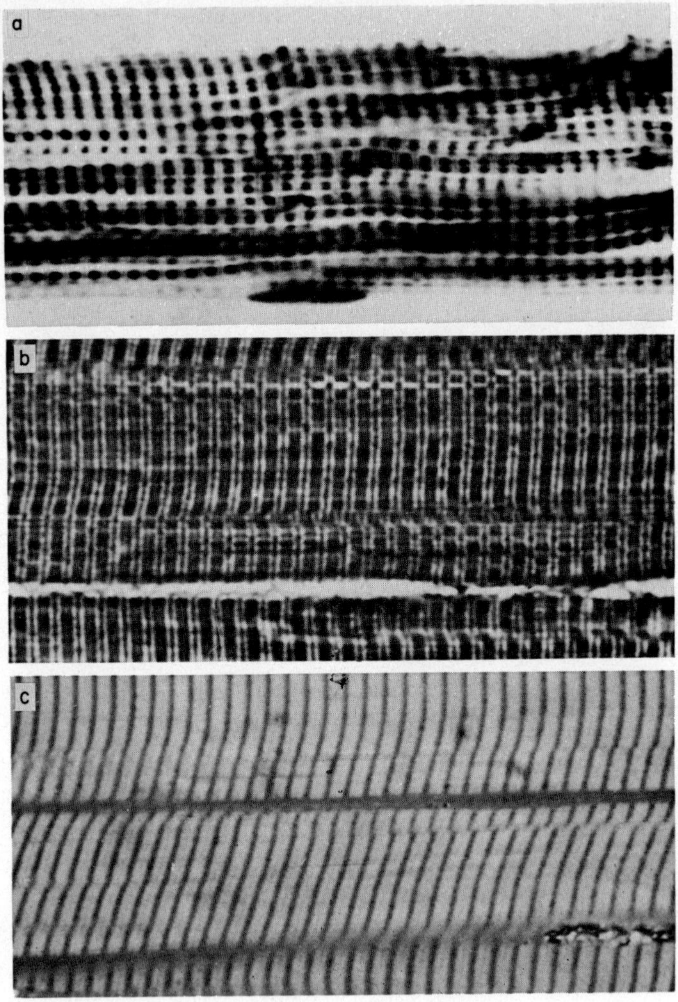

FIG. 7. Striations in muscle fibres demonstrated by: (a) iron haematoxylin stain, ×1160; (b) phase contrast illumination, ×1000; (c) polarized light, ×1160.

FIG. 8. Electron micrograph showing the structure of sarcomeres in longitudinal section, ×10 000.

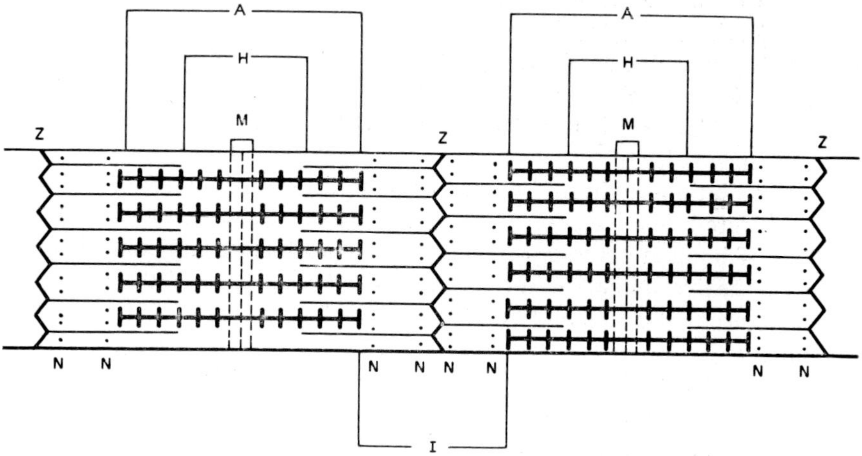

FIG. 9. Diagram of sarcomere structure to show the location of the main features described in the text.

(Huxley, 1969). These have a diameter of about 5·5 nm. In this form the filaments are described as F-actin. Each turn of the helix is comprised of 13 molecules, and the pitch of the helix is about 36·5 nm. In beef muscles the actin filaments are 1·3 μm in length (Bendall and Voyle, 1967). Other proteins associated with the thin filaments are tropomyosin and the troponin complex. The presence of these proteins has been demonstrated microscopically by antibody-labelling techniques (Pepe, 1966) as well as by biochemical methods. The troponin complex 'turns on' the contractile process by sequestering the calcium ions which are released by stimulation of the muscle.

The fine structure of the Z-band is of interest. When observed in profile it presents a lattice structure made up of a series of Y-shaped elements to which actin filaments are attached. A slight thickening of these filaments in the region of their attachment to the Z-band may be seen in longitudinal sections.

2. Dark band (A-band)

The denser region of the sarcomere, sited symmetrically between the Z-bands, is generally known as the A-band. It is comprised of thick filaments and the overlapping parts of thin filaments (Fig. 9). This dark band derives its name from its anisotropic appearance under polarized light, showing considerable birefringence (Fig. 7c). The thick filaments, containing the protein myosin, extend from one end of the A-band to the other. There is a less-dense central zone which represents the absence of overlap between thick and thin filaments. This is known as the H-zone. Like the Z-band it derives its name from the German language; 'heller', meaning 'clearer' or 'brighter', describes this feature. As with the thin actin filaments, the length of the thick filaments is found to be very consistent. In beef animals thick filaments have a length of 1·6 μm (Bendall and Voyle, 1967).

High-resolution studies of thick filaments in very thin sections show a series of cross-bridges spaced uniformly along the filament. These represent the sites of binding between thick and thin filaments during contraction of muscle in the live animal, or in rigor in the dead one. Huxley (1972) has given a detailed account of the molecular structure of thick and thin filaments and their interaction in the contractile process.

In the centre of the H-zone is a narrow, moderately dense transverse structure, the M-line. Near the boundary of the A- and I-bands and on either side of the Z-band are thin broken lines, the N-lines. These are not very conspicuous in routinely stained thin sections but they can be clearly demonstrated by histochemical methods (Yarom and Meiri, 1971). The M-line is made up of a series of cross-bridges between thick filaments in the centre of the H-zone, as described later. The N-lines are possible sites of calcium accumulation associated with the location of troponin. Calcium ions are rapidly released and dispersed when the myofibre receives a nervous stimulus. The loss of precise staining of the N-lines is to be expected in this condition, and this is found to be the case.

3. Overall structure of the sarcomere

As has been shown (Figs 8 and 9), thick and thin filaments are assembled to form a structural unit in which the two types of filament overlap, or interdigitate, with each other in such a way as to produce a transversely striated appearance. Normally there is a high degree of register in adjacent myofibrils between the dark (A) and light (I) bands, and also between Z-bands. In addition there appears to be a high degree of uniformity in the periodic spacing along the myofibrils but, as will be described later, this feature can be subject to considerable variation under certain *post mortem* conditions.

A combination of the observations made by A. F. Huxley, and H. E. Huxley working with Dr Jean Hanson in the early 1950s, using light (interference) microscopy, electron microscopy and X-ray diffraction, gave rise to the now widely accepted sliding filament theory of contraction in muscles. Certain aspects of the texture of meat can also be explained in these terms.

It can now be appreciated that the dimensions of a sarcomere are determined by the contractile state of the myofibre. Contraction results in an increase in the amount of overlap between thick and thin filaments, with a consequent reduction in the width of the H-zone accompanied by a reduction in the distance between adjacent Z-bands. This distance is commonly referred to as the 'sarcomere length' and it is used as an index of the contractile state of the muscle.

The description of myofibrillar structure has so far been based on the appearances observed in a longitudinal orientation, parallel to the long axis of the myofibre, as revealed by electron microscopy. The structural appearance in transverse section, illustrated in Fig. 10, gives further information about the assembly of the myofilaments in each myofibril. The appearance presented will depend on the level of section within the sarcomere. Four typical levels are illustrated in Figs 11a–d. A section through that part of the A-band which includes both thick and thin filaments shows a double hexagonal array (Fig. 11a). Each thick filament is surrounded by six thin filaments. Similarly, each thin filament is associated trigonally with three thick filaments. The thick filaments themselves are arrayed in a hexagonal pattern. The 'end-on' view of the thin filaments to be found in the I-band shows a rather loosely ordered arrangement of the filaments (Fig. 11b). In a section through the H-zone thick filaments only can be observed, again displaying a hexagonal arrangement. A section through the M-line shows thin connecting structures

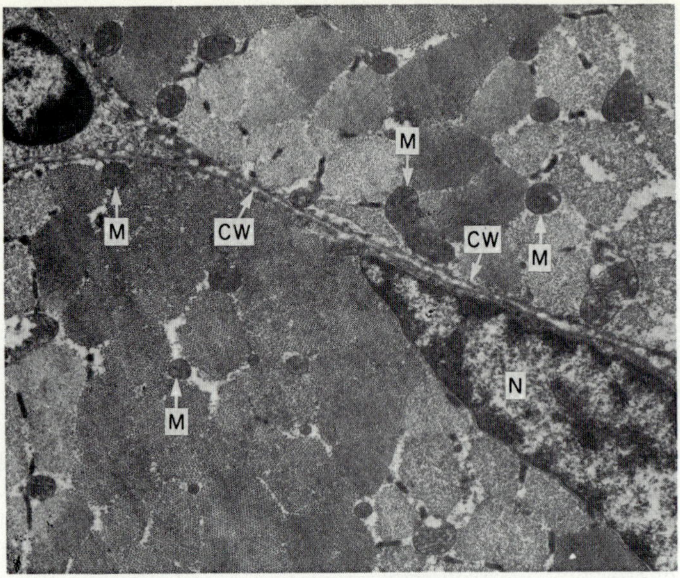

FIG. 10. Electron micrograph of a myofibre in transverse section showing parts of the cell wall (CW), nucleus (N) and mitochondria (M), ×10 000.

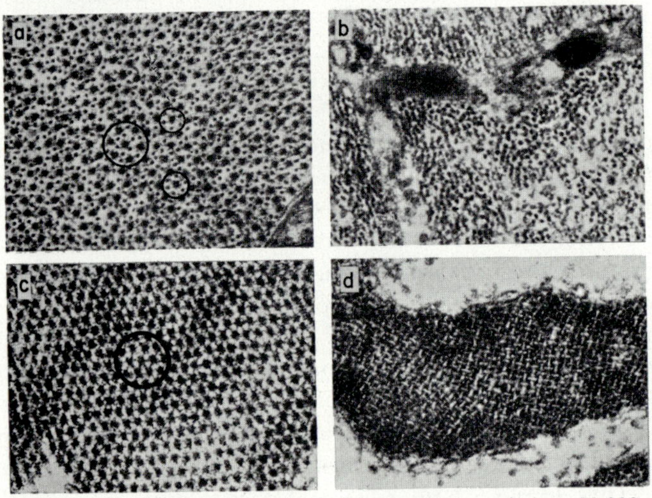

FIG. 11. Different constituents of myofibrils in transverse section, ×50 000. (a) A-band. The large circle marks a hexagonal array of thick filaments surrounding a similar array of thin filaments; small circles mark a hexagonal array of thin filaments around a single thick filament. (b) I-band. (c) M-line. The area within the circle shows connections between myosin filaments. (d) Z-line.

radiating from each thick filament to adjacent ones (Fig. 11c). The net-like structure of the Z-band in transverse section is clearly seen in Fig. 11d.

4. *Other structural features*

There are several other structures which can be observed by either light or electron microscopy in thin sections of muscle tissue. These have their equivalent in cells of other types but vary in morphological detail.

The many nuclei are generally situated just beneath the cell membrane and are usually elongated in shape. They resemble nuclei of other cell types in structure. Nuclei are distributed fairly evenly along the myofibre but tend to be more numerous in the terminal region. This is the region of growth in developing muscle.

Another organelle is the mitochondrion. A number of these are to be found beneath the cell wall and between the myofibrils (Fig. 8). The number and distribution of mitochondria in a myofibre is one of the criteria used to distinguish fibre types. Many systems of classification have been proposed, based on such criteria as the size, colour and cytochemistry of myofibres, shape of myofibrils, sarcomere structure and physiological activity. None of these systems is entirely definitive, but the trend seems to be towards the use of morphological criteria as the basis for classification.

The fine structure of mitochondria is common to almost all cells which contain them. A double-layered membrane forms the boundary, with the inner layer extending internally to form a series of folds or 'cristae'. The layers, which are composed of protein and lipid in the form of a 'unit membrane', are each about 7 nm in thickness. The space between them is about 8 nm wide. The outer surface of each crista is covered with particles, 8–9 nm in diameter, each mounted on a short stalk. These structures are associated with the enzymes which are responsible for much of mitochondrial function through the mechanism of the Krebs or tricarboxylic acid cycle. General mitochondrial morphology can be studied by electron microscopy in thin sections stained with uranyl acetate and lead citrate. Fine detail is frequently better revealed in preparations of isolated mitochondria, using the negative-contrast method of staining.

Mitochondria vary considerably in size and shape. It has also been observed that in 'red' muscle fibres (e.g. diaphragm, part of *M. semitendinosus*, *M. sternomandibularis*) chains of mitochondria are formed between the myofibrils. In 'white' muscle, e.g. *M. psoas*, mitochondria are generally fewer in number and are more widely separated in their distribution.

Granules of glycogen, the carbohydrate which is oxidized to carbon dioxide and water during muscle activity under aerobic conditions, and hydrolysed to lactic acid under anaerobic conditions, may be demonstrated by both light and electron microscopy. At a suitable level of magnification the granules may be seen between the myofibrils. Since this substance, which is a polymer of glucose molecules, is soon broken down by enzymic action in *post mortem* tissue, prompt fixation is necessary for successful demonstration. In particular, the process of glycolysis in *post mortem* muscle, terminating when rigor is complete, means that virtually no residual glycogen will be observed in histological sections of post-rigor muscle.

Intracellular lipid droplets may be observed in sections examined by electron microscopy, but not in large numbers. Much of the lipid which can be demonstrated within muscle cells is associated with membranous structures such as mitochondria. In addition there are considerable amounts of lipid in the perimysium surrounding bundles of myofibres, especially in older animals. This gives rise to the macroscopic appearance known as 'marbling'.

A very important intracellular component is the tubular system known as the sarcoplasmic reticulum (SR). First described by the classical histologists of the nineteenth century from their study of silver-stained preparations, the fine and complex nature of this system was later fully revealed in observations made with the electron microscope (Porter, 1961). This system plays an important part in the mechanism of contraction and relaxation, especially in relation to the release and recapture of calcium ions. Its function is frequently described as that of a calcium pump. Morphologically the SR is analogous to the smooth endoplasmic reticulum of other cells. As seen in thin sections, the network of SR consists of longitudinal and transverse arrays of tubules. The system pervades the entire myofibre by enmeshing each myofibril in the network. Transverse sections of tissue impregnated with silver and examined by light microscopy show the interfibrillar array, while ultrathin longitudinal sections frequently include elongated elements of the tubular system. Figures 12a and b illustrate these features in muscles of the ox in transverse and longitudinal array, respectively (Voyle and Lawrie, 1964).

A particular feature of the SR in mammalian muscle is the 'triad' system. This is to be found in association with each sarcomere. It consists of the expanded ends or 'cisternae' of the longitudinal elements in juxtaposition with a transverse element. The latter is sandwiched between the cisternae, and in the muscles of mammals is situated in the region of the A–I junction. The cisternae are the parts of the longitudinal system which are associated with the capture and release of calcium ions. In muscles of amphibians and birds the transverse element of the SR is often situated at the level of the Z-band, but some variation in location does occur. These transverse elements are continuous with the muscle wall (the plasmalemma). The invaginations on the

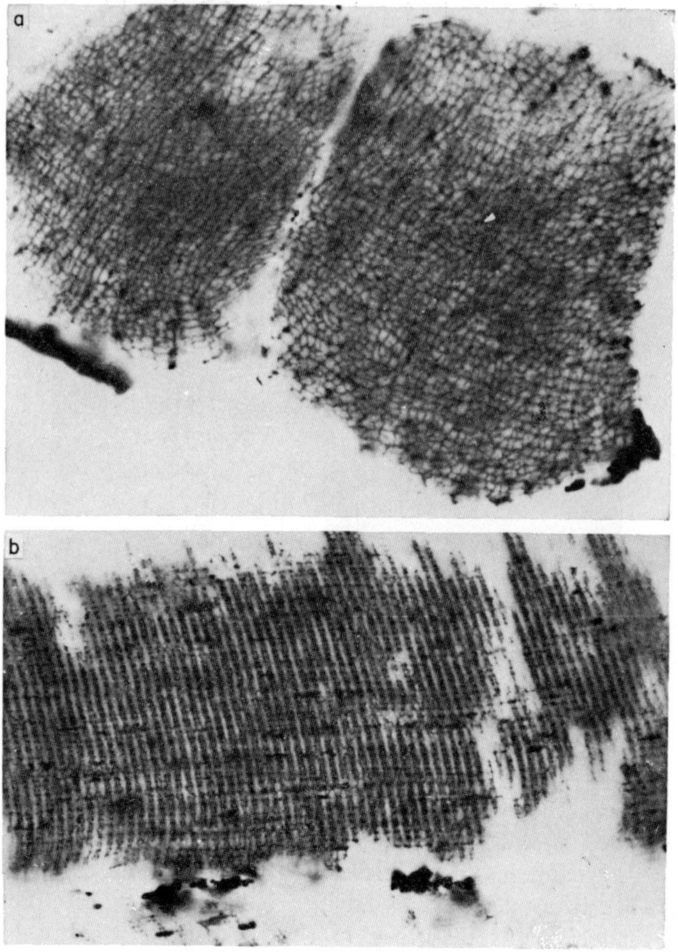

FIG. 12. Sarcoplasmic reticulum shown by silver impregnation, ×1160. (a) Transverse section. (b) Longitudinal section.

surface of the myofibre, seen by scanning electron microscopy and illustrated in Fig. 13, represent the sites of emergence of the transverse elements of the SR. Thus there is a continuous membrane between the surface and the deepest parts of the myofibre. This is the means by which the contractile stimulus in the form of a depolarizing wave is conveyed throughout the myofibre.

As with mitochondria, the amount of SR within each myofibre is related to the type of activity which the fibre performs. So-called 'fast' muscles tend to have a more highly organized and extensive SR than 'slow' muscles. The terms 'fast' and 'slow' refer to the physiological activity of the myofibres.

In the living animal myofibres are under the control of the nervous system which initiates the process of contraction in a muscle. The neuromuscular junction or motor end-plate is a structure which can be observed at the surface of the myofibre. Generally there is one motor end-plate for each myofibre.

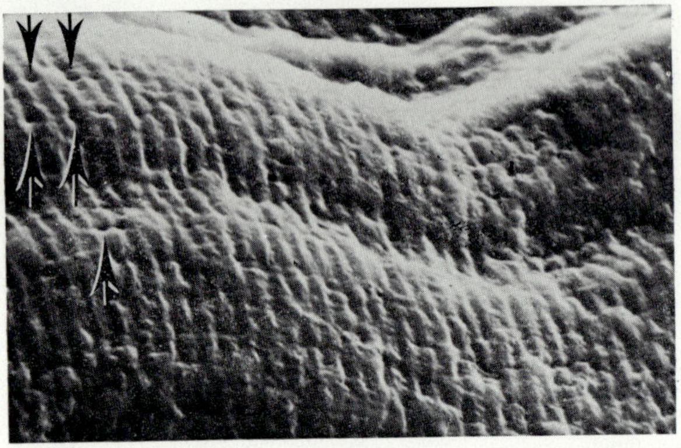

FIG. 13. Surface of myofibrils as seen by scanning electron microscopy, ×8000. Note the openings of transverse elements of SR (arrowed).

III. CHEMICAL COMPOSITION

It is not intended to give an exhaustive discussion of the chemistry of muscle in this chapter. Detailed information on this topic may be found in textbooks of biochemistry and in the published works of Bendall (1969, 1973) and Lawrie (1974). It is relevant, however, to indicate the relation between the morphological elements described in the preceeding paragraphs and the chemical composition of the tissue. It is important to remember also that large changes in composition occur in the transition from pre- to post-rigor muscle.

In general terms post-rigor muscle, prior to the onset of any degradative changes, consists of 75% water, 19% protein, 2·5% fat and 3·5% soluble non-protein substances. The water is divided between the extracellular space (5–12% of total water content) and that which is within the myofibres (88–95% of total water content). A further breakdown of water distribution may be made. Of the total water in the muscle, 22–24% is located between myofibrils in the sarcoplasmic space, while 66–71% is to be found within the myofibrils. This is distributed between the filaments which form a hexagonal lattice. The sarcoplasmic space contains fluid in which many soluble enzymes are to be found. When the rigor process is complete in a muscle at its rest-length, the filament lattice volume decreases and fluid is released into the sarcoplasmic space. About 5% of the total volume of water in muscle is bound onto the myofibrillar proteins. With the fall in pH which results from *post mortem* glycolysis, there is a drop in the water-holding capacity of the proteins in the muscle.

The proteins of muscle may be classified as follows: sarcoplasmic, myofibrillar and stroma. About 50 proteins have been identified in the sarcoplasmic fluid (Scopes, 1970). These proteins are soluble in water or in a salt solution of low ionic strength. Many of them are enzymes involved in the breakdown of glycogen to lactic acid in the rigor process. The sites of activity of some of these enzymes can be demonstrated by histochemical methods. The main myofibrillar proteins are myosin and actin which, in post-rigor muscle, form the complex actomyosin. Myofibrillar proteins are soluble only in salt solutions of higher concentration than those required to extract sarcoplasmic proteins. Filaments containing myosin and actin with other myofibrillar proteins become clearly visible in fixed tissue suitably prepared for microscopic examination. These have already been described in some detail. The stroma proteins include those found in connective tissue, i.e. collagen and elastin, and those found in the structure of intracellular organelles such as mitochondrial membranes and SR. These also can be clearly demonstrated and their morphology has already been described.

Intramuscular fat content varies both in quantity and in composition between muscles and species, as well as with breed, age and plane of nutrition. Sex also influences the fat content, males having less than females. Castrates have more intramuscular fat than entire animals. Lipids also occur within the cell as droplets in the cytoplasm or in association with protein in the structure of membranes.

The remaining components of muscle in the immediate post-rigor condition are carbohydrates and other non-protein substances. The carbohydrates include glycogen and the intermediate products of the glycolytic process. The end-product of anaerobic glycolysis, lactic acid, and the remnants of non-protein nitrogenous compounds are also included. Among the latter are the phosphonucleotides ADP and ATP, and creatine which is formed by the hydrolysis of phosphocreatine during the rephosphorylation of ADP to ATP in active muscle. Also there are small but important amounts of inorganic ions present. Sodium and potassium ions play an important part in the propagation of the action potential which is a consequence of the depolarization of the muscle cell membrane by nervous stimulation. Calcium ions, stored in the cisternae of the SR, are released into the sarcoplasm as a result of the changed permeability of the SR membrane when depolarized. Together with magnesium ions, these initiate the splitting of ATP in the contractile process. Chloride, phosphate, and other anions are present both within the myofibre and in the extracellular space. Finally, there are minute amounts of various fat- and water-soluble vitamins.

IV. *POST MORTEM* CHANGES IN STRUCTURE

A. Effect of Final pH

Myofibres remain sensitive to a stimulus and capable of contraction for some time after slaughter. With the onset of rigor there is a fall in pH due to the production of lactic acid under the anaerobic conditions obtaining after death. Myofibres lose the ability to contract when the pH falls below about 6·3, that is about half-way through the rigor process. Rigor sets in when the ATP is no longer available to keep the actin and myosin filaments apart in a relaxed state. If the increase in acidity occurs rapidly while the carcass temperature remains high, certain muscle proteins are denatured with a concomitant loss of their water-holding capacity. Histological preparations show that the denatured sarcoplasmic proteins are precipitated onto myofibrils in a more or less random manner. Myofibrillar proteins are also denatured under these conditions. Figures 14a and b show, in longitudinal and transverse sections respectively, the deposition of denatured sarcoplasmic protein within a myofibre. Muscle tissue in which this occurs is generally pale in colour, soft in texture and produces considerable exudate. Much interest has been shown in this pale, soft, exudative (PSE) condition in the musculature of certain breeds of pig, especially those which display a marked degree of stress-susceptibility, and which show a very rapid *post mortem* production of lactic acid in their muscles (Cassens *et al.*, 1975).

In animals where the glycogen reserves are severely depleted prior to slaughter because of excessive exercise or muscular tension due to stress, the post-rigor muscle will have a high pH. Values in excess of 6·2 have been reported in *M. longissimus dorsi* of calves subjected to stressful conditions immediately before slaughter. Such muscles have an appearance which contrasts sharply with that of PSE muscle, being dark, firm and dry (DFD) (Callow, 1935; Kidwell, 1952). Because of the high final pH, the water-binding properties of the muscle proteins will be considerably enhanced. This results in the myofibres being tightly packed together and having a mean cross-sectional area which is greater than that found in normal post-rigor tissue.

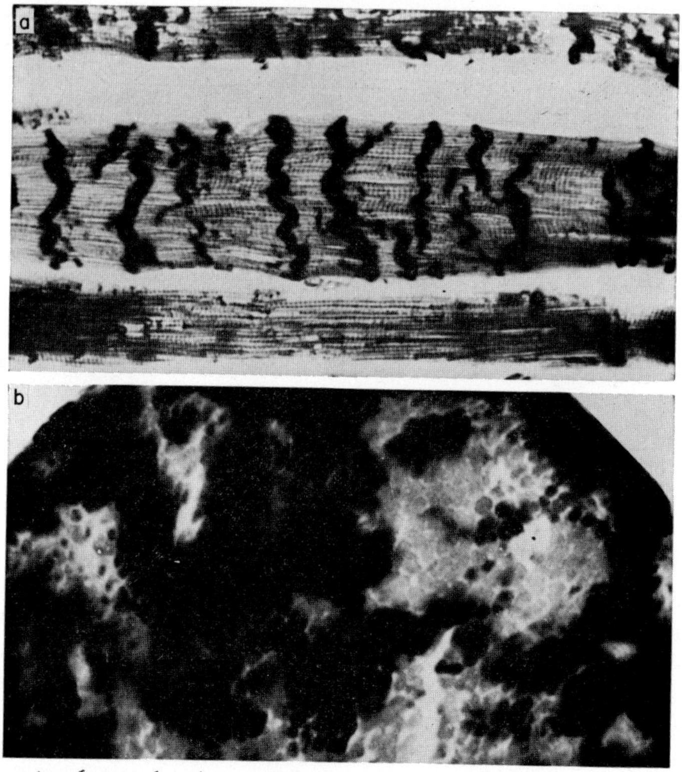

FIG. 14. Denatured sarcoplasmic protein on a myofibre from PSE pig *M. longissimus dorsi*. (a) Longitudinal section, ×380. (b) Transverse section, ×1160.

B. Effect of Posture of the Carcass

Traditionally, a carcass is hung by placing a hook between the tibia and the Achilles tendon very soon after slaughter. This assists drainage of blood as well as allowing the slaughterman easy access for cleaning and dressing the carcass. The posture thus adopted by the carcass, however, represents a considerable departure from the normal position in life, with the result that some muscles, particularly in the hind quarter, are subject to an excessive degree of stretch, e.g. the flexor muscles of the thigh joint. Other muscles, e.g. the extensors of the thigh joint, are somewhat compressed. Since the texture of meat is influenced by the contractile state of the muscle (a state which on completion of the rigor process is irreversible), interest has been shown in the effect of altering the posture of the carcass so that the position of the hind limbs more closely resembles that of the standing animal (Hostetler *et al.*, 1970). With beef carcasses this has been achieved by suspending the carcass from the hip bone by placing a hook through the obturator foramen. Similar results have been achieved with lamb carcasses by passing a smooth rod, such as a broom handle, through the internal cavity (Davey and Gilbert, 1974). As a result of the altered posture an improvement in tenderness has been achieved in such muscles as *M. biceps femoris* and *M. gluteus medius* in the thigh and rump. *M. psoas*, the fillet, suffers an increase in toughness due to shortening in this posture. Even so, it remains one of the most tender muscles because of its very low connective tissue content. These changes in texture are reflected in comparisons of mean sarcomere length in muscles from control and experimental carcasses. It should be remembered, however, that textural quality cannot be explained solely in terms of this particular parameter.

C. Influence of Storage Conditions

1. 'Ageing'

It is a well-known practice to hang meat in order to achieve a degree of tenderization by the process of 'ageing'. During hanging of meat at temperatures above its freezing point (−1·5°C) proteolysis occurs to some extent. Evidence for such action is to be found in the change in the extractability of myofibrillar proteins (Penny, 1970), and in the increased fragility of myofibrils (Davey and Dickson, 1970). The latter condition can be demonstrated experimentally by mechanical disruption of tissue in a blender. Microscopic examination of the slurry shows that aged myofibrils have a marked tendency to break at the Z-band. This suggests a weakening of the attachments between the tropomyosin lattice and the thin actin filaments. It has also been shown that the Z-band fades from muscle held at 15°C for 3 days *post mortem*. A similar effect can be shown by treating post-rigor muscle with a proteolytic enzyme which is present in the sarcoplasm. In the presence of calcium ions this enzyme causes the removal of the Z-band (Penny *et al.*, 1974). Figure 15 shows a thin section of myofibrils which have been treated in this way. The enzyme has been described as a calcium-activated sarcoplasmic factor (CASF). It was also found that the application of CASF had a tenderizing effect on meat when cooked, due at least in part to the removal of the Z-bands and the consequent ease of fragmentation of the myofibrils.

2. Chilling

Whereas it is desirable to reduce the temperature of a carcass at a rate and to a level which minimizes bacterial spoilage, it is also necessary to recognize

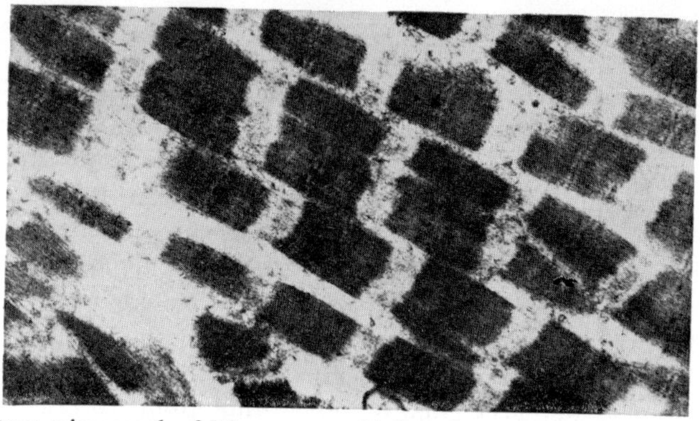

FIG. 15. Electron micrograph of *M. sternomandibularis* from ox after treatment with CASF, ×10 000. Note loss of the Z-band.

that rapid cooling of muscle in the pre-rigor condition serves to stimulate the contractile mechanism of the myofibres. A serious consequence of rapid cooling is the occurrence of increased toughness in some muscles when they are eventually cooked. This is a result of the so-called 'cold-shortening' effect which occurs at temperatures below 10°C (Marsh and Leet, 1966), provided that a supply of ATP is still available in the muscles (Cassens and Newbold, 1967). It happens readily in the musculature of small animals, e.g. lamb, and in the superficial muscles of larger animals. The phenomenon has only been detected in red muscles of mammals and birds. Since the amount of sarcoplasmic reticulum in these muscles is less than that of white muscles it has been suggested that the reduced facility for withdrawing calcium ions from the sarcoplasm makes the red muscles more reactive to the cold stimulus. It is known that the efficiency of the SR as a calcium pump is temperature-dependent, the efficiency being reduced as the temperature falls. The involvement of the cisternae of the longitudinal elements of the SR in the release and recapture of calcium ions has already been described. Furthermore, it has been shown that calcium release from mitochondria in muscle subject to anaerobic conditions at low temperatures is increased. Since red muscles have more mitochondria than white muscles, the availability of calcium ions in the sarcoplasm to initiate contraction by the splitting of ATP will be greater in the red muscles. Coupled with the reduced ability of the SR in these muscles to remove calcium ions at low temperatures, the red muscles would seem to be a prime target for the cold-shortening phenomenon (Beuge and Marsh, 1975).

Extensive investigations of this phenomenon have been made using the neck muscle, *M. sternomandibularis,* from lamb and beef animals. This muscle has the advantage that its myofibres lie parallel to the long axis of the muscle, which is strap-like in shape. Changes in muscle length are, therefore, readily reflected in changes in mean sarcomere length.

If this muscle is excised immediately after slaughter and is suspended vertically under slight load in an environment at 15°C for a period of 24 hours, it will lengthen slightly and then shorten by a small amount during the rigor process. Sections of the tissue examined by light microscopy show a highly uniform sarcomere pattern with a mean sarcomere length of about 2·6 μm. This approximates to the mean sarcomere length of mammalian muscle at rest-length, i.e. when it is not in a stretched condition or subject to a contractile stimulus.

If, by contrast, the muscle is suspended in a chill room at 1–2°C it will show a progressive but slow contraction. Within about 10 hours the muscle will have shortened by about 40% of its initial length. The effect is fairly rapid at first, the muscle shortening by some 30% of its initial length in less than 3 hours. Figure 16 shows a pair of neck muscles, one of which has been cold-shortened in the manner described. The distance between the marker threads was initially the same in each case.

Histological examination of myofibres in sections of cold-shortened muscle shows that the fibre population falls into two groups (Voyle, 1969). Less than

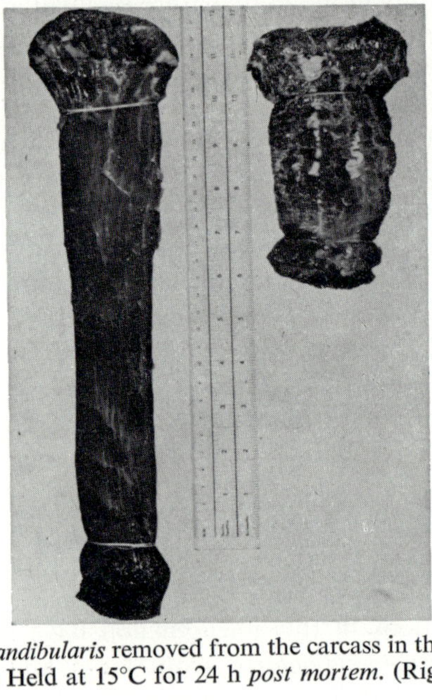

FIG. 16. Ox *M. sternomandibularis* removed from the carcass in the pre-rigor condition and freely suspended. (Left) Held at 15°C for 24 h *post mortem*. (Right) Held at 2°C for 24 h *post mortem*.

half of the fibres undergo 'active' contraction in which there is a marked reduction in sarcomere length. In myofibres which do exhibit active contraction very short sarcomeres of about 1·0 μm in length have been observed; this implies some considerable distortion of the array of myofilaments. The endomysium surrounding these fibres is also grossly distorted. The remainder of the fibres exhibit 'passive' contraction. These shorten only slightly, but because of the overall length change in the muscle are markedly changed in shape and assume a zig-zag form. This can be observed in stained sections

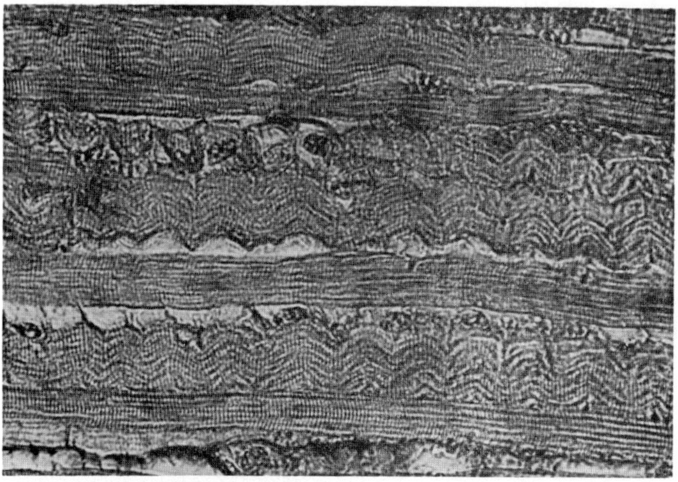

FIG. 17. Longitudinal section of cold-shortened ox muscle, × 500. Straight myofibres are 'actively' shortened; wavy fibres show 'passive' shortening.

examined by light microscopy. Figure 17 shows myofibres of both types. A dramatic representation of the morphology of myofibres affected in the manner described can be observed by scanning electron microscopy, as illustrated in Fig. 18. Preparations examined by transmission electron microscopy show the expected variations in sarcomere structure; these are illustrated in Fig. 19.

The morphology of myofibres from a muscle exposed to the cooling regime described above, while being held at a fixed length as would be the case with a muscle on a carcass, shows even more variation than the freely suspended muscle. In addition to active and passive shortening, there is considerable localization of contraction in the myofibres. These nodes of contraction alternate with groups of stretched sarcomeres, thus compensating for the constraint on the muscle against an overall reduction in length. In some areas the myofibres are stretched to the point of rupture of the myofibrils, while the cell membrane remains intact.

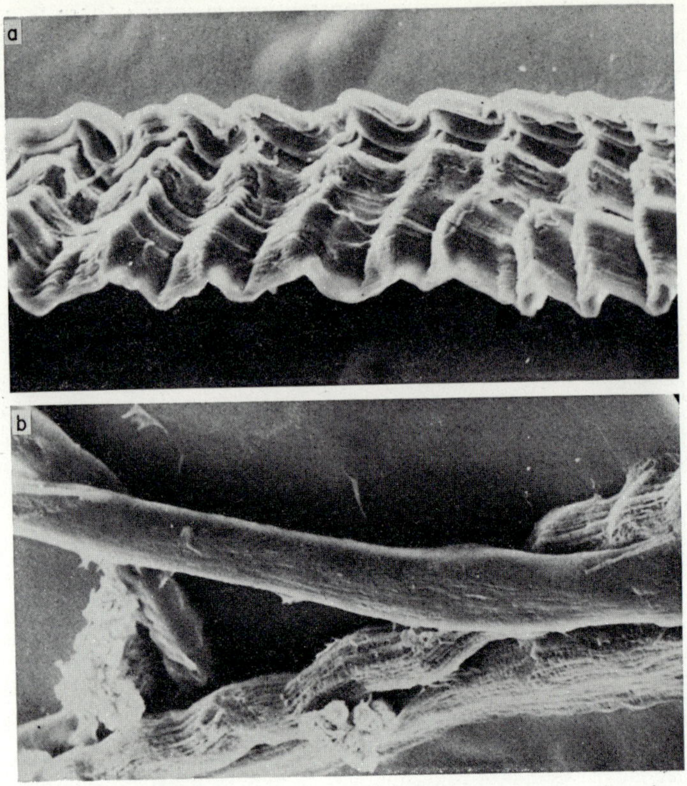

FIG. 18. Myofibres from cold-shortened ox muscles, as seen by scanning electron microscopy. (a) 'Passive' shortening, ×428. (b) 'Active' shortening, ×408.

3. *Freezing*

a. Conditions governing ice crystal formation. The influence of freezing on muscle structure is of interest and importance due to the widespread use of refrigeration for the storage of meat and other foodstuffs. Interest in this aspect of meat preservation goes back to the advent of refrigerated transport of meat over many thousands of miles between producer and consumer. A fundamental study of the formation of ice in biological systems has provided the basic knowledge for the successful storage of living tissue at low temperatures and its subsequent viability. The subject has been reviewed by Meryman (1966).

The rate of cooling to temperatures below the freezing point of the tissue was found to be important in relation to structural damage due to the formation of ice crystals (Moran, 1932). The consequences of so-called 'rapid' and

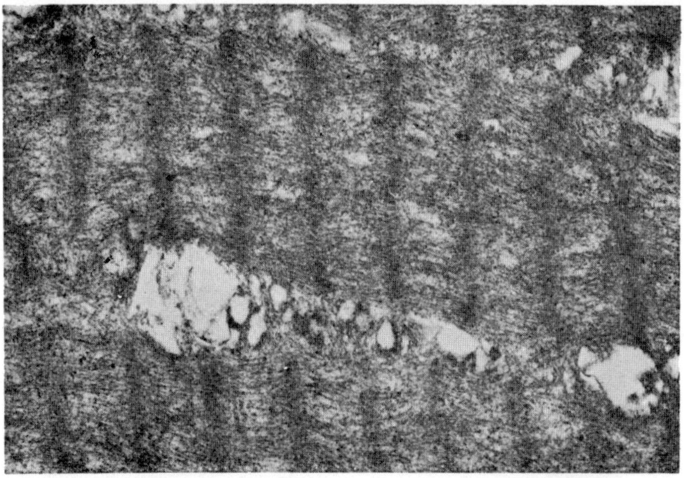

FIG. 19. Electron micrograph of severely contracted myofibrils from cold-shortened muscle, ×20 000. A- and I-bands show total overlap. The dense lines are Z-bands.

'slow' freezing have been assessed, but it is first necessary to define these terms. Simple definitions can be expressed numerically as 'degrees per minute fall in temperature'. Alternatively, the definitions can be related to the respective effects of the rates of cooling on the morphology of the material being frozen. It has become common practice to define rapid freezing as that rate which results in the formation of ice crystals within the cell. Conversely, slow freezing gives rise to the formation of ice crystals in the extracellular space. In circumstances other than those of food preservation, tissue damage resulting from ice formation can be avoided by the use of cryo-protectants such as glycerol or dimethyl sulphoxide. It is important that the concentration of cryo-protectant used does not result in toxic damage to the cells.

Ice crystal formation can be observed microscopically in thin slices of muscle tissue, suitably mounted on a cold-stage. Alternatively, the freeze-drying process can be used, whereby the ice formed is sublimed from the tissue under vacuum. The spaces that remain between the elements of tissue structure may be regarded as representing the location of the ice crystals (Voyle, 1974a). The dried tissue is subsequently embedded in a suitable supporting medium, such as paraffin wax, from which sections are cut and stained. An epoxy resin embedding medium would be used for electron microscopy.

Formation of intracellular ice by rapid freezing leaves the overall appearance of the tissue intact, but microscopic examination of individual cells shows needle-like spaces, the dimensions of which vary with the rate of cooling. This is illustrated in Fig. 20. The location of the spaces suggests that structural elements have been pushed aside by the developing ice crystals. Sections of tissue which has been frozen and thawed look very similar to those of unfrozen tissue, however.

Formation of extracellular ice crystals by slow freezing gives rise to varying amounts of deformation of the overall tissue structure, depending on the rate of cooling. During the transition from the liquid to the solid phase there is a period of thermal arrest. This is represented by a plateau in a typical cooling curve, such as that shown in Fig. 21. It is during this period that the latent

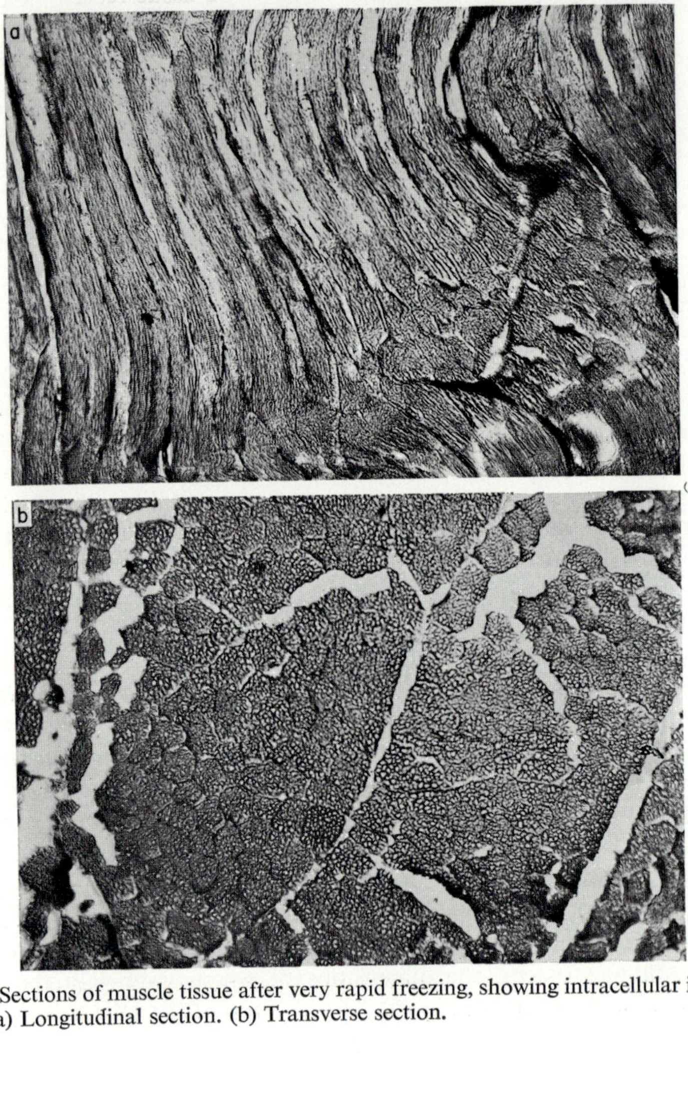

FIG. 20. Sections of muscle tissue after very rapid freezing, showing intracellular ice spaces, ×100. (a) Longitudinal section. (b) Transverse section.

heat of crystallization is extracted from the water which is available for ice formation, and the temperature remains steady. When all the latent heat has been extracted the temperature in the tissue falls further to a point of equilibrium with the environment. It is during the period of thermal arrest that ice crystals are formed. The longer the period, the larger are the crystals. Figure 22 shows ice spaces formed during the slow cooling of a sample of *M. psoas* from a pig.

The amount of disruption due to ice formation during slow freezing will determine how closely the structure of the thawed tissue resembles unfrozen

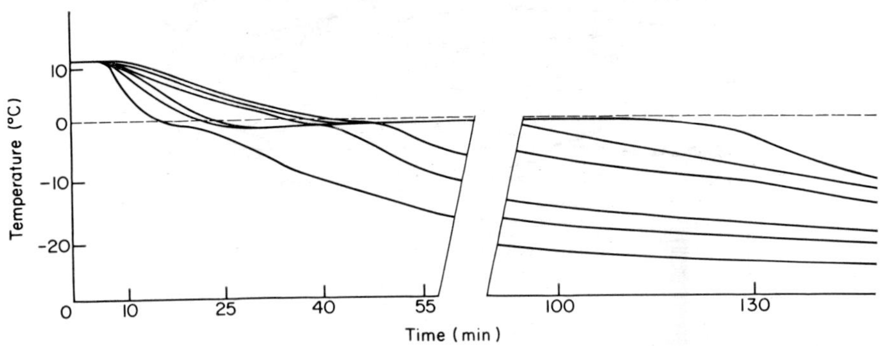

FIG. 21. Cooling curves for different rates of freezing of pig *M. psoas*. Note the varying periods of thermal arrest.

tissue in appearance. Large crystals may damage some of the tissue components and cause fragmentation of the myofibres. Furthermore, during freezing a certain amount of water will be extracted from the cells, by osmosis, into the extracellular space. The increased salt concentration in the sarcoplasmic fluid causes denaturation of protein and a loss of water-binding capacity. The permeability of the cell membrane is also altered. When thawing occurs some of the water in the extracellular space will re-enter the cell, but some of it will be released from the tissue as 'drip' (Moran, 1929). (A certain amount of drip is released as a result of the rigor process and change in tissue pH, quite apart from the effects of freezing.)

b. Freezing of pre-rigor muscle. The effects of freezing on muscle tissue while it is still in the pre-rigor condition are even more marked than those resulting from chilling. In order that freezing should be complete before the supply of ATP is exhausted, a rapid rate of cooling is required. Damage to the myofibres resulting from the formation of intracellular ice crystals can be measured chemically by determining the concentration of DNA in the extracellular fluid (Love, 1955). It is only in the nuclei of the myofibres in unfrozen, undamaged tissue that DNA occurs. If damage is caused to the cell membrane of the myofibre there is an increase in the amount of drip released on thawing. In the absence of cell wall damage, the intracellular water which was separated as ice recombines with the myofibrillar proteins.The microscopical appearance of the thawed tissue is virtually indistinguishable

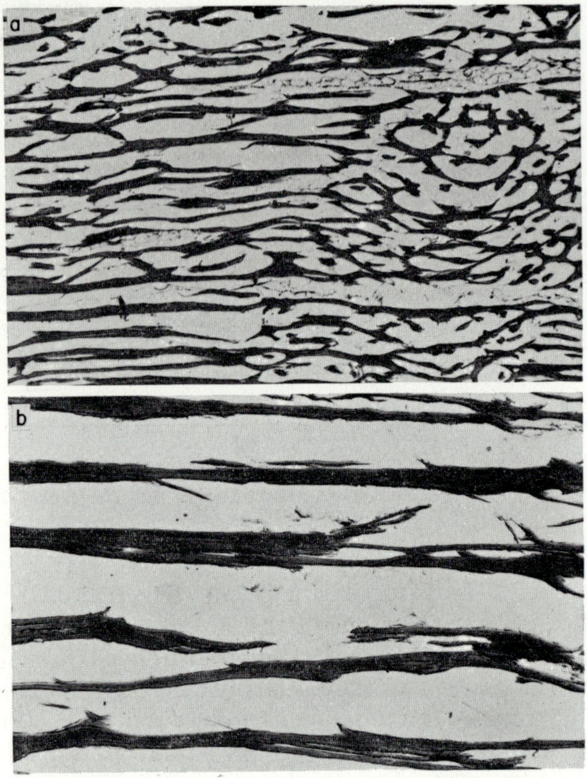

FIG. 22. Sections of muscle tissue after slow freezing, ×30. (a) +5°C to −5°C in 13·8 min. (b) +5°C to −5°C in 109 min.

from that of unfrozen tissue, provided that the following condition is avoided.

If pre-rigor frozen meat is thawed rapidly there is a massive contraction of the myofibres accompanied by a substantial loss of drip (Bendall and Marsh, 1951). This, in turn, gives rise to increased toughness when the meat is cooked, an increase which is more marked than that resulting from cold-shortening. This phenomenon of excessive shortening on thawing is known as 'thaw-rigor' or 'thaw-contracture'. Changes in the sarcomere pattern are similar to those described for cold-shortening, but more extreme. Shortening of up to 70% of the initial length of an excised muscle has been recorded. If ATP is still present in a muscle subject to thaw-rigor it can relax or be extended if a load is applied.

Because of the size of large meat animals and their concomitant high heat capacity, it is unlikely that a sufficiently fast cooling rate would be achieved under commercial conditions of refrigeration to cause thaw-rigor, except in the superficial layers of muscle. The phenomenon does occur, however, in lamb carcasses during blast-freezing, and it is also likely to happen if cooling in liquid nitrogen is practised. Furthermore, the advent of the quick freezing of hot de-boned meat in small packs for supermarket retail outlets heightens the risk of thaw-rigor occurring, with consequent loss of tenderness. If the frozen meat is stored at about −3°C to −5°C for at least 48 hours, it is unlikely that sufficient ATP will remain for this type of shortening to occur.

c. Freezing of post-rigor muscle. Although post-rigor muscle may be frozen rapidly it will not be subject to cold-shortening or thaw-rigor. The microscopic appearance of frozen post-rigor muscle is governed by the same factors that apply to pre-rigor muscle, i.e. that the size, number and location of ice crystals depend on the rate of cooling. A good eating quality is maintained if freezing is delayed for a period of at least 24 hours after slaughter. The rate of thawing prior to cooking has no significant effect on the eating quality of the meat unless it is extremely slow.

4. Freeze-drying

The rate of freezing and the consequent ice crystal size are important considerations in relation to freeze-drying of meat. The ability to form a satisfactory product on rehydration depends on the efficiency of water uptake by the dried material. Ingress of water is aided by large extracellular spaces, as with a sponge, but if the water is not re-bound to myofibrillar proteins the texture of the rehydrated product is characteristically woody. In experiments using pork and beef muscles the effects of low and high final pH on the rehydratability of freeze-dried tissue were compared (Penny *et al.*, 1963, 1964). Dried meat, prepared from muscle which had a final pH of 6·7, gave a more tender and more juicy product than that which had a final pH of about 5·5. Sections of muscle fibres from the reconstituted high-pH muscle showed these fibres to have an increased mean diameter compared with the material having a lower final pH. This reflected the increased water-binding capacity of the former compared with the latter. *In vitro* experiments showed that myofibrillar swelling in reconstituted meat from samples of high final pH was greater than that found in samples of low pH. Figures 23a and b show sections of muscle tissue after dehydration and rehydration respectively.

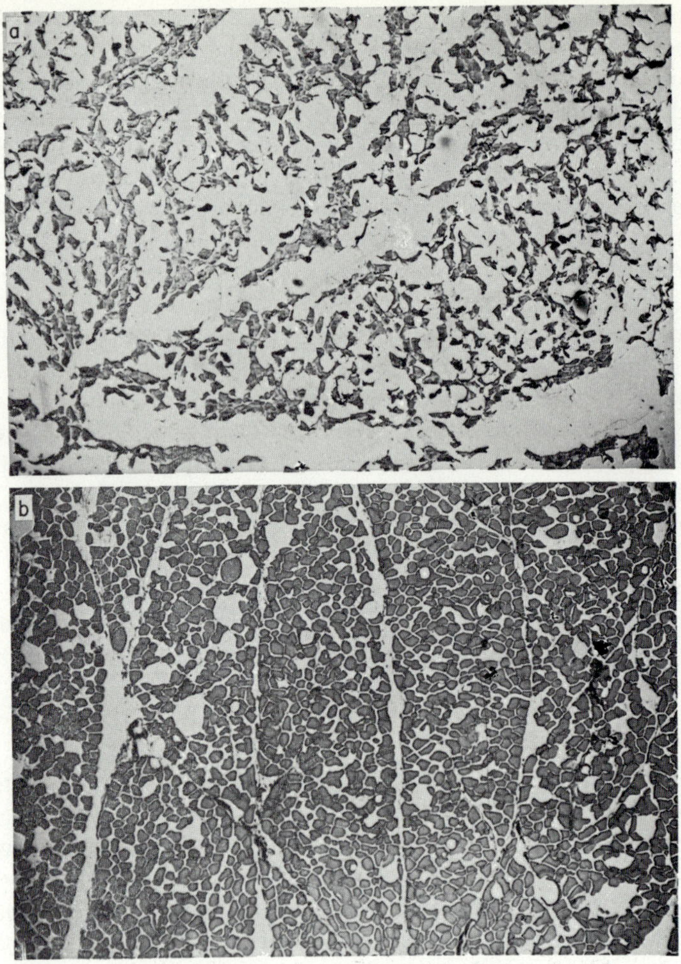

FIG. 23. Sections of muscle tissue, ×30. (a) After dehydration. (b) After rehydration.

D. Effect of Heat on Structure

The primary objective in the cooking of meat is to make it acceptable for eating purposes. Much work has been done on the influence of cooking on texture, measured objectively by instrumental methods, or subjectively by sensory assessment. Instrumental assessment of texture usually involves the measurement of the work done in shearing a sample of standard thickness or estimation of the depth of penetration of a probe applied under standard conditions. Sensory evaluation is carried out by trained panels of tasters or by randomly chosen groups of consumers. The results obtained by these methods provide a gross assessment of texture due to the combined contributions of myofibrillar proteins and connective tissue. Structural studies allow an assessment of the effect of heat on the individual tissue components. Light microscopy has been used to evaluate the damage done to muscle fibres when subjected to various methods of cooking as applied to conventional cuts of meat (Paul, 1963; Reid and Harrison, 1971). Typically, the myofibres display transverse breaks which may be complete, or only partial, giving a zig-zag appearance to the fibre. The optical density of the myofibres in a stained section is increased compared to that of a preparation of raw tissue. This may be explained by the denaturation of the fibrillar proteins which occurs as a result of the heat treatment. Figure 24 shows something of the disruption in myofibres under these conditions.

Studies of the ultrastructure of myofibres, heated singly or in small bundles, indicate that the morphological changes detectable at this level are temperature dependent (Voyle, 1974b). Some of these changes are shown in Fig. 25. The degree of damage sustained by the myofilaments is also limited by the contractile state of the muscle and the consequent overlap between thick and thin filaments.

There are two significant effects of heat on collagen, both of which influence the texture of cooked meat. Collagen fibres undergo a marked shrinkage in

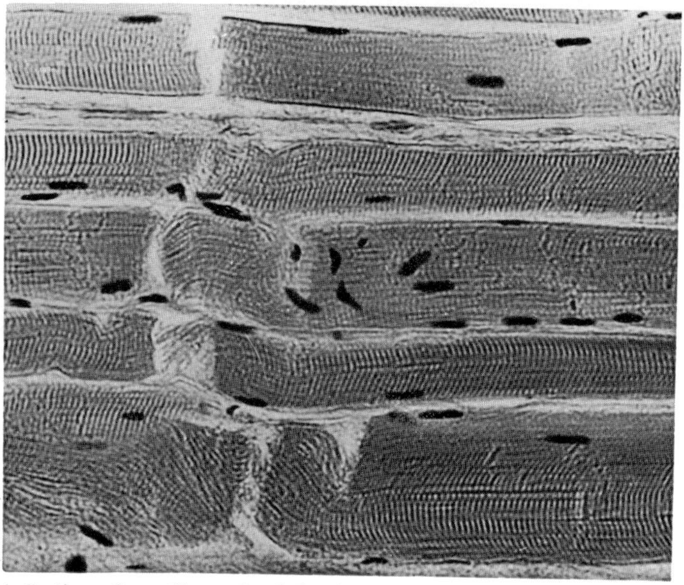

FIG. 24. Section of myofibres after being heated, showing transverse breaks, ×100.

the temperature range 60–70°C. There is a specific shrink-temperature for collagen from any particular source, e.g. tendon collagen from bovine *M. sternomandibularis* begins to shorten at 62°C while intramuscular collagen from the same muscle begins to shorten at 64·5°C (Mohr and Bendall, 1969). As a consequence of this phenomenon myofibres tend to be pulled into a wavy configuration in contrast to the usual straight array, and the entire piece of meat is reduced in size.

At a temperature of about 80°C collagen is solubilized or converted to gelatin. The extent to which this happens depends on the age of the animal which, in turn, determines the amount of cross-linking in the molecular structure of the collagen fibres. The older the animal, the more inter- and intramolecular cross-links are formed (Shimokomaki *et al.*, 1972), and the amount of solubilization which can occur on heating is decreased. In addition, the amount of solubilization which occurs at a given cooking temperature depends on the rate of heating and on the length of time at that temperature. Thus, muscles with a high collagen content need a long, slow, cooking schedule compared with muscles of low collagen content.

Using conventional histological stains for collagen, such as van Gieson's acid picro-fuchsin, there is a marked reduction in staining intensity of heat-denatured collagen compared with its native counterpart.

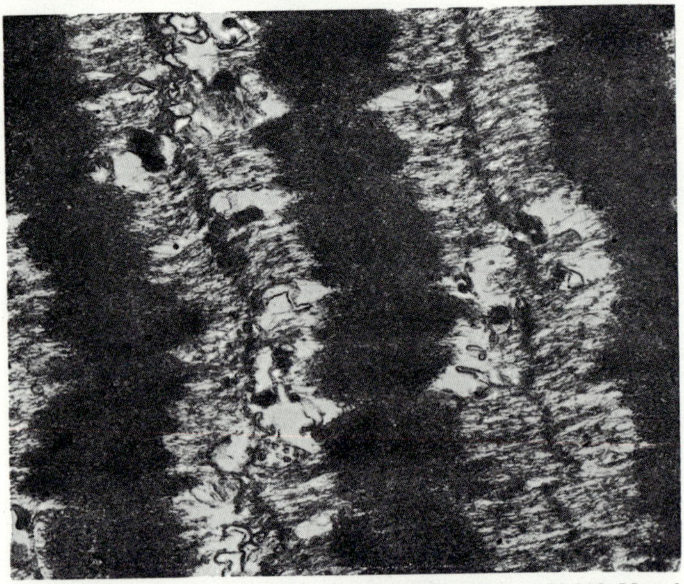

FIG. 25. Electron micrograph of ox *M. psoas*, heated to 70°C, ×20 000. Note loss of filament structure in the A-band and the reduced density of the Z-band.

Relating these structural changes to the texture of heat-treated tissue, it can be postulated that fragmentation of the myofibres will tend to increase the tenderness of the tissue as meat, since it is more easily subdivided by chewing. The denaturation of the myofibrillar proteins, however, tends to increase the hardness of the tissue resulting in a loss of tenderness. Solubilization of collagen also results in an increase in tenderness. Elastic fibres do not contribute significantly to texture. The ultimate texture, therefore, is due to the net effect of all these contributory factors together with such others as have been mentioned, e.g. the effects of *post mortem* treatment.

V. OFFALS

The various anatomical parts of the meat animal which do not come into the classification of skeletal muscle are generally described as offals. This group of organs may be subdivided into three further groups. These are (i) the edible offals, (ii) those used for pharmaceutical purposes and (iii) those used for a wide range of industrial and other purposes. Gerrard (1971) gives a detailed classification of offals and their uses.

Of the edible offals probably the most familiar is the liver. Kidneys are also widely used, generally in conjunction with other meats. Interest in other edible offals tends to be regional, as does the variety of methods of presentation. Some are incorporated into meat products, a procedure which is controlled by legislation.

The gross anatomy of these organs varies between species, showing, for example, differences in shape, in the thickness of a capsule, and in the amount of connective tissue. There are also some differences between species in the histological structure of each organ, although certain major features are common to all. Identification of the various offal tissues should not be difficult given a modicum of knowledge of animal histology. The normal structure of some common edible offals is illustrated in Figs 26, 27, 28 and 29.

The microscopic structure of liver is characterized in most animals by the lobular arrangement of cells with a vein at the centre of each lobule. A septum of connective tissue separates each lobe of cells, forming part of a supporting framework throughout the organ. A collection of blood capillaries and other ducts is found at the intersections of the connective tissue septa. This delineation is most clearly seen in the liver of the pig and other ruminants, but other domestic mammals have very little interlobular connective tissue. The capsule surrounding the liver, known as the capsule of Glisson, is thicker in the pig and large ruminants than in other domestic animals. Figures 26a, b illustrate sections of liver from beef and pig respectively, and show the difference in the appearance of the interlobular septum.

In the kidneys there are macroscopic differences between small ruminants, in which the organ consists of a single lobe or pyramid, and the pig and large ruminants which have a multi-lobed organ. The lobar structure is clearly recognizable in a kidney which is bisected along its major axis. The main features of structural organization in the kidney are the cortex, surrounded by a capsule, and the medulla. These each have a distinctive microscopic structure. The cortex (Fig. 27) is the region where glomeruli are located. Each glomerulus is a cluster of capillaries surrounded by a capsule, Bowman's capsule. The medulla (Fig. 27) is comprised of numerous collecting tubules through which the waste material which is extracted from the blood supply passing through the kidney is drained via the ureter to the bladder.

The tongue, a section of which is shown in Fig. 28, is a highly muscular organ covered by stratified squamous epithelium. The striated myofibres are gathered into bundles surrounded by collagen and elastic fibres. The bundles interlace with one another in three planes. A section cut at right angles to the dorsal surface shows myofibres in both longitudinal and transverse arrays. This unique feature is of considerable assistance in the identification of tongue tissue. As in skeletal muscle, each myofibre is surrounded by endomysium.

The pancreas, sometimes referred to as sweetbreads, is a large gland

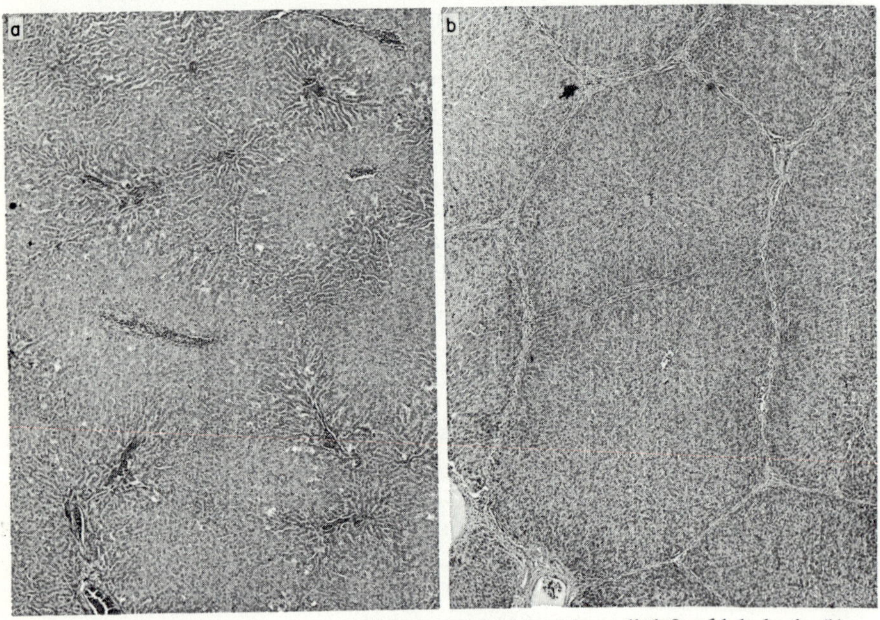

FIG. 26. Sections of liver, ×27. (a) Beef. (b) Pig. Note the well-defined lobules in (b).

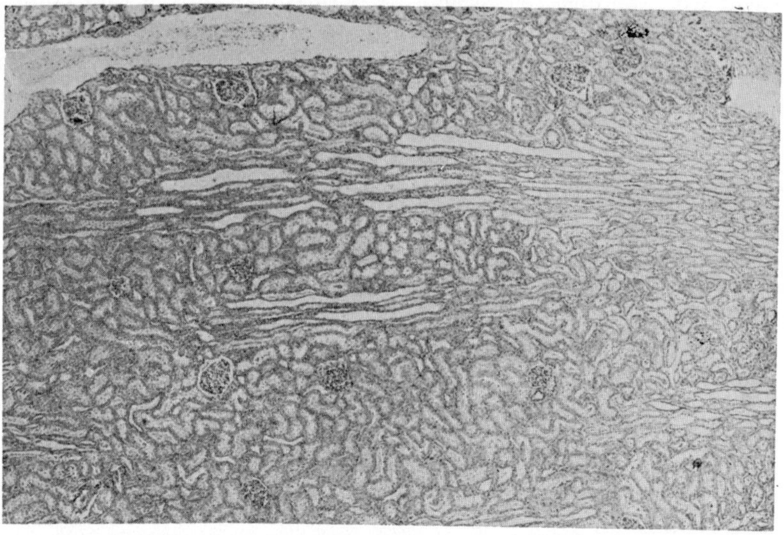

FIG. 27. Section of kidney showing cortex (left) and medulla (right), ×30. Glomeruli are conspicuous in the cortex. Note the rays of medullary tissue protruding into the cortex.

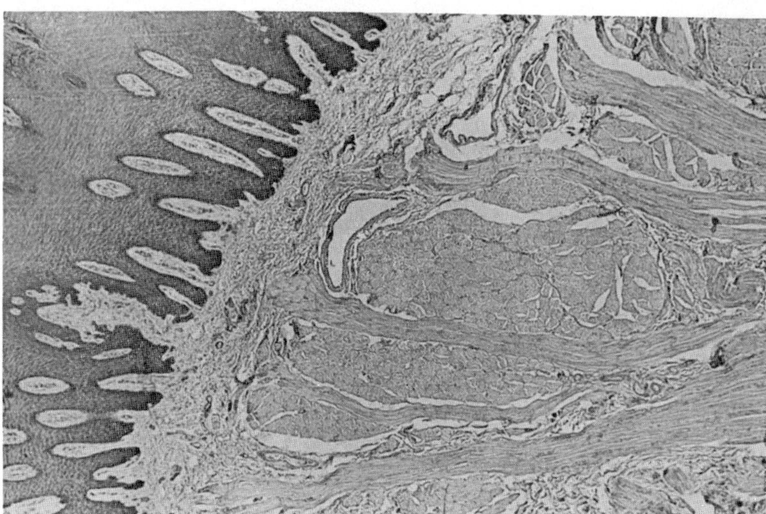

FIG. 28. Section of tongue showing epithelial coat (left) and muscle layers (right) in longi tudinal and transverse orientations, ×30.

associated with the digestive system. It is surrounded by a thin connective tissue capsule which is extended inwards to form septa dividing the organ into lobules. The cell structure falls into two main groups. There are secretory units, formed at the ends of highly branching ducts, which resemble a bunch of grapes in appearance. They are referred to as acini for this reason (*acinus* = grape). Also there are isolated groups or islets of cells having no attachment to the duct system. These are known as Islets of Langerhans and

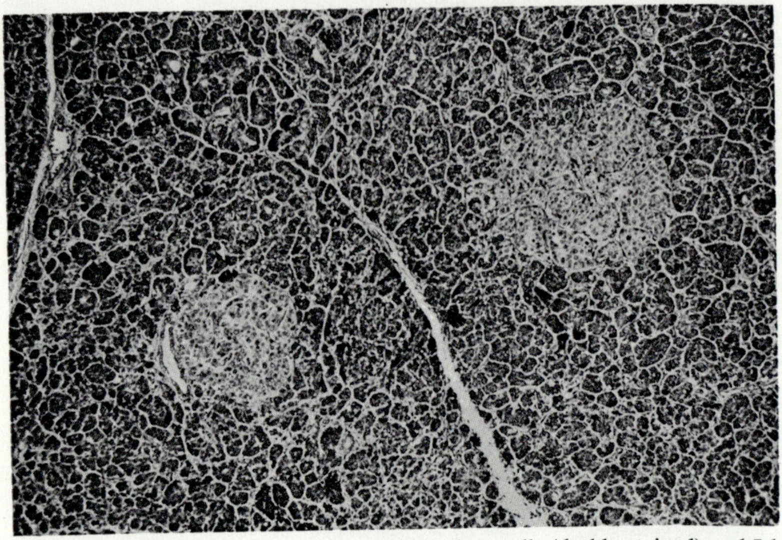

FIG. 29. Section of pancreas showing clusters of acinar cells (darkly stained) and Islets of Langerhans (lightly stained), ×75.

contain several different cell types which are termed alpha, beta and delta cells. The distribution of these cell types varies in different species, and 'C' and 'F' cells have been described uniquely in the guinea pig and dog respectively. Beta cells are involved in the production of insulin, and its absence results in the condition known as diabetes mellitus. It is possible to identify the acinar structures and the Islets of Langerhans in sections stained with haematoxylin and eosin, but more specialized procedures are necessary to identify the cell types in the islets. However, such a refinement is not likely to concern those who are primarily interested in identifying the tissue. Figure 29 shows a low-power view of a section of pancreas, illustrating the lobules of acini and the Islets of Langerhans.

Some difficulty in the identification of tissues arises if the material has been comminuted or subjected to some other form of processing as a result of which the cellular morphology may be substantially impaired. This aspect of meat structure is dealt with in Chapter 6.

ACKNOWLEDGEMENTS

The author wishes to acknowledge the constructive criticism of the typescript by Dr J. R. Bendall, and thanks D. J. Restall and A. Cousins for their help in preparing material for microscopy and photography.

REFERENCES

Annual Report of the National Food Survey Committee (1976). Tables 6 and 7. H.M.S.O., London.

Bendall, J. R. (1967). *J. Sci. Fd Agric.* **18**, 553.

Bendall, J. R. (1969). 'Muscles, Molecules and Movement.' Heinemann Educational London.

Bendall, J. R. (1973). *In* 'The Structure and Function of Muscle' (G. Bourne, ed.), Vol. 2, 2nd Edn, Ch. 5. Academic Press, New York and London.

Bendall, J. R. and Marsh, B. B. (1951). Proc. 8th Int. Cong. of Refrig., London, p. 351.

Bendall, J. R. and Voyle, C. A. (1967). *J. Fd Technol.* **2**, 259.

Beuge, D. R. and Marsh, B. B. (1975). *Biochem. Biophys. Res. Commun.* **65**, 478.

Bloom, W. and Fawcett, D. W. (1975). 'A Textbook of Histology', 10th Edn, Ch. 11. W. B. Saunders, Philadelphia.

Bowman, W. (1840). *Phil. Trans. R. Soc.* **130**, 457.

Callow, E. H. (1935). Ann. Rept. Fd Invest. Bd London, p. 57.

Cassens, R. G. and Newbold, R. P. (1967). *J. Fd Sci.* **32**, 269.

Cassens, R. G., Marple, D. N. and Eikelenboom, G. (1975). *In* 'Advances in Food Research' (E. M. Mrak and G. F. Stewart, eds), Vol. 21, p. 71. Academic Press, New York and London.

Cross, H. R., Carpenter, Z. L. and Smith, G. C. (1973). *J. Fd Sci.* **38**, 998.

Davey, C. L. and Dickson, M. R. (1970). *J. Fd Sci.* **35**, 56.

Davey, C. L. and Gilbert, K. V. (1974). *J. Sci. Fd Agric.* **25**, 923.

Disbrey, B. D. and Rack, J. H. (1970). 'Histological Laboratory Methods.' E. and S. Livingstone, Edinburgh.

Gauthier, G. F. (1969). *Z. Zellforsch. Mikrosk. Anat.* **95**, 462.

Gerrard, F. (1971). 'Meat Technology', 4th Edn. Leonard Hill, London.

Getty, R. (1975). Sisson and Grossman's 'The Anatomy of the Domestic Animals', 5th Edn. W. B. Saunders, Philadelphia.

Hammond, J. (1932). 'Growth and Development of Mutton Qualities in the Sheep.' Oliver and Boyd, London.

Hostetler, R. L., Landmann, W. A., Link, B. A. and Fitzhugh, H. A., Jr (1970). *J. Anim. Sci.* **31**, 47.

Huxley, H. E. (1969). *Science*, **164**, 1356.

Huxley, H. E. (1972). *In* 'The Structure and Function of Muscle' (G. Bourne, ed.), Vol. 1, 2nd Edn, Ch. 7. Academic Press, New York and London.

Joubert, D. M. (1956). *J. Agric. Sci., Camb.* **47**, 59.

Kidwell, J. F. (1952). *J. Hered.* **43**, 157.

Lawrie, R. A. (1974). 'Meat Science', 2nd Edn, Ch. 4. Pergamon Press, Oxford.

Love, R. M. (1955). *J. Sci. Fd Agric.* **6**, 30.

Marsh, B. B. and Leet, N. G. (1966). *J. Fd Sci.* **31**, 450.

Meryman, H. T., ed. (1966). 'Cryobiology'. Academic Press, New York and London.

Mohr, V. and Bendall, J. R. (1969). *Nature, Lond.* **223**, 404.

Moran, T. (1929). *Proc. R. Soc. B*, **105**, 177.

Moran, T. (1932). *J. Soc. Chem. Ind., Lond.* **51**, 16T.

Paul, P. C. (1963). *In* 'Proceedings Meat Tenderness Symposium', p. 225. Campbell Soup Company, Camden, New Jersey.

Penny, I. F. (1970). *J. Sci. Fd Agric.* **21**, 303.

Penny, I. F., Voyle, C. A. and Lawrie, R. A. (1963). *J. Sci. Fd Agric.* **14**, 535.

Penny, I. F., Voyle, C. A. and Lawrie, R. A. (1964). *J. Sci. Fd Agric.* **15**, 559.

Penny, I. F., Voyle, C. A. and Dransfield, E. (1974). *J. Sci. Fd Agric.* **25**, 703.

Pepe, F. A. (1966). *J. Cell Biol.* **28**, 505.

Porter, K. R. (1961). *J. Biophys. Biochem. Cytol.* (Suppl.) **10** (4), Pt 2.

Reid, H. C. and Harrison, D. L. (1971). *J. Fd. Sci.* **36**, 206.

Scopes, R. K. (1970). *In* 'The Physiology and Biochemistry of Muscle as a Food' (E. J. Briskey, R. G. Cassens and B. B. Marsh, eds), Vol. 2, Ch. 22. University of Wisconsin Press, Madison.

Shimokomaki, M., Elsden, D. F. and Bailey, A. J. (1972). *J. Fd Sci.* **37**, 892.

Voyle, C. A. (1969). *J. Fd Technol.* **4**, 275.

Voyle, C. A. (1974a). Proc. M.R.I. Symposium No. 3, 'Meat Freezing: Why and How', 6.1–6.6. Meat Research Institute, Langford, Bristol.

Voyle, C. A. (1974b). Proc. 20th European Meeting of Meat Research Workers, Dublin, p. 233.

Voyle, C. A. and Lawrie, R. A. (1964). *Jl R. Microsc. Soc.* **82**, 173.

Yarom, R. and Meiri, U. (1971). *Nature, Lond.* **234**, 254.

6. Meat Products

D. F. LEWIS

The Leatherhead Food Research Association,
Leatherhead, U.K.

I. INTRODUCTION

Meat is processed in a number of ways. Historically, processes were designed to improve the keeping properties of meat. In the present-day processing of meat, the manufacturer has to consider a number of criteria of acceptability. These criteria may relate to the consumer, e.g. flavour, texture, colour and behaviour on cooking, or to the manufacturer, e.g. cost, best use of materials, avoidance of weight loss during processing and avoidance of deterioration on storage. Many meat processes are designed to improve acceptability without improving keeping quality, e.g. hamburger, sausage.

In order for the acceptance criteria to be attained a comprehensive understanding of the systems is required. This is obtained by using techniques developed in a number of scientific disciplines, and the aim of this chapter is to illustrate how the use of microscopes can assist in the process of acquiring the relevant knowledge. The chemical and biochemical aspects of meat processing are related to the structural changes which occur, and the microscopy of meat systems must be considered against the background of information avilable from other techniques.

In order to understand the microscopy of meat products some consideration must be given to the microscopical procedures which can be used for meat and the usefulness of these procedures for particular problems.

In general, the meat microscopy has concentrated on 'fresh' meat and the changes occurring in the conversion of muscle to meat and the effect of factors such as storage conditions on the structure and quality of meat. The processing of meat to produce such products as hams, sausages, burgers, pastes and bacons has not been so extensively studied and is an area where microscopy offers considerable scope.

Processed meats can be considered in two categories: those sold as intact meat pieces and those in which the meat is in a comminuted or finely divided state. This classification is used in this chapter, and for each type of product an outline of the process involved and the changes produced will be given as well as the microscopical observations that have been made on the products.

II. METHODOLOGY

A. Light Microscopy

The methods used for the examination of meat products are based on those employed by animal histologists although, as the tissue is essentially 'dead' when received by the microscopist, chemical fixation with formalin etc. can sometimes be avoided. The most useful specimen preparation procedure is sectioning, and especially the cutting of frozen sections using a microtome mounted in a refrigerated chamber. The specimens may be fixed or unfixed and are rapidly frozen, normally using liquid nitrogen or pressurized carbon dioxide as a coolant. Once sections have been cut in the cryostat they are collected on glass slides and allowed to dry in air. Most sections will adhere to the slides acceptably well but in cases of difficulty an adhesive such as glycerin albumen or Haupt's medium may be used.

Wax sections can also be used, although this is a much less convenient procedure. Fixation is essential for wax embedding and the samples must be carefully dehydrated in alcohol or acetone and cleared in a suitable medium before being embedded in paraffin wax. For systems containing fat, wax embedding and sectioning is less useful than freeze-sectioning, but the former approach may be used for the examination of lean meat samples where the avoidance of the slight ice crystal damage, which may be produced in freeze-sectioning, is desired. Where very thin sections are desired for light microscopy, the samples can be embedded in a 'plastic' resin before sectioning. The resins used are similar to those employed for electron microscopy, and often samples embedded for electron microscopy can be sectioned first for light microscopy and later for electron microscopy.

Sections of meat products can be usefully studied unstained, using polarized light, phase contrast, interference or dark field methods of illumination to obtain contrast. Examination of sections between crossed polars will give much information on the form of fat in a product and phase contrast allows the striations of meat pieces to be identified.

In most cases, however, the sections will need to be stained in some way before being examined. Again, the microscopist can generally use the methods of the animal histologist, either directly or with slight modification. Toluidine blue and acid fuchsin are good general stains which will colour the non-lipid components of a meat system. Some differentiation between connective tissue and 'cellular' meat protein can be obtained by using acid fuchsin and picric acid together in van Gieson's stain. Triple staining with Mallory's stain, and staining for elastic tissue with Weigert's stain, have been used for staining connective tissue in meat (Ramsbottom *et al.*, 1945).

In addition to its use as a general stain for meat products, toluidine blue can also be used to differentiate between textured non-meat proteins and meat in a product (F. O. Flint, personal communication; Jewell, 1974).

Fat in meat products can be stained using the oil-soluble dyes, such as oil red O or the Sudan stains, or else by using osmium tetroxide. Osmium tetroxide stains fat brown-to-black depending on the duration of staining; it should be used with care, however, as it is highly toxic. It has the advantage of also fixing the fat, but sections should be examined shortly after staining as considerable darkening occurs with time.

It is possible to use many more specialized histochemical procedures to identify the constituents of a product, however the stains mentioned above can be used to give good contrast for studying the morphology of the product.

In addition to sectioning it is possible to prepare specimens by dispersing them; this procedure is of course not suitable for studying the arrangement of the components relative to one another. Dispersion of the specimen can be used for observing the size and shape of pieces in the product, e.g. for measuring the sarcomere lengths in meat (Herring *et al.*, 1965; Locker and Hagyard, 1963), or for the identification of particular constituents in the product, e.g. the detection of soya flour (A.O.A.C., 1975). In general, however, the stained section is the most useful form of specimen preparation for the light microscopy of meat products.

B. Electron Microscopy

Once again the techniques available are those developed for use by microscopists working on the study of biological ultrastructure.

The usual technique for the electron microscopy of meat products is the use of ultrathin sections—in this case the sections must be less than ~100 nm thick to be of any use, and for many purposes sections are required to be only ~30 nm thick. Fixation is an all-important step in the preparation of thin sections, although in the case of meat products the aim of fixation is to hold the materials in place during subsequent processing and the killing of tissue is not relevant. Glutaraldehyde, introduced by Sabatini *et al.* (1963) as a fixative, is almost universally employed for fixing the proteinaceous components of meat and meat products. The glutaraldehyde solutions used for fixation are generally buffered to a pH of approximately 7·2, and this can be achieved with phosphate, cacodylate or veronal acetate buffers according to the choice of the operator. Where the prime object of the investigation is the form of the proteins in the meat product, fixation in glutaraldehyde alone may be sufficient and this can be simply followed by a thorough washing before the specimen is dehydrated.

Where the investigation is required to consider membranous or lipid structures, fixation with glutaraldehyde is generally followed with osmium tetroxide fixation. The osmium tetroxide may be applied buffered, using the same buffer as for the glutaraldehyde, or unbuffered. The period of fixation in osmium tetroxide is important as penetration of this fixative is quite slow. For preservation of membrane systems a fixation time of one or two hours is generally sufficient, however where the retention of bulk lipid is important very long contact times of 2 weeks or more may be required. When long-contact osmium fixation is used two phenomena can be observed; firstly the osmium associates itself with the liquid fat portion which shows up fat crystals as clear spaces in a sea of liquid fat (Fig. 1), and secondly the protein constituents of the meat product are rendered more labile and may be lost during subsequent processing (Figs 2 and 3). Hence, studies involving osmium fixation over a long period of time should be coupled with an examination of systems subjected to little or no osmium treatment in order to obtain an overall picture of the structure of the product.

Following fixation the samples must be dehydrated, and this is usually carried out using a graded series of alcohols, cleared in propylene oxide or acetone and embedded in a suitable resin. In the same way as the buffers used for fixation are dependent on the preferences of the operator, so are the resins chosen for embedding. Methacrylates are preferred by a few workers, although Araldite (Glauert *et al.*, 1956), Epon (Kushida, 1959) or combinations

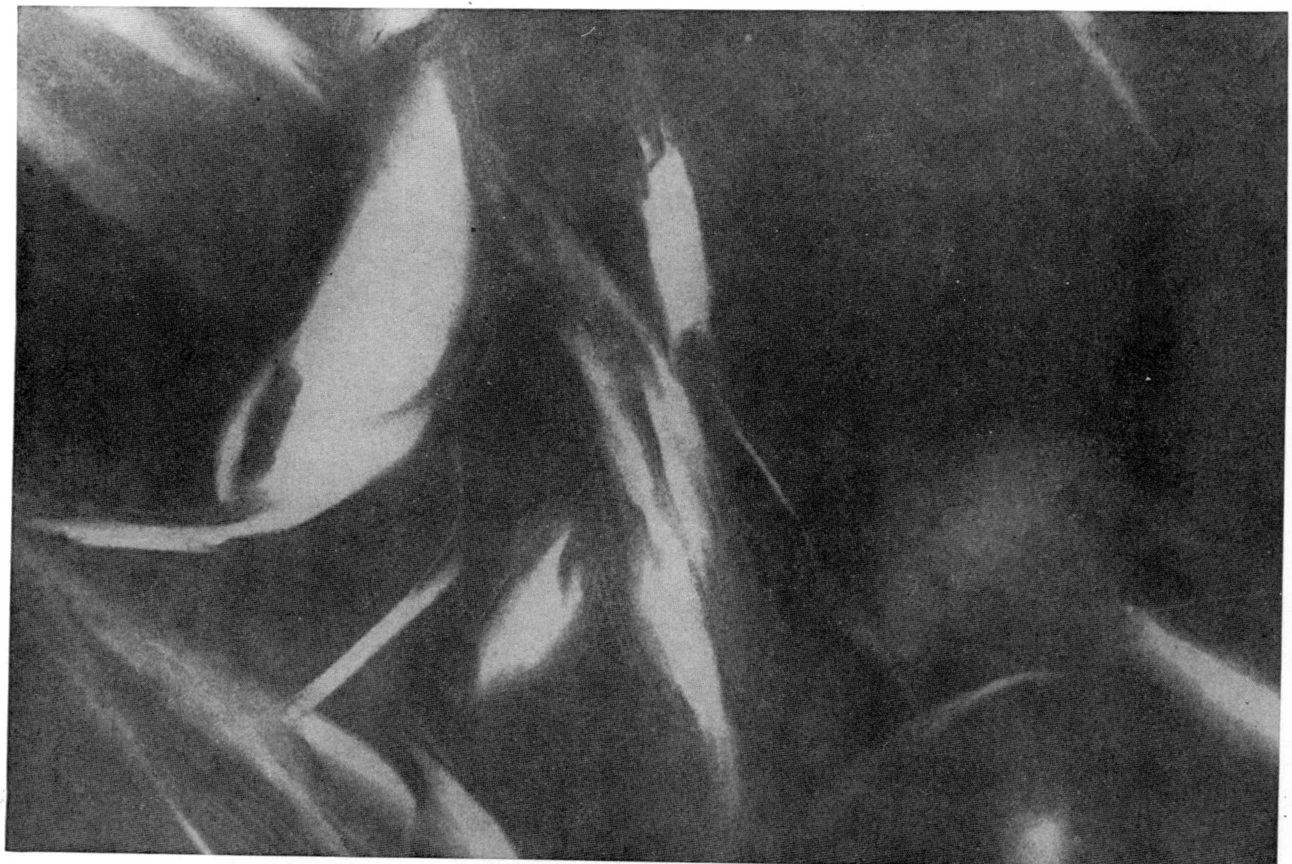

FIG. 1. Sample of fatty tissue prepared by thin-sectioning after fixation in osmium tetroxide for 2 weeks, $\times 70\,000$; shows clear fat crystals in a dark sea of liquid fat.

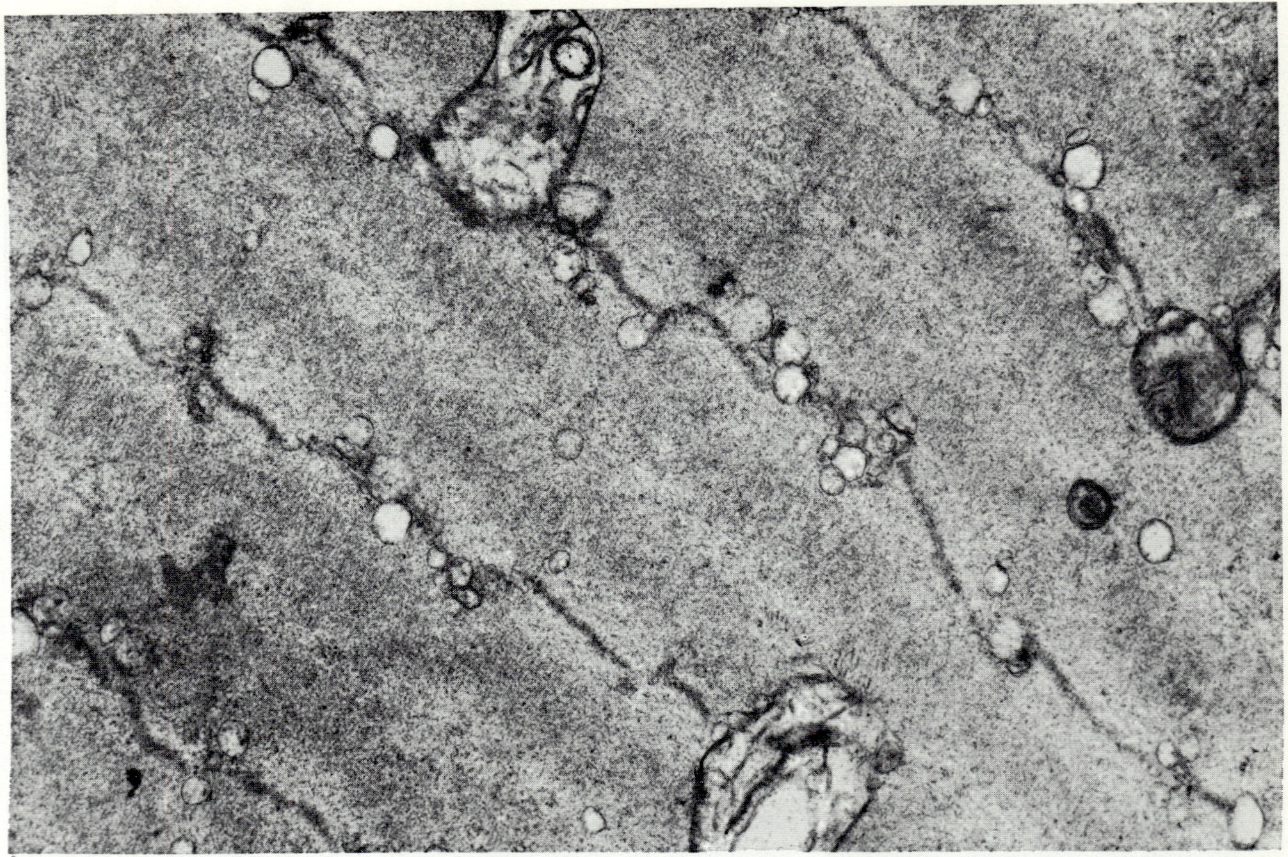

FIG. 2. Sample of salted meat, thin-sectioned after fixation in osmium tetroxide for 1 h, ×15 000; stained with uranyl acetate and lead citrate.

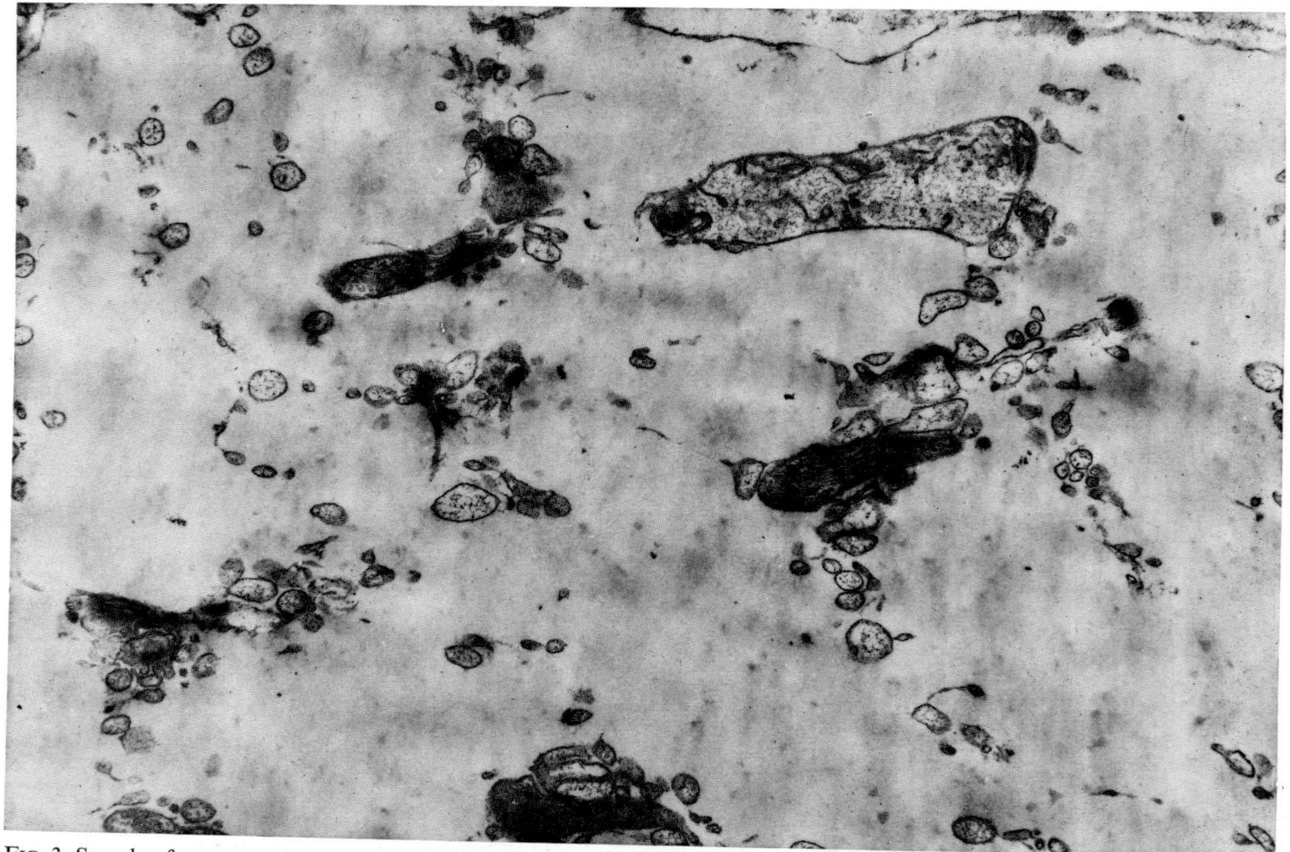

FIG. 3. Sample of same meat as Fig. 2, thin-sectioned after fixation in osmium tetroxide for 2 weeks, $\times 20\,000$; stained with uranyl acetate and lead citrate.

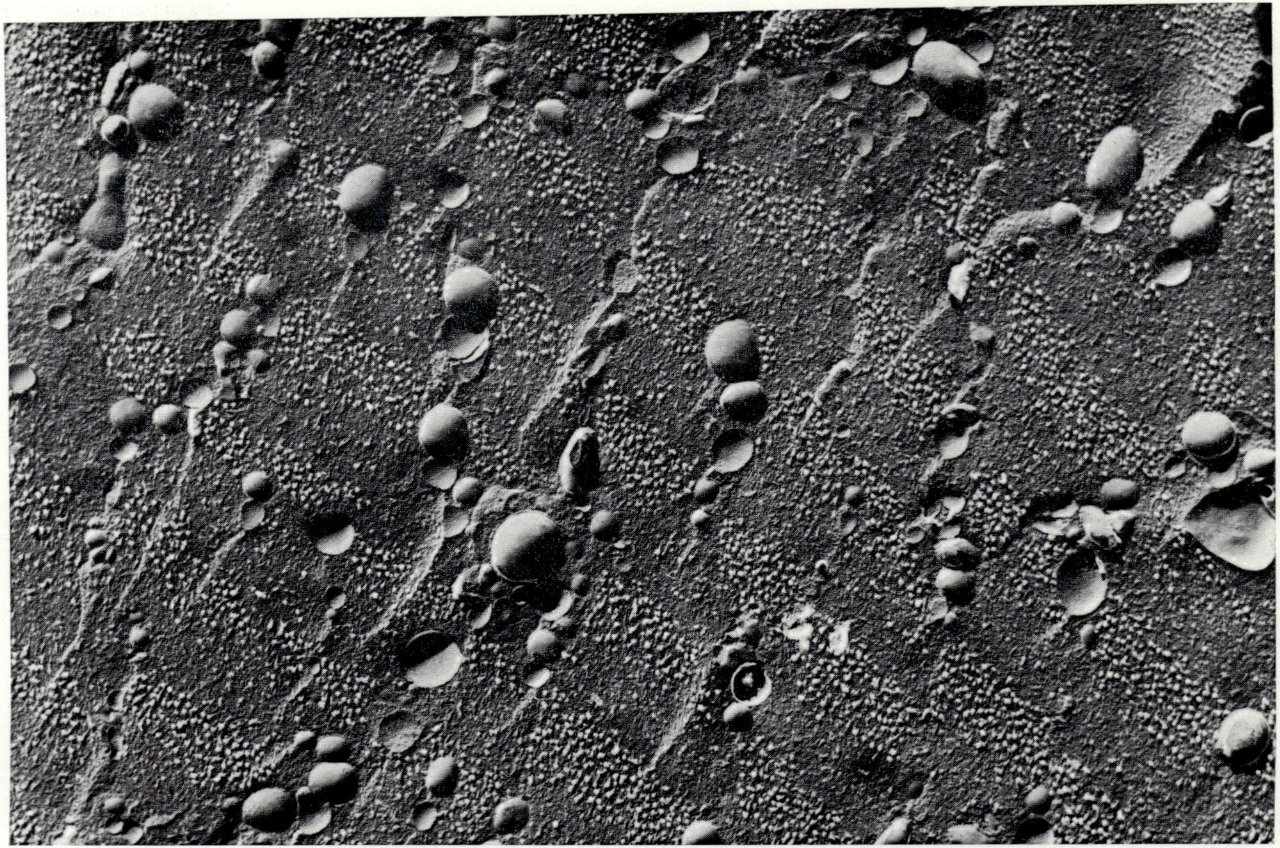

FIG. 4. Fresh meat prepared by freeze-etching after infiltration with 30% glycerol, ×15 000.

FIG. 5. Fresh meat prepared by thin-sectioning after fixation with glutaraldehyde, ×18 500; stained with uranyl acetate.

of the two are probably most widely used. Specially designed resins such as Spurr or maraglass (Spurr, 1969; Freeman and Spurlock, 1962) are also employed. Water-soluble resins, where the resin is also used as the dehydrating agent, may have some advantages for fatty systems, although in most cases with meat products careful dehydration with alcohol is adequate. Durcopan (Staubi, 1963) and glycol methacrylates are the most commonly used water-soluble resins.

After sectioning, sections can be stained, and in this case conventional uranyl acetate and lead citrate methods are adequate. In samples which have undergone long osmium fixation staining may not be necessary and often the staining of protein components is diminished in any case.

Thin-sectioning has many drawbacks in the sense that a sample is subjected to fairly severe chemical treatments and these are likely to produce some structural alterations. Freeze-etching (Steere, 1957; Moor and Mühlethaler, 1963) is an alternative method of specimen preparation which avoids chemical fixation and dehydration and replaces them with the physical process of freezing. In some cases good agreement can be obtained between the freeze-etched and thin-sectioned appearance; but in others freeze-etching produces images which are not immediately compatible with those obtained from thin-sectioning. Meat and muscle tissue are examples of materials for which it is difficult to see an immediate connection between the freeze-etched and thin-sectioned specimens (cf. Figs 4 and 5). These differences have led some workers (Reed *et al.*, 1969; Achatz, 1970) to suggest that the structure of muscle should be viewed quite differently from the sliding filament hypothesis proposed by Hanson and Huxley (1953). Others however, have claimed that the freeze-etch image of muscle can be explained by considering that irregular fracture planes and some deformation of fibres during fracture may account for the appearance without deviation from the sliding filament model (Rayns, 1972; Willison *et al.*, 1972). As will be seen later in the consideration of meat processing, some changes in structure can be observed by both thin-sectioning and freeze-etching, and the use of both techniques can be beneficial.

The formation of ice crystals during freezing can spoil freeze-etch preparations and in most cases the specimen must be infiltrated with glycerol prior to freezing. Freeze-etching has the advantage of retaining fat, and in the area of meat–fat mixtures is most useful for demonstrating the relationship between the fat and non-fat constituents in a single sample.

Another preparative technique which may be of value in the study of meat products is negative staining. Like the dispersion techniques used for light microscopy, negative staining is of little use for showing the relationship between components, but the method will reveal information on the conformation of structural elements isolated from the system. The technique has been applied to the study of sarcoplasmic reticulum (Greaser *et al.*, 1969) in meat, and to myofibrillar structures (Huxley, 1963; Maruyama and Ebashi, 1970). It has potential, albeit largely unexploited, for use in the examination of the structure of extracted proteins in relation to their functional characteristics and their reactions with other proteins.

Thus it can be seen that the microscopist has a wide variety of techniques at his disposal for studying the structure of meat products. This variety can be extended by considering scanning electron microscopy for the examination of surfaces, histochemistry and cytochemistry for the localization of components using both light and electron microscopy, and the possible use of analytical electron microscopy.

Preparation of samples for scanning electron microscopy is generally simpler than for transmission electron microscopy, although the sample still needs to be dried and coated with a conducting layer.

A variety of procedures have been suggested for the dehydration of meat and meat products prior to examination by scanning electron microscopy. These include freezing, freeze-drying (Stanley and Geissinger, 1972), fixation followed by air-drying, freeze-fixation and substitution (Schaller and Powrie, 1971), and critical point drying (Jones *et al.*, 1976). Alcohol freeze-fracture of fixed material, followed by critical point drying or air-drying, was reported by Jones *et al.* (1976) as being the most useful procedure. An example of meat produced by critical point drying is shown in Fig. 6.

III. NON-COMMINUTED PRODUCTS

A. Outline of Processes Involved

Two of the oldest methods employed for preserving meat are drying and salting. Of these, salting (or curing) is the more commonly used in the present day. The main outlet for dehydrated meat is in so-called convenience foods and some use is made of them for expedition rations where their low weight is an advantage.

The processes available for drying meat are air-drying and freeze-drying; of these freeze-drying gives the better quality product and can be used for moderate-sized pieces of meat (Lawrie, 1974). However this process is not very popular for meat, probably because the product does not compete economically with the frozen equivalent in this age of home deep-freezers. Air-dried meats are often pre-cooked and the meat is dried in small pieces.

Curing is a process based on treating meat with salt, although many

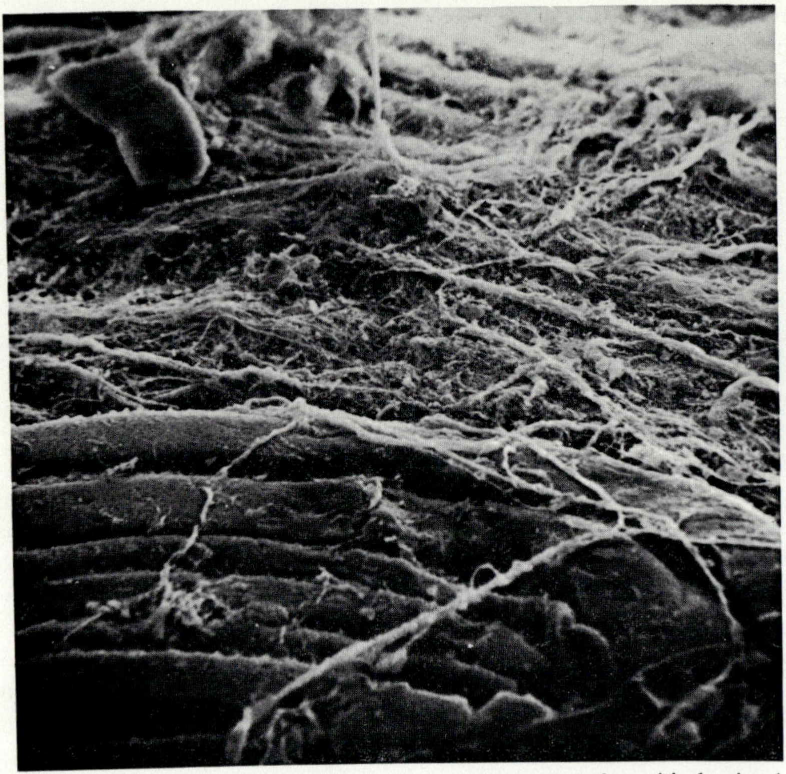

FIG. 6. Fresh meat as revealed by scanning electron microscopy after critical point drying, ×220.

variations on the detail of the procedure are possible depending on the product required. Originally, curing was carried out by packing salt in and around them eat and allowing it to dissolve in the 'meat juices'. However, this process normally brought about a loss of weight, and later meat was cured by placing it in concentrated brine. More recently, injection of the brine into the meat has been adopted, and this may be coupled with a tumbling process where the sample is mechanically agitated to give a more rapid distribution of the brine through the meat. Tumbling is also claimed to cause meat pieces to bind together more effectively. Meat slices may also be cured by passing them through a solution of the appropriate curing salts.

Pork is the main meat processed by curing and the cured product is quite distinctive and different from the fresh meat. In addition to the salt in the brine, other substances may be added to control the properties of the product. These include sodium or potassium nitrate and sodium nitrite, which have important anti-microbial effects and are involved in the development of colour in the product, polyphosphates, which aid water retention in the product (especially on cooking), sugar and potassium sorbate. After treatment with the curing brine the meat will generally be allowed to mature for a period during which time the salts can equilibrate themselves through the sample and the characteristic flavour can develop. The meat may also be smoke derived from hard woods.

The product may be sold uncooked as gammon, bacon, ham steaks or as ham or bacon joints. Where the product is sold as rashers, these may be produced in the factory and vacuum-packed or cut in the shop where they are bought.

Cooked products are most commonly cooked in cans or moulds into which the meat pieces are packed. The heat treatment applied varies according to the size of the pack, but normally either a pasteurizing process, heating to an internal temperature of about 75°C, or a sterilizing process at retort temperatures of 110–120°C is used.

More details of recipes and processes for producing meat products may be found in Hill (1969), Halliday (1971), Treharn (1971) and Gerrard (1976).

B. Microscopical Observations

Dehydration of meat is reported by Wang *et al.* (1953) to be associated with shrinkage, and this shrinkage occurs both within the fibres and in the spaces between the fibres. Shrinkage was found to be more extensive in cases where the meat had been cooked prior to air-drying. The ability of dehydrated meat to rehydrate is an important characteristic, and this is affected by the structure and extent of denaturation of the proteins present. The proteins may be denatured by the build-up of salts in the meat, and Wang *et al.* (1954a) found an accumulation of potassium at the periphery of the cells. Proteins are also denatured by heat, and in this respect the freeze-drying process has the advantage of not affecting the meat so badly. The characteristics of the meat after freeze-drying depend partly on the freezing process and partly on the conditions in the drying chamber.

As with normal freezing of meat, least structural and physical damage is caused by very fast freezing rates when ice crystals may be within the myofibrils (Menz and Luyet, 1961). However, such freezing rates are only achievable using very small meat pieces; probably 2 mm cubes are about the largest size that could be cooled sufficiently rapidly, even using liquefied refrigerant gases such as Freon. Even if such a cooling rate could be achieved with normal size pieces the very fine honeycomb structure that would be formed is likely to make rehydration very difficult. This problem may be encountered even with slower cooling rates (Wang *et al.*, 1954b) and the freezing process for freeze-drying should produce a relatively coarse structure which then rehydrates more easily.

The drying process of accelerated freeze-drying involves plates to enhance heat exchange during the initial stage of ice sublimation, and the rate of heat input into the meat is important to avoid thawing and re-freezing of the meat which could produce problems of inorganic salt concentration or changes in the nature of the honeycomb network (Wang *et al.*, 1954b; Rolfe, 1958). Freeze-drying produces changes in the myofibrils, which tend to become attached to one another, and in the sarcoplasmic reticulum which does not take up stain readily after freeze-drying (Voyle and Lawrie, 1964). In the later stages of drying the structural changes which become marked at higher plate temperatures relate to difficulties in reconstitution (Aitken *et al.*, 1962).

The addition of salt to meat produces benefits in terms of aiding water retention in addition to its anti-microbial properties. This property of increased water retention is aided by the incorporation of polyphosphates into the brine and the improvement of water retention has been the subject of much attention. As early as 1929, Callow found the optimum level of salt for water retention to be associated with the level of salt that produced the most solubilization of the meat proteins. Maximum hydration is achieved when the ionic strength is in the region of 0·6–1·0 (Hamm, 1960). Evans and Ranken (1972) studied the effects of varying salt levels and water-to-meat ratio on weight loss from the meat on cooking. They found that the optimum yield was obtained when the salt concentration, expressed as a percentage of the total water in the system (i.e. added water plus the water normally found in the meat), is in the region of 3–8%. In general those conditions that favoured protein solubilization also favoured water retention. Microscopically the changes in structure on treating meat with salt solutions can be followed and tend to support this view. Examination of salted meats with the light microscope does not reveal very great differences; only occasionally are cells found which show a marked change in appearance. When individual cells are treated with solubilizing salt solutions the effect is much more marked than when meat is treated in bulk; this is probably due to the vastly increased amount of solubilizing solution available in relation to the amount of protein.

At the electron microscope level, changes can be readily seen, even in bulk meat samples which have been treated with salt solutions. Figures 5 and 7 show the changes which occur. Figure 5 shows the typical appearance of meat at 24 hours *post mortem* and illustrates the sliding filament structure, whilst Fig. 7 shows how this alters when the meat is treated with 4% salt and

FIG. 7. Meat treated with 4% salt solution and prepared by thin-sectioning, ×18 500; shows some change in organization of myosin-containing H-zone.

FIG. 8. Meat treated with 1% sodium tripolyphosphate solution and prepared by thin-sectioning, ×41 000.

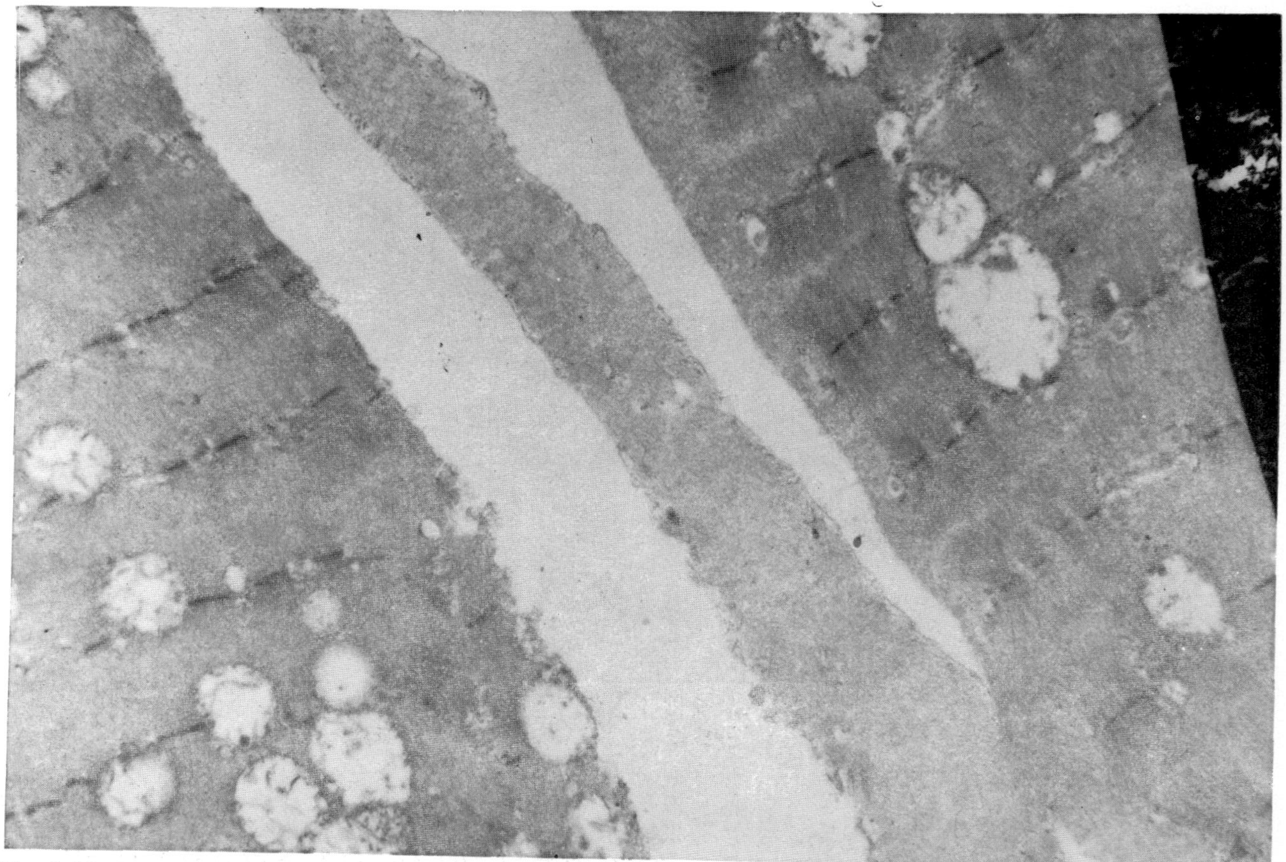

FIG. 9. Meat treated with 4% salt and 1% sodium tripolyphosphate solution and prepared by thin-sectioning, ×9000; shows considerable breakdown of myofibril organization.

a marked breakdown in structure can be seen. The conditions which favour weight increases in meat also tend to favour structural breakdown (Whitworth, 1968; Lewis, 1974). This level tends to be around 6% of the total water in the system (including that in the meat as well as that in the brine). Above this level increasing salt concentrations tend to produce rather more denaturation and precipitation of proteins than solubilization, and consequently the hydration state of the meat is decreased.

Salt and polyphosphate behave synergistically in promoting water retention and this has been observed by many workers (Bendall, 1954; Kotter, 1960; Mahon, 1961; Lewis and Jewell, 1975). The suggested action of polyphosphates is that they mimic ATP in dissociating actomyosin formed after the death of the muscle. This suggestion was first put forward by Bendall (1953) and has been supported by many workers, notably Kotter (1960) who further suggested that the synergism found between salt and polyphosphate was effective because the dissociated actin and myosin were more readily solubilized by salt than undissociated actomyosin. Unfortunately, the microscopic evidence does not offer very much support to this view (Lewis, 1974; Lewis and Jewell, 1975). Figure 8 shows meat treated with polyphosphate alone, and Fig. 9 meat treated with salt and polyphosphate. Polyphosphate does make the individual filaments stand out with greater clarity, which could be due to dissociation of actomyosin, but the combined action of salt and polyphosphate seems to have acted more on those areas where actin and myosin are not combined initially. (See Addendum p. 272.)

One of the features of muscle tissue is that it is generally considered to be made up of more than one type of muscle cell. These types are referred to as red, white and intermediate and have been found to be structurally and biochemically distinct (Ogata, 1958; Gauthier, 1969, 1970). When any meat product is examined a range of structures is found in different cells (Fig. 10). It is tempting to suggest that the cells originated as distinct types in the muscle and have consequently behaved differently on processing. Voyle (1969) found that when muscles cold-shorten, some fibres contract actively whilst others are passively contracted. It is possible that the different responses of the cells to processing may be brought about by their shortening to different extents *post mortem*. Some observations on the changes in muscles containing varying proportions of the different fibre types have been made (Beecher *et al.*, 1965), although the behaviour of individual fibres has not been clearly demonstrated. Fox *et al.* (1970) considered red and white fibres and their behaviour on processing, assessing the behaviour of the fibres by their change in diameter during the production of ham. They found more variaton between red fibres in different 'qualities' of meat than between red and white fibres in the same piece of meat. Fox *et al.* differentiated meat fibres into red

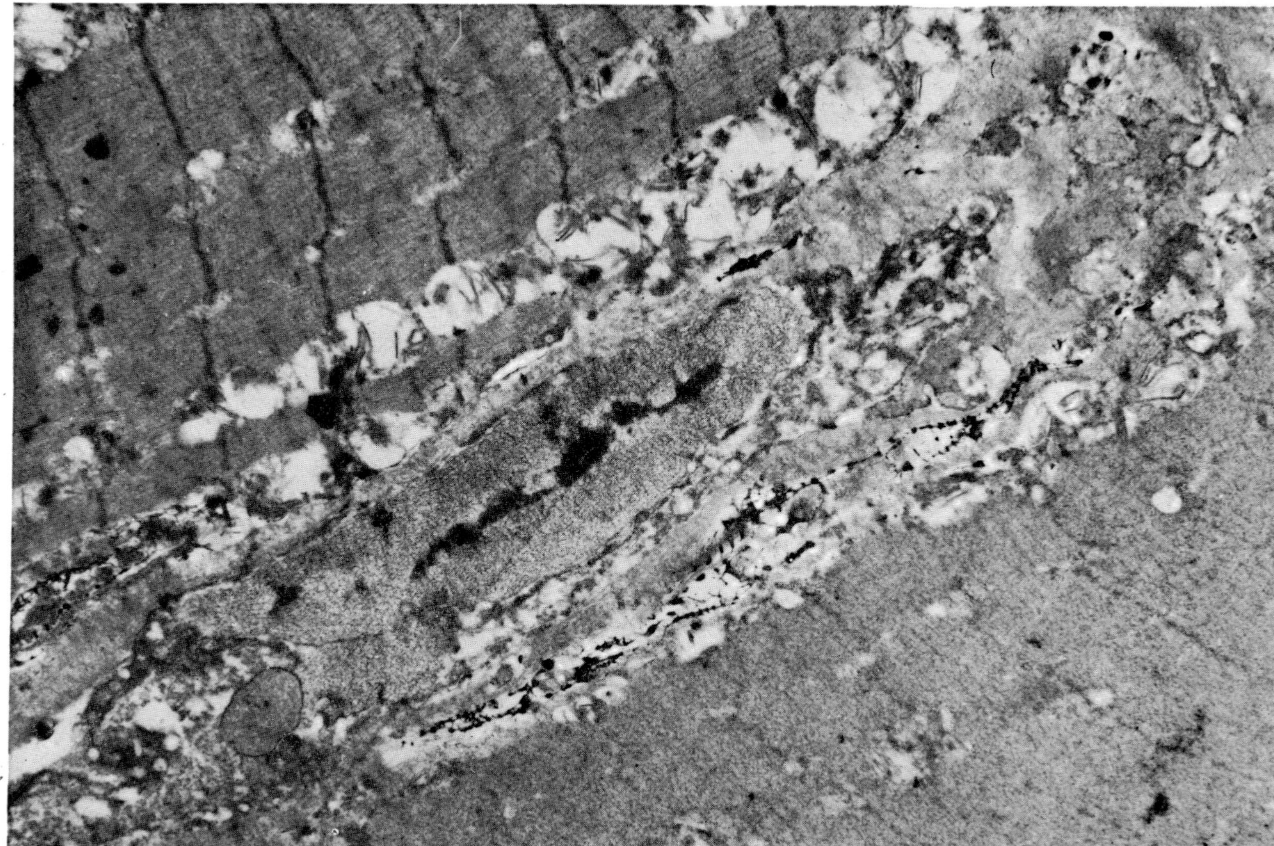

FIG. 10. Sample of commercial bacon prepared by thin-sectioning, $\times 10\,000$; shows two adjacent meat cells which have reacted to give quite different structures.

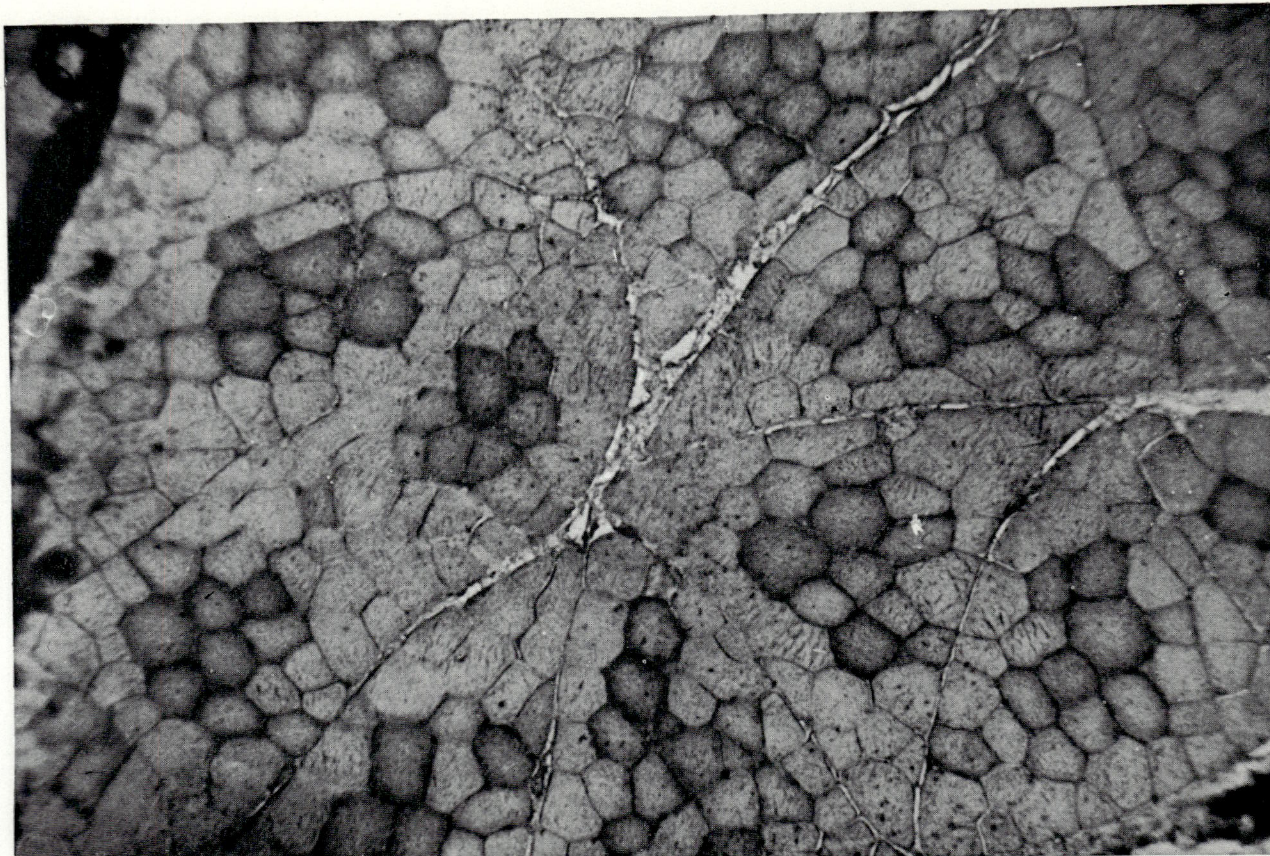

FIG. 11. Light micrograph of pork meat stained with Sudan black B to demonstrate intracellular lipids, $\times 100$. The meat had been treated with salt and polyphosphate.

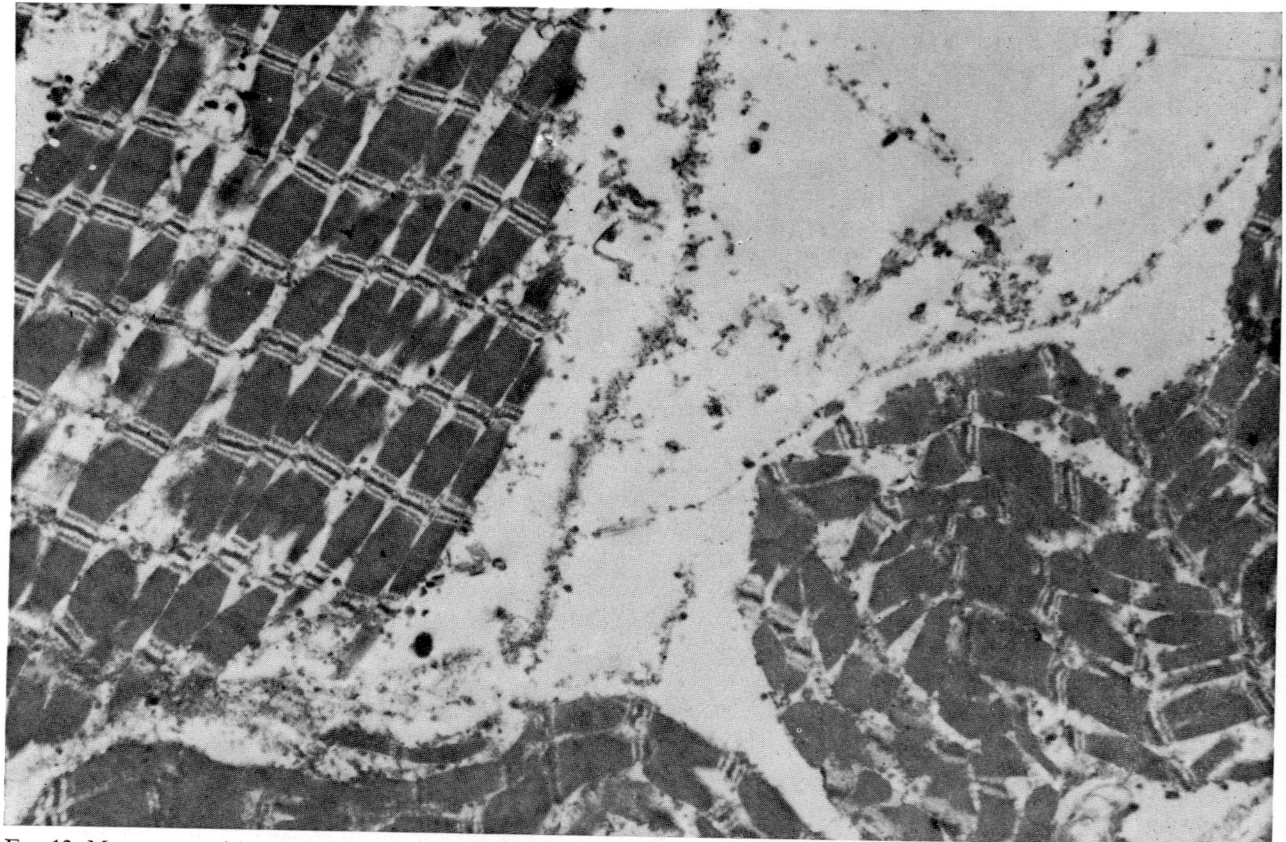

FIG. 12. Meat prepared by thin-sectioning after heating to 75°C in water, ×7500; shows coagulation of proteins in sarcomeres and disruption of connective tissue.

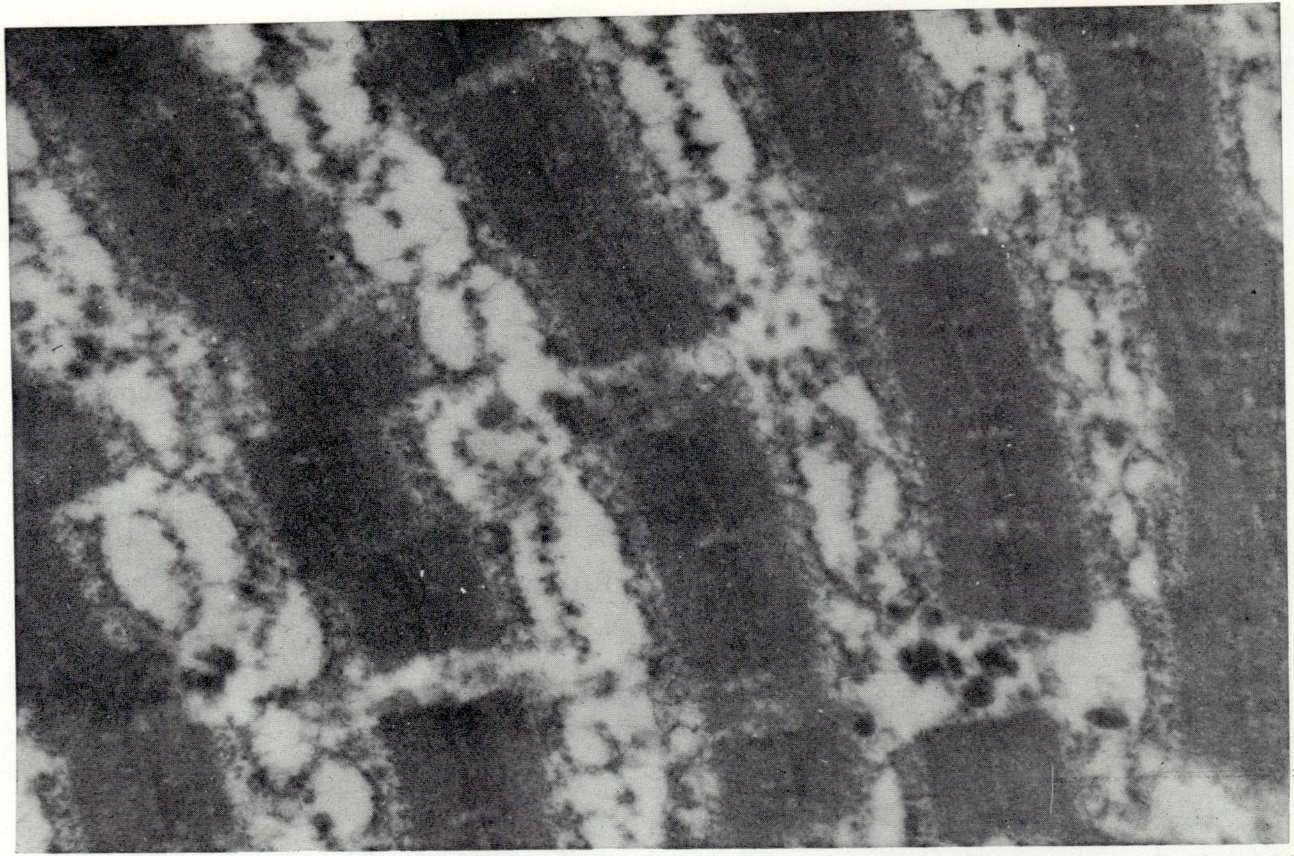

FIG. 13. Meat prepared by thin-sectioning after treatment with 4% salt solution and heating to 75°C, ×17 000; shows loss of material from the actin-rich I-band.

FIG. 14. Meat prepared by thin-sectioning after treatment with 4% salt and 1% sodium tripolyphosphate solution, and heating to 75°C, ×12 000; shows narrow coagulated bands (DB) at each end of the A-band where actin and myosin filaments overlap.

FIG. 15. Sample of commercially manufactured ham prepared by freeze-etching, $\times 25\ 000$; shows densely packed bands (DB) resembling those dark bands seen in thin-section preparations.

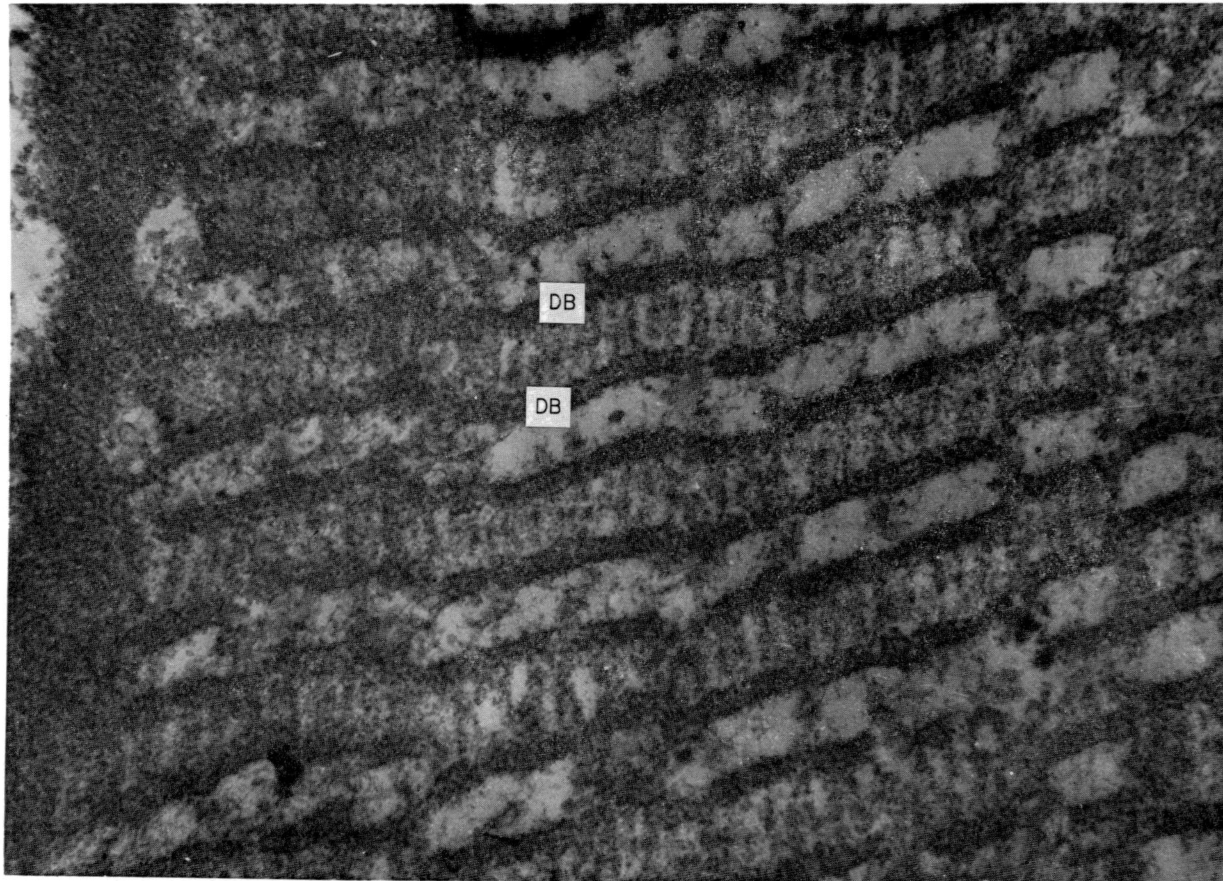

FIG. 16. Sample of commercially manufactured ham prepared by thin-sectioning, ×14 000; shows dark bands (DB) similar to those seen in Fig. 14.

and white types, using Sudan black B stain to illustrate intracellular lipids. This method is simpler but less precise than enzyme histochemistry; however it has the advantage of being applicable to salt-processed meats (Fig. 11). Generally, as meat is processed the differentiation of red and white fibres becomes more difficult.

Heating of meat generally produces shrinkage and weight loss accompanied by denaturation of proteins. These changes are seen microscopically as a shortening in sarcomere length and as a decrease in cell diameter (Paul, 1965; Giles, 1969; Voyle and Restall, 1970–71; Schmidt and Parrish, 1971). The electron microscope reveals the changes as a loss of fine structure; in particular the A-band regions apparently become amorphous, detail is lost from the Z-discs and the I-bands are disrupted (Schmidt and Parrish, 1971; Lewis, 1974). The appearance of heated meat is as shown in Fig. 12 and heat-induced changes in meat are more fully described in Chapter 5.

When meat has been treated with salt and/or polyphosphate, the coagulation pattern on heating is altered. Polyphosphate alone results in a complete loss of the Z-disc on heating whilst salt alone induces the loss of material from the actin-rich I-band (Fig. 13). Salt and polyphosphate together produce a very much more disrupted pattern after heating (Fig. 14); in this case the Z-disc and A- and I-bands are replaced by a series of much thinner bands of apparently coagulated material. A similar effect may be seen in freeze-etched preparations (Fig. 15) and also in commercially manufactured hams (Fig. 16). The narrow coagulation bands seem to originate from the A–I overlap region of the sarcomere where actomyosin is likely to form in the *post mortem* muscle. Thus it seems that dissociation of actomyosin by polyphosphate is not a major effect and that the improved dispersion of 'free' actin and myosin to produce a water-retaining gel on heating is a more important feature of polyphosphate action. The regions where actomyosin is formed seem to be the most resistant areas to dispersion on processing.

IV. COMMINUTED PRODUCTS

A. Outline of Processes Involved

Comminuted products include sausages, burgers, luncheon meats and pastes. The meat may be used fresh or after curing, depending on the type of product; for fresh sausage and burgers the meat is generally used fresh, whilst for dried sausage or luncheon meat the meat is often cured before comminution. Mixtures of fresh and cured meat may also be used. In addition to lean meat, fat, often in the form of pork fat, is included in most comminuted meat products and a number of other additives may be used. These additives may include bread, rusk, farina, soya flour, textured soya protein, soluble proteins (soya isolates, caseinates, blood proteins, etc.), seasoning, colouring and spices.

The normal procedure followed is to mince or flake the meat and then chop the meat together with any added ice or water, salt and additives in a bowl chopper. The fat in the recipe is normally added later in the chopping process in either minced or diced form. The length of the chopping operation depends partly on the type of product desired, coarser products such as 'English-type' sausages not requiring such a prolonged period, whilst luncheon meats and frankfurters which are much finer are more severely comminuted.

Meat pastes and spreads are prepared from meats which are first coarsely minced and cooked in a retort before being finely minced, filled into jars and given a sterilizing heat treatment.

Comminuted meat products may be sold raw, e.g. sausages or burgers, or after heat treatment, e.g. luncheon meat, canned frankfurters or meat pastes; smoking may also be applied.

B. Microscopical Observations

From a microscopical standpoint a comminuted meat product can be thought of as containing three elements. Firstly there is the lean meat portion, secondly the fat content and finally the additives portion. The microscopy of comminuted meats can also be considered from two viewpoints, viz. analytical and functional.

The analytical aspect has been reported extensively in German, notably by Prandl (1961). The techniques do not seem to have received much attention in Britain although some other countries do appear to use them (A. Rantsios, personal communication). The main aim of the German work is to assess the quality of the product by determining the relative amounts of the connective tissue and lean meat components. The detection of extraneous material such as bone fragments and offals can also be carried out by microscopical examination. Some workers claim to be able to detect added proteins histochemically, and several others have been able to demonstrate the presence of soy products by their morphology and staining characteristics (Lewin and Flint, 1976; Linke, 1969; Coomeraswamy and Flint, 1973; Jewell, 1974).

The functional approach to meat microscopy is concerned with relating the structural changes which take place during processing to the handling properties of the product. The behaviour of the lean portion of meat during comminution is such that it breaks down into pieces normally about 200–500 μm in diameter which comprise several cells or parts of cells. Some isolated cell

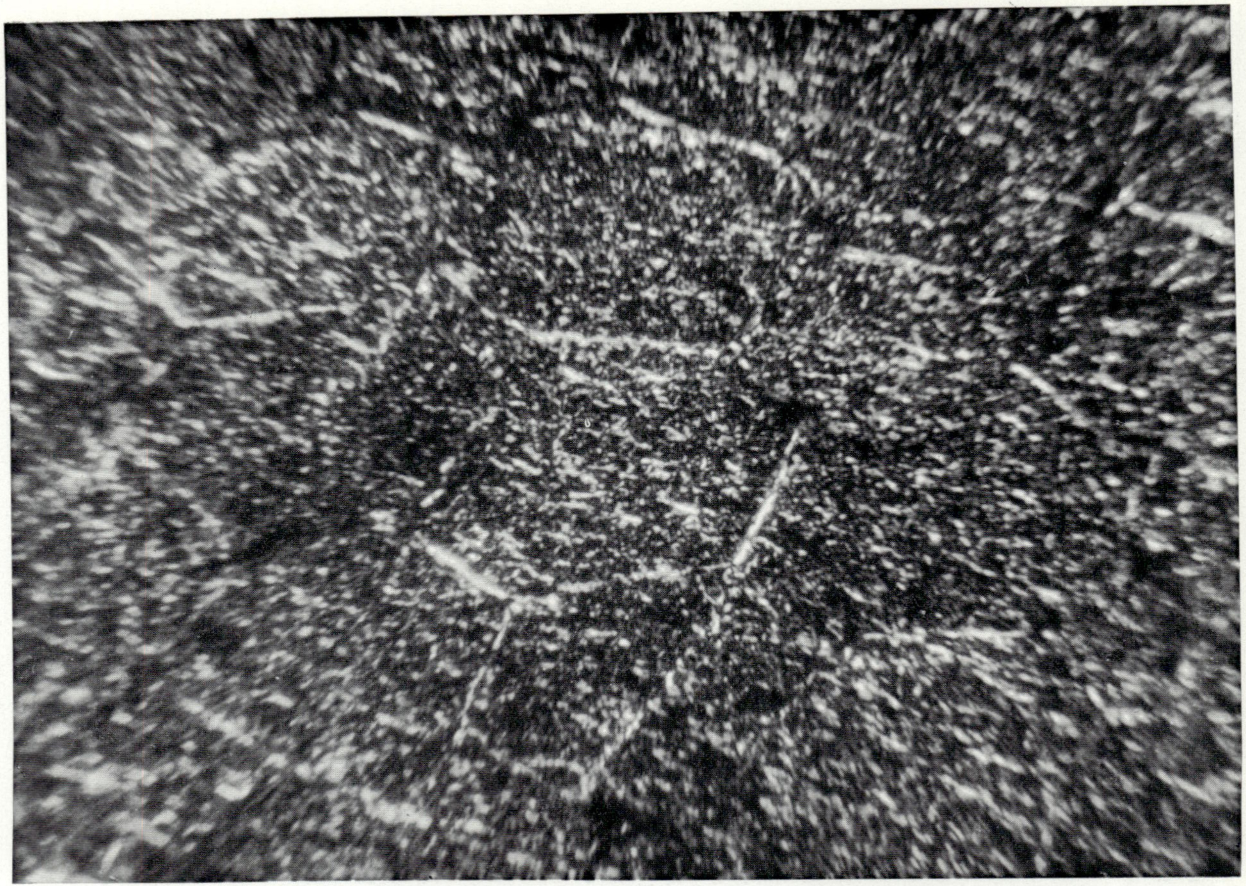

FIG. 17. Light micrograph of frozen section of pork flare fat viewed between crossed polars, $\times 300$.

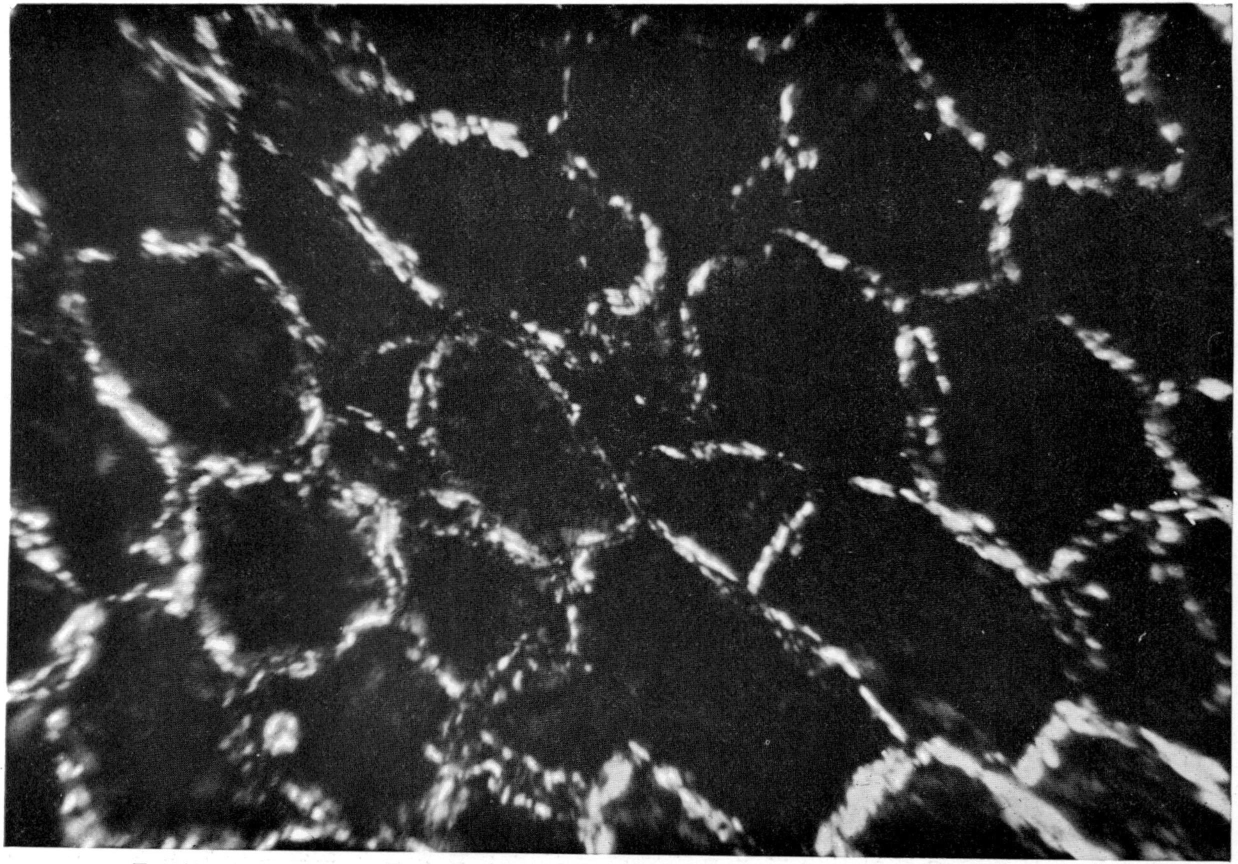

FIG. 18. Light micrograph of frozen section of pork head fat viewed between crossed polars, $\times 300$.

parts are released from the cells and individual myofibrils may be seen. Some proteins are extracted from the cells and these tend to form a continuum which 'cements' the larger pieces together. Protein denaturation seems to occur during chopping and 'clumps' of protein may be found (Evans and Ranken, 1975).

The structure of adipose tissue is essentially that of cells packed with fat (Sheldon, 1969). However this structure varies with the anatomical location of the fat, and this variation in structure has been related to the processing behaviour of the tissue (Barnet and Evans, 1973; Lewis, 1976). Flare or mesenteric fat appears highly crystalline when viewed between crossed polars (Fig. 17), whilst head or jowl fat is less crystalline but often has a well defined and coherent cell structure. The appearance of this tissue (Fig. 18) is similar to that seen in plant parenchyma cells, and the term 'cell wall' does not seem inappropriate for the interstitial connective tissue; botanists may regard this use of a botanical term as profane, however! Electron microscopy using long-term osmium fixation also reveals the crystalline nature of the fats. Of particular interest is the crystalline layer found at the edge of some cells (Fig. 19); this probably accounts for the well-defined cell outlines seen with the polarizing microscope. A similar crystalline region can also be seen after freeze-etching (Fig. 20). Flare fat rarely shows this structure, but electron microscopy generally does reveal more crystalline structure in the cells (Fig. 21).

Electron microscopy of the 'cell wall' material of fatty tissue is best carried out after minimal or no osmium fixation, the retention of fat tends to obscure the detail of the collagen. Flare fat is found to have poorly ordered cell walls, whilst head fat seems to have more ordered and collagenous structures (Figs 22 and 23).

Generally, the structures found in head fat lead to greater stability on processing and in consequence less fat is lost from this tissue than from flare fat. The structures of other fatty tissues tend to be intermediate between these two, as is their stability.

Comminuted meat products are often referred to as 'emulsions'; however this is only partly true. Some emulsification of the fat in the system will generally occur, but the whole structure of comminuted meat products is much more than a protein–lipid emulsion.

Hansen (1960) described the formation of an emulsion in finely comminuted sausage and demonstrated the breakdown of fatty tissue in the early stages of comminution. However, in many comminuted meat products groups of fatty tissue cells can still be found and it has been suggested that these may play an important role in the stability of fat in these products (Van den Oord and Visser, 1973; Evans and Ranken, 1975). Fat which is not

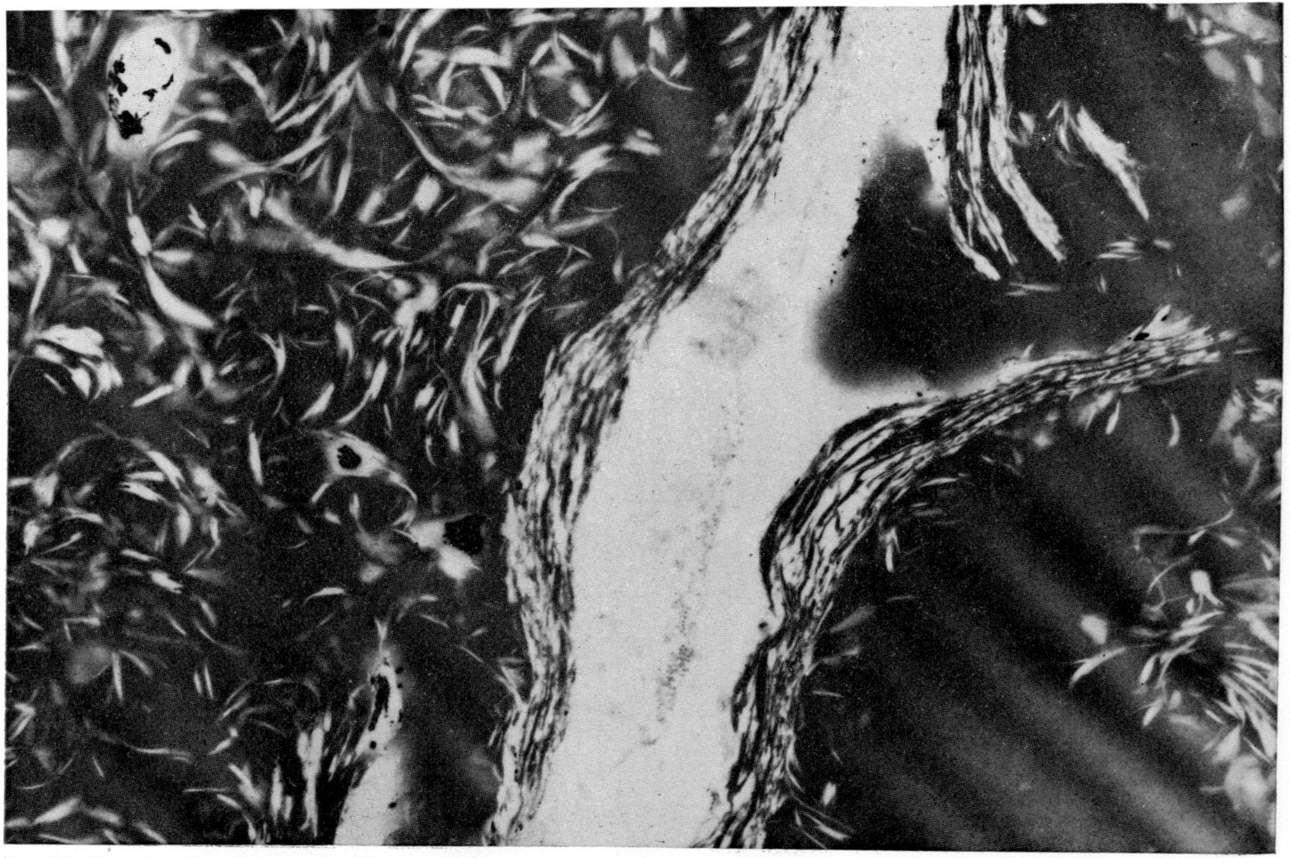

FIG. 19. Sample of head fat prepared by thin-sectioning after fixation in osmium tetroxide for 2 weeks, ×9000; shows crystalline layer at edge of cells.

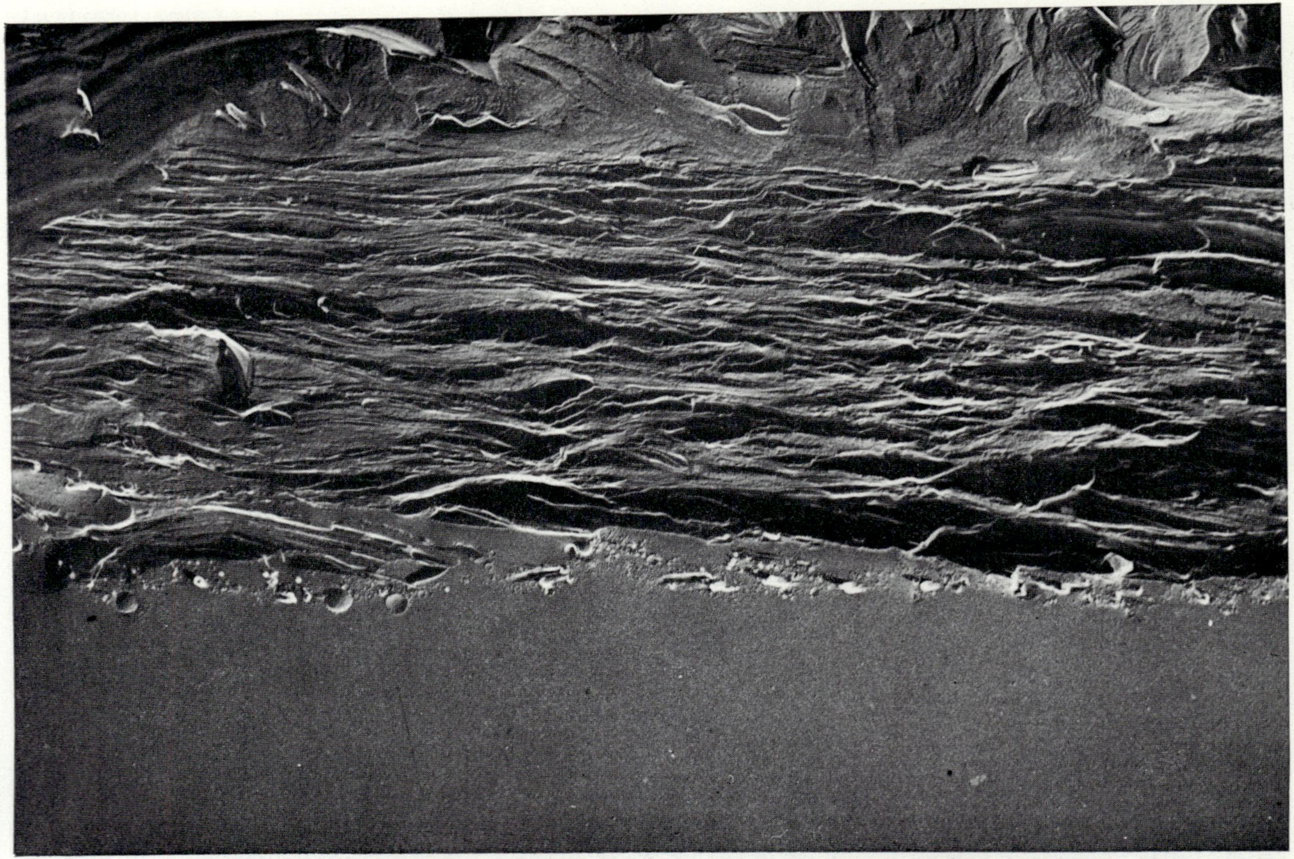

FIG. 20. Crystalline layer at the edge of a fat cell as revealed by freeze-etching, $\times 25\,000$.

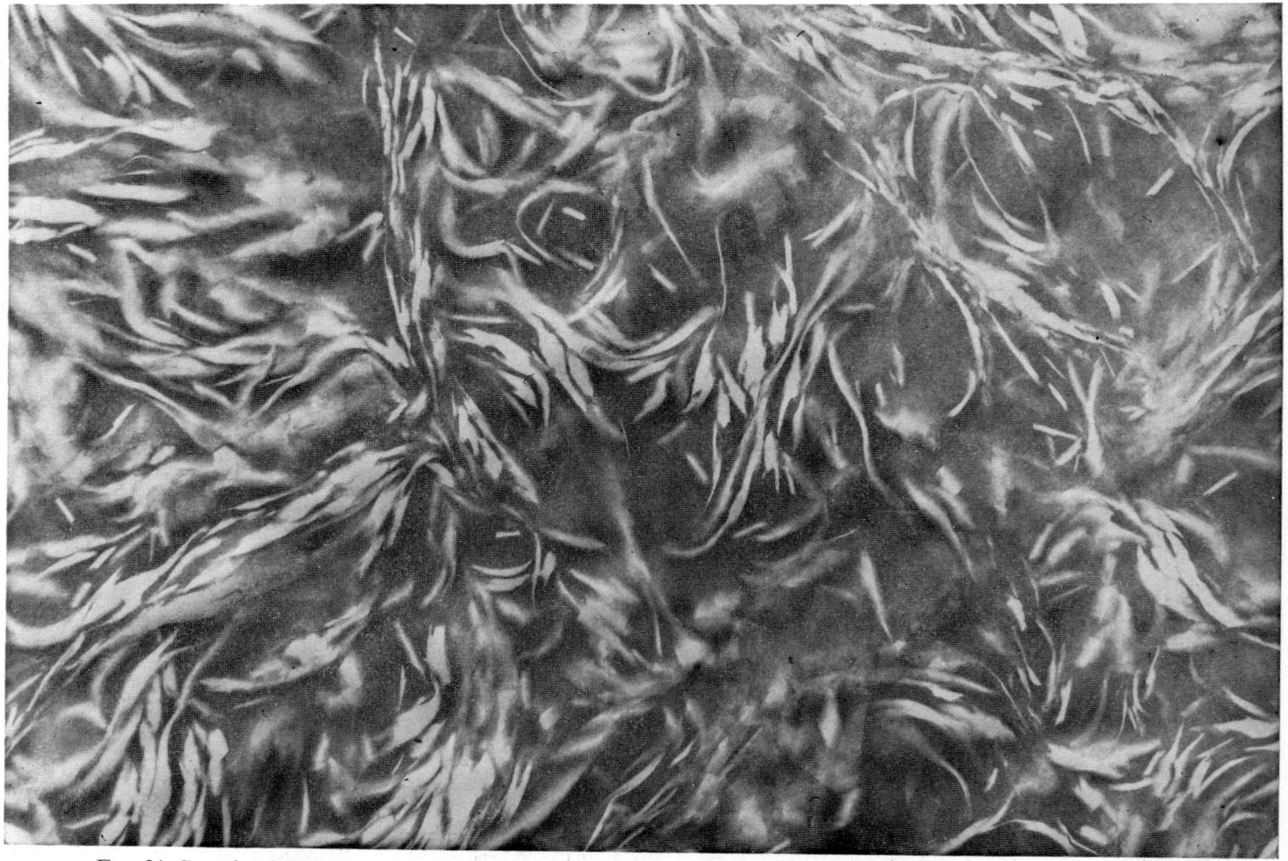

FIG. 21. Sample of flare fat prepared by thin-sectioning after fixation in osmium tetroxide for 2 weeks, $\times 12\,000$.

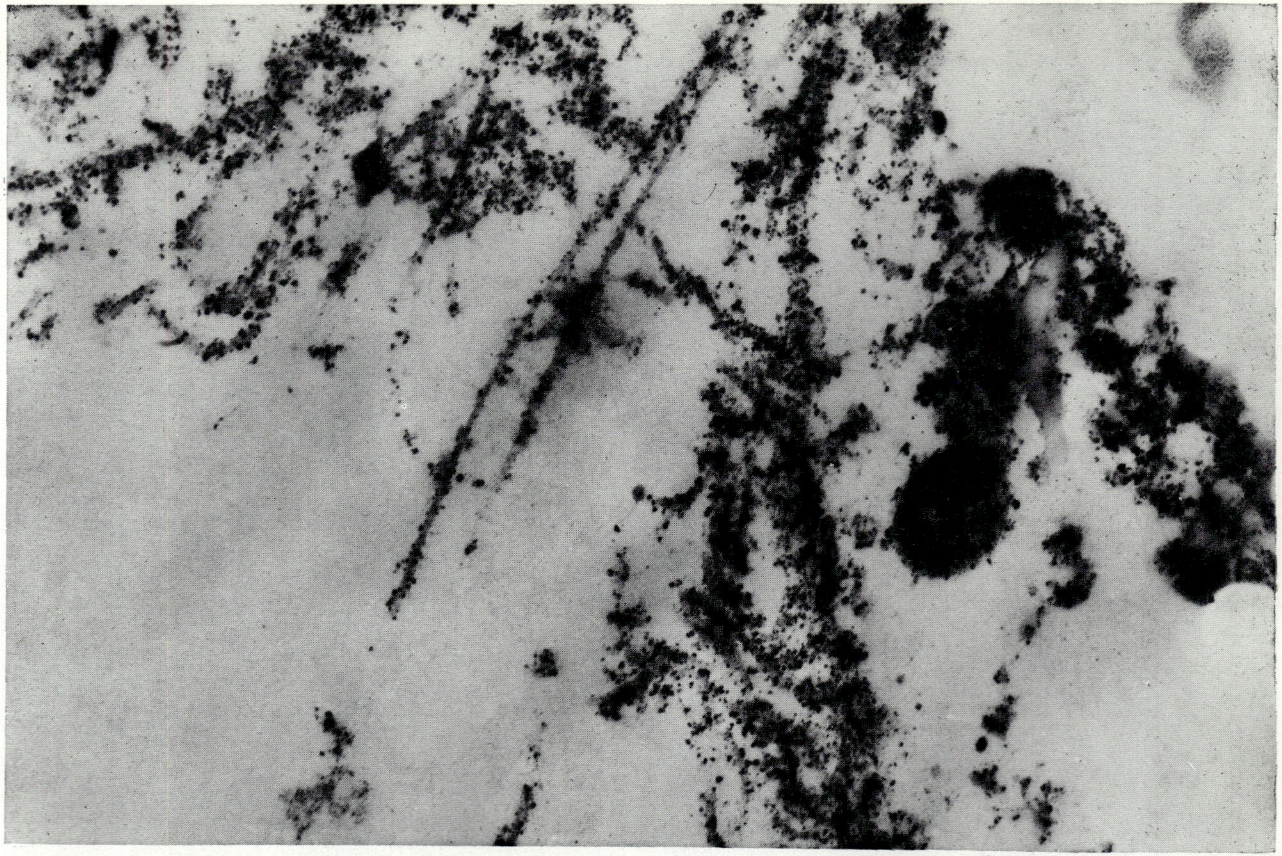

FIG. 22. Sample of flare fat prepared by thin-sectioning after fixation in osmium tetroxide for 1 h, $\times$ 29 000; stained with uranyl acetate and lead citrate, shows disorganized interstitial connective tissue.

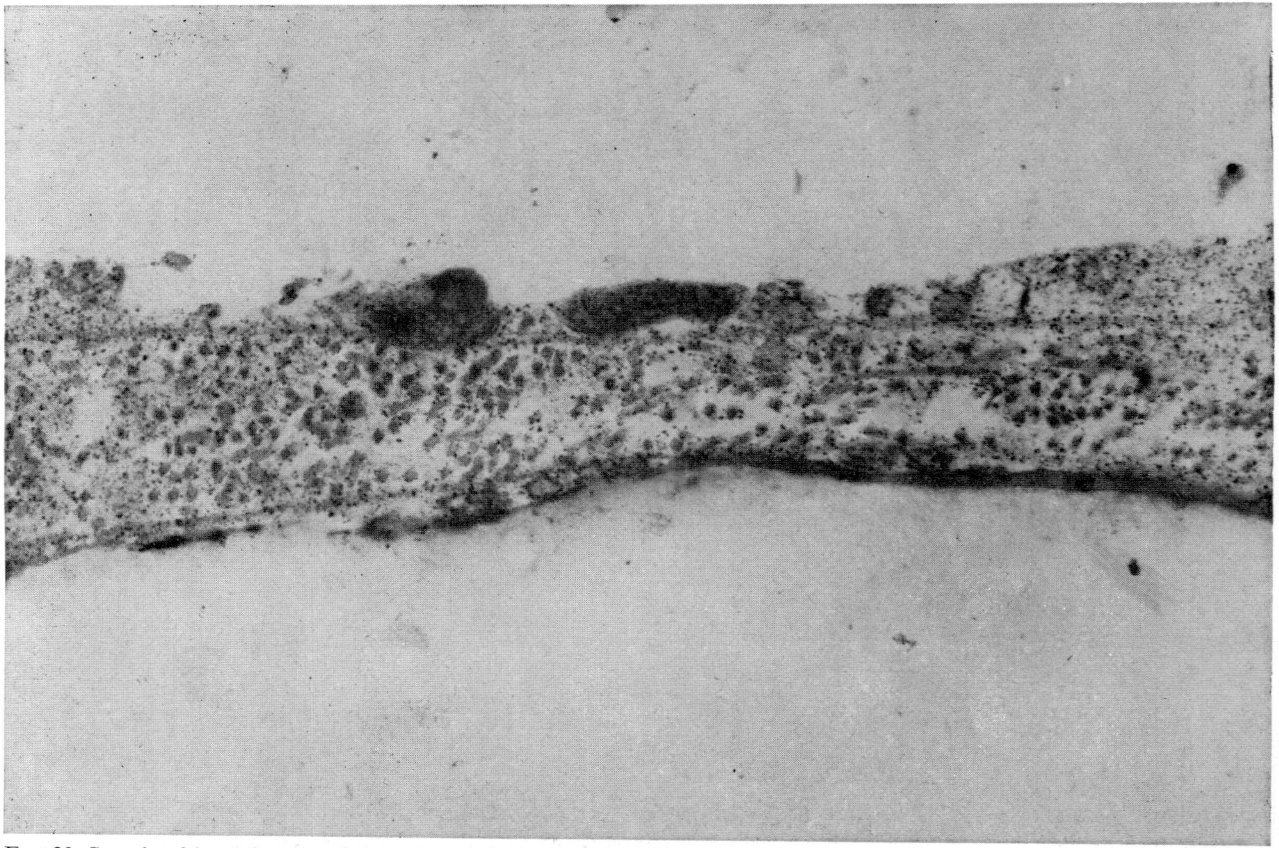

FIG. 23. Sample of head fat prepared by thin sectioning after fixation in osmium tetroxide for 1 h, ×29 000; stained with uranyl acetate and lead citrate, shows well organized interstitial connective tissue.

contained in cells may be retained by emulsification and in this case the conditions for stability are governed by the efficiency of the emulsifying protein. The protein must not only emulsify the fat effectively but must also 'set' to form a rigid network when the system is heated (Schut, 1976).

'Meat emulsions' have been fairly extensively examined microscopically and presumably because of this there is some conflict of views as to the importance of emulsification in the stability of meat products. This conflict partly arises over the difficulty in defining a meat emulsion which contains a large number of different structures. For example, the lipid part of a comminuted product may be present as groups of fat cells, individual cells, 'lakes' of fat, small fat droplets or even as a continuous 'phase' containing aqueous droplets. In a similar manner the myofibrillar material may exist in forms ranging from sizeable groups of muscle fibres to a solubilized protein network enmeshing fat droplets. When it is remembered that the structure of comminuted meat products may be complicated by additives ranging from purified proteins to spices it is easy to see how differences arise. Figure 24 shows some features of a meat emulsion.

Hansen (1960) was content to describe sausage-making in terms of producing and maintaining an emulsion, and many workers since then have considered the stability of comminuted meats in these terms (Swift *et al.*, 1961; Helmer and Saffle, 1963; Meyer *et al.*, 1964; Froning *et al.*, 1970; Sulzbacher, 1970). More recently, however, Van den Oord and Visser (1973) and Evans and Ranken (1975) have suggested that too much emphasis is placed on emulsification. Van den Oord and Visser consider that fat exists largely as fat cells and groups of cells in coarsely comminuted products, and as dispersions or suspensions of fat (mainly as crystals or groups of crystals) in a meat matrix. They criticized the fixation and staining procedures adopted by other workers, claiming that these interfered with the structures found in meat products. Van den Oord and Visser used frozen sections with a simple amaranth stain, determining the presence of fat by its birefringent crystals between crossed polars. These workers consider all fat commonly used in meat manufacture to be present as crystals and largely ignore the form of any associated liquid fat. However their general questioning of the overriding importance of emulsification appears justified and should assist in the understanding of fat stability in meat products.

Electron microscopic investigation of meat emulsions has only been reported to a limited extent (Ackerman *et al.*, 1971) and this has shown the existence of oil droplets which are below the resolution limit for the light microscope. So far these studies have been carried out on model systems and the total form of any emulsion in real meat products still awaits demonstration.

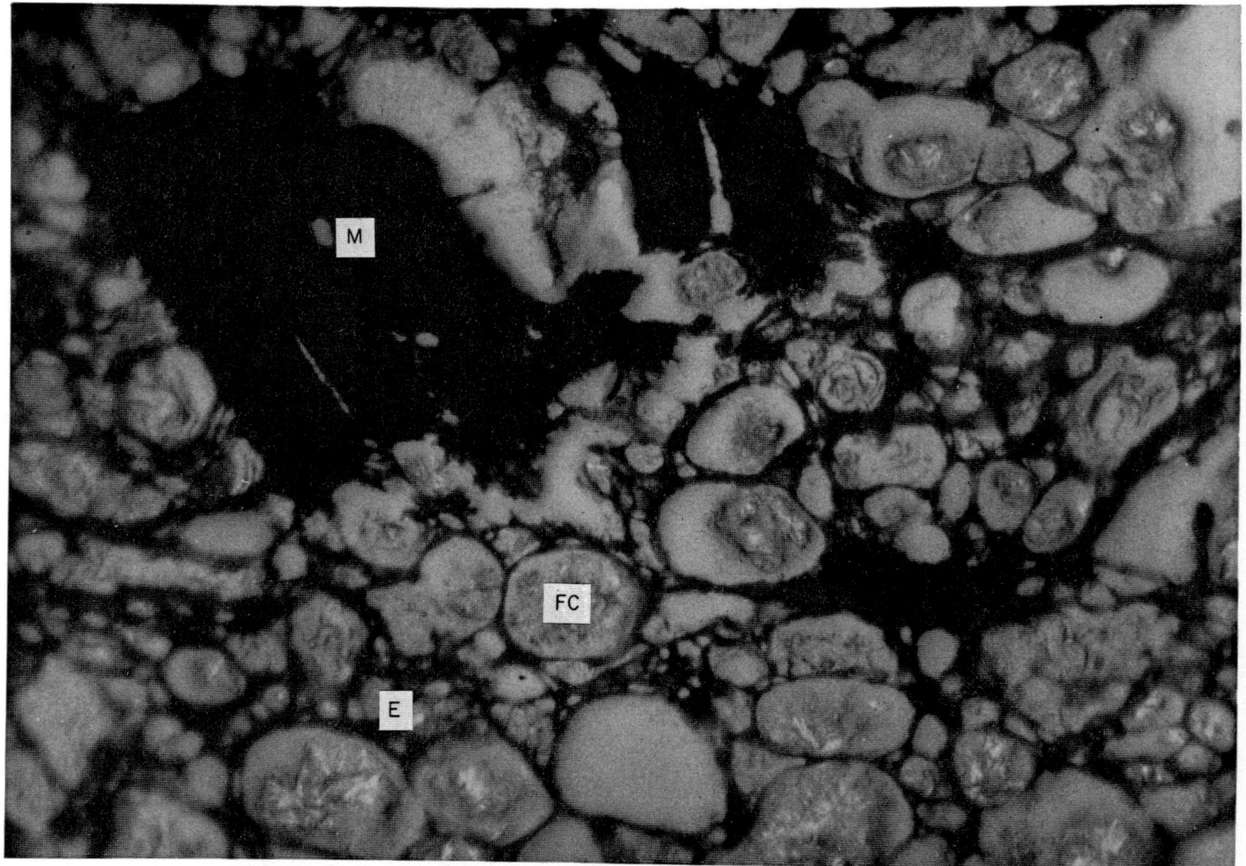

FIG. 24. Sample of finely comminuted 'meat emulsion' prepared for light microscopy, ×100; shows meat pieces (M), dark protein network, fat cells (FC) and emulsified fat (E).

V. CONCLUSION

Microscopes have been shown to contribute in an important way to the study of meat products. The use of the microscope may be analytical, assessing the composition of a product, or it may be as a research tool explaining technological observations and thereby advancing our understanding of these systems. In both roles there is room for further exploitation of the potential of light and electron optical techniques.

ACKNOWLEDGEMENT

All photographs in this chapter are reproduced by permission of the Leatherhead Food Research Association.

REFERENCES

Achatz, J. (1970). *Acta Biochim. Biophys. Acad. Sci. Hung.* **5**, 259.

Ackerman, S. A., Swift, C. E., Carroll, R. J. and Townsend, W. E. (1971). *J. Fd Sci.* **35** (6), 852.

Aitken, A., Casey, J. C., Penny, I. F. and Voyle, C. A. (1962). *J. Sci. Fd Agric.* **13**, 439.

A.O.A.C. (1975). 'Methods of Analysis', 12th Edn, p. 424.

Barnet, P. D. and Evans, G. G. (1973). Leatherhead Food R.A. Technical Circular No. 529.

Beecher, G. R., Cassens, R. G., Hoekstra, W. G. and Briskey, E. J. (1965). *J. Fd Sci.* **30**, 969.

Bendall, J. R. (1953). *Nature, Lond.* **172**, 586.

Bendall, J. R. (1954). *J. Sci. Fd Agric.* **5**, 468.

Callow, E. H. (1929). Ann. Rept Fd Invest. Bd London, p. 70.

Coomeraswamy, M. and Flint, F. O. (1973). *Analyst*, **98**, 542.

Evans, G. G. and Ranken, M. D. (1972). Leatherhead Food R.A. Research Report No. 188.

Evans, G. G. and Ranken, M. D. (1975). Leatherhead Food R.A. Research Report No. 219.

Fox, J. D., Moody, W. G., Kemp, J. D. and Henning, W. (1970). *J. Anim. Sci.* **31** (2), 323.

Freeman, J. A. and Spurlock, B. O. (1962). *J. Cell Biol.* **13**, 437.

Froning, C. W., Anderson, J. and Mebus, C. A. (1970). *Poult. Sci.* **49**, 497.

Gauthier, G. F. (1969). *Z. Zellforsch. Microsk. Anat.* **95**, 462.

Gauthier, G. F. (1970). *In* 'The Physiology and Biochemistry of Muscle as a Food' (E. J. Briskey, R. G. Cassens and B. B. Marsh, eds), Vol. 2, p. 103. University of Wisconsin Press, Madison.

Gerrard, F. (1976). 'Sausage and Small Goods Production', 6th Edn. Northwood Publications, London.
Giles, B. G. (1969). Proc. 15th European Meat Workers Conf., Helsinki, p. 289.
Glauert, A. M., Rogers, G. E. and Glauert, R. H. (1956). *Nature, Lond.* **178**, 803.
Greaser, M. L., Cassens, R. G., Hoekstra, W. G., Briskey, E. J., Schmidt, G. R., Cair, S. D. and Galloway, D. E. (1969). *J. Anim. Sci.* **28**, 589.
Halliday, D. A. (1971). *Process Biochem.* **6** (5), 15.
Hamm, R. (1960). *Adv. Fd Res.* **10**, 355.
Hansen, L. J. (1960). *Fd Technol., Champaign*, **14**, 565.
Hanson, J. and Huxley, H. E. (1953). *Nature, Lond.* **172**, 530.
Helmer, R. L. and Saffle, R. L. (1963). *Fd Technol., Champaign*, **17**, 115.
Herring, H. K., Cassens, R. G. and Briskey, E. J. (1965). *J. Sci. Fd Agric.* **16**, 379.
Hill, F. (1969). *In* 'Food Industries Handbook' (A. H. Woollen, ed.), 20th Edn, p. 280. Leonard Hill, London.
Huxley, H. E. (1963). *J. Molec. Biol.* **7** (3), 281.
Jewell, G. G. (1974). Leatherhead Food R.A. Technical Circular No. 571.
Jones, S. B., Carroll, R. J. and Cavanaugh, J. R. (1976). *J. Fd Sci.* **41**, 867.
Kotter, L. (1960). 'Zur Wirkung Kondensierkers Phosphate und Anderen Salze auf Tierisches Eiweiss.' Verlag M. and H. Schaper, Hanover.
Kushida, H. (1959). *J. Electron Microsc.* **8**, 72.
Lawrie, R. A. (1974). 'Meat Science.' Pergamon Press, Oxford.
Lewin, Y. A. and Flint, F. O. (1976). *J. Fd Technol.* **11**, 137.
Lewis, D. F. (1974). Ph.D. Thesis, University of Leeds.
Lewis, D. F. (1976). Leatherhead Food R.A. Research Report No. 238.
Lewis, D. F. and Jewell, G. G. (1975). Leatherhead Food R.A. Research Report No. 212.
Linke, H. (1969). *Fleischwirtschaft*, **49** (4), 469.
Locker, R. H. and Hagyard, C. J. (1963). *J. Sci. Fd Agric.* **14**, 787.
Mahon, J. H. (1961). Proc. 13th American Meat Industry Res. Conf.
Maruyama, K. and Ebashi, S. (1970). *In* 'The Physiology and Biochemistry of Muscle as a Food' (E. J. Briskey, R. G. Cassens and B. B. Marsh, eds), Vol. 2, p. 373. University of Wisconsin Press, Madison.
Menz, L. and Luyet, B. (1961). *Biodynamica*, **8**, 261.
Meyer, J. A., Brown, W. L., Giltner, N. E. and Guinn, J. R. (1964). *Fd Technol., Champaign*, **18**, 138.
Moor, H. and Mühlethaler, K. (1963). *J. Cell Biol.* **17**, 609.
Ogata, T. (1958). *Acta Med. Okoyama*, **12**, 216.
Paul, P. C. (1965). *J. Fd Sci.* **30**, 960.
Prandl, O. (1961). 'Die Histologische Analyse von Wurstwaren.' Gerhard Rottger Verlag, Munchen.
Ramsbottom, J. M., Strandine, E. J. and Koonz, C. H. (1945). *Fd Res.* **10**, 497.
Rayns, D. G. (1972). *J. Ultrastruct. Res.* **40**, 103.
Reed, R., Houston, T. W., Tood, P. M., Kennelly, J. S., Frazier, J. A., Flawith, S. M. and Boyes, I. D. (1969). *Proc. Leeds Phil. Lit. Soc. Sci. Sec.* **10** (2), 17.

Rolfe, E. J. (1958). *In* 'Fundamental Aspects of the Dehydration of Foodstuffs', p. 211. Society of Chemical Industry, London.
Sabatini, D. D., Bensch, K. and Barnett, R. J. (1963). *J. Cell Biol.* **17**, 19.
Schaller, D. R. and Powrie, W. D. (1971). *J. Fd Sci.* **36**, 552.
Schmidt, J. D. and Parrish, F. C., Jr (1971). *J. Fd Sci.* **36**, 110.
Schut, J. (1976). *In* 'Food Emulsions' (S. Friberg, ed.), p. 385. Marcel Dekker, New York.
Sheldon, H. (1969). Meat Industry Res. Conf. Proc., Chicago, p. 9.
Spurr, A. R. (1969). *J. Ultrastruct. Res.* **26**, 31.
Stanley, D. W. and Geissinger, H. D. (1972). *J. Inst. Can. Sci. Tech. Aliment.* **5** (4), 214.
Staubi, W. (1963). *J. Cell Biol* **33**, 319.
Steere, R. L. (1957). *J. Biophys. Biochem. Cytol.* **3**, 45.
Sulzbacher, W. L. (1970). *J. Sci. Fd Agric.* **24**, 589.
Swift, C. E., Lockett, G. and Fryar, A. J. (1961). *Fd Technol., Champaign*, **15**, 468.
Theno, D. M., Siegel, D. G. and Schmidt, G. R. (1978). *J. Fd Sci.* **43** (2), 483.
Treharn, T. (1971). *Fd Mf.* **46** (10), 35.
Van den Oord, A. H. A. and Visser, P. R. (1973). *Fleischwirtschaft*, **53**, 1427.
Voyle, C. A. (1969). *J. Fd Technol.* **4**, 275.
Voyle, C. A. and Lawrie, R. A. (1964). *Jl R. Microsc. Soc.* **82**, 173.
Voyle, C. A. and Restall, D. J. (1970–71). Meat Research Institute Annual Report.
Wang, H., Andrews, F., Rasch, E., Doty, D. M. and Kraybill, H. R. (1953). *Fd Res.* **18**, 351.
Wang, H., Auerbach, E., Bates, V., Andrews, F., Doty, D. M. and Kraybill, H. R. (1954a). *Fd Res.* **19**, 154.
Wang, H., Auerbach, E., Bates, V., Andrews, F., Doty, D. M. and Kraybill, H. R. (1954b). *Fd Res.* **19**, 543.
Whitworth, V. A. (1968). Ph.D. Thesis, University of Leeds.
Willison, J. H. M., Cocking, E. C. and Lawrie, R. A. (1972). *J. Microscopy*, **95** (3), 511.

Addendum

p. 250

Theno *et al.* (1978) have investigated the effect of massaging meat in the presence of salt and phosphate by light microscopy and scanning electron microscopy. They report that an exudate of extracted protein and cell fragments is formed which acts as a binder for meat pieces. The exudate sets on heating and thus makes it possible to slice the product easily.

7. Milk and Its Products

B. E. BROOKER

*National Institute for Research in Dairying,
Shinfield, Reading, U.K.*

I. INTRODUCTION

Interest in the microscopic structure of milk and milk products has been stimulated by economic problems arising from their production, storage and marketing, and from the realization that such information is central to a proper understanding of their physical and chemical properties. It is customary to think of the cow as the source of the raw material, but the milk of other domesticated animals such as the goat, the buffalo and the ewe is used for human consumption in various parts of the world. Therefore, the following account of the light and electron microscopy of milk and some of its products, which of necessity must draw largely from work using cow's milk as the starting material, will not be restricted by reference only to this species. There are distinct differences in composition between the milks of various species (Ling *et al.*, 1961; Jenness, 1974). Quantitatively, some parameters of milk structure are notoriously variable both between and within species. Intraspecifically, diet, season, stage of lactation and breed are among the factors contributing to such variation. However, experience has shown that qualitative differences in their microscopic structure are not marked.

II. MILK

A. Composition and Secretion

Before considering the structure of milk, it is necessary to provide some basic information concerning its composition and secretion. Comprehensive accounts of these two aspects are given by Jenness (1974) and Saacke and Heald (1974) respectively.

Milk is a heterogeneous liquid which, at the time of its secretion and storage in the mammary gland at body temperature, contains two liquid phases, namely fat and water. Milk fat is composed predominantly of triglycerides of saturated and unsaturated fatty acids. Although the fat occurs in the form of numerous small fat globules, milk is not a simple oil-in-water emulsion since the boundary of each fat globule is marked by a membrane (the milk fat globule membrane or MFGM) of complex composition (Anderson and Cawston, 1975) and ultrastructure. About 90% of the aqueous phase is composed of water in which are found, either in true or colloidal solution, milk serum proteins (lactalbumin, lactoglobulin, blood serum albumin and immunoglobulins), lactose, mineral salts, vitamins and trace elements. The principal protein of milk (about 80% of the total) is casein which exists as three major molecular species denoted as α, β and κ. Most of the casein in milk occurs in association with other, lower molecular weight species, such as calcium, phosphate, citrate and magnesium, as minute particles or micelles dispersed throughout the aqueous phase.

Milk is secreted by epithelial cells which line bulbous alveoli in the mammary gland. Since fat globules and casein micelles are the only particulate components of milk synthesized by the mammary gland, they are the only elements whose secretion can be followed in detail by light and electron microscopy. The rough endoplasmic reticulum in which casein is synthesized is a prominent feature of these secretory cells during lactation and is largely responsible for the pronounced basophilia of their cytoplasm. The protein is transported to the Golgi apparatus in the cisternae of which it first becomes visible by electron microscopy as dispersed, 15–20 nm diameter particles (Fig. 1). These particles have been referred to by Buchheim and Welsch

(1973) as submicelles, since they apparently give rise to much larger granules, the casein micelles, by a process of aggregation in the Golgi vesicles (Fig. 1). Movement of the Golgi vesicles to the apical cytoplasm and the fusion of their limiting membrane with the plasmalemma of the cell results in the release of casein micelles into the lumen of the alveolus.

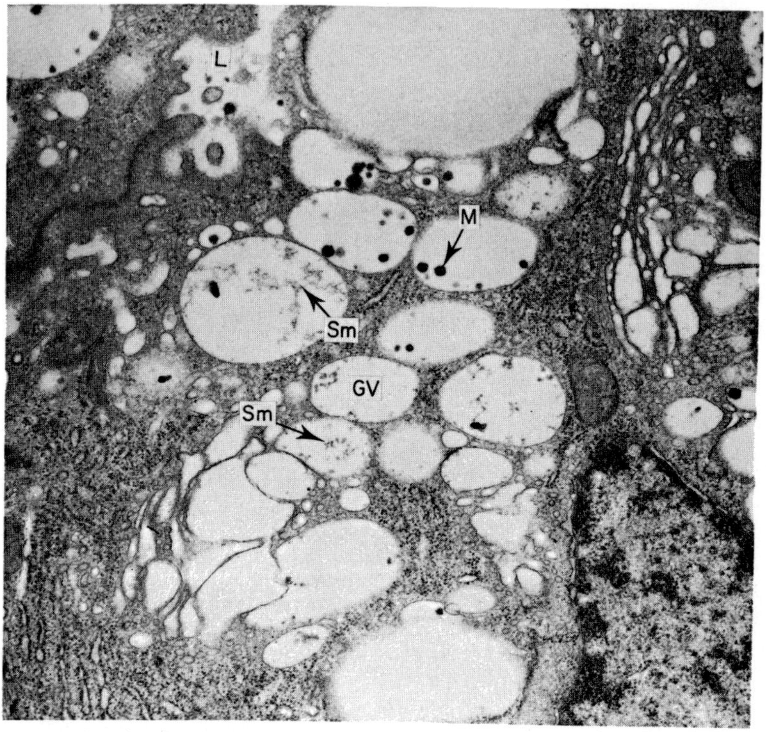

FIG. 1. Lactating bovine mammary gland, ×17 000. Golgi vesicles (GV) in the apical cytoplasm contain micellar (M) and submicellar casein (Sm). L = alveolar lumen.

The mechanism of secretion of milk fat globules is quite different and does not directly involve the Golgi apparatus. Fat globules usually first appear as small lipid droplets in the basal cytoplasm of the mammary gland. These droplets are not bounded by a unit membrane but a single electron-dense layer is visible at their interface with the cytoplasmic matrix. The fat globules grow in size, move through the apical cytoplasm and closely approach, but do not make direct contact with, the plasmalemma. At this and subsequent stages of its secretion, the surface of the fat globule is separated from the plasmalemma by a thin (10–20 nm wide) layer of cytoplasm. As the fat globule emerges from the apex of the cell it becomes totally enveloped by the plasmalemma (Fig. 2). The free fat globule immediately after secretion is therefore surrounded by an outer, unit membrane and an inner, interfacial layer adjacent to the lipid (Fig. 3). Between the two there is an intermediate layer derived from the cytoplasm (Fig. 4). The outer two layers have been referred to collectively as the initial MFGM, and the inner, interfacial layer, as the

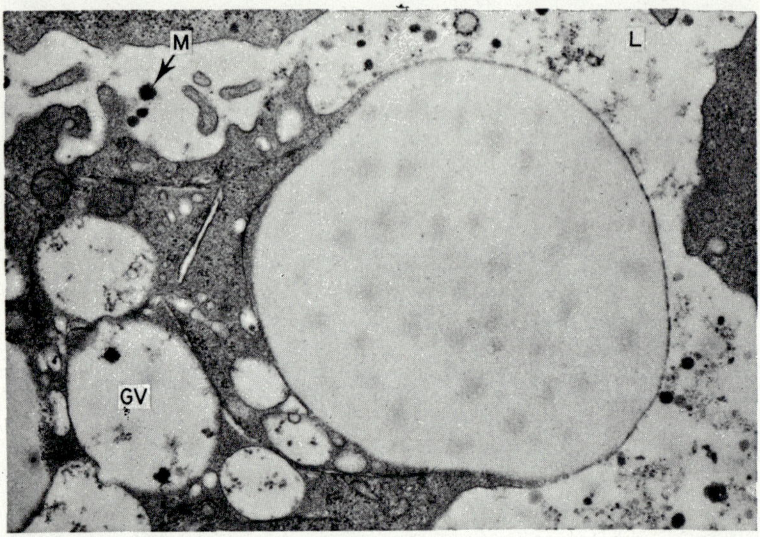

FIG. 2. Lactating bovine mammary gland, ×17 000. Emergence of a milk fat globule. Labels as Fig. 1.

secondary MFGM (Wooding, 1971a) (Fig. 5). This nomenclature has been widely adopted and will be followed here.

An alternative sequence of events involved in fat secretion has been proposed by Wooding (1971b). This author observed that fat globules in the apical cytoplasm were surrounded by vesicles which were believed to be of Golgi origin since some contained casein micelles. It was suggested that the progressive fusion of these vesicles with the plasmalemma would result in the gradual emergence of the globule from the cell. In this mechanism, the cytoplasmic layer of the initial MFGM arises because, in the pre-secretory phase, the Golgi vesicles remain separated from the interfacial layer of the globule by a distance of 10–20 nm. Although this theory is attractive, it has not been widely accepted as the sole mechanism for fat secretion in the mammary gland since, as noted by Saacke and Heald (1974), the necessary lipid droplet–Golgi vesicle association is not always observed during secretion.

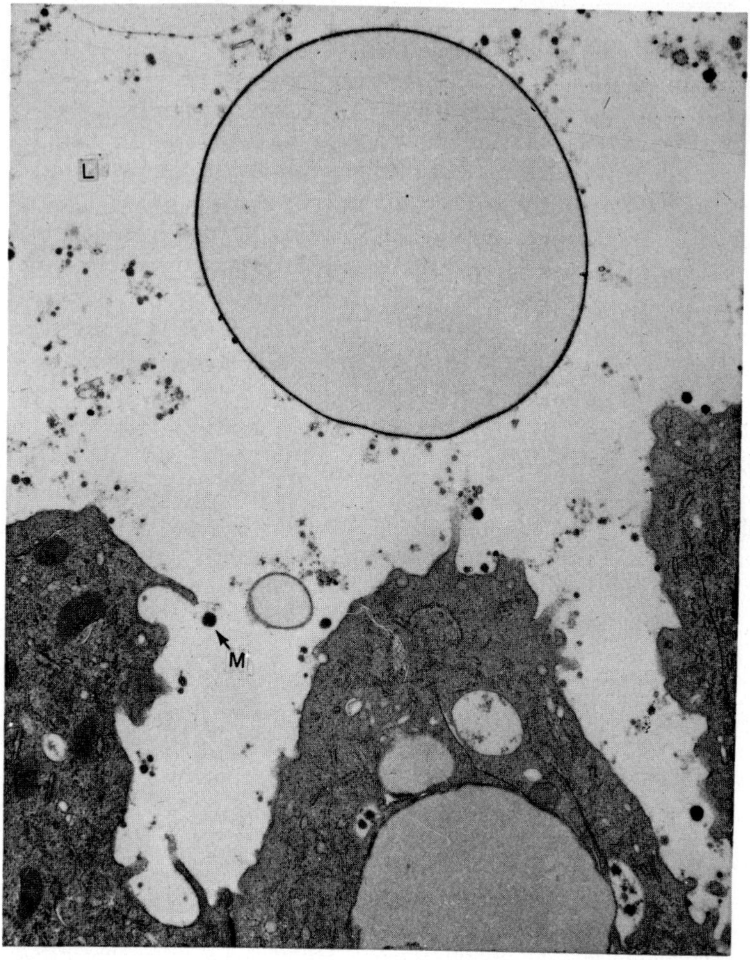

FIG. 3. Bovine mammary gland, ×12 500. A fat globule soon after its secretion; the uniformly electron-dense MFGM entirely surrounds the globule. Labels as Fig. 1.

B. Light Microscopy

When a wet preparation of milk is examined by phase contrast light microscopy, the most numerous particulate elements visible are casein micelles. They appear as small, phase-dense granules. All show some degree of Brownian movement and, although some variation in their diameter can be detected, the full particle size range can only be appreciated by the use of electron microscopy (see below). The fat globules vary considerably in diameter from those (<0·2 μm) which are beyond the resolution of the light microscope to about 30 μm (Figs 6 and 7). Although the mean diameter for the globules in milk is usually given as 3·0 μm for the cow and 2·0 μm for the goat, there is considerable variation in size distribution between breeds and stages of lactation, and even during the milking process itself (Kernohan and Lepherd, 1969). It is interesting to note that whereas globules in the size range 2–6 μm contain over 90% of the fat, those <1 μm represent about 80% of the total number of globules but contain only 1–2% of the fat (Walstra, 1969). Although the fat phase of milk is liquid at body temperature, there is a change in its physical state when expressed milk is cooled. This process is complex and involves the gradual crystallization of the milk fat rather than its simple solidification. Its rate is affected by such factors as temperature, fat composition and globule size. After a number of hours at room temperature, globules contain a mixture of crystalline and liquid fat. When examined in polarized light therefore, some of the fat globules contain birefringent inclusions which represent regions of crystalline fat. Three types of milk fat globules, differing in their patterns of birefringence, were distinguished by Walstra (1967). In one of these, elongate crystals of variable size were distributed throughout the globule, and in another birefringence was limited to an outer, but not necessarily continuous, layer. The third type was intermediate

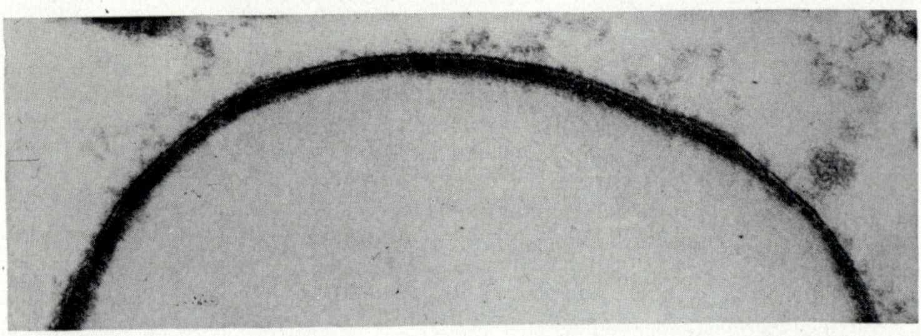

FIG. 4. Profile of the intact MFGM soon after secretion, ×100 000. The initial MFGM is clearly visible.

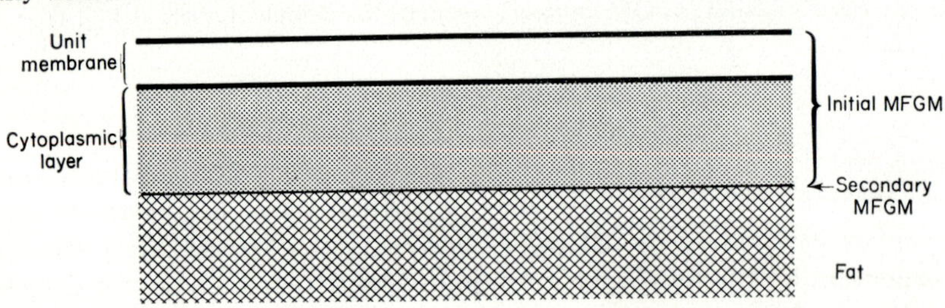

FIG. 5. Diagrammatic profile of the milk fat globule membrane (MFGM) immediately after secretion.

in appearance between these two. Not surprisingly, cooling milk to lower temperatures leads to a higher proportion of birefringent globules of all three types.

Careful examination of a wet preparation of milk shows that a small number of fat globules ($<1\%$) have a crescent of phase-dense material attached to their surface (Fig. 11). These globules are referred to as 'signets' and are formed when a small portion of alveolar cell cytoplasm remains attached to the globule after secretion. They will be discussed in more detail below.

Somatic cells and fragments of cellular debris are always present in milk. The term 'somatic cells' refers to epithelial cells derived from the alveoli and ducts of the gland itself and to white blood cells, principally polymorphonuclear leucocytes (polymorphs), lymphocytes and macrophages which are migrants from blood vessels and the connective tissue. Epithelial cells of different varieties and occasionally large squamous cells form the major cellular structures in milk throughout lactation (Zlotnik, 1947). However, some of the vacuolated mononuclear cells in bovine milk previously identified as epithelial cells have recently been shown to be lipid-filled macrophages (Jensen and Eberhart, 1975). In milk from a normal mammary gland, the total number of cells is very variable but may range from 20 000–500 000 cells ml^{-1}. Inflammation of the mammary gland (mastitis) induced by microbial infection or by trauma is accompanied by a large rise in the number of polymorphs; this is one of the criteria for the diagnosis of mastitis. The light microscopy of cells found in milk from the normal and mastitic bovine mammary gland is extensively dealt with by Schalm *et al.* (1971). Cell fragments greatly outnumber the somatic cells in normal milk (Paape and Tucker, 1966; Wooding *et al.*, 1970). As will be seen below, available evidence suggests that the fragments are mainly derived from intact secretory cells of the glandular epithelium. In the goat, unlike the cow, these fragments frequently contain one or more fat droplets (inset, Fig. 14). In all species, however, lipid droplets are often found in the cytoplasm of the phagocytic cells of milk. Most of the somatic cells and cell fragments in milk can be conveniently examined in the pellet that is obtained when milk is gently centrifuged, e.g. at 1000 **g** for 10 minutes.

C. Electron Microscopy

It has already been mentioned above that casein micelles in the Golgi vesicles of secretory cells appear to arise by the aggregation of 15–20 nm diameter subunits or submicelles (Buchheim and Welsch, 1973). In favourable thin sections of secreted micelles, what are regarded as individual subunits can be distinguished with some ease (Knoop *et al.*, 1973). The electron-dense lines which delineate the subunits and produce the characteristic mosaic-like appearance of the micelles (Fig. 8) are believed to represent the sites of calcium in calcium–phosphate stabilizing bridges between subunits (Knoop *et al.*, 1973). Freeze-fractured and -etched casein micelles are quite different in appearance and are apparently composed of a number of loosely packed 10 nm diameter subunits. To account for this difference in appearance between sectioned and fractured micelles, Buchheim and Welsch (1973) have suggested that each subunit consists of a nucleus of low water content, 10 nm in diameter, surrounded by a highly hydrated protein shell. In sectioned material it is suggested, both nucleus and shell would react with the fixative and stains to give subunits 15–20 nm in diameter but the sublimation of water which follows freeze-fracturing would leave only the nuclear (10 nm diameter) region in relief. The size range of casein particles in milk extends from 10 nm (subunits) to about 300 nm (large micelles). Differences between authors in the size range and distribution given for the casein micelles in cow's milk is, to some extent, attributable to differences in the techniques used to prepare milk for examination by electron microscopy. Most of these studies have used

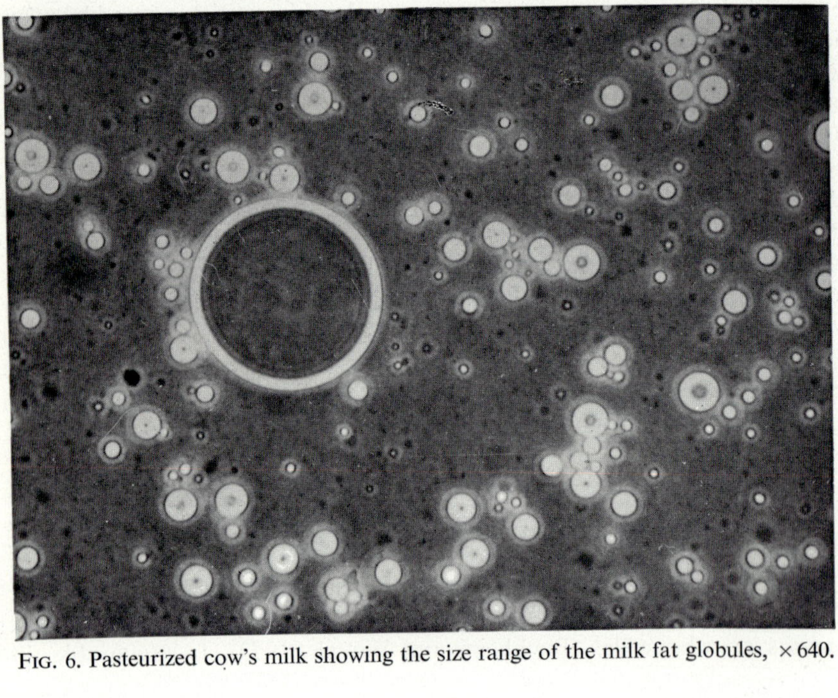

FIG. 6. Pasteurized cow's milk showing the size range of the milk fat globules, ×640.

metal-shadowed or negatively stained preparations (Nitschmann, 1949; Hostettler and Imhoff, 1952; Carroll *et al.*, 1968), but Knoop and Wortmann (1960) derived correction formulae which enabled them to calculate micelle diameters from thin sections of embedded milk. By these methods, which do not easily permit the detection of submicelles and must, by their very nature, produce some distortion and/or shrinkage of the micelles, the most numerous casein particles in cow's milk are those with a diameter of about 100 nm. Quite different and probably more reliable results have been obtained by Schmidt *et al.* (1973) using freeze-etched preparations. They concluded that submicelles greatly outnumber the larger micelles (although they comprise less than 3% of the total micelle volume) and that there is a gradually decreasing number of particles with increasing diameter.

Although immediately after secretion each fat globule is surrounded by an initial MFGM, post-secretory changes occur in the appearance of this membrane as the globule moves away from the cell from which it originated. According to Wooding (1971a), the cytoplasmic layer beneath the outer, unit membrane increases both in electron density and in thickness and this is followed by loss of a variable amount of the initial MFGM. In this process, which exposes areas of the underlying secondary MFGM, some of the initial

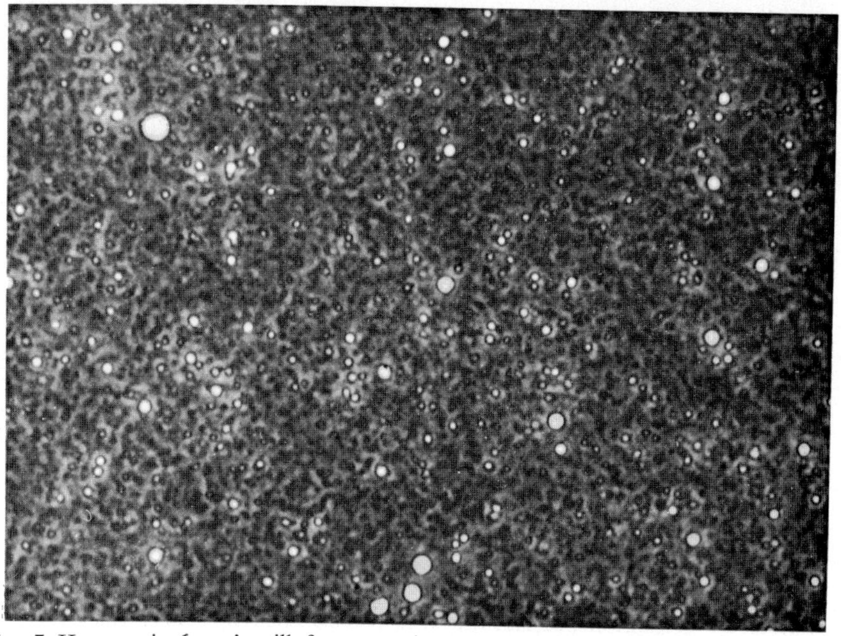

FIG. 7. Homogenized cow's milk for comparison of the fat globule size range with that in Fig. 6, ×640.

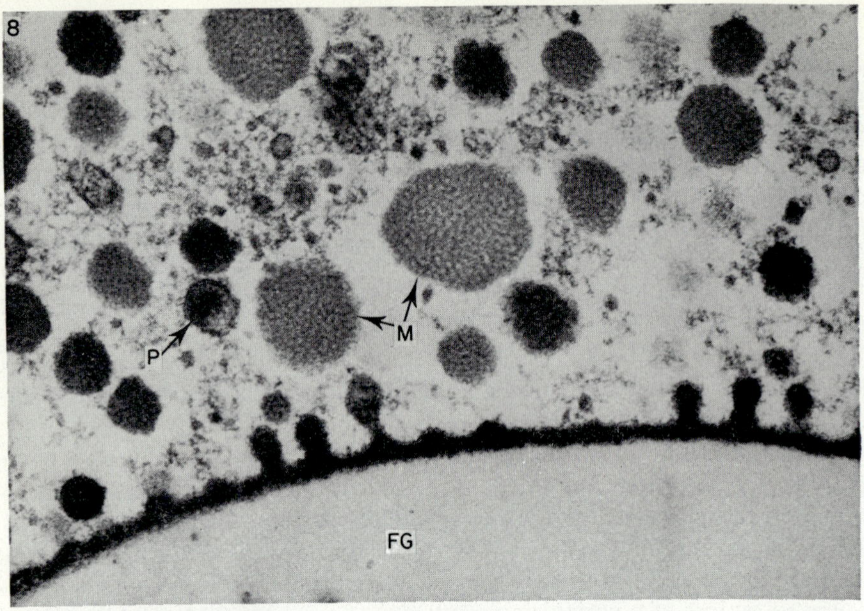

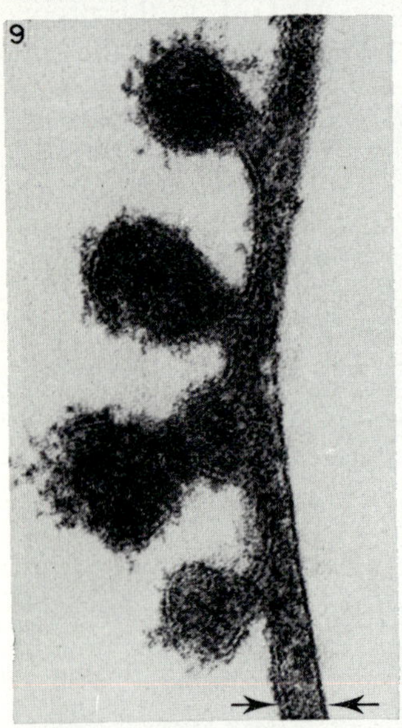

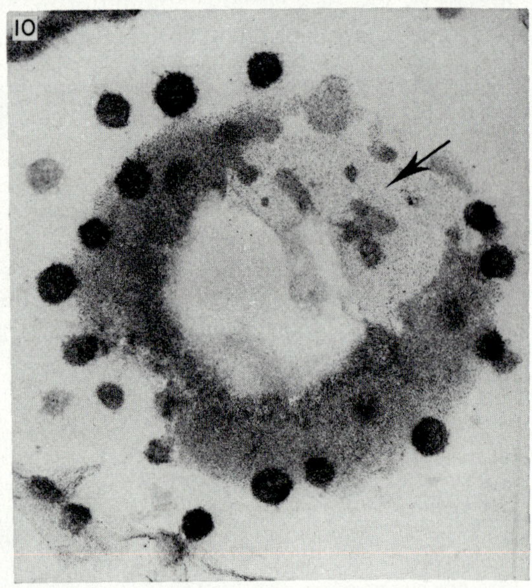

FIG. 8. Lactating mouse mammary gland, ×65 000. The typical mosaic-like appearance of sectioned casein micelles (M) can be clearly seen and there is post-secretory 'blebbing' of the MFGM. A free, unit membrane-bound plasmalemmasome (P) is also present. FG = fat globule.

FIG. 9. Blebbing of the MFGM in expressed porcine milk, ×112 000. The initial and secondary MFGM are both visible between the arrows. (Cf. Fig. 5.)

FIG. 10. The surface of a milk fat globule from porcine milk seen in tangential section, ×39 000. The many circular 'blebs' are probably still attached to the underlying MFGM. The pale area (arrowed) represents secondary MFGM denuded of the overlying initial MFGM.

MFGM is shed in sheets but most is lost by 'blebbing' in which small vesicles or microvilli formed by evagination of the unit membrane are pinched off by a constriction at their base (Fig. 9). When this process is complete, the fat globules still possess patches of initial MFGM, but the thin secondary MFGM remains as the only continuous boundary of the fat globule. This is the usual appearance of fat globules in expressed milk although evidence of blebbing still in progress is common (Fig. 10). The amount of initial MFGM retained by the globule appears to be species-variable, more being retained in the case of the goat for example (Fig. 19) than in that of the cow (Wooding, 1971a).

The free vesicles or plasmalemmasomes (Wooding, 1972) originating from

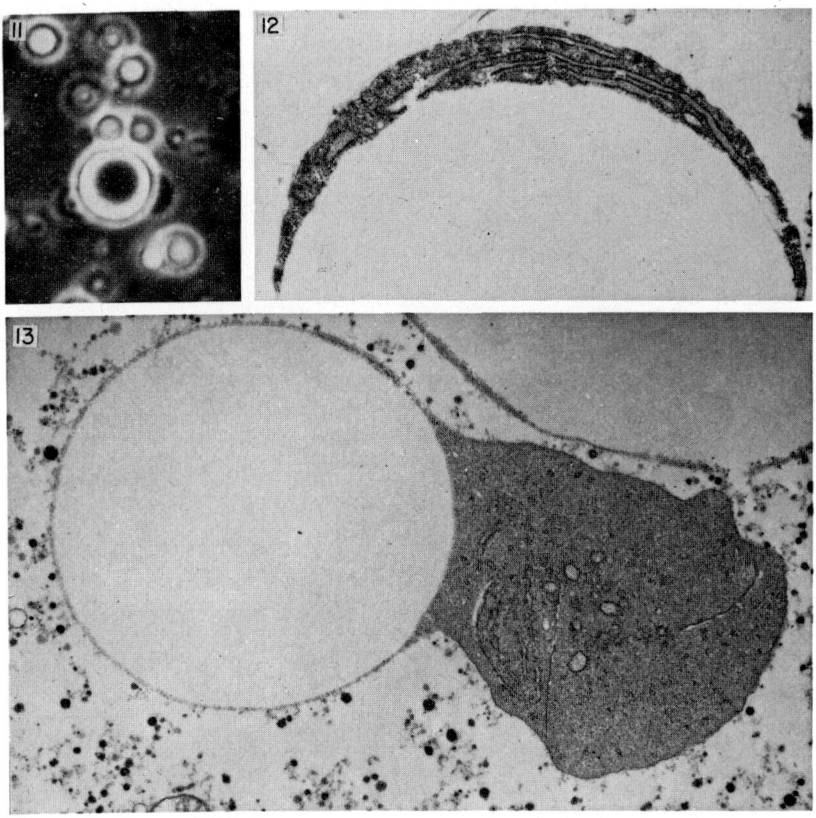

FIG. 11. A signet in goat's milk, ×1400. Phase contrast.
FIG. 12. Part of a signet from goat's milk, ×12 500. The crescent of cytoplasm contains prominent cisternae of rough endoplasmic reticulum.
FIG. 13. A signet from cow's milk, ×12 500.

the initial MFGM (Figs 8 and 10) are clearly normal components of milk and are probably equivalent to the particles which were isolated by Morton (1954) and called by him 'milk microsomes'. Wooding (1972) has drawn attention to the similarity of plasmalemmasomes and some virus particles reported from milk, and has emphasized the danger of attempting to identify the presence of such viruses on purely morphological grounds. Other membrane fragments, whose origin cannot be identified with such certainty, are always present in milk and can be recovered after high-speed centrifugation as a thin layer adjacent to the resulting casein pellet (Anderson and Cawston, 1975). Leucocytes, mammary gland epithelial cells and the MFGM all seem likely sources, but the quantitative contribution made by each in normal milk has not been determined.

The crescent of material associated with a small percentage of the fat globules in milk has been shown by Wooding *et al.* (1970) to be a portion of mammary gland cytoplasm. They are more numerous in goat's milk than in cow's milk. During the extrusion of the fat globule from the secretory cell, there is a transient stage during which it remains attached to the cell only by

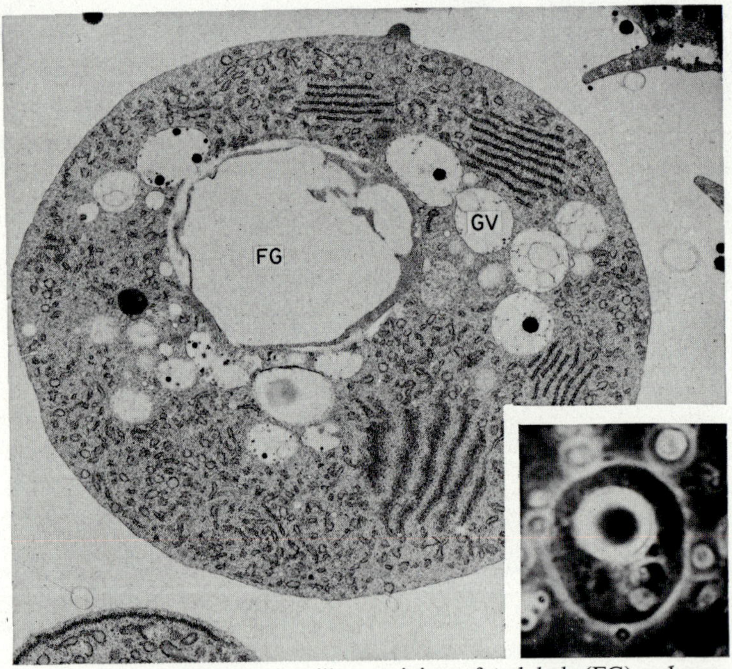

FIG. 14. A cell fragment from goat's milk containing a fat globule (FG) and several Golgi vesicles (GV), ×7800. Note the extensive rough endoplasmic reticulum. Inset shows a similar fragment seen by phase contrast light microscopy, ×1100.

a narrow neck of cytoplasm. Normally, the fat globule is released from this cytoplasmic neck, but when both leave the cell together a signet is the result. The cytoplasmic crescents usually contain rough endoplasmic reticulum, but mitochondria and vesicles believed to be of Golgi origin are also encountered (Figs 11, 12 and 13). Wooding *et al.* have estimated that 1–5% of fat globules in goat's milk are signets. This figure is higher than that obtained by light microscopy because some crescents are so thin that they can only be resolved by electron microscopy.

Particles that are essentially similar in appearance to signets, but in which the ratio of cytoplasm to lipid is much higher, have recently been described from goat's milk by Christie and Wooding (1975). These are of considerable interest since, like the mammary gland cells from which they are believed to arise, they are able to synthesize triglyceride from fatty acid precursors. Their origin from alveolar secretory cells is suggested not only by their extensive rough endoplasmic reticulum but also by the occasional presence of Golgi vesicles containing casein micelles. Christie and Wooding suggest that they represent portions of apical cytoplasm which have been released by the sequential fusion of adjacent Golgi vesicles in essentially the same way as has

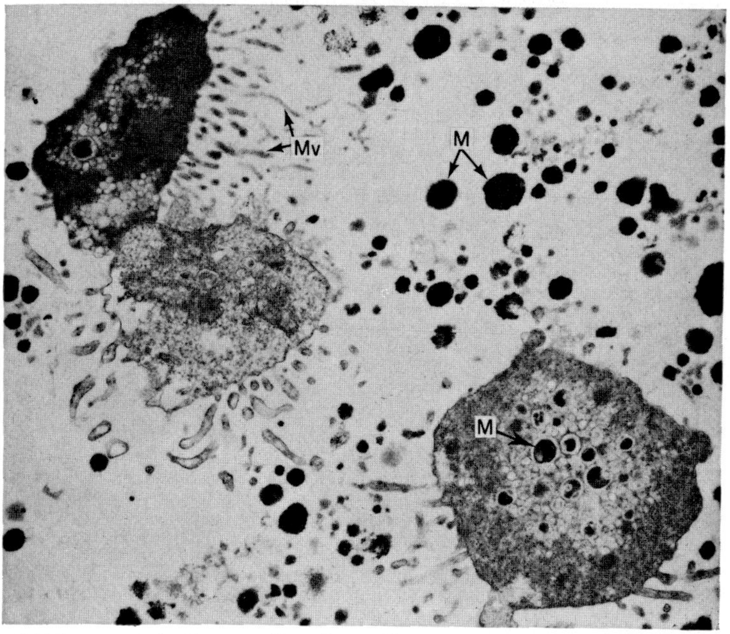

FIG. 15. Cell fragments in the centrifugation pellet of cow's milk, ×12 500. They possess microvilli (Mv) and contain micellar casein (M).

been suggested by Wooding (1971b) for milk fat globule secretion. Although these cell fragments are large (up to 5 μm in diameter), they do not sediment when milk is centrifuged at high speed (40 000 **g**) for short periods (less than 15 minutes) because of the buoyancy conferred on them by their lipid inclusions. In the goat, cell fragments that do sediment easily differ from the particles described by Christie and Wooding (1975) in containing less free lipid (Wooding *et al.*, 1970). Apart from this difference in free lipid content

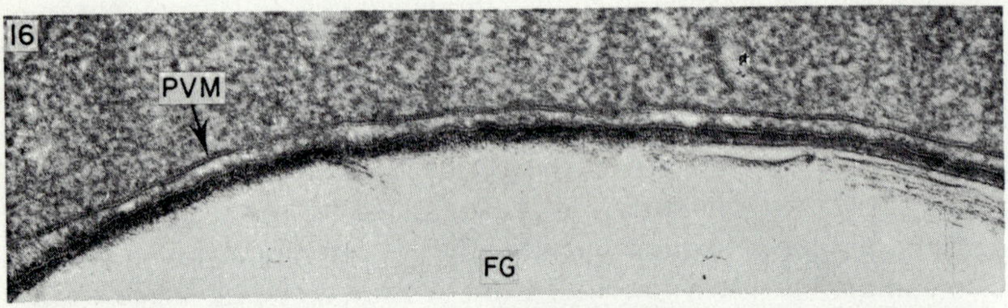

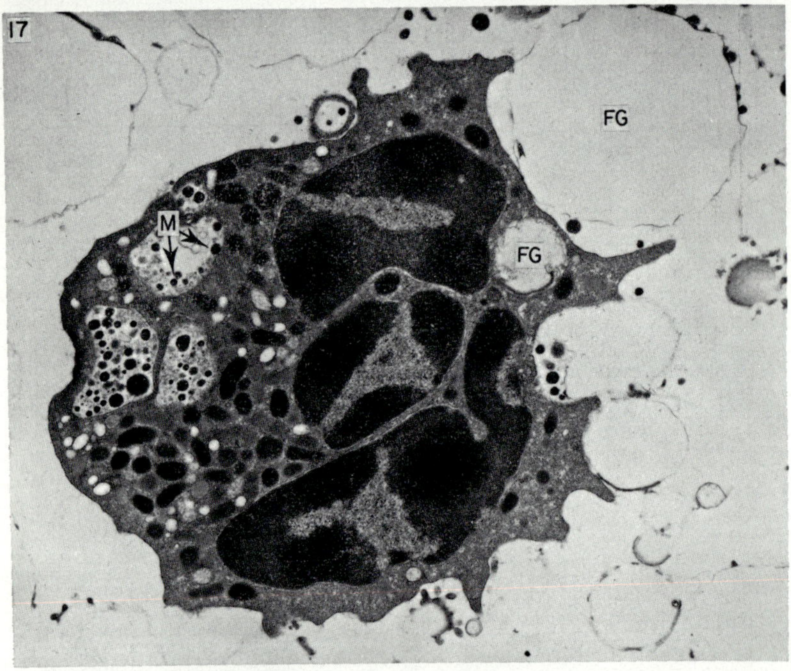

FIG. 16. Section through the phagocytic vacuole of a bovine polymorph containing an ingested milk fat globule, ×81 000. The initial MFGM of the globule is clearly visible. PVM = phagocytic vacuole membrane, FG = fat globule.
FIG. 17. A polymorph in the cream layer of porcine milk, ×9200. Ingested casein micelles and a fat globule are located in separate phagocytic vacuoles.

which allows their separation into two populations by centrifugation, they are similar both in their ultrastructure and probable origin from the secretory cells of the mammary gland (Fig. 14). Cytoplasmic fragments, probably of secretory cell origin, are also found in the low-speed centrifugation pellet derived from cow's milk, but these are quite different in appearance from the corresponding particles from goat's milk since they do not contain large amounts of rough endoplasmic reticulum but possess abundant microvilli (Fig. 15).

One of the features of the phagocytic cells in milk is that they often contain lipid droplets. With few exceptions, close examination shows that these droplets do not represent endogenous fat but rather milk fat globules lying within phagocytic vacuoles. Thus, the lipid is separated from the cytoplasmic matrix of the cell not only by the limiting membrane of the food vacuole but also by remnants of the MFGM (Fig. 16). Casein micelles are also actively ingested, although it is interesting to note that fat globules and micelles are usually taken into separate vacuoles (Fig. 17). The reported decrease in the phagocytic or bactericidal capability of polymorphs isolated from milk compared with that of polymorphs from blood (Wisniowski *et al.*, 1965; Kent and Newbould, 1969; Russell *et al.*, 1976) has been attributed to their prior ingestion of fat globules (Paape *et al.*, 1975) or casein micelles (Russell *et al.*, 1976). The ingestion of these particles has therefore been claimed to be a factor contributing to the establishment of intramammary infection by invading pathogens.

III. CREAM

Cream is the separated fat phase of milk. Its production therefore involves concentration of the fat globules. Because milk fat has a lower density than the aqueous phase (milk plasma), fat globules rise under the influence of gravity. In fresh milk, this 'natural creaming' is accelerated by cooling since the globules are then agglutinated (by a process involving cryoglobulins in the milk plasma) and rise to the surface much more rapidly than they would as individual globules (Fig. 18). Since the effect of the differential density of the fat and aqueous phases of milk is dependent on the strength of the gravitational field, creaming is greatly accelerated by centrifugation. This is the principle used in the milk separator for the commercial preparation of cream, and in the laboratory centrifugation can be used to produce a semisolid cream layer which can be easily handled for examination by light and electron microscopy.

The fat globules, of which cream is principally composed (Fig. 19), are

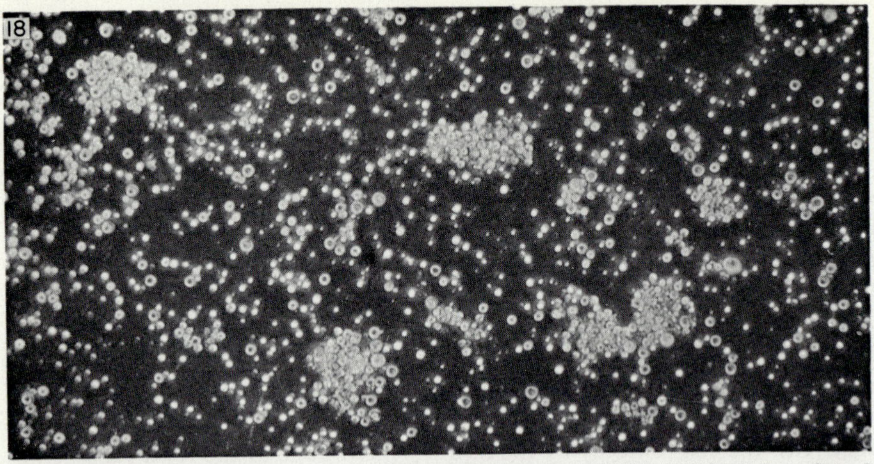

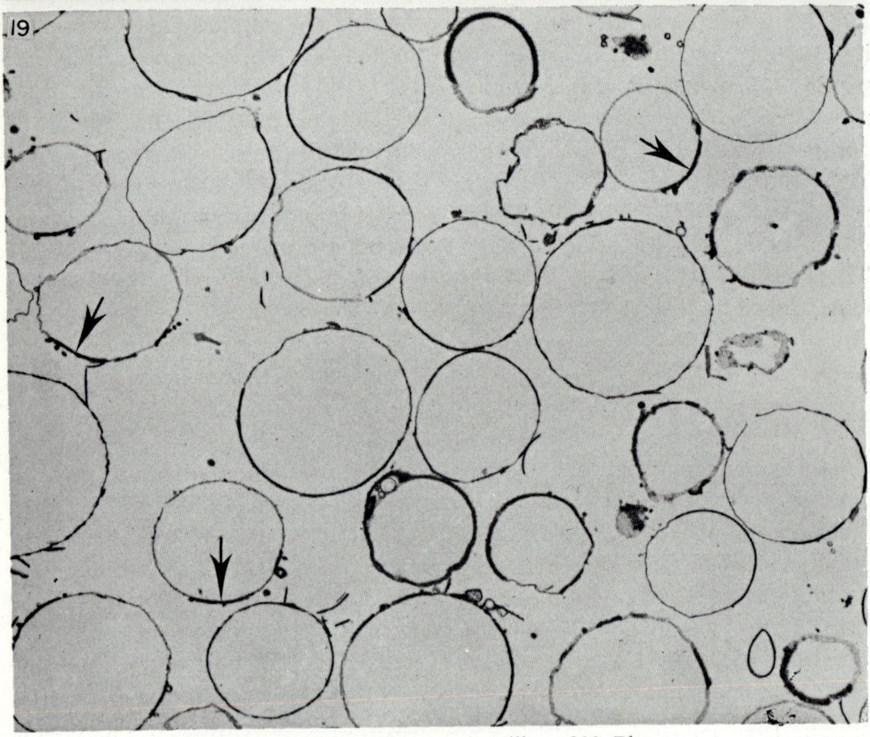

FIG. 18. Fat globule clusters in cold (10°C) cow's milk, ×200. Phase contrast

FIG. 19. Fat globules in the cream layer of goat's milk, ×5600. The surface of each globule appears more electron dense where initial MFGM is retained (arrowed).

identical to those more widely dispersed in whole milk. Some casein micelles become trapped between the globules during their ascent but they are a relatively minor component. White blood cells which have ingested so many fat globules that they become buoyant are sometimes found in cream but they are more common when it has been derived from abnormal milk with an elevated cell count.

IV. SKIM MILK

Strictly, this term is applied to milk from which most of the fat globules have been skimmed after they have risen naturally to the surface. Microscopically, skim milk still contains a number of small fat globules, but in every other respect it is identical to whole milk. The product obtained when cream has been removed from milk after centrifugal separation is called separated milk, although it too is often referred to as skim milk. This can be misleading because separated milk is deficient in the cells, cell fragments and other sedimentable particles (separator slime) retained in skim milk. Skim milk and separated milk therefore differ microscopically in the numbers of cells and particles of debris that they contain.

V. HOMOGENIZED MILK

Milk is homogenized to prevent separation of the fat and aqueous phases, i.e. creaming, and to improve the texture and body of some of the products (ice cream, soft cheese) derived from it. Essentially, the process involves forcing milk at high pressure through a fine slit (homogenization valve) to break the fat globules into much smaller ones. The resulting globules are not of uniform diameter but their size variation is much less than in untreated milk (Fig. 7). The size distribution of the fat globules after homogenization is dependent on the processing conditions, but most globules are usually less than 2 μm in diameter. The immediate result of homogenization is a large increase in the fat globule surface area which, naturally, cannot be covered by the existing MFGM. Instead, the newly formed surface is stabilized by the adsorption of surface-active proteins of which casein is probably the most important. Thus, electron micrographs of freeze-etched specimens show the fat globules to be partly covered by small particles which have been tentatively identified as casein micelle subunits (Eggmann, 1969; Buchheim and Knoop, 1970). In thin sections of homogenized milk, these individual subunits cannot be resolved and most of the fat globule surface is bounded

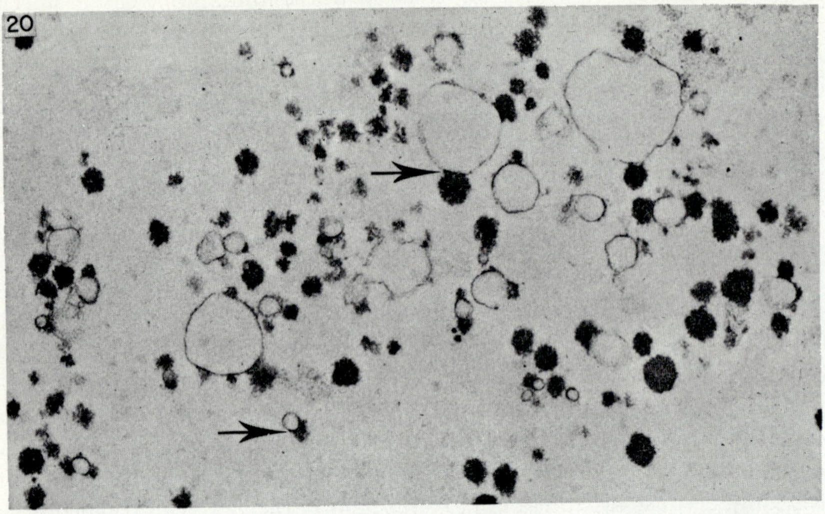

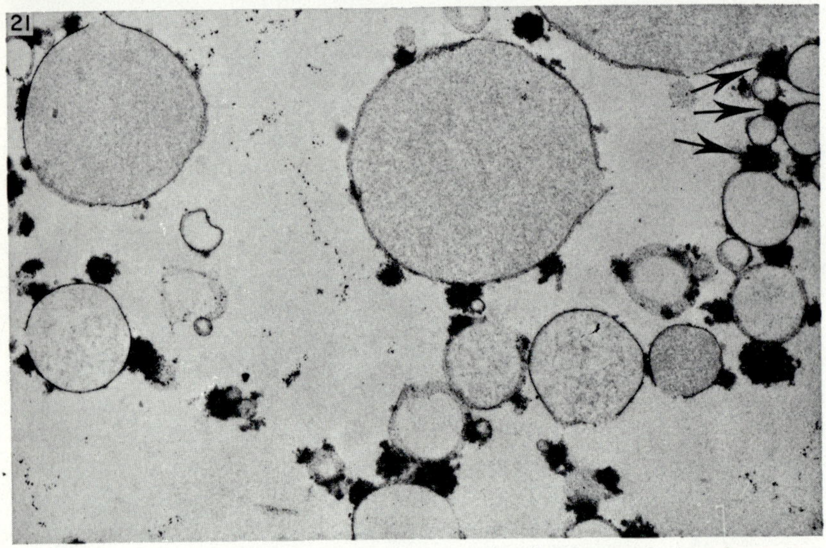

FIG. 20. Homogenized cow's milk, ×25 000. Casein micelles are attached to the surface of many fat globules (arrowed). The number of small globules is much greater than that in untreated milk.

FIG. 21. Homogenized cream, ×27 500. In this homogenization cluster, many fat globules are joined together by the casein micelles shared by their surfaces. This is very clear on the right of the micrograph (arrowed).

by a uniform membrane about 4 nm thick which appears as a single dense line (Figs 20 and 21). Only small patches of initial MFGM are encountered. Some of the adsorbed casein is in the form of micelles (Fig. 20), many of which appear to be partially spread over the surface of the globule (Henstra and Schmidt, 1970; Schmidt *et al.*, 1971). There are large differences between globules in the amount of micellar casein bound to their surface; in some cases none is visible, whilst some small globules appear totally enveloped by a single micelle. The observed ability of a single micelle to be shared by the surface of more than one fat globule must contribute to the formation of the stable fat globule clusters sometimes found in high-fat products such as homogenized cream (Fig. 21). These, unlike the globule aggregates which form during natural creaming in cold milk, cannot be disrupted by gentle agitation or by heating, and are not dependent on the presence of agglutinins (which are destroyed by homogenization) for their formation. Thus, reduction in globule size and increase in globule density caused by their absorption of protein are two of the factors which prevent phase separation in homogenized milk.

As might be expected, homogenization also has a marked effect on the total cell count of milk. Peters and Trout (1945) found that homogenization of milk at 2500 p.s.i. (17 237 kNm^{-2}) reduced the cell count to 60% of the original value, and that the use of 5000 p.s.i. (34 474 kNm^{-2}) for 10 minutes destroyed almost all of the cells. Similar results were obtained by Peterson and Keyser (1963) who found that the effect of homogenization on cell count was reduced if the milk was first heated. In both cases, disruption of cells resulted in a corresponding increase in the number of cell fragments.

VI. CHEESE

Although numerous different types of cheese are produced throughout the world, comments here will be restricted to the relatively few whose development and structure have been studied in some detail. Detailed accounts of the methodology in cheesemaking are given by Davis (1966) and Kosikowski (1966), but essentially the process involves clotting milk with rennet, cutting the resultant curd into small pieces and driving out some of the whey by the combined effect of heating (scalding) and acid development, before finally packing the curd into a mould to allow fusion of the curd particles. Generally, a heated, acid curd produces a hard, dry cheese with good keeping qualities, whereas more gentle treatment yields a soft, moist cheese with a limited shelf-life.

The ultrastructural changes that occur during the formation of cheese curd from milk are best described by brief reference to the results obtained by Kimber *et al.* (1974) with Cheddar cheese, although similar studies have been reported in the case of Camembert (Knoop and Peters, 1971; Peters and Knoop, 1974, 1975) and Emmental (Rüegg and Blanc, 1972). The addition of lactic acid-producing starter bacteria (e.g. *Streptococcus cremoris*) to milk produces little change in its ultrastructural appearance. As in untreated milk, the casein micelles appear as discrete particles and never make permanent contact because they are mutually repelled by their high negative surface charge (Fig. 22). Rennet, whose primary effect is the cleavage of a highly acidic macropeptide from κ-casein (Mackinlay and Wake, 1971), reduces the electrical repulsive forces between micelles and allows them to make contact (Green, 1973). Thus, within minutes of the addition of rennet, the micelles come together to form small aggregates and chains of various lengths (Fig. 23). Knoop and Peters (1976) have gained further insight into the formation of milk coagula by observing the effect of calcium, phosphate and citrate ions on the ultrastructure of rennet and acid gels. By the time the milk forms a soft gel, adjacent micelles in large, interconnecting aggregates have undergone partial fusion and have formed a three-dimensional network, the spaces of which are occupied by widely dispersed fat globules and starter bacteria. When the curd is cut into small pieces and scalded, syneresis of the milk gel is accompanied by several changes in curd structure. The fusion of micelles becomes complete so that the contours of individual particles can no longer be detected and adjacent portions of the convoluted casein network fuse (Fig. 24). The interstitial spaces are correspondingly reduced and the enclosed bacteria and fat globules are brought into closer contact, both with each other and with the casein network. The overall structure may be visualized as a whey-filled casein sponge in which fat globules and bacteria are trapped. During subsequent stages (pitching and cheddaring), the whey is drained off and the curd is compressed under its own weight. This leads not only to further fusion of the casein matrix but also to the coalescence of many fat globules and the progressive elimination of the interstitial spaces (Fig. 25). These events continue during the subsequent milling, salting and pressing of the curd, after which the young cheese is left to mature. It is very probable, from the similarity of this description to that given by Rüegg and Blanc (1972) for Emmental cheese, that the curd of all hard cheeses undergoes a similar sequence of developmental changes. Several alterations in the ultrastructural appearance of Cheddar cheese take place during maturation. The granular casein matrix becomes coarser and in some areas appears disrupted and dispersed (Fig. 26). In sectioned material the protein matrix is in close contact with, or closely apposed to, the surface of fat globules, but there is no structure equivalent to the water-rich layer around the fat which was

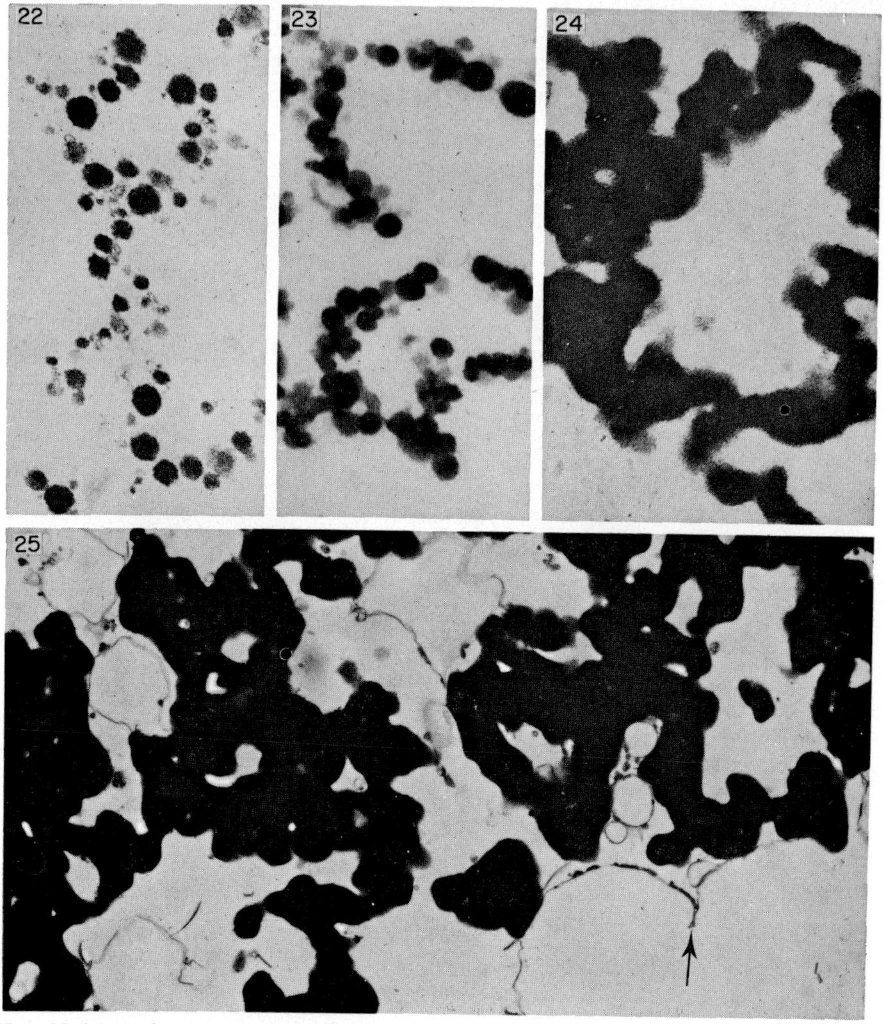

FIGS 22–25. Early stages in the formation of cheese curd.

FIG. 22. Appearance of casein micelles 15 min after the addition of starter bacteria, ×28 500.

FIG. 23. Milk gel just before cutting, ×28 500. Micelles are in chains and aggregates.

FIG. 24. Fusion of micelles at maximum scald, ×20 000.

FIG. 25. Cheese curd at pitching, ×7800. Some of the trapped fat globules have fused (arrowed) and the casein matrix is a dense, convoluted network.

discovered by Kimber *et al.* (1974) by deeply etching fractured Cheddar cheese. Using a similar technique, Hall and Creamer (1972) found differences between mature Cheddar, Cheshire and Gouda cheeses in the size of the globular units composing the protein matrix. Although they cannot be

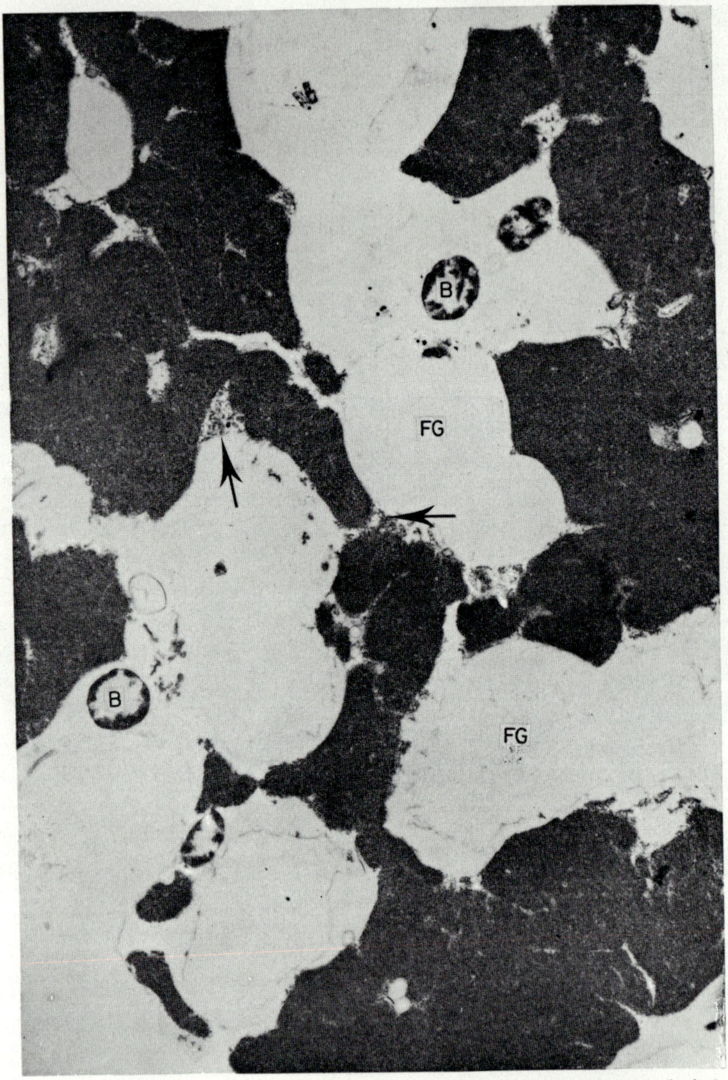

FIG. 26. Mature Cheddar cheese (6 months old), ×8300. Note that the matrix is less dense than that in Fig. 25 and that it is breaking down in some places (arrowed). Between the tightly packed fat globules (FG), the remains of starter bacteria are visible (B).

positively identified as streptococci, profiles of what are presumed to be lysed starter bacteria are common (Fig. 26). The crystalline inclusions dealt with in more detail below appear soon after pressing and increase in size and/or number with time.

During pitching and cheddaring the curd particles, produced when the milk gel is cut, are allowed to fuse; the curd so-formed is later broken into relatively large pieces by milling and these, in their turn, fuse during pressing. After cutting and milling, the surface of the free curd particles develops a distinct surface film or 'skin' which, because it contains fewer fat globules and has a denser protein matrix than the enclosed curd, can be stained preferentially for light microscopy (King, 1956, 1958; Rammell, 1960). Thus, after suitable staining, curd junctions in cheese can be recognized as fine lines which represent the fused 'skin' of adjacent particles (Fig. 27). Rammell (1960) has shown that it is possible to distinguish milled curd junctions from those of the more numerous cut curd particles. Thus, by reference to Fig. 27, it can be seen that many curd junctions terminate abruptly on either side of a milled curd junction.

Most hard cheeses contain microscopic cyrstalline inclusions which, although widely distributed in the body of the cheese, are concentrated along curd junctions (Laxa, 1926, 1927; Swiatek and Jaworski, 1959; Rüegg and Blanc, 1972; Brooker *et al.*, 1975). In Cheddar cheese, two distinct types of inclusion are found. They are clearly seen as birefringent particles in frozen sections examined by polarized light (Fig. 28). One of these is irregular in shape, may measure up to 80 μm in diameter, and consists of randomly arranged bundles of slightly curved, needle-like crystals (Fig. 29). These appear within 1 month of pressing and have properties which have led to their identification as calcium lactate (Brooker *et al.*, 1975). The other type of inclusion appears by light microscopy as a small aggregate (20–30 μm in diameter) of triangular crystals and occurs in very large numbers (Fig. 28). There is good histochemical and circumstantial evidence to suggest that they are principally composed of calcium phosphate (Swiatek and Jaworski, 1959; Brooker *et al.*, 1975) which has crystallized from pockets of residual whey in small spaces between curd particles. Thus, when examined by electron microscopy, most of them occupy the featureless spaces between the fat and protein phases of the cheese, but part of their surface is frequently embedded in the casein matrix. Each is composed of numerous, closely packed, needle-like crystals (25 nm in diameter and up to 4·5 μm long) surrounded by electron-dense material (Fig. 30). It appears from the work of Conochie and Sutherland (1965) that the defect in Cheddar cheese known as 'seaminess' is due to the excessive formation of these calcium phosphate crystalline inclusions at the junctions of the milled curd particles. All of these

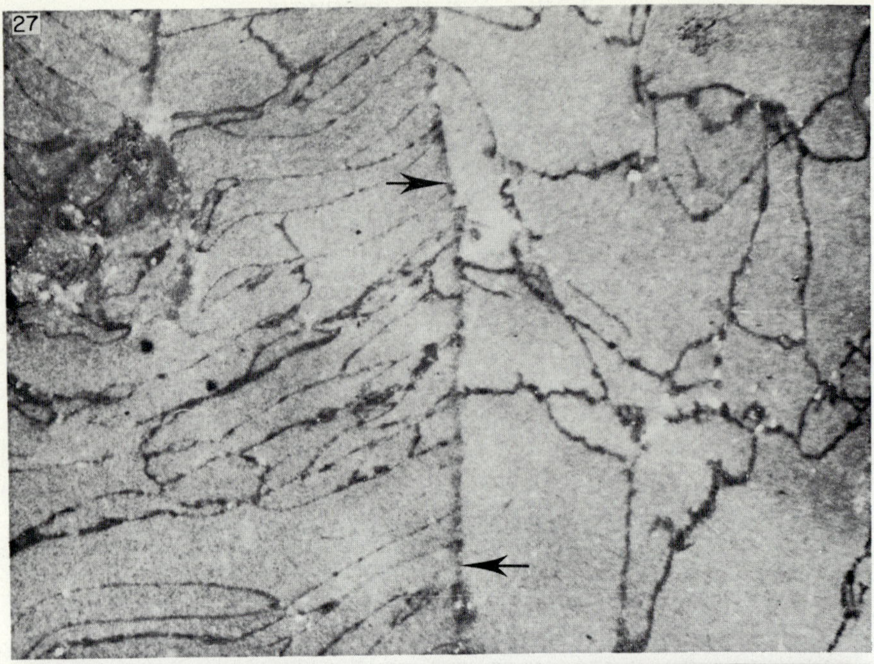

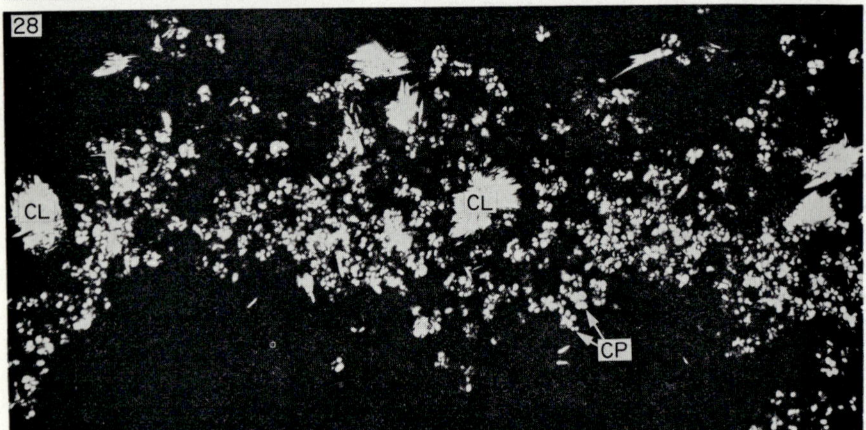

FIG. 27. Frozen section of mature Cheddar cheese stained with methyl violet, ×18. The cut curd junctions appear as lines. A milled curd junction (arrowed) runs from top to bottom.
FIG. 28. Frozen section of 4-month-old Cheddar cheese examined with polarized light, ×170. Aggregates of calcium lactate (CL) and calcium phosphate (CP) crystals are visible along this curd junction.

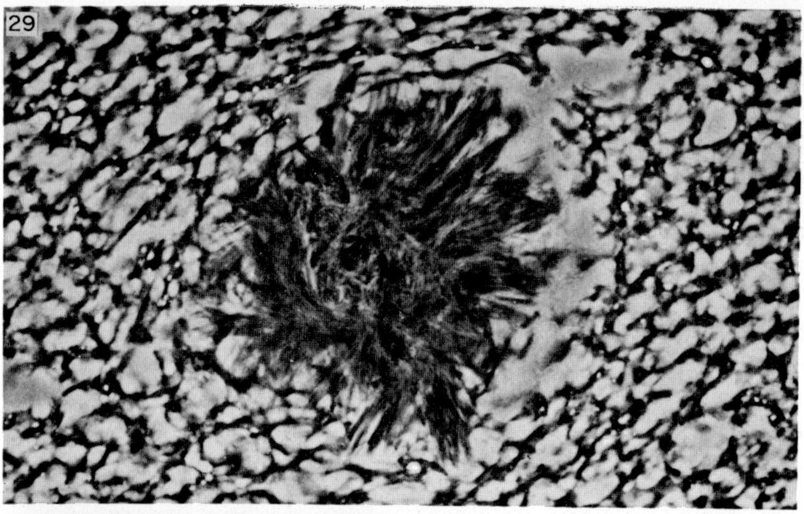

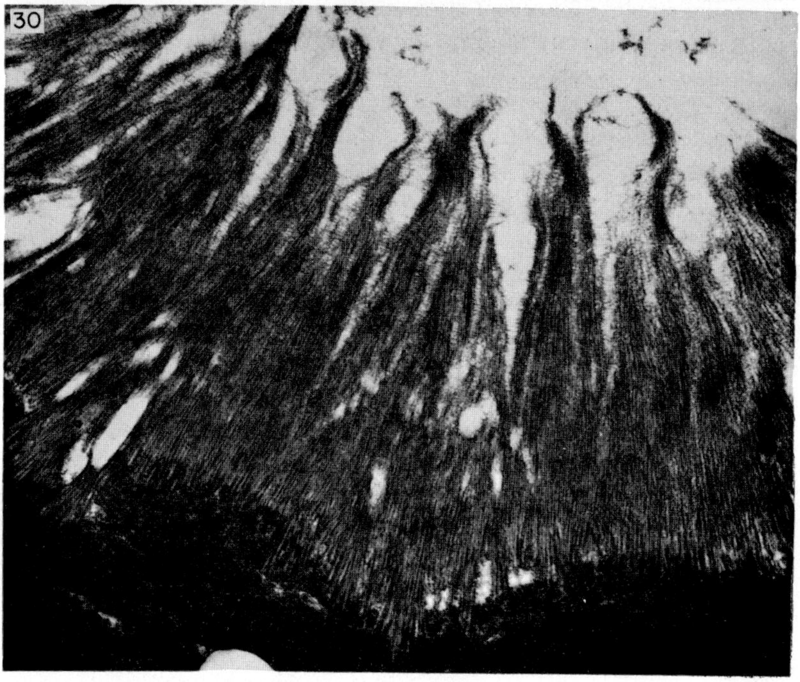

FIG. 29. An aggregate of calcium lactate crystals in 6-month-old Cheddar cheese, ×640. Frozen section examined by phase contrast microscopy.

FIG. 30. Part of a small aggregate of needle-like calcium phosphate crystals in Cheddar cheese, ×19 500.

microscopic inclusions are distinct from the white, macroscopic crystals that are sometimes visible in the body and on the surface of cheese. Depending on the age of the cheese, these have been identified as tyrosine derived from the proteolysis of casein (Flückiger and Schilt, 1963), calcium lactate (Tuckey *et al.*, 1938; McDowall and McDowell, 1939; Conochie *et al.*, 1960; Farrer and Hollberg, 1960; Pearce *et al.*, 1973) or mixtures of both (Harper *et al.*, 1953; Conochie *et al.*, 1960).

In addition to producing acid to accelerate syneresis, the retention of starter bacteria by the curd is important for the later development of good cheese flavour (Reiter *et al.*, 1967, 1969). It is apparent that the loss of streptococci from the curd with the escaping whey is prevented by their entrapment as the pore size of the shrinking curd diminishes. Although this is undoubtedly true by the time maximum scald has been reached, the more open structure of the curd at earlier stages appears to offer no such restraint. However, the results of a recent study suggest that bacterial adhesion to the casein matrix, mediated by an extracellular carbohydrate, may be responsible for retention of the starter bacteria immediately after the curd has been cut (Brooker, 1976). The attachment of bacteria to the casein matrix by filamentous material is seen throughout curd development (Fig. 31).

The number of starter bacteria increases progressively during curd development, but decreases after pressing; thus in mature Cheddar cheese,

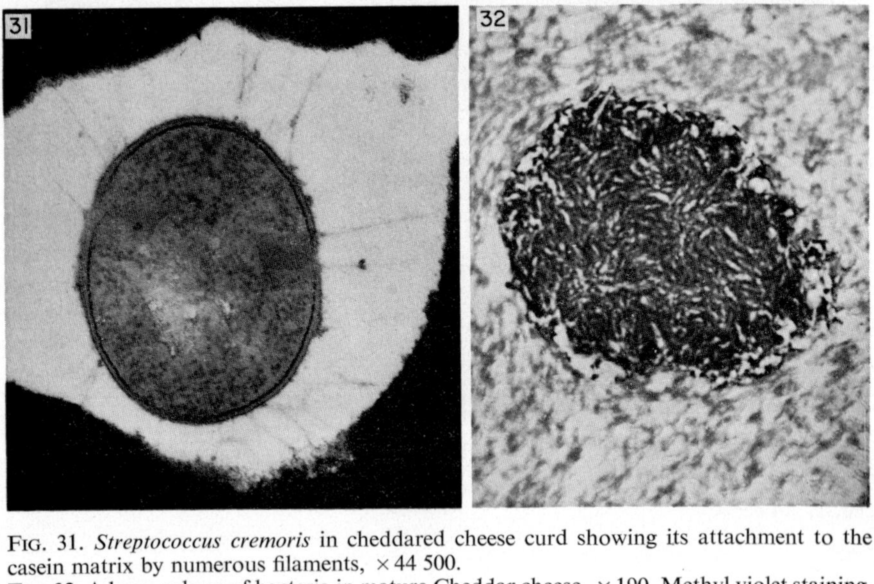

FIG. 31. *Streptococcus cremoris* in cheddared cheese curd showing its attachment to the casein matrix by numerous filaments, ×44 500.
FIG. 32. A large colony of bacteria in mature Cheddar cheese, ×190. Methyl violet staining.

although numerous, widely distributed streptococci can be demonstrated in stained preparations (Rammell, 1960), it is probable that very few of these cells are viable and that many of them correspond to the lysed bacteria observed by electron microscopy. The decline of the starter bacteria after pressing is accompanied by a gradual rise in the numbers of lactobacilli and other adventitious bacteria which are collectively referred to as the 'reference flora'. These reach maximum numbers after several months and, in contrast to the starter bacteria, form large, well-defined colonies (Fig. 32) located mainly in cracks (Dean *et al.*, 1959) or along the curd junctions (Rammell, 1960; Rüegg and Blanc, 1972).

The development of flavour in the cheeses referred to so far depends on the changes brought about by starter streptococci and adventitious bacteria distributed throughout the cheese. In some cheeses, however, the development of characteristic texture and flavour is dependent not so much on the presence of starter bacteria as on the activity of enzymes released from a mould growing in or on the cheese. Camembert is probably the best known example of a cheese ripened by surface mould. Microscopically, at least four zones may be distinguished in this cheese (Seeler, 1968; Knoop and Peters, 1971) (Fig. 33). The white surface layer is entirely composed of a mass of branching hyphae of the mould *Penicillium camemberti*. Some of the hyphae penetrate into the next layer or rind (Fig. 34), the greater part of which consists of sparse fat globules embedded in the digested remains of a casein matrix; in deeper parts of the rind, evidence of digestion is less marked and more localized and the integrity of the protein matrix is largely maintained. A detailed electron microscope study of digestion in the rind has been reported by Knoop and Peters (1971). Beneath the rind is the ripe zone into which the free enzymes, but few hyphae, of the mould have penetrated. It consists of a remarkably uniform casein matrix enclosing numerous large fat globules and there is little evidence of interstitial spaces (Fig. 35). Starter steptococci occur throughout this zone, either as widely distributed chains of cells or as occasional small colonies (Fig. 35). A similar distribution of starter bacteria occurs in the inner, unripened nucleus of the cheese, but the structure and texture of the curd is quite different. Both by light and electron microscopy, the protein phase is seen to be a reticulum of convoluted strands (Knoop and Peters, 1971) closely resembling the scalded curd of Cheddar cheese described above (Fig. 36). In blue-veined cheeses such as Roquefort, Stilton and Gorgonzola, ripening is mediated by the growth of *P. roqueforti* or *P. glaucum* within the cheese. Since this growth depends on the amount of air that can penetrate the cheese, airways are usually made with needles. This allows the germination of mould spores introduced into the milk or curds, and thus the formation of the characteristic greenish-blue mycelium which ultimately

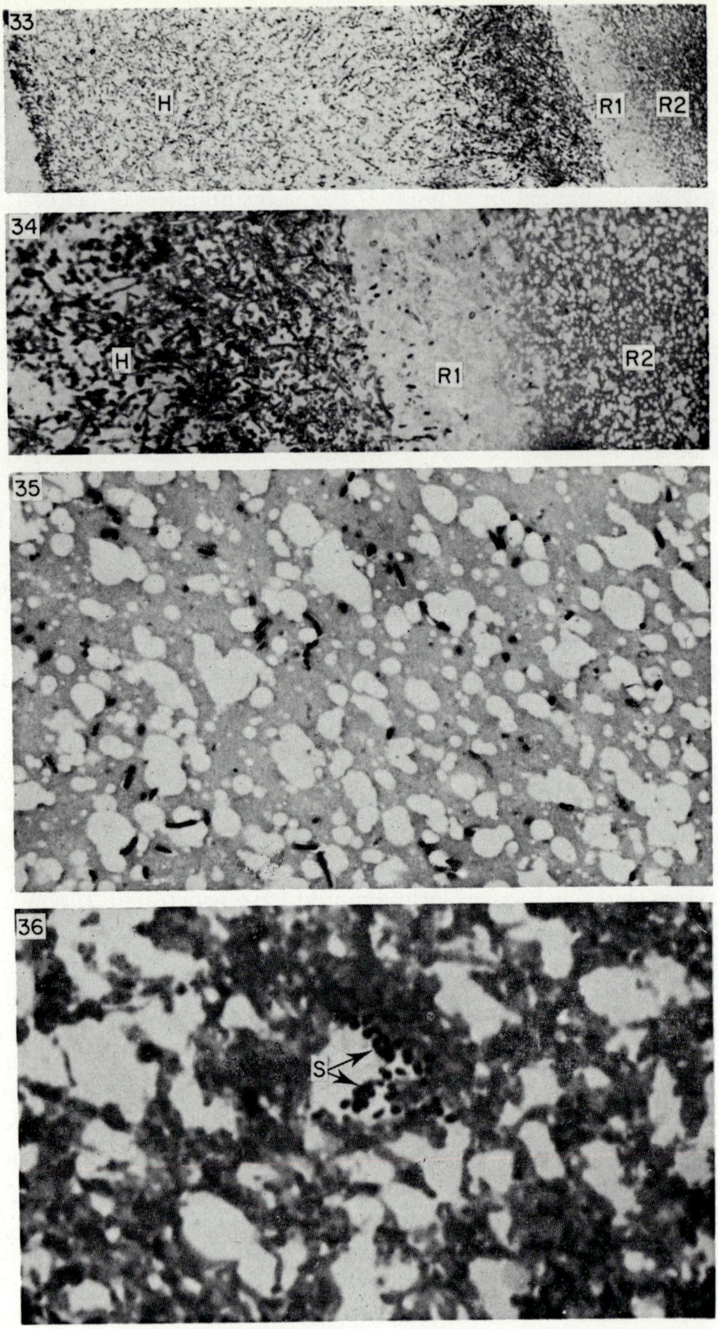

FIGS 33–36 (*Captions on opposite page*)

lines the cracks that develop between the loosely packed curd particles. Growth of the mould is not as luxuriant as that found on surface-ripened varieties, but penetration of the cheese by large numbers of hyphae and the release of proteolytic enzymes produces similar degenerative changes in the appearance of its protein matrix.

VII. YOGHURT

Yoghurt is a gel resulting from the acidification of homogenized or separated milk by mixed starter cultures of *Lactobacillus bulgaricus* and *Streptococcus thermophilus*. In sections examined by light microscopy, yoghurt appears as a three-dimensional network of casein micelle chains and aggregates, the whey-filled spaces of which contain numerous lactobacilli and streptococci (Fig. 37). However, Kalab *et al.* (1976) have shown that the development and final structure of the gel observed by electron microscopy is variable and depends on the way the milk has been treated before fermentation. Although it is possible to make youghurt from raw or pasteurized (heated to 60–70°C) milk, normally milk is heated to 90°C and then cooled before the addition of starter bacteria. Heat treatment in this way has several advantages. Not only is bacterial growth rate and gelation more rapid, but the final gel is firmer and exhibits less syneresis. It is, perhaps, not surprising therefore that the ultrastructure of yoghurts prepared from milk with and without heat treatment is quite different. Kalab *et al.* (1976) showed that during the fermentation of heat-treated milk the micelles lose their smooth outline, grow in size and, by the time gelation occurs, are closely associated or partially fused in clusters and chains (Figs 38 and 39). It is probable that the characteristic appearance of the gel so-formed is dependent on heat-induced changes that occur not only in the micelles but also in the soluble plasma proteins. In the case of pasteurized milk, the micelles enlarge considerably and their surface becomes covered with short projections which, according to Kalab *et*

FIGS 33–36. Camembert cheese.
FIG. 33. A section showing the two superficial layers of Camembert cheese, ×64. The thick layer of mould hyphae (H) overlies the largely digested (R1) and partly digested (R2) zones of the rind.
FIG. 34. Part of Fig. 33 enlarged, ×190. Note that the largely digested casein matrix of R1 stains very little.
FIG. 35. The ripe zone, ×640. The uniform casein matrix is typical. Starter bacteria are scattered throughout.
FIG. 36. Unripe nucleus of Camembert cheese, ×1400. The matrix is reticular in appearance. A small colony of starter organisms (S) is visible.

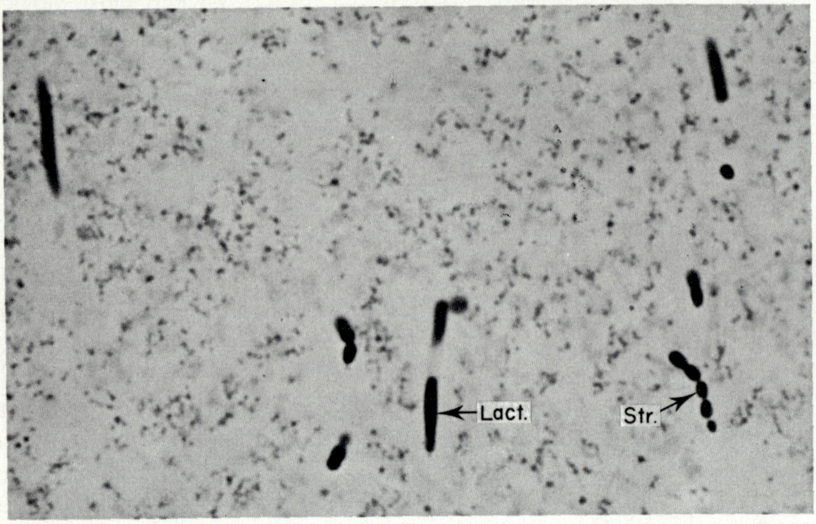

FIG. 37. A 2 μm thick section of yoghurt prepared from heat-treated separated milk showing the granular casein matrix and starter organisms, ×1600. Lact. = lactobacilli, Str. = streptococci.

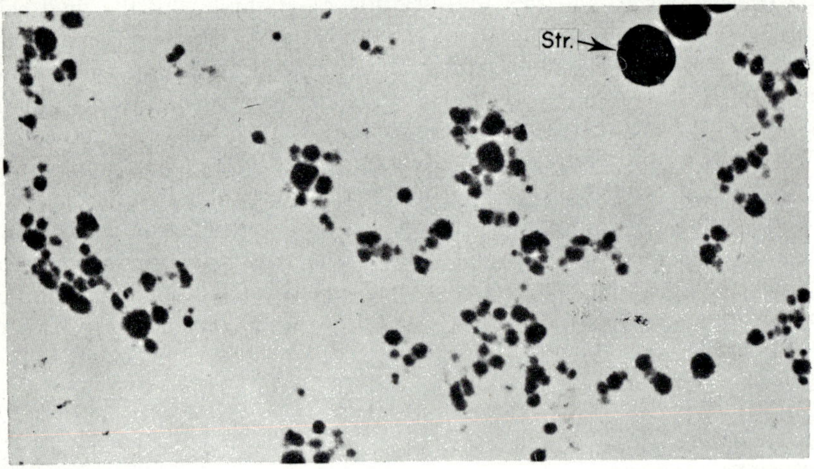

FIG. 38. Chains and aggregates of casein micelles in yoghurt prepared from heat-treated milk, ×7800. Str. = streptococci.

al. (1976), probably represent adsorbed non-micellar protein. The spinous micelles then fuse into large units, and at gelation have formed a three-dimensional network of smooth-surfaced casein aggregates which are much larger than those formed in heat-treated milk. Yoghurt made from raw milk is very similar to this but the casein micelles are smaller (Fig. 40).

Thickening agents such as carrageenan and gelatin are sometimes added to

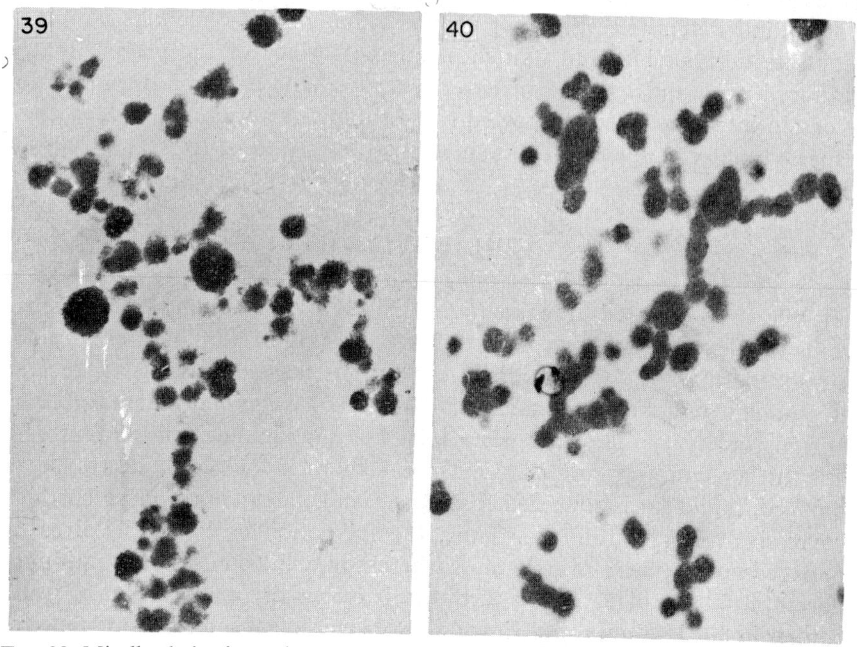

FIG. 39. Micelle chains in yoghurt derived from heat-treated milk, × 12 900. (Cf. Fig. 40.)
FIG. 40. Fused micelles in yoghurt prepared from raw, unheated milk, × 12 900.

milk before fermentation in order to improve the texture, increase firmness and reduce syneresis in the yoghurt. The effect of these additives on the ultrastructure of yoghurt has been studied by Kalab *et al.* (1975) using scanning and transmission electron microscopy. Whereas gelatin has little or no effect, carrageenan produces a gel in which the casein micelles form large, discrete aggregates connected by carrageenan fibres. It is of some interest to note that the cell wall of *Str. thermophilus* is covered with a thin layer of extracellular carbohydrate morphologically identical to that seen on *Str. cremoris* in cheese curd. In yoghurt, casein micelles are often seen attached directly to this surface layer or to the filaments which extend from it.

Ultrastructurally, acid gels, produced by the addition of gluconic acid lactone to milk, resemble yoghurt made from raw or pasteurized milk. The structure of these gels is considerably influenced by the conditions under which they are made. In a comparative study of rennet and acid gels, Knoop and Peters (1975a) have shown that whereas gels produced at 25°C are composed of micelle chains or aggregates, at higher temperatures (37°C and 44°C) the micelles fuse before a gel is formed from the resulting small clumps of casein. The effect of milk acid phosphatase on acid gels has been studied by Knoop and Peters (1975b). They reported that this enzyme was unable to penetrate micelles but that its action on the surface of the protein matrix produced profound changes in its appearance. The enzymic disintegration of casein into what were purported to be submicelles was manifest as finely dispersed material adjacent to the matrix.

VIII. BUTTER

When cream is churned to make butter, the fat globules clump to form small grains which grow in size with time. The butter grains are collected, drained of milk plasma and worked into a continuous mass consisting of free fat, fat globules and their remnants, and droplets of trapped plasma. During churning, many globules fuse when their limiting membranes are damaged; with further churning and working many more globules are destroyed or damaged, liquid fat is squeezed from them and the aqueous phase becomes progressively less continuous. Thus, the formation of butter involves an almost complete phase inversion, for whereas in milk, water is the continuous phase and fat the disperse phase, butter is essentially an emulsion of water droplets in fat. However, experimental evidence suggesting the coexistence of continuous fat and aqueous phases is provided by the ability of both fat-soluble and water-soluble materials to diffuse into butter. This forms the basis of the theory of butter structure proposed by Mulder (1947), according to which the water-containing membranes of many fat globules in the continuous fat phase lie so close to moisture droplets that a continuous network of water channels is formed. Earlier theories which considered either the lipid or the aqueous phase to be continuous have been reviewed by King (1964).

The main structural elements of butter can be seen when a smear is examined by bright field or phase contrast microscopy. Fat globules and water droplets are surrounded by free fat which is plainly visible as a clear zone on the edge of the specimen when pressure is exerted on the coverslip (Fig. 42). In polarized light (between crossed polarizer and analyser) the fat globules show a bright peripheral layer (Fig. 41). This birefringent layer is

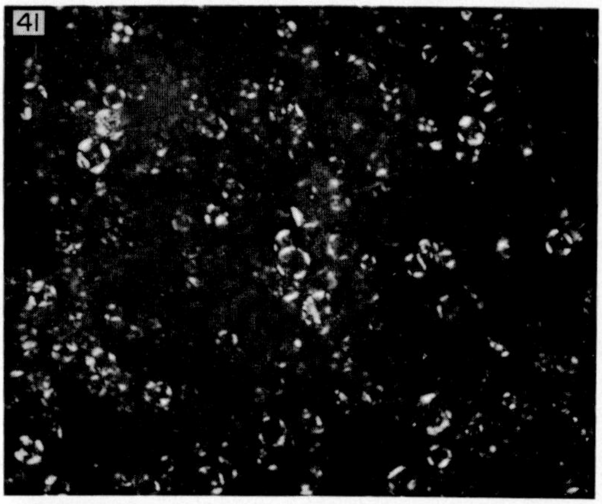

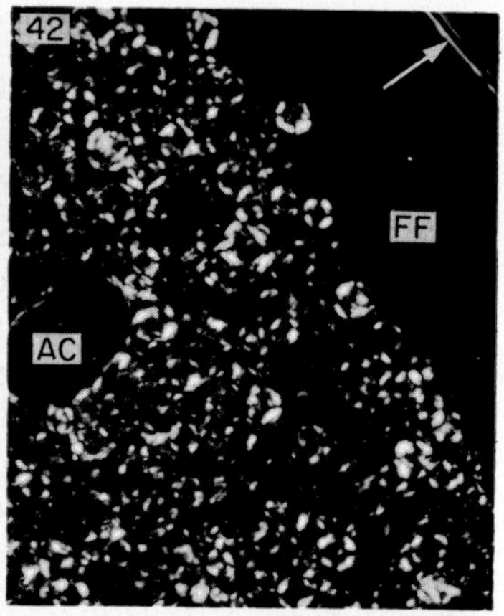

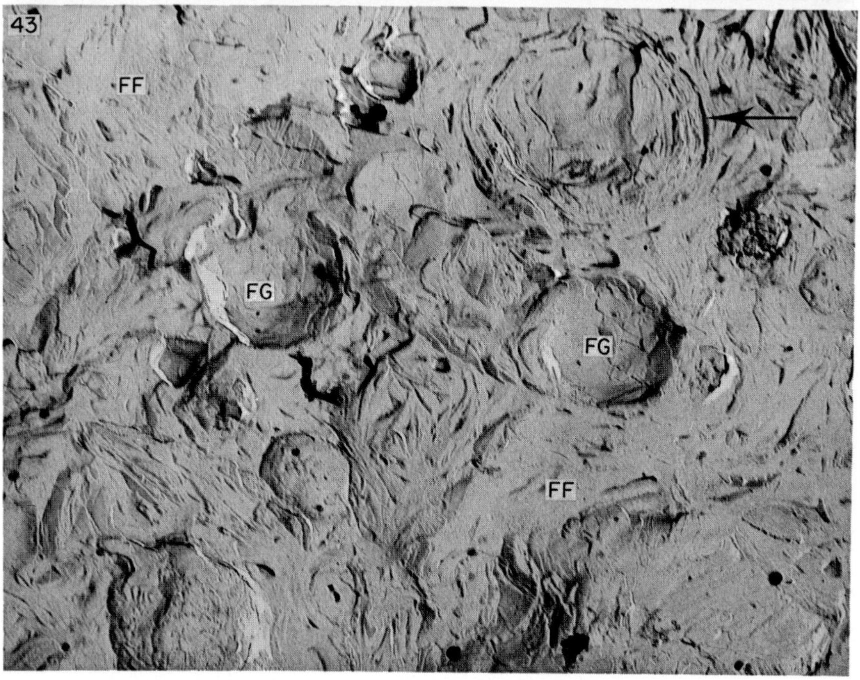

FIG. 41. Squash smear of butter examined in polarized light showing the peripheral birefringence of the fat globules, ×640.

FIG. 42. The edge of a squash preparation of butter in polarized light, showing a clear zone of liquid free fat (FF) and a fat–air interface (arrowed), ×640. Numerous birefringent free fat crystals are visible in the background and surrounding an air cell (AC).

FIG. 43. Freeze-etched butter, ×14 500. Fat globules (FG) are embedded in a matrix of free fat (FF). The typical 'onion skin' appearance of peripheral crystalline fat is clearly visible in one globule (arrowed). (Courtesy A. M. Kimber.)

interrupted in the four places where the polarization planes of polarizer and analyser intersect the globule surface. According to Walstra (1962, 1967), this layer is formed by birefringent, tangentially orientated, needle-like fat crsytals along the globule boundary. It is apparent that the properties of fat globules in butter are slightly different to those in milk, for whereas most fat globules in butter are birefringent, in milk at the same temperature relatively few globules contain birefringent material. The work of King (1950) suggests that the exertion of pressure on fat globules and their loss of some liquid fat during churning and working is important for the formation of the outer birefringent layer. He observed that a mounted sample of cooled cream contained a higher proportion of birefringent globules directly after application of pressure to the coverslip.

Using a 'dilution' method in which butter was dispersed in butter oil and examined microscopically (King, 1947a), King (1947b) obtained volume-average diameters of 3·6–4·2 μm for the fat globules. A similar modal diameter was obtained by Gavél (1956) who measured and counted fat globules in microtome sections of butter which had been cut in a cold room at 0–3°C. This information supports the conclusion reached by Gavél (1956) that the fat globules in butter correspond to the original globules in milk or cream. Similarly, the ultrastructural studies of Knoop and his co-workers (Knoop and Schulz, 1960; Knoop and Knoop, 1961; Knoop and Knoop, 1962; Knoop and Wortmann, 1962; Knoop, 1965; Knoop *et al.*, 1966), in which replicas of etched butter and/or thin sections of butter fixed for 5–6 months in osmium tetroxide were examined, have revealed more or less numerous globular bodies of the size of the original milk fat globules (Fig. 43). Knoop and Schulz (1960) referred to these globules as 'micro-fat granules' and showed that only fragments of the original MFGM were retained by them. Knoop and Knoop (1962) distinguished between two extreme cases of butter structure which differed in the conspicuousness of their micro-fat granules. In one of these the butterfat was markedly granular in appearance, but in the other it was relatively featureless and homogeneous. Which of these structures predominates in any given butter apparently depends on the treatment given to the cream, the churning temperature and the intensity of the mechanical treatment during the various stages of butter-making.

The free fat phase of butter contains a crystalline fraction which is visible as numerous birefringent particles when examined microscopically in polarized light (Fig. 42). These small ($< 1\mu$m), needle-like fat crystals originate from the fat globules that are broken during working, and from partial crystallization in the free fat. Their probable importance in determining the firmness and thixotropic properties of butter has been discussed by Prentice (1972).

The aqueous phase of butter is present mainly in the form of small droplets dispersed throughout the free fat; these droplets originate partly from the milk plasma and partly from the wash-water added after the butter grains have been strained. However, Mulder (1957) has distinguished five forms of moisture in butter.

(i) Internal moisture of butter grains. This is cream plasma in the spaces enclosed by aggregates of globules and it occurs as droplets that are smaller than the surrounding fat globules.
(ii) External moisture of butter grains. This is buttermilk and wash-water retained on the surface of and between butter grains. These droplets are generally larger than those of internal moisture.
(iii) Water in the fat globule membrane. This is present in very small amounts but is important because it contains a number of substances, e.g. proteins, enzymes, salts and phospholipids, in high concentrations.
(iv) Water dissolved in the fat ($<0{\cdot}25\%$).
(v) Water added during working to adjust the moisture content of the butter. This, like the external moisture of butter grains, occurs as relatively large droplets.

Whilst the diameter of the smallest water droplets is somewhat less than 1 μm, the upper limit depends on the intensity and amount of working but should not greatly exceed 30 μm. Droplets with a diameter less than 2 μm greatly outnumber those larger than this size. Electron microscopy has shown that a membrane is formed at the interface between milk plasma droplets and the surrounding butterfat (Knoop and Wortmann, 1962; Knoop, 1965). This interfacial layer, which does not have unit membrane structure (Robertson, 1959), appears to be formed from components in the water droplet and probably represents the major component of the abundant membraneous material that can be isolated from lipid-extracted butter (Fig. 44).

In addition to these major structural elements of butter, curd granules and air cells are constantly present. Curd granules of varying size have been observed in sections of butter examined by light microscopy (Gavél, 1956; Mohr *et al.*, 1957). They are particularly conspicuous in butter prepared from ripened cream and are composed of casein micelles connected by fine filaments (Fig. 45). Butter contains more than 5% air by volume. Some of this is in solution in the butterfat but the remainder exists as a free gas in air cells (Fig. 42). Gavél (1956, 1957) has found that their size, shape and uniformity of distribution is dependent on the manufacturing method used to prepare the butter.

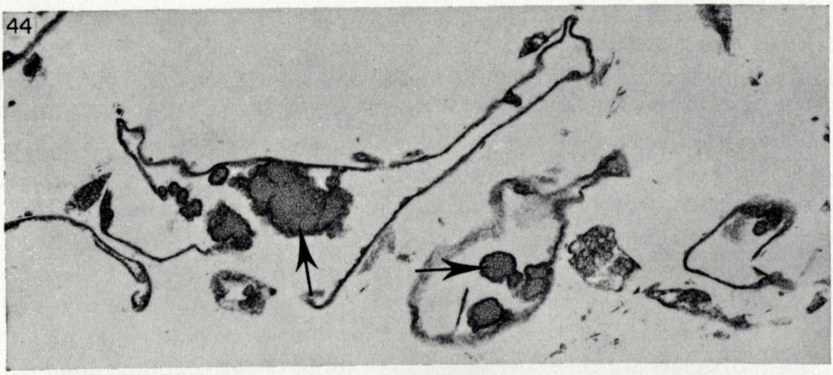

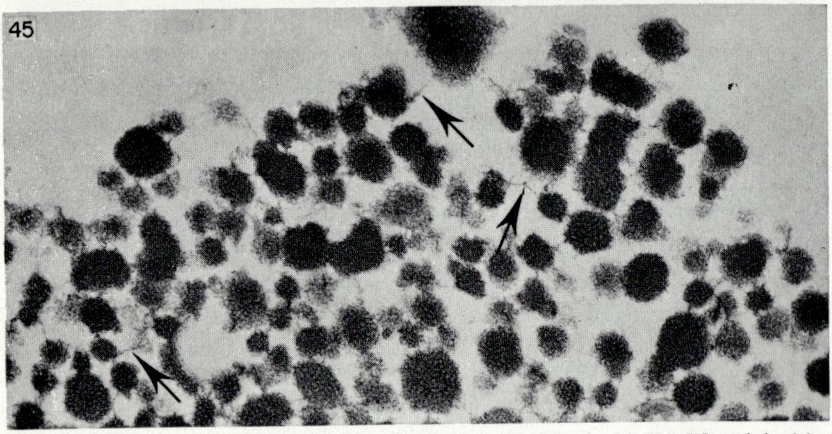

FIG. 44. Membrane material isolated from fat-extracted butter, ×25 500. Material with the mosaic-like appearance of casein (arrowed) is usually closely associated with it.
FIG. 45. Part of a curd granule in butter, ×43 000. Fine filaments (arrowed) connect neighbouring micelles.

REFERENCES

Anderson, M. and Cawston, T. E. (1975). *J. Dairy Res.* **42**, 459.
Brooker, B. E. (1976). *J. Dairy Res.* **43**, 283.
Brooker, B. E., Hobbs, D. G. and Turvey, A. (1975). *J. Dairy Res.* **42**, 341.
Buchheim, W. and Knoop, E. (1970). *Kieler Milchw. ForschBer.* **22**, 323.
Buchheim, W. and Welsch, U. (1973). *Ned. Melk- en Zuiveltijdschr.* **27**, 163.
Carroll, R. J., Thompson, M. P. and Nutting, G. C. (1968). *J. Dairy Sci.* **51**, 1903.
Christie, W. W. and Wooding, F. B. P. (1975). *Experientia*, **31**, 1445.
Conochie, J. and Sutherland, B. J. (1965). *J. Dairy Res.* **32**, 35.
Conochie, J., Czulak, J., Lawrence, A. J. and Cole, W. F. (1960). *Aust. J. Dairy Technol.* **15**, 120.

Davis, J. G. (1966). 'A Dictionary of Dairying.' Leonard Hill, London.
Dean, M. R., Berridge, N. J. and Mabbitt, L. A. (1959). *J. Dairy Res.* **26**, 77.
Eggmann, H. (1969). *Milchwissenschaft*, **24**, 479.
Farrer, K. T. H. and Hollberg, W. C. J. (1960). *Aust. J. Dairy Technol.* **15**, 151.
Flückiger, E. and Schilt, P. (1963). *Milchwissenschaft*, **18**, 437.
Gavél, L., von (1956). *Z. Lebensmittelunters. u.-Forsch.* **104**, 1.
Gavél, L., von (1957). *Dt. Molk. -Ztg.* **78**, 1489.
Green, M. L. (1973). *Ned. Melk- en Zuiveltijdschr.* **27**, 278.
Hall, D. M. and Creamer, L. K. (1972). *N.Z. Jl Dairy Sci. Technol.* **7**, 95.
Harper, W. J., Swanson, A. M. and Sommer, H. H. (1953). *J. Dairy Sci.* **36**, 368.
Henstra, S. and Schmidt, D. G. (1970). *Ned. Melk- en Zuiveltijdschr.* **24**, 45.
Hostettler, H. and Imhof, K. (1952). *Landw. Jb. Schweiz.* **66**, 307.
Jenness, R. (1974). *In* 'Lactation' (B. L. Larson and V. R. Smith, eds), Vol. 3, p. 3. Academic Press, New York and London.
Jensen, D. L. and Eberhart, R. J. (1975). *Am. J. Vet. Res.* **36**, 619.
Kalab, M., Emmons, D. B. and Sargant, A. G. (1975). *J. Dairy Res.* **42**, 453.
Kalab, M., Emmons, D. B. and Sargant, A. G. (1976). *Milchwissenschaft*, **31**, 402.
Kent, G. M. and Newbould, F. H. S. (1969). *Can. J. Comp. Med.* **33**, 214.
Kernohan, E. A. and Lepherd, E. E. (1969). *J. Dairy Res.* **36**, 177.
Kimber, A. M., Brooker, B. E., Hobbs, D. G. and Prentice, J. H. (1974). *J. Dairy Res.* **41**, 389.
King, N. (1947a). *Ned. Melk- en Zuiveltijdschr.* **1**, 19.
King, N. (1947b). *Ned. Melk- en Zuiveltijdschr.* **1**, 114.
King, N. (1950). *Ned. Melk- en Zuiveltijdschr.* **4**, 30.
King, N. (1956). *Aust. J. Dairy Technol.* **5**, 3.
King, N. (1958). *J. Dairy Res.* **25**, 312.
King, N. (1964). *Dairy Sci. Abstr.* **26**, 151.
Knoop, A.-M. and Knoop, E. (1962). *Milchwissenschaft*, **17**, 604.
Knoop, A.-M. and Peters, K. H. (1971). *Milchwissenschaft*, **26**, 193.
Knoop, A.-M. and Peters, K. H. (1975a). *Kieler Milchw. ForschBer.* **27**, 227.
Knoop, A.-M. and Peters, K. H. (1975b). *Milchwissenschaft*, **30**, 674.
Knoop, A.-M. and Peters, K. H. (1976). *Milchwissenschaft*, **31**, 338.
Knoop, A.-M., Knoop, E. and Wiechen, A. (1973). *Ned. Melk- en Zuiveltijdschr.* **27**, 121.
Knoop, A.-M., Knoop, E. and Wortmann, A. (1966). 17th Int. Dairy Congr., C, 253.
Knoop, E. (1965). *Kieler Milchw. ForschBer.* **17**, 73.
Knoop, E. and Knoop, A.-M. (1961). *Milchwissenschaft*, **16**, 449.
Knoop, E. and Schulz, M. E. (1960). *Milchwissenschaft*, **15**, 439.
Knoop, E. and Wortmann, A. (1960). *Milchwissenschaft*, **15**, 273.
Knoop, E. and Wortmann, A. (1962). 16th Int. Dairy Congr., B, 104.
Kosikowski, F. (1966). 'Cheese and Fermented Milk Foods.' Edwards Bros., Ann Arbor, Michigan.
Laxa, O. (1926). *Lait*, **6**, 885.

Laxa, O. (1927). *Lait*, **7**, 521.

Ling, E. R., Kon, S. K. and Porter, J. W. G. (1961). *In* 'Milk: the Mammary Gland and its Secretion' (S. K. Kon and A. T. Cowie, eds), p. 195. Academic Press, New York and London.

Mackinlay, A. G. and Wake, R. G. (1971). *In* 'Milk Proteins' (H. A. McKenzie, ed.), Vol. 2, p. 175. Academic Press, New York and London.

McDowall, F. H. and McDowell, A. K. R. (1939). *J. Dairy Res.* **10**, 118.

Mohr, W., Mohr, E. and Drachenfels, H. J., von (1957). *Milchwissenschaft*, **12**, 246.

Morton, R. K. (1954). *Biochem. J.* **57**, 231.

Mulder, H. (1947). 'Problemen en Resultaten bij Weternschappelijk Zuivelonderzoek', Vol. 2. Algemeene Nederlandsche Zuivelbond (F.N.Z.), The Hague.

Mulder, H. (1957). *Ned. Melk- en Zuiveltijdschr.* **11**, 25.

Nitschmann, H. (1949). *Helv. Chim. Acta* **32**, 1258.

Paape, M. J. and Tucker, H. A. (1966). *J. Dairy Sci.* **49**, 265.

Paape, M. J., Guidry, A. J., Kirk, S. T. and Bolt, D. J. (1975). *Am. J. Vet. Res.* **36**, 1737.

Pearce, K. N., Creamer, L. K. and Gilles, J. (1973). *N.Z. Jl Dairy Sci. Technol.* **8**, 3.

Peters, I. I. and Trout, G. M. (1945). *J. Dairy Sci.* **28**, 251.

Peters, K. H. and Knoop, A.-M. (1974). *Milchwissenschaft*, **29**, 193.

Peters, K. H. and Knoop, A.-M. (1975). *Milchwissenschaft*, **30**, 205.

Peterson, R. D. and Keyser, W. L. (1963). *J. Dairy Sci.* **46**, 988.

Prentice, J. H. (1972). *J. Text. Stud.* **3**, 415.

Rammell, C. G. (1960). *J. Dairy Res.* **27**, 341.

Reiter, B., Fryer, T. F., Pickering, A., Chapman, H. R., Lawrence, R. C. and Sharpe, M. E. (1967). *J. Dairy Res.* **34**, 257.

Reiter, B., Sorokin, Y., Pickering, A. and Hall, A. J. (1969). *J. Dairy Res.* **36**, 65.

Robertson, J. D. (1959). *Biochem. Soc. Symp.* **16**, 3.

Rüegg, M. and Blanc, B. (1972). *Schweiz. Milchwirt. Forsch.* **1**, 1.

Russell, M. W., Brooker, B. E. and Reiter, B. (1976). *Res. Vet. Sci.* **20**, 30.

Saacke, R. G. and Heald, C. W. (1974). *In* 'Lactation' (B. L. Larson and V. R. Smith, eds), Vol. 2, p. 147. Academic Press, New York and London.

Schalm, O. W., Carroll, E. J. and Jain, N. C. (1971). 'Bovine Mastitis.' Lea and Febiger, Philadelphia.

Schmidt, D. G., Buchheim, W. and Koops, J. (1971). *Ned. Melk- en Zuiveltijdschr.* **25**, 200.

Schmidt, D. G., Walstra, P. and Buchheim, W. (1973). *Ned. Melk- en Zuiveltijdschr.* **27**, 128.

Seeler, G. (1968). *Milchwissenschaft*, **23**, 661.

Swiatek, A. and Jaworski, J. (1959). 15th Int. Dairy Congr. **3**, 1509.

Tuckey, S. L., Ruehe, H. A. and Clark, G. L. (1938). *J. Dairy Sci.* **21**, A161.

Walstra, P. (1962). *Ned. Melk- en Zuiveltijdschr.* **16**, 273.

Walstra, P. (1967). *Ned. Melk- en Zuiveltijdschr.* **21**, 166.

Walstra, P. (1969). *Ned. Melk- en Zuiveltijdschr.* **23**, 99.

Wisniowski, J., Romaniukowa, K. and Grajewski, H. (1965). *Bull. Vet. Inst. Pulaway*, **9**, 140.
Wooding, F. B. P. (1971a). *J. Ultrastruct. Res.* **37**, 388.
Wooding, F. B. P. (1971b). *J. Cell Sci.* **9**, 805.
Wooding, F. B. P. (1972). *Experientia*, **28**, 1077.
Wooding, F. B. P., Peaker, M. and Linzell, J. L. (1970). *Nature, Lond.* **226**, 762.
Zlotnik, I. (1947). *J. Comp. Path.* **57**, 196.

8. The Domestic Hen's Egg

D. S. ROBINSON

Procter Department of Food Science,
University of Leeds, U.K.

I. INTRODUCTION

It is believed that primitive man-like creatures began to appear in different parts of the world some 4–500 000 years ago. From fossilized remains we know that these very early men were food gatherers living like most wild animals catching, trapping or collecting whatever was acceptable. It is likely that the eggs of wild birds may have formed one of the most readily available, but unfortunately seasonal, sources of protein, minerals and vitamins. Such is the composition of the egg, containing protein, carbohydrate, fat and not least water in addition to the essential nutrients mentioned above, that it provides a complete food for the developing embryo; thus it also affords an almost complete food for man.

Today the need for eggs, both in the home for direct nourishment and in the food processing industry as components of many convenience foods, is probably greater than it has ever been. Fortunately, concomitant with this increased demand, major advances have been achieved both by genetic selection of the most efficient strains of hens and by the development of efficient animal husbandry techniques for intensive rearing of poultry. Also, some of the infectious diseases prevalent in large, confined poultry units are gradually being eliminated by the development of effective vaccines. These advances have meant that in economic terms the egg and poultry industry is now one of the major producers of food in the Western world. Because of the size of the industry, and above all because the egg itself is an unstable product, much research in recent years has been directed towards improving the physical properties of the egg. For example, losses due to shell faults and cracks may be approaching approximately £10 million per year in the United Kingdom alone. Also there are equally important losses due to the spontaneous liquefaction of thick egg white during the storage of eggs at ambient temperatures, but these are more difficult to quantify in financial terms. Consequently, much research during the last two decades has been devoted to studying the structure and the mechanical properties of egg shells, egg white and egg yolk. It is in this context that both the light microscope and more recently the electron microscope have proved to be the most valuable tools for fundamental structural studies of the domestic hen's egg.

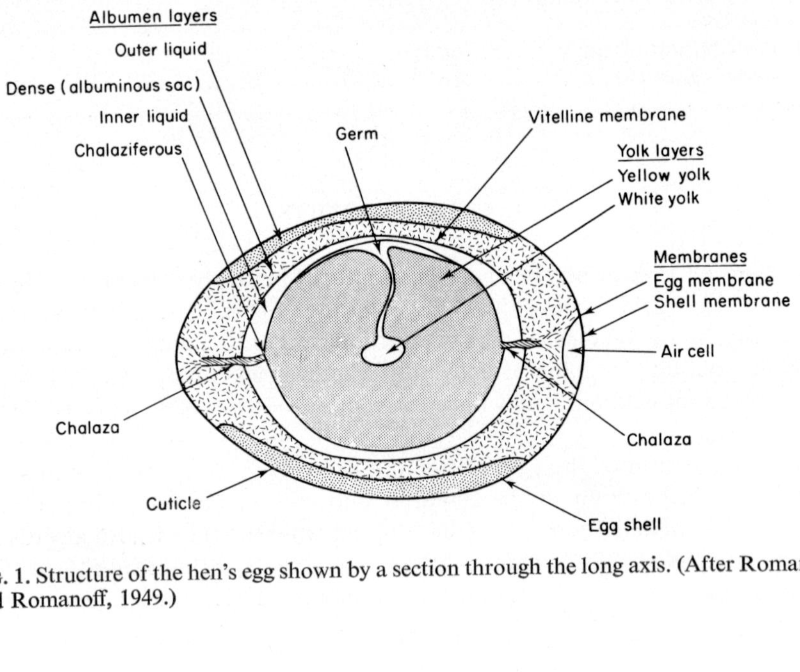

FIG. 1. Structure of the hen's egg shown by a section through the long axis. (After Romanoff and Romanoff, 1949.)

II. GROSS STRUCTURE OF THE EGG

A diagram of the structure of the whole egg is shown in Fig. 1. The yolk is held in a central position by the two chalazea which are embedded in both the small amount of thick egg white surrounding the vitelline membrane and the bulk of the thick egg albumen gel. The liquid content of the egg is surrounded by layers of inner and outer egg shell membranes. The outer and inner membrane separate at the broad pole of the egg to form the air sac; this offers a readily available source of air for a developing embryo, or alternatively may function as an expansion–contraction vessel to minimize any possible deleterious effects on a developing embryo brought about by changes in air pressure. The shell surrounds the membranes and has been the subject of considerable microscopical research in recent years. It is composed mainly of calcium carbonate in the calcite form and is covered on the outside by an insoluble organic cuticle.

III. MICROSCOPICAL STRUCTURE OF THE SHELL

The first substantial microscopic study of the shells of numerous avian species was carried out by Nathusius about a century ago; the first paper on the subject appeared in 1868. A translated and edited version of Nathusius's work has been published by Tyler (1964). Although many of Nathusius's views on general biology were doubted even at the time of his work, his description, his technique, his drawing and his industry for studying so many different kinds of egg shells is to be admired. Since the time of Nathusius and until quite recently the study of egg shells has been a neglected subject and it is only during the last 15 years, partly because of their economic importance in maintaining the integrity of the egg contents and partly due to the more widespread use of electron microscopes, that a detailed understanding of the ultrastructure of egg shells has been obtained.

The egg shell is approximately 300–350 μm thick and consists of approximately 97% (w : w) calcium carbonate together with small amounts of other minerals and approximately 2% (w : w) of protein. Observations of both radial and tangential ground sections of egg shell under the light microscope and recent studies using the scanning electron microscope (Simons and Wiertz, 1970; King and Robinson, 1972; Fujii, 1974) have established that the egg shell can be considered to consist of two morphologically distinct layers of calcium carbonate. With the advent of scanning electron microscopy, much of the structure deduced painstakingly by light microscopy was quickly confirmed, and it is now well established that the shell of the domestic hen's egg is composed of the mammillary layer, made of cone-shaped blocks of calcite firmly attached at their bases to the outer egg shell membrane, and the more diffuse and less well-defined crystalline outer palisade layer (Fig. 2).

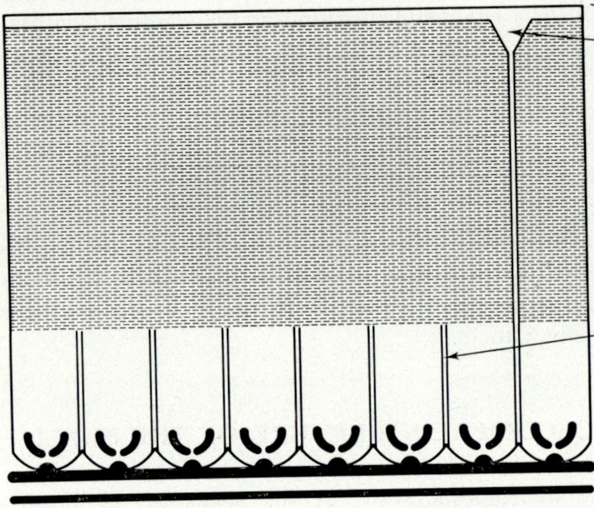

FIG. 2. Diagram of the structure of a radial section of the hen's egg shell.

A. The Palisade Layer

This layer of the shell represents approximately two-thirds of the total shell and is covered on the outer surface by an organic cuticle. The surface view of a transmission electron micrograph of a carboned replica of the cuticle is shown in Fig. 3. The cuticle is seen as a structureless layer of material overlying the outer surface of the egg shell, and the cracks in the surface as shown in Fig. 3 probably arose during the drying of the egg shell surface after laying or during the handling of the prepared replicas. Because of the clear absence of any cellular or fibrous structure, and in the light of results of chemical analysis, it is thought that the cuticle is mainly an amorphous film of highly polymerized and probably oxidized protein.

For the shell proper, electron micrographs (Figs 4, 5, 6, 7 and 8) of the replicas of a series of radial fracture surfaces, ranging from the outer edge through the central part to ultimately the inner mammillary layer of the shell, show that the microstructure of the shell is heterogeneous. Close to the cuticle, the crystalline matrix (Fig. 4) is granular with a number of pit-holes and irregularly shaped and spaced aggregates of unknown material. Towards the centre of the shell, radial fracture surfaces (Figs 5, 6 and 7) show a more defined crystalline structure with obvious cleavage planes within the crystallites. In some parts of the shell a very large number of overlying globules of material, which was perhaps protein or mucopolysaccharide, can be seen (Fig. 5). The globules seem to be arranged in parallel rows and in many cases accumulate at the edges or cleavage planes of the underlying shell structure. The ultrastructure of the mammillary layer (Fig. 8) is more rock-like in appearance, and is characterized by the almost complete absence of pit-holes and extraneous globules. Although it is difficult to unequivocally identify organic substances by electron microscopy, the results of the studies shown in Figs 4–7 indicate that the organic material is dispersed through the shell and does not form discrete layers. This is supported by similar observations on replicas of both whole shell and decalcified shells (Simons, 1971). Further, earlier histological and histochemical studies by light microscopy of sections of decalcified shell (Simkiss and Tyler, 1957, 1958) showed the presence of protein and polysaccharide generally distributed throughout the thickness of the palisade layer. Chemical analysis of the organic matrix has shown the presence of protein and the mucopolysaccharides choindroitin sulphate (Baker and Balch, 1962) and hyaluronic acid (Heaney and Robinson, 1976).

While scanning and transmission electron microscopy have shown the topography and the microstructure of the shell respectively, these techniques do not provide information on the crystallographic nature of the shell. To date, the crystallographic orientation of the columns of calcite has been studied by polarized light microscopy and X-ray diffraction. Polarized light microscopical studies of radial sections by Schmidt (1962) and Terepka (1963a, b) have shown that the gross structure of shell is comprised mainly of adjacent vertical columns of calcite, each of definite crystallographic orientation but with no obvious relationship between the different crystallographic orientations for the various separate aggregate columns of calcite. However this has been a matter for dispute, and Cain and Heyn (1964) from studies of X-ray diffraction patterns, while verifying that the egg shell is composed of calcium carbonate in the rhombohedral calcite form and not the orthorhombic aragonite form, have claimed that the main portion of shell contains calcite orientated with the hexagonal axis inclined at $28 \pm 16°$ from the normal of the shell surface. However the latter conclusion, which is only based on the visually estimated intensities of diffraction rings, has been disputed by Favejee *et al.* (1965) who used an X-ray goniometer. For the few specimens examined it was claimed that the c-axis was randomly arranged with the rhombohedral crystal face parallel to egg's surface. It is not known how many samples of shell from either the same egg or different eggs were studied by Cain and Heyn (1964). To my knowledge no further X-ray diffraction studies of the domestic hen's egg shell have been reported. Tangential

FIG. 3. Surface of the cuticle, obtained by the replica technique, ×10 000.

FIG. 4. Radially fractured surface of the palisade layer near the cuticle; obtained by the replica technique, ×25 000.

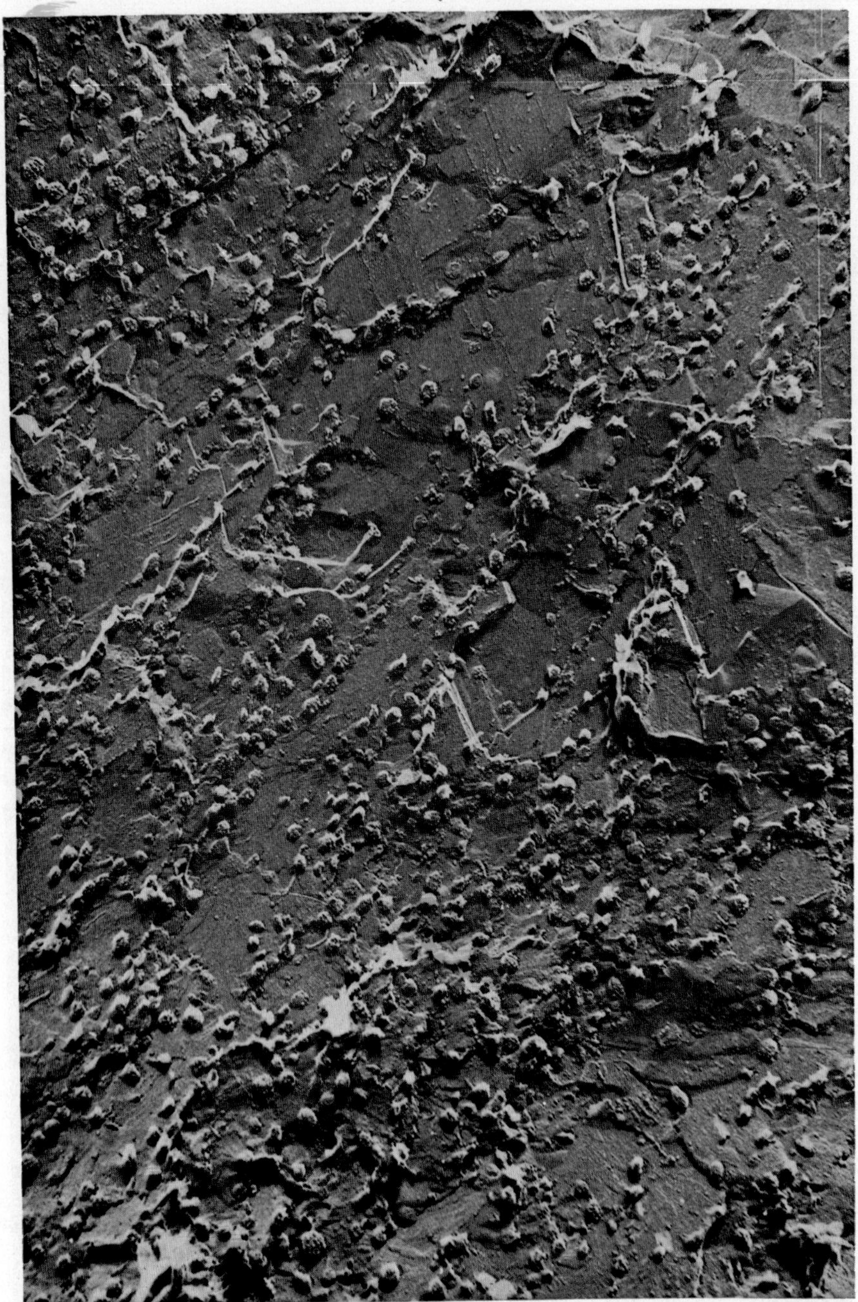

FIG. 5. Radially fractured surface of the palisade layer from the centre of the shell; obtained by the replica technique, ×10 000.

FIG. 6. Radially fractured surface of the palisade layer from the lower centre of the shell; obtained by the replica technique, $\times 10\,000$.

FIG. 7. Radially fractured surface of the palisade layer near the mammillary layer; obtained by the replica technique, ×25 000.

FIG. 8. Radially fractured surface of the mammillary layer; obtained by the replica technique, ×10 000.

sections of the egg shell have shown that the crystalline columns of calcite interlock like pieces of a jigsaw puzzle, but these sections do not, and neither could they be expected to, reveal any ordered and preferred crystallographic orientation for the individual aggregate columns of calcite. From the studies of Schmidt (1957, 1962), later supported by electron microscopy (Heyn, 1963), it now seems that the egg shell is composed of large crystal columns exhibiting spherulitic growth, i.e. crystal growth from a common centre, in some cases with associated twinning of the crystallites.

B. The Mammillary Layer

Light microscopy of ground, radial sections of the egg shell first indicated that the mammillary layer is composed of discrete blocks of calcite. The blocks of calcite, or mammillae, are arranged in juxtaposition to each other (Fig. 2) like the foundation stones of a building, although, of course, on a microscopic scale. The use of the scanning electron microscope (Simons, 1971; King and Robinson, 1972) has resulted in the production of photographs of much greater clarity in detail and definition than could previously be obtained through the use of the light microscope. The mammillae form the 'foundation stones' for the bulk of the egg shell, the palisade layer being formed by presumably a more random and diffuse crystallization during the continued growth of the columns of calcite while still in the shell gland of the hen. The mammillae are closely packed together, each being attached to the outer surface of the egg shell membrane. A small number of fibres from the outer surface of the egg shell membrane penetrate the base of each well-formed block of calcite (Fig. 9a). The study of tangential surfaces by either light microscopy or scanning electron microscopy has shown that the mammillae are randomly distributed on the outer surface of the egg shell membrane, and it is claimed by Cain and Heyn (1964) that the inner surface of the shell, unlike the outer surface, contains crystallites in entirely random orientation. The most reliable method for confirmation of this proposal could involve polarized light studies of the mammillae as they are being formed on the shell membrane and are thus free of other contaminating crystalline and organic material. Eggs where the shell is just being formed can be obtained from the dissected reproductive tract of the hen. The first crystals seem to be formed while the egg is in the isthmus region of the reproductive tract (Robinson *et al.*, 1968), while the main bulk of the shell is secreted later by the uterus or, as is more commonly known, the shell gland—part of the reproductive tract.

Decalcification of the inner surface of the egg shell with solutions of ethylene-diamintetraacetic acid results in the release of the shell membrane fibres which penetrated the mammillae. A small residue of organic material which represents the centre of the mammillae can be found attached to the membrane fibres. Such aggregates of organic material attached to the outermost layer of the outer shell membrane are known as organic mammillary cores. By very carefully controlled decalcification of the inner surface of the shell, from which the bulk of the outer and inner shell membranes have been removed, preparations of a thin layer (17 μm) of mammillary membrane have been obtained (Robinson and King, 1963, 1968, 1970; Robinson, 1970). Although Simkiss and Tyler (1957) reported that the organic composition of the mammillary cores and the main part of the shell were basically similar and contained a protein acidic-polysaccharide complex, later studies on the cleaner, washed mammillary membranes have shown that the core itself is composed of a neutral mucin surrounded by weakly acidic substances such as sialomucins (Robinson and King, 1968). Further it was shown that the cores of neutral mucin were firmly bound to the membrane fibres through disulphide bonds, whereas the weakly acidic substances which surround the mammillary cores could be easily washed off to leave the thin mammillary membranes containing the covalently linked true organic mammillary core material. The importance of the organic mammillary core during calcification was emphasized by the findings of Schmidt (1962) and Robinson and King (1963); these workers identified the organic mammillary core as the initial site of calcification during the formation of the egg shell by demonstrating that these cores lie within the first-formed small crystals present on eggs removed from the uteri of laying hens. Although the role played by the neutral mucin of the organic mammillary core in the initiation of nucleation of calcification on the outer shell membrane is unknown, it has been demonstrated histochemically by Robinson and King (1963) that the organic mammillary cores also contain the enzyme carbonic anhydrase. The presence of carbonic anhydrase in egg shell has recently been verified by Krampitz *et al.* (1974a, b). Therefore, although the overall mechanism of shell formation involving the transfer of calcium ion from the uterine gland to the egg shell is unknown, it seems likely that carbonic anhydrase situated at the organic mammillary cores aids shell deposition at these special sites by catalysing the hydration of carbon dioxide and thus increasing the amount of available carbonate anion at the calcification sites.

While much research (Robinson and King, 1970; Carter 1971, 1974, 1976) has shown that the durability of the hen's egg shell as a natural package for the protection of the foodstuff is approximately related to the thickness and curvature of the shell, it is equally obvious on grounds of common sense alone that the tensile strength, and the resistance to impact and snapping forces, must also be dependent on the intrinsic composition and microstructure of the shell. Studies with the scanning electron microscope on the structure of a large number of weak- and strong-shelled eggs obtained from a flock of 20 000 Shaver starcross 288 hybrid hens have shown that the mammillae are poorly formed in weak and thin shells (King and Robinson, 1972). The mammillae of weak egg shells were of irregular shape and very porous and frequently were not attached to the outer shell membrane fibres (Figs 9b and 10b). In weak egg shells the bases of the mammillae were often pointed, and scanning electron microscopy of the radial fracture surfaces of such shells showed that the planes of fracture passed through the mammillae rather than between them as in stronger and thicker shells. In strong egg shells (Figs 9a and 10a) the mammillary layer was found to contain well-defined compact mammillae (average width 70 μm), with approximately 196 mammillae per square millimetre of shell surface. Therefore, in strong and thick egg shells it seems that the individual mammillae are more compact and hence more resistant to fracture. Detailed examination of the mammillae of weak egg shells has shown them to be very porous (Fig. 9b) and it seems likely that such high microporosity may be a cause of the inherent weakness of the material which is analogous to weakness caused in any solid material by natural or man-made holes. It has been proposed that the unknown physiological factors responsible for the formation of holes and small pores in the mammillae may be fundamental to the oviposition of weak and thin egg shells. Thus it is advised that future workers studying the strength of egg shells take into consideration the microstructure of the shell material as revealed by scanning electron microscopy as well as more easily determined factors such as the thickness and the curvature of the shell.

C. Egg Shell Membranes

Although the egg shell membrane forms part of the inner surface of the egg shell and forms a base structure on which the shell is crystallized, only the thin, outermost mammillary layer of the shell-membrane-containing organic mammillary cores is calcified. The remainder of the membrane, which represents approximately 80% (w : w) of the shell membrane substances, has no part in the structure of the shell, and thus the bulk of the shell membranes which can be easily removed by hand is made up of the distinct inner and outer shell membranes which separate at the broad pole of the egg. There is no clearly understood function for the egg shell membranes, but like membranes associated with other calcified tissues, such as the periosteum of bone and the periosteum of mollusc shells, they may be considered to be an essential accompaniment to the calcite layer. The normal thickness of the outer and inner membranes according to Romanoff and Romanoff (1949) is 0·07 mm. The membranes are highly fibrous and completely insoluble, and it is this

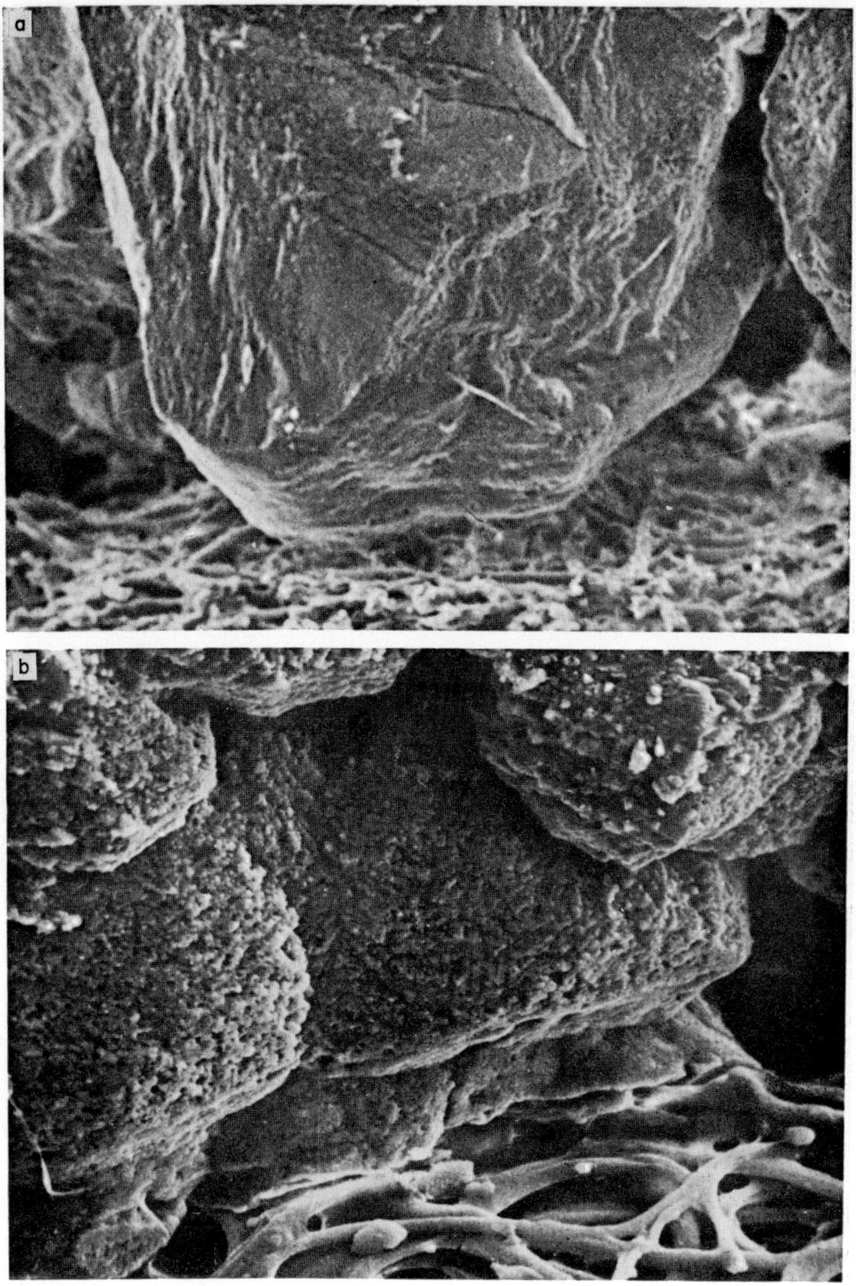

FIG. 9. Scanning electron micrographs of the mammillary cores from (a) strong egg shells, ×1300, and (b) weak egg shells, ×4500.

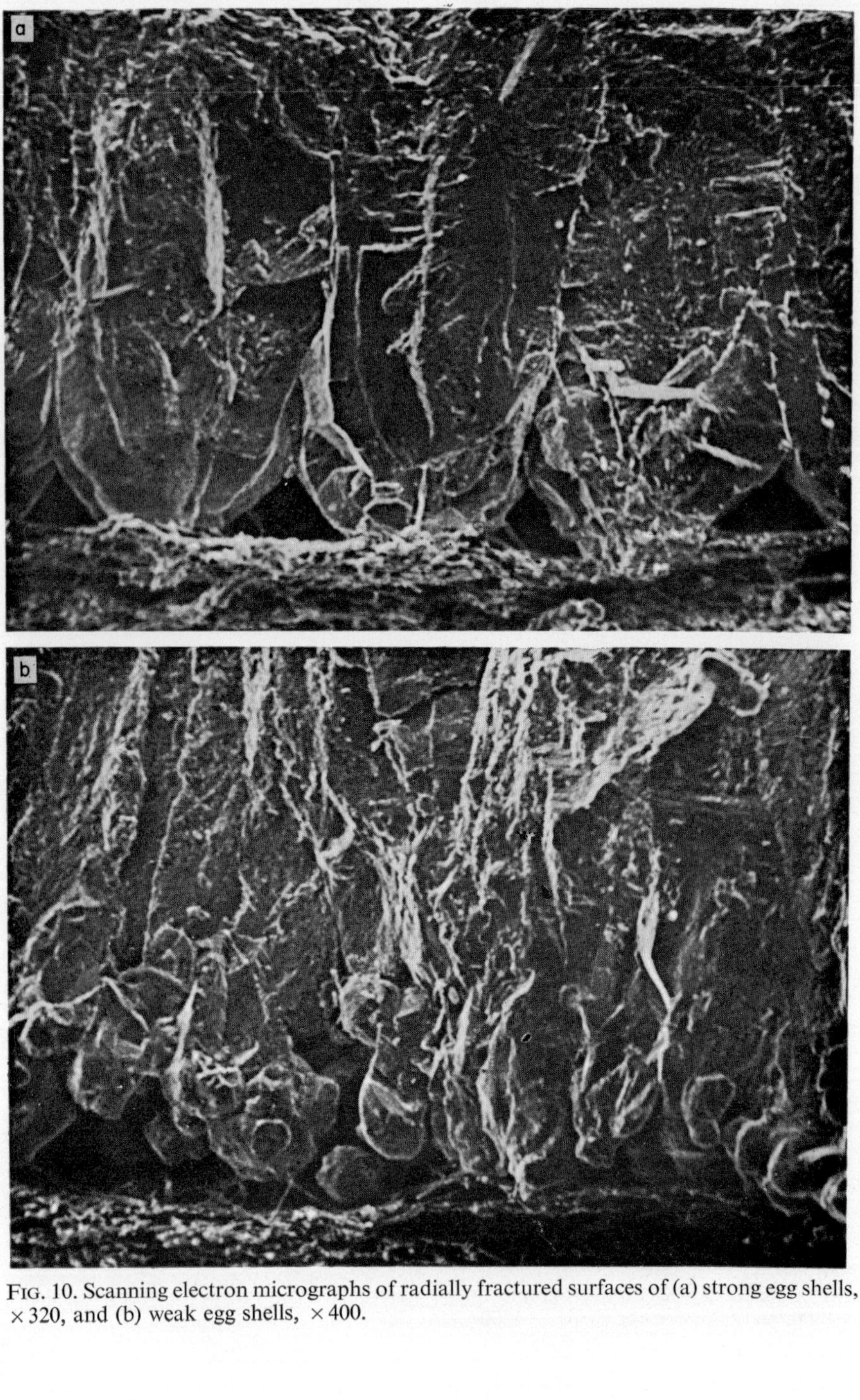

FIG. 10. Scanning electron micrographs of radially fractured surfaces of (a) strong egg shells, $\times 320$, and (b) weak egg shells, $\times 400$.

lack of solubility which is responsible for our very poor understanding of the molecular structure of the membrane fibres. For instance, although it seems probable that the membrane fibre is composed of a basic repeating protein unit, the resistance of the membranes to dissolution without destruction of the chemical structure has prevented the possible identification of such a subunit in terms of either molecular weight or chemical composition.

When the outer shell membrane is bulked such preparations always appear pink; this is due to the presence of ooporphyrin (Klose and Almquist, 1937). The outer shell membrane (48 μm) studied by Moran and Hale (1936) can be teased into three separate layers of approximately equal thickness, each layer being composed of an intricate mesh of fibres with a maximum diameter of 3 μm. The inner shell membrane (22 μm) studied by Simons (1971) was composed of fibres with a maximum diameter of 1·5 μm which seemed to be arranged in three sheets parallel to the shell, the fibres in each sheet lying at right angles to those present in adjacent sheets. However, Masshoff and Stolpmann (1961) observed fibres for the inner and outer membranes with diameters of 0·5–0·1 μm and 0·8–1·0 μm, respectively, which are small when compared with other results of Simons and Wiertz (1963), Moran and Hale (1936), Sajner (1955) and Wolken (1951). Spaces between the fibres of 1–2 μm and up to 28 μm have been measured for inner and outer shell membranes respectively (Wolken, 1951; Von Wittich, 1851). Although the shell membrane fibres are obviously composed of bundles of relatively large fibrils like muscle tissue, it is possible that the fibres might be composed of very small, coiled fibrils which may not yet have been detected in the presence of large amounts of overlying cementing material. Indeed, Masshoff and Stolpmann (1961) claimed that fibrils with a diameter of 3–4 nm were present, but subsequent work by Simons (1971) and Candlish (1972) has not substantiated this. The fibres of the outer shell membrane layers have been shown to be composed mainly of protein, but like the organic mammillary cores they react positively to the periodate–Schiff stain but not to the basophilic dyes such as alcian blue or toluidine blue. Therefore it is thought that the fibres also contain small amounts of neutral polysaccharides, although the location of these latter substances within the fibres themselves has not been identified. The fibre itself has been thought to be composed largely of keratin because of its obvious fibrous structure, insolubility, the absence of significant amounts of hydroxyproline and the presence of substantial amounts of cystine (Baker and Balch, 1962). Histochemically, disulphide groups have been shown to be present in the membrane fibres by Robinson and King (1968). However, in the absence of convincing X-ray data to support the presence of keratin, and due to the known presence of other substances containing glycoproteins and small amounts of hydroxylysine, it can only be assumed that the egg shell membrane fibres contain a keratin-like substance, but one which has a molecular structure different from that of the keratin of skin, hair or feathers. Just as a better understanding of the morphological structure of the egg shell was obtained by the use of electron microscopy, a much greater knowledge of the microstructure of egg shell membrane fibres has been similarly achieved, particularly through the use of transmission electron microscopy. Masshoff and Stolpmann (1961) and Simons and Wiertz (1963) have shown that the outer shell membrane fibres consist of an electron-dense core and a lighter mantle layer.

The inner shell membrane fibres have been shown by transmission electron microscopy to be finer than those of the outer membrane. They are bounded internally by a smooth sheet of material continuous with the fibre mantles which is called the inner layer of the inner membrane. The clear separation of the core material from the outer mantle substances seen in fixed material by electron microscopy has led to the assumption that the core and mantle materials are chemically distinct. However there is little evidence to support this view, particularly as studies to date have been carried out on chemically fixed material, and it seems likely that some of the differences observed may be artefacts introduced during chemical fixation. Simons (1971) found that the cleft between the cores and mantles was approximately 0·1 μm thick, which is less than the cleft thickness mentioned by Masshoff and Stolpmann (1961). Also, the thickness of the mantle varied in the different parts of the membranes between 0·1 and 1·0 μm, being thinner for the inner membrane fibres.

Using a ruthenium red staining technique prior to electron microscopy, Kobayashi and Asboe-Hanson (1970) and Candlish (1972) have concluded that the mantle substances contain mucopolysaccharide. Also, ionic staining of the mantles (Candlish, 1972) rendered them more electron dense, thus indicating the presence of ionic groups possibly associated with mucopolysaccharides. However other studies by Candlish (1970, 1972), using lead citrate and uranium acetate stains prior to electron microscopy, are difficult to interpret as the heavy metal treatments were reported to have been used in reverse sequence. In order to detect the presence of such substances as hyaluronic acid, collagen and amylose by negative means, attempts have been made to digest these specific components by incubation with the appropriate enzymes. From such studies Masshoff and Stolpmann (1961) claimed that the fibre mantles contained hyaluronic acid, whereas Candlish observed no apparent digestion of the fibres by hyaluronidase, trypsin or papain. However the results of such experiments are very difficult to interpret, firstly because the purity of many of the enzyme preparations used is not defined and secondly because their action on well-known model compounds is not described. Despite all these studies which have been carried out on fixed material, we have little understanding of the microstructure of the egg shell membrane fibres; further work must await the development of improved techniques not requiring chemical fixation of the samples, and substantially new staining methods, or even physical methods, for the identification of proteins and specific carbohydrates at the molecular level.

IV. MICROSTRUCTURE OF EGG ALBUMEN

Egg white, which can easily be separated by sieving through a wide-mesh gauze into thin and thick albumen fractions, is known to be a solution containing up to 12% (w : w) of glycoproteins which are mainly soluble (Robinson, 1972). The chemical changes which accompany the gel-to-sol transformation during the liquefaction of the thick albumen fraction have been described by Donovan *et al.* (1970), Kato *et al.* (1970) and Robinson and Monsey (1975).

The thick fraction of albumen which surrounds the yolk has well-defined physical properties, such as rigidity and elasticity, which have been measured using microscopic methods (Brooks and Hale, 1959). When examined under the light microscope a distinct impression of the presence of fibres is obtained, and indeed Brooks and Hale claimed to have measured the rigidity of the gel in and between albumen fibres. However, although albumen does appear at first sight to contain parallel fibres, it seems likely that this may be an optical effect due entirely to the high viscosity of the gel rather than to the presence of distinct fibres. This view is supported by the work of Forsythe and Berquist (1951) who were unable to locate individual fibres in unaltered thick egg white. However disturbance of the gel structure by stirring results in entanglement of the gel, and fibres can then be collected which resemble the chalazea. Thus it does seem that the native gel may not contain fibres as such, but, due to the unusual rheological properties of the gel, fibres may be easily spun from the natural thick albumen which is initially secreted in concentric layers. Although it is generally agreed that the gel property of the thick albumen is due to the glycoprotein ovomucin (Shenstone, 1968), which after concentration by precipitation or centrifugation can be isolated in a fibrous form (Robinson and Monsey, 1975), this is no reason or evidence for supposing that ovomucin fibres exist naturally in the thick egg white gel.

During the period of the last four years transmission electron microscopy has been used to study the ultrastructure of egg albumen for a rather different purpose. Using transmission electron microscopy, Meyer and co-workers have demonstrated the effect of heat on the ultrastructure of albumen. Meyer and Hood (1973) showed that all raw albumens, that is both thin and thick albumens at either pH 7·4 or pH 9·0, possessed a similar ultrastructure in which heterogenous protein matrix was characterized by a continuous protein network (Fig. 11a). Not surprisingly, since the albumen contains 88% water, open spaces were evident within the network. However, in contrast, cooked albumens showed different ultrastructures at pH 7·4 and 9·0. Electron-dense, symmetrical, spherical structures with a diameter of 100–200 μm were present in the samples of albumen cooked at pH 7·4 (Fig. 11b), whereas slightly smaller asymmetrical aggregates consisting of dense particles were observed in cooked albumens at the higher pH value of 9·0. The technique developed by Meyer and Hood (1973) has been extended and applied by Meyer and Potter (1974) to study the effect of various commercially-used additives on the ultrastructural changes induced by heat denaturation during the pasteurization of egg albumen. It is claimed that treatment with lactic acid and aluminium sulphate gives rise to the least denaturation, and the assumption is made that this is due to an increased stability of lysozyme at pH 7·4 and the protection by the aluminium salt of the iron-binding protein, conalbumin. These rather simple but very direct studies of the ultrastructure of albumens have now been extended to investigation of the effects of freezing and thawing on the ultrastructure of pasteurized egg albumen treated with various commonly used additives such as sodium hexametaphosphate, triethyl citrate and trisodium citrate. Meyer and Potter (1975a, b) have suggested that sodium hexametaphosphate and trisodium citrate increase the foam stability of albumen and strengthen the protein matrix of egg albumen against freeze–thaw stress by cross-linking the protein matrix.

V. THE VITELLINE MEMBRANE

The yolk of the egg is contained within the vitelline membrane and it is the ability of this membrane to withstand rupture that is responsible for the maintenance of the yolk as a separate body within the egg. The vitelline membrane is permeable to water and some of the solutes of the egg. However during the storage of eggs there is a slow transfer of water-soluble substances between the yolk and the white. It is claimed that the thickness of the membrane varies between 0·012 and 0·025 mm (Moran and Hale, 1936); however, the membrane is probably more even and smoother than these values suggest since much of the variation can be accounted for by the adsorbtion of varying amounts of the thick egg white gel which surrounds the outer surface of the membrane. Fromm (1967) described the gel-like amorphous material which encircles the vitelline membrane as the fibrous chalaziferous layer of the

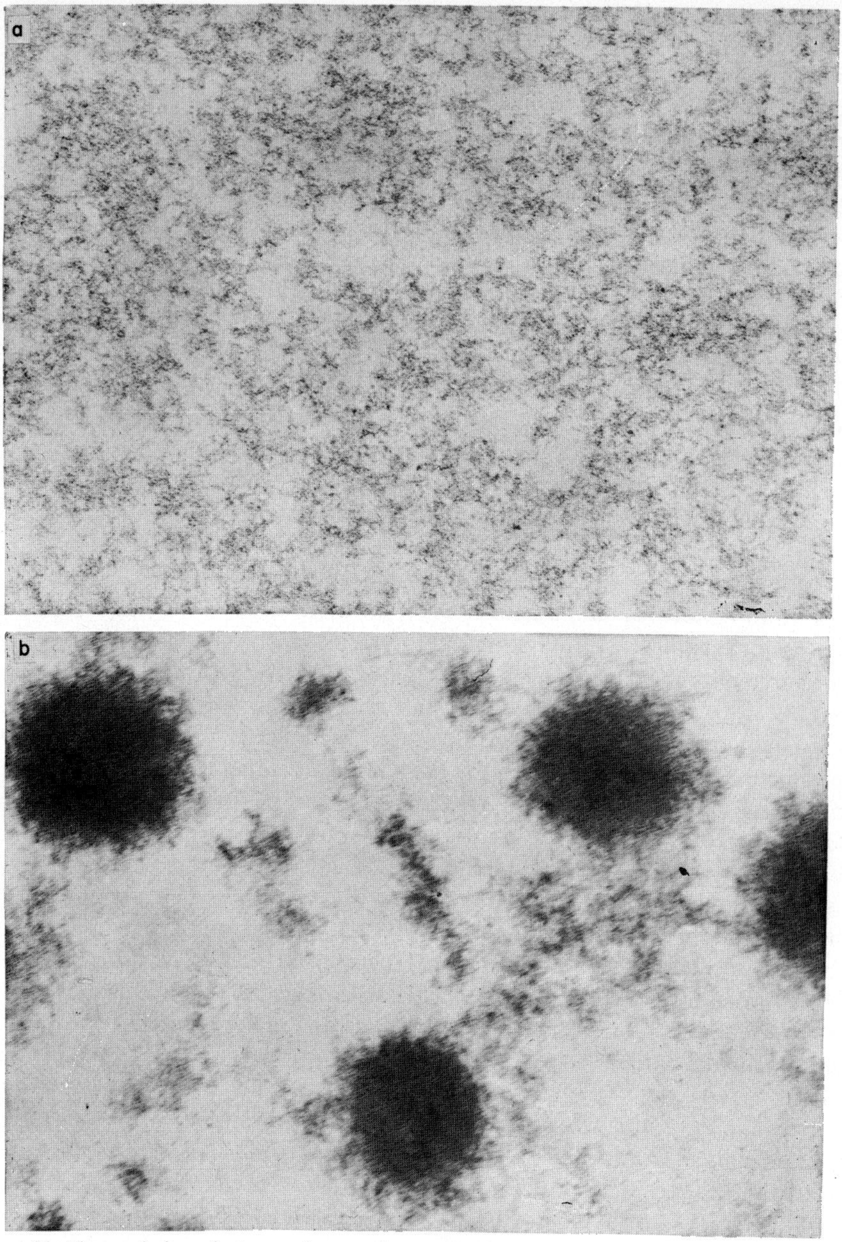

FIG. 11. Transmission electron micrographs, ×90 300, of (a) raw thick egg albumen and (b) cooked thick egg albumen. (Meyer and Hood, 1973.)

membrane and supported the theories of Feeny *et al.* (1956) who claimed that the strength of the vitelline membrane was to a great extent dependent on the quality and nature of the surrounding gel-layer. Consequently, it is now thought that those factors which affect the gelatinous nature of the thick egg white also affect the chalaziferous layer of the vitelline membrane, and thus, in turn, the physical properties of the membrane itself. For as the egg ages and the thick egg white gel liquefies, the vitelline membrane stretches and becomes weak so that the yolk tends to lie flat when the egg is broken out. Thus, the vitelline membrane plays a very important role in the maintenance of egg quality. Additionally, it can become almost impossible to open eggs without rupturing the vitelline membrane and this causes problems for the egg products processor concerned with the preparation of yolk-free albumen.

Until the microscopical studies of Bellairs *et al.* (1963) it was thought that the vitelline membrane was composed of either collagen, elastin (McNally, 1943; Shalumovitch, 1955) or keratin (Moran and Hale, 1936; Wolken, 1951). Bellairs *et al.* showed that the vitelline membrane was composed of two quite distinct layers, an inner layer formed in the ovary and an outer layer added as the yolk enters the oviduct prior to secretion of egg albumen. Measurements made on electron micrographs of fixed membranes indicated that the inner layer (2·7 μm) consists of a meshwork of cylindrical fibres 0·2–0·6 μm in diameter. However, no studies have been carried out on unfixed specimens to date, and in some of the samples studied the fibres were seen to consist of small fibrils with diameters ranging from 15 to 120 nm. The outer layer of vitelline membrane described was approximately four-times thicker (3–8·5 μm) than the inner membrane and was comprised of a variable number of sublayers lying on top of each other. Tangential sections showed that each sublayer of the outer vitelline membrane is composed of fibrils with diameters of approximately 15 nm. No striations were observed in the fibrils, thus suggesting the absence of collagen which was also thought to be absent due to failure of the fixed membrane material to react positively with Verhoeff's stain.

Other studies of the vitelline membrane have either been connected with embryonic development in fertilized eggs or concerned with studies of mottled yolks. Doran and Mueller (1961) have shown in cross-sections of mottled membranes that the chalaziferous layer, which encircles the true vitelline membrane layers of Bellairs, is separated by up to 240 μm from the vitelline membrane at the mottled sites. It is now generally agreed that this separation of the surrounding chalaziferous layer is brought about by increased permeability of the vitelline membrane which, during the later stages of egg formation or during storage of eggs, promotes a more rapid movement of water from the albumen to the yolk and ultimately forces the membranes apart. Generally yolk mottling is associated with an increased moisture content of the yolk, which suggests an increased permeability of the vitelline membrane (Polin, 1957). All factors commonly associated with mottled yolks have been reviewed by Cunningham and Sanford (1973, 1974).

VI. EGG YOLK

Although the intact yolk appears to the naked eye to be an homogeneous liquid contained within the vitelline membrane, it in fact consists of an internal structure made up of alternate layers of yellow and white yolk. The layering is thought to be due to the diurnal rhythm occurring during the formation of the yolk in the ovary of the hen (Riddle, 1911). The layers of yellow and white yolk are 2·0 mm and 0·25–0·40 mm thick respectively (Romanoff and Romanoff, 1949) and each layer has been shown to contain numerous spheres of yolk and granules suspended in a continuous plasma (Bellairs, 1961; Bellairs *et al.*, 1972). It has been thought for many years (Hale and Hardy, 1933; Moran and Hale, 1936) that this organized microstructure of the yolk is partly responsbile for the restricted diffusion of water and salts from the albumen to the yolk during the storage of eggs. For although the difference in osmotic pressure between the yolk and the albumen is of the order of 1·8 atm, as calculated from the freezing points of the white and yolk, the transfer of water to the yolk is still relatively slow at a rate of approximately 10 mg d^{-1} at ambient temperature (Smith, 1931). Further, this slow rate of diffusion cannot be due to the presence of an impermeable vitelline membrane, since the isolated vitelline membrane offers little or no resistance to the passage of water or salts when it is used to separate simple salt solutions (Needham, 1931). Also, Grodzinski (1951) has suggested that each yolk sphere is surrounded by a semi-permeable membrane which he claimed would account for the observed slowness of diffusion of water and salts from the albumen to the yolk.

The morphology of the yolk has been studied by Bellairs (1961) and Bellairs *et al.* (1972). One of the most interesting conclusions to emerge from these studies is that although a large number of different lipoproteins have been isolated from the yolk (Tattrie and Cook, 1968) only a very limited number of discrete types of structural components have been seen by the microscopist. Indeed, it is now thought (Bellairs *et al.*, 1972) that the whole yolk consists of a continuous phase containing ring-like structures measuring 25–30 nm in diameter and numerous suspended white and yellow yolk spheres and smaller granules. The white yolk spheres vary in diameter from 4 to 75 μm (whereas that of the yellow spheres ranges from 25 to 150 μm (Bellairs, 1961)), and are composed of the same basic structure as the rest of yolk in so far as they contain a number of granules suspended in a continuous fluid. Fujii *et al.* (1973) have studied the ultrastructure of isolated yolk spheres by scanning and transmission electron microscopy (Fig. 12). The yolk spheres are the most conspicuous structures found in natural, whole yolk, but little is known of the mechanism of their formation. Also, opinions vary concerning the presence of a membrane surrounding each sphere. The studies of Fujii *et al.* (1973), who used scanning electron microscopy, indicate that a wall made of viscous substances surrounds each sphere, and therefore it is still possible that yolk spheres are contained by something like a membranous structure. However, to date, no preparations of such membranes have been isolated by ultracentrifugation of yolk. A photomicrograph of the granules described by Bellairs is shown in Fig. 13.

During the last 20 years a considerable amount of biochemical research relating to the molecular structure and chemical composition of purified yolk lipoproteins has been carried out (Parkinson, 1966). However, until the combined studies of Bellairs *et al.* (1972) no attempt was made to correlate the results of biochemical investigations with the various distinct morphological structures present in the yolk. The free-floating lipid drops of Bellairs (1961) and the contents of the yolk spheres are now seen to correspond with the granule fraction of the biochemists and are assumed to be rich in lipovitellin and phosvitin. It is also suggested that only the protein–phospholipid part of lipovitellin, together with phosvitin, may correspond to the observed granule particles. Bellairs *et al.* (1972) have additionally claimed that the continuous-phase fluid of the microscopists is the supernatant of the biochemists which consists mainly of low density lipoproteins. The latter are normally isolated by floation during ultracentrifugation and the ring-like structures observed by the microscopist may represent the outer peptide–phospholipid covering of the low density lipoprotein molecule. However, many of these microscopic studies have been carried out on stained, fixed and subsequently dehydrated samples of yolk and the various preparations of purified lipoproteins. Thus it seems likely that during at least some of these procedures the free, neutral lipid parts of the original aggregated molecules may have been extracted. Although some microscopical studies on unfixed samples of negatively stained samples of low density lipoprotein have been reported (Nichols *et al.*, 1969; Kamat *et al.*, 1972; Fujii *et al.*, 1973; Evans *et al.*, 1974), further studies of the various constituents of the yolk are required.

Just as microscopy is being employed to assess the effect of physical treatments and additives on the ultrastructure of albumen, so microscopy, particularly scanning electron microscopy, will be increasingly used to study

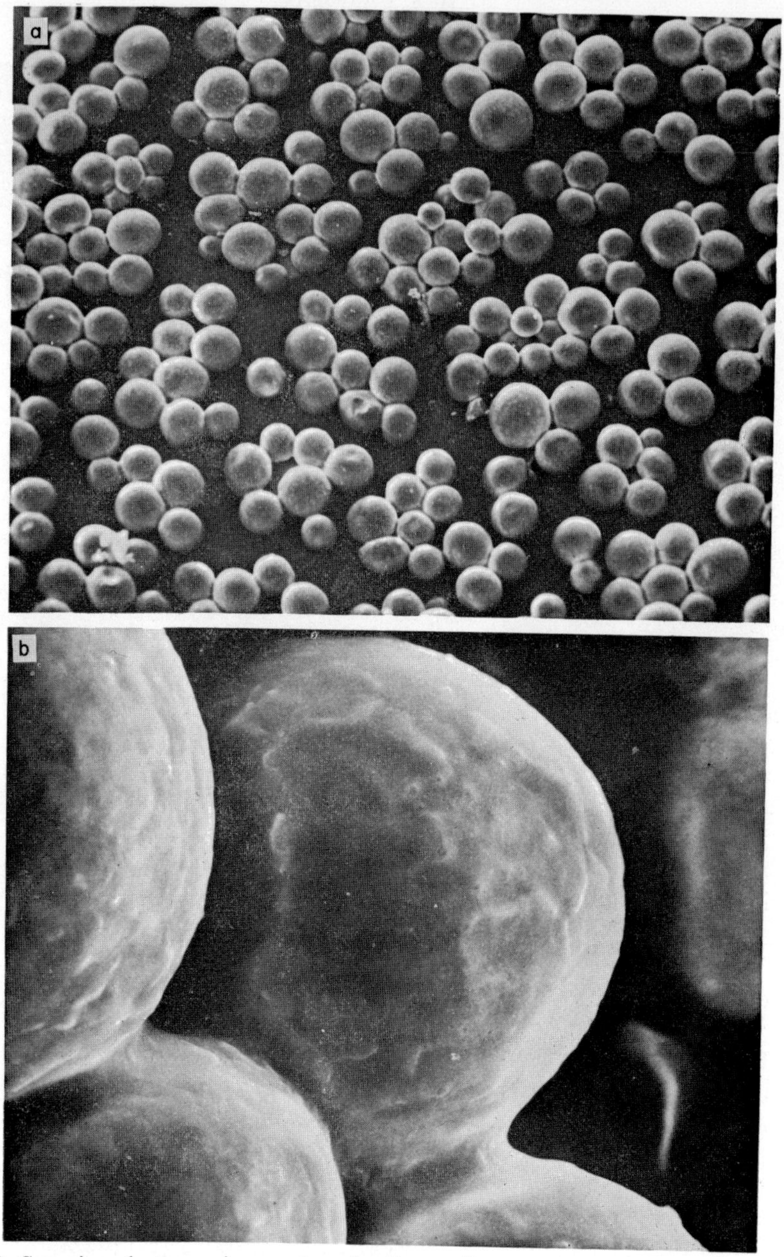

FIG. 12. Scanning electron micrographs of yellow yolk spheres (a) ×100, (b) ×2000. (Fujii *et al*., 1973.)

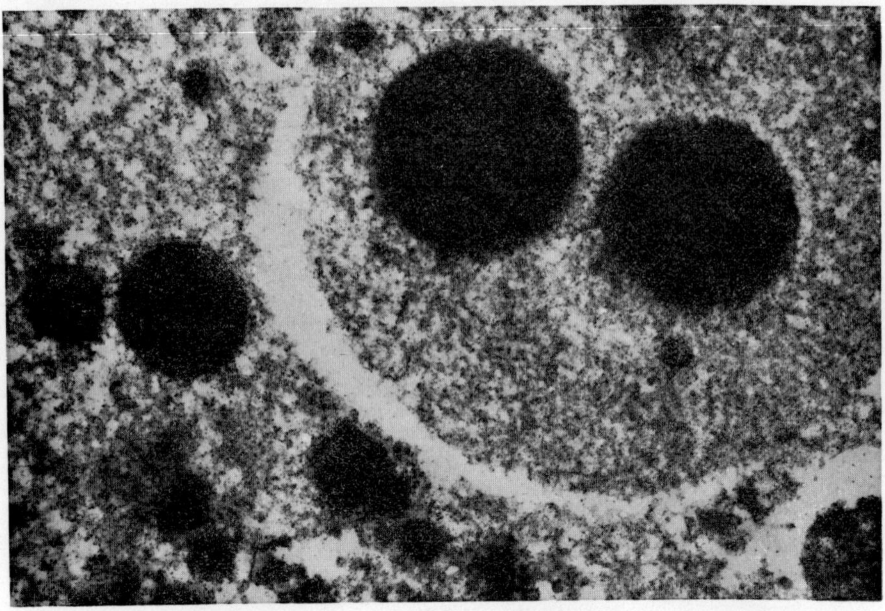

FIG. 13. Transmission electron micrograph of centrifugally sedimented egg yolk showing a yolk sphere containing granules and the continuous phase, ×20 000. (Bellairs *et al.*, 1972.)

the effect of similar physical treatments on the ultrastructure of yolk. Those studies which have been carried out have indicated that yolk lipoproteins may aggregate during freezing processes in such a way that the viscosity of the whole yolk is increased; this might account for the phenomenon of yolk gelation which occurs during freezing (Powrie, 1968; Kamat *et al.*, 1972).

VII. DISTRIBUTION OF EGG IN MIXED FOODS

When whole egg or the separated parts such as albumen or yolk are mixed into savoury foods or confectionery products it might generally be thought that the egg will be evenly distributed throughout the product. However this is not necessarily the case, since the distribution of the various ingredients will depend to a large extent on the efficiency of mixing and on the physical nature of the various ingredients; the latter may vary from a natural liquid form to that of reconstituted powders. Few workers (Fooste and Mackey, 1952; Flint *et al.*, 1970), have studied the microstructure of such prepared foods, and indeed it has been assumed that various proteins and fats of different origin cannot be differentiated from each other (Greethead, 1969;

Rees, 1971). However, the overall composition of proteins from different sources such as egg, milk, cereals and legumes is dissimilar. For example, egg proteins are rich in the sulphur amino acids whereas leguminous proteins are deficient of such amino acids but contain greater quantities of lysine than cereal proteins. Similarly, plant materials contain a high proportion of unsaturated fats whereas foodstuffs of animal origin contain a higher proportion of saturated fats due to hydrogenation in the rumen of ruminant animals such as the cow or the sheep. Nevertheless, since fats are easily extracted from plant and animal products and can be used separately in the baking industry, it is unlikely that specific histochemical methods for the specific identification of animal and plant fats will find application. However, the distribution and location of neutral saturated fats, phospholipids, unsaturated fats and even synthetic emulsifying agents in prepared food mixes during food processing, and ultimately in frozen and cooked products, should add knowledge to our understanding of mixing affinity reactions between ingredients and the desirable physical and chemical compositions required for the production of high quality foods. A study of the distribution of fats and proteins, and in particular egg protein, either in the liquid form or as a reconstituted freeze-dried product, has recently been reported (Passmore, 1977), and many of the matters related to this subject are more fully described in Chapter 14.

ACKNOWLEDGEMENTS

Thanks are due to Mr N. R. King for the photographs of egg shells from work previously carried out with the author at the Agricultural Research Council's Food Research Institute, Norwich.

Appreciation is also expressed to Drs R. Bellairs, R. Meyer and T. Tamura for their generous gifts of photomicrographs of their research on egg yolk and egg white Further, I am grateful to Pergamon Press Ltd. for permission to reprint Fig. 13.

REFERENCES

Baker, J. R. and Balch, D. A. (1962). *Biochem. J.* **82**, 352.
Bellairs, R. (1961). *J. Biophys. Biochem. Cytol.* **11**, 207.
Bellairs, R., Backhouse, M. and Evans, R. J. (1972). *Micron*, **3**, 328.
Bellairs, R., Harkness, M. and Harkness, R. D. (1963). *J. Ultrastruct. Res.* **8**, 339.
Brooks, J. and Hale, H. P. (1959). *Biochim. Biophys. Acta*, **46**, 289.
Cain, C. J. and Heyn, A. N. J. (1964). *Biophys. J.* **4**, 23.

Candlish, J. K. (1970). *Br. Poult. Sci.* **11**, 341.
Candlish, J. K. (1972). *In* 'Egg Formation and Production' (B. M. Freeman and P. E. Lake, eds), Vol. 8, p. 87. Poult. Sci. Symp. No. 8, Br. Poult. Sci. Ltd.
Carter, T. C. (1971). *Br. Poult. Sci.* **12**, 57.
Carter, T. C. (1974). *Br. Poult Sci.* **15**, 507.
Carter, T. C. (1976). *Br. Poult. Sci.* **17**, 199.
Cunningham, F. E. and Sanford, P. E. (1973). *Feedstuffs*, **45**, 27.
Cunningham, F. E. and Sanford, P. E. (1974). *Wld's Poult. Sci. J.* **30**, 103.
Donovan, J. W., Davis, J. G. and White, L. M. (1970). *Biochim. Biophys. Acta*, **207**, 190.
Doran, B. M. and Mueller, W. J. (1961). *Poult. Sci.* **40**, 474.
Evans, R. J., Baker, D. H. and Flegal, C. J. (1974). *Poult. Sci.* **53**, 645.
Favejee, J. Ch. L., van der Plas, L., Schoorl, R. and Floor, P. (1965). *Biophys. J.* **5**, 359.
Feeney, R. E., Weaver, J. M., Jones, J. M. and Rhodes, M. B. (1956). *Poult. Sci.* **35**, 1061.
Flint, F. O., Moss, R. and Wade, P. (1970). *Fd Trade Rev.* **40** (4), 32.
Fooste, M. E. and Mackey, A. L. (1952). *Stain Technol.* **18**, 27.
Forsythe, R. H. and Berquist, D. H. (1951). *Poult. Sci.* **30**, 302.
Fromm, D. (1967). *J. Fd. Sci.* **32**, 52.
Fujii, S. (1974). *J. Fac. Fish. Anim. Husb.* **13**, 29.
Fujii, S., Tamura, T. and Okamoto, T. (1973). *J. Fac. Fish. Anim. Husb.* **12**, 1.
Greethead, G. F. (1969). *Fd Technol., Aust.* **21**, 228.
Grodzinski, Z. (1951). *Biol. Rev.* **26**, 253.
Hale, H. P. and Hardy, W. B. (1933). *Proc. R. Soc. B*, **112**, 473.
Heaney, R. K. and Robinson, D. S. (1976). *Biochim. Biophys. Acta*, **451**, 133.
Heyn, A. N. J. (1963). *J. Ultrastruct. Res.* **8**, 176.
Kamat, V. B., Lawrence, G. A., Barratt, M. D., Darke, A., Leslie, R. B., Shipley, G. G. and Stubbs, J. M. (1972). *Chem. Phys. Lipids*, **9**, 1.
Kato, A., Nakamura, R. and Sato, Y. (1970). *Agric. Biol. Chem.* **34**, 1009.
King, N. R. and Robinson, D. S. (1972). *J. Microsc.* **95**, 437.
Klose, A. A. and Almquist, H. J. (1937). *Poult. Sci.* **18**, 173.
Kobayashi, T. and Asboe-Hanson, G. C. (1970). Micro 70 Abstr., p. 231. Royal Microscopical Society, London.
Krampitz, G., Engels, J. and Helfgen, I. (1974a). *Experientia*, **30**, 228.
Krampitz, G., Engels, J., Heindl, I., Heinrich, A., Hamm, M. and Faust, R. (1974b). *Arch. Gefluq.* **6**, 197.
McNally, E. H. (1943). *Poult. Sci.* **22**, 40.
Masshoff, W. and Stolpmann, H. J. (1961). *Z. Zellforsch. Mikrosk. Anat.* **55**, 818.
Meyer, R. and Hood, L. F. (1973). *Poult. Sci.* **52**, 1814.
Meyer, R. and Potter, N. N. (1974). *Poult. Sci.* **53**, 761.
Meyer, R. and Potter, N. N. (1975a). *Poult. Sci.* **54**, 241.
Meyer, R. and Potter, N. N. (1975b). *Poult. Sci.* **54**, 727.
Moran, T. and Hale, H. P. (1936). *J. Exp. Biol.* **13**, 35.
Needham, J. (1931). *J. Exp. Biol.* **8**, 330.

Nichols, A. V., Forte, G. M. and Coggiola, E. L. (1969). *Biochim. Biophys. Acta,* **175**, 451.
Parkinson, T. L. (1966). *J. Sci. Fd Agric.* **17**, 101.
Passmore, N. (1977). 'The Microstructure of Cakes and their Batters.' Ph.D. Thesis, University of Leeds.
Polin, D. (1957). *Poult. Sci.* **36**, 831.
Powrie, W. D. (1968). *In* 'Recent Advances in Food Science' (J. Hawthorne and E. J. Rolfe, eds), Vol. 4, p. 319. Pergamon Press, Oxford.
Rees, R. F. (1971). *Bak. Ind. J.* **4**, 27.
Riddle, O. (1911). *J. Morph.* **22**, 455.
Robinson, D. S. (1970). *Ann. Biol. Anui. Bioch. Biophys.* **10**, 27.
Robinson, D. S. (1972). *In* 'Egg Formation and Production' (B. M. Freeman and P. E. Lake, eds), Vol. 8, p. 65. Poult. Sci. Symp. No. 8, Br. Poult. Sci. Ltd.
Robinson, D. S. and King, N. R. (1963). *Nature, Lond.* **199**, 497.
Robinson, D. S. and King, N. R. (1968). *Jl R. Microsc. Soc.* **88**, 13.
Robinson, D. S. and King, N. R. (1970). *Br. Poult. Sci.* **11**, 39.
Robinson, D. S. and Monsey, J. B. (1975). *Biochem. J.* **147**, 55.
Robinson, D. S., King, N. R. and Bowen, D. J. (1968). *Histochemie,* **13**, 97.
Romanoff, A. L. and Romanoff, A. J. (1949). 'The Avian Egg.' John Wiley and Sons, New York.
Sajner, J. (1955). *Acta Anat.* **25**, 141.
Schmidt, W. J. (1957). *J. Orn., Lpz.* **98**, 432.
Schmidt, W. J. (1962). *Z. Zellforsch. Mikrosk. Anat.* **57**, 848.
Shalumovitch, V. D. (1955). *Dokl. Acad. Nauk. SSSR,* **105**, 584.
Shenstone, F. S. (1968). *In* 'Egg Quality: A Study of the Hen's Egg' (T. C. Carter, ed.), p. 41. Oliver and Boyd, Edinburgh.
Simons, P. C. M. (1971). Agricultural Research Report No. 758, Centre for Agric. Pub. and Doc., Wageningen, Holland.
Simons, P. C. M. and Wiertz, G. (1963). *Z. Zellforsch. Mikrosk. Anat.* **59**, 555.
Simons, P. C. M. and Wiertz, G. (1970). *Ann. Biol. Anim. Biochim. Biophys.* **10**, 31.
Smith, A. J. M. (1931). Ann. Rept. Fd Invest. Bd, London, p. 151.
Simkiss, K. and Tyler, C. (1957). *Q. Jl Microsc. Sci.* **98**, 19.
Simkiss, K. and Tyler, C. (1958). *Q. Jl Microsc. Sci.* **99**, 5.
Tattrie, N. H. and Cook, W. H. (1968). *In* 'Surface-active Lipids in Foods', p. 92. Monogr. No. 32, Society of Chemical Industry, London.
Terepka, A. R. (1963a). *Expl Cell Res.* **30**, 171.
Terepka, A. R. (1963b). *Expl Cell Res.* **30**, 183.
Tyler, C. (1964). 'Wilhelm Von Nathusius (1821–1899) on Avian Eggshells.' University of Reading.
Wittich, Von (1851). *Z. Wiss. Zool.* **3**, 213.
Wolken, J. J. (1951). *Anat. Rec.* **111**, 79.

9. Fish

P. HOWGATE

*Torry Research Station,
Aberdeen, U.K.*

I. INTRODUCTION

This chapter is concerned with the structure of fish muscle and how it is affected by processing. The word 'fish' is used here in the sense it has in the fishing and fish-processing industries, that is, to include all aquatic animals, vertebrate and invertebrate, marine and freshwater, but excluding warm-blooded creatures like seals and whales. The review is directed to food scientists and technologists, and hence is restricted to the part of the animal that is usually and selectively eaten, i.e. the muscle tissue. Other organs and tissues are used as food (skin and gonads, for example, quite commonly so) but to cover all parts that could possibly be eaten would considerably lengthen this chapter with very little profit to the food scientist.

Before dealing with the main subject, the status of fish as a food commodity and the special features that distinguish fishing from the production of other animal foods will be briefly described.

A. Fish as Food

Fish is an important food commodity providing first-class protein and, from some species, lipids. The world catch in 1975 was 68·5 million tonnes, though only 48 million tonnes were used directly for human food, almost all the rest being reduced to fish meal for compounding into animal feed (Food and Agriculture Organization, 1976a). Globally, fish is not as important a source of animal protein as are meat, poultry and dairy products. For example, in 1974 the world production of meat, including poultry, was 116 million tonnes (Food and Agriculture Organization, 1976b). Altogether fish provides about 15% of the total world supply of animal proteins used for food. This overall figure obscures a wide variation in the relative importance of the various protein sources amongst countries and communities. In Western societies with a high standard of living, meat and dairy products provide most of the animal protein and fish plays a relatively minor part in the diet. In many developing countries, particularly in Asia, and also in some advanced countries like Japan, fish may be the main source of first-class protein.

One feature that distinguishes fish as a commodity from other sources of animal foods is the wide range of species that are consumed around the world. A complete list, were it possible to prepare one, would number several thousands. This is in sharp contrast to the mere handful utilized by the meat and poultry industries. Two main factors account for this disparity: the fact that in the main fish are not farmed but are hunted or collected, and the richness of the fauna in the aquatic environment.

So far as fishing is concerned man is still predominantly a hunter–gatherer. The same basic techniques of fishing, the net, the hook and line, and the trap, are now used as they were in the past. What characterizes modern fishing is the use of power to handle heavier gear, the use of powered craft to exploit more distant fishing grounds and the use of electronic aids to find the fish. The amount of fish that is deliberately farmed is still a small proportion of total consumption and as yet there has been almost no selective breeding for characteristics other than size. Man is an opportunist hunter and gatherer of fish, willing to consume most of what he gathers; fortunately few aquatic animals are intrinsically inedible, and truly poisonous species are rare.

All the zoological phyla are represented in the aquatic environment, though three of them, the Chordata, Arthropoda and Mollusca, contain virtually all the species used for food. Food-fish from other phyla are of little relative importance and will not be included in this review. About 90% of the world catch consists of chordates, which in the rest of this chapter will be referred to less pedantically as vertebrates to distinguish them from invertebrates, animals from the other phyla. Once again certain classes of vertebrates like frogs and turtles, though coming within my original definition of fish, are of minor importance so far as food is concerned and will not be considered further. Almost all the rest of the vertebrate food animals come from the Super-order Teleostei, the true bony fishes. About 2% come from the Class Elasmobranchii, cartilaginous fishes like sharks, dogfish, skates and rays (Food and Agriculture Organization, 1976a). (Classification taken from Love (1970).)

Members of the phyla Arthropoda and Mollusca, though providing only 10% by weight of the food catch, are commercially very important as many are highly prized and have a high value. The arthropods include such varieties as crabs, lobsters, crayfish, prawns and shrimps. The food molluscs come principally from three classes: Gastropoda, whelks and winkles; Lamellibranchia, oysters, mussels, cockles and other bivalve shellfish; and Cephalopoda, squids and octopuses.

II. MORPHOLOGY OF FISH MUSCLE

The body muscle of vertebrate fish is divided into segments, or myotomes, separated by sheets of connective tissue, or myocommata. When seen at the surface of the body after the skin has been stripped off, the myocommata have the shape of broad Ws set across the long axis of the body with their central points directed towards the head. This simple shape is only superficial; in three dimensions a myotome can be visualized as consisting of two hollow cones, one on each side of a plane bisecting the fillet at the level of the backbone, with the apices of the cones pointing towards the tail. The cones of one myotome nest into those of the next (Le Danois, 1952). (This arrangement can be readily seen in many species of fish where the myocommata are not stable to heating. In these cases the myocommata break down leaving the myotomes, still in their original shapes, separated into the flakes so characteristic of cooked fish.) The muscle fibres extend across the myotomes approximately parallel to the backbone with their ends inserted into the myocommata. Alexander (1969) has described the orientation of fibres in the myotomes and its influence on the mechanics of the swimming motions of fish.

Two types of muscle tissue, differing in colour, can be distinguished in the body muscle of fish. In most species the bulk of the flesh is white or nearly so; in migratory species such as herring, mackerel and salmon, it is light brown or pink. At the surface of the muscle and lying in a fairly broad band along the mid-line is a strip of darker muscle; in white-fleshed species it is reddish, in darker-fleshed species it is brown. Conventionally, in all varieties, the lighter coloured flesh is referred to as white muscle, the darker as red. The proportion of red muscle compared with white varies among species. In many the former is not confined to a layer just beneath the skin but also penetrates inwards as a broadly-based wedge towards the backbone (Love, 1970). There are differences in the chemistry and physiology of the two types of muscle (Love, 1970) and though there has been some controversy over their respective functions the relative amounts appear to be related to the activity of the fish (Boddeke *et al.*, 1959; Bone, 1966). There are also differences in the microscopical structure which will be dealt with in a later section.

Crustacean shellfish have an exoskeleton and the muscles are attached to the inner walls of the shells. The muscle of the abdomen is divided into segments, or somites, and is only loosely attached to the shell. This muscle is well developed in shellfish like lobsters, prawns and shrimps but in crabs the abdomen is much reduced and most of the edible flesh comes from the legs and claws. The muscle fibres in the legs and claws are in the form of sheets or bundles attached to the shell on either side of the joints.

The muscles of molluscs are quite different from the previous examples. In bivalve shellfish the main muscle tissue is the adductor muscle. Its morphology and function vary amongst species. In most of the shellfish used for food, such as mussels, clams and oysters, its function is to hold the valves closed against the spring of the shell. In others, cockles and razor shells for example, the muscle is also adapted for burrowing (Morrison and Odense, 1973). Two portions are often visible to the eye, one translucent, the other more opaque. There are differences in physiology, function and microscopical anatomy between the two which will be dealt with in a later section. In cephalopods the mantle that surrounds the viscera of the animal consists of muscle tissue which in squids is stiffened in part by an endoskeleton, the cuttlebone or pen. The mantle appears homogeneous and without any obvious structure to the naked eye, but it consists of two layers of muscle, one circular, the other radial (Ward and Wainwright, 1972). This enables the mantle to contract and expand and drive the animal along by a form of jet propulsion (Ward, 1972). In gastropod molluscs most of the muscle tissue is in the head–foot, the part of the animal that protrudes from the shell when it is fully extended, though there are some muscles in the visceral mass within the shell. The anterior portion is the head region which contains the mouth and carries the eyes on stalks. The head–foot is the locomotory organ and the animal moves by a succession of waves passing along the sole of the foot region. The morphology of the musculature is complex (Plesch *et al.*, 1975) and the way the muscles are disposed to produce the pedal waves has been described by Gainey (1976). The foot extends dorsally as a body wall enclosing some of the organs.

III. STRUCTURE OF VERTEBRATE MUSCLE FIBRE

A. White Muscle

Vertebrate fish muscle is composed of striated muscle fibres. When viewed in longitudinal section by light microscopy they are seen to be divided into alternate A-(anisotropic) and I-(isotropic) bands, the latter being narrower than the former. The I-bands are divided by thin Z-lines which also delimit the sarcomeres, the regularly repeating units of the fibre. The A-band is divided by a clearer region, the H-zone, which in its turn is divided by a diffuse line, the M-line. In these respects vertebrate fish muscle fibres are similar to striated muscle fibres in other animals.

Within the fibre and running the length of it are fibrils surrounded by sarcoplasm. The arrangement in cod is shown in transverse section in Figs 1, 2, 11 and 17. At the periphery of the fibre the fibrils are elongated in section, with the long axes approximately normal to the surface of the cell. The entire fibril then has a ribbon-like shape. Fibrils in the interior of the cell are roughly polygonal in section and have a rod-like shape. This arrangement of peripheral ribbon-like fibrils and inner rod-like fibrils seems to be a common pattern in fish muscle and has been observed in many species by light microscopy (Veratti, 1902; Bühn, 1940; Barets, 1952; Nakajima, 1969; Walker, 1970) and electron microscopy (Buttkus, 1963; Bishop and Odense, 1967; Kilarski, 1967; Nishihara, 1967; Liljemark, 1969; Nakajima, 1969; Patterson and Goldspink, 1972). However, a few examples have been reported where the ribbon-like fibrils are centrally disposed rather than at the periphery (Hamoir, 1955). Krüger (1952) describes two types of fibrils in vertebrate muscle which he called ‘Felderstruktur’ and ‘Fibrillenstruktur’ for ribbon-like and rod-like respectively, but in higher vertebrates a muscle fibre will contain one type or the other. He associated the Felderstruktur type with tonic, slow-acting muscle fibres, and Fibrillenstruktur with tetanic, fast-acting fibres. These conclusions were based on examples of mammalian and avian muscle and cannot be extended to the situation in fish where both types of fibril are present in a fibre.

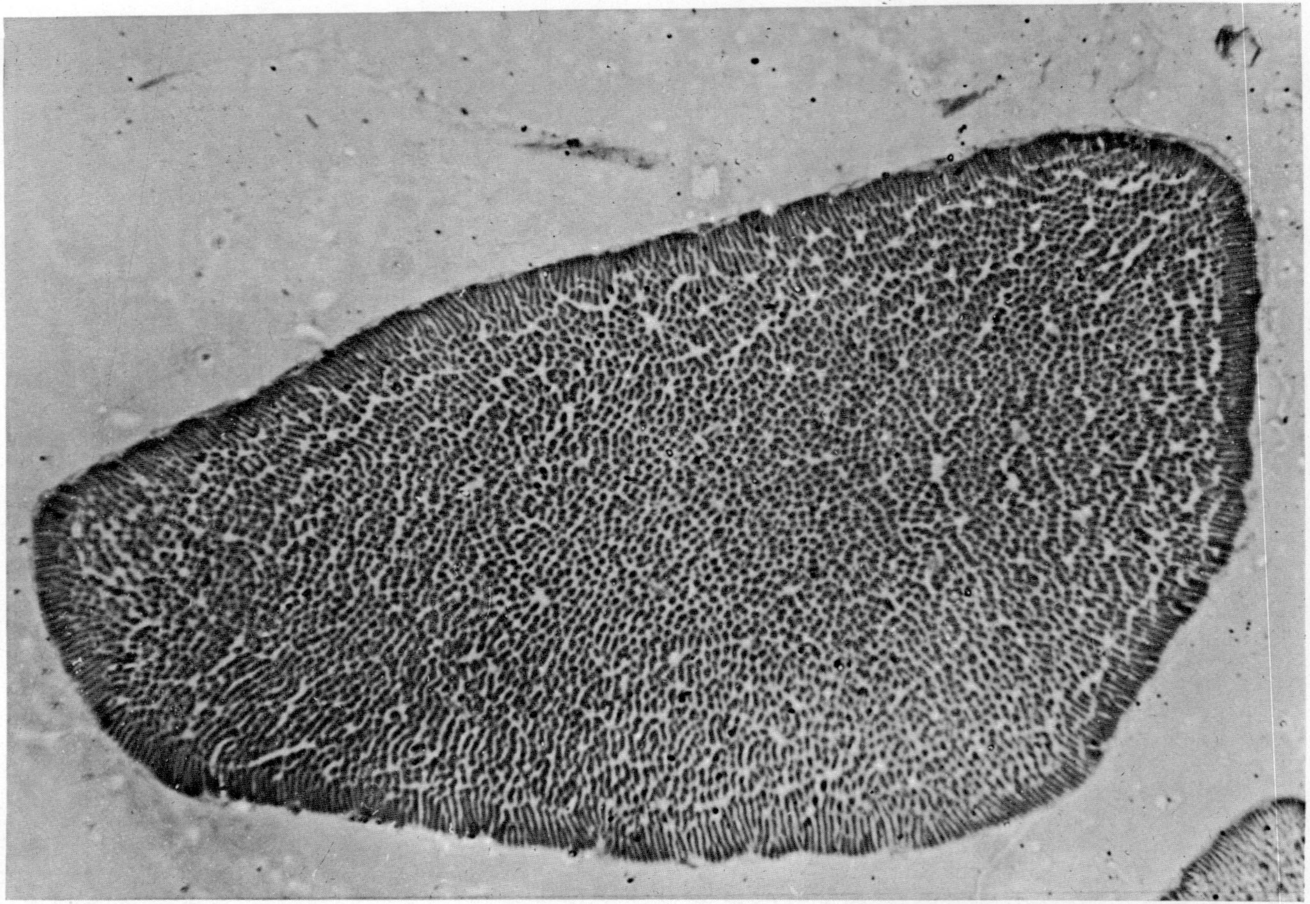

FIG. 1. Transverse section of cod muscle fibre, ×1000. Specimen prepared for electron microscopy, embedded in Epon and a section viewed by phase contrast microscopy.

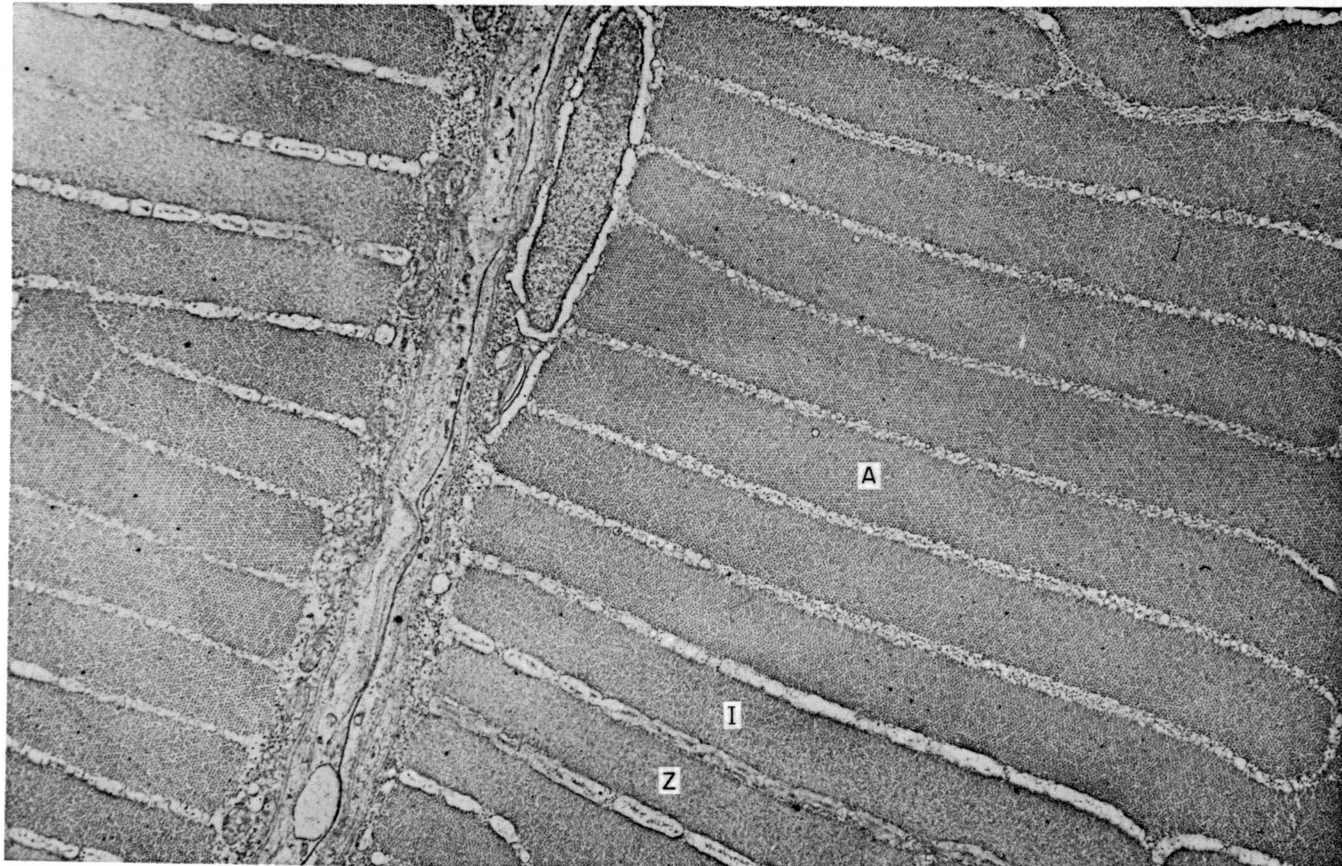

FIG. 2. Electron microscope photograph of a transverse section of two muscle fibres of cod, ×1020. The fibrils are elongated near the edges of the fibres. A nucleus is present just beneath the sarcolemma of the right-hand fibre and mitochondria are similarly situated in the left-hand fibre. The section is cut slightly oblique to the long axis of the fibre and passes through the regions marked in the right-hand cell. The specimen was fixed in glutaraldehyde, post-fixed in osmium tetroxide and embedded in Epon. A = A-band; I = I-band; Z = Z-line.

A muscle fibre is a single cell and the cell wall is referred to as the sarcolemma. In electron microscope sections this can be seen to consist of the plasmalemma, the semi-permeable membrane proper, and a basement layer. In glutaraldehyde-fixed specimens of 'live' muscle the plasmalemma can be seen to have frequent invaginations that are pinocytotic vesicles (Liljemark, 1969) (Fig. 3). The basement membrane is diffuse and contains collagen fibrils. It is continuous with the myocommata (Love *et al.*, 1969). The muscle fibre is a large multi-nucleate cell and the nuclei lie in the sub-sarcolemmal space (Fig. 2). White muscle cells contain few mitochondria most of which lie in the sub-sarcolemmal space and only occasionally among the fibrils. They are small with rather poorly developed cristae.

The fine structure of the fibrils can be seen by electron microscopy (Figs 4 and 5). The sarcomere has two sets of interdigitating filaments, thin and thick. Thin filaments are composed predominantly of the proteins actin, tropomyosin and troponin and are connected to the Z-line and extend into the sarcomere. In the resting muscle, however, they do not meet those filaments originating from adjacent Z-lines. Thick filaments, composed of the protein myosin, run from the centre of the sarcomere towards the Z-lines, but they do not meet them when the muscle is in the normal relaxed state. The I-band is the region containing only thin filaments, and the A-band contains both. The H-zone is the region between the ends of the thin filaments and contains only thick filaments interconnected by protein links which give rise to the denser M-line. When seen in cross-section the thick filaments are arranged in a regular hexagonal array. In the overlap region each thick filament is surrounded by six thin filaments which are arranged in a triangular array. The unit of the pattern is formed of an equilateral triangle with thick filaments at the vertices and a thin filament at the intersection of the altitudes (Fig. 14). Six triangles fit together to form a regular hexagon. Since each thick filament contributes to six adjacent triangles, there are twice as many thin filaments as thick. Transverse sections of the various regions are illustrated in Figs 7b, c.

The arrangement of filamentous structure described above was first reported by Huxley (1957) in rabbit muscle but has been confirmed in other vertebrate animals including fish (Peachey, 1961; Buttkus, 1963; Slautterback, 1966; Bishop and Odense, 1967; Kilarski, 1967; Nishihara, 1967; Liljemark, 1969; Nakajima, 1969; Gas, 1972; Patterson and Goldspink, 1972; Schattenberg, 1973; Jarenbäck and Liljemark, 1975a). Jarenbäck and Liljemark (1975a) have published excellent photographs of the ultrastructure of cod myofibrils using freeze-etching techniques. Partmann (1967) has reviewed the fine structure of vertebrate fish muscle.

Within the I-band the thin filaments are not regularly arranged except where they are attached to the Z-line. In favourably orientated longitudinal sections of a fibril the Z-line can be seen to have a zig-zag configuration with thin filaments attached at the angles. In transverse section the Z-line is seen to be a disc showing a tetragonal pattern (Fig. 7a). The structure of the Z-disc in frog muscle has been described by Knappeis and Carlsen (1962) and can be visualized as being like the trays in an egg box. Thin filaments are attached to the protruding peaks of the disc and extend into the sarcomeres at each side.

As the muscle contracts and relaxes so will the sarcomeres decrease and increase in length. My own measurements on cod muscle fixed immediately after death give an average value of 1·46 μm for the sarcomere length. Measurements made on the illustrations in the papers by Bishop and Odense (1967), Liljemark (1969) and Jarenbäck and Liljemaɪk (1975a, b) on the electron microscopy of cod muscle give values of 1·51 μm, 1·63 μm and 1·65 μm respectively. Walker (1970), using light microscopy, measured sarcomere lengths of fibres, fixed in formalin, from a large number of cod and found the average value for fish over 20 cm long to be 2·1 μm; this is considerably greater than the electron microscopy values. The sarcomere lengths of other species of fish measured from plates in published papers or quoted by the authors range from 1·4 μm to 2·4 μm.

The estimated lengths of the thick filaments, i.e. the A-band, are more consistent. My own data from about 40 measurements on six cod give an average value of 1·28 μm which agrees very well with the value quoted by Bishop and Odense (1969) of 1·3 μm and with measurements made on plates in published papers on cod (Liljemark, 1969; Jarenbäck and Liljemark, 1975a, b). The values for other species, once again measured on published photographs, fall mostly in the range 1·20–1·32 μm (Peachey, 1961; Slautterback, 1966; Nishihara, 1967; Gas, 1972). These values are less than that usually quoted for other vertebrate skeletal muscle ,1·5 μm, and it may be a feature of vertebrate fish muscle that the thick filaments are shorter than those in higher animals.

The fibrils are surrounded by sarcoplasm and a well-structured reticulum. The reticulum has two parts; the transverse tubular (T) system and the sarcoplasmic reticulum (SR) proper. These are best seen in the electron miscroscope in longitudinal sections. Where the section passes through the fibrils the reticulum has a triad structure at the level of the Z-disc (Figs 4 and 5). The central element is a section of the tubes making up the T system. These tubes are interconnected, forming a network across the fibre at the level of the Z-disc and surrounding each fibril. The tubes open out through the sarcolemma into the extracellular space. The membrane of the T system

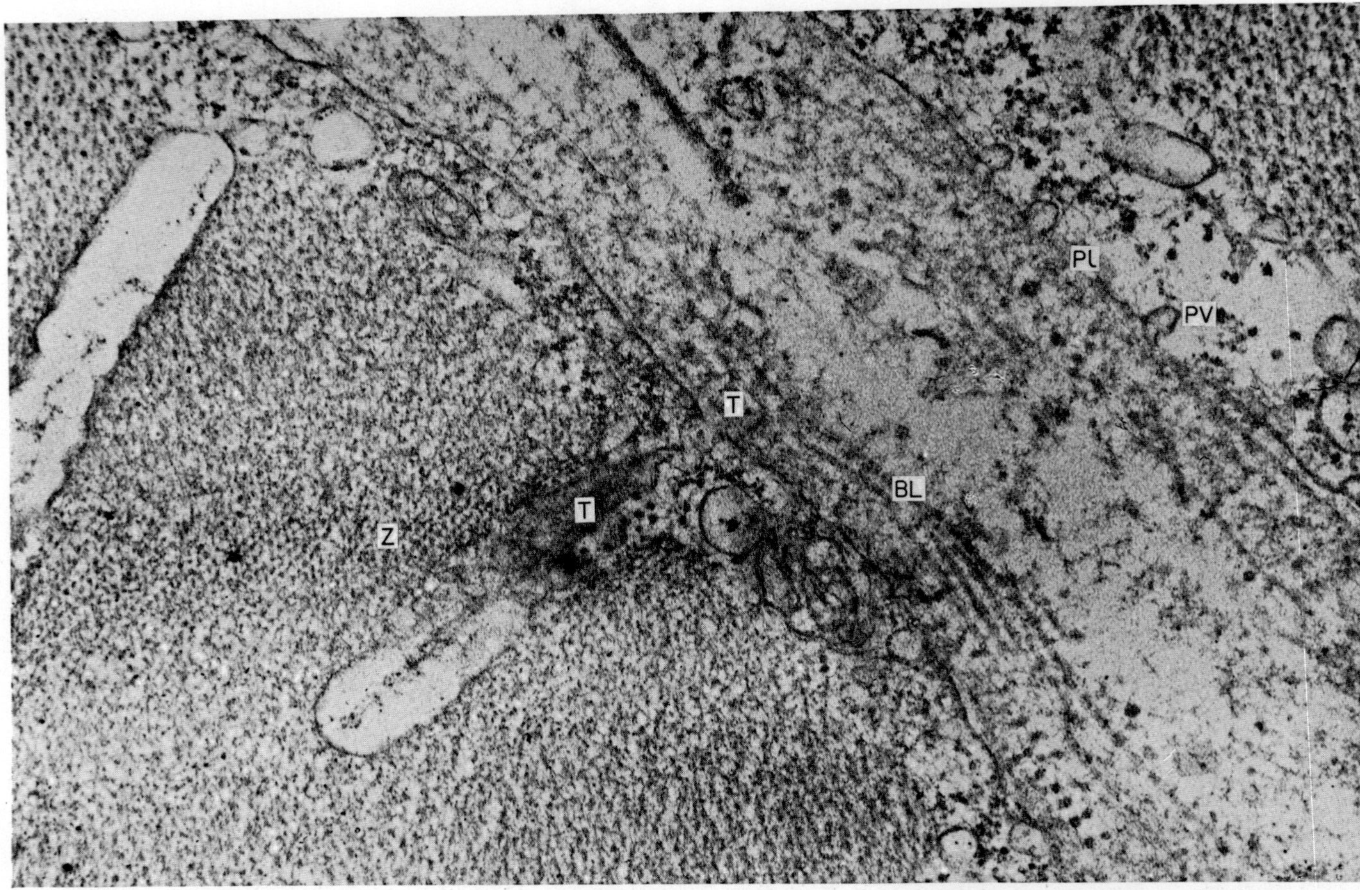

FIG. 3. Approximately transverse section of cod fibres illustrating the structure of the sarcolemma, × 51 000. The section is at an angle to the long axis of the fibre, hence the filaments have an elongated cross-section. The T system (T) at the level of the Z-line (Z) is shown opening into the extracellular space. The sarcolemma consists of the plasmalemma (Pl) with pinocytotic vesicles (PV) and basement layer (BL).

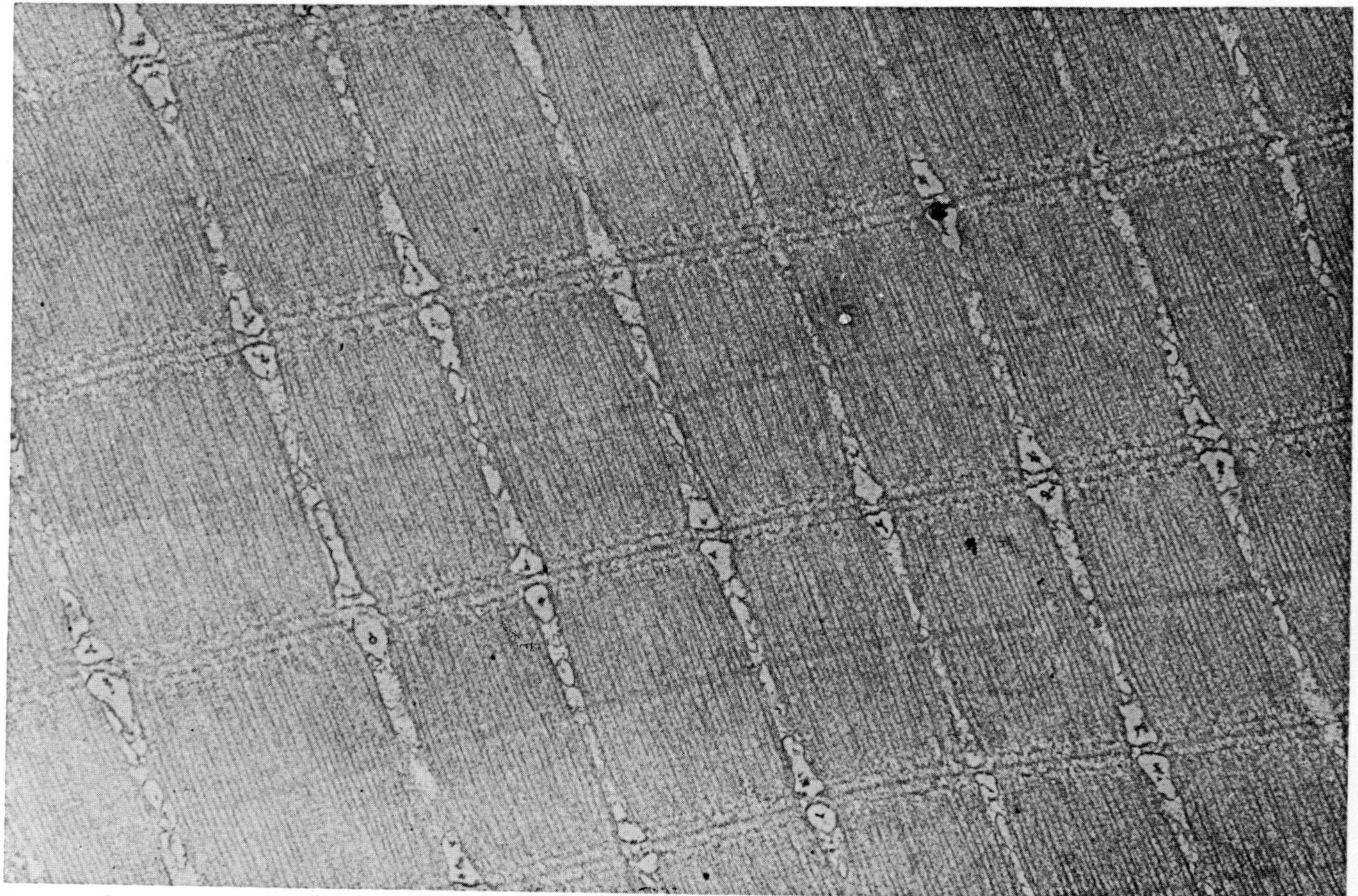

FIG. 4. Longitudinal section of cod muscle fibre away from the edge, $\times 25\,800$. The fibrils are narrow and there are no mitochondria or other cell inclusions other than the reticulum.

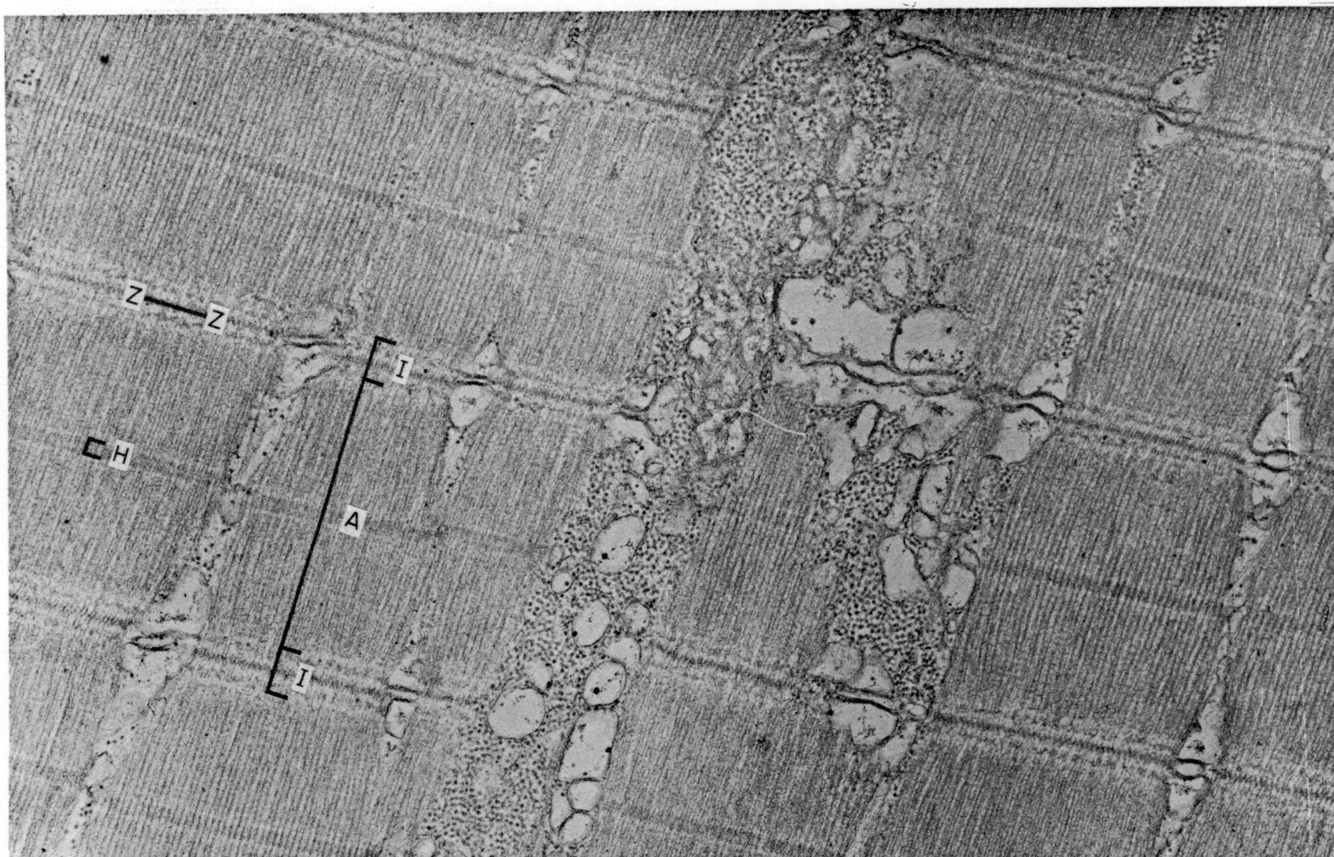

FIG. 5. Longitudinal section of cod muscle fibrils, ×25 500. The triad structure consisting of a central T tubule between enlarged terminal cisternae of the SR system can be seen at the level of the Z-lines. The section grazes one fibril and shows part of the SR system. There are numerous glycogen granules in the sarcoplasmic space. A = A-band; H = H-zone; Z = Z-line; I = I-band.

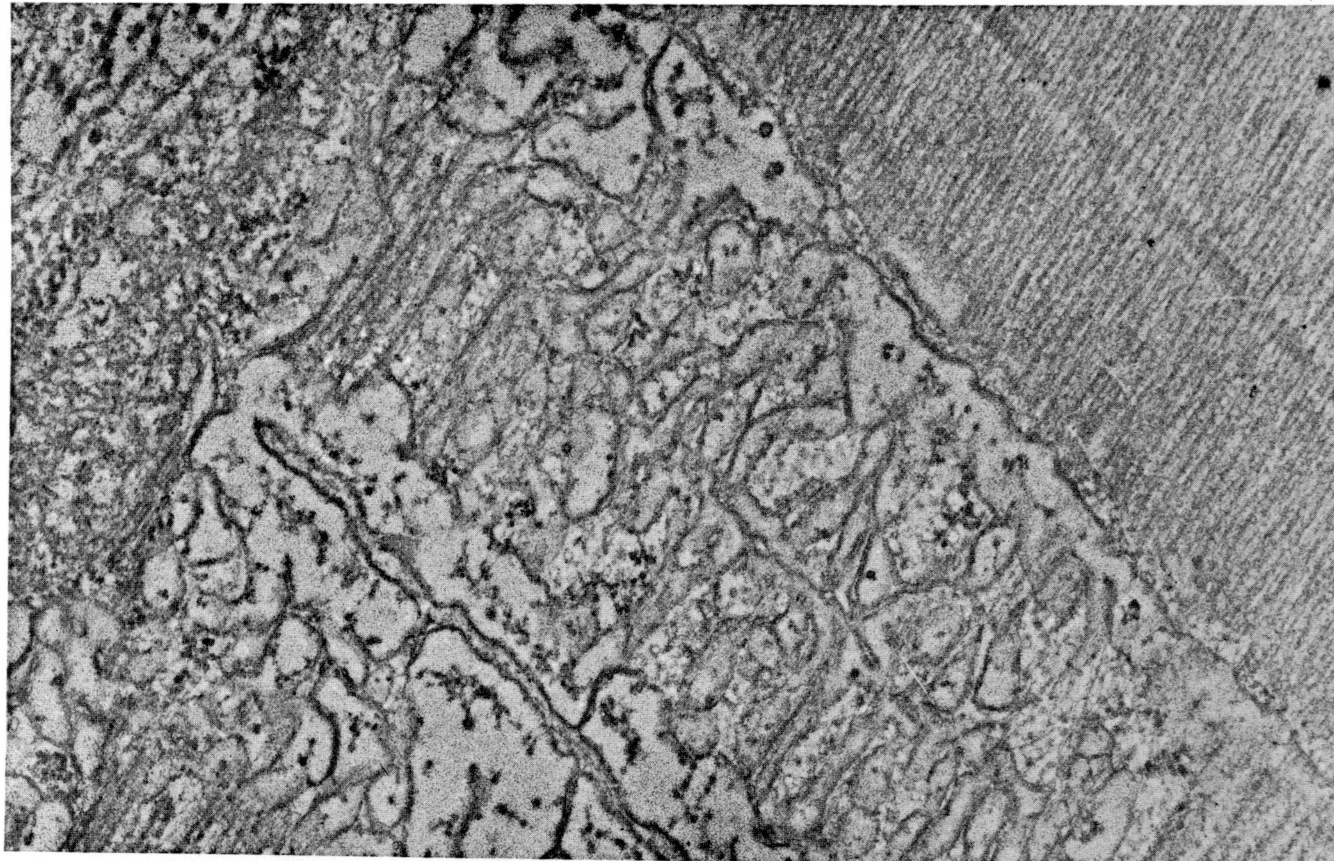

FIG. 6. Longitudinal section grazing a fibril at the periphery of the fibre showing the structure of the SR system, ×51 000. The sarcolemma is at the top left corner.

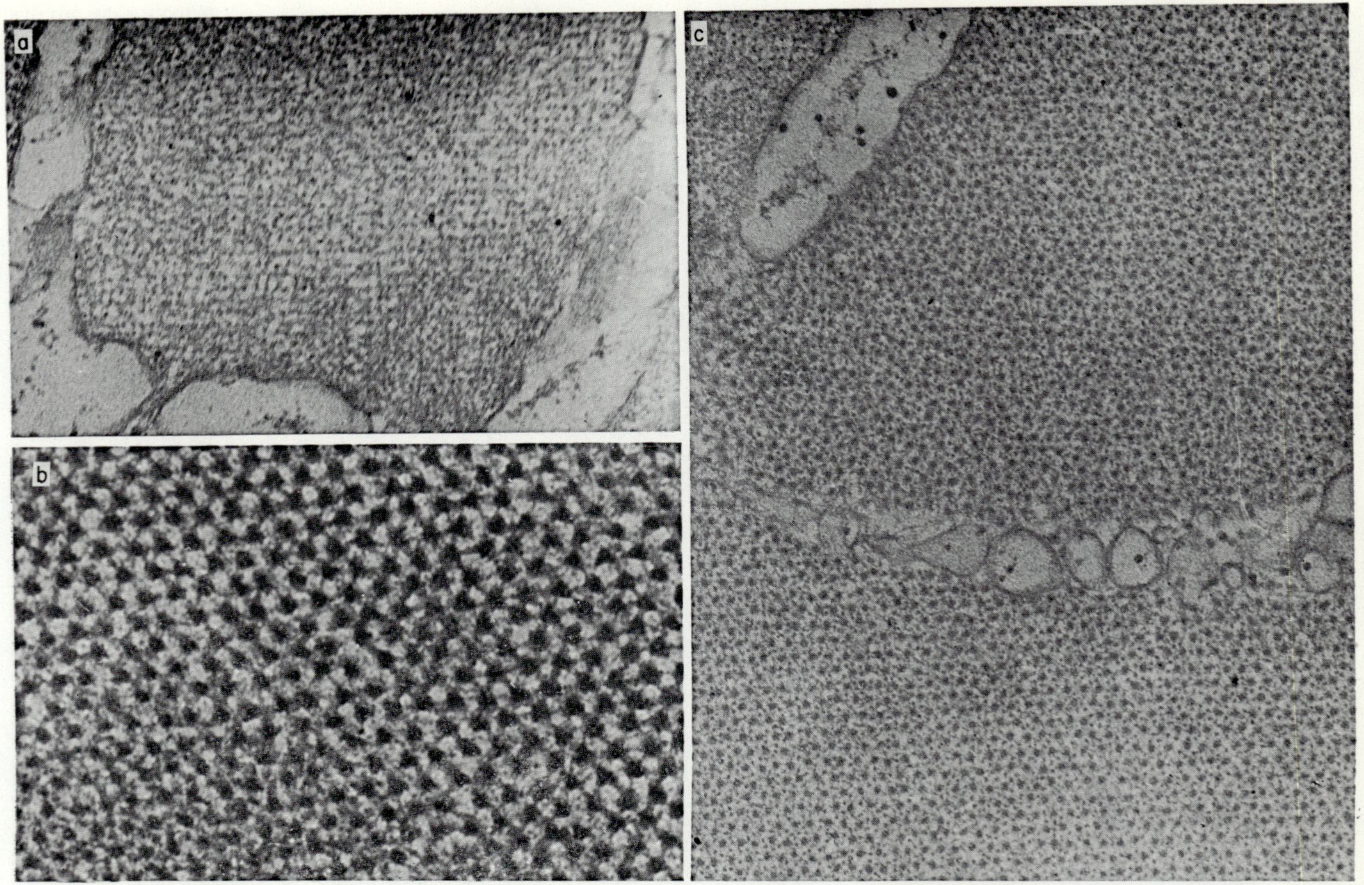

FIG. 7. Transverse sections of cod fibrils through (a) Z-line and I-band, × 51 000; (b) H-zone and M-line, × 102 000; (c) A-band, × 51 000.

FIG. 8. Longitudinal section of trout red muscle fibre, × 13 000. There are numerous mitochondria and some oil globules between the fibrils. (By permission of A. Davison and G. Goldspink.)

is continuous with the plasmalemma and indeed can be considered as a complex invagination of it.

The SR-system, however, is discontinuous, each sarcomere having its own portion of reticulum surrounding it like a hollow collar. The collar is not complete and is perforated by holes elongated along the long axis of the fibre. This is well seen where a longitudinal section grazes the periphery of a fibril (Fig. 6). Where the reticulum approaches the Z-disc the terminal cisternae are slightly swollen and cisternae from adjacent sarcomeres form the outer elements of the triad structure. The structure of the SR and T systems and their function in promoting contraction and relaxation of muscle are well illustrated and described in reviews of muscle structure and function (Peachey, 1970; Bourne, 1972; Bendall, 1974). It is not always easy to demonstrate that the T system opens into the extracellular space (Peachey, 1970) and the first evidence that this is so was obtained using fish muscle (Franzini-Armstrong and Porter, 1964). Kilarski (1967) has published a photograph of the continuity of the T-system with the extracellular space in the extrinsic eye muscle of the tench. Other workers (Bishop and Odense, 1967; Nishihara, 1967; Nakajima, 1969) show the T system opening in the sarcoplasmic space just beneath the sarcolemma. However this may just be a consequence of an unfavourably orientated section, since current models for the induction of the contraction of a muscle fibre assume the T system is continuous with the plasmalemma. In Fig. 3 a tubule of the T system can be seen opening out into the extracellular space. All the published reports on vertebrate fish skeletal muscle show the T system traversing the fibre at the level of the Z-discs; in higher vertebrates it usually lies at the junction of the A- and I-bands.

Fish in temperate or arctic waters may not eat for long periods during winter when food is scarce. During this time muscle tissue will be metabolized as an energy source when lipid reserves have been exhausted, and consequently the fish become emaciated (Love, 1970). The effects of starvation on muscle structure have been reported by Walker (1971) and Gas (1972).

B. Red Muscle

Many workers have compared the microstructures of red and white muscles by light and electron microscopy in a wide range of vertebrate fish species and have found a consistent pattern of differences (Barets, 1952; Boddeke *et al.*, 1959; George, 1962; Buttkus, 1963; George and Bokdawala, 1964; Bone, 1966; Kilarski, 1967; Nishihara, 1967; Nakajima, 1969; Walker, 1970, 1971; Patterson and Goldspink, 1972; Best and Bone, 1973). Many of the differences can be seen at the light microscope level. The diameters of muscle fibres are rather variable amongst fish species but red muscle fibres are appreciably smaller than those of white muscle. Red muscle has more extracellular space and it is better supplied with blood capillaries. The sarcomere lengths of red muscle fibres are about 0·2–0·3 μm greater than those of white muscle. White muscle fibres are innervated only at the end where they are inserted into the myocommata, whereas red muscle fibres are innervated along their entire length. These differences are similar to those that have been reported between the red and white muscles of higher vertebrates (Cassens and Cooper, 1971).

Specific staining with Sudan black and carmine, for lipid and glycogen respectively, demonstrates that red muscle is richer in both these energy reserves than is white muscle. This differentiation is also borne out by the results of chemical analysis (Love, 1970). Associated with this is the observation from histochemical studies that lipase and succinic dehydrogenase activities are also greater in red muscle (George and Bokdawala, 1964). There are no differences in the structures of the sarcomeres of red and white fibres that can be detected by electron microscopy apart from a possible thickening of the Z-line in red muscle fibres. The SR and T systems also appear to be the same, though in red muscle the sarcoplasmic space is greater and more densely packed with glycogen granules. The most striking difference is in the mitochondria; these are more numerous in red muscle, and are dispersed throughout the fibre and not confined to the sub-sarcolemmal space (Fig. 8). They are also larger in red muscle and have better developed cristae.

These features of the structure of red muscle are consistent with what is known from biochemical and physiological studies. Red muscle is capable of continuous aerobic metabolism and can sustain prolonged exercise; white muscle, on the other hand, soon incurs an oxygen debt on being exercised and can operate only in short bursts. This has led to the concept of 'stayers' (fish with high contents of red muscle that can cruise continuously over long periods of time) and 'sprinters' (those that do not move very far but can dart swiftly to catch prey or escape predators) (Boddeke *et al.*, 1959). Walker (1971) has shown that at cruising speeds in an exercise tank only the red muscle of coalfish hypertrophies but at faster speeds both red and white muscle do so. Best and Bone (1973) have pointed out that the power required to propel a fish through water varies approximately as the cube of the velocity and that the bulk, if not all, of the white muscle will be used only at high speeds. The red muscle can then be considered as normal muscle tissue and the white muscle as specialized. The particular features of the structure of white muscle (large-diameter fibres, low extracellular volume, reduced vascularization, very low mitochondrial volume (in red muscle mitochondria can occupy one-fifth of the volume) and innervation at the fibre ends only) are adaptations to ensure a very high density of contractile tissues (Best and Bone, 1973). These adaptations in their turn attract a penalty in that the muscle is unable to maintain aerobic metabolism except at a low level of activity.

As usual, in nature there is not such a clear distinction between red and white muscle as has just been made out. The two types are anatomically quite distinct, with a well-defined septum separating them, and the red muscle is quite homogeneous in respect of the type of fibre. In the white muscle, however, there is an intermediate zone at its border with red muscle which is a mosaic of red and white fibres (Bone, 1966; Johnston *et al.*, 1974, 1975).

IV. EFFECTS OF PROCESSING

A. Icing

Fish is a rather perishable food commodity and it is commonly, though not invariably, held in ice or at chill temperatures during storage and distribution in the wet state. No comprehensive account of the effects of storage on the structure of fish muscle is available, and the following description is based almost entirely on published (Liljemark, 1969; Jarenbäck and Liljemark, 1975a) and unpublished observations on one species, cod.

Rigor mortis in cod stored at 0°C sets in about 2–12 hours after death and is fully resolved after 2–3 days. The fish in rigor is rigid and the flesh is hard to the touch, but there is no shortening of the myotomes while the muscle is still 'on the bone'. (If a fillet is removed before the onset of rigor and not restrained it may contract to about two-thirds of its original length.) Though there is no change in the overall length of a fibre, there is slippage within it so that the Z-lines no longer remain in register. There are also localized regions of contraction, extension and distortion of the fibrils (Fig. 9). (This figure is illustrative only and should not be taken as representative; it is not possible to estimate the average degree of distortion from electron microscopic studies alone.) In those regions where there is little distortion of the fibril it can be seen that the filamentous structure of the sarcomere has not been affected. The filaments are clearly separated and retain their hexagonal packing. The sarcoplasmic reticulum, however, is altered as the muscle goes through rigor mortis. In the post-rigor state the ordered structures of the T and SR systems are replaced by a conglomeration of approximately spherical vesicles situated between the fibrils. Their shapes can be visualized from an examination of longitudinal and transverse sections (Liljemark, 1969) (Figs 9 and 10) where the vesicles appear as unbroken circles or ovals. This is a particularly interesting phenomenon considering the role the reticulum plays in initiating contraction and relaxation. It prompts the question as to which comes first; disintegration of the reticulum leading to rigor mortis, or rigor mortis leading to disintegration of the reticulum? Unfortunately there seem to be no published observations of the structure of muscle fibres in rigor. No further changes in the structure of the fibres during storage in ice for up to 17 days have been noted, except posssibly a reduction in the amount of reticulum present.

Another major change occurring soon after the death of the fish is in the structure of the sarcolemma. The plasmalemma degrades, losing its integrity and allowing the cell contents to be shed into the extracellular space (Fig. 11). The basement membrane becomes more diffuse, but the collagen fibres are still visible.

B. Freezing and Frozen Storage

Fish flesh is somewhat unstable, even in the frozen state, and its stability is highly temperature dependent. When stored at −30°C it remains virtually unchanged for several years, but the shelf-life of a lean species like cod at −20°C is about 9 months and at −15°C about 3 months. Fatty fish tend to deteriorate at a faster rate. There are changes in flavour during prolonged storage under unsuitable conditions, but of more relevance to the subject of this chapter are the changes in texture and in physical properties. The flesh of frozen fish stored under unsuitable conditions before thawing is firm to the touch and the tissue is unable to retain water to the same extent as the original wet fish, the excess fluid being released as 'drip'. When cooked, badly stored frozen fish has a tough, chewy, dry and stringy texture.

There is considerable evidence that alterations in the myofibrillar proteins are associated with these phenomena and indeed, in the opinions of many workers, go a long way towards explaining them. The review by Sikorski *et al.* (1976) summarizes the evidence and arguments very well. In my view chemical changes can provide only a partial explanation of alterations in the ultrastructure of the fibre; for example, changes in the sarcoplasmic reticulum (described below) may play a more important role than is generally realized.

On freezing, ice crystals are formed in the tissue. Their size, shape and position, extra- or intracellular, all depend on rate of freezing and whether or not the flesh is frozen pre- or post-rigor (Partmann and Schlaszus, 1973). The subject has been well reviewed (Love, 1966, 1971) and need not be discussed in detail here. Suffice it to record that in fish frozen under commercial conditions the ice will form large intracellular, columnar crystals. The spaces occupied by the crystals can be clearly seen in sections of tissue that

FIG. 9. Longitudinal section of a muscle fibre from cod that had been stored at 0°C for 3 d, i.e. after the fish had passed through rigor mortis, ×15 300. The fibrils are distorted and the reticulum has lost its integrity.

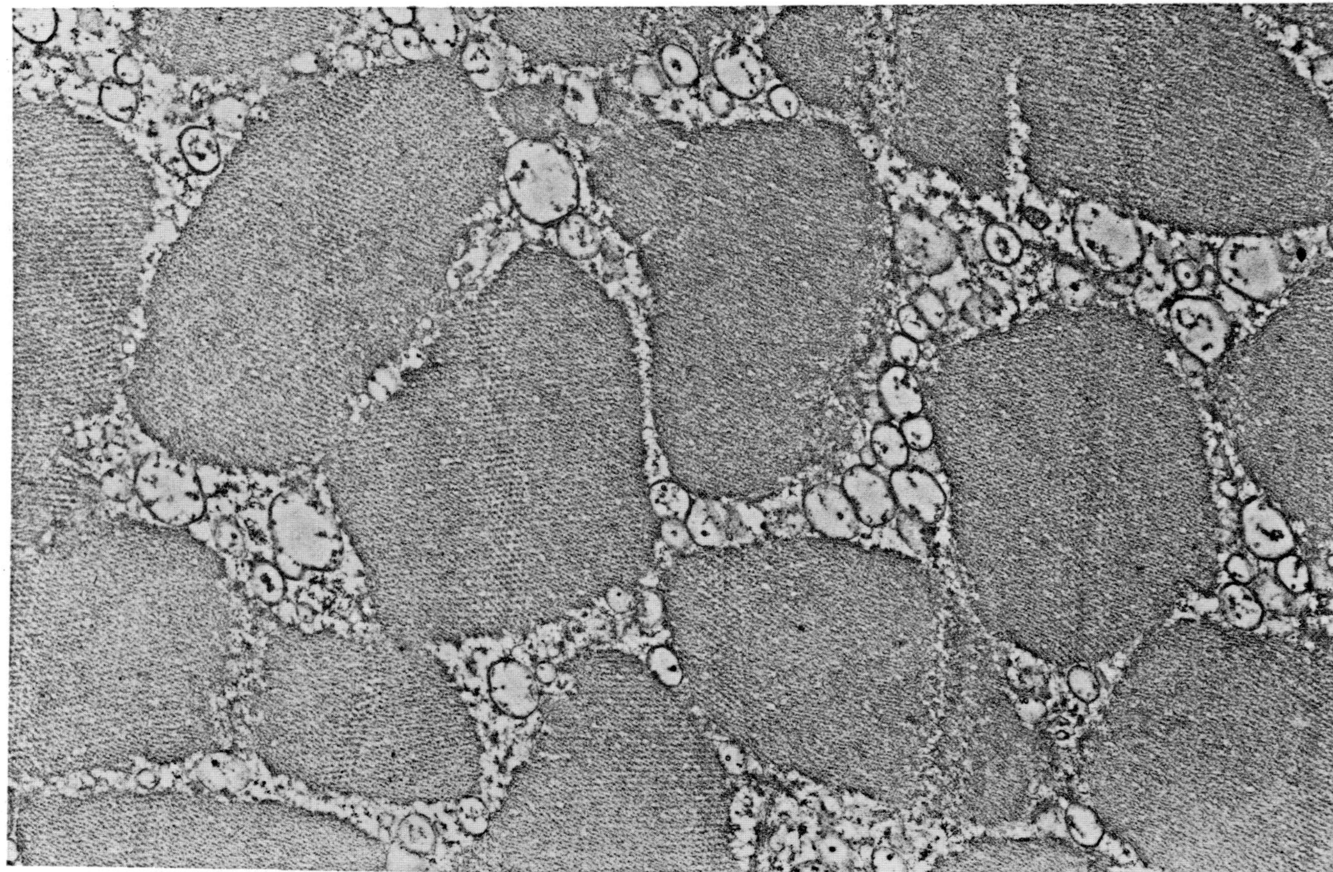

FIG. 10. Transverse section of post-rigor cod muscle fibre, ×25 500. The reticulum appears as small, approximately spherical, membranous vesicles. On account of the distortion of the fibril the filaments are not normal to the plane of the section and appear elongated.

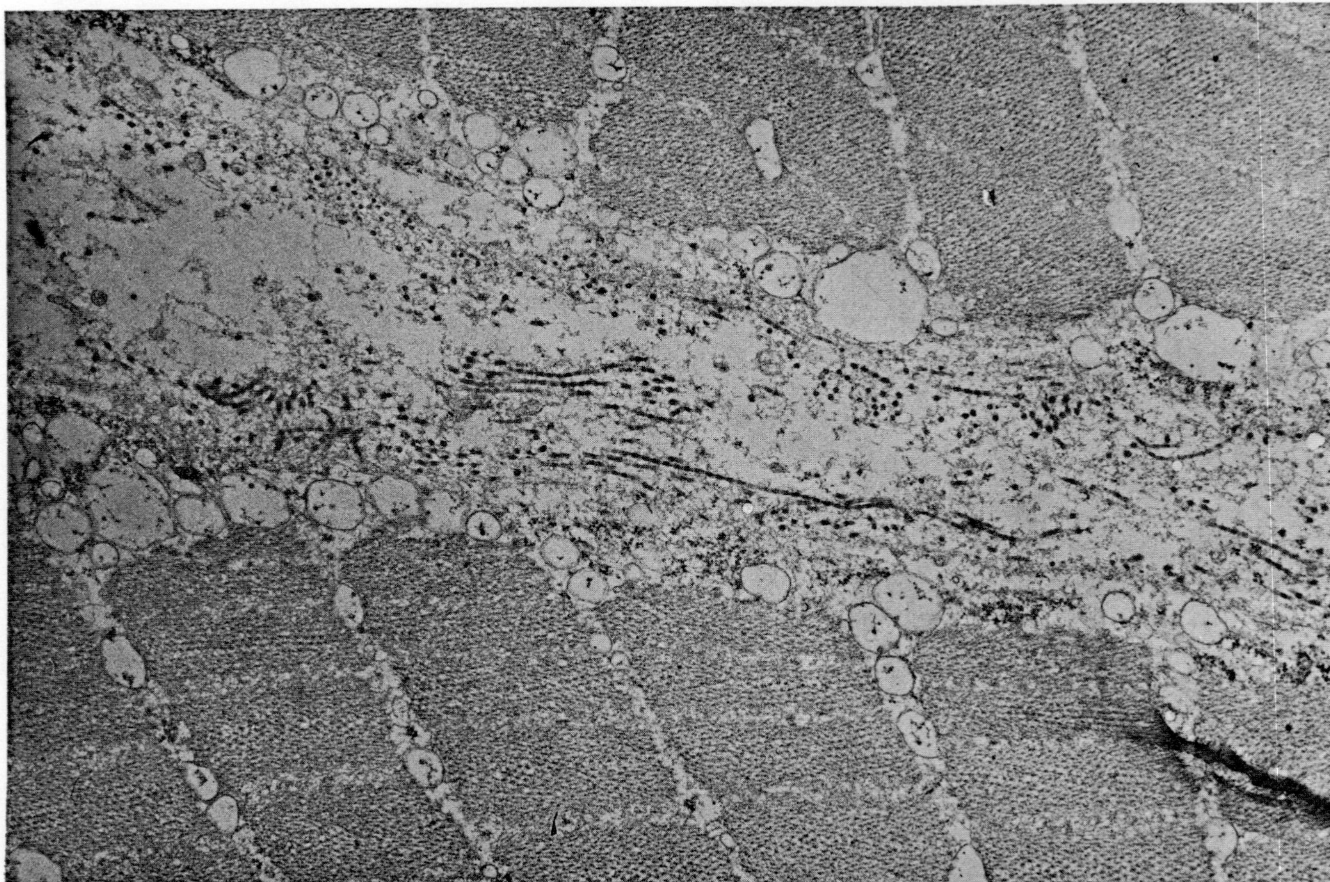

FIG. 11. Transverse section of post-rigor cod muscle fibres illustrating the breakdown of the sarcolemma, ×25 500. The fibrils have distorted on going through rigor and the section passes through a succession of bands.

have been prepared to retain the structure of the frozen material (Figs 12a, b).

On thawing cod muscle tissue immediately after freezing or after storage under good conditions, the fibres resorb the melt water, completely or nearly so, regaining their original shapes though fissures are left as evidence of the previous existence of ice crystals (Fig. 12c). However, in fish that has been badly stored the water is only partially resorbed and the fibres retain the structure of frozen muscle (Fig. 12d).

When fibres are teased from freshly frozen muscle and allowed to fix in glutaraldehyde solutions at 0°C, they retain their frozen shapes while they thaw and the voids occupied by the ice crystals remain. Figure 13 shows a transverse section of such a preparation as seen in the electron microscope. The fibrils are distorted and compressed along a line approximately normal to the edge of the ice crystal. The filament pattern is disrupted in the region close to the ice crystal; elsewhere the filaments are visible but the hexagonal array is modified. The spacing between thick filaments is reduced along the line of compression and in places is reduced sufficiently for them to touch, forming an unbroken line on either side of which are rows of still-separate thin filaments. The way in which this comes about is shown diagrammatically in Fig. 14. Though the interfilament distance has been reduced in one dimension, it does not necessarily mean that the cross-sectional area of the fibril has decreased in relative proportion; examination of sections of several fibrils suggests that as the interfilament spacing decreases along the lines, so does the spacing between the lines increase. On thawing freshly frozen tissue, the fibril regains its original structure and filament spacing.

As already mentioned, during frozen storage under unfavourable conditions the compacted nature of the contents of the fibre is somehow fixed and the original shape is not fully restored on thawing. This setting of the shape appears to be due to changes in the sarcoplasm, the fibril structure remaining apparently unchanged. Figures 15a, b are longitudinal and transverse sections of a muscle fibre from cod that has been stored for 5 months at −15°C before thawing and fixing. This sample would have the typical properties of badly stored frozen fish. The fibrils are still clearly visible and distinct, and in transverse sections they can be seen to have the typical hexagonal packing. There are no signs of any compression lines. However, there is virtually no space between the fibrils and the sarcoplasmic reticulum is compressed into a thin layer. This effect has been noted by Connell and Howgate (1969), Liljemark (1969) and Jarenbäck and Liljemark (1975a, b) in cod, and by Tanaka (1965) in Alaska pollack. The reduction in sarcoplasmic space and the compression of the reticulum is progressive during storage, and Fig. 16, a transverse section from cod held for 2 months at −15°C, shows an intermediate stage. These changes do not occur at low temperatures or do so only

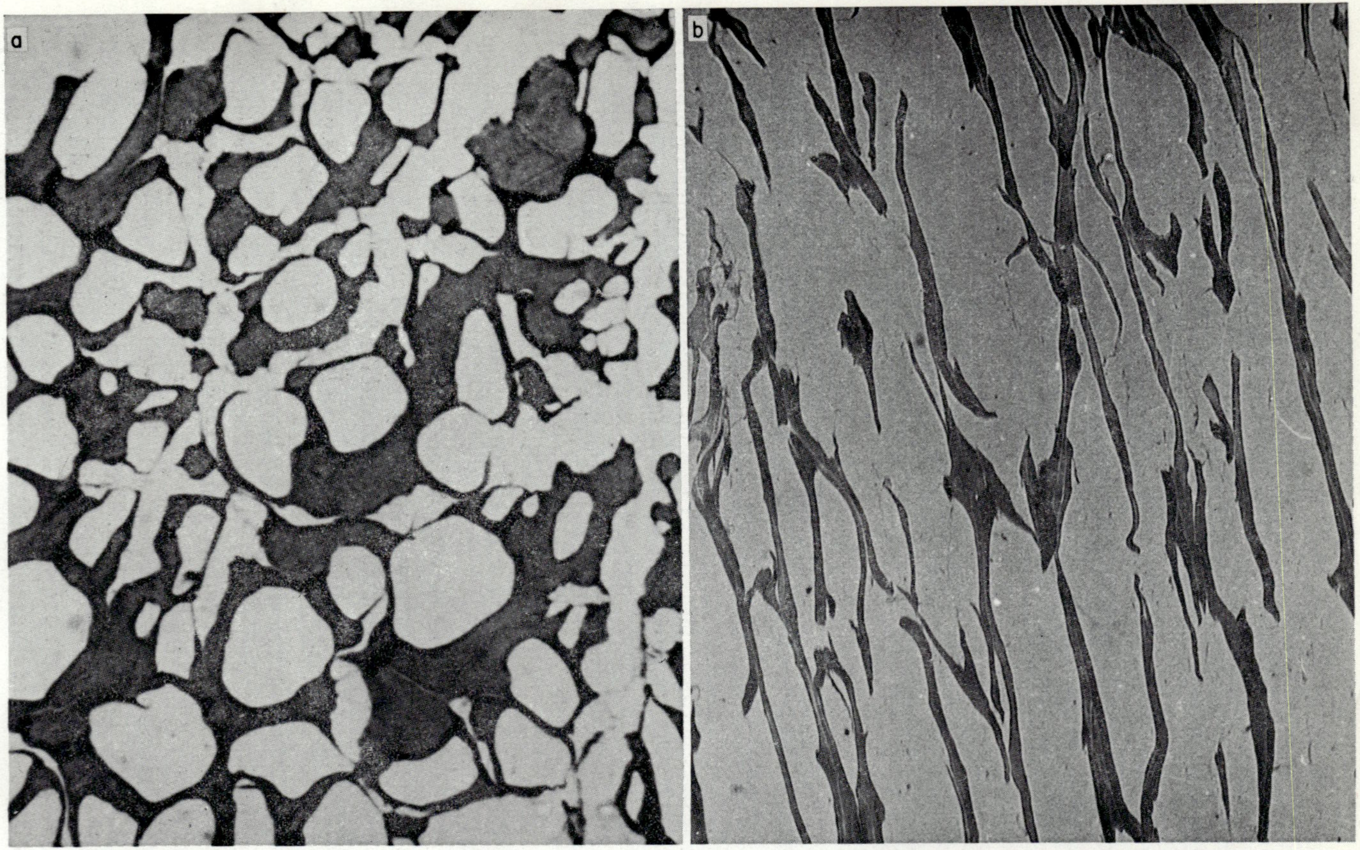

FIG. 12. Light microscope photographs of the effect of freezing on post-rigor cod muscle fibres. (a) Transverse and (b) longitudinal sections of frozen muscle, $\times 85$. The frozen tissue was freeze-dried, impregnated with paraffin wax and sectioned. Ice crystals have formed within the cells. (By permission of Love (1966).)

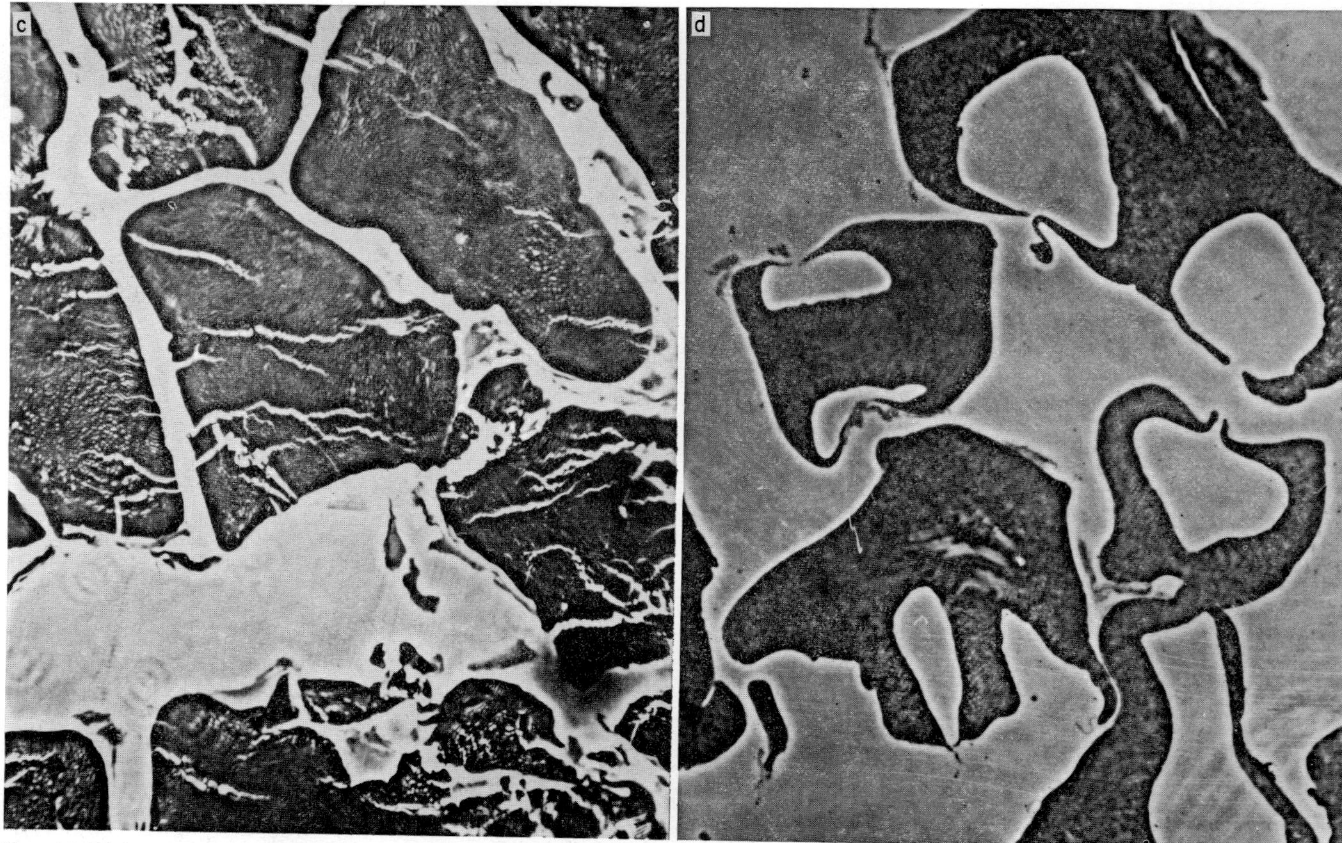

FIG. 12. Light microscope photographs of the effect of freezing on post-rigor cod muscle fibres. (c) Transverse section of post-rigor cod muscle frozen to $-30°C$ then thawed, $\times 213$. Tissue was prepared for electron microscopy, a thick section cut and viewed by phase contrast microscopy. The ice crystals have melted, water being resorbed, but fissures remain where the crystals have been. (d) Transverse section of fibres from cod that had been stored at $-15°C$ for 5 months, $\times 213$. Tissue prepared as in (c). The ice crystals have melted but all the water has not been resorbed and the fibre retains the shape it had in the frozen state.

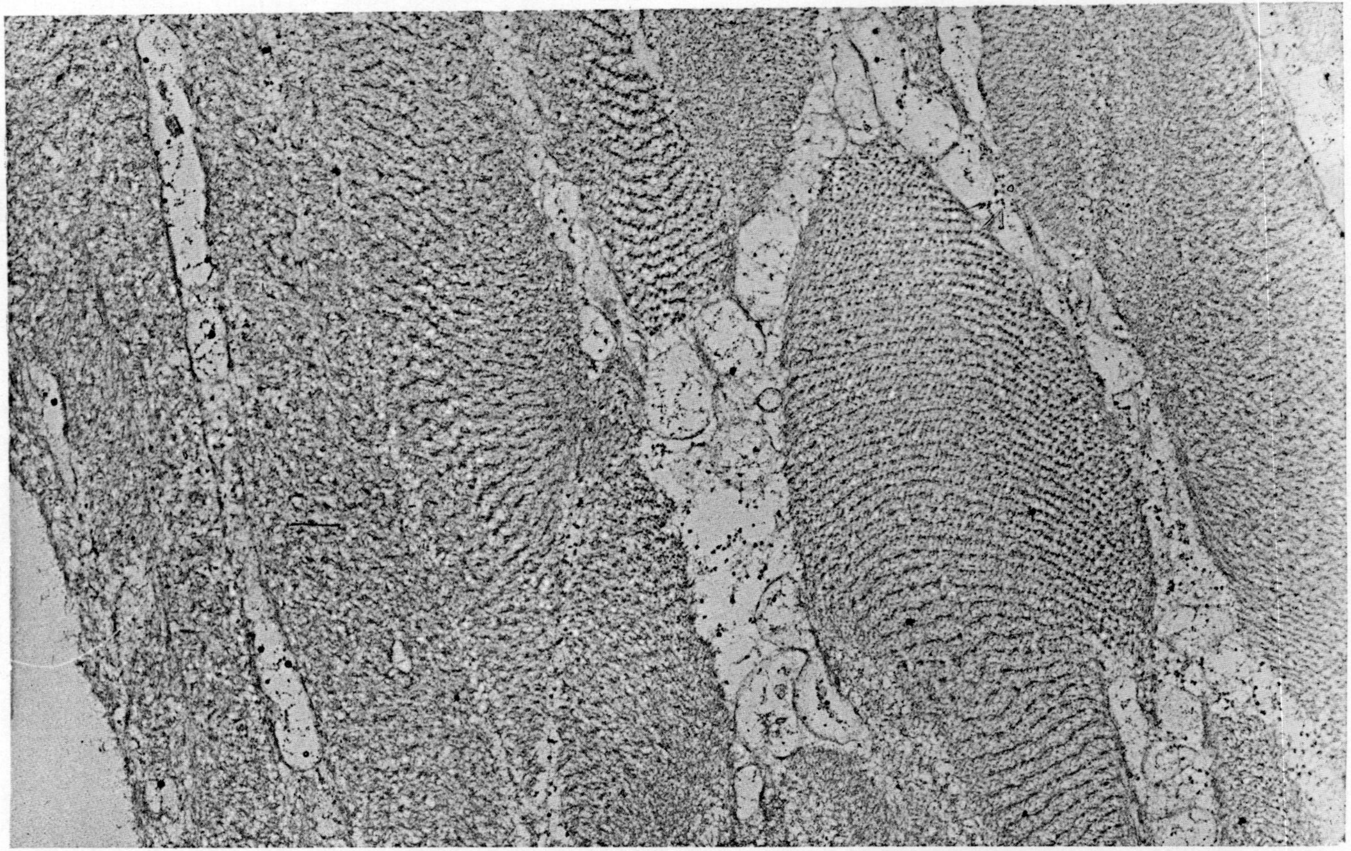

FIG. 13. Transverse section of a fibre from frozen cod, × 25 500. The frozen fibre was allowed to fix in glutaraldehyde solution at 0°C. Thick sections viewed by light microscopy showed a structure similar to that illustrated in Figs 12a, b and d, i.e. still retaining the voids previously occupied by the ice crystal. The edge of the void can be seen (bottom left).

very slowly. A transverse section of a fibre from cod stored at −29°C for 9 months (Fig. 17) is indistinguishable from a corresponding section of iced, post-rigor fish.

It appears as if remnants of the sarcoplasmic reticulum, as it degrades during frozen storage, are acting as a glue cementing the fibrils together. Fibrils bound together like this will produce a stiffer, firmer fibre and hence a tougher product. Such a model was proposed by Connell (1962) to account for the changes taking place in fish flesh during frozen storage. Love (1967) concluded from a study of the microscopy of macerates of cod muscle that

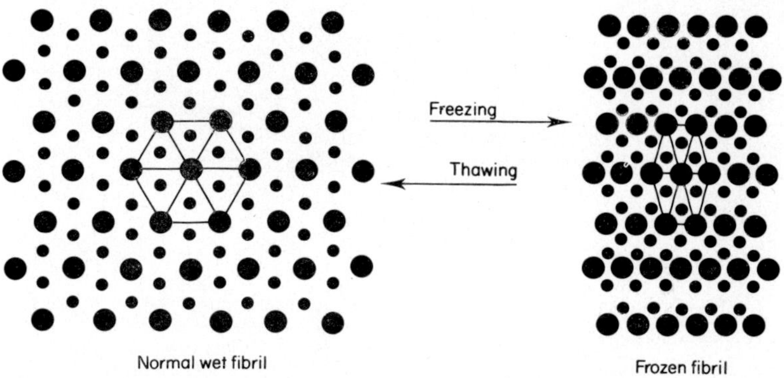

FIG. 14. Diagram illustrating the distortion of the regular hexagonal arrays of thick and thin filaments on freezing.

toughening of the flesh during frozen storage is associated with an increased resistance of the fibre to disruption into its constituent fibrils. He also postulated that the sarcoplasmic reticulum could play a part in this. The results from electron microscope observations appear to support these ideas.

A mechanism that has been suggested as a contributing cause of toughening is the cross-linking of protein molecules from adjacent filaments when these are closely approximated in the frozen state. If this does happen, then one might expect a reduction of interfilament distance during frozen storage as the cross-links form. Two groups of workers, Connell and Howgate (1969) and Liljemark (1969), have published results for spacing between thick filaments for cod based on sectioned material. Both found a decrease in spacing in post-rigor fish compared with pre-rigor, but observed no effect due to frozen storage even in material that was very tough. Both sets of results were obtained on tissue that had been fixed in glutaraldehyde, post-fixed in osmium tetroxide, dehydrated in ethanol and embedded in Epon. Subsequently, Jarenbäck and Liljemark (1975a) published data based on cod muscle

FIG. 15a. Longitudinal section of fibrils of cod muscle stored at $-15°C$ for 5 months, $\times 25\,500$. Cf. Fig. 12d. The sarcoplasmic space has been considerably reduced and the fibrils appear to be 'cemented' together by the remnants of the reticulum.

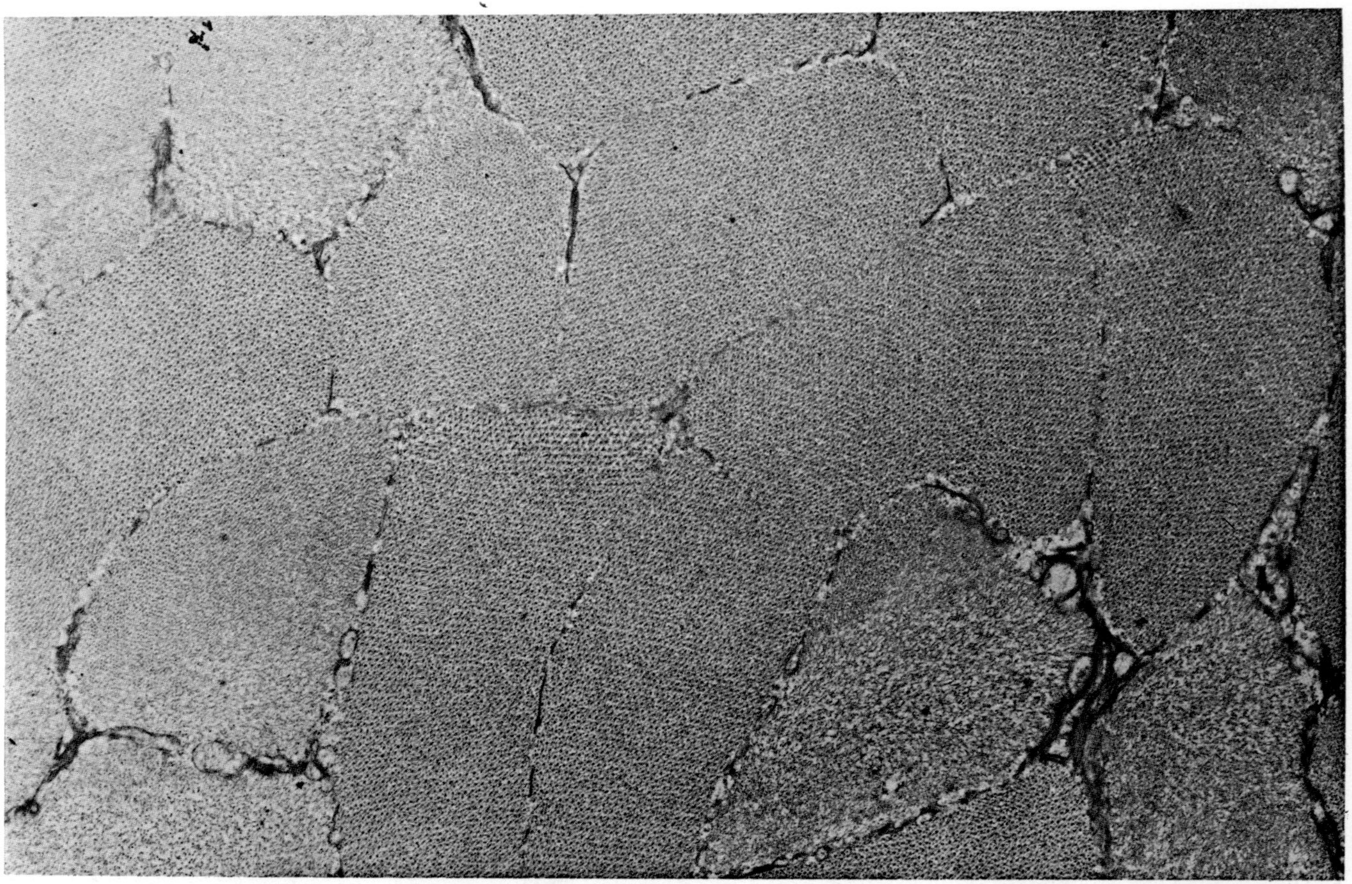

FIG. 15b. Transverse section of the same specimen, $\times$ 25 500. Note that the filaments have a regular packing apparently identical with that of unfrozen fish (Fig. 10).

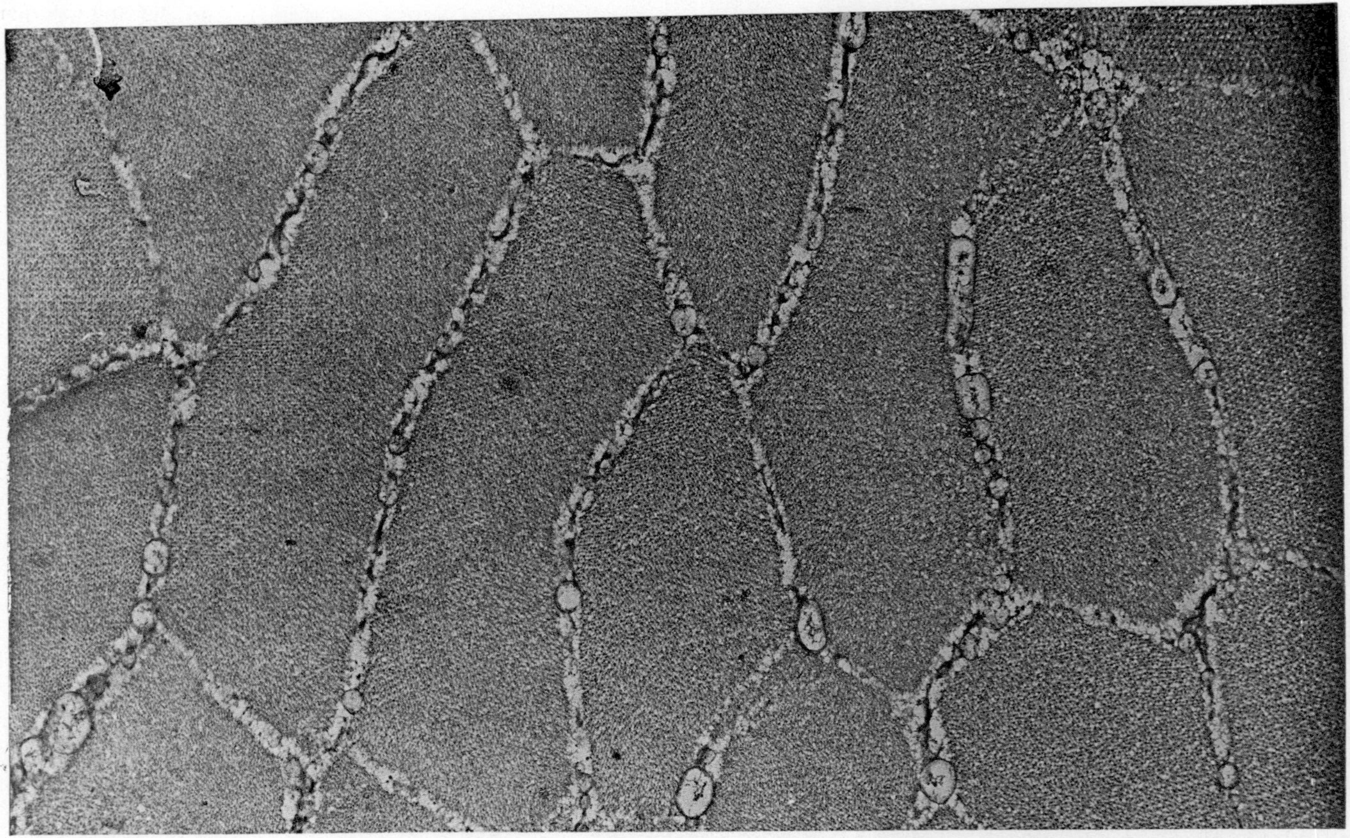

FIG. 16. Transverse section of fibrils from cod muscle stored at $-15°C$ for 2 months, $\times 25\ 500$. It shows a stage in the reduction of the sarcoplasmic space intermediate between those illustrated in Figs 10 and 15.

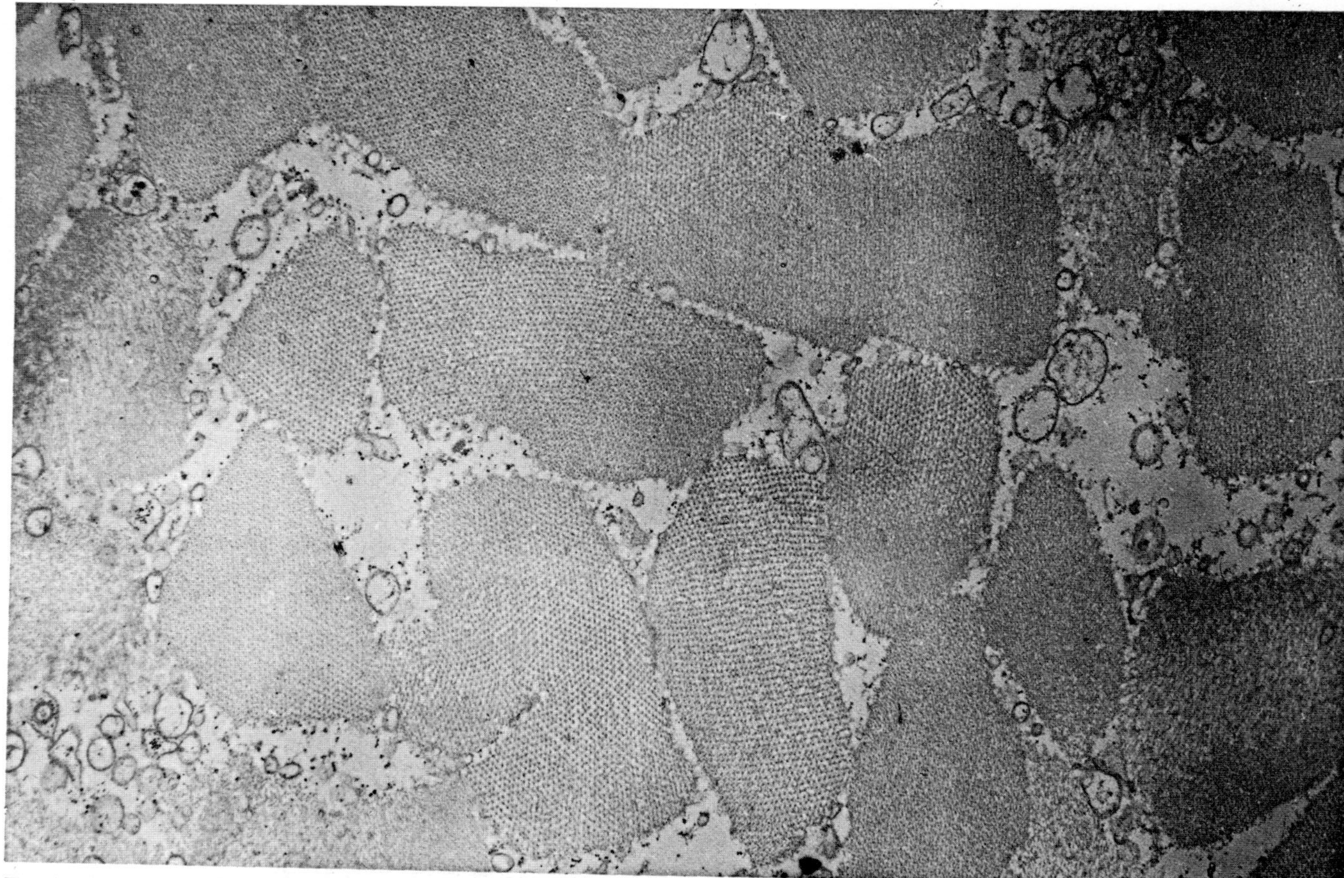

FIG. 17. Transverse section of fibrils from cod muscle stored at $-29°C$ for 9 months, $\times 25\,500$. There is no reduction in sarcoplasmic space.

that had been cryo-fractured and freeze-etched. Data from the three reports are summarized in Table 1. The freeze-etched preparations show an appreciably greater interfilament spacing compared with the embedded specimens. Jarenbäck and Liljemark (1975a) did not examine pre-rigor cod to confirm a decrease in spacing on going through rigor mortis, but they did demonstrate a statistically significant smaller spacing in tough, frozen-stored fish.

A noticeable feature of all of the results is their very high variance compared with the size of the effects observed. For instance Liljemark (1969) quotes a range of 33–44 nm for the interfilament spacing of post-rigor tissue.

TABLE 1. Distance in nm between thick filaments in fibrils of cod muscle

Treatment	Liljemark (1969)[a]	Jarenbäck and Liljemark (1975a)[b]	Connell and Howgate (1969)[a]
Pre-rigor	41·4	—	39·2
Post-rigor	38·0	43·0	33·2
Frozen, not tough	37·0	42·1	37·4
Frozen, tough	38·8	36·7	35·2

[a] Data obtained from tissue fixed in glutaraldehyde, embedded in Epon and sectioned.
[b] Data obtained from cryo-fractured and freeze-etched samples.

Two effects can contribute to this variance. It is known from X-ray studies on living muscle that the sarcomere volume remains constant during contraction and relaxation and that interfilament spacing varies inversely with the square root of the sarcomere length. In post-rigor cod muscle sarcomere lengths are very variable, and if the same principle holds then a similarly variable interfilament distance would be expected. In addition, it is very difficult to orientate a fibre so that the plane of sectioning or fracturing is normal to the axis of the fibre, and in an oblique section the regular hexagonal pattern is distorted. In such a section the interfilament distances measured along the three principal axes of the hexagon may be different and greater than or equal to the true value. The data of Connell and Howgate (1969) were corrected for this effect, whereas those of Liljemark (1969) were obtained from 'well-oriented sections'. Jarenbäck and Liljemark (1975a) quote both the average distance and the minimum, the latter being incorporated in Table 1.

A further difficulty in evaluating measurements from microscope preparations is the effect of fixation and preparation. In the freeze-etching technique used by Jarenbäck and Liljemark (1975a) the tissue was treated with cryoprotective agent, 20% glycerol, in a physiological buffer before freezing. This treatment could cause swelling of the muscle fibres (glycerol is known to do so under some conditions) and may account for the greater interfilament spacing found by this technique. Conversely, fixation, dehydration and embedding may cause contraction, and Jarenbäck and Liljemark (1975a) did observe that post-rigor cod tissue fixed in glutaraldehyde before treatment with the cryoprotective agent gave a spacing 2·7 nm less than the non-fixed control.

There must then be some doubt as to whether or not interfilament spacing does decrease as fish toughens in the cold store. The cod used by Jarenbäck and Liljemark (1975a) to demonstrate a reduction had been stored for 2–3½ years at −20°C, a storage treatment that would produce an extremely tough product; storage times of 6 months or less at this temperature would cause appreciable toughening. If cross-linking of proteins with associated changes in interfilament spacing is a factor in the toughening of fish muscle during frozen storage, then it would appear that electron microscopy is not a sensitive technique for detecting it. (Colleagues at the Torry Research Station (Aitken and Connell, 1977) have recently examined frozen-stored cod muscle by X-ray diffraction and found no change in average interfilament spacing.)

C. Other Processes

Drying has been used as a means for the long-term preservation of fish and other foods since ancient times, and although its importance has been superseded in the more technologically advanced countries by canning and freezing it is still practised in many parts of the world. The most commonly used technique is still to expose the fish, usually split to present an increased surface area for evaporation, to sun and wind at ambient temperatures, although artificial methods are also used. The air-dried product is compact and does not rehydrate fully on soaking in water. The cooked, reconstituted flesh is tough, chewy and dry. Freeze-drying produces a better product; it resorbs water well, almost to its original moisture content, although much of the water can be expelled with moderate pressure, and after cooking it is less tough and dry than air-dried products but still very different from wet fish.

Connell (1957) examined the microscopical structure of dried cod muscle after rehydration. The air-dried sample had a compact, coherent structure with hardly any spaces between the fibres; the freeze-dried samples had an open, cellular structure with the spaces originally occupied by the ice crystals still being retained. It appears that though the freeze-dried samples regained virtually their original weight on rehydration, the water was not fully resorbed

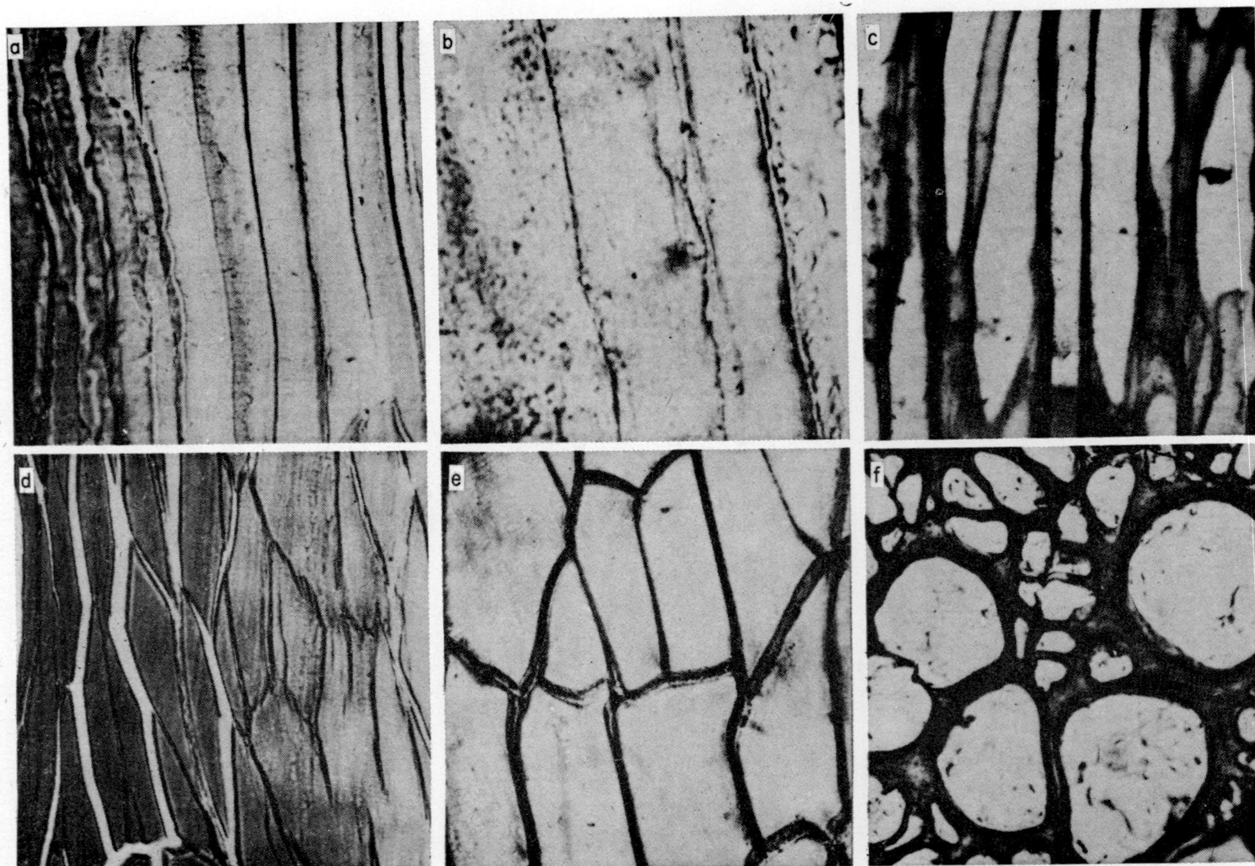

FIG. 18. Light microscope photographs of dried fish muscle, ×180. Photographs (a)–(c) are longitudinal sections, (d)–(f) are corresponding transverse sections: (a) and (d), air-dried at 18–20°C; (b) and (e), air-dried over silica gel at 0°C; (c) and (f), freeze-dried. Voids previously occupied by the ice crystals are retained. (By permission of Kanna *et al.* (1971).)

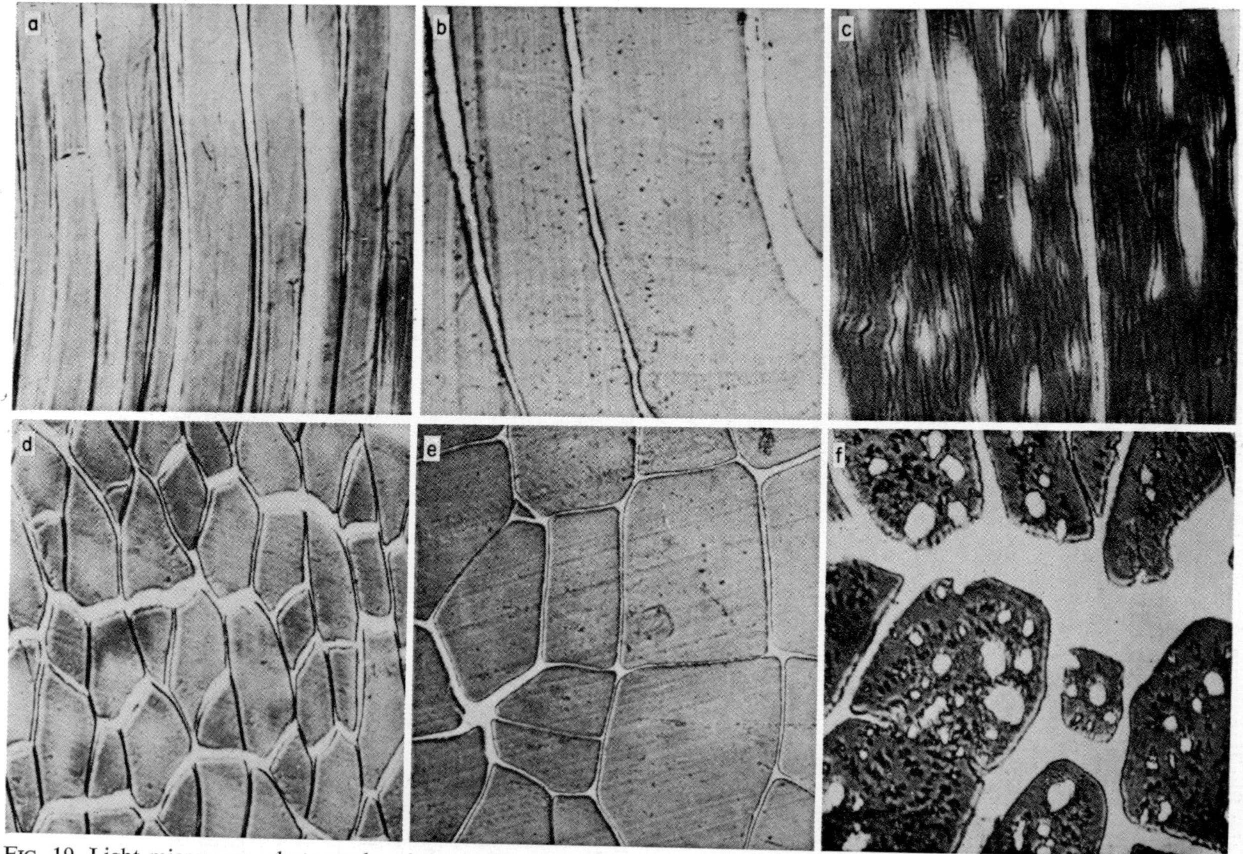

FIG. 19. Light microscope photographs of the dried muscle specimens illustrated in Fig. 18 after rehydration, ×180. Lettering corresponds to that in Fig. 18. (By permission of Kanna *et al.* (1971).)

into the fibrillar tissue. Similar results were obtained by Kanna *et al.* (1971) using sea bass. Their micrographs are reproduced in Figs 18 and 19. They show that temperature is a factor controlling the resorption of water in air-dried fish. In the sample dried at 0°C the fibres have regained virtually all their original volume, whereas those in the sample dried at 18–20°C remain partially shrunken. This is confirmed by the water contents of the reconstituted products which were 51% and 72% for the samples dried at 18–20°C and 0°C respectively, compared with 76% in the original. These workers also confirmed the observations of Connell (1957) that freeze-dried fish does not fully resorb water into the tissue and voids remain where ice crystals have been. There are parallels here with the situation in badly-stored frozen fish.

It is noticeable that in the reconstituted flesh (Fig. 19), the fibres are separate even in the most affected case. The changes occurring during dehydration that affect the physical and sensory properties of dried fish must lie within the fibres; there are no connections between the fibres that can contribute to them. Unfortunately, there do not appear to be any published studies on the ultrastructure of dried fish muscle.

Smoke-curing is another traditional way of preserving fish, although nowadays the products consumed in most westernized countries are smoked primarily to give them an attractive flavour and texture rather than a long shelf-life. Fish is usually brined before smoking, and smoking can be carried out at temperatures near ambient (cold-smoking) or at temperatures where the product is cooked (hot-smoking). There is a loss of weight due to dehydration in the smoking process of around 10–25% depending on the product.

Bromlei (1949) studied the microscopical changes in both hot- and cold-smoked fish. There was no gelatinization of the connective tissue during cold-smoking but there were signs of desiccation with diminution of extracellular space. During hot-smoking there was complete gelatinization of the connective tissue of the sarcolemma and myocommata.

Yuditskaya (1959) made a histological and histochemical study of smoke-cured fish to investigate the deposition of smoke constituents in the skin and flesh. He described a dark brown pellicle, 10–20 μm thick, which contained carbon particles covering the skin of the fish. The brown colour also diffused into the subcutaneous tissue. In sections stained with acid haematoxylin and eosin there was an intense bluish coloration of the basement membrane of the epidermis and of the subcutaneous muscular tissue. This colour was not seen in the control, unsmoked fish.

Charley and Goertz (1958) studied the cooking of salmon by baking at several temperatures and the effect of this on the microscopical structure of the flesh. Granular material was precipitated on the fibres during cooking and the connective tissue was gelatinized. There was also precipitation of granular material in the connective tissue areas. They also reported that the width of the fibres is greater in the cooked material but that there was less 'inter-fibrillar' space; probably extracellular space is what was meant by the latter term.

The effect of heating trout muscle at 60°C and 97°C has been studied by Schaller and Powrie (1972) using scanning electron microscopy. The fibrillar structure of the fibres was retained though there was disintegration of the filaments in the A-band and H-zone.

V. INVERTEBRATE MUSCLE

A. Crustacea

Atwood (1967, 1972) has reviewed the structure and physiology of crustacean muscle. The muscle fibres are generally of large diameter and readily visible to the naked eye. Several kinds of fibres can be distinguished physiologically in crustacean muscles but there are two main types. One is slow to contract and relax but is capable of sustaining a prolonged contracture—this is referred to as a 'tonic' fibre; the other is fast acting but is easily fatigued—this is a 'phasic' fibre. This physiological distinction is similar to that between red and white muscle fibres in vertebrate animals but with the important distinction that in crustaceans both types are white and cannot be distinguished by eye. A muscle can consist predominantly of one type or another (for example the crushing claw of the lobster is mainly comprised of 'tonic' muscle) or can be mixed. Fibres with properties intermediate between the two main types are known.

At the microscopical level crustacean muscles are striated with prominent A- and I-bands and the sarcomeres are delimited by Z-lines as in vertebrate muscle fibres. The H-zone is not well resolved. The two physiological fibre types exhibit structural differences. The 'tonic' fibres have long sarcomeres, typically about 10 μm in length, whereas 'phasic' fibres have sarcomeres 2–3 μm long, similar in length to those in vertebrate muscle.

The fibril has a filamentous structure with thin filaments originating at the Z-disc and extending into the A-band where they interdigitate with the thick filaments. The filaments are correspondingly long in fibres with long sarcomeres. In cross-sections the two fibre types can be seen to have different filament packings. In the short-sarcomere fibrils the thick and thin filaments are arrayed in the regular hexagonal structure already described for vetebrate fibrils. In the long-sarcomere fibrils the thick filaments are more loosely arrayed, though the pattern is still approximately hexagonal, and each thick filament is surrounded by 12 thin filaments rather than six as in the other type and in vertebrate muscle. The thin filaments do not form a regular array of themselves.

There is a well-developed reticulum between the fibrils containing a transverse tubular system and a sarcoplasmic reticulum proper. The SR system surrounds the fibrils without a break, with the T system tubules lying outside of it. In the short-sarcomere fibres there is a pair of tubules in each sarcomere lying at the junctions of the A- and I-bands. In the long-sarcomere fibres there are several tubules to a sarcomere.

Crab is often cooked before sale or further processing. Giddings and Hill (1976) have published striking scanning electron microscope pictures of cooked, frozen crab muscles. The illustrations are reproduced in Fig. 20. They demonstrate that in rapidly frozen muscle the ice crystals are initially very small (Fig. 20b) but grow during storage even at −29°C (Fig. 20c). On the other hand, slow freezing induces the formation of large, extracellular ice crystals (Fig. 20d).

B. Cephalopoda

Cephalopod mantle muscle is often stated to be an example of smooth muscle. Hanson and Lowy (1960) in their review of the structure of invertebrate muscle classified it as 'helical smooth' muscle. This classification derives from the presence in the muscle fibres when seen in the light microscope of helically wound fibrils and the absence of bands typical of striated muscle. Nevertheless, the electron microscope reveals that the structure of cephalopod muscle has more in common with striated muscle than smooth and should more properly be referred to as obliquely striated muscle. The structure of this type of muscle has been reviewed by Rosenbluth (1972) but with hardly any reference to the cephalopods.

The fibres are slender and in them the fibrils are helically wound along their lengths. Tanaka (1958) has described the microscopical structure of the mantle and tentacle muscle of a few species of squid and octopus. The text of the paper is in Japanese, a language inaccessible to most western readers, but the article is well illustrated. The absence of striations in the muscle is obvious in the micrographs (illustrations of striated fish muscle are shown for comparison) but the helical structure is not visible in these fixed and embedded specimens.

Hanson and Lowy (1957, 1960), Kawaguti (1962), Gonzalez-Santander and Garcia-Blanco (1972), Ward and Wainwright (1972) and Moon and Hulbert (1975) have published pictures of the ultrastructure of squid mantle muscle as seen in the electron microscope. Transverse sections show that the fibre contains only a few fibrils, roughly rectangular in cross-section and well separated by sarcoplasm. The fibrils contain both thick and thin filaments the disposition of which is seen in longitudinal sections (Fig. 21a). The thick filaments are regularly stepped, forming the oblique pattern. Thin filaments lie between the thick but are attached in small bundles to nodes rather than to Z-discs. The ratio of thin to thick filaments is 6 : 1 (Gonzalez-Santander and Garcia-Blanco, 1972). The overlap region of thick and thin filaments is similar to the A-band of striated muscle. There is even an H-zone where the filaments do not meet. This is seen clearly in sagittal sections (Fig. 21b) where the sarcomeres are aligned. This similarity to striated muscle is confirmed by biochemical and physiological studies (Hanson and Lowy, 1960).

Though Hanson and Lowy (1957) report that squid muscle has a highly ordered reticulum, this is not evident in their published plates; rather their figures reveal a reticulum of globular vesicles similar to the material in post-rigor vertebrate muscle. However, they used glycerol-washed fibres where it might be expected that the reticulum would not be in the 'living' state. The more recent papers by Moon and Hulbert (1975) and by Ward and Wainwright (1972) better illustrate the structure of the reticulum and other cell inclusions.

C. Lamellibranchia

Lamellibranchs are bivalve shellfish, e.g. mussels, cockles, clams and oysters, in which the valves of the shell are held closed by one or two adductor muscles. There are two parts to an adductor muscle. One, translucent in appearance, is capable of fast contraction and relaxation but cannot maintain tension for long periods, the other, more opaque in appearance, is slow acting but can maintain tension over long periods. The relative amounts of the two types in an animal vary among species depending on its mode of life. Sessile species such as oysters and mussels have only small proportions of translucent muscle which are sufficient to close the shell rapidly before the opaque muscle holds it firm. Clams are motile, propelling themselves by vigorously flapping their shells; they have high proportions of translucent muscle tissue. Lamellibranch muscle tissue is unusual in that it contains the protein paramyosin as part of the contractile apparatus as well as myosin and actin. It is found in both the opaque and translucent portions but its concentration is higher in the former (Philpott *et al.*, 1960). In some species the translucent, fast-acting portion is composed of cross-striated muscle fibres, in others, of oblique-striated fibres (Bowden, 1958; Morrison and Odense, 1974). The structure of the cross-striated fibres is very similar to that in vertebrate fish muscle fibres, the main difference being that in lamellibranchs the ratio of thin

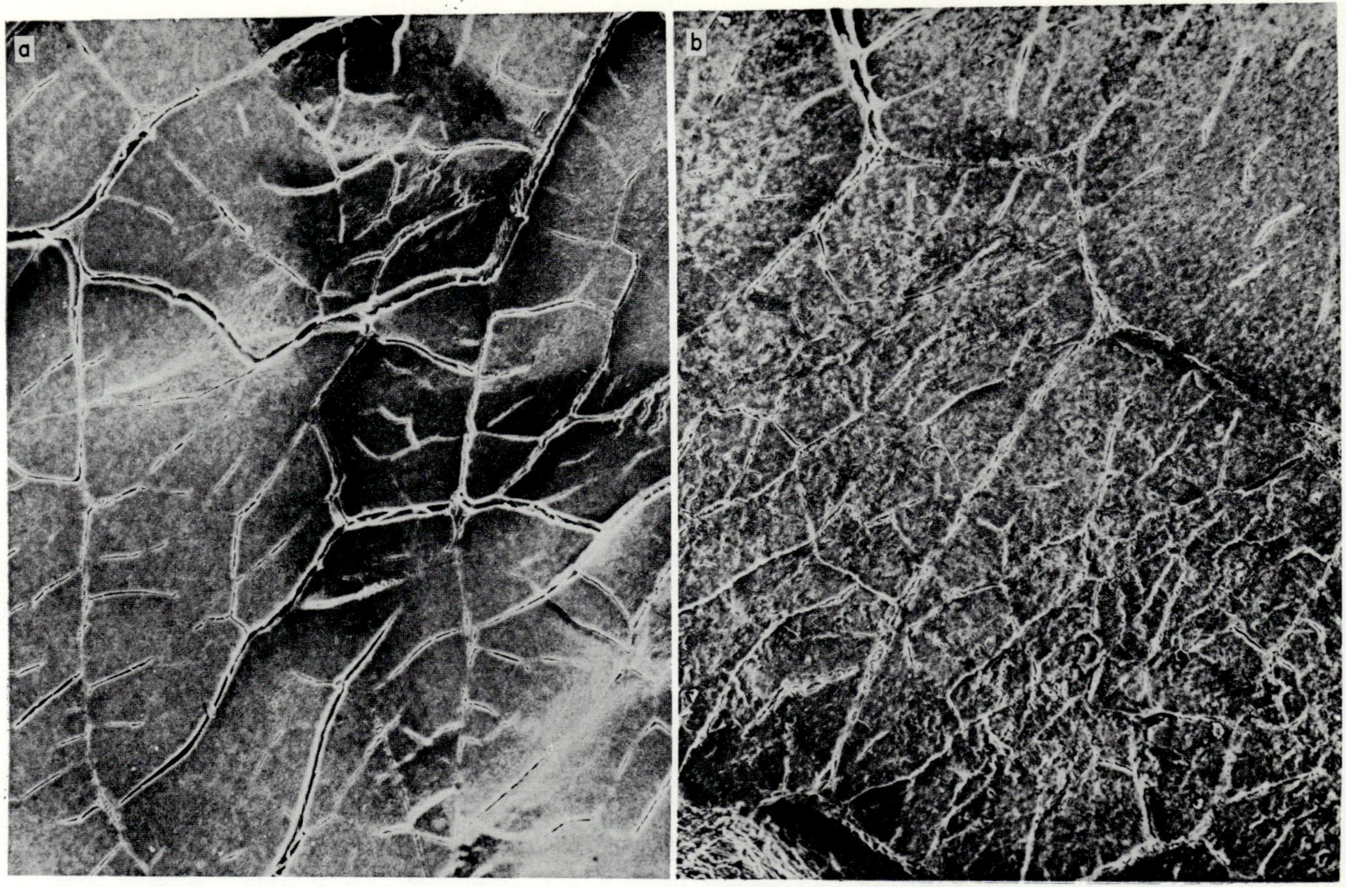

FIG. 20. Scanning electron microscope photographs of cooked crab muscle, $\times 160$. (a) Not frozen. (b) Frozen in liquid nitrogen and stored at $-29°C$ for 3 d.

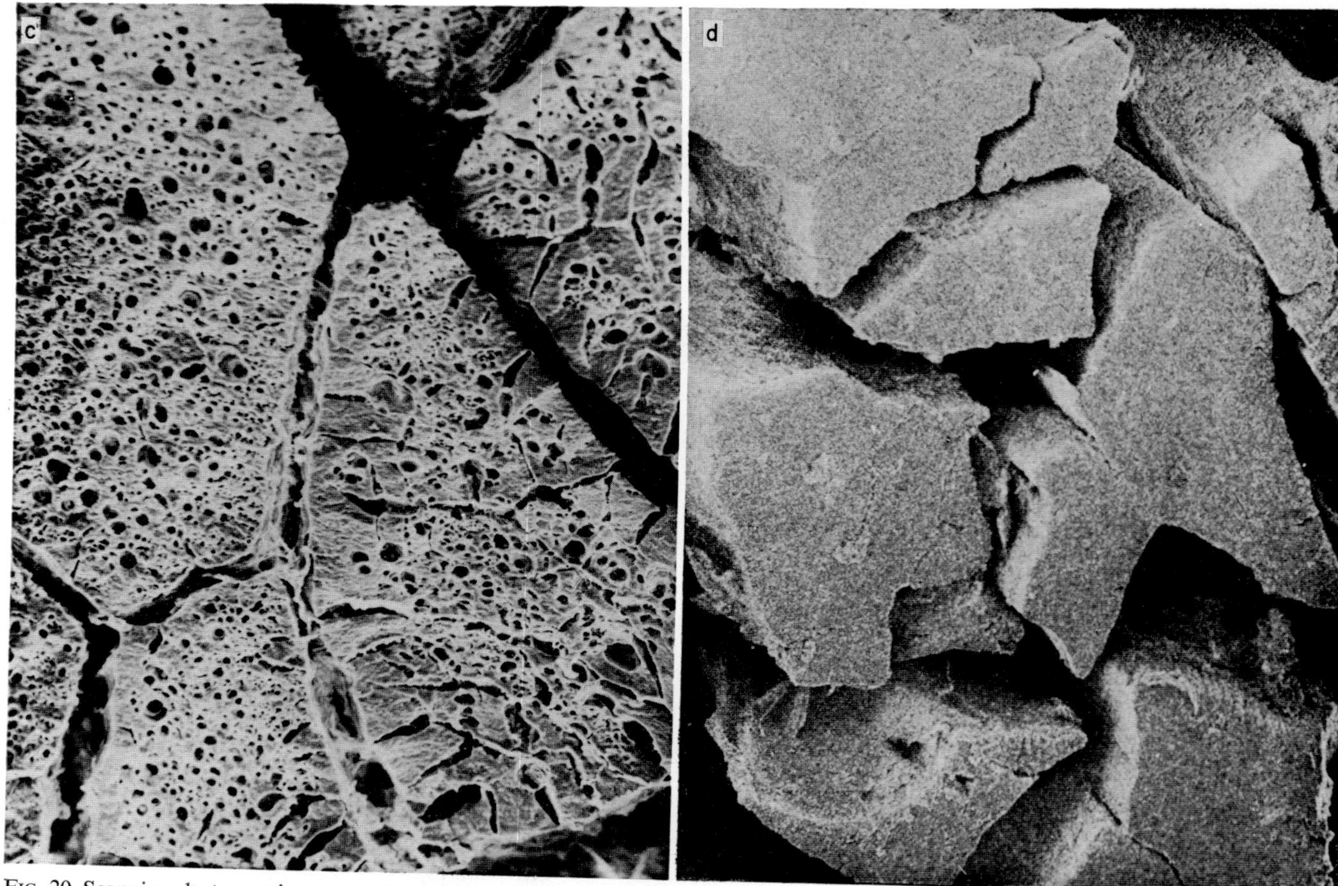

FIG. 20. Scanning electron microscope photographs of cooked crab muscle, ×200. (c) Frozen in liquid nitrogen and stored at −29°C for 1 yr illustrating the growth of ice crystals during frozen storage. (d) Frozen slowly at −29°C and stored for 1 yr at −29°C. Ice is formed between the fibres. (By permission of Giddings and Hill (1976) and reproduced from the *Journal of Food Science*.)

to thick filaments, about 6 : 1, is much greater and the Z-disc is more diffuse (Philpott *et al.*, 1960; Morrison and Odense, 1968; Nisbet and Plummer, 1968; Millman and Bennett, 1976). The oblique-striated muscle is similar to that already described for cephalopods. The fibre contains thick and thin filaments, with the thin filaments attached to dense bodies obliquely aligned in the fibre (Morrison and Odense, 1974).

The opaque, slow-acting muscle tissue is smooth muscle, i.e. it shows no striations under the microscope, and possesses some interesting physiological and biochemical properties (Hanson and Lowy, 1960, 1964; Ruegg, 1964; Needham, 1971). It is able to sustain a powerful contraction for considerable periods without fatigue. Although this gives an impression of the muscles being locked in a contracted state, the maintenance of tension is an active, not a passive process, requiring periodic excitation. Many of the studies of the structure of lamellibranch smooth muscle have been conducted not on adductor muscle but on the anterior byssus retractor muscle (ABRM) of the

FIG. 21a. Electron microscope photograph of squid muscle, longitudinal section, ×92 400. (By permission of Kawaguti (1962).)

marine mussel. (The mussel is attached to its substrate by a clump of protein strands, the byssus. The ABRM manipulates the muscle that is connected to the byssus.) It has the same properties and structure as smooth adductor muscle and can be used as a model of it (Hanson and Lowy, 1964).

Sobieszek (1973) has given a detailed and comprehensive description of the structure of the contractile apparatus of this muscle. The fibre contains two kinds of filament (Fig. 22a), a thin type containing actin and a thick one containing myosin and paramyosin. The ratio of thin to thick filaments is 17 : 1. The thick filaments are tapered from the centre to the ends, and since they are not aligned a cross-section will show various diameters (Fig. 22b). They have a maximum diameter of about 70 nm and a length of 25 μm. (In comparison, the thick filaments of striated muscle are about 10 nm in diameter and 1·3 μm in length.) The filaments are thought to consist of a core of paramyosin with an outer layer of myosin. The actin filaments are about 11 μm long.

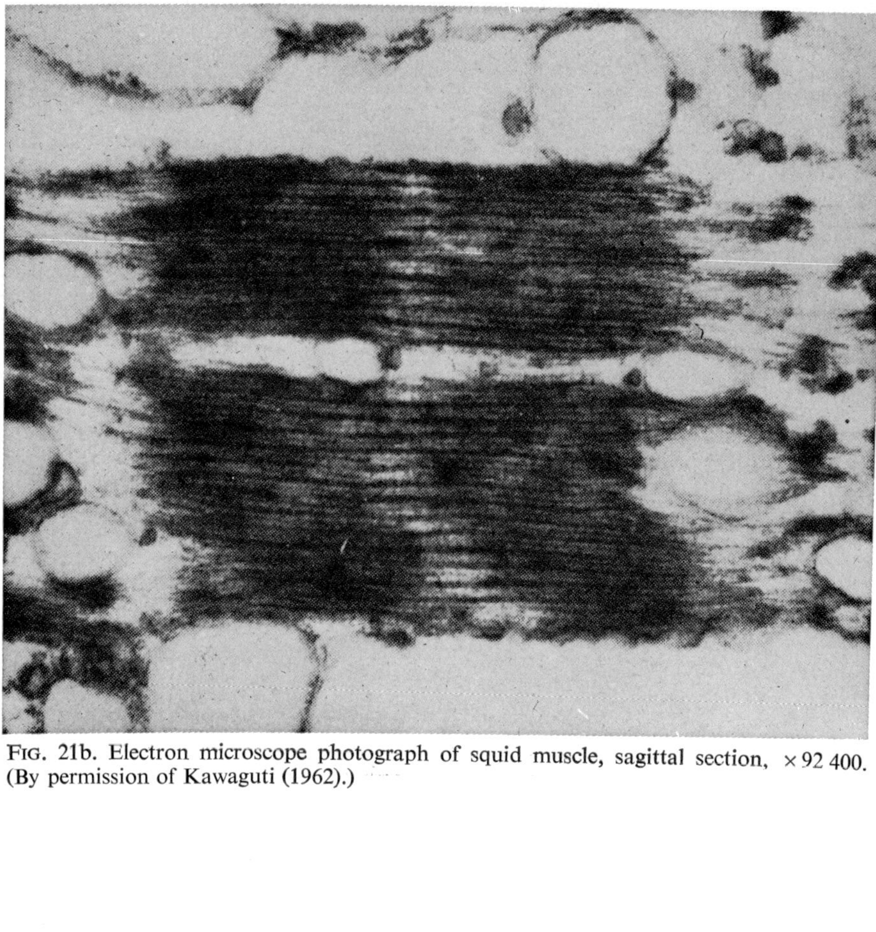

FIG. 21b. Electron microscope photograph of squid muscle, sagittal section, ×92 400. (By permission of Kawaguti (1962).)

Fig. 22a. Electron microscope photograph of anterior byssus retractor muscle of the marine mussel, longitudinal section, ×61 200. (By permission of Sobieszek (1973).)

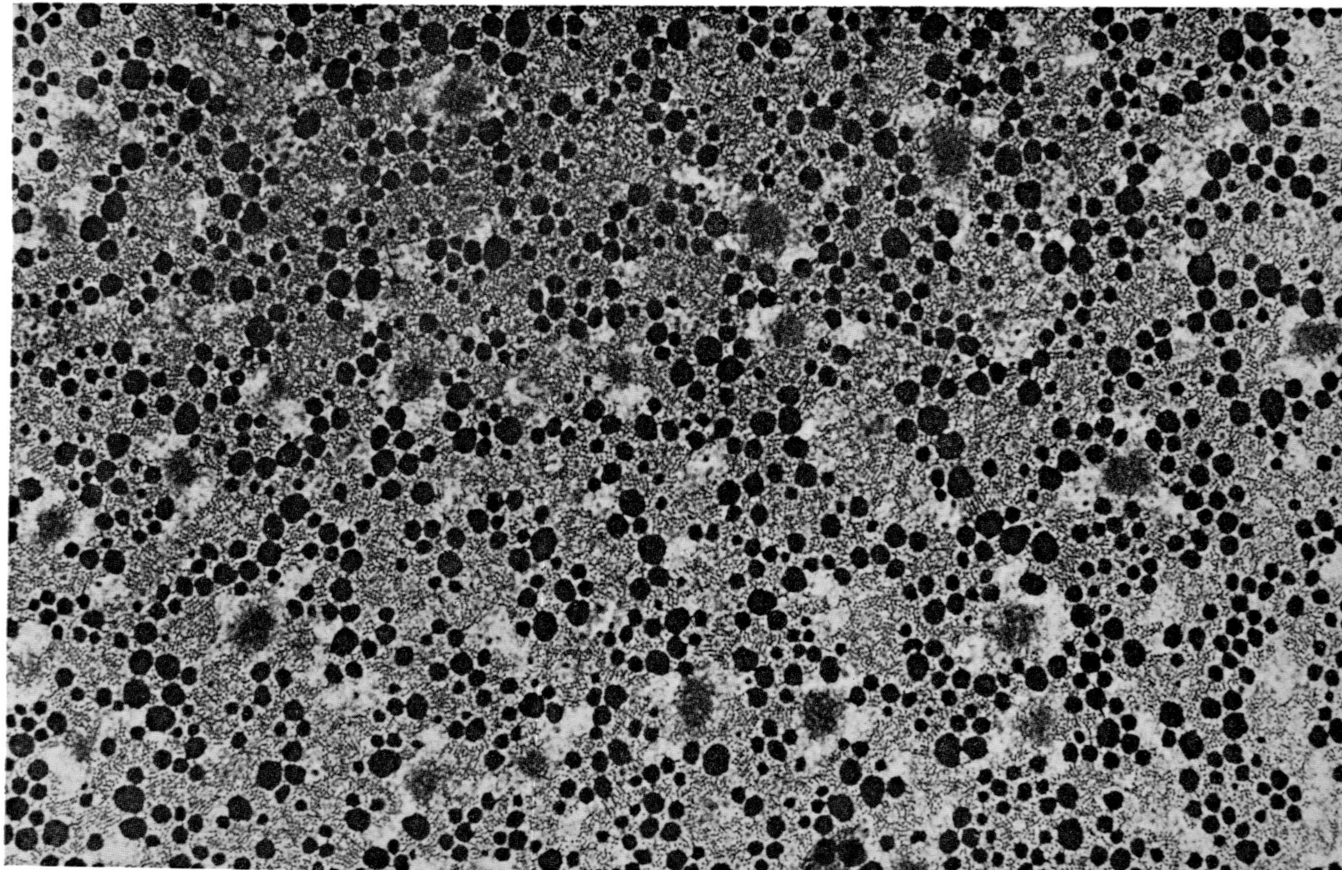

Fig. 22b. Electron microscope photograph of anterior byssus retractor muscle of the marine mussel, transverse section, × 61 200. (By permission of Sobieszek (1973).)

There are no Z-discs in the fibre but there are dense bodies distributed throughout the cell and attached to the cell wall. Sobieszek (1973) describes thin filaments attached to these bodies and argues that they are equivalent to Z-discs in cross-striated muscle and the nodes in oblique-striated muscle. From this conclusion he deduced a length of 50 μm for the sarcomere.

The muscle fibre is surrounded by a sarcolemma with pinocytotic vesicles. There is a well-developed sarcoplasmic reticulum with a transverse tubular system (Gilloteaux, 1976).

D. Gastropoda

Plesch *et al.* (1975) examined the structure of the musculature of a freshwater snail by light microscopy and identified the three types of muscle already described—cross-striated, oblique-striated and smooth. The first was found only in the heart, the second in a few visceral organs. Smooth muscle was by far the most common and was found in the head–foot and body wall. The ultrastructure of this last type is very similar to that in lamellibranchs which was described in the previous section (Nisbet and Plummer, 1968; Sanger and Hill, 1973).

VI. CONCLUSION

In this chapter I have surveyed the structure of the muscle tissue of the main types of aquatic animals used for food. It has been written from the viewpoint of a fish technologist, and a biologist may find the treatment somewhat superficial. Nevertheless, I hope it will be of some interest to marine biologists as I am not aware of any recent review of the subject.

As a result of the work of microscopists, and particularly of electron microscopists over the last 15 years or so, we now have a very good picture of the structure of fish muscle. Certainly our knowledge of 'normal' muscle is more than adequate for the needs of the food scientist and technologist. Unfortunately, from their point of view our knowledge of the effects of processing on structure is, in my view, inadequate. It might be asked, quite fairly, to what extent does a study of the structure of muscle contribute to an understanding of the effects of processing on fish flesh? I would argue that some awareness of the structure of muscle tissue should be part of the basic knowledge of anyone concerned with the science and technology of fish processing, but in addition a more detailed study of the subject can give a valuable insight into factors influencing the texture and physical properties of fish and fish products.

Fish and fish products exhibit a wide range of textural properties, even before any effects of processing are considered (Howgate, 1977), and undoubtedly the differences in structure already described must have some influence. For instance, vertebrate fish flesh is generally soft to firm, tender and fibrous; molluscan flesh is very firm, tough and chewy. Their structures are also very different, the former consisting of cross-striated muscle fibres, the latter of smooth or oblique-striated fibres. It is hard to avoid the conclusion that there is some causal relationship between these observations. Even within a species texture may vary due to a number of biological factors. The effect of some of these factors on the association between texture and structure have been investigated in meat, but so far there have been few corresponding studies on fish.

Some storage and processing conditions affect the texture of fish and have already been mentioned along with the associated structural changes. At present the picture is very incomplete and the links between structural and textural changes are only broadly understood. Even a fundamental process like cooking, which has a profound effect on texture in converting a soft, gelatinous material into a firmer, fibrous product, has not been adequately studied by microscopical methods.

In preparing this review I have been struck by the lack of information we have about the structure of processed fish and by the fact that more recently developed techniques like scanning electron microscopy, cryo-fracturing and freeze-etching have hardly been exploited. I hope that this review will stimulate microscopists to further investigate this field of study.

REFERENCES

Aitken, A. and Connell, J. J. (1977). *In* 'Freezing, Frozen Storage and Freeze-drying', p. 187. International Institute of Refrigeration, Paris.

Alexander, R. McN. (1969). *J. Mar. Biol. Ass. U.K.* **49**, 263.

Atwood, H. L. (1967). *Am. Zool.* **7**, 527.

Atwood, H. L. (1972). *In* 'The Structure and Function of Muscle' (G. H. Bourne, ed.), Vol. 1, 2nd Edn, p. 421. Academic Press, New York and London.

Barets, A. (1952). *Archs Anat. Microsc. Morph. Exp.* **41**, 305.

Bendall, J. R. (1974). 'Muscles, Molecules and Movement.' Heinemann, London.

Best, A. C. G. and Bone, Q. (1973). *Z. Zellforsch. Mikrosk. Anat.* **143**, 495.

Bishop, C. M. and Odense, P. H. (1967). *J. Fish. Res. Bd Can.* **4**, 2549.

Boddeke, R., Slijper, E. J. and van der Stelt, A. (1959). *Proc. K. Ned. Akad. Wet. C* **62**, 576.

Bone, Q. (1966). *J. Mar. Biol. Ass. U.K.* **46**, 321.

Bourne, G. H., ed. (1972). 'The Structure and Function of Muscle', Vol. 1, 2nd Edn. Academic Press, New York and London.

Bowden, J. (1958). *Int. Rev. Cytol.* **7**, 295.

Bromlei, G. F. (1949). *Izv. Tikhookean. Nauchno-issled. Inst. Ryb. Khoz. Okeanogr.* **31**, 158.
Bühn, W. (1940). *Z. Zellforsch. Mikrosk. Anat.* **30**, 323.
Buttkus, H. (1963). *J. Fish. Res. Bd Can.* **20**, 45.
Cassens, R. G. and Cooper, C. C. (1971). *Adv. Fd Res.* **19**, 1.
Charley, H. and Goertz, G. E. (1958). *Fd Res.* **23**, 17.
Connell, J. J. (1957). *J. Sci. Fd Agric.* **9**, 526.
Connell, J. J. (1962). *In* 'Recent Advances in Food Science' (J. Hawthorn and J. M. Leitch, eds). Butterworth, London.
Connell, J. J. and Howgate, P. (1969). *In* 'Freezing and Irradiation of Fish' (R. Kreuzer, ed.), p. 145. Fishing News (Books) Ltd., London.
Food and Agriculture Organization of the United Nations (1976a). 'Catches and Landings, 1975.' *Yb. Fish. Statist.* **40**. F.A.O., Rome.
Food and Agriculture Organization of the United Nations (1976b). 'The State of Food and Agriculture, 1975.' F.A.O., Rome.
Franzini-Armstrong, C. and Porter, K. R. (1964). *J. Cell. Biol.* **22**, 675.
Gainey, L. F. (1976). *Malacologia,* **15**, 411.
Gas, N. (1972). *C.R. Hebd. Séanc. Acad. Sci., Paris, D*, **275**, 1403.
George, J. C. (1962). *Am. Midl. Nat.* **68**, 487.
George, J. C. and Bokdawala, F. D. (1964). *J. Anim. Morph. Physiol.* **11**, 124.
Giddings, G. G. and Hill, L. H. (1976). *J. Fd Sci.* **41**, 455.
Gilloteaux, J. (1976). *Cytobiologie,* **12**, 440.
Gonzalez-Santander, R. and Garcia-Blanco, E. S. (1972). *J. Submicr. Cytol.* **4**, 233.
Hamoir, G. (1955). *Adv. Protein Chem.* **10**, 227.
Hanson, J. and Lowy, J. (1957). *Nature, Lond.* **180**, 906.
Hanson, J. and Lowy, J. (1960). *In* 'The Structure and Function of Muscle' (G. H. Bourne, ed.), Vol. 1, 1st Edn, p. 265. Academic Press, New York and London.
Hanson, J. and Lowy, J. (1964). *In* 'Biochemistry of Muscle Contraction' (J. Gergely, ed.), p. 400. J. A. Churchill, London.
Howgate, P. (1977). *In* 'Sensory Properties of Food' (G. G. Birch, J. G. Brennan and K. J. Parker, eds), p. 249. Applied Science Publishers, London.
Huxley, H. E. (1957). *J. Biophys. Biochem. Cytol.* **3**, 631.
Jarenbäck, L. and Liljemark, A. (1975a). *J. Fd Technol.* **10**, 229.
Jarenbäck, L. and Liljemark, A. (1975b). *J. Fd Technol.* **10**, 309.
Johnston, I. A., Patterson, S., Ward, P. S. and Goldspink, G. (1974). *Can. J. Zool.* **52**, 871.
Johnston, I. A., Ward, P. S. and Goldspink, G. (1975). *J. Fish. Biol.* **7**, 451.
Kanna, K., Tanaka, T., Kakadu, K. and Shimizu, T. (1971). *Bull. Tokai Reg. Fish. Res. Lab.* No. 68, 51.
Kawaguti, S. (1962). *In* 'Electron Microscopy: Proceedings of the 5th International Congress for Microscopy' (S. S. Breese, ed.), Abstr. M-11. Academic Press New York and London.
Kilarski, W. (1967). *Z. Zellforsch. Mikrosk. Anat.* **79**, 562.
Knappeis, G. G. and Carlsen, F. (1962). *J. Cell Biol.* **13**, 323.

Krüger, P. (1952). 'Tetanus and Tonus der quergestreiften Skelettmuskeln der Wirbeltiere und des Menschen.' Akademische Verlags-gesellschaft Geest und Portig K.-G., Leipzig.

Le Danois, Y. (1952). *In* 'Traite de Zoologie' (P. P. Grasse, ed.), Vol. 13, p. 783. Masson et Cie, Paris.

Liljemark, A. (1969). *In* 'Freezing and Irradiation of Fish' (R. Kreuzer, ed.), p. 140. Fishing News (Books) Ltd., London.

Love, R. M. (1966). *In* 'Cryobiology' (H. T. Meryman, ed.), p. 317. Academic Press, London and New York.

Love, R. M. (1967). *In* 'Rheology and Texture of Foodstuffs', Monogr. 27, p. 120. Society of Chemical Industry, London.

Love, R. M. (1970). 'Chemical Biology of Fishes.' Academic Press, London and New York.

Love, R. M. (1971). *In* 'Recent Advances in Quantitative Histo- and Cytochemistry', p. 17. Hans Huber Publishers, Bern.

Love, R. M., Lavety, J. and Steel, P. J. (1969). *J. Fd Technol.* **4**, 39.

Millman, B. M. and Bennett, P. M. (1976). *J. Molec. Biol.* **103**, 439.

Moon, T. W. and Hulbert, W. C. (1975). *Comp. Biochem. Physiol.* **52B**, 145.

Morrison, C. M. and Odense, P. H. (1968). *J. Fish. Res. Bd Can.* **25**, 1339.

Morrison, C. M. and Odense, P. H. (1973). *J. Fish. Res. Bd Can.* **30**, 1583.

Morrison, C. M. and Odense, P. H. (1974). *J. Ultrastruct. Res.* **49**, 228.

Nakajima, Y. (1969). *Tiss. Cell*, **1**, 229.

Needham, D. (1971). 'Machina Carnis.' Cambridge University Press.

Nisbet, R. H. and Plummer, J. M. (1968). *Symp. Zool. Soc. Lond.* No. 22, 193.

Nishihara, H. (1967). *Arch. Histol. Jap.* **28**, 425.

Partmann, W. (1967). *Arch. Fischwiss.* **19**, 64.

Partmann, W. and Schlaszus, H. (1973). *Kältetech.-Klim.* **25**, 3.

Patterson, S. and Goldspink, G. (1972). *Z. Zellforsch. Mikrosk. Anat.* **133**, 463.

Peachey, L. D. (1961). *In* 'The Sarcoplasmic Reticulum', Suppl. to *J. Biophys. Biochem. Cytol.* **10**, No. 4, Part 2, 159.

Peachey, L. D. (1970). *In* 'The Physiology and Biochemistry of Muscle as a Food' (E. J. Briskey, R. G. Cassens and J. C. Trautmann, eds.), Vol. 2, p. 273. University of Wisconsin Press, Madison.

Philpott, D. E., Kahlbrock, M. and Szent-Györgi, A. G. (1960). *J. Ultrastruct. Res.* **3**, 254.

Plesch, B., Janse, C. and Boer, H. H. (1975). *Neth. J. Zool.* **25**, 332.

Rosenbluth, J. (1972). *In* 'The Structure and Function of Muscle' (G. H. Bourne, ed.), Vol. 1, 2nd Edn, p. 389. Academic Press, New York and London.

Ruegg, J. C. (1964). *In* 'Biochemistry of Muscle Contraction' (J. Gergely, ed.), p. 412. J. A. Churchill, London.

Sanger, J. W. and Hill, R. B. (1973). *Proc. Malac. Soc. Lond.* **40**, 335.

Schaller, D. R. and Powrie, W. D. (1972). *Can. Inst. Fd Sci. Technol. J.* **5**, 184.

Schattenberg, P. J. (1973). *Z. Zellforsch. Mikrosk. Anat.* **143**, 569.

Sikorski, Z., Olley, J. and Kostuch, S. (1976). *CRC Crit. Rev. Fd Sci. Nutr.* **8**, 97.

Slautterback, D. B. (1966). *In* 'The Physiology and Biochemistry of Muscle as Food' (E. J. Briskey, R. G. Cassens and J. C. Trautmann, eds), Vol. 1, p. 39. University of Wisconsin Press, Madison.

Sobieszek, A. (1973). *J. Ultrastruct. Res.* **43**, 313.

Tanaka, T. (1958). *Bull. Tokai Reg. Fish. Res. Lab.* No. 20, 77.

Tanaka, T. (1965). *In* 'The Technology of Fish Utilisation' (R. Kreuzer, ed.), p. 121. Fishing News (Books) Ltd., London.

Veratti, E. (1902), *Translated in* 'The Sarcoplasmic Reticulum', Suppl. to *J. Biophys. Biochem. Cytol.* **10**, No. 4, Part 2, 1.

Walker, M. G. (1970). *J. Cons. Perm. Int. Explor. Mer*, **33**, 228.

Walker, M. G. (1971). *J. Cons. Perm. Int. Explor. Mer*, **33**, 421.

Ward, D. V. (1972). *J. Zool. Lond.* **167**, 487.

Ward, D. V. and Wainwright, S. A. (1972). *J. Zool. Lond.* **167**, 437.

Yuditskaya, A. I. (1959). *Ryb. Khoz.* **35**, 65.

10. Animal Feeds—Plant Constituents

J. G. VAUGHAN[1] and JANE A. STUBBS[2]

[1] *Biology Department, Queen Elizabeth College, London, U.K.*
[2] *Royal Botanic Gardens, Kew, Richmond, U.K.*

I. INTRODUCTION

Microscopy may be employed for several reasons in the quality control of animal feeds. When parts of animals and plants are processed for inclusion in feeds, they are often reduced to such small fragments that identification by the naked eye or low magnification is impossible. However, the arrangement of cells and tissues, also the cell contents, can often characterize a species and therefore microscopic examination of feed materials can lead to identification of the plant and animal constituents present.

Such a technique is most important because it is the only means of establishing the composition of a feed in terms of its plant and animal constituents. Chemical composition is sometimes taken as a guide to the authenticity of a feed material but this can be unreliable. Feed microscopy is therefore an essential technique for the detection of adulteration or admixture. In addition, it is most important for the detection of certain poisonous plant materials and also materials which have a stimulant drug action or affect the flavour of the feed. The various problems that can be investigated by microscopy are to be found in raw materials, straight feeds, concentrates and compounds. The technique can also be used to check the composition of a compound as produced at the end of the manufacturing process. Feed microscopy is probably used mainly in qualitative analysis but has also been adapted for quantitative or semi-quantitative analysis.

II. HISTORY AND PRESENT PRACTICE

For the analysis of animal feeds, normally only light microscopy is employed. As this type of microscopy has been available for many years, a number of reference works on feed microscopy have appeared. These, in the main, have been concerned with plant constituents. One problem encountered by the present-day microscopist is the difficulty of obtaining many of these reference works.

The following books are of value to the feed microscopist. *U.S.A.*: Winton (1906), Winton and Winton (1932, 1935, 1937, 1939) and American Association of Feed Microscopists (1961). *U.K.*: Morris (1928), Parkinson and Fielding (1930) and Vaughan (1970). *France*: Collin and Perrot (1904), Bussard and Brioux (1925), Juillet *et al.* (1955) and Ferrando and Henry (1966). *Germany*: Moeller and Griebel (1928), Czaja (1971), Gassner (1973) and Mészáros and Deutschmann (1975). Of special interest is a recent review of methods and literature (Huss, 1976).

As far as the authors of this chapter are aware, there are two organizations for the promotion of feed microscopy. The American Association of Feed Microscopists, with a membership drawn mainly, but not entirely, from North America, holds annual meetings. Official proceedings of this meeting are made available to its members. The Association also organizes short courses of instruction in feed microscopy. Annual meetings are also held in various countries on the continent of Europe by the 'Sektionen Futtermittelmikroskopie und Futtermittelmikrobiologie der Internationalen Arbeitsgemeinschaft fur Futtermitteluntersuchung'. This organization produces annual proceedings (Protokoll). In the U.K., occasional short courses on feed microscopy are given at Queen Elizabeth College, London University.

III. METHODS

This section deals only with the methods of microscopy that are particularly applicable to the analysis of animal feeds (plant constituents). Most of the references in the previous section give detailed information on these methods.

A. Microscopes

Only the light microscope is normally employed in feed analysis. In the present circumstances it is unlikely that transmission or scanning electron microscopy would be used extensively although, in recent years, the technique has been utilized in plant systematics (Cole and Behnke, 1975).

Both the stereomicroscope (capable of magnifications $\times 5$ to $\times 50$) using incident light and the compound microscope (capable of magnifications $\times 50$ to $\times 500$) using transmitted light are essential. The compound microscope should be equipped with a mechanical stage, an inclined binocular body tube and a built-in lighting system. Preliminary examination of a feed material with the stereomicroscope is always recommended.

B. Preparation of Material for Microscopy

Most plant materials used in animal feeds are of course derived from storage organs (such as seeds and tubers) and are therefore rich in oil, starch and protein. Microscopic analysis is based to some extent on the characters of the cell contents, but the arrangement of cells is probably more important and to observe this it is often necessary to remove the cell contents.

However, before any clearing procedures are carried out, it is strongly suggested (Huss, 1976) that the material be investigated with the stereomicroscope. Sieving into fractions may facilitate examination with the stereomicroscope. Pellets must be given special treatment prior to examination with the stereomicroscope (Huss, 1961). Because of differences in specific gravity, plant materials may be separated from fish scales, bone fragments and minerals by flotation in chloroform or carbon tetrachloride.

A variety of procedures exists for clearing plant materials before examination with the compound microscope. It should be emphasized, however, that there is no one set system of procedures to be adopted. The type of animal feed under examination determines the treatment.

Oil or fat in plant materials in animal feeds does not normally hinder microscopic examination but any difficulty is easily obviated by removing the oil or fat with benzene or some similar organic solvent. Similarly, the presence of molasses does not normally cause difficulty but, again, a method is available for dealing with the situation (Huss, 1976).

To examine plant tissues in animal feeds it is normal to remove the starch and protein contents and a variety of techniques exist for this purpose (Wallis, 1965). Heating in chloral hydrate solution is particularly useful for expanding delicate and compressed tissues. Another advantage is that calcium oxalate crystals are not affected. Treatment with 2–5% caustic soda or potash has a strong softening and disintegrating effect upon cellulose and therefore is useful for separating tissues, although it does cause cell walls to swell. Crystals are also affected. Crude fibre or digestion methods are vigorous ways of removing starch and protein by alternatively boiling with acid and alkali. Various modifications exist involving caustic soda and nitric or sulphuric acid. Crude fibres may be preserved for reference in formalin. To effect identification, it is sometimes necessary to macerate plant fragments or soften them prior to sectioning. Nitric acid, chromic acid and hydrofluoric acid can be used for these purposes (Wallis, 1965).

Because of heavy pigmentation, it is sometimes difficult to observe the arrangement of cells in plant fragments. Under these conditions, it is necessary to bleach the tissues with Javelle water, hydrogen peroxide (Gassner, 1973) or commercial domestic bleach. In addition to bleaching, these reagents may also act as macerating agents.

C. Sections

For the sake of speed in quality control work, identification should be based as far as possible on the surface features of plant fragments. Occasionally it is necessary to cut sections and the authors have found the Reichert sliding-type microtome most effective. Prior to sectioning it is frequently necessary to soften the fragments by soaking in water or, in cases of difficulty, hydrofluoric acid. For this type of sectioning, material is held between two pieces of cork, balsa wood or polystyrene.

Sections may also be prepared with the freezing-type microtome (Mészáros and Deutschmann, 1975). It is also possible to cut sections by hand with a single-edged razor blade. In routine work, it is hardly likely that embedding in paraffin wax will be required. Details of these techniques are given by Sass (1958).

D. Mounting Media

Huss (1976) discusses the effects of mounting media in terms of refractive index. He points out that the refractive index of most animal and plant material is about 1·52–1·54. Water ($n_D = 1{\cdot}333$) is a commonly used mountant in feed microscopy and is an excellent medium for demonstrating important diagnostic features such as the pericarp ‘cross cells’ of wheat and rye and the internal features of starch grains. Other commonly used mountants are glycerol ($n_D \approx 1{\cdot}47$) and phenol–glycerol ($n_D \approx 1{\cdot}525$). As the refractive index of the mounting medium approaches that of the object, the outlines of the cells are less clear but there is an increase in the transparency of the particles and the brightening of the visual field.

In dealing with the different plant fractions of the same compound feed, a variety of mounting media may be used. The authors find that water is satisfactory as a starch mountant while a crude fibre or material cleared in chloral hydrate can be mounted in a mixture of glycerol and water (1:1).

E. Types of Microscopy

Most observations in feed microscopy are made under bright field conditions. Although, of course, the phase contrast method is well established in modern biology, it is rarely used in the routine analysis of plant constituents. However, it could be used with advantage for the identification of micro-organisms and yeasts. Similarly, phase contrast microscopy would be useful for dealing with any delicate cell layers of plant fragments. In contrast, polarized light microscopy is a commonly used and highly advantageous technique. With crossed nicols or polarizing plates, birefringent (anisotropic) objects become illuminated. Many plant materials incorporated into feeds contain starch and the microscopic structure of starch grains can facilitate identification. With crossed nicols or polarizing plates, starch grains show an interference cross (Maltese cross or Brewster’s cross) where the arms intersect at the hilum. Most of the general works on feed microscopy, already referred to, give information on starch grain identification. More detailed references to starch microscopy are to be found in Reichert (1913), Seidemann (1966) and Whistler and Paschall (1967). The contemporary German feed microscopists use, in addition to the polarizing system, a Red I Compensator which gives a distinct play of colours without staining (Huss, 1960). Polarized light microscopy can also assist in the identification of plant species with a characteristic crystal distribution (e.g. sesame seed (Vaughan, 1970), cruciferous seed residues (Huss, 1956) and leguminous leaf meals (Mäckel, 1943)).

The possibility of identifying plant constituents with fluorescence (both primary and secondary) microscopy has been investigated on a number of occasions (Huss, 1976). Perhaps the most comprehensive account is given by Juillet *et al.* (1955). An interesting method for the identification of *Datura* seed fragments with the stereomicroscope is available (Huss, 1976). Although fluorescence microscopy might well occupy a significant role in the routine analysis of animal feeds, most investigators nevertheless agree that confirmation of any identification by fluorescence microscopy should be carried out with regard to the histological features of the plant fragments.

F. Staining and Histochemical Tests

For efficient quality control, identification of plant fragments should be carried out as quickly as possible. Consequently, the minium of staining should be employed. The following are the most useful reagents: (i) iodine/potassium iodide solution for starch and possibly protein, (ii) Sudan stains for fats and (iii) ferric chloride solution for tannins. The general reference works on feed microscopy give details of these tests.

Some special methods are available. Parkinson and Fielding (1930) state that fragments of charlock (*Sinapis arvensis*) testa turn bright red when heated in dilute sulphuric acid or chloral hydrate. Dilute suspensions of Indian ink in water have been used for identifying cocoa shells and distinguishing between rye and wheat flour (Czaja, 1971; Huss, 1976).

The methods of staining plant tissues usually employed by the botanist are not required in feed microscopy. However, plant materials not previously utilized occasionally appear in animal feeds. Frequently, no information is available in the literature concerning the microscopy of these materials, either in the entire or fragmented states. Original investigations are therefore required. To understand the structure of fragments, it is best to first investigate the structure of the entire organ. In this situation, the normal staining methods are usually employed (Vaughan, 1960; Jensen, 1962).

G. Reference Materials

The investigator normally bases his identification of plant fragments on descriptions, photomicrographs and drawings in the general reference works on feed microscopy. It must be emphasized, particularly in the case of the inexperienced microscopist, that such identification should be checked against authentic material. The authentic material should include a museum of seeds, other plant organs and the products of these structures used in feeds. Reference slides of fragments may also be made, using glycerine jelly as the mountant.

H. Quantitative Microscopy

Without a doubt, the most difficult aspect of feed microscopy is the quantitative assessment of constituents. This topic has lately been reviewed in an excellent manner by Huss (1976). He records five methods.

(i) Visual estimations. This is a subjective assessment of the constituents of a feed, using the stereomicroscope and/or the compound microscope. The accuracy of the method relies heavily on the experience of the analyst. Reference mixtures, with a range of composition ratios, may be used in conjunction with the method. However, as far as possible, the fragments of the constituents of the reference mixture must be of the same size and have the same characteristics as the fragments of the sample under examination.

(ii) Sorting and weighing. This involves picking out the individual constituents of a feed under the stereomicroscope and subsequently weighing them. According to Huss (1976), this method is applicable to particles greater than 0·2 mm in size (those that can be clearly identified at magnifications between $\times 5$ and $\times 50$). The relative proportions of particles less than 0·2 mm must be estimated. Obviously this method of sorting and weighing is only valid if the fine fraction (particles less than 0·2 mm) forms a small part of the whole. This method is laborious but probably the most accurate of the quantitative methods.

(iii) Flotation techniques. When constituents of a feed differ in density, they may be separated by sedimentation in liquids of appropriate specific gravity. A quantitative method, using carbon tetrachloride and petroleum ether, is given by the American Association of Feed Microscopists (1961). Under this method of flotation techniques, elutriation might be mentioned. This technique will be described later in the chapter, in conjunction with castor determination.

(iv) Counting methods. Various counting methods have been employed in the quantitative analysis of animal feeds. There is the Dutch method (Bussard and Brioux, 1925), and a technique for the determination of rice hulls is given by the American Association of Feed Microscopists (1961). Wallis (1965) developed a counting method with lycopodium powder (spores) but, as far is known, this has not been used widely for animal feeds.

(v) Measurements of area. Area measurements of plant fragments have been used by some workers for quantitative determinations. The Nebeltau method (De Jong, personal communication) is currently used in Holland.

Another technique, using lobelia seeds, has been developed by Hamence (1966).

Most analysts will agree that the quantitative methods available in microscopy cannot match quantitative chemical methods for accuracy. However, microscopy is the only method available for obtaining information on the quantities of plant and animal materials in feeds and the techniques described are currently being used with success. It is possible that other methods may appear when more progress is made with automatic image recognition and scanning (Rosenfeld, 1976).

IV. PLANT CONSTITUENTS OF ANIMAL FEEDS

The range of plant constituents that may appear in animal feeds is enormous (Morrison, 1961). Amongst others, these include leaf meals, cereals, oil seed residues, roots, tubers, legume seeds and the by-products from the manufacture of soft and alcoholic drinks. Descriptions of the microscopy of the common plant constituents are to be found in the manuals referred to in Section II.

New plant constituents are constantly being suggested for or appearing in animal feeds. Two comprehensive surveys of oil seeds and their residues have recently appeared (Vaughan, 1970; Mészáros and Deutschmann, 1975). These works include new information on important groups such as cruciferous and palm seed residues. Descriptions of the microscopy of new plant constituents appear quite regularly in the *Protokoll* (available annually) of the Sektionen Futtermittelmikroskopie und Futtermittelmikrobiologie der Internationalen Arbeitsgemeinschaft für Futtermitteluntersuchung, also in the official proceedings (*Jahresbericht*) of the Institut für Angewandte Botanik der Universitat Hamburg, West Germany. Isolated descriptions appear in other journals and publications from time to time but, because of the widespread nature of these descriptions, it is advisable to have access to an efficient abstracting system if the literature on any particular plant constituent is required.

To produce a description of the microscopy of a new plant constituent, it is sensible in the first instance to investigate the normal structure of the plant organ from which the constituent has been produced. Most plant constituents of animal feeds are the result of some form of industrial processing. This processing may affect the microscopic structure, but the situation becomes clearer if the untreated plant organ is first investigated. With regard to seeds and fruits, the major constituents of animal feeds, the recent work of Corner (1976) has surveyed in a most comprehensive manner the structure of dicotyledonous seeds. No modern general work is available for monocotyledonous seeds but useful information can be obtained from Netolitzky (1926) and Barton (1967). As regards stems, roots and leaves of dicotyledonous plants, the best reference work is Metcalfe and Chalk (1950). Information on the vegetative organs of monocotyledons is being produced at the Jodrell Laboratory of the Royal Botanic Gardens, Kew, U.K. (Metcalfe, 1960, 1971; Tomlinson, 1961, 1969; Cutler, 1969; Ayensu, 1972).

V. A CONTEMPORARY PROBLEM OF PLANT CONSTITUENT IDENTIFICATION

Feed microscopy is an essential technique in quality control, both within commercial organizations and at the level of official legislation. A striking example of the latter has recently appeared in connexion with the European Economic Community (E.E.C.). Certain plant constituents have been designated as 'undesirable substances' to be included in feeding stuffs within certain prescribed limits. The following descriptions include information on the nomenclature, commercial significance and microscopy of these constituents. There is no detailed discussion in the following sections concerning the toxicological effects of these materials, although it must be pointed out that it is most difficult to find relevant information in the literature and one wonders how the prescribed limits for the undesirable constituents were arrived at in some cases.

A. **Mustard Seeds** (Figs 1–8)

The toxicity of mustard seeds and their residues is related to the production of isothiocyanate and oxazolidinethione compounds from glucosinolates (Juillet *et al.*, 1955; Vaughan, 1970). The microscopic identification of these residues is based on characters of the testa or seed coat, the embryo tissue being of little value in this respect. As stated earlier in this chapter, under conditions of commercial quality control the microscopist seeks to identify plant fragments on surface features. The production of sections of plant fragments is a time-consuming process and should be avoided, if possible, in feed analysis. However, identification of mustard seed residues often relies on the production of sections, although the necessary information can sometimes be deduced from fragments when the edge of the fragment is so orientated as to produce a sectional effect. Although the present work deals only with the undesirable mustards, a full understanding of the situation

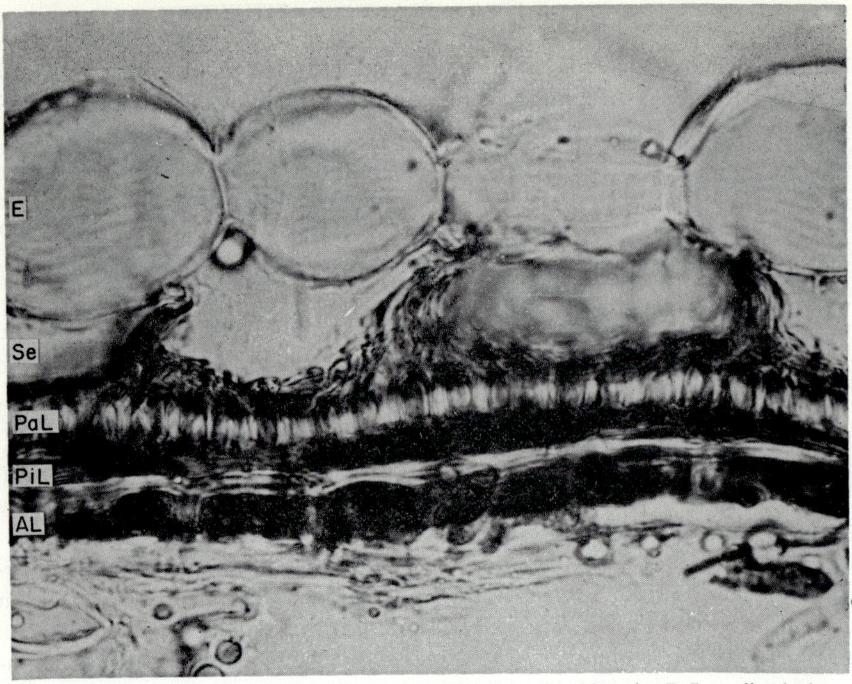

FIG. 1. TS of *B. nigra* testa, ×540. E, epidermis; Se, subepidermis; PaL, palisade layer; PiL, pigment layer; AL, aleurone layer.

FIG. 2. TS of *B. carinata* testa, ×540.

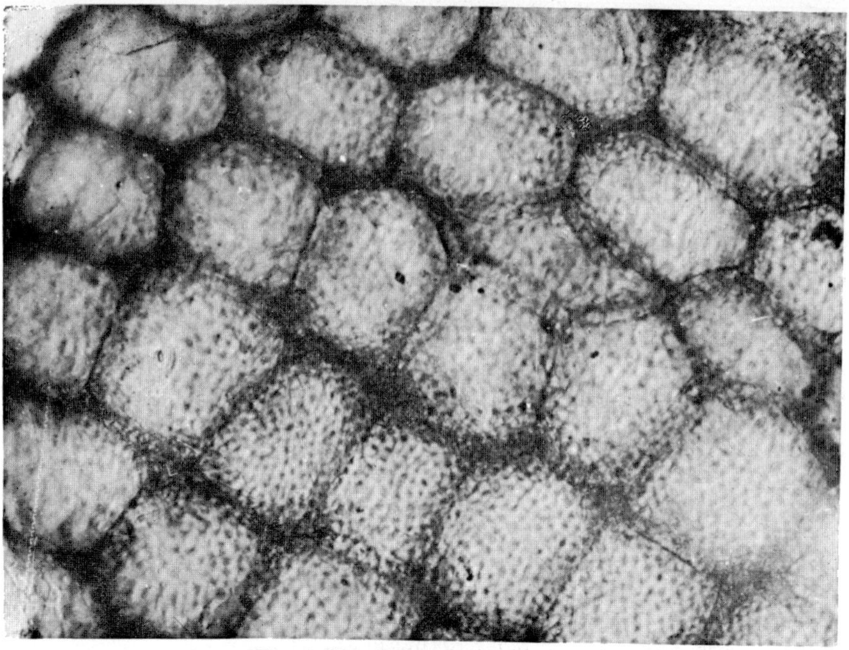

FIG. 3. SV of *B. nigra* testa, ×165.

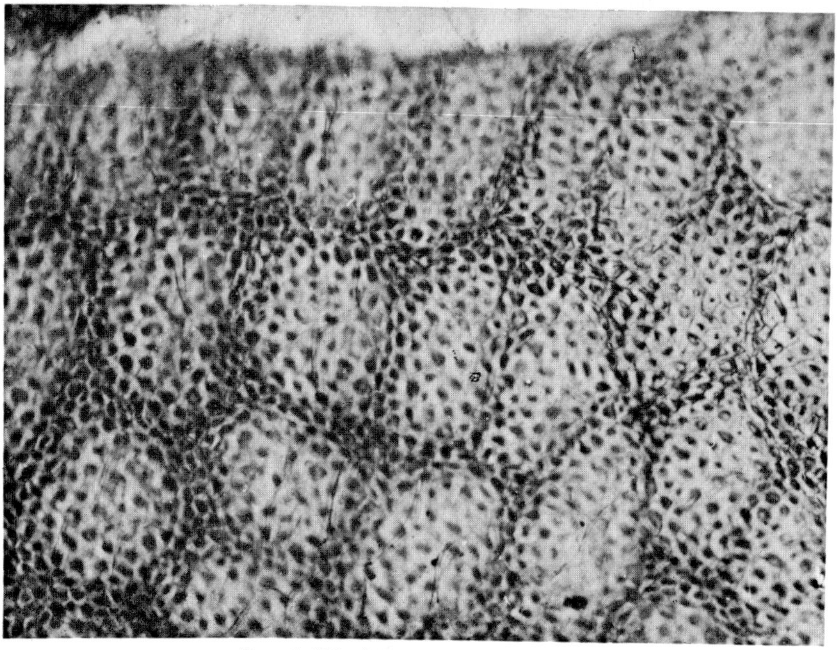

FIG. 4. SV of *B. juncea* testa, ×165.

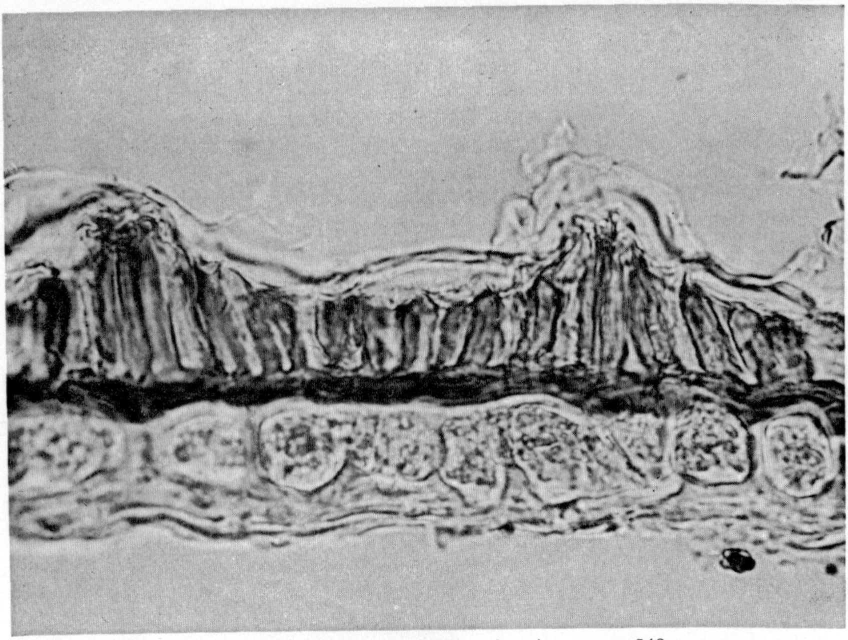

FIG. 5. TS of *B. juncea* (Indian form) testa, × 540.

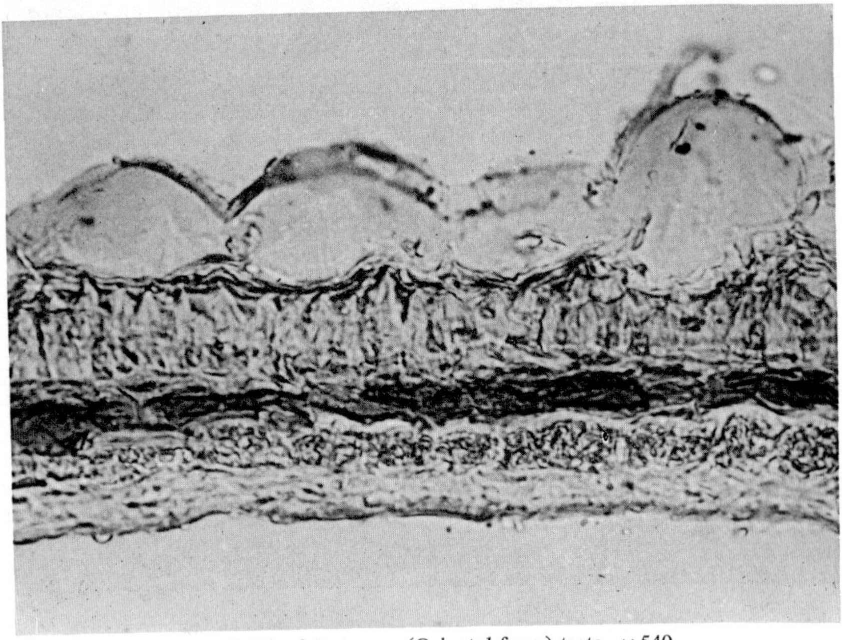

FIG. 6. TS of *B. juncea* (Oriental form) testa, × 540.

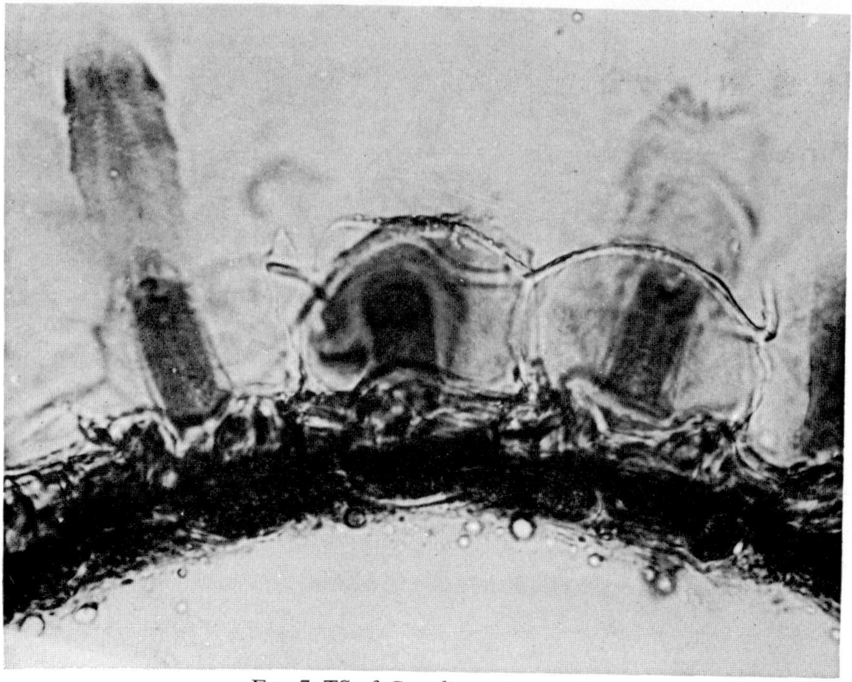

FIG. 7. TS of *Camelina* testa, ×540.

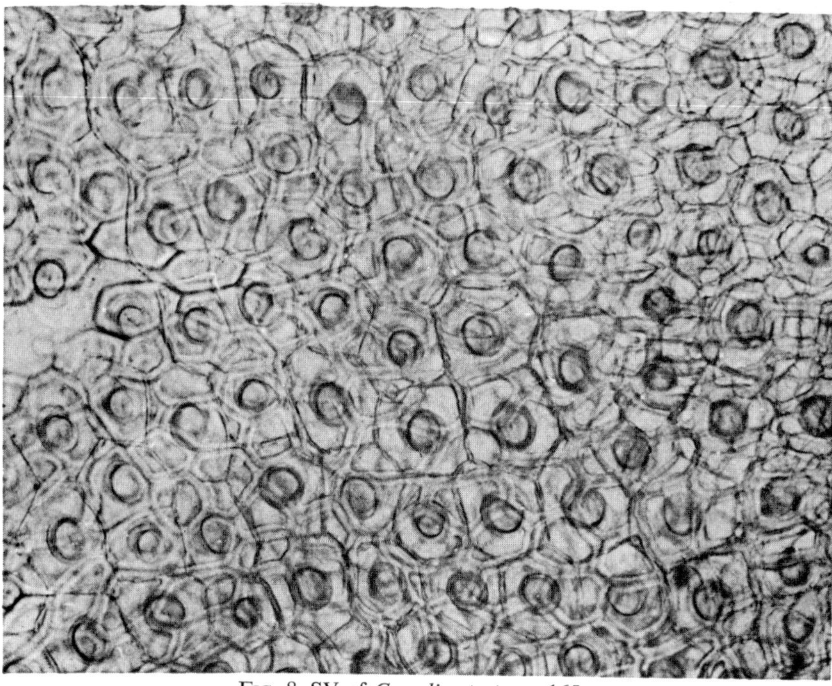

FIG. 8. SV of *Camelina* testa, ×165.

can only be obtained by reference to the microscopy of the rapes and other cruciferous seeds (Vaughan, 1970; Mészáros and Deutschmann, 1975).

1. Brassica nigra *(L) Koch (black mustard)*

B. nigra seed used to be one of the major constituents of condiment mustard but it has now been largely replaced by *B. juncea* (Vaughan and Hemingway, 1959). The testa of *B. nigra* shows all the usual cruciferous features: (i) a mucilaginous epidermal layer, one cell thick, (ii) a subepidermal layer, (iii) a palisade layer, (iv) a layer of pigmented parenchyma cells and (v) a single layer of aleurone cells. As far as the other mustards are concerned, *B. nigra* is most likely to be confused with *B. juncea* and *B. carinata*. In section, the distinguishing feature of *B. nigra* is that the radial walls of some of the palisade cells are drawn out into threads and this is the most important feature for identification. Fragments of the palisade layer in surface view show a marked reticulation but this feature is common to some other *Brassica* species. However, the diameter of the individual palisade cells is between 8 μm and 14 μm. This distinguishes *B. nigra* from other *Brassica* species.

2. Brassica juncea

Of the commercial mustards, *B. juncea* is likely to be most confusing to the analyst, not only because of its microscopy, but also because of nomenclature. Over the years *B. juncea* has been known under a number of synonyms (Vaughan *et al.*, 1963; Vaughan and Gordon, 1972). In the northern parts of the Indian subcontinent, the plant (rai) is grown for oil seeds. It is also grown elsewhere (e.g. China, Russia, Europe and North America) for oil seeds, also condiment mustard. Probably the crop cultivated in Russia, Europe and North America originated in China. Seed from the Indian subcontinent produces mainly 3-butenyl isothiocyanate; seed from other regions, with a presumed origin in China, produces mainly allyl isothiocyanate. It has been suggested (Vaughan *et al.*, 1963; Vaughan and Gordon, 1972) that there are two 'geographical races' of *B. juncea*, Indian and Oriental.

B. juncea seed, normally brown, from India and Pakistan, is sometimes described as *Brassica integrifolia* (West) O. E. Schulz. Seed, which may be brown or yellow, from other regions can be described as *Brassica juncea* (L) Czern. and Coss. subsp. *juncea* var. *lutea* Batalin, *Brassica juncea* (L) Czern. and Coss. subsp. *juncea*, *Brassica cernua* (Thunb.) Forb. and Hems., *Brassica besseriana* Andrz.

As regards microscopy, the palisade layer of all forms in surface view shows a marked reticulation. The diameter of the palisade cells varies between 10 μm and 24 μm. In section, the testa of Indian subcontinent material shows a flattened epidermis; the epidermis of material from other regions is well marked and mucilaginous. The radial walls of the palisade cells are not drawn out into threads as in *B. nigra.*

3. Brassica carinata *A. Braun* (*Ethiopian or Abyssinian mustard*)

This plant is cultivated only in Ethiopia and neighbouring territories but the seed and seed residue appear in international commerce. The microscopy of *B. carinata* seed residues is virtually the same as that of the mucilaginous epidermis form of *B. juncea.*

4. Camelina sativa (*L*) *Crantz.* (*false flax*)

This is a minor oil seed crop of Russia (Crosfield, 1963). Although in the same family as the mustards and rapes, the microscopy of *Camelina* is quite distinctive. The testa epidermis is mucilaginous and, in water, the mucilage escapes as characteristic 'fingers'. Also, the diameter of the palisade cells is up to 60 μm.

B. *Ricinus* and Related Species (Figs 9 and 10)

Ricinus communis L. (castor) is a major oil seed but its residue is highly unsuitable as a feed constituent because it contains at least three toxic substances (Vaughan, 1970). In tropical and subtropical countries, castor may occur as a weed. As far as recent E.E.C. regulations are concerned, the maximum content for castor seed (expressed in husks) is 10 mg kg^{-1} of feeding stuff. The family Euphorbiaceae, in which *Ricinus* is placed, contains other species which are also toxic. Two of these are mentioned in the E.E.C. regulations, namely *Jatropha curcas* L. (physic nut) and *Croton tiglium* L. (croton). These have been cultivated for oil seeds but on a minor scale.

The toxic constituents of castor are actually located in the seed endosperm but microscopic identification in feeding stuffs is based on character of the testa (husk). The complete microscopy of *Ricinus* seed is recorded by Vaughan (1970). For identification, the following features of the husk are of importance: (i) the pitted epidermal cells, (ii) the layer, 300–350 μm in width, of curved pitted fibres and (iii) the spongy parenchyma of the testa with anastomosing vascular tissue. In commercial analysis, the curved pitted fibres are most important.

Various methods (e.g. Robertson Dodd, 1932; Huss, 1976) have been described for estimating the amount of castor in a sample. A common method relies on the fact that the castor husk is heavier than other plant fragments and is not easily bleached. Basically, a crude fibre of the sample is subjected to decantation or elutriation to remove the lighter fragments. The residue is bleached and the castor husks are picked out for weighing. To compensate for the previous drastic treatment, the weight of the husks is multiplied by a factor of 1·2.

Details of the methods employed can be obtained from the literature but microscopy is essential for the identification of the husk. The fact that fragments do not bleach and are hard is not sufficient evidence of their being castor. Fragments of grape seed residue are often mistaken for castor. Castor husk fragments, under low magnification, show pointed edges. However, the best method of confirmation is to macerate the fragments and to examine microscopically for the testa fibres. In castor, the fibres constitute only one

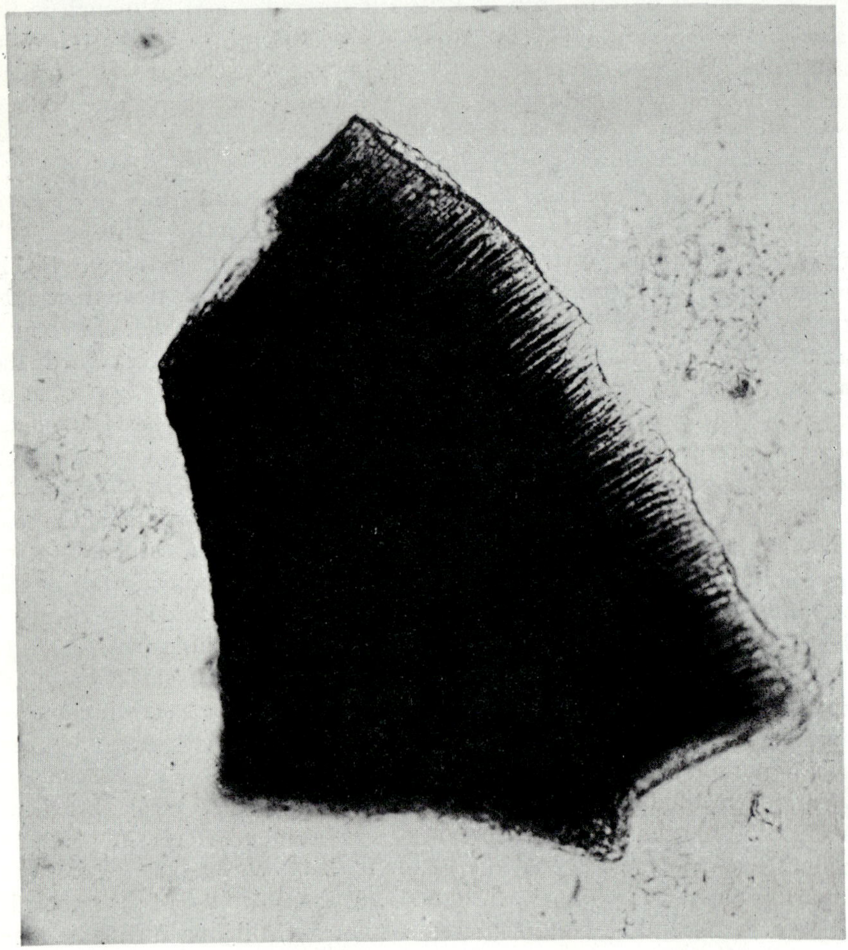

FIG. 9. SV of castor fragment, ×138.

layer. These fibres are relatively long and curved. In grape, the fibres or stone cells are straight and constitute a region, some two to three cell layers thick (Vaughan, 1970).

Although *Jatropha* and croton are included in the E.E.C. regulations, their appearance in feeding stuffs is likely to be a rare event. There are some differences in microscopic structure between these two species and *Ricinus* (Vaughan, 1970), but the main feature of identification, namely the curved fibres, is the same as that for *Ricinus*. Consequently, the commercial method of analysis normally employed would cover the situation for *Jatropha* and croton.

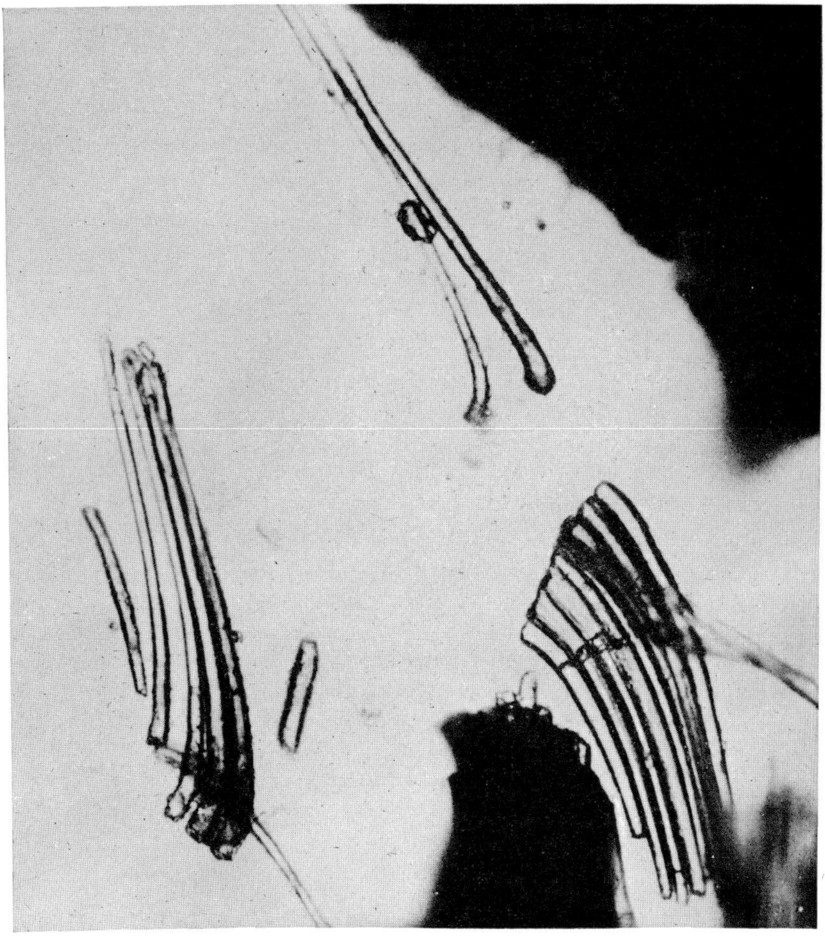

FIG. 10. Castor fibres, after maceration in nitric acid, ×223.

C. Apricot and Bitter Almond (Figs 11–14)

Apricot (*Prunus armeniaca* L.) and bitter almond (*Prunus amygdalus* Batsch var. *amara* (DC.) Focke = *Prunus dulcis* (Mill.) D. A. Webb) seeds and their processing derivatives are considered undesirable because of the production of hydrocyanic acid from amygdalin (Watt and Breyer-Brandwijk, 1962). Apricot stones or pits may be available as a by-product from dried apricot fruits or the canning industry, and bitter almond seeds from the production of 'oil of bitter almond'.

The 'seed', stone or pit of these two commodities is botanically speaking

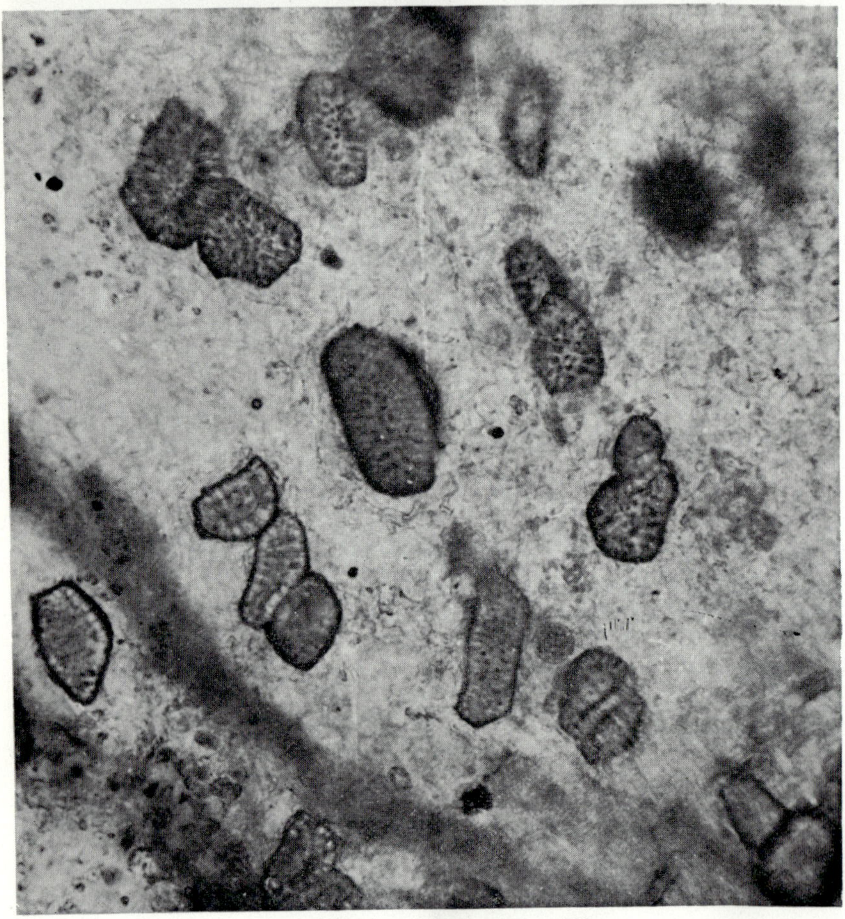

FIG. 11. SV of apricot testa, ×280.

the endocarp enclosing the true seed. Their microscopy has been described by Vaughan (1970). The endocarp consists of stone cells but there are no good features for identification. The seed or seed residue is also difficult to identify and, in fact, the only distinguishing character is the microscopic structure of the outer epidermis of the testa. In both apricot and bitter almond the outer epidermis of the testa in surface view shows thin-walled cells and characteristic stone cells. To distinguish between the two species, it is necessary to examine the stone cells both in surface view and in transverse section. Apricot: stone cells 40–50 μm in diameter, but some, in transverse section, are elongated tangentially and attain a length of 70 μm. Their walls are

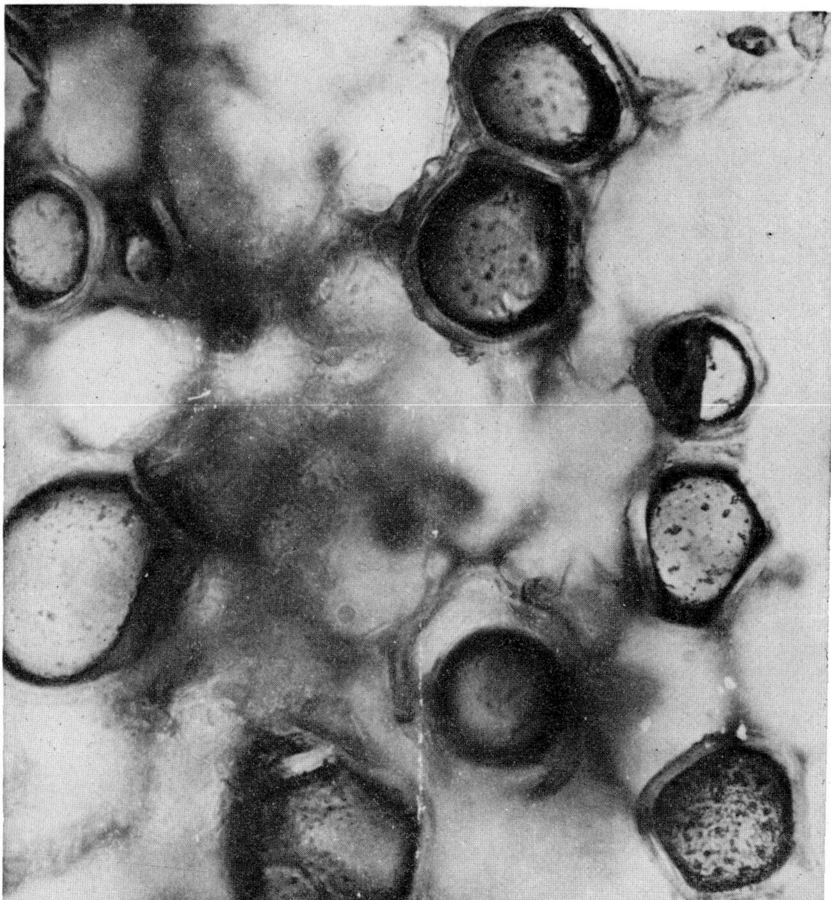

FIG. 12. SV of bitter almond testa, ×280.

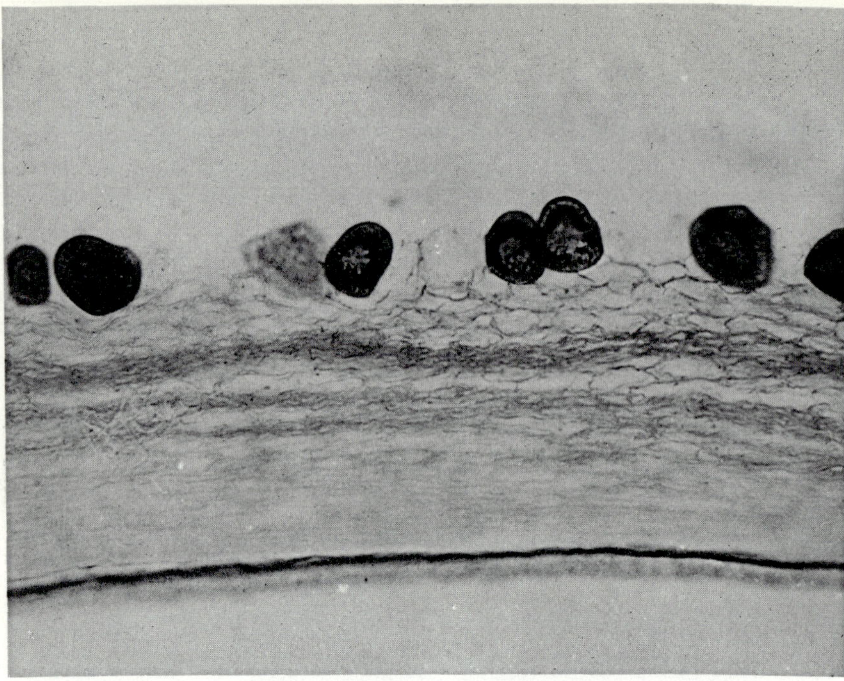

FIG. 13. TS of apricot testa, ×200.

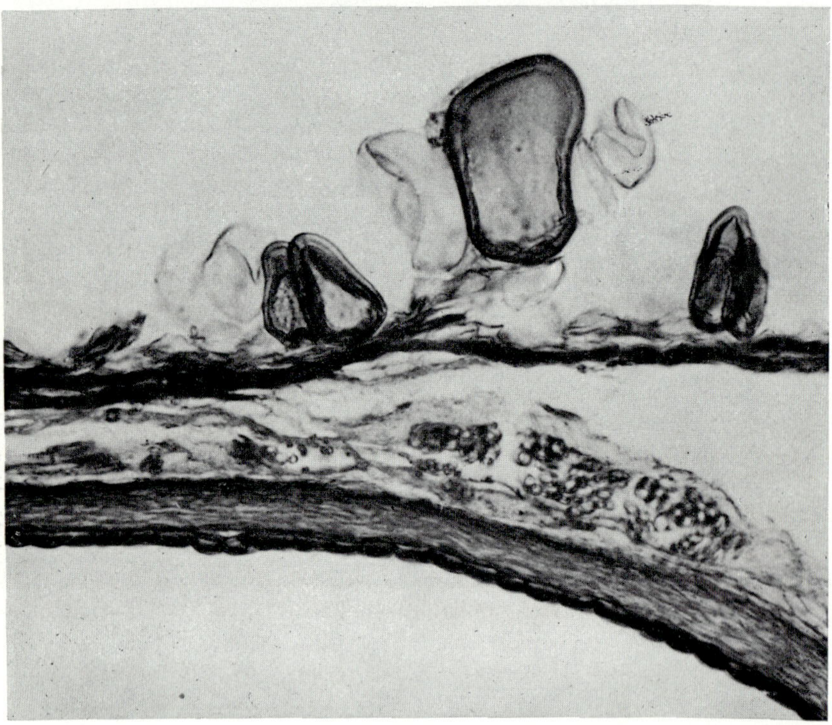

FIG. 14. TS of bitter almond testa, ×200.

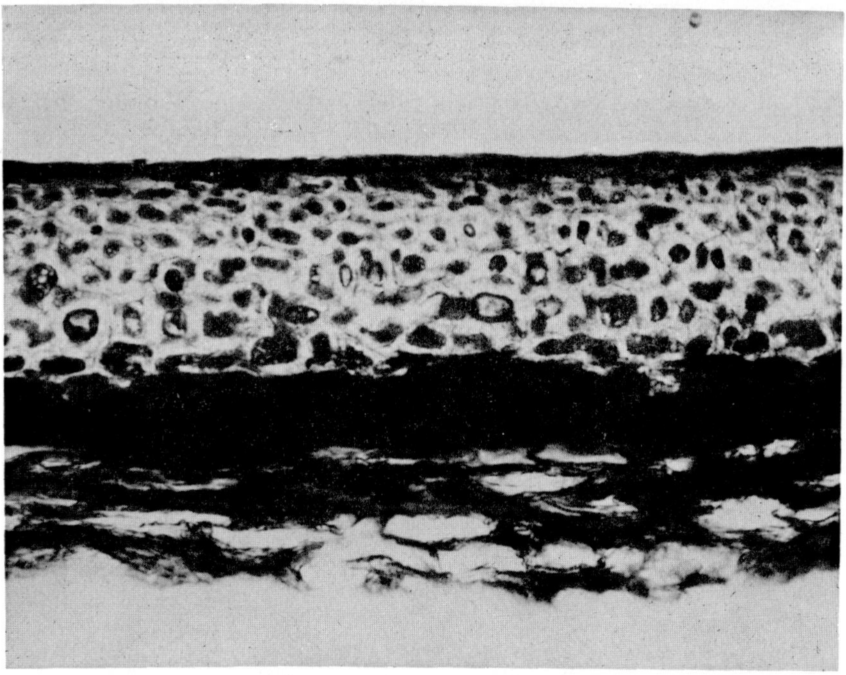

FIG. 15. TS of beech pericarp, ×190.

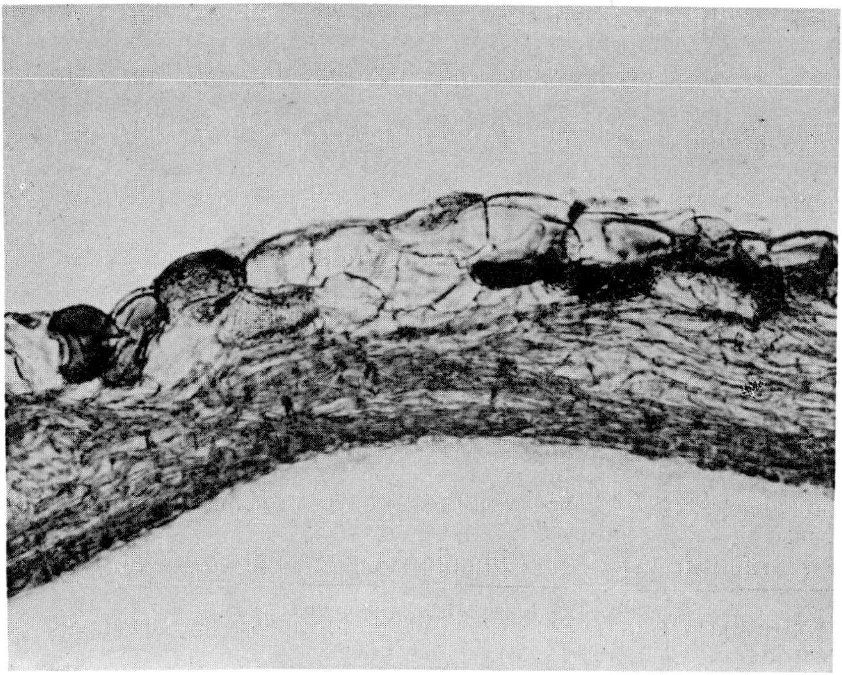

FIG. 16. TS of beech testa, ×280.

pitted throughout. Bitter almond: stone cells have an average radial diameter of 140 μm and an average tangential diameter of 100 μm. Normally they are truncated and their walls are pitted only in the inner halves.

D. Unhusked Beech (*Fagus sylvatica* L.) (Figs 15–17)

Although beech can be fed to many types of animals, it is said to be somewhat toxic to horses and asses because of the presence of an alkaloid, fagine.

The microscopy of beech has been described by Vaughan (1970) and Mészáros and Deutschmann (1975). The husk or pericarp has an epicarp

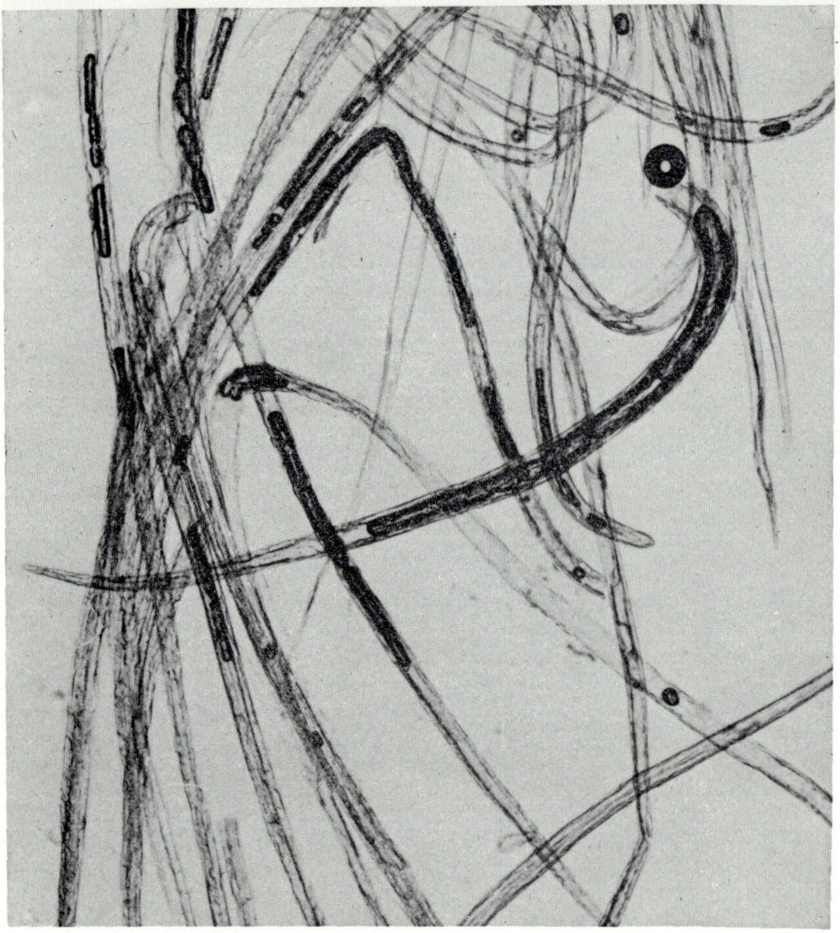

FIG. 17. Beech hairs from pericarp, ×143.

consisting of thin-walled cells, a mesocarp of stone cells and an endocarp of thin-walled parenchymatous cells. Vascular bundles are present in the endocarp. There are hairs, up to 400 μm in length, on the outer surface of the husk and hairs, up to 3 mm in length, on the inner surface of the husk. The distinguishing feature of the testa is the rather bulbous cells of the outer epidermis.

E. Rye Ergot (*Claviceps purpurea* (Fr.) Tul.) (Fig. 18)

Rye ergot is the resting stage or sclerotium of the fungus *Claviceps purpurea* which attacks rye, some other cereals and a variety of wild grasses. The

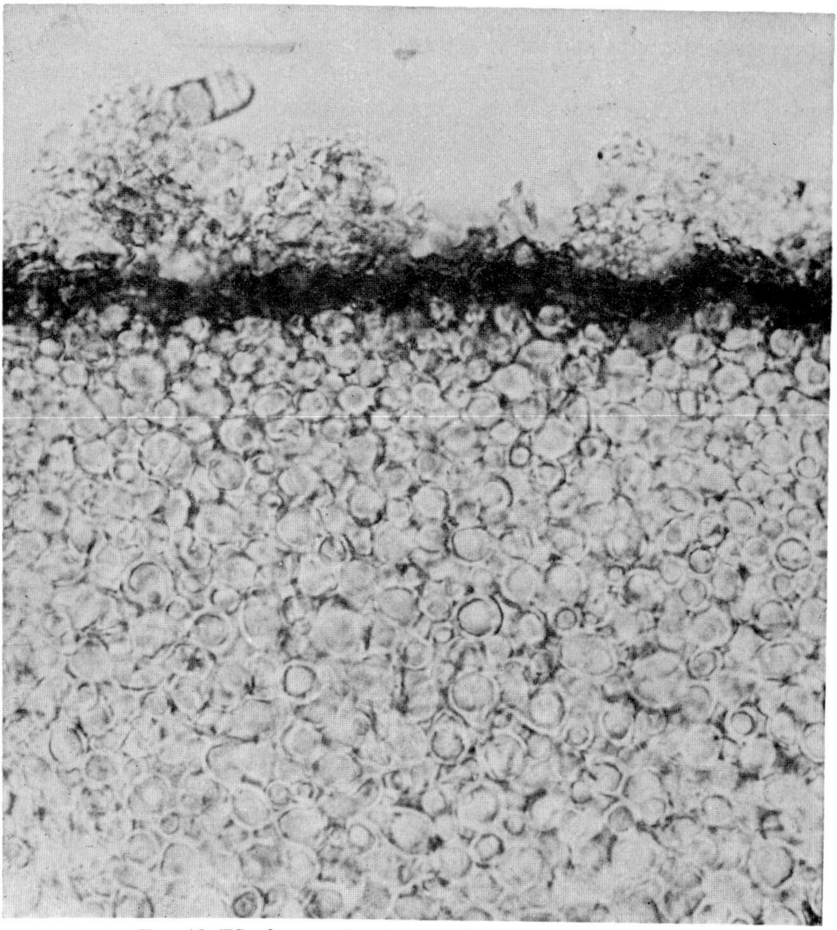

FIG. 18. TS of ergot showing pseudoparenchyma, ×655.

poisonous principles are alkaloids (Kingsbury, 1964). Ergots are hard, greyish or purplish structures. Microscopically, ergot consists of pseudoparenchyma made up of rounded cells, 3–12 μm in diameter. Wallis (1965) states that heating ergot in chloral hydrate liberates oil globules.

F. *Crotalaria* L. spp. (Figs 19 and 20)

Miller (1967) has given an excellent account of the seed morphology, microscopy and toxicology of *Crotalaria* seeds. He points out that the toxic principle is the alkaloid, monocrotaline, but that not all species are toxic. However, the E.E.C. regulations do not name definite species and presumably, therefore, identification is required only at the generic level.

Microscopic identification of leguminous seed residues is based mainly on characters of the testa and the starch grains, when present, of the cotyledons. Starch is not present in *Crotalaria*. The epidermis of the *Crotalaria* testa consists of the usual rod-like cells or macrosclereids (according to species, 35–109 μm in length).

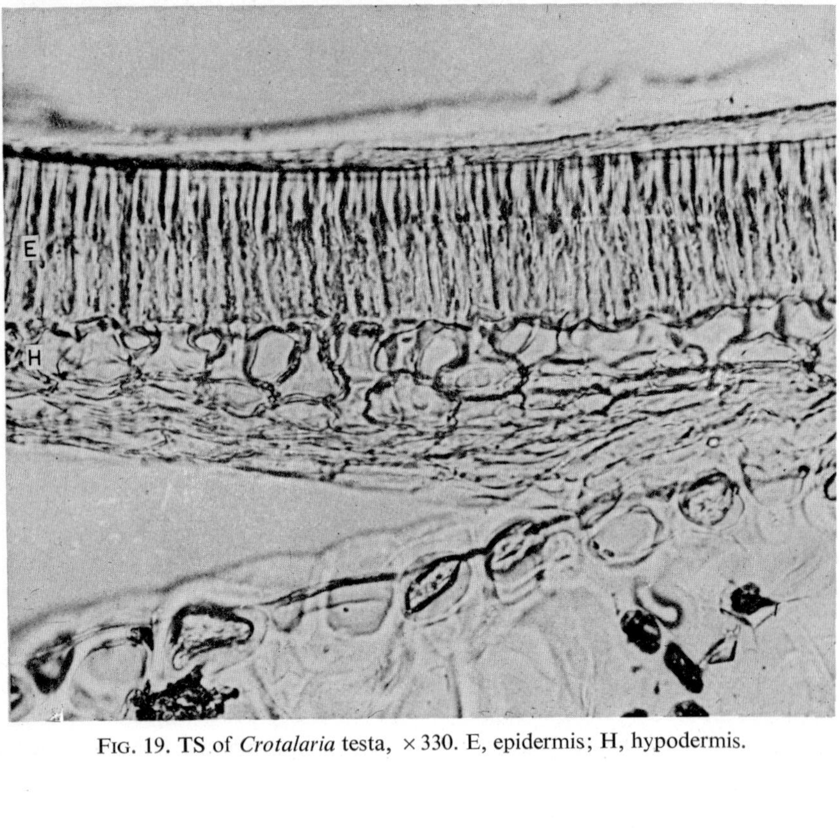

FIG. 19. TS of *Crotalaria* testa, ×330. E, epidermis; H, hypodermis.

Covering the epidermis is a distinct cuticle (according to species, 2–17·5 μm in width) which is said to be principally hemicellulosic and pectinaceous rather than fatty. The hypodermis consists of flask-shaped cells or lagenosclereids which, depending on the species, vary in length from 12 μm to 56 μm and in width of their basal portions from 30 μm to 70 μm. *Crotalaria* differs from a number of leguminous food seeds in that it possesses endosperm tissue which is mucilaginous when moistened.

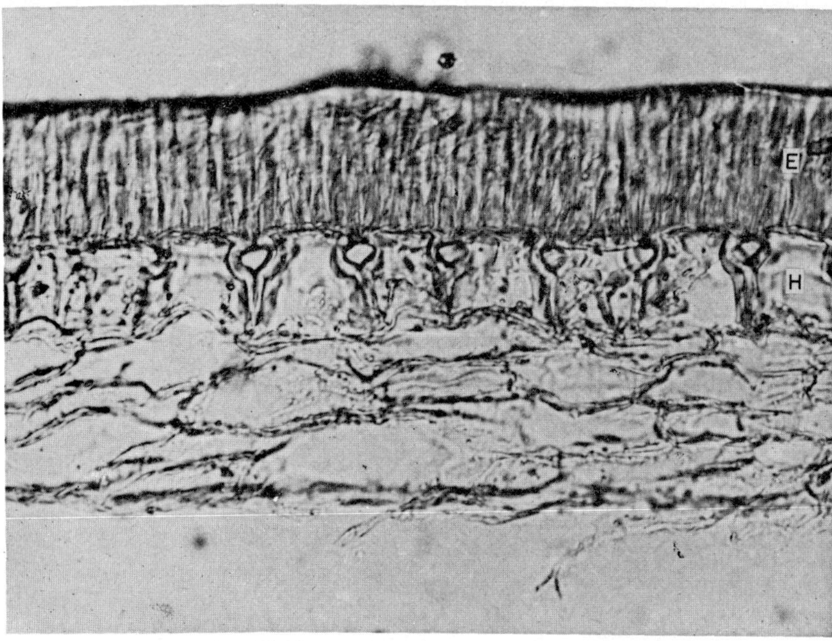

FIG. 20. TS of guar testa, ×330. Labels as in Fig. 19.

To identify *Crotalaria* seed fragments in feeding stuffs, attention should be focussed on the lagenosclereids. In most other leguminous food seeds the hypodermis consists of hourglass cells or osteosclereids. However, a reasonably common constituent of feeding stuffs is the leguminous seed, guar (*Cyamopsis*). This has lagenosclereids and, incidentally, a mucilaginous endosperm.

G. Illipe, Mowrah, Bassia, Madhuca (Figs 21 and 22)

For reasons of nomenclature, this product can be confusing to the analyst. It is the residue left after the extraction of fat from the seeds of Indian trees belonging to the family Sapotaceae. The botanical nomenclature is confusing with varying combinations given both at the generic and specific levels. Some combinations, which are given in the E.E.C. regulations and the reference books on feed microscopy are *Madhuca longifolia* (L.) Macbride, *Bassia longifolia* L., *Illipe malabarorum* Engl., *Madhuca indica* J. F. Gmel., *Bassia latifolia* Roxb., *Madhuca latifolia* (Roxb.) Macbride, *Illipe latifolia* (Roxb.)

FIG. 21. TS of mowrah cotyledon showing dark tannin cells, × 350.

F. Muell. This material is undesirable because of the production of saponins. Another and quite different material found in commerce is the seed residue of various species of *Shorea* (Dipterocarpaceae). Unfortunately this product is also referred to as illipe but, as it is non-toxic, the analyst must be able to distinguish between the two quite different products. The microscopy of both mowrah and *Shorea* has been fully described by Vaughan (1970) and Mészáros and Deutschmann (1975). The mowrah product is essentially cotyledon tissue and shows abundant tannin cells, pitted cotyledon cells and chains of latex cells. *Shorea* meal also possesses tannin cells but can be distinguished easily from mowrah by the presence of starch grains, 4–10 μm in diameter.

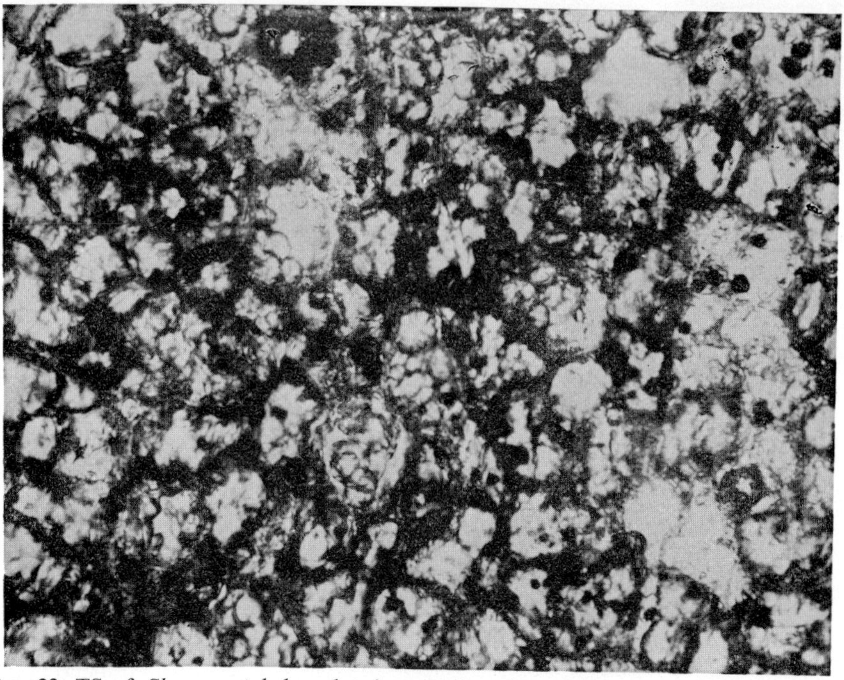

FIG. 22. TS of *Shorea* cotyledon showing starch grains which have been stained with iodine, × 350.

H. *Datura stramonium* L. (Figs 23 and 24)

Much has been written about the toxicity of the alkaloids of *Datura* (e.g. Kingsbury, 1964). The microscopic features of the seed useful for identification have been described by Gassner (1973) and Czaja (1971). In surface view the walls of the epidermal cells, containing dark contents, present a characteristic sinous appearance. In section the endosperm cells are angular and contain aleurone grains. Also, Czaja (1971) describes large oil drops, 250 μm in diameter, in the endosperm. According to Huss (1976), *Datura* seed fragments shown an intense bluish-green fluorescence in chloral hydrate or water. This changes to golden yellow after the addition of alkali.

I. Darnel (*Lolium temulentum* L.) (Figs 25 and 26)

The toxicity of darnel seed, botanically a fruit, possibly due to the alkaloids loliine and temulentine, has been the subject of much investigation and discussion (Watt and Breyer-Brandwijk, 1962; Kingsbury, 1964). One point at

FIG. 23. SV of *Datura* testa, ×153.

FIG. 24. TS of *Datura* testa, ×245.

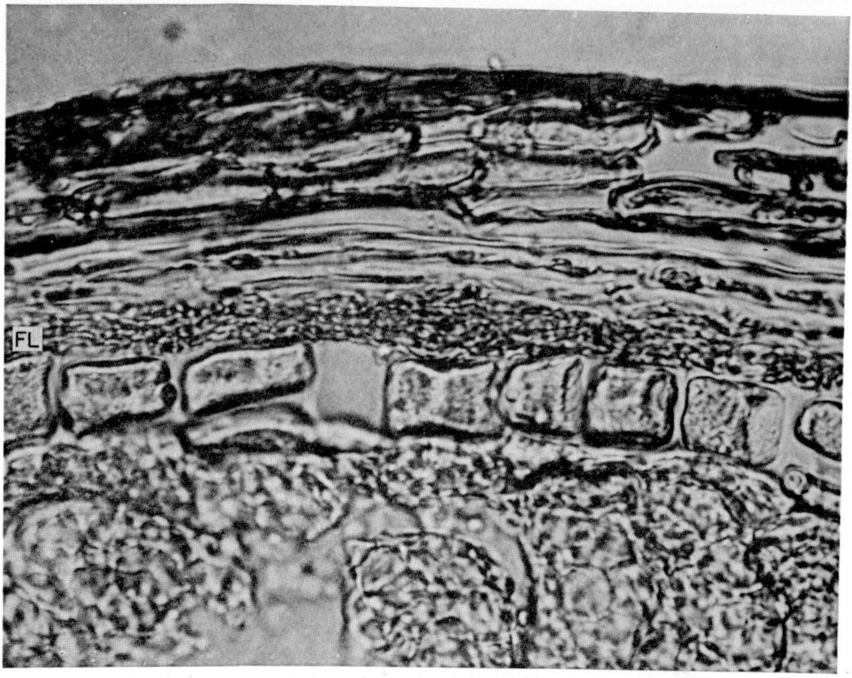

FIG. 25. TS of darnel showing fungus layer (FL), ×484.

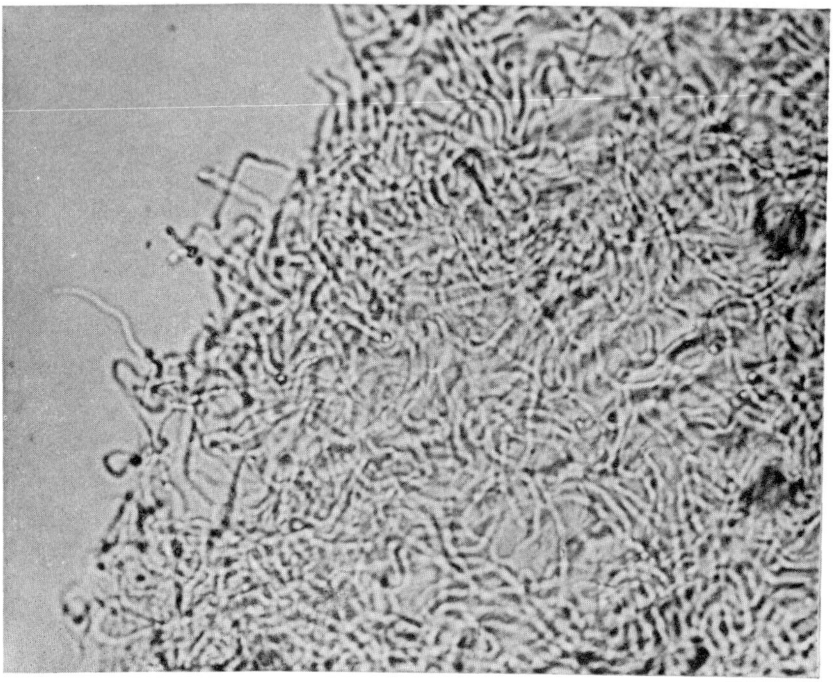

FIG. 26. SV of fungus layer, ×484.

issue has been the possible connexion between toxicity and the presence of a fungus (*Endoconidium temulentum* Prill. and Del.) in the seed.

The microscopy of darnel has been described in detail by Winton and Winton (1932). In general terms it resembles that of the chaffy cereals. Distinctive features for identification are: (i) chaff with numerous round cells and thorn-like hairs, (ii) epicarp without hairs, (iii) one layer of thin-walled cross cells, (iv) some of the tube cells form spongy parenchyma, (v) a fungal zone is sometimes present between the nucellus or perisperm and the aleurone layer and (vi) the small starch grains of the endosperm vary in diameter between 2·5 μm and 8 μm and form spherical-to-ovoid aggregates, about 25 μm to 30 μm in diameter.

Lolium remotum Schrank is mentioned in the E.E.C. regulations. This has been described as a weed of flax (Hubbard, 1954) but there appears to be no information concerning its toxicology.

REFERENCES

American Association of Feed Microscopists (1961). 'Manual of Microscopic Analysis of Feeding Stuffs.' (Revised.)

Ayensu, E. S. (1972). *In* 'Anatomy of the Monocotyledons' (C. R. Metcalfe, ed.), Vol. VI. Clarendon Press, Oxford.

Barton, L. V. (1967). 'Bibliography of Seeds.' Columbia University Press, New York.

Bussard, L. and Brioux, C. (1925). 'Torteaux.' Libraire Polytechnique Ch. Beranger, Paris.

Cole, G. T. and Behnke, H. D. (1975). *Taxon*, **24**, 3.

Collin, E. and Perrot, E. (1904). 'Les Residues Industriels.' A. Joanin, Paris.

Corner, E. J. H. (1976). 'The Seeds of Dicotyledons', Vols I and II. Cambridge University Press, Cambridge.

Crosfield, J. and Sons Ltd. (1963). 'Oil Seeds in the Soviet Seven Year Plan (1958–65).' Warrington, U. K.

Cutler, D. F. (1969). *In* 'Anatomy of the Monocotyledons' (C. R. Metcalfe, ed.), Vol. IV. Clarendon Press, Oxford.

Czaja, A. Th. (1971). 'Methoden der Lebensmittel-mikroskopie und Lebensmittel-uberwachung.' Umschau Verlag, Frankfurt am Main.

Ferrando, R. and Henry, N. (1966). 'Determination Microscopique des Ailments du Betail.' Vigot Frères, Paris.

Gassner, G. (1973). 'Mikroskopische Untersuchung Pflanzlicher Lebensmittel', 4th Edn. G. Fischer Verlag, Stuttgart.

Hamence, J. H. (1966). *J. Ass. Public Anal.* **4**, 15.

Hubbard, C. E. (1954). 'Grasses.' Penguin Books, U.K.

Huss, W. (1956). *Landw. Forsch.* **9**, 110.

Huss, W. (1960). *Zeiss Mitt. Fortschr. Tech. Opt.* **2**, 92.

Huss, W. (1961). *Z. Landw. Vers-u. UntersWes.* **7**, 11.

Huss, W. (1976). 'Microscopy and Quality Control in the Manufacture of Animal Feeds.' Roche Information Service. (Also available in German.)

Jensen, W. A. (1962). 'Botanical Histochemistry.' W. H. Freeman, San Francisco.

Juillet, A., Susplugas, J. and Courp, J. (1955). 'Les Oleagineaux et Leurs Tourteaux.' Paul Lechevalier, Paris.

Kingsbury, J. M. (1964). 'Poisonous Plants of the United States and Canada.' Prentice-Hall, New Jersey.

Mäckel, H. G. (1943). *Z. Tierernähr. Futtermittelk.* **6**, 131.

Mészáros, L. and Deutschmann, F. (1975). 'Atlas für die Mikroskopie von Nahrungsgrundstoffen und Futtermitteln', Vol. I. Verlag J. Neumann-Neudamm, Melsungen, Berlin.

Metcalfe, C. R., ed. (1960). *In* 'Anatomy of the Monocotyledons', Vol. I. Clarendon Press, Oxford.

Metcalfe, C. R., ed. (1971). *In* 'Anatomy of the Monocotyledons', Vol. V. Clarendon Press, Oxford.

Metcalfe, C. R. and Chalk, L. (1950). 'Anatomy of the Dicotyledons', Vols I and II. Clarendon Press, Oxford.

Miller, R. H. (1967). '*Crotalaria* Seed Morphology, Anatomy and Identification.' Tech. Bull. No. 1373, U.S. Department of Agriculture, Washington D.C.

Moeller, J. and Griebel, C. (1928). 'Mikroskopie der Nahrungs und Genussmittel ans dem Pflanzenreich', 3rd Edn. Springer-Verlag, Berlin.

Morris, T. N. (1928). 'Microscopic Analysis of Cattle-Foods.' Cambridge University Press, London.

Morrison, F. B. (1961). 'Feeds and Feeding', 9th Edn. Morrison Publishing, Clinton, Iowa.

Netolitzky, F. (1926). 'Anatomie der Angiospermen Samen.' *Handb. PflAnat.* (K. Linsbauer, ed.), **10**, (14).

Parkinson, S. T. and Fielding, W. L. (1930). 'The Microscopic Examination of Cattle Foods.' Headley Brothers, Ashford and London.

Reichert, E. T. (1913). 'The Differentiation and Specificity of Starches in Relation to Genera, Species, etc.' Pub. 173, Part 1, Carnegie Institution of Washington, Washington D.C.

Robertson Dodd, F. (1932). *Analyst.* **57**, 488.

Rosenfeld, A., ed. (1976). 'Digital Picture Analysis.' Springer-Verlag, Berlin.

Sass, J. E. (1958). 'Botanical Microtechnique', 3rd Edn. Iowa State University Press, Ames, Iowa.

Seidemann, J. (1966). 'Starkeatlas. Grundlagen der Stark-mikroskopie und Beschreibung der wichtigsten Starke-arten.' Parey-Verlag, Berlin.

Tomlinson, P. B. (1961). *In* 'Anatomy of the Monocotyledons' (C. R. Metcalfe, ed.), Vol. II. Clarendon Press, Oxford.

Tomlinson, P. B. (1969). *In* 'Anatomy of the Monocotyledons' (C. R. Metcalfe, ed.), Vol. II. Clarendon Press, Oxford.

Vaughan, J. G. (1960). *Stain Technol.* **35**, 229.

Vaughan, J. G. (1970). 'The Structure and Utilization of Oil Seeds.' Chapman and Hall, London.

Vaughan, J. G. and Gordon, E. I. (1972). *Ann. Bot. N.S.* **37**, 167.
Vaughan, J. G. and Hemingway, J. S. (1959). *Econ. Bot.* **13**, 196.
Vaughan, J. G., Hemingway, J. S. and Schofield, H. (1963). *J. Linn. Soc.* (*Bot.*) **58**, 435.
Wallis, T. E. (1965). 'Analytical Microscopy', 3rd Edn. J. A. Churchill, London.
Watt, J. M. and Breyer-Brandwijk, M. G. (1962). 'The Medicinal and Poisonous Plants of Southern and Eastern Africa', 2nd Edn. E. and S. Livingstone, Edinburgh.
Whistler, R. L. and Paschall, E. F., eds (1967). 'Starch: Chemistry and Technology', Vol II. Academic Press, New York and London.
Winton, A. L. (1906). 'The Microscopy of Vegetable Foods.' John Wiley, New York.
Winton, A. L. and Winton, K. B. (1932). 'The Structure and Composition of Foods', Vol. 1. John Wiley and Sons, New York.
Winton, A. L. and Winton, K. B. (1935). 'The Structure and Composition of Foods', Vol. 2. John Wiley and Sons, New York.
Winton, A. L. and Winton, K. B. (1937). 'The Structure and Composition of Foods', Vol. 3. John Wiley and Sons, New York.
Winton, A. L. and Winton, K. B. (1939). 'The Structure and Composition of Foods', Vol. 4. John Wiley and Sons, New York.

11. Animal Feeds—Animal Constituents

H. VOEHRINGER

Landw. Untersuchungs- und Forschungsanstalt,
Bonn, West Germany

I. INTRODUCTION

Feeds of animal origin are, in general, products of marine animals (fish, sea mammals, crustaceans, molluscs, echinoderms), of terrestrial mammals (mainly cattle and pigs) and of poultry; milk products are also added. They may be produced from the whole animal (e.g. anchovy meal, tankage). or from parts of animals (e.g. bone scraps, blood meal, feather meal, liver meal). Generally these products have been dried and ground after an initial process of heating/steaming to sterilize the material.

Most of these feedstuffs are rich in phosphorus and calcium, and in protein of high biological value. They are of great importance in the formulation of complete rations for livestock. Quality and price of these animal products are related to composition, authenticity and purity of the raw materials and careful processing. Identification and control of the products in straight feeds and compounds are often required for commercial and legal reasons. Microscopy is used for this purpose, either as the sole method or in conjunction with chemical analysis.

II. HISTORY AND PRESENT PRACTICE

Due to the relatively late introduction of animal constituents into commercially produced feeds, and to the small quota of these constituents as compared with plant constituents in feeds, little has been published on the subject. Boehmer (1900) and Barnstein (1920) deal briefly with tankage and meat meal products. Other early papers were written because of the possible danger of anthrax contamination in cases of adulteration of fish meal by tankage (Lucks, 1915) and by bone scraps of mammals (Bartschat, 1927; Goerttler and Muenster, 1929; Claussen, 1930, 1931, 1932; Groening, 1932; Heyck, 1934a). Claus (1932) and Mach and Claus (1933) deal primarily with the problems of authenticity and purity. The comprehensive monograph of Schweitzer (1931) contains not only 21 tables with explanations, but also gives directions for mounting media, staining and other microscopic methods. The short summary of Bartschat (1936) is rather restricted. Voehringer (1958a) deals with all aspects referring to fish meal microscopy and to some extent with bone scraps. Fish and meat meals were also investigated by Lafond-Grellety (1965). New information appears in the Official Proceedings of the Meetings of the American Association of Feed Microscopists (e.g. Brown, 1964, 1968), produced annually, and in the 'Protokoll der Sektionen Futtermittelmikroskopie und Futtermittelmikrobiologie der Internationalen Arbeitsgemeinschaft für Futtermitteluntersuchung', mostly written in German and also produced annually. General works on feed microscopy which contain some information on animal constituents are the German *Handbuch* (Anon, 1951), the Dutch 'Methods of Analysis' (Anon, 1954), the Manual of the American Association of Feed Microscopists (1961), Ferrando and Henry (1966) and Huss (1977a).

III. METHODS

The microscopic identification of animal material in feeds is based upon fundamental histological knowledge. A substantial difference to normal histology is that only the so-called 'hard constituents' such as bones, teeth, scales and horn maintain their microscopic structure more or less unaltered during processing. The 'smooth constituents', e.g. muscle fibres, connective tissues and various organs, have a microscopic structure which is often markedly changed during processing (Ingling, 1975). In the following account emphasis is laid upon the most characteristic features for identification, according to the literature and the personal experience of the author.

Microscopic techniques for the identification of animal and plant constituents in feeds are essentially the same and reference may be made to Chapter 10.

Three points, however, are emphasized as regards animal constituents:

(i) Preparation by flotation is indispensable. In liquids of high specific gravity, such as carbon tetrachloride or chloroform, the heavier, mainly inorganic particles, e.g. bones, teeth, shells and minerals, will sink. After separation and drying, the sediment may be sieved. The sediment contains the very characteristic hard constituents and the sediment percentage often provides information regarding the amount of ash and protein. Brown (1964) refers to a refinement of the simple flotation test.

(ii) Polarizing microscopy facilitates the identification of calcified and other anisotropic materials (Schmidt, 1924; Stoves, 1957; Voehringer, 1958a, b; Huss, 1977a). A revolving stage, as well as the use of the mountant phenol–glycerol (saturated, $n_D \approx 1{\cdot}525$), is recommended.

(iii) Due to the effect of the processing method on the microscopic appearance of the finished product, and because of the development of new sources of raw materials, a permanent collection of authentic samples for reference is indispensable.

IV. MARINE ANIMALS

A. Fish (Teleosteans and Elasmobranchs)

The most important features for identification are bones, scales, teeth and otoliths (ear-stones). They cannot be confused with mammalian fragments.

The bones of teleosteans have a yellowish, glassy appearance, as seen by the naked eye and under the low power of the microscope, while slender and blade-shaped fragments are frequent. The amphicoel vertebrae are also characteristic. As regards microscopic structure and feed analysis, two main types may be recognized: (i) bone with, and (ii) bone without lacunae containing osteocytes (Koelliker, 1859; Stephan, 1900). Type (i) is found in the Clupeoideae (e.g. herring, anchovy), Cyprinoideae (e.g. carp) and Salmonoideae (e.g. salmon, trout); type (ii) in the Gadeoideae (e.g. cod, hake) and other groups. In herring or anchovy bone, the lacunae are seen as flat, hollow structures with long, ramified canals (canaliculi) which remain filled with air in microscopic preparations (Fig. 1). Tuna bones have filiform lacunae. The bones of fish such as cod show only the weak striation caused by the fibrous collagen ground substance. The penetrating crossing Sharpey's fibres (Schmid-Monnard, 1883), occurring in all fish bones, are often conspicuous (Fig. 2).

In elasmobranchs, the bone–cartilage is interspersed with small, globular lacunae without any canals and, in processed material, the lacunae appear compressed. The surface of the cartilage skeleton of sharks and rays is formed from polygonal calcified plates.

Fish bones with a striking structure are gill-rakers, fin rays and, characteristic of Clupeoideae skulls, bullae penetrated by canals (Voehringer, 1958a).

The scales of teleosteans vary considerably and in a characteristic way regarding number, size and structure. Early papers on this matter were presented by Baudelot (1873), Klaatsch (1890) and Tims (1905). The scales consist of a fibrous basal layer and an often highly mineralized superficial layer. The latter shows an arrangement of concentric rings and grooves which are characteristic for families, genera and, in some cases, species (Fig. 3). Even scale fragments may be recognized in fish meals (Voehringer, 1958a, 1959). The placoid scales of sharks (Fig. 4) and rays may also be identified in commercial products.

The teeth of fish are usually conical in shape and possess a pulp. An enamel tip may occur and, in large specimens, dentine tubules (Voehringer, 1958a).

Otoliths are pearl-white structures consisting of calcium carbonate, with a length varying from 1 mm to 20 mm. Even small fragments are easily recognizable. The distinguishing microscopic feature is the striated crystalline structure, indicated by alternate light and dark rings of growth (Fig. 5).

With regard to the soft constituents of fish, striated muscle tissue is present in the greatest amount. In processed fish meal, fibres of 20–200 μm thickness are to be found (Fig. 6). The cross striations have mainly faded and often only the longitudinal striations are to be seen. Connective tissues deform, dissolve and coagulate during processing and act as a glue in forming small conglomerates of the finer portions of the material.

Special diagnostic value is attached to the melanophores (Fig. 7) and guanin. The guanin may occur in thin layers and bands but in processed material will be divided into a large number of minute crystals, recognized by their strong birefringence in polarized light. The guanin crystals, soluble in dilute acids and alkalis, are found in conglomerates as well as on the surface of muscle, scales, etc. (Fig. 8).

Other fragments with a distinctive shape and of some diagnostic value are eye lenses, eggs and portions of the stomach and heart (Voehringer, 1965).

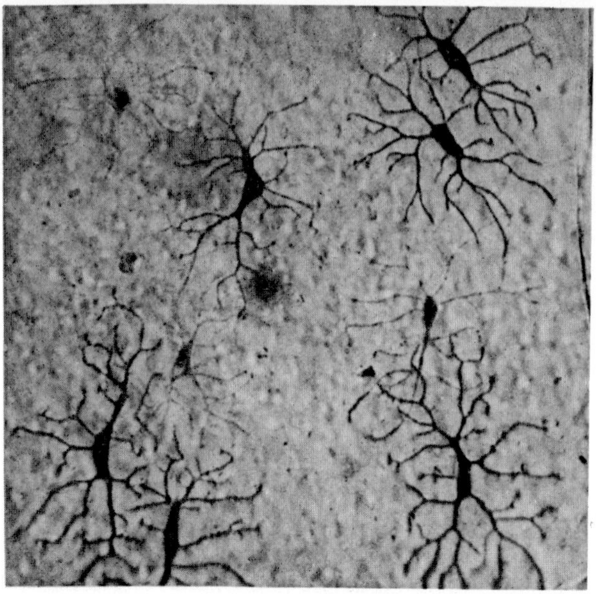

FIG. 1. Fish bone. Herring (*Clupea harengus*), ×300.

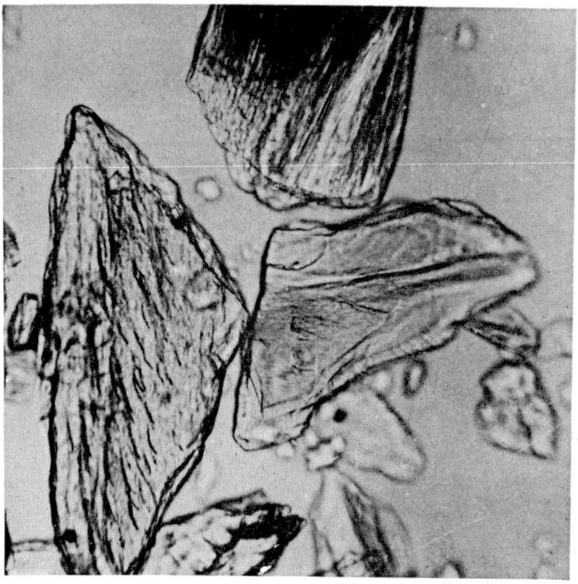

FIG. 2. Fish bone. Cod (*Gadus morrhua*), ×120.

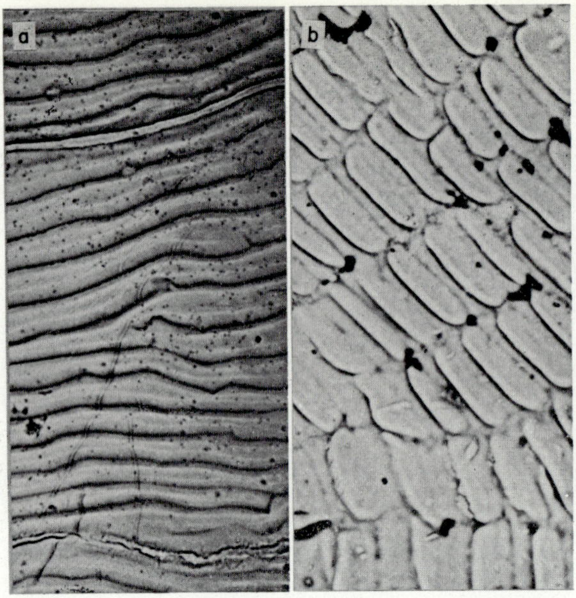

FIG. 3. Parts of fish scales, ×200. (a) Sardine (*Sardinops sagax*); (b) haddock (*Melanogrammus aeglefinus*).

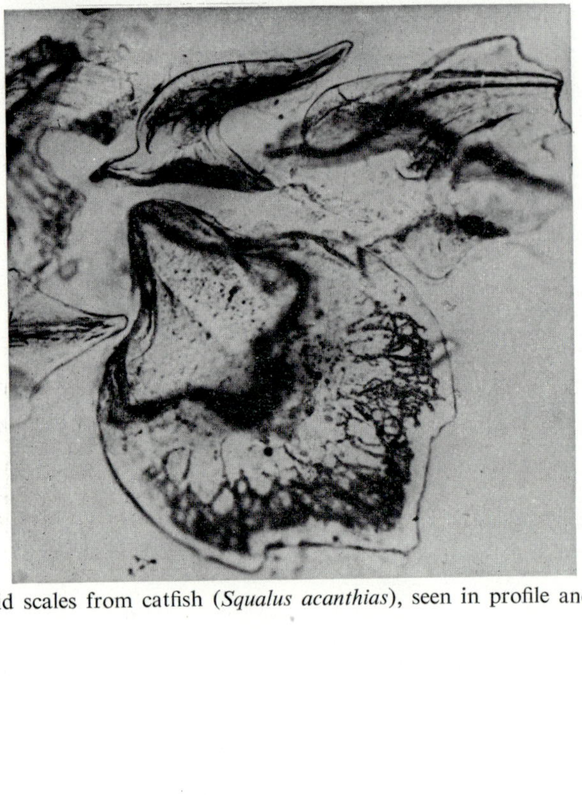

FIG. 4. Placoid scales from catfish (*Squalus acanthias*), seen in profile and from above, ×200.

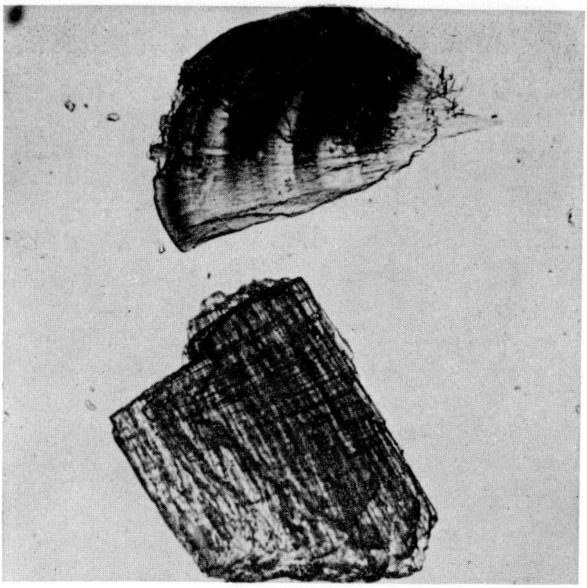

FIG. 5. Otolith fragments, ×70.

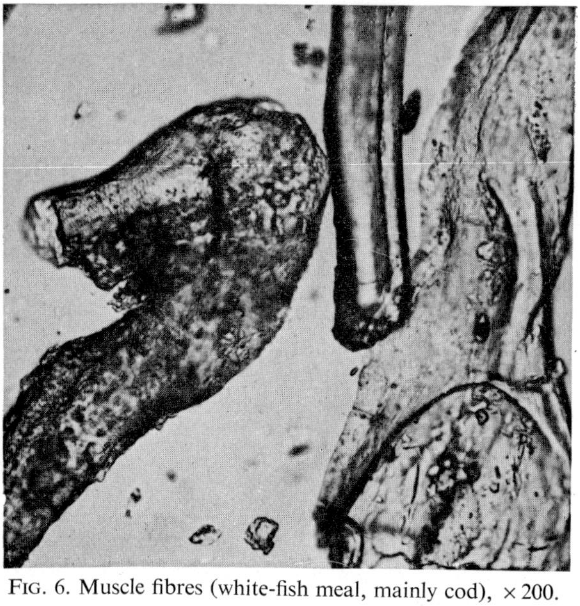

FIG. 6. Muscle fibres (white-fish meal, mainly cod), ×200.

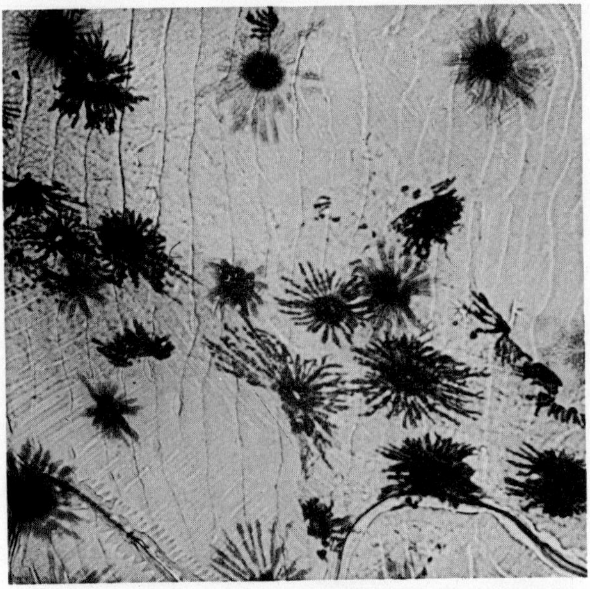

FIG. 7. Melanophores on a sardine scale, ×200.

FIG. 8. Guanin crystals on a redfish (*Sebastes marinus*) scale (polarized light), ×200.

B. Other Marine Animals

The most important identification feature for crabs and shrimps is the calcified exoskeleton, as spheroid or flat, rosette-shaped concretions (Fig. 9). The exoskeleton surface of crayfish resembles cellular plant tissue, often showing a reddish colour. Crab muscle fibres are characterized by a prominent cross-striation. It is not difficult to identify crustacean shell fragments because of the crystalline calcified exoskeleton and the colour. Further detail may be found in Schweitzer (1931) and Voehringer (1958a).

Starfish and sea urchins possess a characteristic spongy skeleton consisting of calcium carbonate (Fig. 10).

C. Miscellaneous Information

Some investigations deal with fish meals of a particular geographical origin, e.g. North Europe (Voehringer, 1959), Japan (Voehringer, 1961), Chile (Hirsch-Reinshagen *et al.*, 1966) and Pakistan (Voehringer, 1974).

An extremely fibrous structure of the product indicates previous open-air drying; browned and burnt fragments occur in the case of unsuitable gas-fire drying, with deterioration of protein digestibility. Microscopic analysis may also demonstrate the presence of salt or spoilage.

Fish liver meals have no characteristic microscopic features. Fish solubles, condensed or dried, may sometimes be identified by a high guanin content, and traces of muscle fibres, bones and scales, which are usually present, facilitate identification.

D. Marine Mammals

Whale meal is similar to ordinary meat meal. A good description of this material is available (American Association of Feed Microscopists, 1961). Guanin and hairs are absent and bone fragments may be encountered, but there are no real differences in microscopic features between marine and terrestrial mammals (Heyck, 1934a, b; Hemmert-Halswick, 1937).

V. TERRESTRIAL MAMMALS

A. Diagnostic Microscopic Features

The diagnostic hard constituents are bones, teeth and, to some extent, calcified cartilage and horn.

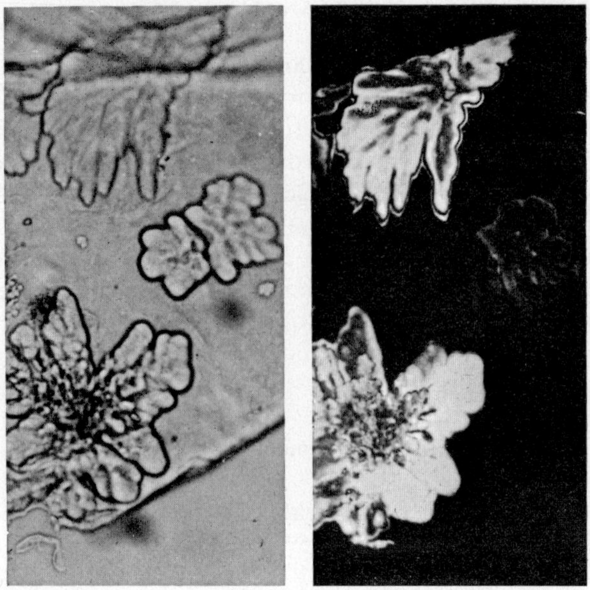

FIG. 9. Calcium carbonate concretions in a shrimp shell, seen in bright field and in polarized light, ×200.

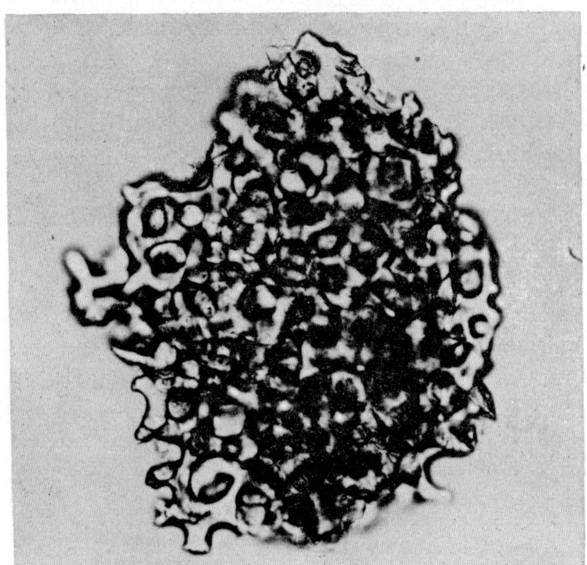

FIG. 10. Fragment of a starfish skeleton, ×300.

The microscopic features of bone are unmistakable. Lacunae, containing osteocytes, have very small, non-ramified canaliculi running in all directions (Fig. 11). The direction and shape of the lacunae relate to the overall bone structure, which is either compact with prominent Haversian canals or spongy. The birefringence is positive as to the direction of the fibrous collagen basic substance (Schmidt, 1924). This optical character will change to low-negative when, by autoclaving, the collagen has at least partly been removed leaving the apatite crystalline structure ('steamed bone meal'); concomitant with this the refractive index rises from $n_D \approx 1{\cdot}55$ to $1{\cdot}60$ (Voehringer, 1958b). Under the low power of the microscope, bone is seen as opaque, white or yellowish particles with really no distinct shapes.

Teeth occur only as extremely hard, white fragments. Parallel-running, small, wavy, dentine tubules are significant (Fig. 12). The birefringence of dentine is low.

The original cross-striations of the muscle fibres can rarely be seen, as described by Schweitzer (1931) and others. Although not an absolutely reliable diagnostic feature, the diameter of mammal muscle fibres is generally smaller than that of fish fibres. Cartilage, sinews and other connective tissues coagulate during processing and may form glassy, yellow or brownish particles.

Hairs, bristles, horn, claws, hooves (and feathers) consist of very resistant keratin which becomes somewhat plastic during the usual cooking and drying processes but does not lose the original structure. Hairs are particularly significant for identification. A histochemical test for keratin with alkaline lead acetate (a covering of black lead sulphide forms) was suggested by Voehringer (1972). The microscopy of the hair regions (medulla, cortex and cuticle with characteristic overlapping scales) has been well described by zoologists (e.g. Litterscheidt and Lambardt, 1921), fibre specialists (e.g. Stoves, 1957) and others (e.g. Harris *et al.*, 1960). The most accurate identification can be made when guard-, fur- and curly overhairs are found together (Fig. 13).

Red blood cells cannot be observed in a microscopic preparation of a feed but large amounts of blood in a tissue will give it a dark colour. Chemical tests for blood may be employed, such as benzedine (Anon, 1951) and a fluorescence test; concentrated sulphuric acid converts haemoglobin to red fluorescent haematoporphyrine.

B. Miscellaneous Information

Tankage will contain all parts of the animal, but large quantities of hair, horn and hooves as well as plant residues from the stomach and intestines are to be avoided.

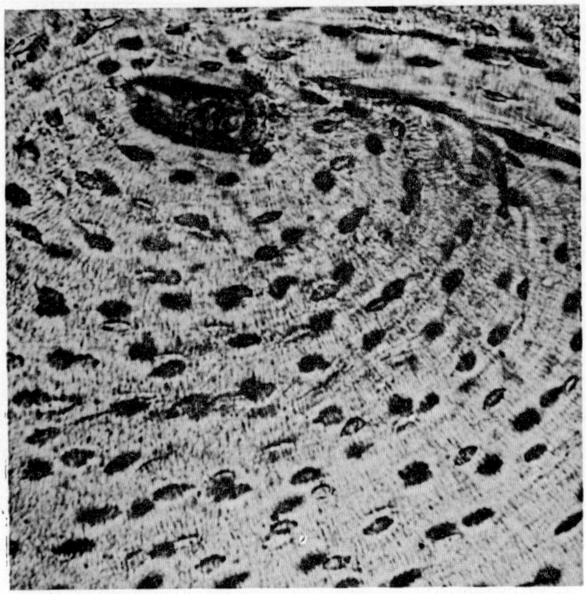

FIG. 11. Mammal bone (most lacunae are filled with air), ×300.

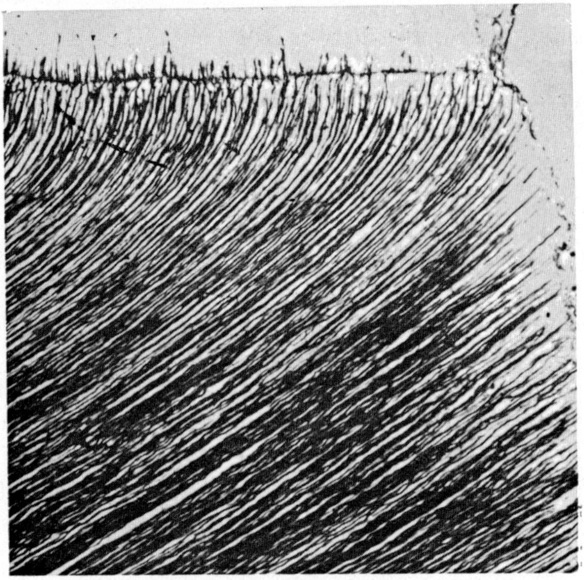

FIG. 12. Dentine from cattle tooth, ×140.

FIG. 13. Guard- and fur-hairs of cattle (medullae are partly filled with air), × 200.

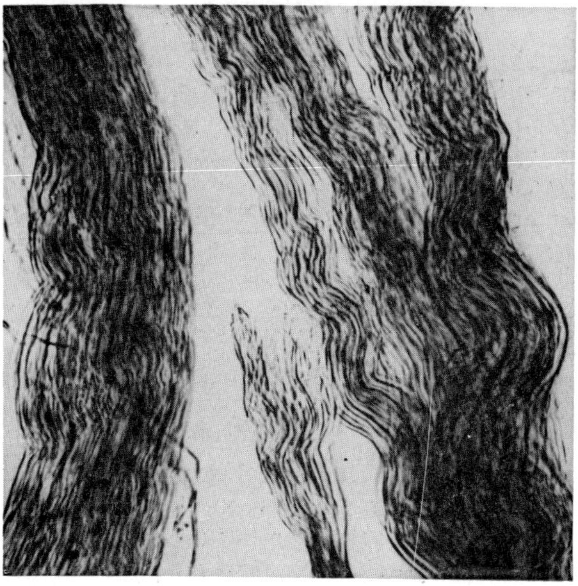

FIG. 14. Fibre bundles from the corium (leather fragments), × 200.

Meat meal in general is understood to be a product with a low amount of bone, containing plant remains only in traces. The approximate percentage of bone can be determined by the separation flotation technique. Steamed bone meal is characterized by finely ground bone particles with negative birefringence and high refraction.

Blood meal, spray- or drum-dried, consists of brown–red particles.

Details of these products are given by the American Association of Feed Microscopists (1961), Brown (1964), Wilder (1967) and Herrera (1970).

Other high-quality products are liver meal and spray-dried meat solubles and protein hydrolysates (Huss, 1973). Except for the high protein content, there are no reliable identification features.

Unhydrolysed leather meal and horn offals are undesirable. Characteristic of leather are the coarse fibre-bundles of the corium (Anon, 1957) (Fig. 14); particles with a treated and/or a coloured surface, as well as a typical odour, may aid identification. Hydrolysed leather fragments have lost almost all of these features. In the case of chromium-tanned leather, a spot test for chromium may facilitate identification (Heckman, 1964).

VI. POULTRY

The offal of poultry slaughter consists of heads, feet, gizzards and intestines; whole birds may also be processed for poultry by-product meal.

The microscopic structure of the bone, muscle fibres and connective tissues is similar to that found in mammals; the lacunae in the wing and long leg bones appear as conspicuous slits.

The feathers are the best identification feature, especially the straight barbules of the down- and pin-feathers with their nodes and bamboo-like branching which may be easily recognized under the stereomicroscope or transmission microscope (American Association of Feed Microscopists, 1961) (Fig. 15).

Raw feathers are almost indigestible. Sometimes they are collected and subjected to autoclave hydrolysation. After such a treatment, the dried feathers look wholly deformed; the barbules crumble and cluster (Brown, 1959). A test for hydrolysed feather meal is to soak the material in 80% sulphuric acid which largely destroys most other animal and plant tissues but leaves the feather as recognizable fragments (Voehringer, 1970, 1971) (Fig. 16). As feathers undergo hydrolysation, there is a loss of birefringence. Huss (1977b) has commented on the relationship between the structure and digestibility of feather material.

Egg shells may be found in poultry by-products. Their identification is not difficult.

FIG. 15. Barbules of a pin-feather (raw), ×200.

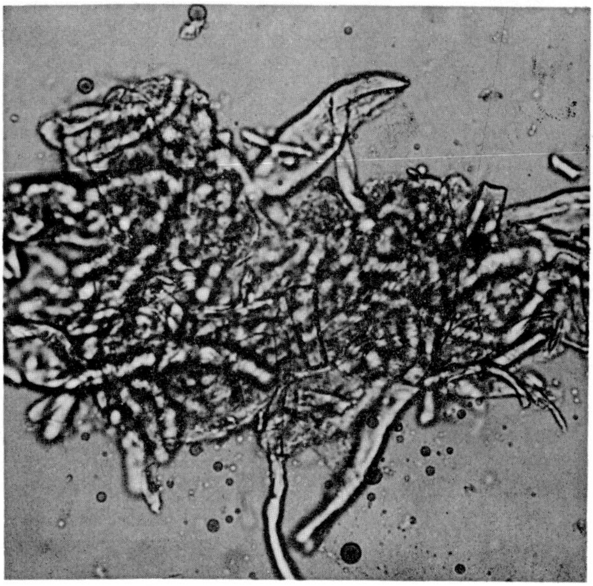

FIG. 16. Clustered barbules from hydrolysed feather meal, after treatment with sulphuric acid, ×200.

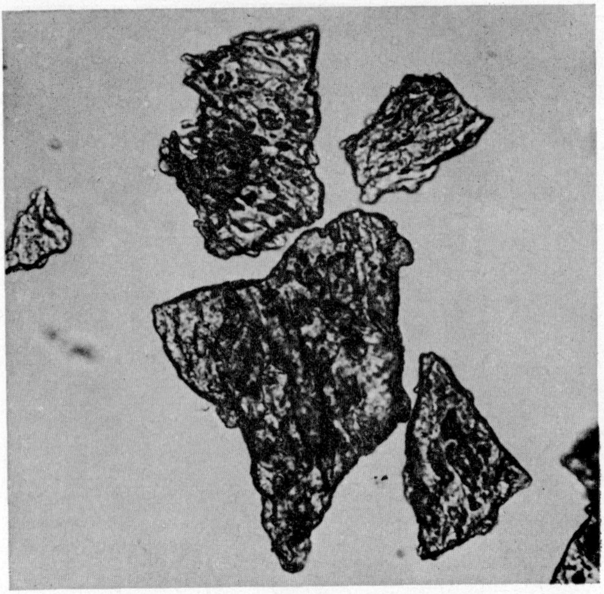

FIG. 17. Roller-dried skim milk, × 200.

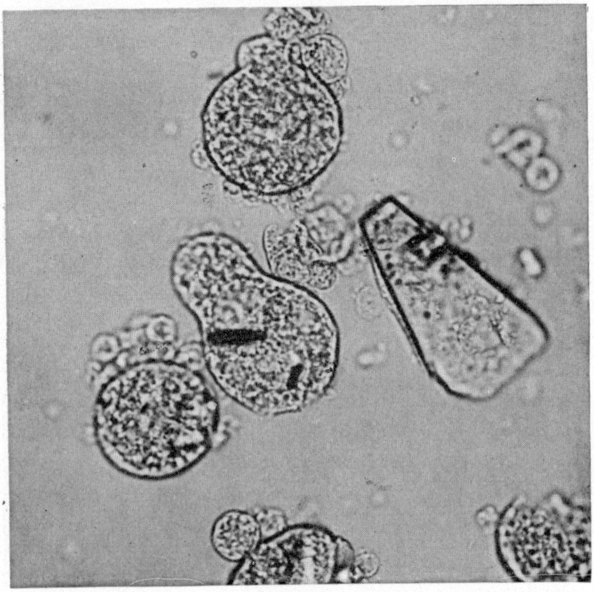

FIG. 18. Spray-dried whey (with a large lactose crystal), × 200.

VII. MILK PRODUCTS

In general, only dried-milk products are used as feed constituents. Roller-dried products consist of crumbled flakes (Fig. 17), spray-dried material of minute spheres, often containing cavities (Fig. 18). Besides the protein component, microscopic analysis is concerned with the identification of fat globules and lactose crystals in order to differentiate between dried whole, skimmed milk and whey products (American Association of Feed Microscopists, 1961; Huss, 1970a, c; Gillet, 1973). Whey products often contain relatively large, characteristic lactose crystals, easily found in polarized light (Huss, 1971) (Fig. 18). They are not to be confused with the small sickle-shaped lactose crystals which are seen in heated phenol–glycerol mounts (Huss, 1970a). The connection between lactose crystallization and loss of lysine in stored milk products is discussed by Huss (1970b). A useful chemical test for milk products in mashes is the use of Fehling's reagent.

REFERENCES

American Association of Feed Microscopists (1961). 'Manual of Microscopic Analysis of Feeding Stuffs.' (Revised.)

Anon. (1951). *In* 'Handbuch der Landw. Versuchs- und Untersuchungsmethodik Methodenbuch' (R. Herrmann, ed.), Vol. III, p. 142. Neumann Verlag, Radebeul and Berlin.

Anon. (1954). 'Methods of Analysis', Series MVS. State Agricultural Test and Research Station, Maastricht.

Anon. (1957). 'Hides, Skins and Leather under the Microscope.' British Leather Manufacturing Research Association, Egham, Surrey.

Barnstein, F. (1920). 'Anleitung zur Mikroskopischen Prüfung und zur Begutachtung der Kraftfuttermittel.' Gebr. Bornträger, Berlin.

Bartschat, F. (1927). *Chemikerzeitung*, **51**, 518.

Bartschat, F. (1936). *In* 'Handbuch der Lebensmittelchemie' (A. Böhmer, A. Juckenack and J. Tillmann, eds), Vol. III, p. 726. Julius Springer, Berlin.

Baudelot, M. E. (1873). *Archs Zool. Exp. Gén.* **2**.

Boehmer, C. (1900). 'Die Kraftfuttermittel.' Paul Parey, Berlin.

Brown, E. E. (1959). 'Microscopic Identification of Hydrolyzed Feather Meal.' Misc. Publ. 363, Texas Agricultural Experimental Station.

Brown, E. E. (1964). 'Off. Proc. 12th Ann. Meet. A.A.F.M.', p. 43. (Unpublished.)

Brown, E. E. (1968). 'Off. Proc. 16th Ann. Meet. A.A.F.M.', p. 81. (Unpublished.)

Claus, G. (1932). *Z. Tierzücht. ZüchtBiol.* **25** (2), 194.

Claussen, E. (1930). *Z. Fleisch- u. Milchhyg.* **40**, 265.

Claussen, E. (1931). *Z. Fleisch- u. Milchhyg.* **41**, 187, 344.

Claussen, E. (1932). *Dt. Tierärztl. Wschr.* **40** (6/7).
Ferrando, R. and Henry, N. (1966). 'Determination Microscopique des Composants des Aliments du Bétail.' Vigot Frères, Paris.
Gillet, H. (1973). 'Protokoll d. Arbeitstagg. d. Sektion Futtermittelmikroskopie', p. 12, Luxemburg. (Unpublished.)
Goerttler, V. and Muenster, E. (1929). *Berl. Tierärztl. Wschr.* **45** (48).
Groening, W. (1932). *Berl. Tierärztl. Wschr.* **48** (59).
Harris, K. L., Smith, F. R. and Scott, D. B. (1960). *In* 'Microscopic-analytical Methods in Food and Drug Control', p. 187. Food and Drug Tech. Bull. No. 1, U.S. Department of Health, Education and Welfare, Washington D.C.
Heckman, M. (1964). 'Off. Proc. 12th Ann. Meet. A.A.F.M.', p. 22. (Unpublished.)
Hemmert-Halswick, K. (1937). *Z. Fleisch- u. Milchhyg.* **48**, 48.
Herrera, M. (1970). 'Off. Proc. 18th Ann. Meet. A.A.F.M.', p. 28. (Unpublished.)
Heyck, A. (1934a). *Z. Fleisch- u. Milchhyg.* **44**, 444.
Heyck, A. (1934b). *Z. Fleisch- u. Milchhyg.* **44**, 461.
Hirsch-Reinshagen, P., Garcia, F. and Tocornal, G. (1966). *Nutricion, Bromatologia, Toxicologia*, **5**, 90.
Huss, W. (1964). 'Protokoll d. Arbeitstagg. d. Sektion Futtermittelmikroskopie', p. 9, Liebefeld/Bern. (Unpublished.)
Huss, W. (1970a). 'Protokoll d. Arbeitstagg. d. Sektion Futtermittelmikroskopie', p. 32, Maastricht. (Unpublished.)
Huss, W. (1970b). *Landw. Forsch.* **23**, 275.
Huss, W. (1970c). *Kraftfutter*, **53**, 430.
Huss, W. (1971). *Landw. Forsch.* **26**/11, Sonderheft 141.
Huss, W. (1973). 'Protokoll d. Arbeitstagg. d. Sektion Futtermittelmikroskopie', p. 31, Luxemburg. (Unpublished.)
Huss, W. (1977a). *In* 'Handbuch der Mikroskopie in der Technik' (H. Freund and J. Grehn, eds), Vol. VIII, p. 383. Umschau Verlag, Frankfurt am Main.
Huss, W. (1977b). 'Protokoll d. Arbeitstagg. d. Sektion Futtermittelmikroskopie', Hameln. (Unpublished.)
Ingling, A. L. (1975). 'Off. Proc. 23rd Ann. Meet. A.A.F.M.', p. 90. (Unpublished.)
Klaatsch, H. (1890). *Morph. Jb.* **16**, 97.
Koelliker, A. (1859). Verhandlg. der Physik.-mediz. Gesellschaft, Würzburg.
Lafond-Grellety, J. (1965). 'Farine de Viande et Farine de Poisson.' Institute Municipal de Recherches sur l'Alimentation Humaine et Animale, Bordeaux. (Unpublished.)
Litterscheidt, F. and Lambardt, H. (1921). 'Die Erkennung der Haare unserer Haussäugetiere und einiger Wildarten.' Reimann, Hamm (Westf.).
Lucks, R. (1915). *Landw. VersStnen*, **86**, 289.
Mach, F. and Claus, G. (1933). *Landw. VersStnen*, **115** (3–4).
Schmid-Monnard, C. (1883). *Z. Wiss. Zool.* **39**.
Schmidt, W. J. (1924). 'Die Bausteine des Tierkörpers im Polarisierten Licht.' Friedrich Cohen, Berlin.
Schweitzer, G. (1931). 'Mikroskopische Bilder der wichtigsten Futtermittel tierischer Herkunft.' Eugen Ulmer, Stuttgart.

Stephan, P. (1900). 'Recherches Histologiques sur la Structure du Tissue Osseux des Poissons.' Dissertation, Paris.

Stoves, J. L. (1957). 'Fibre Microscopy.' National Trade Press, London.

Tims, H. W. (1905). *Q. Jl Microsc. Sci.* **194**, 39.

Voehringer, H. (1958a). 'Die mikroskopische Untersuchung von Fischmehlen auf Echtheit und Reinheit.' Dissertation, Hohenheim.

Voehringer, H. (1958b). *Landw. Forsch.* **11**, 108.

Voehringer, H. (1959). *Landw. Forsch.* **12**, 177.

Voehringer, H. (1961). *Landw. Forsch.* **14**, 201.

Voehringer, H. (1965). 'Protokoll d. Arbeitstagg. d. Sektion Futtermittelmikroskopie', p. 13, Speyer. (Unpublished.)

Voehringer, H. (1970). 'Protokoll d. Arbeitstagg. d. Sektion Futtermittelmikroskopie', p. 38, Maastricht. (Unpublished.)

Voehringer, H. (1971). *Landw. Forsch.* **26**/11, Sonderheft 137.

Voehringer, H. (1972). 'Protokoll d. Arbeitstagg. d. Sektion Futtermittelmikroskopie', p. 31, Oldenburg. (Unpublished.)

Voehringer, H. (1974). 'Protokoll d. Arbeitstagg. d. Sektion Futtermittelmikroskopie', p. 26, Karlsruhe/Augustenberg. (Unpublished.)

Wilder, O. H. (1967). 'Off. Proc. 15th Ann. Meet. A.A.F.M.', p. 28. (Unpublished.)

12. Oils and Fats

K. G. BERGER[1], G. G. JEWELL[2] and R. J. M. POLLITT[1]

[1] *Lyons Central Laboratories, West Kensington, London, U.K.*
[2] *The Leatherhead Food, Research Association, Leatherhead, U.K.*

I. INTRODUCTION

Early applications of the microscope to oils and fats were mainly of an analytical nature. The identification of typical crystal shapes has long been used to detect adulteration of lard with beef or hydrogenated fats (Sutton *et al.* 1940; Williams, 1966). The typical crystal forms of sterols prepared from fats have been used to distinguish between those of animal and those of vegetable origin (Butcher, 1934a, b). More recently, characterization and identification of a wide range of lipids has been made by measurement of critical solution temperatures with the aid of the microscope (Schmid *et al.*, 1965).

Oils which are completely liquid when used or consumed have no microscopical structure. The term fat is conventionally used for materials which appear solid at room temperature. Such materials consist of a liquid phase held in a crystal matrix. In the last 30 years attention has focused on the nature of this matrix as affected by processing or storage conditions, and on its significance in the functions performed by the fat. Many issues of texture, texture change and functionality have been resolved with the aid of the microscope, often in combination with rheological and crystallographic techniques. Relationships are being worked out between microscopic appearance and the functionality of fats and of surface-active lipids. Such knowledge becomes increasingly important as food manufacture becomes more highly mechanized, since efficient operation depends on the use of ingredients with a predictable behaviour. The purpose of this chapter is to bring together recent information in this field.

II. METHODS

A. Light Microscopy

1. Illumination

Bright field illumination is only useful when the sample has been suitably stained, for example with osmic acid or Sudan red (Pearse, 1968).

Phase contrast can be used with unstained samples, for example to yield information on air cells and water droplets in margarines and emulsions, but is of very limited value for the crystal phase.

Polarized light enables the solid and liquid phases to be distinguished, because the solid crystalline material is usually anisotropic and therefore birefringent, whereas liquid oil is isotropic. No pretreatment, with its possible effects on structure, is required. Most optical microscopy of fats has been carried out under polarizing conditions, and indeed all the optical micrographs presented in this chapter were obtained in this way.

The auto-fluorescent properties of lipids have been little used in microscopy. The principle employs illumination of the sample with short wavelengths, which cause usually faint radiation of a longer wavelength to be emitted. The method uses no pretreatment such as the staining with fluorochromes required in the usual fluorescence procedures (Pearse, 1968) and should therefore be of value.

2. Sample preparation

The nature of fats presents special difficulties. For example, conventional embedding in paraffin wax is not feasible, since hot wax will melt and dissolve fats. The use of a cryostat is open to doubt since new crystal structures will be formed during cooling to sectioning temperature. These artefacts will not necessarily disappear when the specimen returns to room temperature. The shearing action of section cutting will itself deform the surface structures. A possible approach would be to shock-cool the specimen to $-100°C$, a procedure that is used in electron microscopy and is known to change the oily phase into a structureless vitreous form. If the specimen is then warmed back to the cryostat sectioning temperature artefacts may be avoided.

Probably the simplest and most useful method of preparation is a modified smear method. In order to observe crystal aggregation, it is necessary to view a relatively thick smear of material. Much information can be gained by focusing up and down; however, for photography a single focus must be selected and only some parts of the image are sharp. Clusters therefore appear as rather vaguely outlined masses and individual crystals are not clearly visible unless large. Useful magnifications are $\times$ 100 and $\times$400 and closer examination may be made at $\times$1000 with oil immersion.

It is necessary to standardize the thickness of the smear for effective comparison of samples. One procedure involves the use of a special jig and template (Meara *et al.*, 1974). More simply, strips of paper may be stuck on to a microscope slide leaving a space about 18 mm wide in the middle. A portion of fat is deposited centrally between the strips by pressing gently through a hole of diameter 2·5 mm in a plate resting across the paper strips. Finally, a 22 $\times$ 22 mm coverslip is placed over the fat, straddling the paper strips, and pressed down slowly with a flat glass plate. The fat is thus squeezed out to a standard thickness, but the effect is not so harsh as to disintegrate the crystal aggregation. A convenient thickness is 30 μm or 60 μm obtained with carbon paper.

Only those crystals in a certain orientation are visible under polarized light. If the microscope stage is rotated, clusters of crystals will be seen to disappear and others appear where none were formerly visible. The original clusters will reappear as a rotation of 90° is approached. An attempt has been made to photograph the whole crystal structure by making six exposures on the same film, rotating both polars 9° between each exposure, so covering a range of 45° (Pollitt, 1974) (Figs 1a–c).

Direct observation of a standard thickness of shortening was made by Herb *et al.* (1956). A weighed portion was pressed out on a slide, just to the edge of the coverslip. An alternative procedure which gave better definition of individual crystal shape was to dilute the fat with eight parts of oil filtered from the fat. Both polarized light and phase contrast were used, the latter showing more clearly the long crystals of lard.

B. Electron Microscopy

The sample preparation methods for the electron microscope are more rigorous than those for light microscopy, and the low melting point and ready

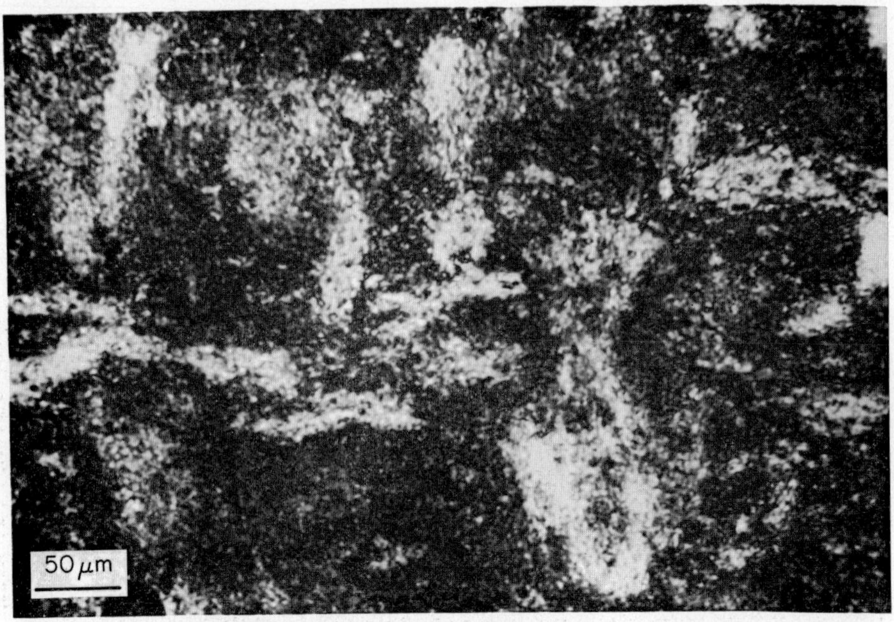

FIG. 1a. Crystal clusters of texturized lard viewed by polarized light. Polars set at 0°.

FIG. 1b. Crystal clusters of texturized lard viewed by polarized light. Same field as Fig. 1a with polars set at 45°.

solubility of many fats in numerous solvents has caused many problems. For transmission electron microscopy (TEM) the four main types of specimen preparation that have been used with fats are ultrathin-sectioning, replicas, freeze-etching and negative staining.

The major problem in the preparation of thin sections (20–50 nm thick) is the solvent action of the embedding materials (normally epoxy or methacrylate resins), coupled with the requirement to polymerize these resins at

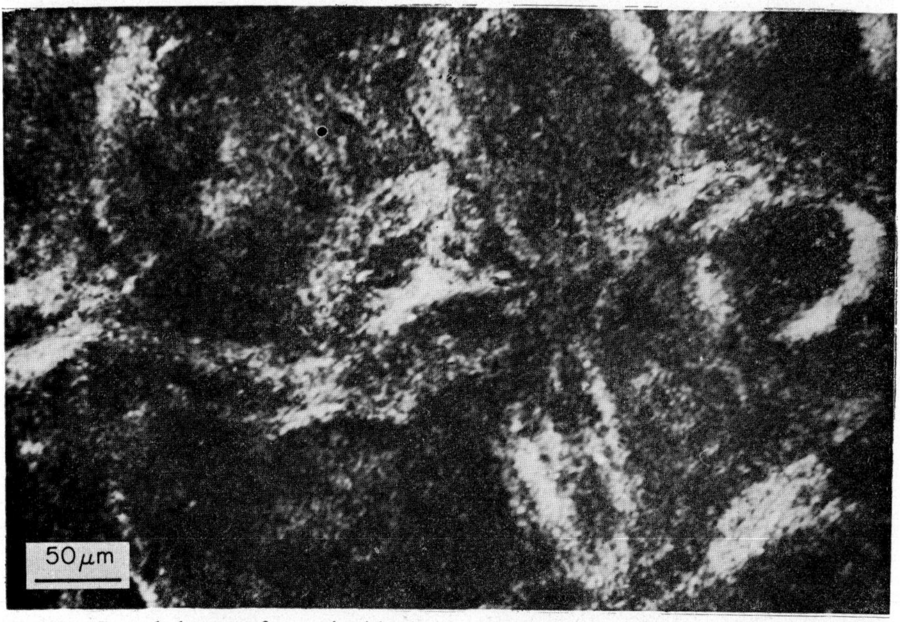

FIG. 1c. Crystal clusters of texturized lard viewed by polarized light. Successive exposures on the same field as Fig. 1a with polars set at 0°, 9°, 18°, 27°, 36° and 45°.

relatively high temperatures (about 60°C). There is also the lack of a really effective fixative for lipids. Several workers have used prolonged fixation with osmic acid as both fixative and stain, thus Wortmann (1965) fixed butter for 6 months whilst Cruickshank (1976) and Lewis (1975) have used fixation periods of between 2 and 4 weeks on chocolate and fatty tissues. Figure 2 shows a thin section of cocoa butter fixed with osmic acid in this way. The crystalline fat shows up white, with the liquid phase dark. The fine-feathered edges to the crystals may be noted at the right-hand end of the large crystal. The embedding problem can be minimized using water-miscible epoxy resins and polymerizing with u.v. light. Great care must be taken in interpretation of thin sections of fats as Cruickshank (1976) has shown that the appearance of cocoa butter differs with various types of embedding procedure. Mouwen

FIG. 2. Electron micrograph of cocoa butter fixed by prolonged exposure to osmic acid. Liquid portion is dark.

(1974) has used thin-sectioning to study the relationship of the components in margarine.

Probably the most widely adopted method of specimen preparation has been the production of replicas, either directly or using the freeze-etching procedure. Two problems have to be overcome during the replica process: firstly, the thermal energy produced during the sputtering of the carbon and heavy metals, and secondly, the location of the liquid phases in the sample which may obscure the structure of the crystalline materials. In general, replicas of fatty materials are prepared while the specimen is maintained at low temperature (about $-100°C$). This can be achieved by a modified specimen support in a conventional vacuum coating unit (Jewell and Meara, 1970) or directly in a commercial freeze-etching apparatus. However, using such an approach, the liquid phase will be frozen and may obscure the detail of the solid phase (Jewell and Meara, 1970). An alternative is to pretreat the sample with detergent to remove the liquid phase and then replicate the exposed solid phase (Jewell and Meara, 1970; Poot *et al.*, 1975).

Another method is to use the freeze-fracture technique and study the fractured solid phase embedded in the surrounding vitrified liquid phase (Berger, 1970; Jewell and Meara, 1970). Of the two approaches the detergent method gives the better information on crystal shape and has been shown to correspond well with crystal size as measured by light microscopy (Meara *et al.*, 1974). Freeze-fracture gives a better insight into the interrelationship and possible networks between the crystals.

The negative staining procedure has had only limited use in the field of fats. Bauer (1972) employed the technique to investigate the morphology of pure triglycerides. A far wider application has been in the study of phospholipid and lipoprotein membranes which falls outside the immediate scope of this chapter.

For scanning electron microscopy (SEM) the samples usually have to be prepared so that they will resist the thermal damage of the electron beam and have a conducting surface. This is normally achieved by sputter coating with a layer of conducting heavy metal, such as gold or palladium, and the use of a cold-stage in the SEM. In this way structural studies have been made of margarine (Mouwen, 1974), chocolate (Cruickshank and Jewell, 1976), compound chocolate (Kawada, 1974), fat in cakes (Shepherd and Yoell, 1976) and freeze-dried monoglycerides (Fig. 3).

Of the two techniques of TEM and SEM, the latter has the advantage of enabling larger specimens to be examined with a greater depth of focus than can be achieved using TEM. However, TEM offers a very much improved resolution that permits examination of the lattice growth steps within the fat crystals.

FIG. 3. Scanning electron micrograph of freeze-dried β crystals of commercially distilled monoglycerides. (By courtesy of Mr N. Krog, Grindsted.)

III. STUDY OF LIPID SYSTEMS

A. Shortening and Margarine

1. Composition and structure

Shortenings consist of mixtures of triglyceride fats processed from a variety of vegetable, animal and marine sources. The fatty phase of margarines is similar in composition, but they contain in addition about 15% emulsified water and usually also some salt and milk protein. The physical properties of both types of product, on which their functionality depends, are mainly governed by the amount of solid fat present and by the shape, size and interrelationship of the fat crystals. Particular rheological properties are required. The products must have a certain rigidity for convenient storage and transport; on the application of stress, such as in mixing or spreading processes, they must flow. Subsequently, they must retain some rigidity in order to perform their shortening and aerating properties. These rheological properties have been studied by means of penetrometry (Haighton, 1963, 1965) and under more dynamic conditions by means of the Ferranti–Shirley cone and plate viscometer (Pollitt, 1974; Stern and Cmolik, 1976).

To achieve the required rheological properties, blends of oils are plasticized by rapidly cooling the melted blend to produce solid components in the form of small crystals. The blend of oils is chosen such that 15–25% of the triglycerides are solid at usage temperature. Modern plant uses a scraped-surface heat exchanger, such as the Votator, which also incorporates a unit in which the shortening or margarine is worked while most of the crystallization develops (Joyner, 1953; Berger, 1976). Some 10–20% air by volume may be dispersed in shortenings.

2. General microscopical nature

The description of the crystalline structure of shortening was initially made without recourse to the microscope. X-ray studies have been used to show a 'tuning fork' structure of the triglyceride molecule in space in which the three fatty acid chains are more-or-less parallel, one pointing in the opposite direction to the others. Van den Tempel (1968) has published an illustration of a typical molecule. In crystal growth, such molecules will pack more easily side-to-side than end-to-end. The resulting individual crystal appears needle-shaped under the microscope, but is probably more like the blade of a knife. Individual crystals may grow in bundles (Fig. 4). In a well-texturized shortening the length of the majority of crystals is less than 1 μm, difficult to measure under the light microscope. Some crystals will be up to 2–3 μm long, and a few perhaps up to 5 μm.

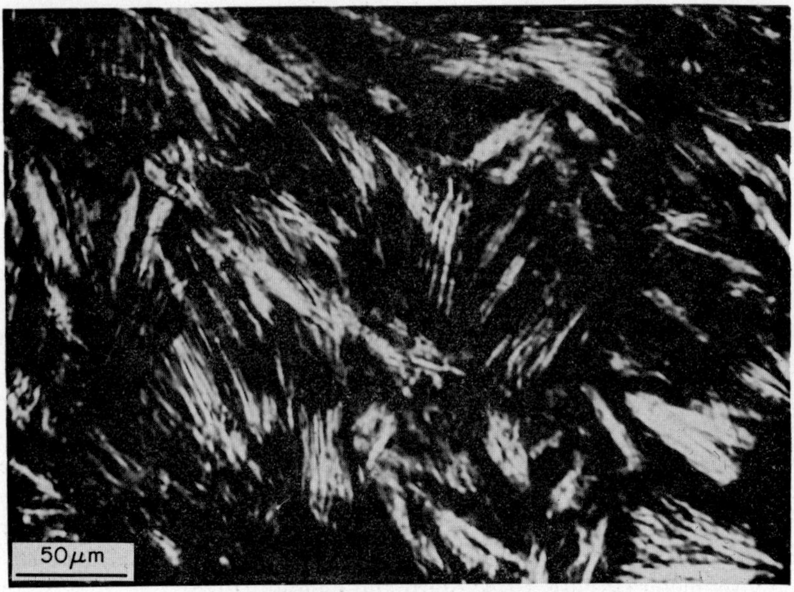

FIG. 4. Bundles of large crystals of lard.

Haighton (1963) calculated that, if all the crystals measured $1 \times 0{\cdot}5 \times 0{\cdot}25$ μm, there would be $1{\cdot}6 \times 10^{12}$ crystals in a shortening of 20% solids content and the average distance apart would be 0·32 μm. In such circumstances, the crystals are attracted to each other by Van der Waal's–London forces and some form of aggregation occurs. The rheological behaviour has been interpreted by Van den Tempel (1961) as indicating that there is a network structure held together by strong primary forces and weaker secondary forces. Haighton (1963) gave a diagrammatic representation of this network structure.

In the use of shortening and margarine in the production of baked goods, large shearing forces are experienced. Haighton (1965) studied the effect of kneading shortening and margarine, measuring the consistency before and after by means of a penetrometer, and confirmed Van den Tempel's network structure. Margarine has essentially the same structure because the many small (up to 10 μm) water droplets present contribute little to the consistency.

Microscopical evidence in support of this network structure was not forthcoming until Berger (1970) published an electron micrograph of a shortening (Fig. 5). The sample was prepared by the freeze-fracture technique which ensures that the liquid components of the fat are set into an amorphous glass and that only those crystals that are part of the permanent structure are shown. The photograph shows a loosely composed string of small crystals

FIG. 5. Electron micrograph of shortening, showing a string of small crystals lying diagonally from the top right-hand corner.

lying diagonally from the top right-hand corner. Individual crystals vary in size from 0·3 μm to 2 μm and have a platelet form. The crystals are not very regular in shape and show many steps where crystallization has started and was interrupted before a symmetrical shape was achieved. The steps are 4–5 nm in height, corresponding to a layer of triglyceride molecules standing on end. There is indication of other chains of crystals above and below the main chain.

Attempts to demonstrate clearly the three-dimensional network under the light microscope have not been entirely successful to date because of the compactness of the network and the small depth of focus of the microscope. To avoid the appearance of a confused matrix, it is necessary to view with polarized light, which reveals only those crystals in the structure which are in a particular orientation (see Section II. A.2), and even then to limit the thickness of material viewed. Nevertheless, very significant information can be gained from observation of the form in which the crystals aggregate together. Photomicrographs illustrating the difference in appearance between a vegetable and a lard shortening were published by Bailey (1964). Joyner (1953) illustrated differences between lard and rearranged lard. Hoerr (1955) drew attention to the possible uses of the microscope in the examination of shortening and margarine.

In 1974, following some years work, the first detailed description of types of crystal aggregates appeared (Meara *et al.*, 1974). The following were identified.

(i) *Spherulite type A*: central core or nucleus of tightly-packed, rather broad, needles surrounded by radially-orientated, long, narrow, needle-like crystals.

(ii) *Spherulite type B*: central core surrounded by small, closely-bound, randomly-orientated crystals.

(iii) *Clusters*: small, more-or-less spherical groups of closely-bound crystals, randomly arranged.

(iv) *Bundles*: crystals arranged more-or-less parallel to each other, each bundle orientated randomly with respect to other bundles, although with some tendency to form an open network. (This type of structure is shown in Fig. 1.)

(v) *Aggregates*: aggregates of spherulites and of clusters were also observed. Typical structures are shown in Figs 6a–e, from photomicrographs prepared by one of us (R. J. M. P.).

3. Specific applications

The early applications of microscopy concerned the illustration of the differences between lard and interesterified lard (in which the fatty acid distribution has been randomized). Joyner (1953) first illustrated the much smaller crystal habit of rearranged lard when crystallized on a microscope slide. Under similar plant chilling (Votator) conditions, he noted that the crystals in vegetable shortenings outnumbered those in lard by 100 : 1. Hoerr (1955) pointed out that the available analytical methods failed to show significant differences between ordinary and rearranged lard; but, when crystallized on a microscope slide, lard contained numerous clusters of large, needle crystals whereas the rearranged lard consisted of a lacy network of uniform, delicate, almost submicroscopic crystals. Air incorporated in Votator-plasticized fats was much more thoroughly dispersed in the rearranged lard. Photomicrographs of cake batters showed that those made with lard contained relatively little air dispersed in large bubbles, whereas those made with rearranged lard contained a relatively greater number of smaller air bubbles. The former produced a low-volume, coarse-grained cake, the latter a high-volume, fine-textured cake. Similar illustrations were published by Herb *et al.* (1956).

Differences found in the texture of blends of hydrogenated cotton seed oil and palm kernel oil from different sources were explained by Landmann and Feuge (1956) on the basis of different crystal forms viewed under the microscope. De Castro and Vazquez (1971) published photomicrographs which indicated, in particular, the increased number and decreased size of lard crystals obtained by the addition of hydrogenated lard.

Variations in processing conditions also alter the crystal network structure. Sambuc and Naudet (1965), using bench-scale texturizing apparatus, showed by means of photomicrographs that longer residence times with lower crystallizing temperatures resulted in a better dispersion of the aqueous phase and a softer texture in margarine. After processing in the Votator, it is usual to submit shortenings to storage for some days at a controlled temperature a few degrees above ambient. This 'tempering' process is considered to lead to some rearrangement of the mixed crystals originally formed, resulting in more ordered and stable crystals. Hellman *et al.* (1955) published photomicrographs and other evidence to show that such a process occurs in 'global' spreads (fatty products intended to remain spreadable over a wide range of temperatures).

A second scraped-surface cooler, operating after the worker unit, has been proposed as a rapid means of tempering which would render the shortening suitable for immediate use (Reigel and McMichael, 1966). Photomicrographs indicated that, with this post-cooling, finer and more uniform crystals were obtained. After conventional processing, the average crystal aggregate size was stated to be about 75 μm and the maximum crystal aggregate size about 300 μm. The same shortening processed with the post-cooler had an average crystal aggregate size of about 15 μm and a maximum of almost 150 μm.

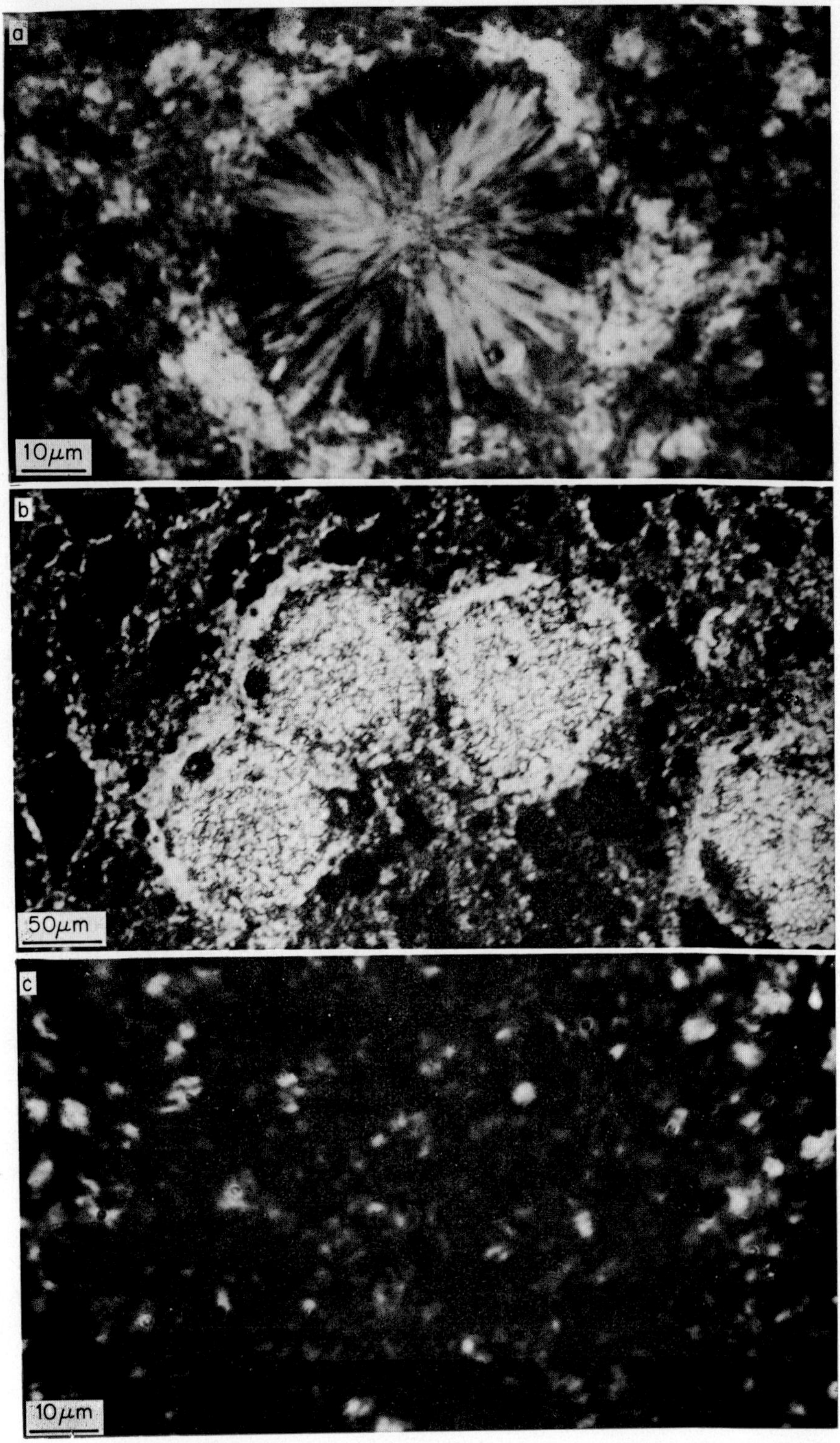
a
10μm
b
50μm
c
10μm

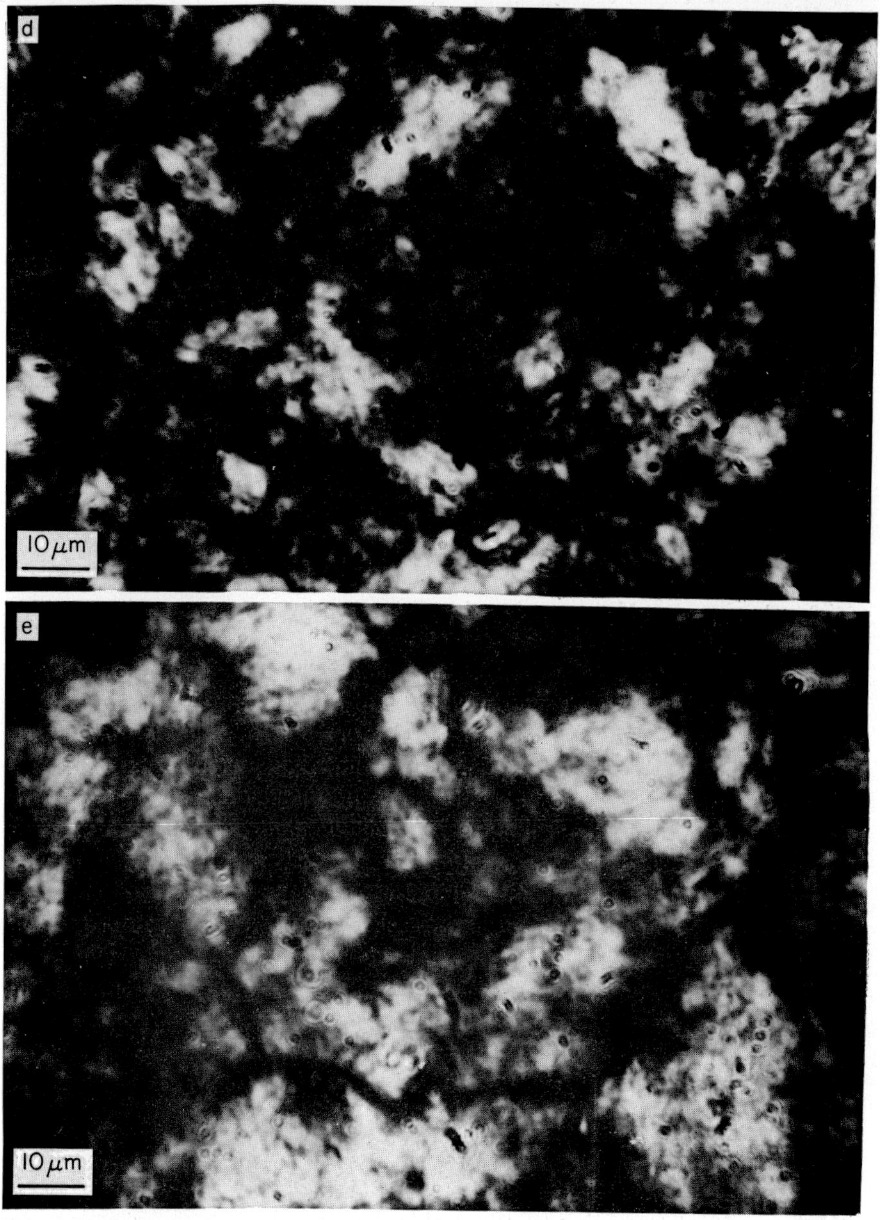

FIG. 6. Typical crystal structures in shortening. (a) A-Type spherulites; (b) B-type spherulites; (c) small clusters; (d) medium clusters; (e) large aggregates.

Sometimes two chilling 'A' units are used in tandem or the product is passed through a second time. Krog and Vrang (1964) recycled part of their margarine emulsion and obtained a somewhat different crystal structure. In polarized light 'crystal caps' could be seen around some of the larger water droplets, similar to those present in butter (Fig. 7). Phase contrast was used to photograph water droplets, starch and mould growth, and fluorescence to indicate protein.

Microscopy was employed by Madsen and Als (1971) in comparative studies on pastry margarine, a stiff but plastic material particularly used to obtain flakiness in pastry, manufactured on a traditional chilling roll and in a pressure cooler similar to the Votator. Microphotographs of the margarines after three weeks at 25°C showed that the chilling roll produced finer crystals, which accounted for the product being smooth. The crystals obtained in the pressure cooler were in the form of clumps and entangled needles. The product softened more on working and gave the poorer baking performance (Fig. 8).

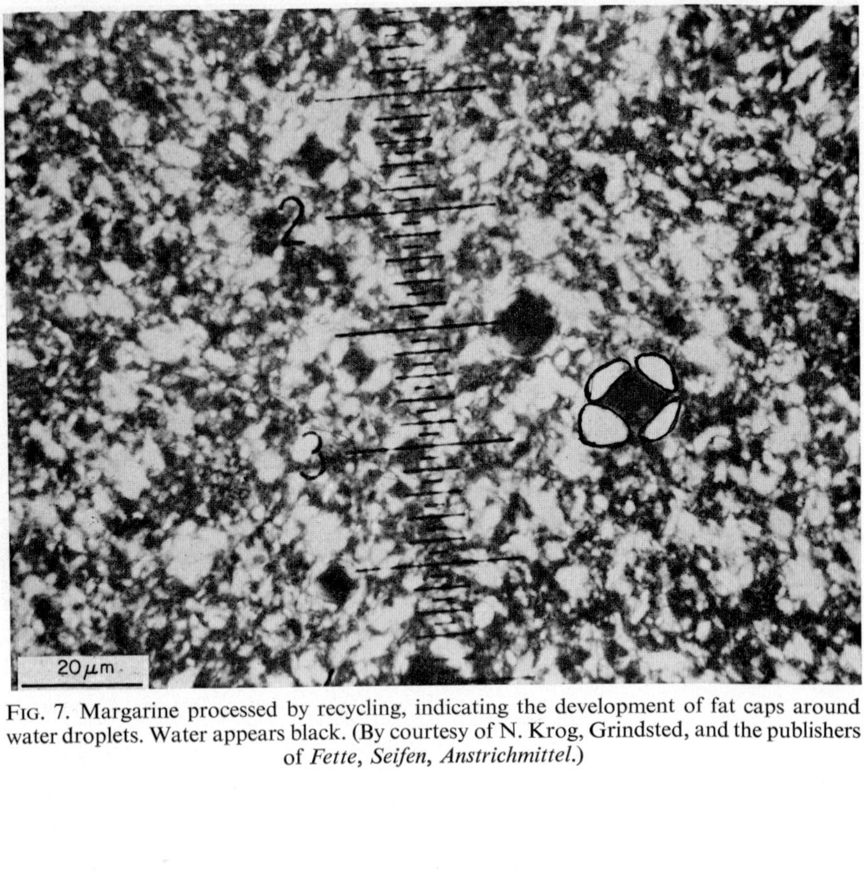

FIG. 7. Margarine processed by recycling, indicating the development of fat caps around water droplets. Water appears black. (By courtesy of N. Krog, Grindsted, and the publishers of *Fette, Seifen, Anstrichmittel*.)

The desired form of crystal aggregation in shortenings when assessed by aerating performance in a cake filling was described by Meara *et al.* (1974). The bulk of the solid material should be present as clusters of more-or-less uniform size (3–5 μm), and individual crystals should be small and fairly uniform (up to 1 μm). A few larger clusters and crystals may be present. Meara *et al.* found that where clusters were more variable in size (4–10 μm), and individual crystals showed a wider scatter in size (up to 6 μm), aerating performance was only moderate. Fats containing aggregated clusters with a tightly packed appearance and those containing bundles had poor aerating performance.

Considerable experience in the commercial production of shortenings has confirmed these observations. Some shortenings show very little crystal aggregation, having a uniform general appearance under the microscope, and these generally have inferior aerating properties. Compact, small clusters are apparently required to surround and stabilize the air cells produced in a whipping process.

Under the electron microscope a shortening with good aerating properties (Fig. 9a) has many small crystals, in contrast to the thick crystals demonstrable in lard (Fig. 9b), which has poor aerating capacity. Interesterified lard also was found to have long crystals (Fig. 9c) which showed wave-like periodic surface deformations (Meara *et al.*, 1974).

The significance of the above observations becomes apparent when one examines the structure of an air cell in a shortening (Fig. 10). It is seen to be completely lined with small individual crystals lying tangentially to the surface and obviously orientated differently from the crystals in the continuous phase. Figure 11 shows an air cell obtained after beating the fat–sugar mixture for a cake batter, and indicates that in this more complicated system the air cell also has the same structure. The small air cell exists in the oil phase and is stabilized by the crystals of the solid fat. Clearly lard cannot perform this function effectively, because its long crystals cannot accommodate themselves to the curved surface. Jewell postulated (Meara *et al.*, 1974) that the improvement in performance obtained after interesterification is due to a loss of mechanical strength by the crystals, so that they can break up easily into the smaller units required. This is indicated by the surface irregularities observed in Fig. 9c.

Wiedermann *et al.* (1961) compared the behaviour of lard, modified to differing extents by interesterification, under the microscope and in baking tests. Their micrographs appear to indicate that the best performance in cakes was obtained with a treated lard which had a moderate degree of crystal aggregation and which did not change on storage. Other samples developed no aggregates, or showed instability with time, and produced poor cakes.

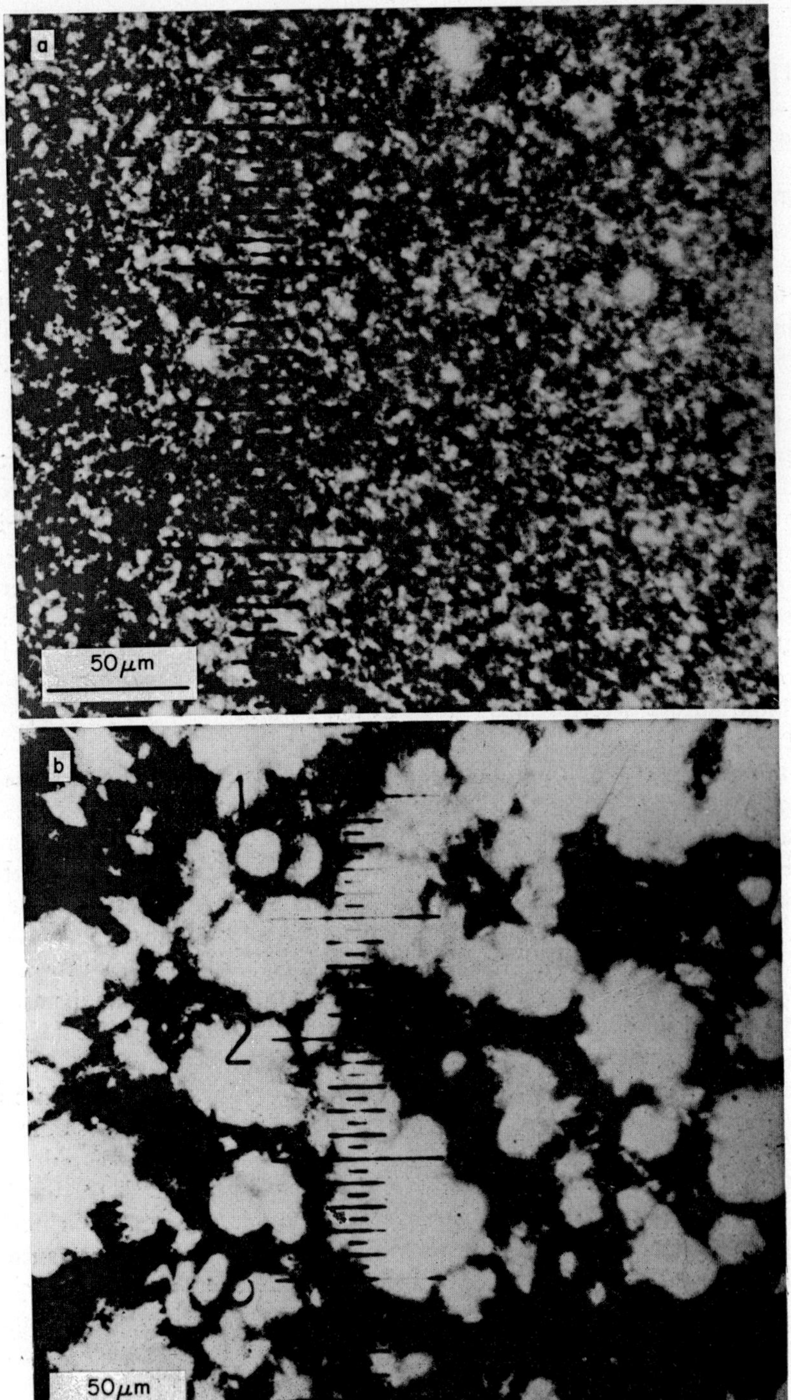
a
50 μm
b
1
2
50 μm

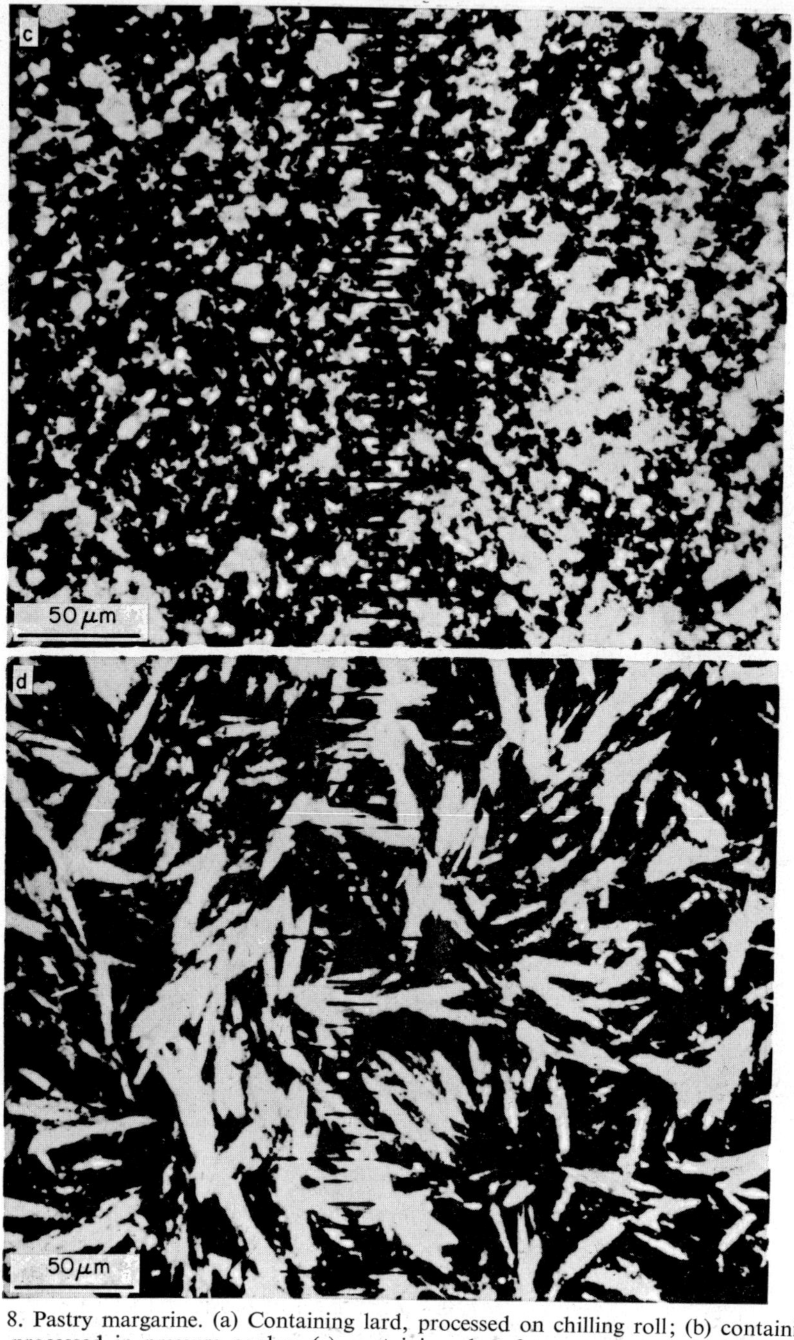

FIG. 8. Pastry margarine. (a) Containing lard, processed on chilling roll; (b) containing lard, processed in pressure cooler; (c) containing shea fat, processed on chilling roll; (d) containing shea fat, processed in pressure cooler. (By courtesy of J. Madsen and G. Als, Grindsted, and the publishers of *Fette, Seifen, Anstrichmittel.*)

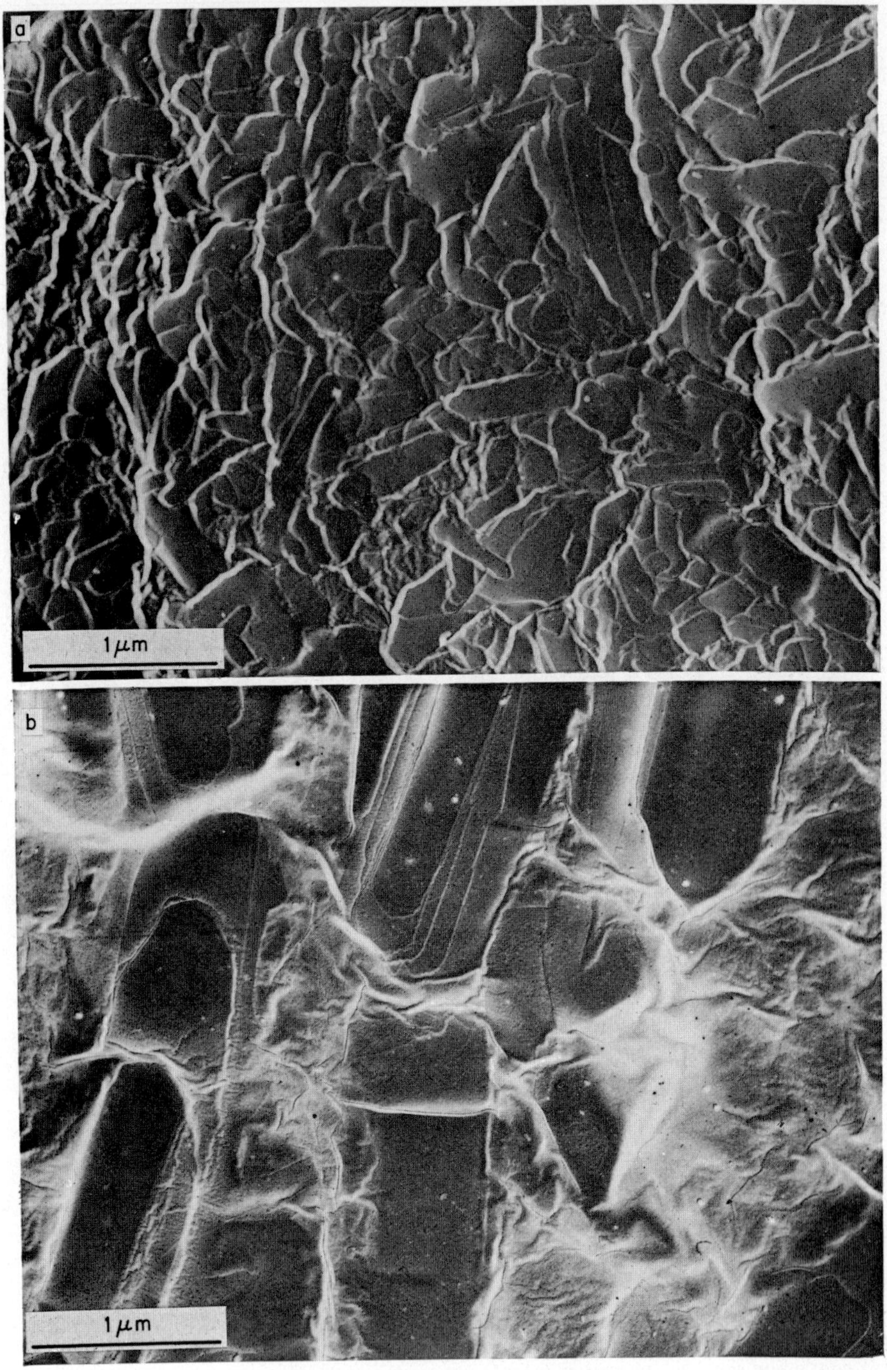
a
1 μm
b
1 μm

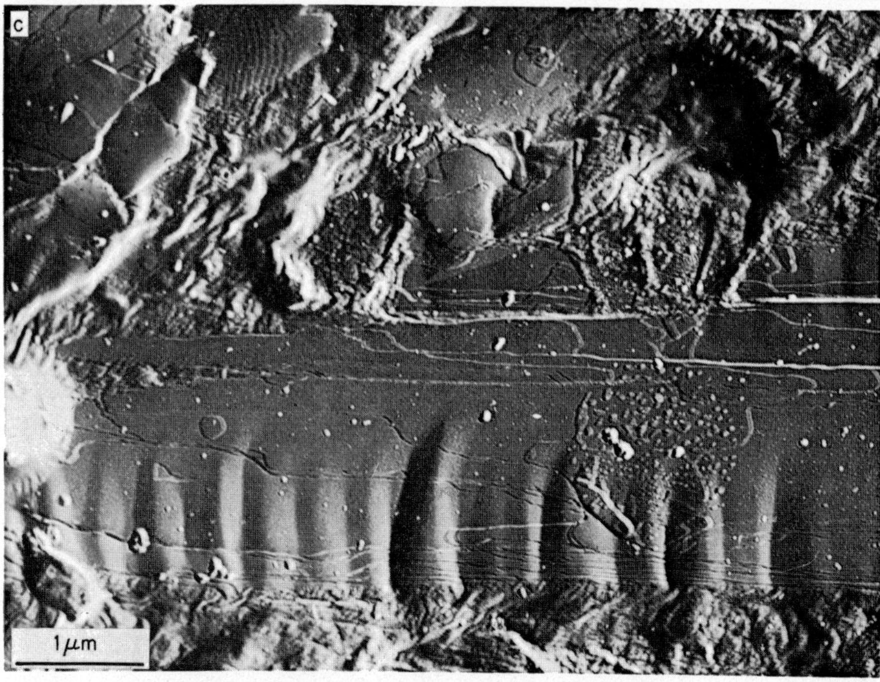

FIG. 9. Electron micrographs of (a) vegetable shortening, (b) lard and (c) interesterified lard. The liquid phase was removed with Teepol.

Cine photomicrography has been used by Bell *et al.* (1975) to extend the study of air cells during baking into the dynamic situation. Cake batters were placed on a Kofler hot-stage which was programmed to simulate the temperature time-cycle in the centre of a baking cake. Time-lapse photography was used to record the development of the cake structure. The shortening used in the cake batter had been stained with oil-soluble dye, and it was demonstrated (i) that all the air initially in the batter was associated with fat, and (ii) that the air cells remained surrounded by fat at least until the fat melted. Electron micrographic evidence (Shepherd and Yoell, 1976) shows that in the finished cake each air cell is still coated by a very thin layer of fat.

The importance of cluster size, and of changes in it which may take place during storage of shortenings texturized on a Votator incorporating a post-cooler, was demonstrated by Berger (1976). Photomicrographs showed increased cluster size in samples stored respectively at 15°C, 21°C and 25°C (Fig. 12). Conclusions from this, supported by rheological measurements, were that at 15°C there is a high solid content but insufficient mobility for crystal aggregation to increase. Aerating performance was poor. At the higher temperatures, some aggregation can occur and the aeration was satisfactory.

The effect of work on a shortening, standardized in terms of shearing at a given rate for a given time on a cone and plate viscometer, was studied by Pollitt (1974). Photomicrographs showed that clusters of crystals were broken up to some extent only and were made to appear more compact (Fig. 13).

To summarize, therefore, two characteristics of shortenings are important in applications requiring aeration. Firstly, enough crystals of a small size must be available to line the air cell surface. Secondly, crystal aggregates or

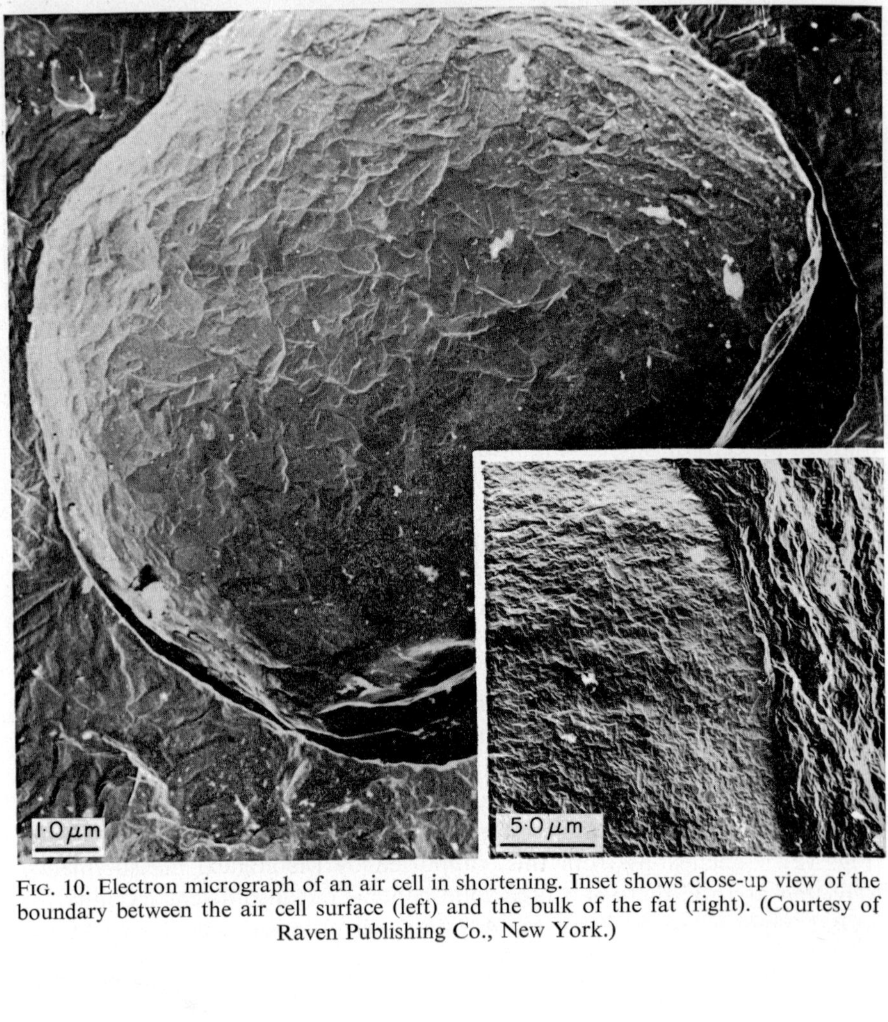

Fig. 10. Electron micrograph of an air cell in shortening. Inset shows close-up view of the boundary between the air cell surface (left) and the bulk of the fat (right). (Courtesy of Raven Publishing Co., New York.)

clusters are required which are not associated with the air cell surface but which remain in the continuous phase of liquid oil and increase its effective viscosity, thus preventing drainage of the oil from between the air cells.

B. Polymorphism

1. Definition and general studies

As early as 1852 Duffy showed that tristearin has three melting points. Much later Malkin (1934) and his associates established that multiple melting was due to polymorphism, that is, the occurrence for a given compound of different crystalline forms with distinct X-ray diffraction patterns. A detailed discussion is beyond the scope of this chapter and the interested reader is referred to the review by Chapman (1962).

It is now generally recognised that purified triglycerides exhibit three main

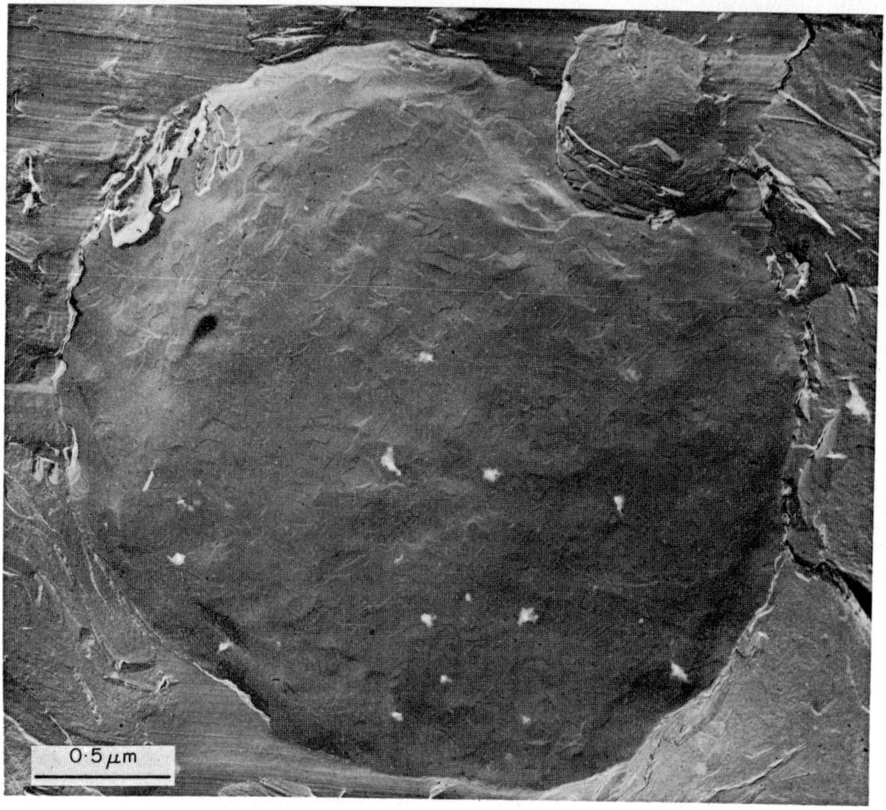

FIG. 11. Air cell in fat–sugar batter.

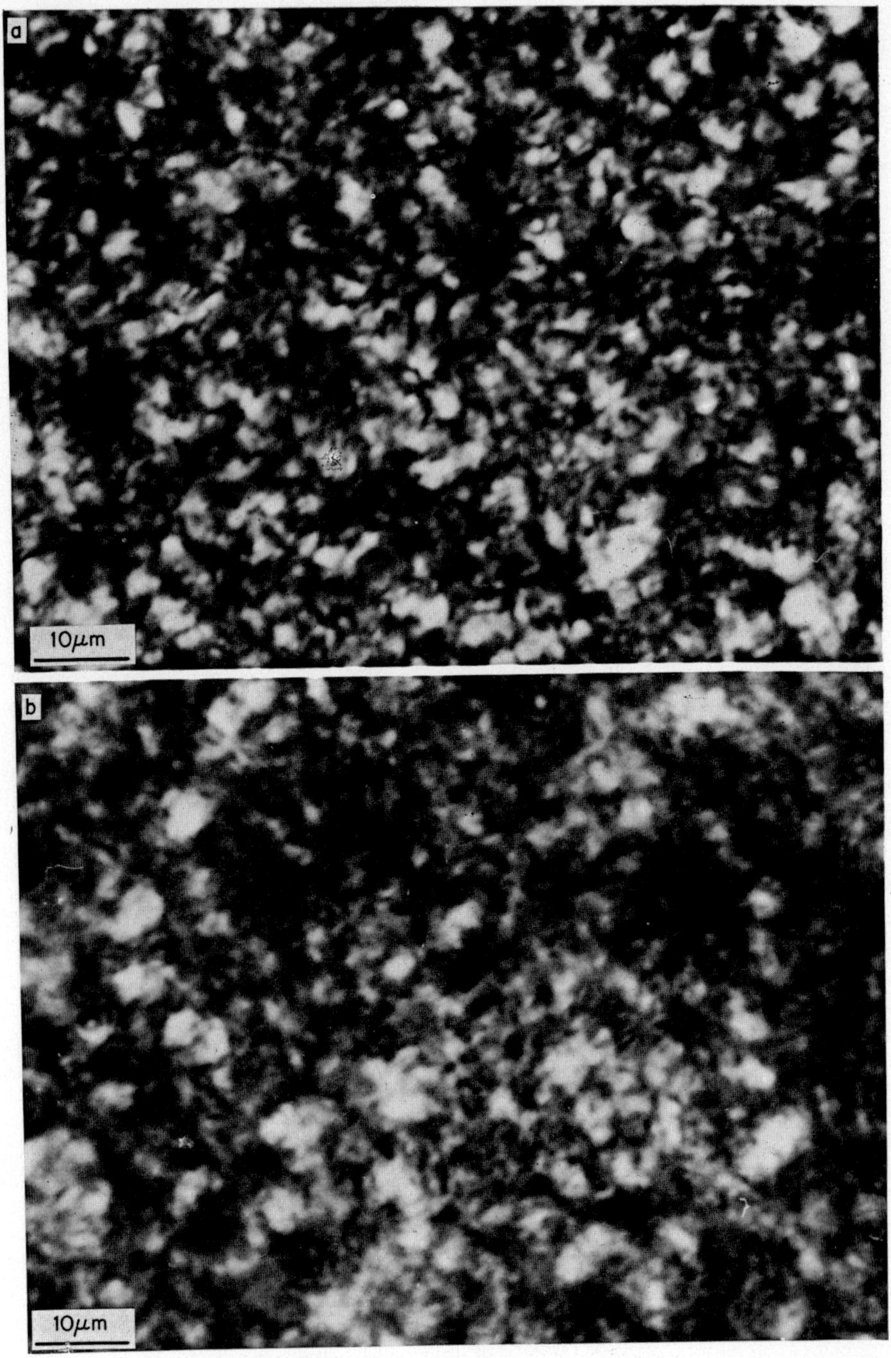
a
10μm
b
10μm

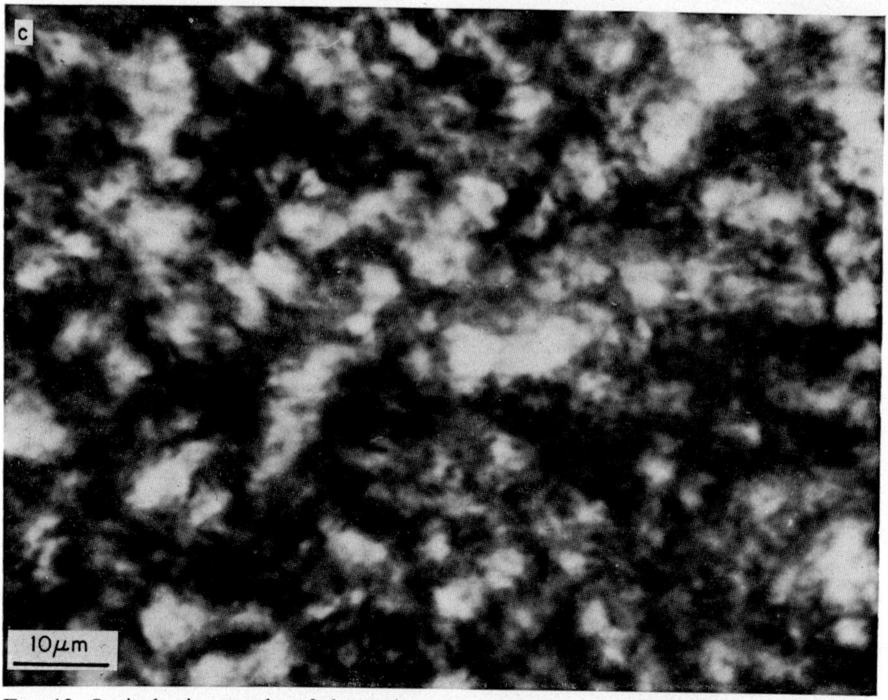

FIG. 12. Optical micrographs of shortening after storage. (a) At 15°C; (b) at 21°C; (c) at 25°C. (Courtesy of Raven Publishing Co., New York.)

forms, namely α, β-prime (β') and β, the α form having the lowest melting point and the β the highest melting point. Some fats develop intermediate forms in addition to those described above, and cocoa butter is recognized as having six polymorphic forms (Wille and Lutton, 1966).

Hoerr (1960), using a combination of X-ray diffraction and photomicrography, attempted to establish some general parameters about the crystal morphology of the various forms. He concluded that the α-form crystals were fragile transparent platelets which readily transformed into the β' type. Beta-prime crystals were tiny delicate needles (about 1 μm long), whilst the β forms were large, coarse crystals about 25–50 μm in length. Although Hoerr's description of appearance has been substantiated by other workers, it would be dangerous to base identification of the polymorphic form of an unknown sample on appearance alone, as this should be done in conjunction with X-ray diffraction.

Whittam and Rosano (1974) used a polarizing microscope equipped with a variable temperature gradient stage. A single photomicrograph illustrated

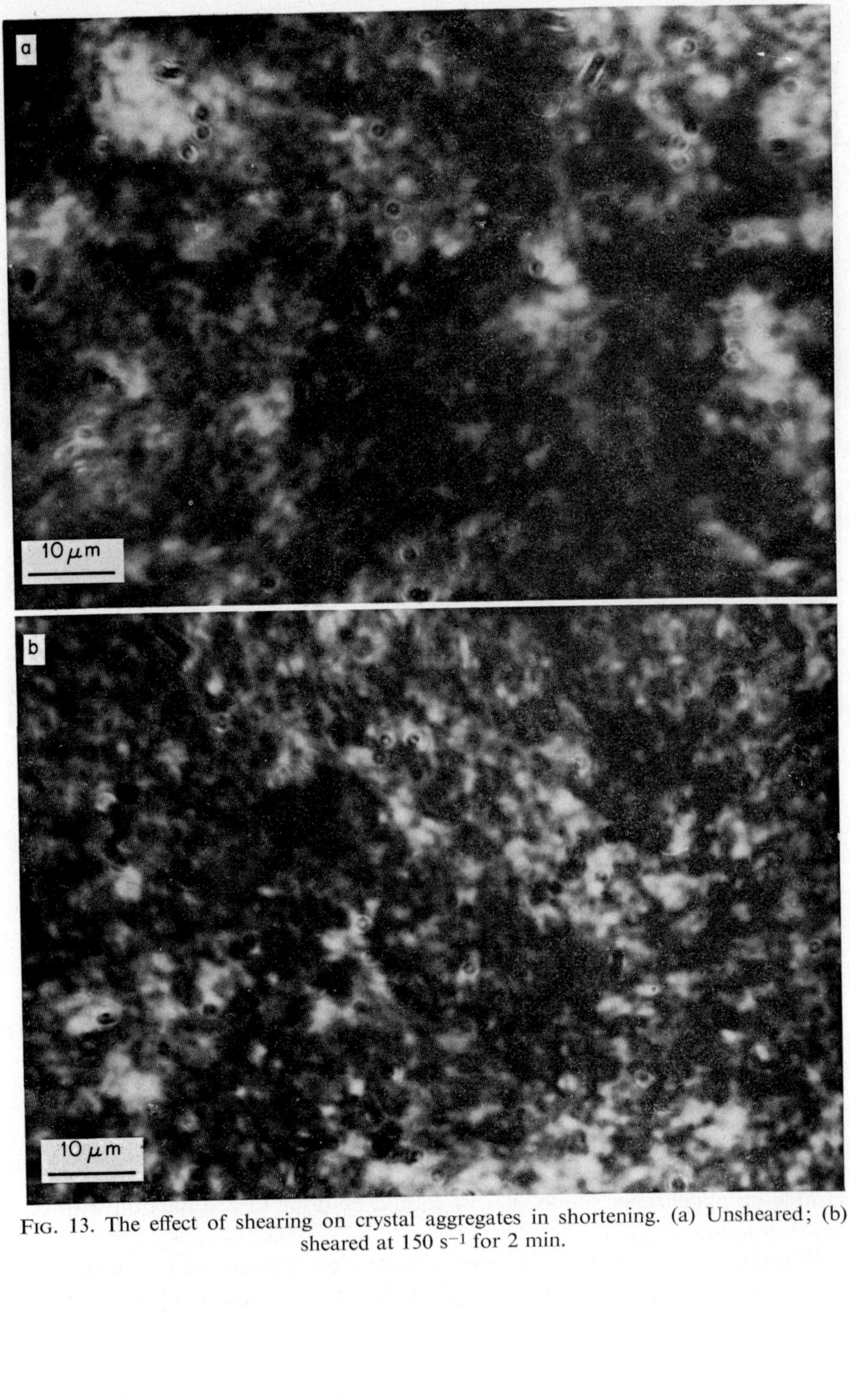

FIG. 13. The effect of shearing on crystal aggregates in shortening. (a) Unsheared; (b) sheared at 150 s^{-1} for 2 min.

the transformation of α to β' to β-tripalmitin crystals with increasing temperature. An investigation of other trisaturated glycerides is described in a later paper by the same authors (Whittam and Rosano, 1975).

2. *Specific problems*

a. Shortenings. In the study mentioned above, Hoerr (1960) established the importance of the various crystal forms in the functionality of shortenings. Their effectiveness in incorporating air bubbles in cake batters was greatest for the β' form and less for the intermediate type, while the β form was quite unsatisfactory.

b. Margarines. Wilton and Wode (1963) found that a long α to β' transition time was unsatisfactory in the production of margarine because it increased the time before the product was firm enough to pack. Their determinations were made by observation under the microscope and in their paper photomicrographs are given in illustration.

Wilton (1965) also prepared colour cine photomicrographs of the polymorphic forms of crystals of fully hardened soybean oil under various conditions. She concluded that crystal form varied according to time and temperature factors.

Margarines sometimes develop a texture fault described as 'sandiness'. This was shown to be due to the formation of large β crystals, identified by melting point determinations under the microscope and confirmed by X-ray crystallography (Madsen, 1971). Blends of sunflower seed oil and hardened sunflower seed oil were particularly prone to this defect, whereas a blend based on soybean oil maintained a satisfactory texture (Figs 14a, b). Merker *et al.* (1958) similarly demonstrated with the microscope the formation of large crystals of the high-melting β polymorph in shortenings and margarines made with hydrogenated soybean or with hydrogenated cotton seed oil blends.

Van der Hoek (1970) proposed a microscopical procedure for predicting the development of structural defects in margarine during storage. A sample of the oil blend or of the solid phase separated from the blend is placed on a microscope slide, heated to 80°C and then brought into contact with a metal plate at the temperature of liquid nitrogen. A uniform appearance of very fine crystals is obtained; but storage at 33°C for two or three days produces marked recrystallization, with large needles or aggregates in some instances, which is invariably associated with structural changes in the margarine made from that blend. Differential thermal analysis indicated that a polymorphic change is involved.

c. Cocoa butter and chocolate. Cocoa butter is recognized as exhibiting six

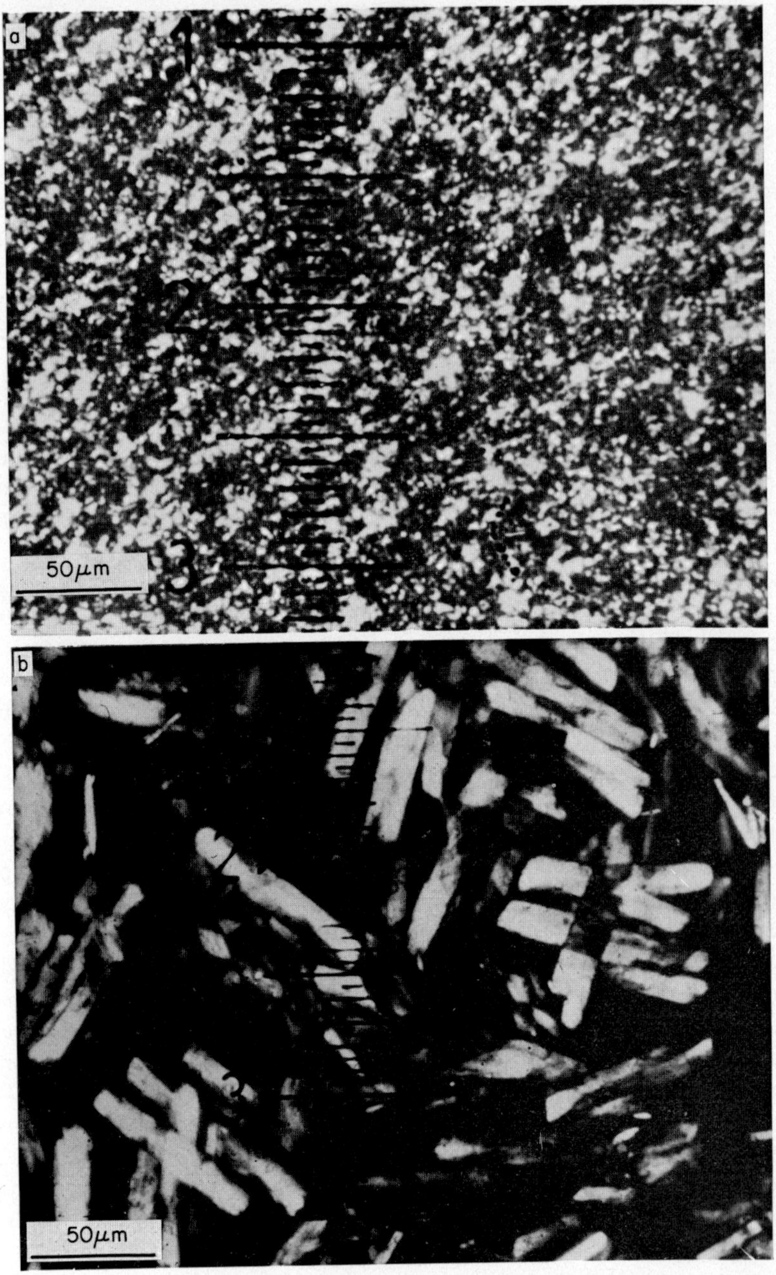

FIG. 14. Margarines after 4 weeks storage at 25°C. (a) Blend based on soybean oil; (b) blend based on sunflower oil. (Courtesy of J. Madsen, Grindsted.)

polymorphic forms (Wille and Lutton, 1966). Their different thermal stabilities have profound effects on the handling properties and shelf-life of products containing cocoa butter. Hence the structures of cocoa butter and chocolate have been studied by numerous authors. On storage many chocolate products may develop a dusty grey–white surface film known as 'bloom'. As early as 1921 Whymper had used the light microscope to show that the bloom consisted of minute fat crystals projecting from the surface of the product. It was quickly realized that if chocolate was subjected to a suitable thermal tempering regime the seed crystal produced was of the more stable β' or β type, and that the product was more resistant to bloom. The light microscope, with polarized light and a hot-stage, can be employed to study the size and type of crystals produced in cocoa butter and chocolate subjected to various tempering regimes. This approach has been applied by Kleinert (1961, 1965, 1970, 1971, 1972) and an interesting paper relating bloom to crystal types was presented by Drachenfels *et al.* (1962).

Chocolate consists of a continuous phase of cocoa butter, substantially crystalline at room temperature and comprising 30–35% of the whole, in which sugar and cocoa solids are suspended. Milk chocolate also contains milk fat and solids. More recently, electron microscopy has been applied to the study of cocoa butter and chocolate. Buchheim and Knoop (1971) used freeze-fracture replicas to study the gross morphology of milk chocolate and were able to identify the cocoa butter, sugar and milk proteins.

Jewell (1972) compared the appearance of stable cocoa butter and chocolate (Figs 15a, b) with chocolate that had developed bloom. The latter shows not only an outgrowth of large needle-shaped crystals above the surface (Fig. 16), but also abnormally large crystal developments throughout the interior (Fig. 16b).

In a later study of cocoa butter tempered by appropriate treatments, Jewell (1974) demonstrated that each of the six forms had a distinct morphology (Figs 17a–e). The similarity of Figs 16a and 17e is particularly striking. The forms were also characterized by X-ray crystallography, and, taken together, the work provides a clear demonstration that the desirable form of cocoa butter is the form V, a β form, and that the bloom defect is associated with a transformation into form VI, a second β form.

Chocolate, containing fatty centres such as nuts, which is particularly prone to bloom, was also studied and the bloom crystals were found to be generally similar to those on plain chocolate. Cruickshank and Jewell (1976) also used thin-sectioning to study the cocoa butter crystals and milk crumb particles in milk chocolate.

Products which contain mixtures of cocoa butter and certain other fats rich in lauric glycerides, such as hardened palm kernel oil (HPKO), are

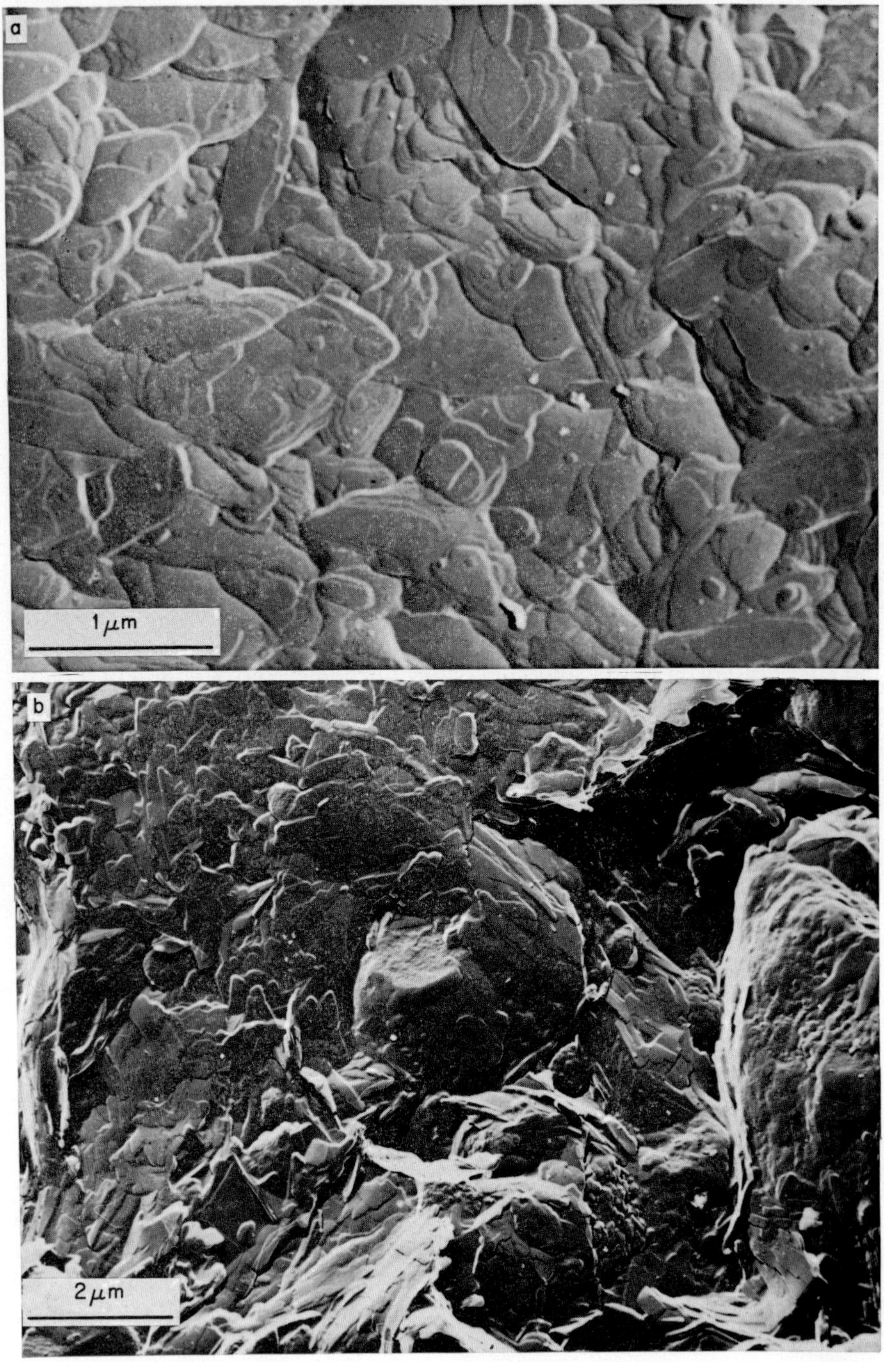

FIG. 15. Electron micrographs of (a) cocoa butter (stable form) and (b) chocolate (fat in the normal state).

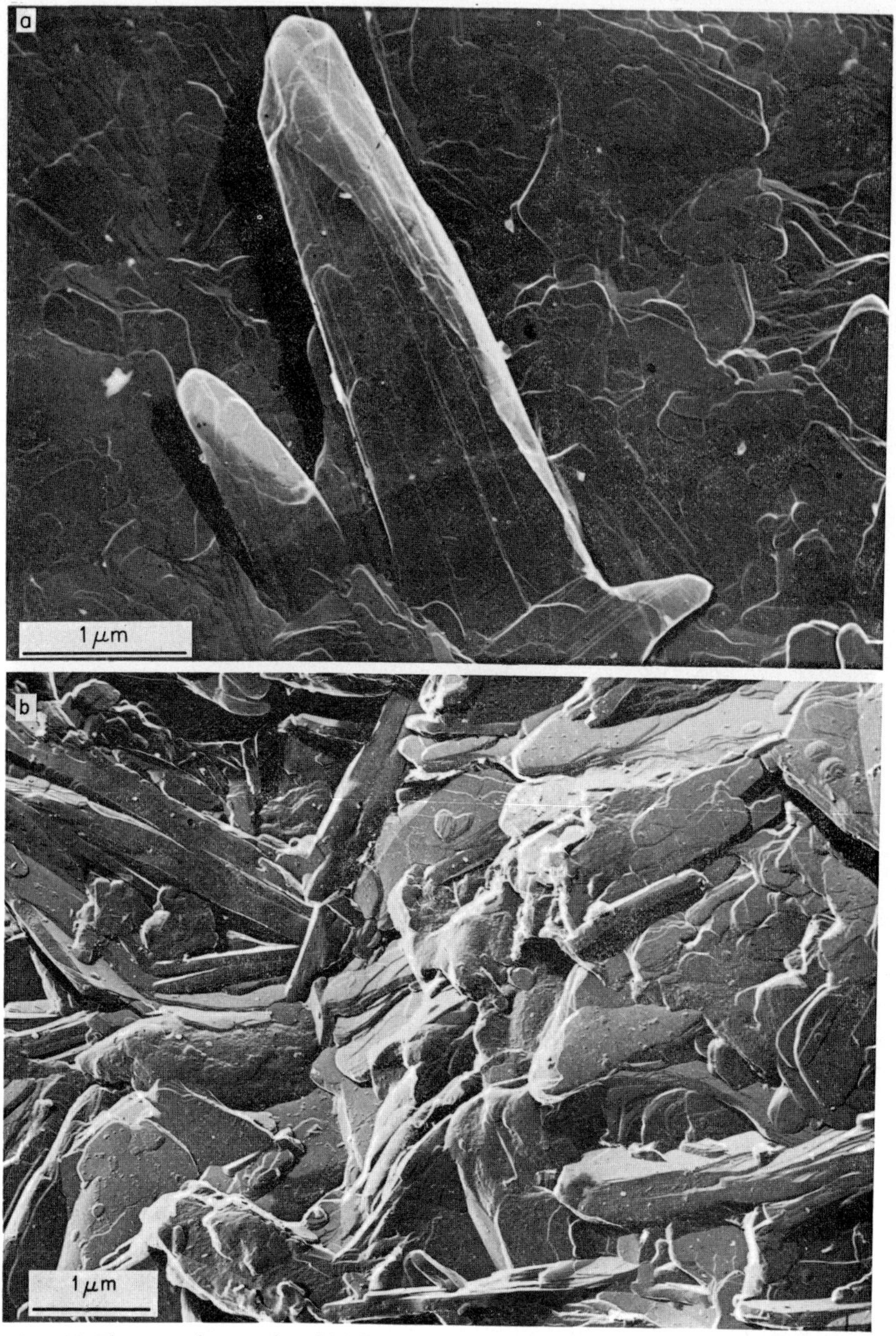

FIG. 16. Electron micrographs of (a) fat crystal on surface of bloomed plain chocolate and (b) fractured edge of bloomed plain chocolate, showing abnormally large crystals.

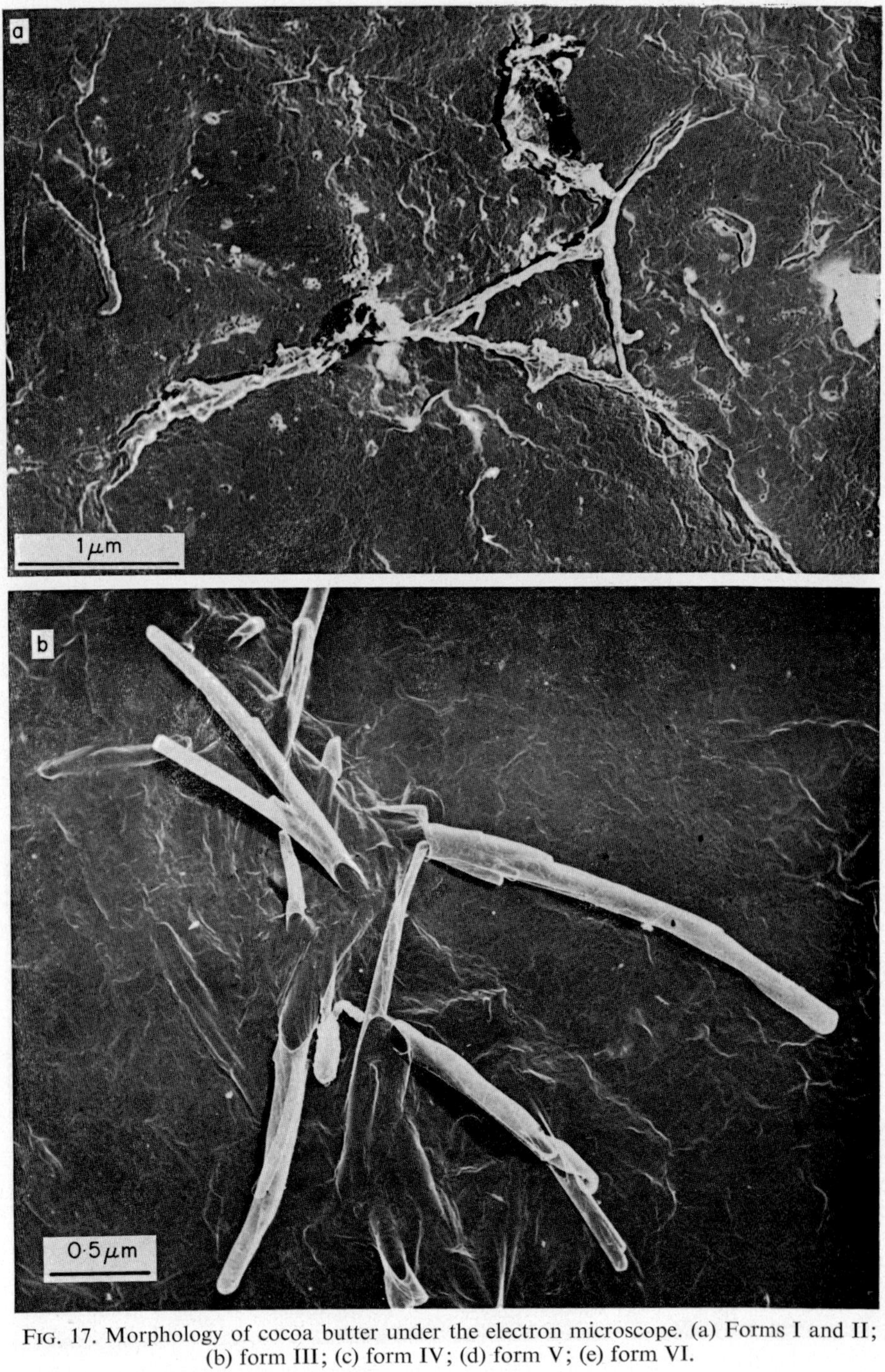

FIG. 17. Morphology of cocoa butter under the electron microscope. (a) Forms I and II; (b) form III; (c) form IV; (d) form V; (e) form VI.

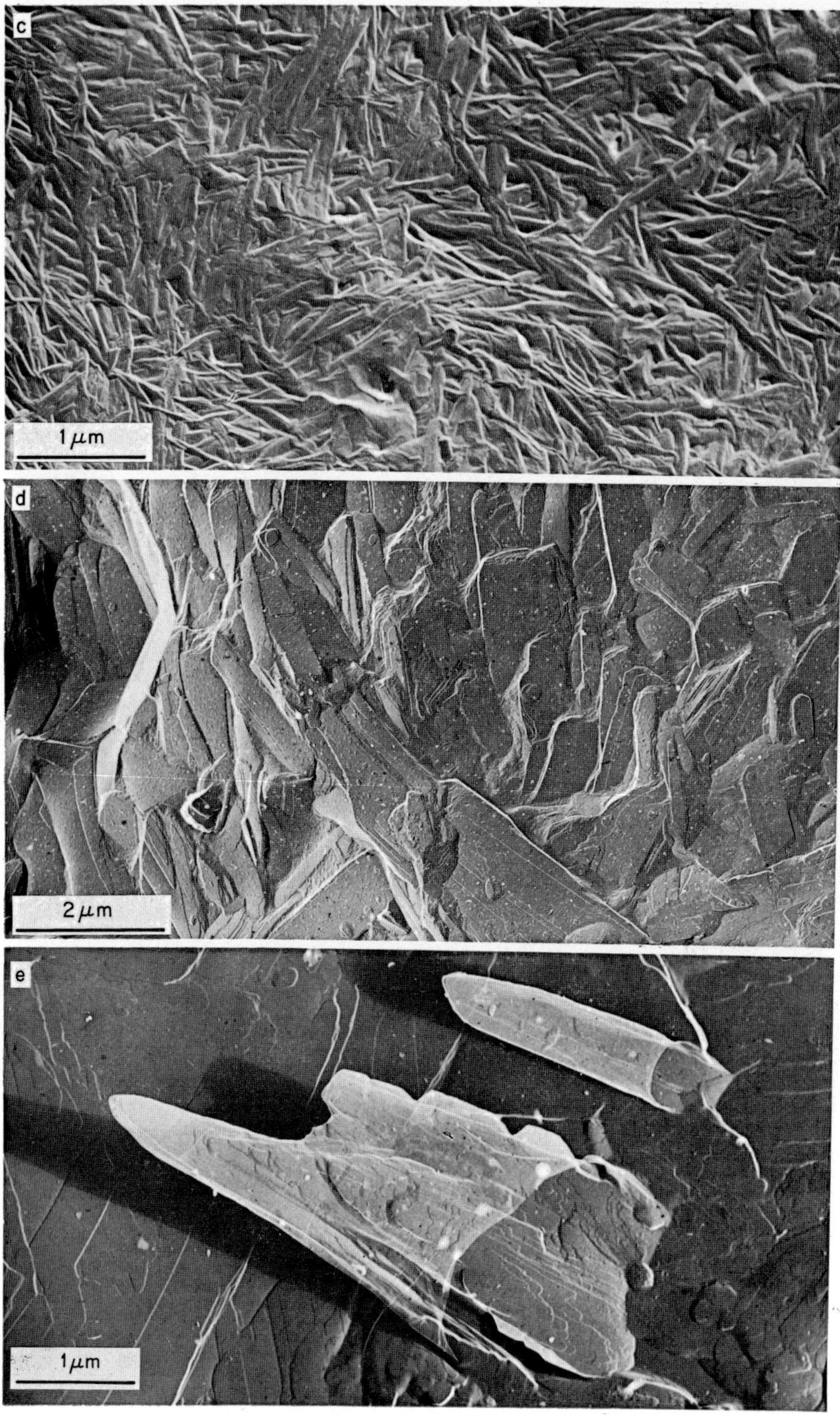
c
1 μm
d
2 μm
e
1 μm

FIG. 18. Electron micrograph of sugar crystal in chocolate, showing a coating of lecithin on the surface.

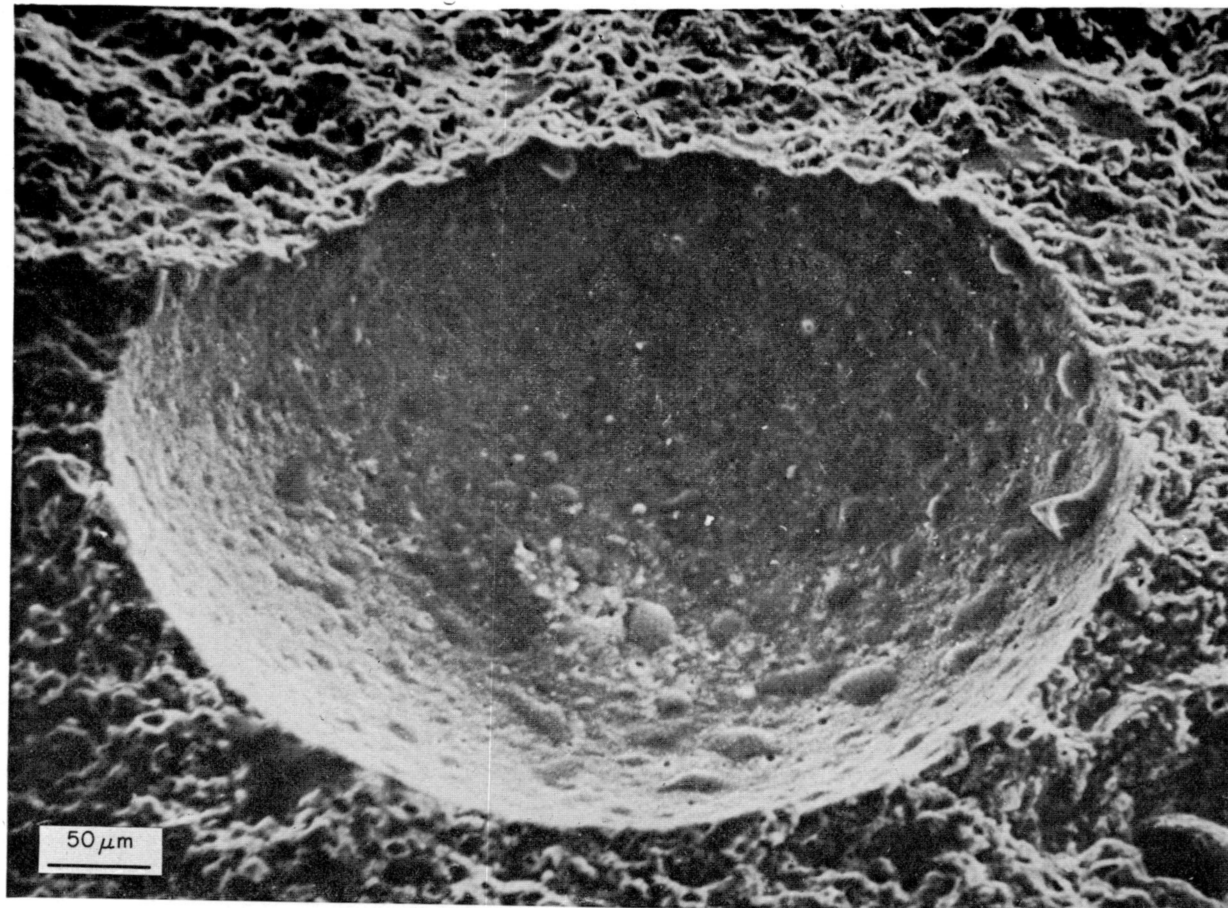

FIG. 19. Scanning electron micrograph of an air cell in chocolate.

usually very susceptible to bloom formation, and the structure of bloom on such products has been studied by Kawada (1974) using the electron microscope.

Other structural features of chocolate have also been elucidated by electron microscopy. The viscosity of molten chocolate is a critical parameter in many processes, for example when it is being applied to food products in an enrober. Control of viscosity is obtained by the addition of lecithin or other surface-active lipids in small amounts. The mode of action is illustrated in Fig. 18, showing a sugar crystal in chocolate with its surface coated with a thin layer of lecithin which has developed some crystallinity. Figure 19 shows an air cell present in chocolate, at low magnification under the scanning electron microscope. The contrasting orientation of fat crystals on the air surface and in the continuous phase presents an interesting analogy with Fig. 10.

C. Liquid Crystals and Mesophases

Monoglycerides and a number of other amphiphilic substances form a series of liquid crystalline mesophases with water depending on concentration, temperature and their chemical structure. These mesophases influence the stability of emulsions and are of great interest in relation to the membranes of biological cells. They have been described in a number of papers by Luzatti, Lutton, Larsson, Krog and others and the available information is effectively summarized by Krog and Lauridsen (1976).

The interrelationship of the various phases is most easily understood by reference to Fig. 20. For example, if we follow a vertical path upwards through Fig. 20a at 35% water, then on reaching about 60°C an apparently clear solution is obtained, consisting of the neat lamellar phase (Fig. 21a). On further heating this transforms into a viscous isotropic phase, after passing through a region of mixed phases (the shaded portion). If, on the other hand, the temperature is held near 60°C and the neat phase is diluted (i.e. a horizontal path to the right is followed), a dispersion phase is obtained. On cooling the dispersion phase (vertical path downwards in Fig. 20a), an α-gel phase is obtained which eventually transforms into β crystals and water. When a neat phase of low water content is heated, a fluid isotropic phase is obtained (top left-hand corner of Fig. 20a).

Under the microscope this phase presents a striking rod-like appearance when a slight pressure is put on the coverslip. The microscopic appearance of the phases described above is shown in Figs 21a–e. The underlying molecular arrangements have been investigated by X-ray crystallography. Thus the neat phase (Fig. 21a) (the name derives from the Greek work for a thread) consists of bilayers of monoglyceride alternating with layers of water

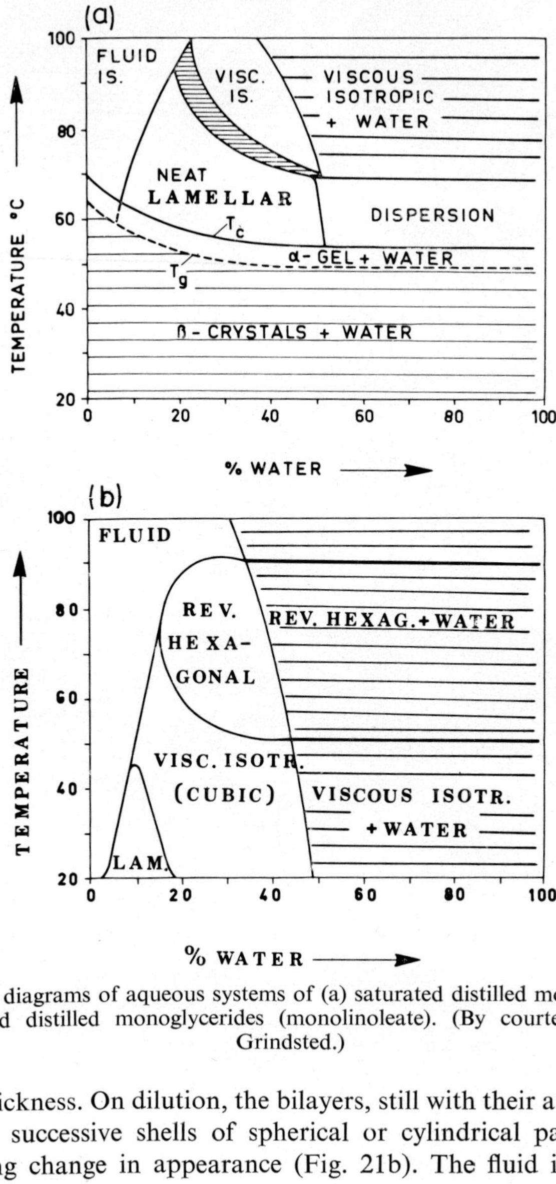

FIG. 20. Phase diagrams of aqueous systems of (a) saturated distilled monoglycerides and (b) unsaturated distilled monoglycerides (monolinoleate). (By courtesy of N. Krog, Grindsted.)

of defined thickness. On dilution, the bilayers, still with their associated water layers, form successive shells of spherical or cylindrical particles, with a corresponding change in appearance (Fig. 21b). The fluid isotropic phase consists of disc-shaped lamellae formed from the lipid–water layers of the neat phase. These are shown in Fig. 21e. Larsson has suggested that when crystallization is allowed to occur by cooling the dispersion phase, β crystals with unusual properties are obtained; this is because the crystals will have a

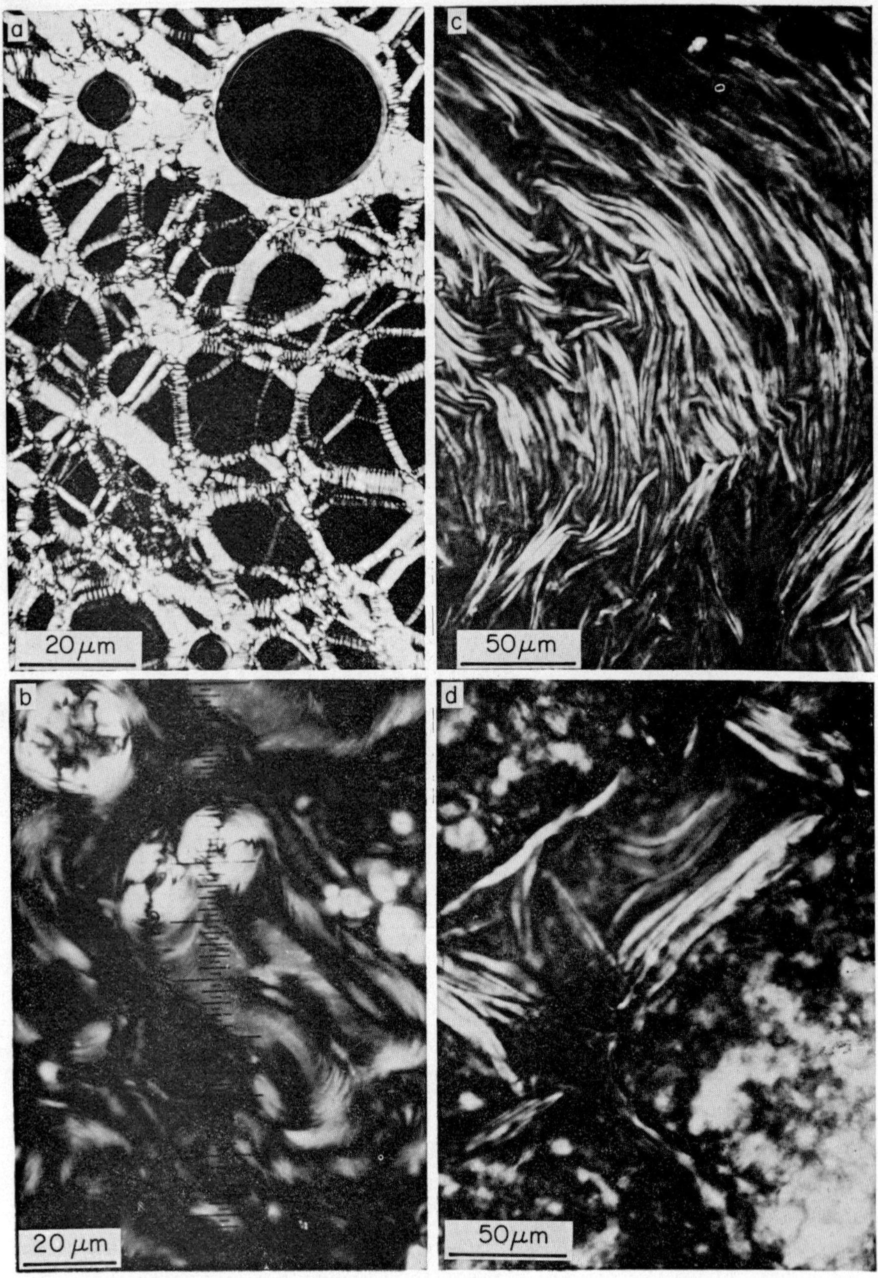
a
20 μm
c
50 μm
b
20 μm
d
50 μm

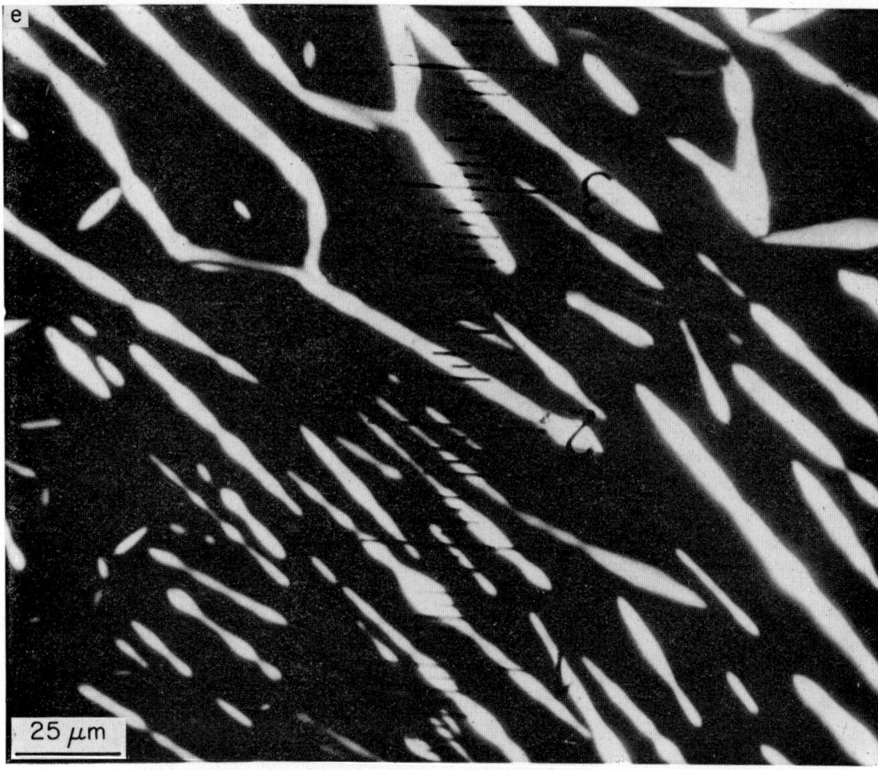

FIG. 21. Emulsifier–water mesophases. (a) Lamellar neat phase of GMS–water; (b) dispersion phase of GMS–water; (c) α-gel phase of GMS–water; (d) α-gel phase of GMS–water partly transforming into β-crystals (see right-hand corner); (e) batonnets of the fluid isotropic phase of polyglycerol ester–water at 60°C. GMS is glycerol mono stearate. (By courtesy of N. Krog, Grindsted, and Marcel Dekker, New York.)

---

surface of polar groups, in contrast with the normal hydrophobic surface. Such special β crystals have high water-binding capacity. The crystallization process is illustrated in Fig. 22 which shows a β-phase crystal with spherical aggregates of the dispersion in the process of adsorbing to the crystal surface. Figure 3 shows a scanning electron micrograph of such β crystals which were formed after freeze-drying the dispersion phase from 90% pure commercial monoglyceride obtained by molecular distillation. One technical advantage of the β crystals is their exceptionally high specific surface of 20 $m^2 g^{-1}$, compared with only 0·03 $m^2 g^{-1}$ of the normal commercial product obtained from the melt (Krog and Lauridsen, 1976).

Molecularly distilled monoglyceride preparations containing more than

90% monoglyceride are widely used in the food industry and their phase behaviour is of great practical importance. Emulsifying activity is extremely high in the lamellar form which is therefore useful in cake systems. A stabilized form of the α-gel phase is commercially available. The β crystals have a high degree of functionality. On the other hand, if during preparation of the neat phase the temperature is allowed to rise too high, the resultant viscous isotropic form can cause severe handling problems on a factory scale. Another phase occurring in some systems is the reversed middle (hexagonal II) phase in which cylinders of lipid molecules, containing a core of water, are arranged in a hexagonal pattern. The appearance of this phase is illustrated

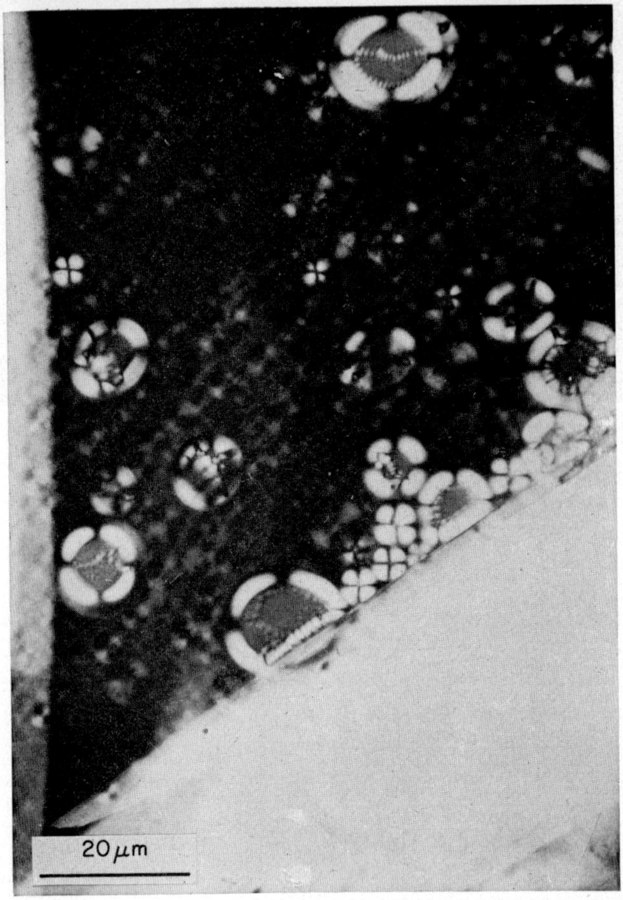

FIG. 22. Dispersion of monocaprin in water adsorbing to a β crystal on the lower right-hand side. (By courtesy of N. Krog, Grindsted, and Marcel Dekker, New York.)

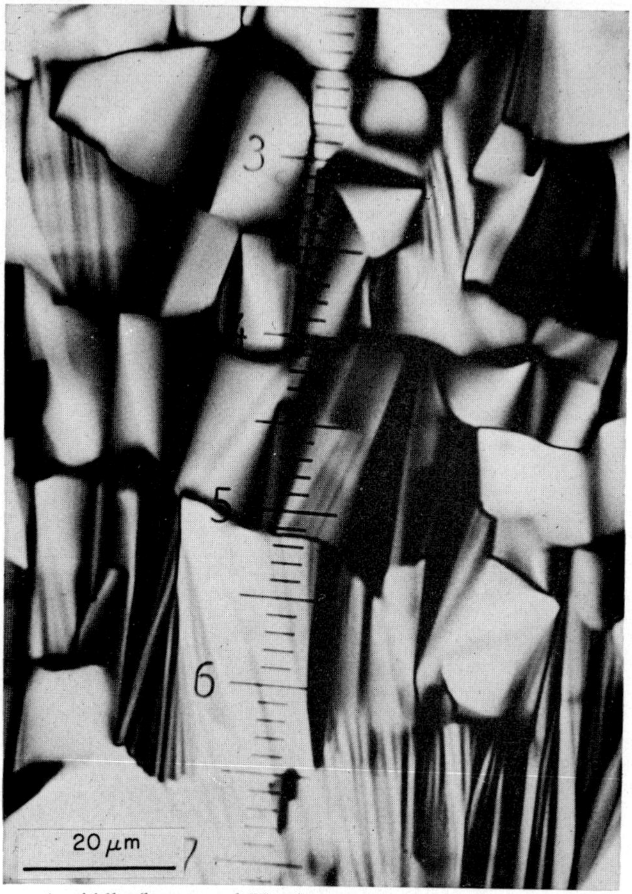

FIG. 23. Reversed middle (hexagonal II) phase of mono-olein–water. (By courtesy of N. Krog, Grindsted, and Marcel Dekker, New York.)

in Fig. 23. Since the hydrocarbon chains of this cylindrical formation point outwards, the reversed middle phase cannot be diluted with water.

The greater magnification available with the electron microscope is of particular value in studying, at the molecular level, associations of the type presented by the various lipid mesophases. Tinker and Pinteric (1971) examined hydrated phosphatidyl ethanolamine from a natural source and were able to demonstrate lamellar, hexagonal I (hydrophilic) and hexagonal II (lipophilic) mesophases. Their preparative technique was negative staining with phosphotungstic acid. The dimensions of the structures found correlated well with those expected from geometrical projections of model structures (Figs 24a–c).

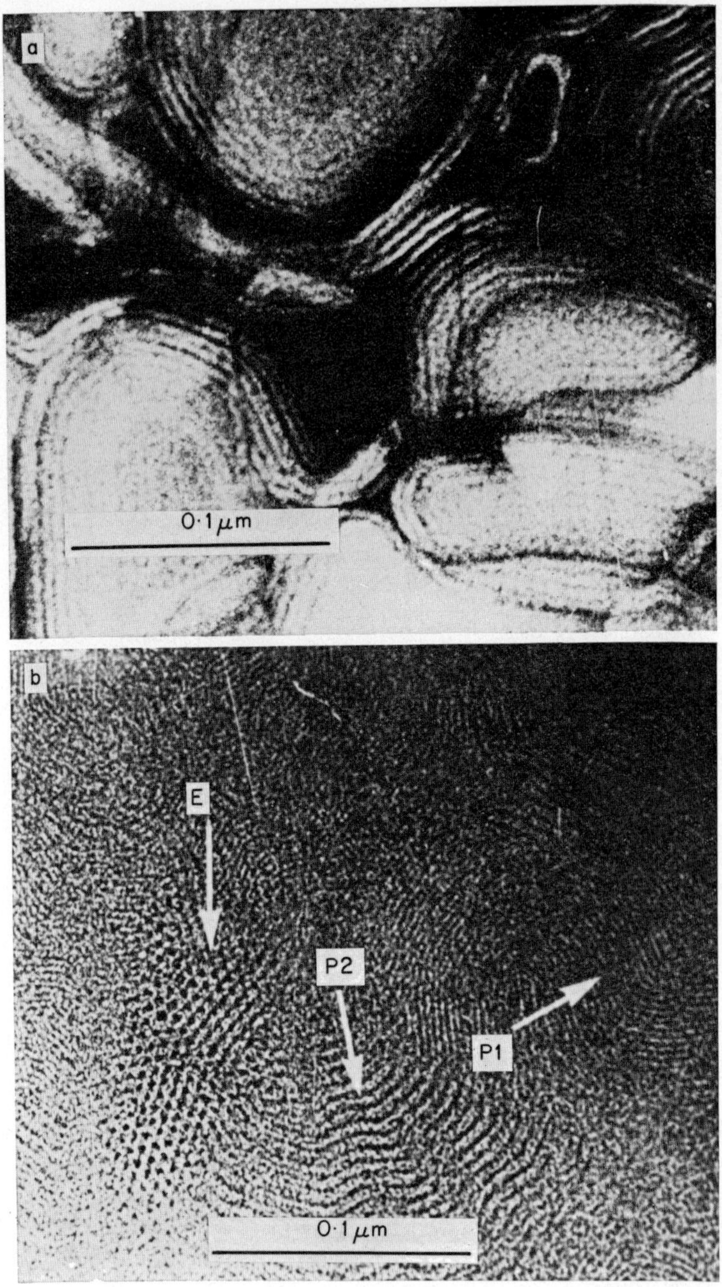
a
0·1 μm
b
E
P2
P1
0·1 μm

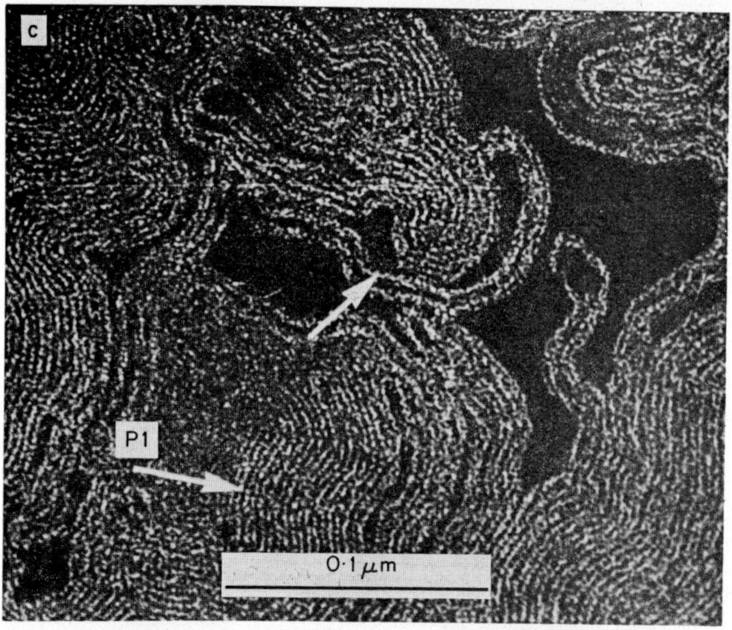

FIG. 24. Phosphatidyl ethanolamine–water mesophases. (a) Lamellar phase. (b) Hexagonal II, reverse middle, phase; arrow E indicates end-on view of lipid cylinders, arrows P1 and P2 indicate projections of side view of cylinder arrays placed at different angles. (c) Hexagonal I phases; arrow P1 indicates cylinders in regular hexagonal array, unlettered arrow indicates unordered cylinders. (Reprinted with permission of D. O. Tinker. Copyright by the American Chemical Society.)

Deamer *et al.* (1970) used freeze-etching to demonstrate similar structures from the same three mesophases formed with a variety of pure polar lipids.

Freeze-etching was also applied by Lundberg (1973) to the mono-olein–water system at 25°C. He was able to demonstrate a lamellar phase at water contents of 4–22%, and a cubic (viscous isotropic) phase at water contents of 25–41%. Figure 25a shows the somewhat uneven fracture face given by the viscous isotropic phase. The spherical nature of the structural unit can be best seen at the arrow, and the cubic packing is evident in Fig. 25b, an increased magnification of a similar surface.

D. Crystallinity in Emulsions

The significance of liquid crystalline phases in emulsions is demonstrated in Fig. 26, showing soybean oil in water emulsion stabilized by the polar emulsifier sodium stearoyl 2-lactylate, and photographed in polarized light. There is a clear indication of crystalline surface layers.

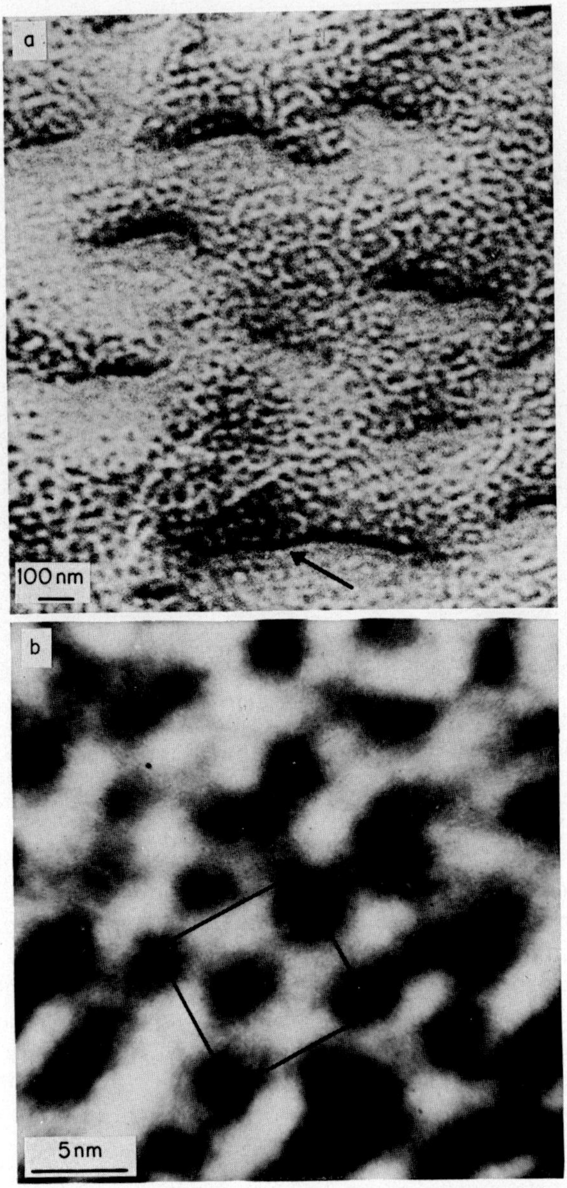

FIG. 25. Viscous isotropic phase of mono-olein–water, 35 : 65 at 25°C. (a) The spherical nature of the unit is best seen at the arrow; (b) at higher magnification the cubic packing becomes apparent. (By courtesy of B. Lundberg and the North Holland Publishing Co.)

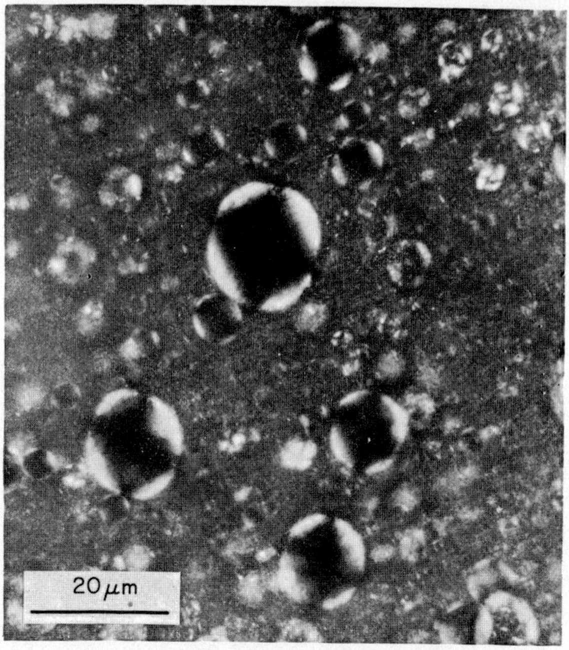

FIG. 26. Emulsion of soybean-sodium stearoyl-2-lactylate–water, viewed by polarized light, and showing liquid crystals at the surface of the droplets. (By courtesy of N. Krog, Grindsted, and Marcel Dekker, New York.)

Temperature-induced phase transitions in relation to the stability of monoglyceride–vegetable oil–water emulsions are discussed by Wilton and Friberg (1971). The paper contains exceptionally beautiful colour photomicrographs demonstrating the neat, viscous isotropic and reverse middle (hexagonal II) phases.

Crystallization of the triglycerides in an emulsion requires a much higher degree of supercooling than in the bulk phase. Skoda and Van den Tempel (1963) measured the droplet number and size distribution of pure triglyceride emulsions microscopically and related them to the crystallization rate at various temperatures measured by dilatometer. They showed that small droplets had to be supercooled to a low temperature for crystallization because only homogenous nucleation was operative. Larger droplets, on the other hand, contained a sufficient number of foreign particles to enable heterogeneous nucleation to take place at a higher temperature. A number of emulsifiers promoted nucleation, particularly if they contained hydrocarbon chains of similar length to those of the triglycerides being examined.

These observations were extended by Phipps (1964) using an elegant

a

b

c

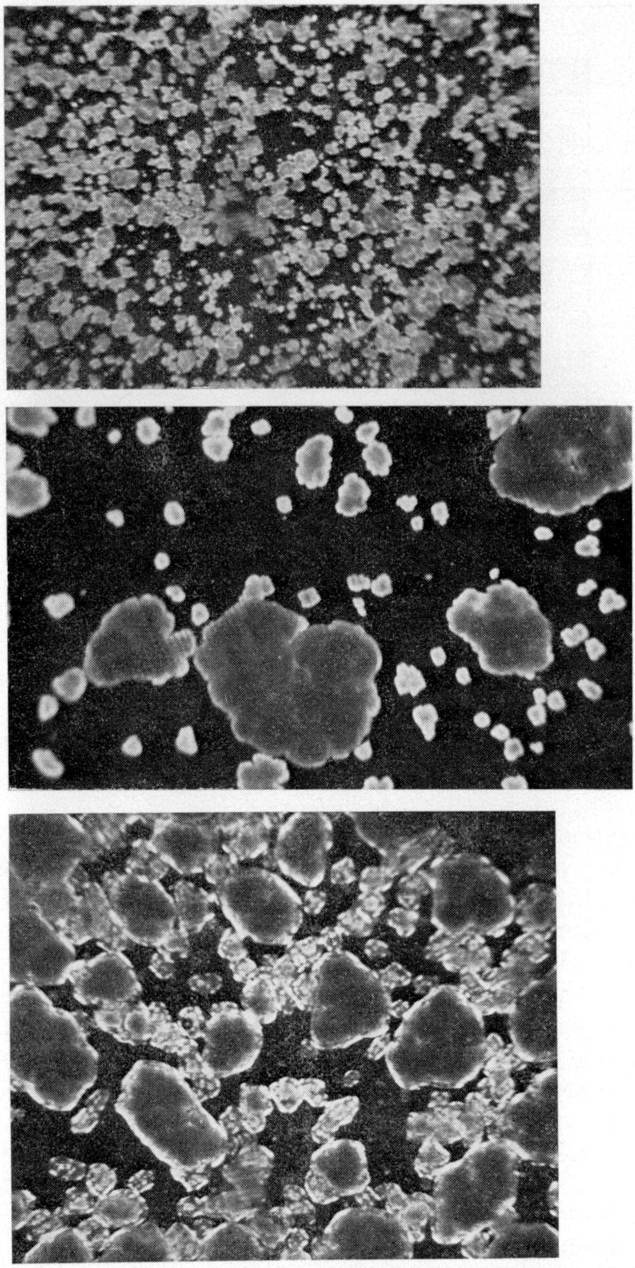

FIG. 27. Crystals for fractionation. (a) β Crystals of palm oil, not filterable, $\times 15$; (b) β' crystals of palm oil, suitable for filtration, $\times 15$; (c) crystals of palm olein, $\times 20$; (d) crystals of beef fat, ideal for filtration, $\times 20$; (e) stearin crystals from hydrogenated soybean oil, rather fragile, filterable with low vacuum, $\times 20$; (f) stearin crystals of hydrogenated fish oil, rather fragile, $\times 20$. (By courtesy of F. Tirtiaux.)

d

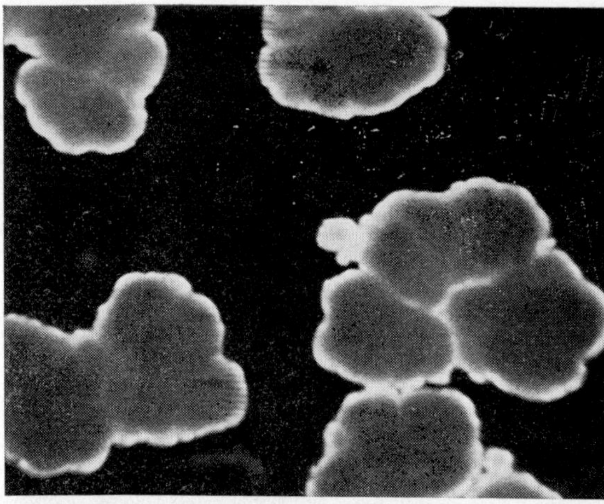

e

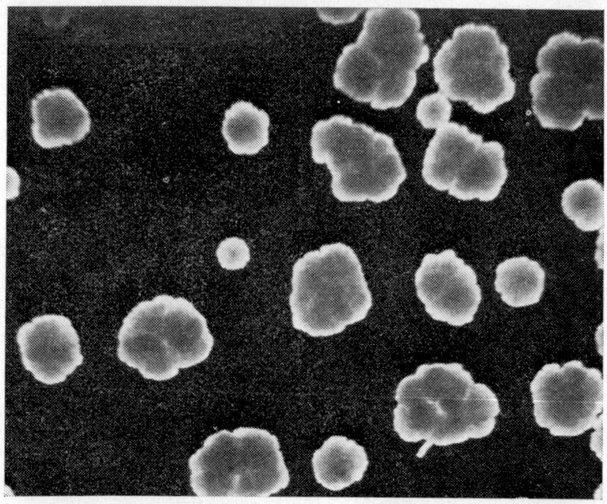

f

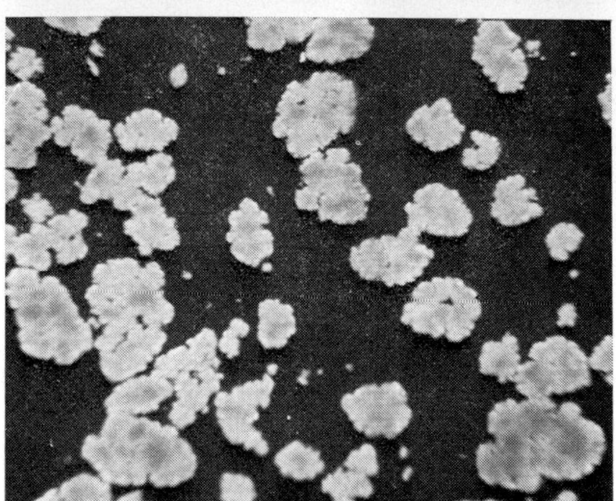

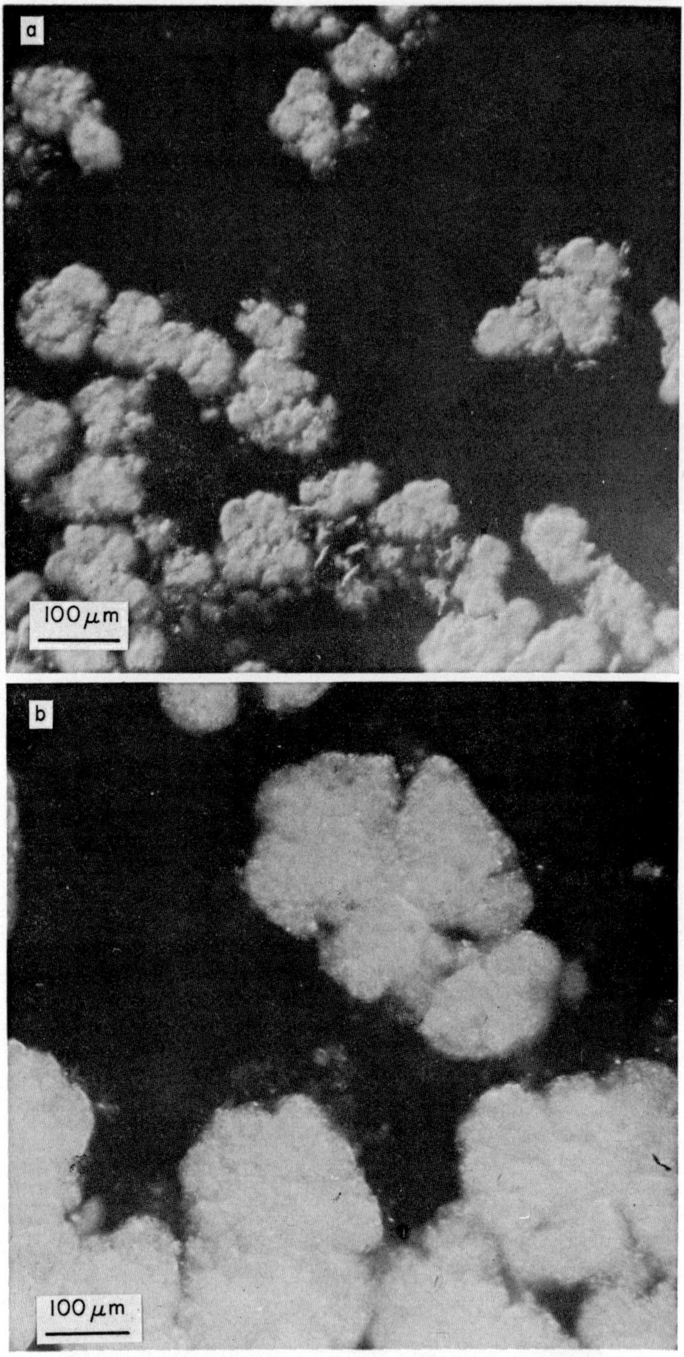
a
100 μm
b
100 μm

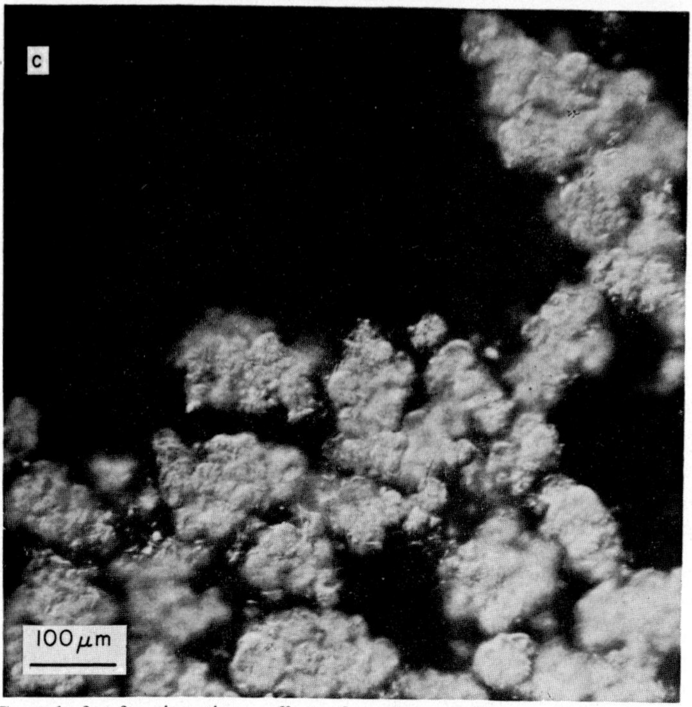

FIG. 28. Crystals for fractionation—effect of cooling conditions. (a) Cooling interrupted at 32°C; (b) continuous cooling to 28°C; (c) cooling interrupted at 30°C. (By courtesy of A. M. Taylor, van den Berg and Jurgens.)

photomicrographic technique. Under partly crossed polars, he photographed emulsions undergoing cooling at 1°C intervals of temperature. He then measured the diameter of individual particles on the photograph in which they last appeared liquid. By plotting the lowest freezing point for each size group against particle size, he quantitatively demonstrated the diameter for each triglyceride studied beneath which homogenous nucleation can be observed.

E. Fractionation of Fats

Fractionation consists, in its simplest form, of the separation of a fat into its more solid and its more liquid glyceride components, respectively called stearin and olein. The process takes place essentially in two steps: (i) the formation of crystals and (ii) the separation of the liquid portion from the crystals. It may be carried out with or without the use of a solvent. Especially in the latter case, and when the separation is by filtration, the size and shape of the crystals is critical for a successful separation step.

Microscopic observation of the crystallizing process conducted under various controlled conditions has been used to determine the optimum conditions. In model studies on palm oil using a pilot plant crystallizer, Deroanne (1976) has shown that large, spherical, crystal aggregates of good mechanical strength are required. These are obtained by ensuring that crystallinity is in the β' form. If residual nuclei are not removed (by heating to 70°C) before the crystallization is started, or if a high temperature is held during crystallization, then crystals of the intermediate or the β form are obtained as flat needles or platelets which do not aggregate, but pack tightly to block the filters. When palm olein is crystallized for a second fractionation, soft aggregates of a vitreous appearance are obtained (Figs 27a–c).

Tirtiaux (1976) extends the discussion to other fats. In the case of beef fat, crystal aggregates of good strength and size are easily obtained. A diameter of 0·3–0·6 mm is preferred. Partially hydrogenated soybean and fish oils, on the other hand, produce rather fragile aggregates which require very careful filtration (Figs 27d–f).

Taylor (1976) demonstrated microscopically the connection between cooling rate, aggregate size and subsequent filtration efficiency. He found that continuous cooling from 70°C to 28°C produced large, rather loose aggregates (Fig. 28b). Although a high filtration rate was obtained, the filter cake retained a high proportion of olein. Somewhat smaller, firmer aggregates were obtained by interrupting the cooling at 32°–33°C (Fig. 28a) or at 30°C (Fig. 28c) for about 30 minutes. Filtration was slower but the separation was more effective.

Beyerlein and Voss (1973) investigated the crystallization of butter fat, with a view to determining the conditions for obtaining large spherulites with good filtration characteristics. Careful microscopic observation of aggregates and fragments indicated that the individual crystal consisted of a bundle of flat strips, each somewhat twisted about its long axis. The growth rate of the crystals at 26°–30°C was 12–35 $\mu m\ h^{-1}$ in length.

IV. FUTURE PROSPECTS

There is no doubt that the trend in the food industry towards larger, more highly mechanized and eventually automated processes will continue in the Western world and will spread to developing countries. This trend inevitably requires a tighter specification of ingredients and operating conditions. Such standardization is most successful when it is based on an intimate knowledge of the physical properties and the structures of ingredients and end-products. The present chapter contains some recent examples where optical and electron microscopy, sometimes backed by X-ray diffraction at the molecular level, have provided a sound rationale for the behaviour of fats.

The authors believe that many more problems await solution and hope that the reader will be tempted to apply some of these powerful techniques in his own work programme. Two areas in particular where one may expect progress in the structural study of fats are the use of the scanning electron microscope, with its capacity to provide vivid three-dimensional impressions, and the correlation of quantitative measurements of microscopical features with other characteristics of the material being studied.

ACKNOWLEDGEMENTS

It is more true of the microscopist than of most that seeing is believing. The authors are therefore most grateful to all the scientists who have made photographic material available, and wish to express their thanks particularly to Mr N. Krog and his colleagues of Grinstedvoerket Aarhus, Denmark, and to Dr A. W. Holmes, Director of the Food Research Association, Leatherhead, for their co-operation.

REFERENCES

Bailey, A. E. (1964). 'Bailey's Industrial Oil and Fat Products' (D. Swern, ed.), p. 272. Interscience Publishers, New York.

Bauer, H. (1972). *J. Dairy Sci.* **55**, 1375.

Bell, A. V., Berger, K. G., Russo, J. V., White, G. W. and Weathers, T. L. (1975). *J. Fd Technol.* **10**, 147.

Berger, K. G. (1970). *Fd Mf.* **45**, 60.

Berger, K. G. (1976). *In* 'Lipids' (R. Paoletti, ed.), Vol. 2, p. 323. Raven Press, New York.

Beyerlein, U. and Voss, E. (1973). *Kieler Milchw. ForschBer.* **25**, 49.

Buchheim, W. and Knoop, E. (1971). *Susswaren*, **24**, 1026.

Butcher, C. H. (1934a). *Food*, **3**, 176.

Butcher, C. H. (1934b). *Ind. Chemist*, **10**, 495.

Chapman, D. (1962). *Chem. Rev.* **62**, 433.

Chapman, D. (1965). 'The Structure of Lipids.' Methuen London.

Cruickshank, D. A. (1976). British Food Manufacturing Industries Research Association Tech. Circ. 611.

Cruickshank, D. A. and Jewell, G. G. (1976). British Food Manufacturing Industries Research Association Res. Rep. 254.

Deamer, D. W., Leonard, R., Tardieu, A. and Branton, D. (1970). *Biochim. Biophys. Acta*, **219**, 47.

De Castro, R. and Vazquez, R. (1971). *Grasas y Aceites*, **22**, 463.

Deroanne, C. (1976). *Proc. Belg. Filt. Soc., Brussels*, p. 173.

Drachenfels, H. J., Kleinert, J. and Hansen, E. (1962). *Int. Choc. Rev.* **17**, 409.
Duffy, P. J. (1852). *J. Chem. Soc.* **5**, 197.
Haighton, A. J. (1963). *Fette Seifen AnstrMittel*, **65**, 179.
Haighton, A. J. (1965). *J. Am. Oil Chem. Soc.* **42**, 27.
Hellman, N. N., Zobel, H. F. and Senti, F. R. (1955). *J. Am. Oil Chem. Soc.* **32**, 110.
Herb, S. F., Audsley, M. C. and Riemenschneider, R. W. (1956). *J. Am. Oil Chem. Soc.* **33**, 189.
Hoerr, C. W. (1955). *J. Am. Oil Chem. Soc.* **32**, 372, 659.
Hoerr, C. W. (1960). *J. Am. Oil Chem. Soc.* **37**, 539.
Jewell, G. G. (1972). *Int. Choc. Rev.* **27**, 161.
Jewell, G. G. (1974). British Food Manufacturing Industries Research Association Res. Rep. 202.
Jewell, G. G. and Meara, M. L. (1970). *J. Am. Oil Chem. Soc.* **47**, 235.
Joyner, N. T. (1953). *J. Am. Oil Chem. Soc.* **30**, 526.
Kawada, T. (1974). Proc. A.O.C.S. Philadelphia Convention No. 178.
Kleinert, J. (1961). *Int. Choc. Rev.* **16**, 201.
Kleinert, J. (1965). *Int. Choc. Rev.* **20**, 86.
Kleinert, J. (1970a). *Int. Choc. Rev.* **25**, 115.
Kleinert, J. (1970b). *Int. Choc. Rev.* **25**, 356.
Kleinert, J. (1971). *Int. Choc. Rev.* **26**, 2.
Kleinert, J. (1972). *Int. Choc. Rev.* **27**, 183.
Krog, N. and Lauridsen, J. B. (1976). *In* 'Food Emulsions' (S. Friberg, ed.), p. 67. Marcel Dekker, New York.
Krog, N. and Vrang, C. (1964). *Fette Seifen AnstrMittel*, **66**, 279.
Landmann, W. and Feuge, R. O. (1956). *J. Am. Oil Chem. Soc.* **33**, 308.
Lewis, D. F. (1975). British Food Manufacturing Industries Research Association Res. Rep. 238.
Lundberg, B. (1973). *Chem. Phys. Lipids*, **11**, 219.
Madsen, J. (1971). Proc. Int. Symp. Deterioration of Lipids, Gdansk, p. 147.
Madsen, J. and Als, G. (1971). *Fette Seifen AnstrMittel* **73**, 403.
Malkin, T. (1934). *J. Chem. Soc.* **137**, 666.
Meara, M. L., Evans, G. G., Jewell, G. G. and Davies, D. P. (1974). British Food Manufacturing Industries Research Association Res. Rep. 204.
Merker, D. R., Brown, I. C. and Wiedermann, L. H. (1958). *J. Am. Oil Chem. Soc.* **35**, 130.
Mouwen, D. W. C. (1974). Proc. A.O.C.S. Philadelphia Convention No. 181.
Pearse, A. G. E. (1968). 'Histochemistry, Theoretical and Applied', Vol. 1, p. 410. Churchill Livingstone, London.
Phipps, L. W. (1964). *Trans. Faraday Soc.* **60**, 1863.
Pollitt, R. J. M. (1974). *Dechema-Monogr.* **77**, 363.
Poot, C., Dijkshoorn, W., Haighton, A. J. and Verburg, C. C. (1975). *J. Am. Oil Chem. Soc.* **52**, 69.
Reigel, G. W. C. and McMichael, C. E. (1966). *J. Am. Oil Chem. Soc.* **43**, 687.
Sambuc, E. and Naudet, M. (1965). *Rev. Fr. Cps Gras*, **12**, 239.

Schmid, H. H. D., Mangold, H. K. and Lundberg, W. O. (1965). *J. Am. Oil Chem. Soc.* **42**, 372.
Shepherd, I. S. and Yoell, R. W. (1976). *In* 'Food Emulsions' (S. Friberg, ed.), p. 215. Marcel Dekker, New York.
Skoda, W. and Van den Tempel, M. (1963). *J. Colloid Sci.* **18**, 568.
Stern, C. and Cmolik, J. (1976). *J. Am. Oil Chem. Soc.* **53**, 644.
Sutton, R. W., Barraclough, A., Mallinder, R. and Hitchin, D. (1940). *Analyst*, **65**, 623.
Taylor, A. M. (1976). *Oléagineaux*, **31**, 73.
Tinker, D. O. and Pinteric, L. (1971). *Biochemistry, N.Y.* **10**, 860.
Tirtiaux, F. (1976). *Oléagineaux*, **31**, 283.
Van den Tempel, M. (1961). *J. Colloid Sci.* **16**, 284.
Van den Tempel, M. (1968). *S.C.I. Monogr.* **32**, 22.
Van der Hoek, W. (1970). *Fette Seifen AnstrMittel*, **72**, 898.
Whittam, J. H. and Rosano, H. L. (1974). *Rev. Fr. Cps Gras*, **21**, 605.
Whittam, J. H. and Rosano, H. L. (1975). *J. Am. Oil Chem. Soc.* **52**, 128.
Whymper, R. (1921). 'Cocoa and Chocolate, Their Chemistry and Manufacture.' J. A. Churchill, London.
Wiedermann, L. H., Weiss, J. W., Jackson, G. A. and Mattil, K. F. (1961). *J. Am. Oil Chem. Soc.* **38**, 389.
Wille, R. L. and Lutton, E. S. (1966). *J. Am. Oil Chem. Soc.* **43**, 491.
Williams, K. A. (1966). 'Oils, Fats and Fatty Foods.' J. A. Churchill, London.
Wilton, I. (1965). Fourth Scandinavian Symposium of Fats and Oils. *In* 'Fat and Oil Chemistry' (1968). Gordon and Breach, New York.
Wilton, I. and Friberg, S. (1971). *J. Am. Oil Chem. Soc.* **48**, 774.
Wilton, I. and Wode, G. (1963). *J. Am. Oil Chem. Soc.* **40**, 707.
Wortmann, A. (1965). *Fette Seifen AnstrMittel*, **67**, 279.

13. Ice Cream

K. G. BERGER and G. W. WHITE

Lyons Central Laboratories, West Kensington, London, U.K.

I. INTRODUCTION

Historically, interest in the microscopical structure of ice cream grew out of early work on the prevention of the sandy texture which arises from the presence of lactose crystals. The light microscope, and later the electron microscope, have proved to be powerful tools for the investigation of many aspects of ice cream, particularly when used in conjunction with other chemical and physical methods of analysis.

This chapter reviews the literature and presents some previously unpublished data on the microscopy of ice cream, describing where it may be of interest some of the methods of specimen preparation.

We can define ice cream as a frozen foam containing, by volume, 50–55% air in the form of air cells, and by weight, 6–12% fat, 7·5–11·5% milk solids-not-fat (MSNF), 13–18% sugars, small amounts of emulsifier, stabilizer and other ingredients, and approximately 64% water, mainly as ice crystals. One of the achievements of microscopists is to tell us how these ingredients are dispersed and arranged in the very complex structure of ice cream.

II. THE MICROSCOPY OF PROCESSING

A. Processing Steps

1. General description

A résumé of the normal processing sequence and the conditions employed for ice cream manufacture is as follows.

(i) Mixing of all ingredients by high-speed stirrer at 50–60°C. A coarse emulsion (about 15 μm in diameter) is obtained.
(ii) Pasteurizing, typically at 80°C for 20 seconds.
(iii) Homogenizing at 176 kg cm^{-2} and 35 kg cm^{-2}.
(iv) Rapid cooling to 4°C.
(v) Ageing at 4°C for 3 hours or more.
(vi) Freezing and aerating in a scraped-surface heat exchanger to a temperature of -5°C to -8°C.
(vii) Hardening in a blast freezer or plate hardener to approximately -35°C.
(viii) Storage at -30°C.

Microscopy has been used extensively for the study of homogenizing and ageing effects and of the final structure after freezing.

2. Homogenizing

The light microscope has been used in studies of homogenization to (i) ensure that fat globules in mixes are sufficiently small to produce a stable emulsion, (ii) compare the efficiency of different homogenizers and (iii) study the effect of operating conditions in the homogenizer.

It has been shown elsewhere (Berger *et al.*, 1972) that the true distribution of fat globules in ice cream mix can be determined only by electron microscopy, as many globules are below the limit of resolution of the light microscope.

Nevertheless, the light microscope can be used to measure the coarse fraction of the distribution. Farrall *et al.* (1941) defined a homogenization index, H, which, for a given volume of mix, is the volume of fat present in globules >2 μm in diameter, expressed as an equivalent number of 2 μm fat globules, or

$$H = \frac{\sum_{2}^{\infty} \frac{\pi}{6} d_x^3 n_x}{\frac{\pi}{6} 2^3} = \tfrac{1}{8} \sum_{2}^{\infty} d_x^3 n_x, \qquad (1)$$

where n_x is the number of fat globules, of diameter d_x, in a given volume of undiluted mix.

The homogenization index is proportional not only to the percentage of fat >2 μm, but also to the fat content of the mix.

TABLE 1. Homogenization indices of various mixes

Mix	Type of fat[a]	Fat content (%)	Homogenizing pressures (p.s.i.)	H
Vanilla 1	—	8·40	—	31
2	100% PKO	9·20	—	49
3	100% PKO	11·16	—	111
4	—	11·25	—	122
5	100% PKO	11·32	2500 500	50
6	50% PKO 50% PO	11·50	1500 500	104
7	80% PKO 20% PO	11·70	—	16
Vanilla after freezing, hardening, thawing	—	11·10	—	470
Vanilla after freezing, hardening, thawing	—	11·90	—	537

[a]PKO = palm kernel oil; PO = palm oil.

We have measured H for a number of ice cream mixes, using the equivalent of 50 000 μm^3 of undiluted mix in each case. The results are shown in Table 1.

If all the fat globules are 2 μm or less in diameter, H will be zero, and for other mixes the nearer H is to zero the more effective has been the homogenization.

Although the electron microscope gives the most detailed information about the fat globule size distribution produced by homogenization, many workers have used spectro-turbidimetry to measure the volume : surface diameter (d_{vs}) in homogenization studies. In an important series of papers on the homogenization process (Goulden and Phipps, 1962, 1964; Phipps, 1971, 1974a, b, 1975) it has been shown that the mean diameter of the fat globules produced by a homogenizer is given by the empirical equation

$$d_{vs} = \frac{a}{p^n}, \qquad (2)$$

where

$$d_{vs} = \frac{\sum nd^3}{\sum nd^2} = \text{mean volume : surface diameter} \quad (3)$$

and p = homogenizer pressure, n = 0·5–0·6 and a = a constant.

It was also shown that neither turbulence (Phipps, 1971) nor cavitation (Phipps, 1974a) are responsible for fat globule disintegration, but direct confirmation was obtained (Phipps, 1974b) of the conclusion, first established microscopically by Loo and Carleton (1953), that fat globules are disrupted just inside the valve slit and close to the inlet boundary. The viscous stresses operating at the inlet to the valve slit primarily control the dispersion process (Phipps, 1975), and it has been shown that the fat globule diameter (d_{vs}) decreases with decreasing valve dimensions and interfacial tensions, and with increasing volume flow rate, and density and kinematic viscosity of the continuous phase.

Multiple homogenization continues to decrease the fat globule diameter, but best results are obtained with two passes; more than this is said to give an ice cream with a buttery texture (Stistrup and Andreasen, 1966).

3. *Cooling and ageing*

The many changes that take place during ageing have been discussed in detail elsewhere (Berger and White, 1976).

Of particular interest here is the slow coalescence of fat globules during the ageing of ice cream mix, and this has been investigated by Sherman (1962, 1965, 1969). The rate of coalescence is governed by the equation

$$N_t = N_0 e^{-kt}, \quad (4)$$

where N_t = number of globules at time $t = t$, N_0 = number of globules at time $t = 0$ and k = rate of particle coalescence.

This can be re-arranged to give

$$\log N_t = \log N_0 - kt, \quad (5)$$

so that $k = -d/dt(\log N_t)$ can be determined from the slope of the straight line plot of $\log N_t$ versus t.

The parameter N_t can be determined microscopically or by other means. Sherman found that if the initial globule diameter was below a certain size, then coalescence proceeded in two stages. After the interfacial area had decreased to a critical value, a phase of relatively rapid coalescence was followed by a phase of slow coalescence. This was interpreted as being due to the monoglyceride–protein film around the globules becoming more condensed and rigid, or that the energy barrier to coalescence increased. It was found that coalescence proceeded in two stages in mixes aged at 20°C, when the initial globule diameter was <0·95 μm. In mixes aged at 4·5°C, however, there was no change in N_t for approximately 5 hours, after which coalescence commenced, so that at this temperature slow flocculation was the rate-determining factor. Investigation showed that milk protein is the main contributor to stability during slow coalescence. The results of Graham and Phillips (1974) suggest that to minimize coalescence in protein-stabilized emulsions, the interfacial layers need to be as thick as possible, and heavily hydrated and charged.

In the above work, the total change in volume mean globule diameter was from approximately 0·8 μm to 1·0–1·1 μm as measured with the light microscope.

More recent studies have shown that the phase inversion temperature (PIT) of emulsion systems, as determined by DTA, is related to the rate of coalescence (k), and that the higher the PIT value the lower is k, so that PIT values can also be used as a measure of emulsion stability (Parkinson and Sherman, 1972; Parkinson, 1976).

4. Freezing, hardening and storage

An important change that occurs on aeration–freezing is a partial breakdown of the fat emulsion, to give a limited amount of clumped and coalesced fat; this confers some desirable properties to ice cream (Berger and White, 1971). The effect will be further discussed in Section III.

B. Processing Problems

When problems occur during homogenization, cooling, ageing or freezing some guidance as to their nature can often be obtained by light and electron microscopy of samples taken at various points.

An example from our experience was associated with a seasonal variation in milk protein composition during the spring period. A very high viscosity was reported in the mix from the pasteurizer, which severely restricted flow through the plate cooler. The problem did not occur with milk given a low heat treatment (88°C for 10 minutes) but was very marked with high heat treatment milk (91°C for 30 minutes). In both cases, samples of ice cream mix were taken, from the pasteurizer, the cooler, and after ageing, and samples of ice cream were taken after freezing and hardening. The mixes or melted samples were diluted 1 : 10 with distilled water, then 1 : 1 with a warm 6% gelatin solution, and allowed to set in haemocytometer cells. Photomicrographs were taken at a magnification of ×100, and are shown in Fig. 1. The phenomenon appeared to be excessive fat clumping associated

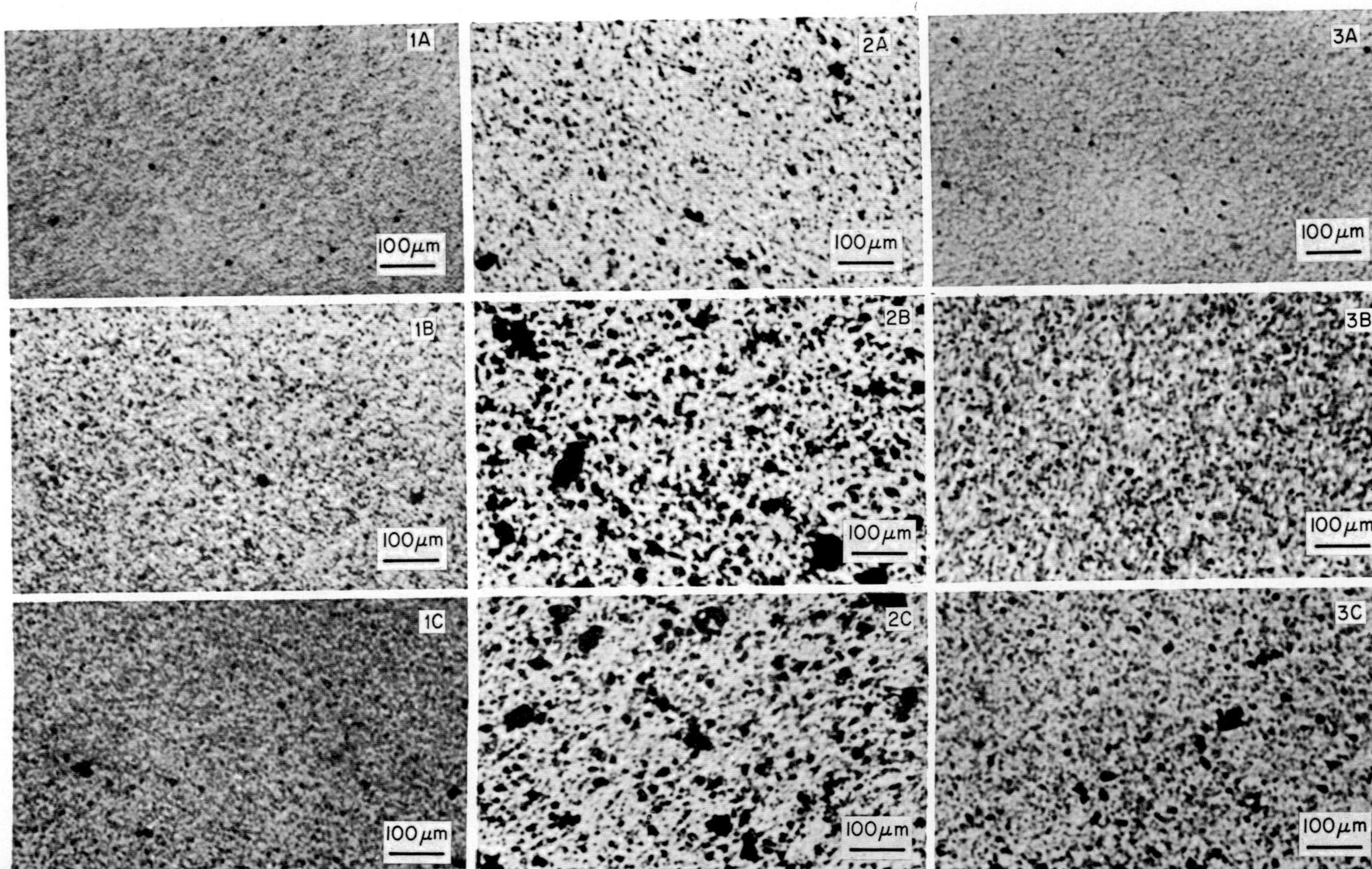
1A
100μm
2A
100μm
3A
100μm
1B
100μm
2B
100μm
3B
100μm
1C
100μm
2C
100μm
3C
100μm

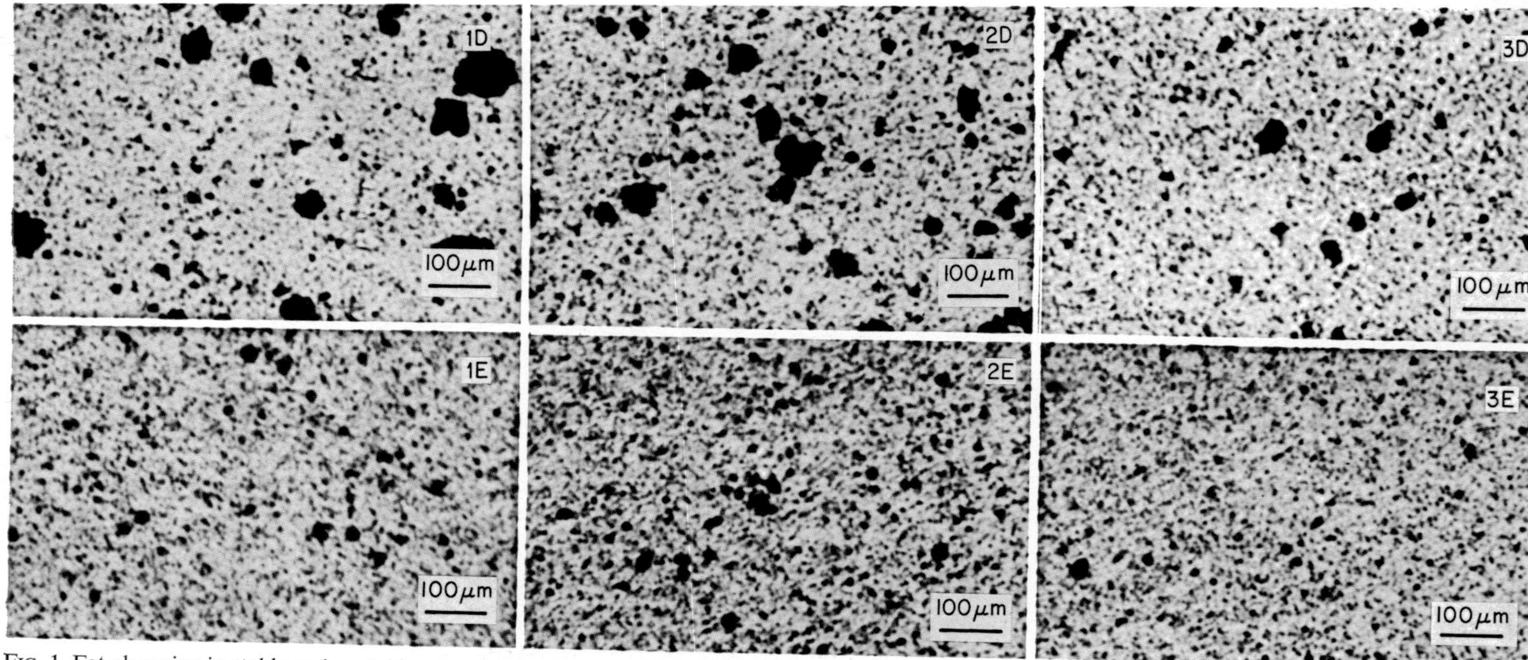

FIG. 1. Fat clumping in stable and unstable mixes. *Column 1*. Ice cream mix made with low heat treatment spring milk (88°C for 10 min); *Column 2*. Ice cream mix made with high heat treatment spring milk (91°C for 30 min); *Column 3*. Same mix as in *Column 2*, but with 0·1% sodium citrate added. *Row A*, pasteurized mix; *Row B*, cooled mix; *Row C*, aged mix; *Row D*, melt from frozen ice cream; *Row E*, melt from hardened ice cream.

with protein instability. It can be seen that ice cream mix made with low heat treatment milk showed very little fat clumping until the freezing stage. Mix made with the high heat treatment milk, on the other hand, showed severe fat clumping in the cooler, causing a high viscosity and restricted flow. Addition of 0·1% sodium citrate to the mix resulted in the micrographs shown in Column 3 of Fig. 1. The trouble at the cooler was eliminated, but some desirable clumping still occurs at the freezer.

An interesting point is that the fat clumps appear to be broken up by hardening, suggesting that the clumps may entrap some water which subsequently expands to ice on hardening.

III. THE STRUCTURE OF ICE CREAM

Much of the early light microscopical work on the structure of ice cream has been reviewed by King (1950) and Arbuckle (1960), while more recently electron microscopical work has been reviewed by Berger *et al.* (1972) and Berger and White (1976).

Early light microscopical work involved the cutting of free-hand sections of ~10 μm at about −23°C to −26°C in the hardening rooms. The sections were mounted with a liquid of refractive index (n) of 1·42, photographed at ×100 in the hardening room, and the micrographs were then measured for air cell diameter, distance between air cells, ice crystal size and thickness of unfrozen material between ice crystals (Arbuckle, 1940). Later improvements in technique involved the use of lower cutting temperatures approaching −30°C, various mountants to show the compounds more clearly, and polarized light. The refractive indices of the elements to be studied are air cells (n = 1·000), ice crystals (n_ω = 1·309, n_ε = 1·313), fat globules (n = 1·430–1·456) and unfrozen material (n = 1·460). For observing fat globules, mountants such as acetone (n = 1·357) and ethyl acetate (n = 1·379) were used, and for observing ice crystals a mixture of 10 : 15 amyl alcohol to kerosene (mixture n = 1·446) was used (Arbuckle and Cremers, 1954). These methods have been superseded by more rapid techniques which are described below.

A. Air Cells and Air Cell Walls

To measure the size distribution and mean size of the air cells in ice cream four Chance No. 1½ coverslips are arranged on a microslide, two on each side of the centre, to form a cell of depth 350 μm. A small piece of frozen ice cream is placed in the well and a fifth coverslip is gently laid on top but not pressed down. As the ice cream melts and warms up, the coverslip slowly settles to give a film of thickness of 350 μm. About 1000 air cells must be sized for a reasonable distribution curve, but only 100–200 are necessary if all that is required is the mean air cell size. A magnification of ×100 is employed. A similar method has been described by Fil'chakova (1972).

Mean air cell diameters for ice cream reported in the literature vary from 30 μm up to 186 μm, and we have previously found a mean diameter of approximately 60 μm, based on a limited number of measurements (Berger *et al.*, 1972). We have since made more extensive measurements on some 10 000 air cells over 47 samples of vanilla ice creams, and found a normal range from 5 μm to 250 μm with a mean value of 30 μm.

A typical distribution curve for air cells in ice cream is shown in Fig. 2.

It has been claimed that the mean air cell wall thickness in ice cream is 100–149 μm (Arbuckle, 1960), but simple calculation shows that the mean value must be much smaller than this. It can be shown that for equidiameter air cells, equally spaced in rhombohedral array, the minimum distance between air cells is given by

$$t = d_a\left[\left(\frac{0{\cdot}7405\,(100+\phi)}{\phi}\right)^{\frac{1}{3}} - 1\right],$$

where d_a is the air cell diameter and ϕ is the percentage overrun. This expression can also be rearranged to give

$$t = d_a\left[\left(\frac{0{\cdot}7405}{\mu}\right)^{\frac{1}{3}} - 1\right],$$

where μ is the volume fraction of air in the mix.

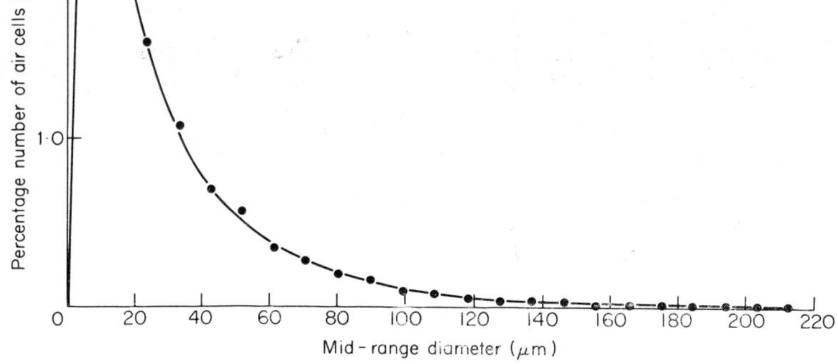

FIG. 2. Mean air cell size distribution in ice cream (based on 10 000 counts).

For 100% overrun and an air cell diameter of 30 μm, the minimum cell wall thickness is approximately 4·2 μm.

Figure 3 shows an electron micrograph of air cells in ice cream, and it can be seen that the air cell is lined with innumerable fat globules. It is also believed that a thin layer of oil from fractured oil globules additionally lines the air cells (Berger and White, 1971). Figure 4 shows a scanning electron micrograph of the air cell walls in a fractured, frozen ice cream. On this micrograph the mean air cell diameter is 25 μm, and the mean cell wall thickness is 5·6 μm.

B. Ice Crystals

A simple method of measuring ice crystal sizes in ice cream, which we use, is to place microslides ($1\frac{1}{2}'' \times 1''$), No. $1\frac{1}{2}$ coverslips, forceps, microspatula and

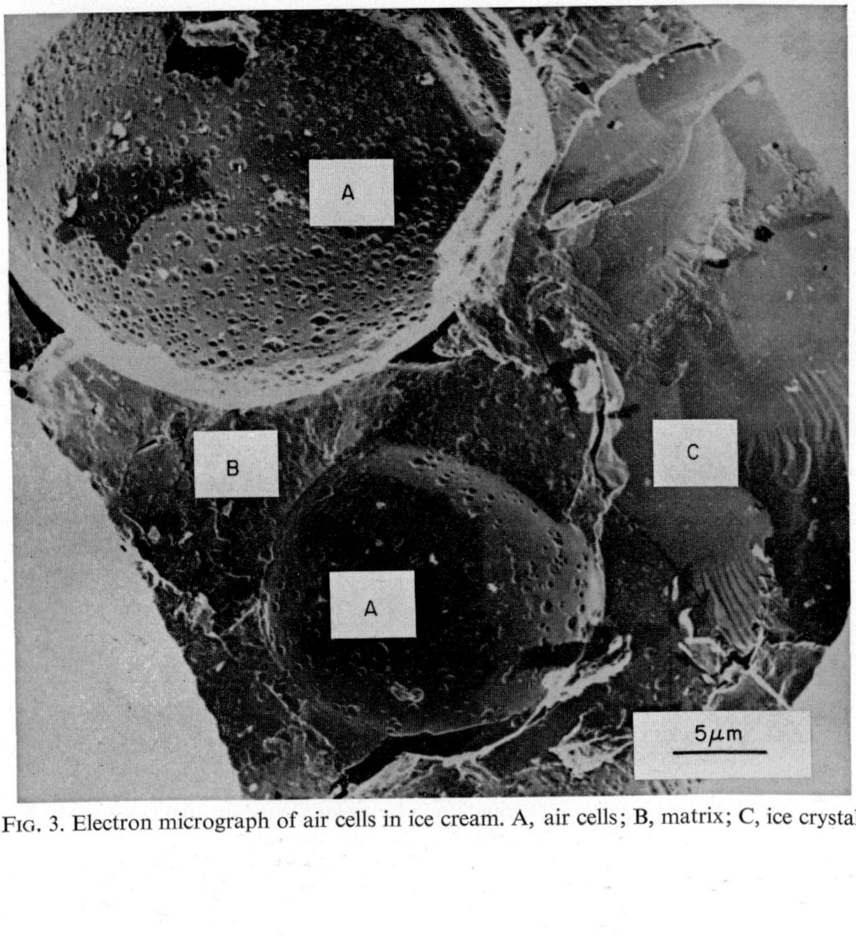

FIG. 3. Electron micrograph of air cells in ice cream. A, air cells; B, matrix; C, ice crystal.

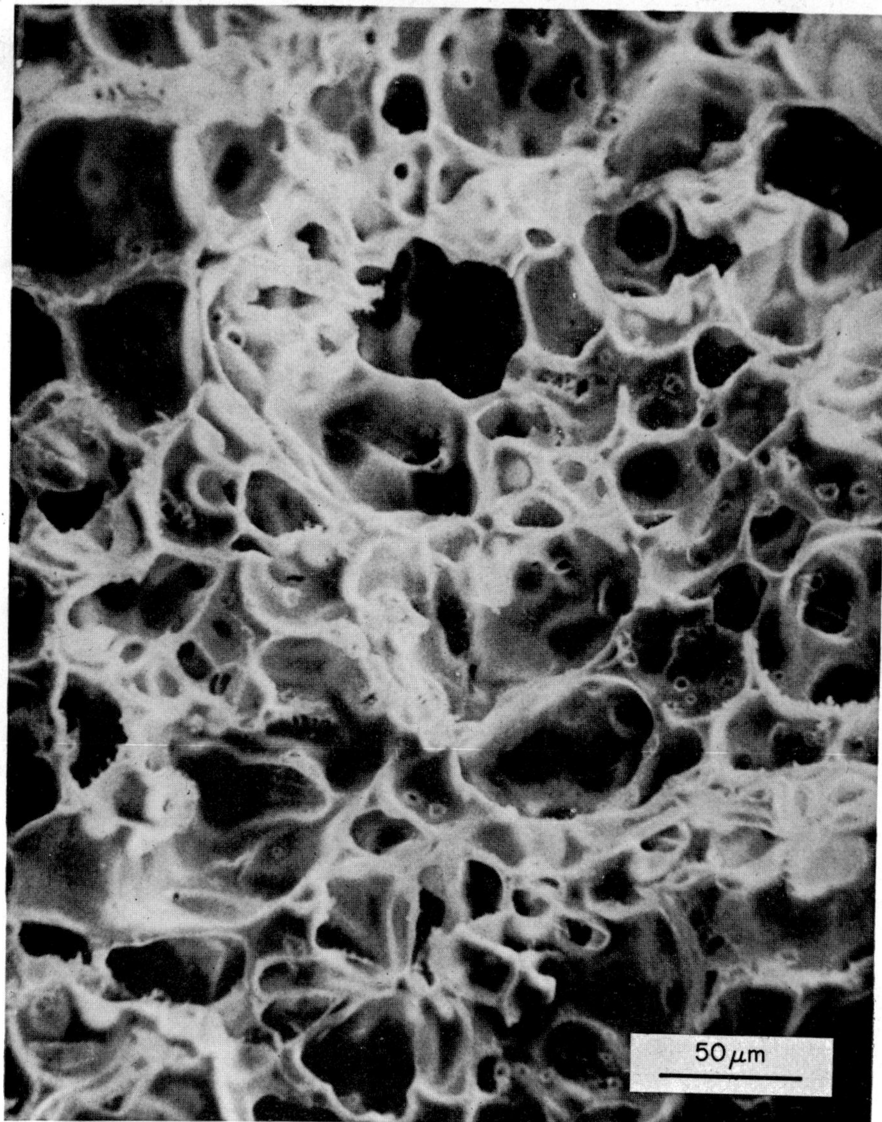

FIG. 4. Scanning electron micrograph showing the air cells and air cell walls in a fractured, frozen ice cream.

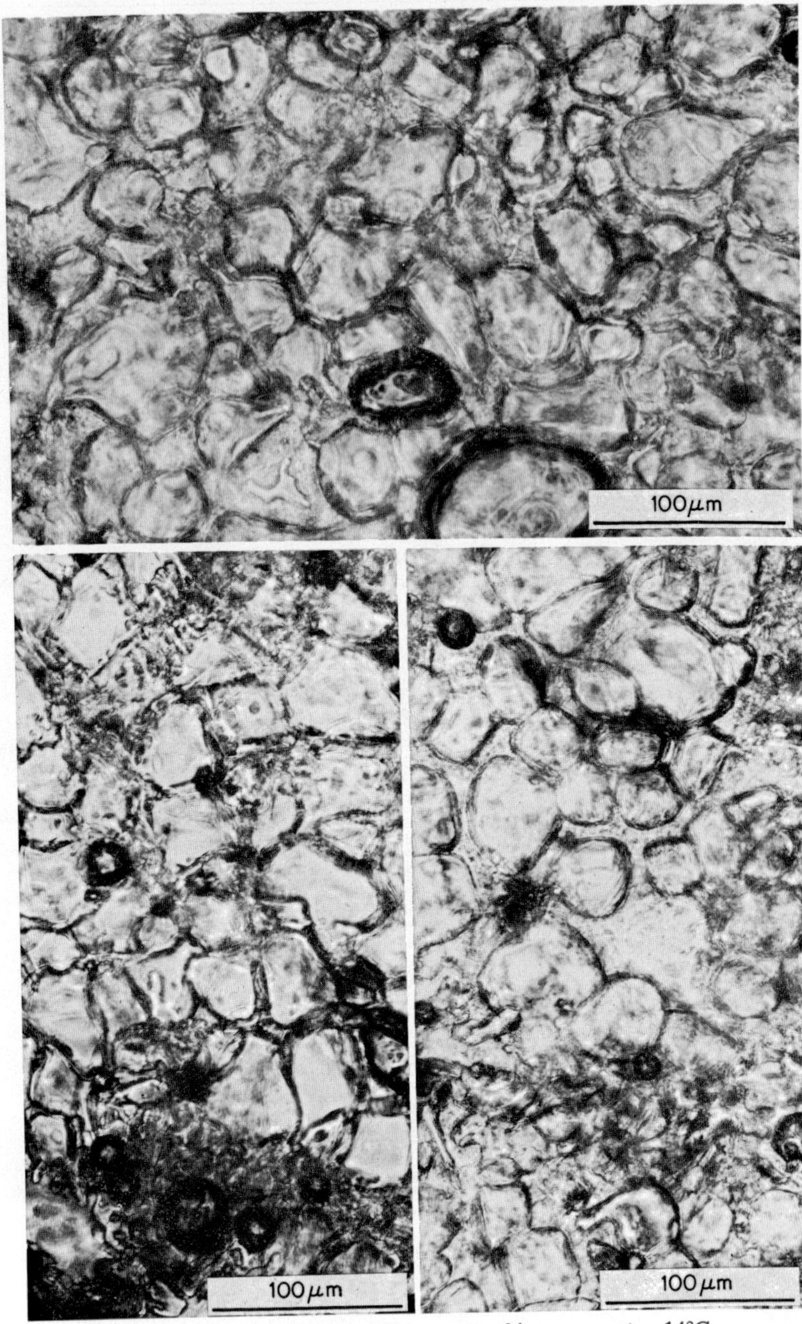

FIG. 5. Ice crystals in squash mounts of ice cream at −14°C.

a suitable mounting medium (50 : 50 amyl alcohol to kerosene) in a refrigerator at −14·4°C (+6°F). The ice cream is also kept in the refrigerator, and a simple squash mount is made with ice cream, the mountant (n = 1·43) and a coverslip. The squash mount is then quickly transferred in a cold box to a Kofler cold-stage, also maintained at −14°C on the microscope, and 100 ice crystals sized at this (serving) temperature, using 3–4 slides. The use of crossed polars or semi-crossed polars is helpful in the identification of ice crystals. The appearance of the ice crystals in a squash mount is shown in Fig. 5.

Our previous review of the literature showed that the mean ice crystal sizes reported for ice cream range from 20–75 μm, with a mean value of about 40 μm (Berger *et al.*, 1972). Re-examination of our experimental data on 37 standard vanilla ice creams showed that the mean ice crystal size, based on 3700 counts, is 43·3 μm. Another interesting feature of our results is that about 70% of the ice crystals are below 55 μm in size, the point at which coarseness becomes apparent; it would appear that coarseness depends on the proportion of ice crystals above 55 μm, i.e. coarseness is a matter of degree.

Arbuckle (1949) has presented microscopical data on the ice crystal and air cell sizes of ice creams frozen and hardened by various procedures; finer ice crystal sizes are produced by continuous (as opposed to batch) freezers, by lower drawing temperatures and by rapid blast hardening, these finer crystals giving smoother ice creams with higher scores for body and texture.

The mean ice crystal size distribution, based on our results, is shown in Fig. 6 and is clearly asymmetric. It should be remembered that these are the ice crystal sizes at the serving temperature (−14°C), when approximately 79% of the water exists as ice. When ice cream is first extruded from the freezer, the ice crystal size is smaller and only about 47% of the water exists as ice. Calculation shows that if the ice crystal size at refrigerator temperature (−14°C) is 43·4 μm, then from the relative amounts of water as ice at this temperature and at the extrusion temperature (−5°C), the ice crystal size at extrusion is probably of the order of 36 μm. The small sizes of ice crystals and air cells at this stage are the main reasons why ice cream drawn straight from the freezer has a smoother, more velvety texture than ice cream taken from refrigerated storage, unless the latter has been very carefully tempered.

Typical electron micrographs of ice crystals in ice cream are shown in Figs 7a and b.

C. Fat Globules, Clumps and Coalesced Fat

Using the light microscope, a rough assessment of the fat globule size distribution in ice cream mixes can be made. The mix is first diluted 1 : 50 with distilled water, then 1 : 2 with a warm solution of 10% gelatin, and a drop is placed in a haemacytometer cell, covered with a coverslip and allowed

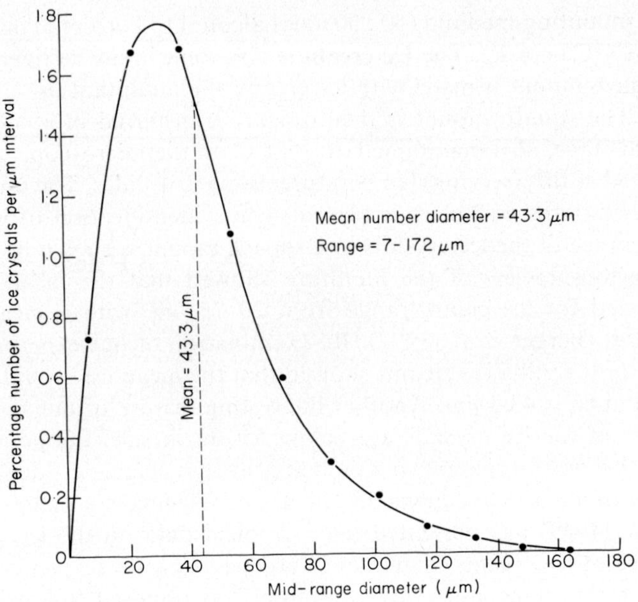

FIG. 6. Mean ice crystal size distribution in ice cream at −14·4°C (based on 3700 counts).

to set. The gelatin inhibits Brownian movement and allows sizing to be done at a magnification of × 1000. However, the lower limit of accurate measurement under the light microscope is about 1 μm (i.e. four times the resolution limit) but the fat globule diameters extend below 0·1 μm (Berger *et al.*, 1972). In these circumstances the only accurate method of obtaining the size distribution is by electron microscopy, and preliminary results showed a range of fat globule diameters of 0·04–3·0 μm, with a mean diameter of 0·45 μm (Berger *et al.*, 1972). We have since collated many more fat globule size distributions in ice cream mixes, measured from electron micrographs prepared by the freeze-etching method (Berger and White, 1971). The fat globule diameter in standard ice cream mixes, based on 2400 measurements, was found to range from 0·05 μm to 2·0 μm, with a mean value of 0·42 μm.

A typical fat globule size distribution curve is shown in Fig. 8, and electron micrographs showing fat globules in ice cream mix and ice cream are shown in Figs 9a and b.

When ice cream mix undergoes aeration–freezing, the combined effects of the low temperature and mechanical agitation cause the membranes around the fat globules to fracture in some cases, releasing still-liquid fat from the centre of these globules. This leads to the appearance of clumped fat globules and coalesced fat in frozen ice cream. Figure 10 shows a fat globule fractured in the freezer, and it can be seen that the liquid fat has run out and cemented together a clump of three fat globules. Valaer and Arbuckle (1961) classified the state of dispersion of butter fat in ice cream into six groups: (i) small, homogenized globules (<2 μm); (ii) improperly homogenized globules (6–8 μm); (iii) small clumps (about 20 μm); (iv) agglomerated fat (up to 25 μm); (v) coalesced globules ($>$ 25 μm); (vi) butter granules. King (1962) described the application of anoptral contrast, fluorescence and incident light microscopy to the study of the dispersion state of milk fat in ice cream. Reid and his co-workers (Reid, 1934, 1938; Reid and Hales, 1934; Reid and Decker, 1942) found that as the proportion of butter fat was raised from 8% to 15%, an increasingly smooth texture resulted, because the fat mechanically obstructed crystal growth and reduced the ice crystal size. This was confirmed by Arbuckle (1940) and Arbuckle and Cremers (1954). Cole and Boulware (1940) observed that the lubricating effect of the fat in the mouth also plays a part in imparting a smooth texture to ice cream.

The partial destabilization of the fat emulsion in ice cream has been extensively described, and an electron microscopical study of the phenomenon has been reported (Berger and White, 1971). Destabilization has both desirable and undesirable effects on ice cream, and as a result most manufacturers aim to produce some emulsion breakdown in the freezer to obtain optimum balance of properties.

D. Emulsifiers and Stabilizers

A number of workers have used the microscope to study the effects of emulsifiers and stabilizers on the structure and properties of ice cream. Reid (1938) found that the addition of egg yolk solids, rather than whole egg solids, produced smaller air cells and gave smoother texture and better body; he also found that as the gelatin content was increased to 0·2–0·3%, the texture became smooth, and at 0·5% the texture became exceedingly smooth and close. Arbuckle (1950) showed that the use of emulsifers produced smaller air cells and ice crystals, more evenly distributed, and gave a smoother ice cream. King (1966) observed that in emulsions of milk fat in skim milk, addition of GMS to the fat caused exudation of liquid fat from the large globules. Shama and Sherman (1966) showed, by cold-stage microscopy, that freeze–thaw cycles on ice cream emulsions caused coalescence, and that applied pressure at low temperatures caused rupture of the protein–emulsifier film around the fat globules. Using the electron microscope, Alsafar and Wood (1968) found that the fat globule membrane in ice creams made with Tween 80 was thinner than in those made with Span 60, and that the amount of fat destabilization on freezing was much greater in ice cream containing Tween 80.

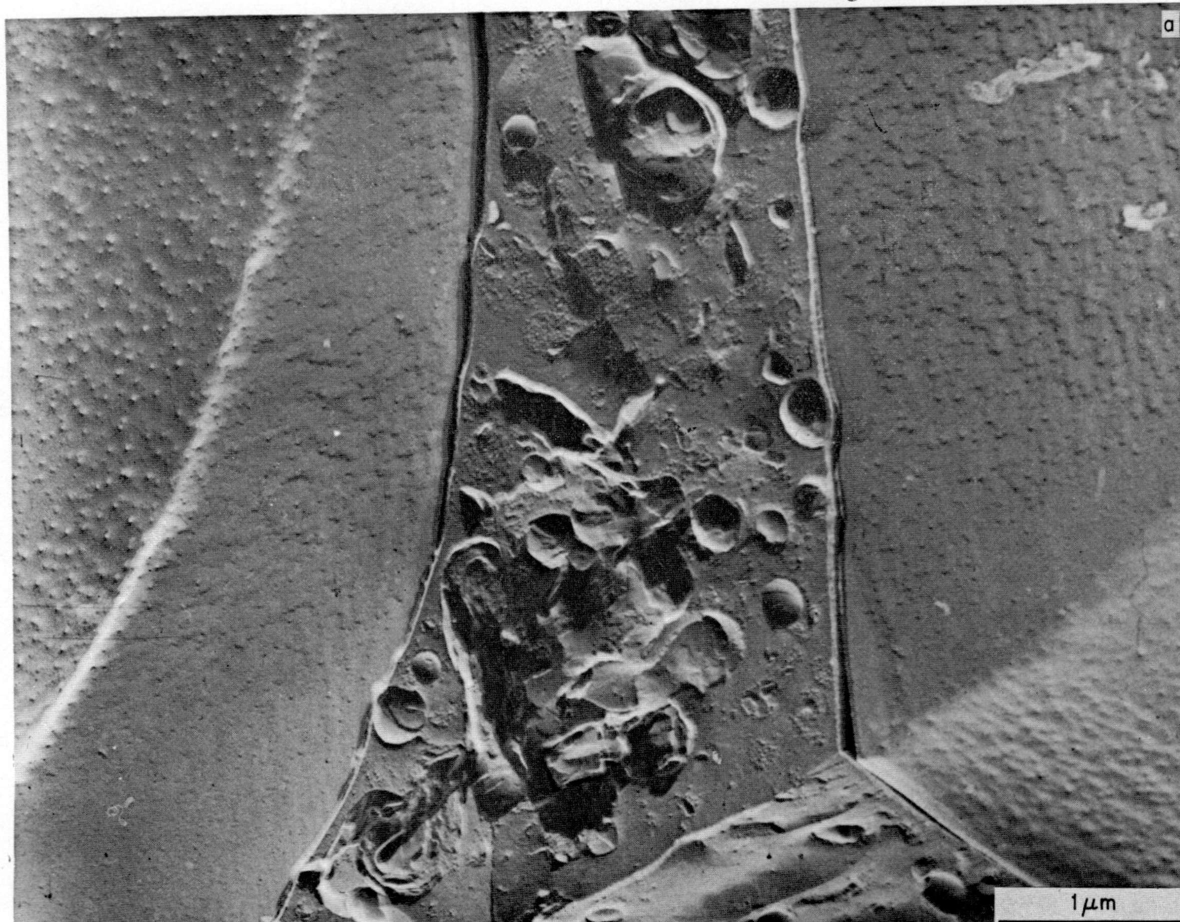

FIG. 7a. Electron micrograph of ice crystals in ice cream.

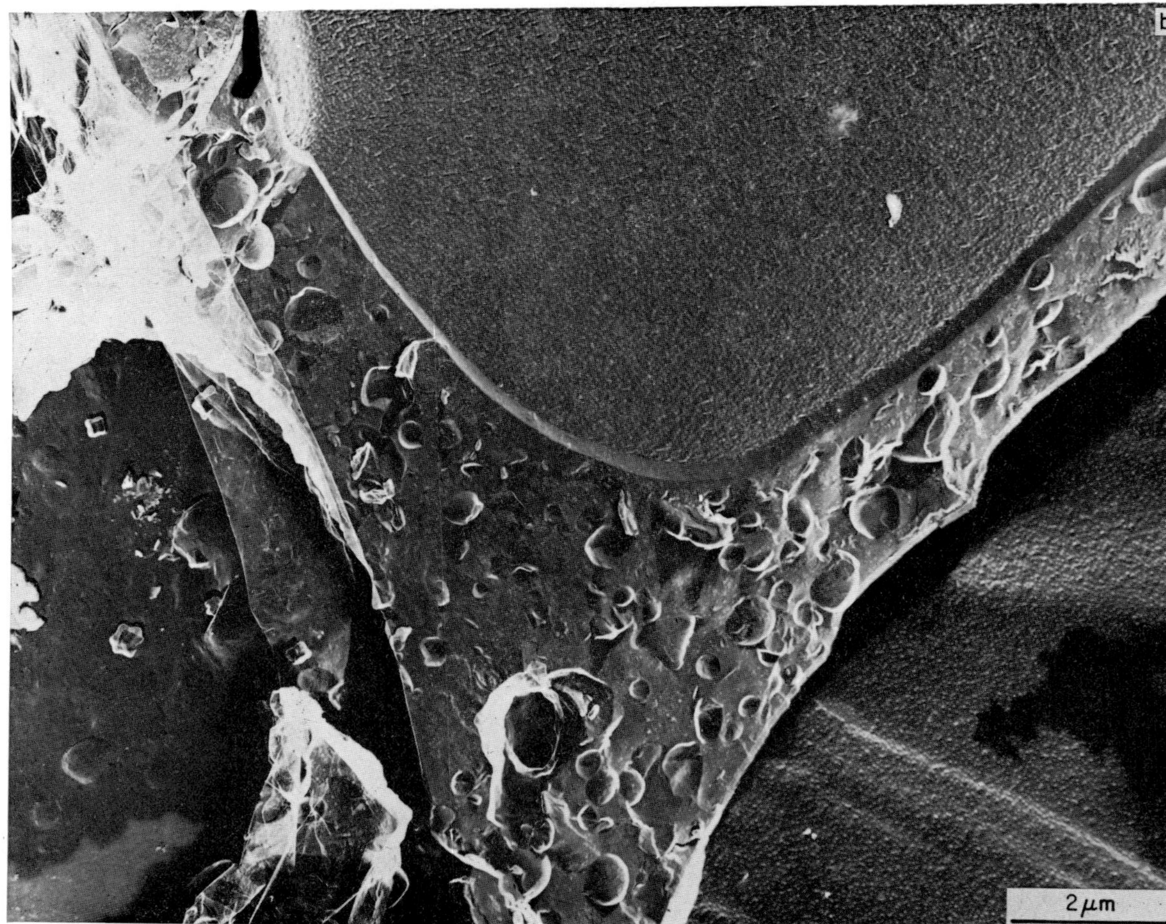

FIG. 7b. Electron micrograph of ice crystals in ice cream.

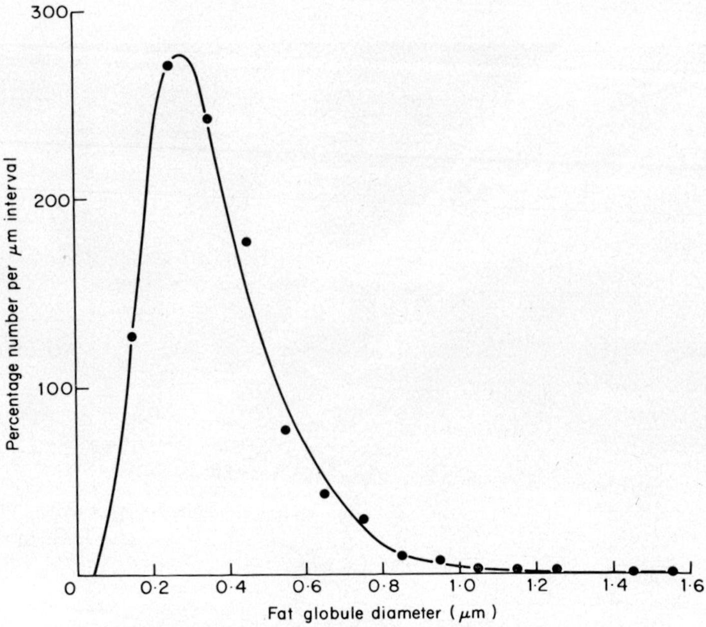

FIG. 8. Mean fat globule size distribution in ice cream mixes.

E. Casein Micelles

Approximately 80% of the milk proteins consists of the caseins, which exist mainly in the form of colloidal particles or micelles. A typical electron micrograph of skim milk (obtained after freeze-etching) is shown in Fig. 11. The structure of the casein micelle is not yet fully established, but electron microscopical studies reported in the literature show that the casein micelles in milk are approximately spherical and range from about 30 nm to 300 nm in diameter, with a mean value of 80–100 nm. Electron microscopy has also shown that the casein micelles have a granular internal structure, consisting of spherical subunits about 10 nm in diameter and separated by a centre-to-centre distance of 14–15 nm (Shimmin and Hill, 1964, 1965).

These findings were confirmed by Buchheim (1969), using the freeze-etching method of specimen preparation, and he assigned a diameter of 10–20 nm to the subunits. Knoop (1972) estimated the subunit diameter as 15–20 nm, and found evidence suggesting that calcium was located in the intermediate spaces between the subunits.

A modern view is that the casein micelles are complex particles containing the caseins, calcium and phosphate, together with lesser amounts of magnesium and citrate, and 60–80% water. The subunits, containing the caseins, have a low water content and are probably held together in the micelles by calcium phosphate bridges.

The effects of heat treating skim milk or whole milk prior to concentration are complex (Sawyer, 1969; Parry, 1974).

Electron microscopical studies of HTST sterilization of concentrated skim milk have shown that heat treatment approximately doubles the mean size of the casein micelles (Schmidt, 1968; Carroll *et al.*, 1971). The enlarged micelles are probably complexed with heat-denatured whey proteins and milk salts.

We have studied the casein micelles in replicas of freeze-etched ice cream. Figure 12 shows a histogram of the micelle size distribution found in ice cream. The micelles range from small aggregates, about 20 nm in diameter and consisting of about four subunits, up to aggregates of about 350 nm in diameter, consisting of many subunits. The mean diameter of some 400 casein micelles that we have sized in ice cream is 120 nm, which corresponds closely to the mean size of about 135 nm found by Carroll *et al.* (1971) for casein micelles in single-strength skim milk. The casein micelles in milk and ice cream are affected by certain factors such as pH, the mineral salt balance (ratio of calcium and magnesium to citrate and phosphate), and heat treatment and freezing; there is a need for more microscopical studies of these effects, which are of importance as the milk proteins, especially casein, are adsorbed at the fat globule interface in ice cream and affect the stability of the emulsion on aeration–freezing.

F. Sugars

During the freezing operation about 50% of the water present is turned into ice, and this proportion increases during hardening and storage.

The unfrozen phase is thus a very concentrated syrup of mixed sugars (sucrose, glucose, lactose, etc.) which sets to a rigid glass or an amorphous solid below about −23°C.

From a microscopical point of view, the interest in these sugars lies in their physical state in ice cream, and whether under certain conditions they can crystallize out, causing undesirable effects. In the latter case, it is important to be able to identify these sugars by means of their optical crystallographic properties. The characteristics of the main forms of sucrose and lactose are shown in Table 2.

The only crystalline forms very occasionally found in ice cream are α-lactose monohydrate, β-lactose anhydride and sucrose. A number of papers have dealt with the effect of the type and amount of sugar on the texture of

FIG. 9a. Electron micrograph of fat globules in ice cream mix.

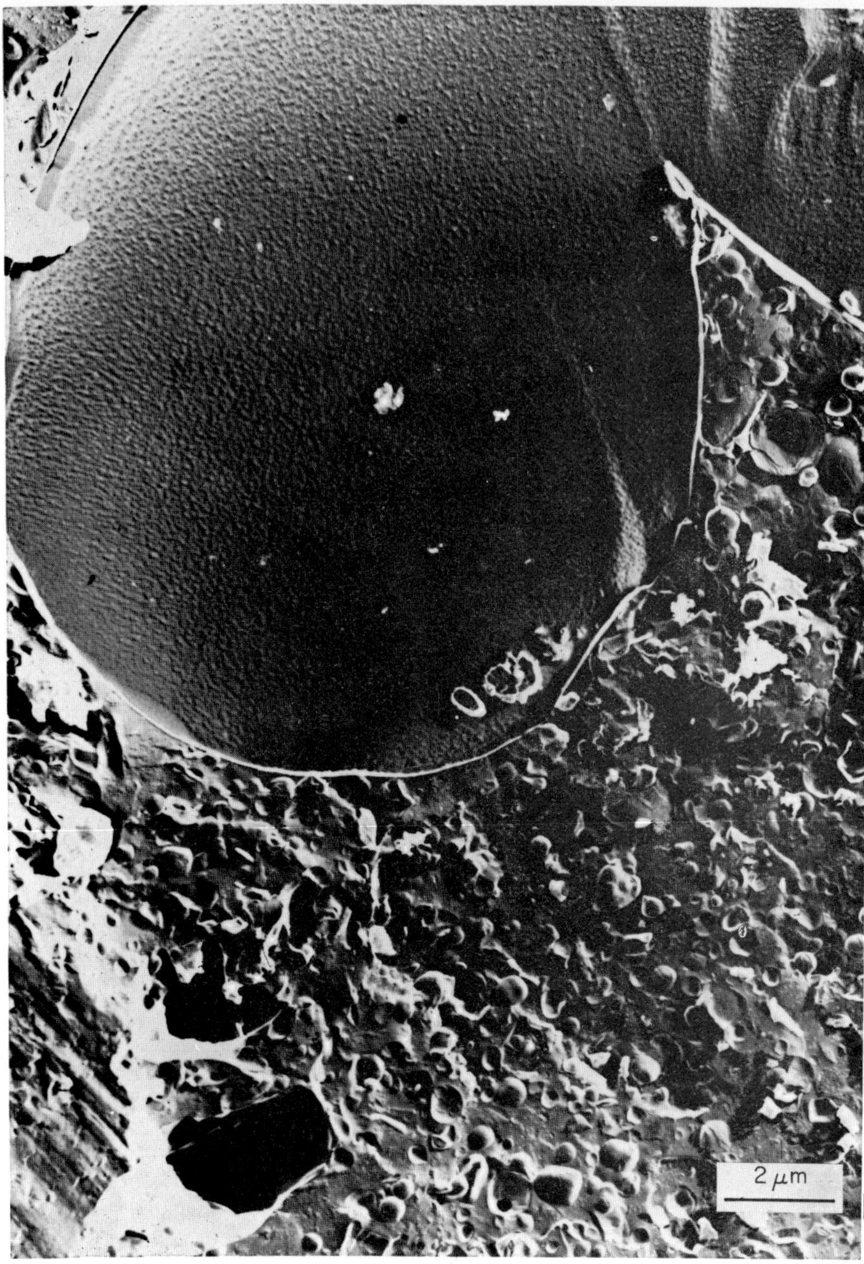

FIG. 9b. Electron micrograph of fat globules in ice cream.

TABLE 2. Characteristics of the crystalline forms of lactose and sucrose

Sugar	α-Lactose anhydride[b]	β-Lactose anhydride[c,d]	α-Lactose monohydrate[c,d]	5α-Lactose, 3β-lactose[b]	Sucrose[d]	Sucrose hemipentahydrate ($C_{12}H_{22}O_{11}2.5H_2O$)[e]	Sucrose hemiheptahydrate ($C_{12}H_{22}O_{11}3.5H_2O$)[e]
m.p. (°C)	225	252·2	201·6	—	185–6	45·7	27·8
Crystal system	—	Monoclinic	Monoclinic	—	—	—	—
Crystal habit	Thin, elongated plates	Tomahawk-shaped, rounded edges	Tomahawk-shaped	Needle-like crystals	Prismatic tablets	Prismatic tablets	—
n_α	1·535	1·542	1·517	1·539	1·538	1·517	—
n_β	—[a]	1·572	1·553	—[a]	1·565	1·526	—
n_γ	1·577	1·585	1·555	1·581	1·571	1·528	—

[a] Not easy to determine as crystals difficult to stand on end.
References: [b]Bushill *et al.* (1965); [c]Wherry (1928); [d]Winchell (1943); [e]Young and Jones (1949).

ice cream. In general, low sugar content gives a coarse texture and crumbly body, and a high sugar content imparts a soft texture, but also a tendency towards a soggy body. Increasing the sugar content replaces water, binds more water, gives smaller ice crystals, and also lowers the freezing point of the mix so that there is less ice at a given serving temperature (Reid, 1934, 1938; Reid and Hales, 1934; Reid *et al.*, 1939; Arbuckle, 1940). These effects improve body and texture, and lower the melting resistance. The replacement of up to 40% of the sucrose by dextrose or starch conversion products is claimed to improve the flavour, body, texture and melt-down of ice cream. The effects of such replacement on the structure and properties of ice cream have been studied by Reid *et al.* (1940), Reid and Decker (1942, 1943), Reid and Minert (1942), and Reid and Hensley (1959).

The gritty condition of ice cream known as sandiness is caused by the development of lactose crystals >25 μm in size (Figs 13a–d) and is encouraged

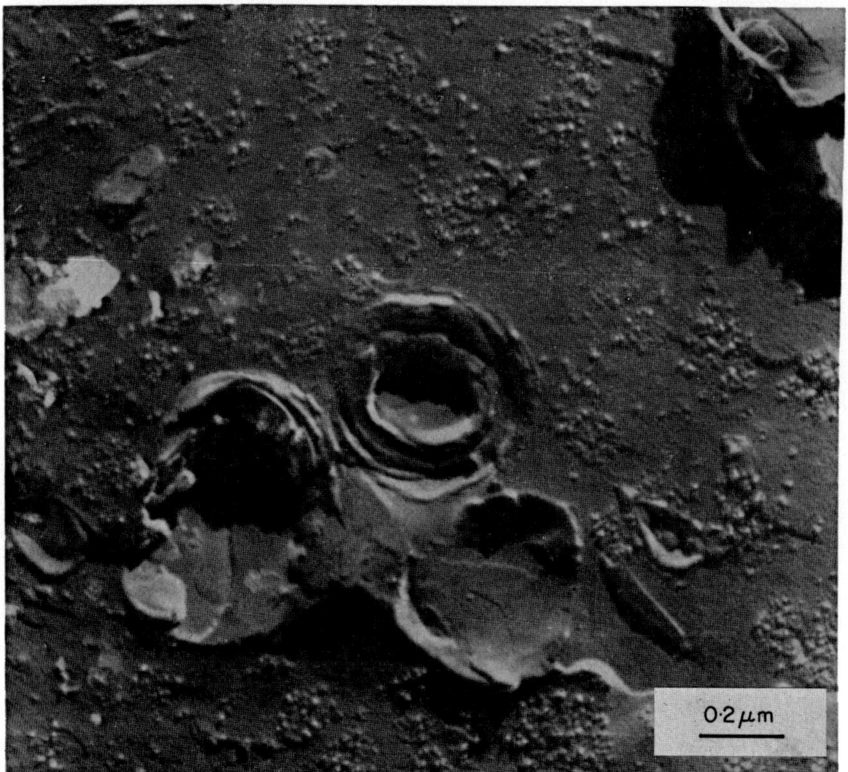

FIG. 10. Electron micrograph of fat clump.

by a high MSNF content, prolonged storage, and by fluctuating storage temperatures. Nickerson (1954, 1956, 1957) found that (i) below $-23°C$ sandiness did not develop, (ii) between $-23°C$ and $-17°C$ lactose crystallization is greatly favoured, and (iii) between $-17°C$ and $-11°C$, lactose crystallization proceeded rapidly only when nuclei were present. The sandy condition is rare in ice cream produced in the U.K., where the MSNF content is usually below the danger level for lactose crystallization ($\not> 11\%$).

The simplest technique for studying the crystals is the melted-drop technique, in which a small piece of frozen ice cream is placed on a microslide, covered with a coverslip, and allowed to melt. Immediate examination between crossed polars then shows up any anisotropic crystalline material present. Generally speaking, both α-lactose monohydrate and β-lactose anhydride crystals may be found in sandy ice cream. Both may be tomahawk-

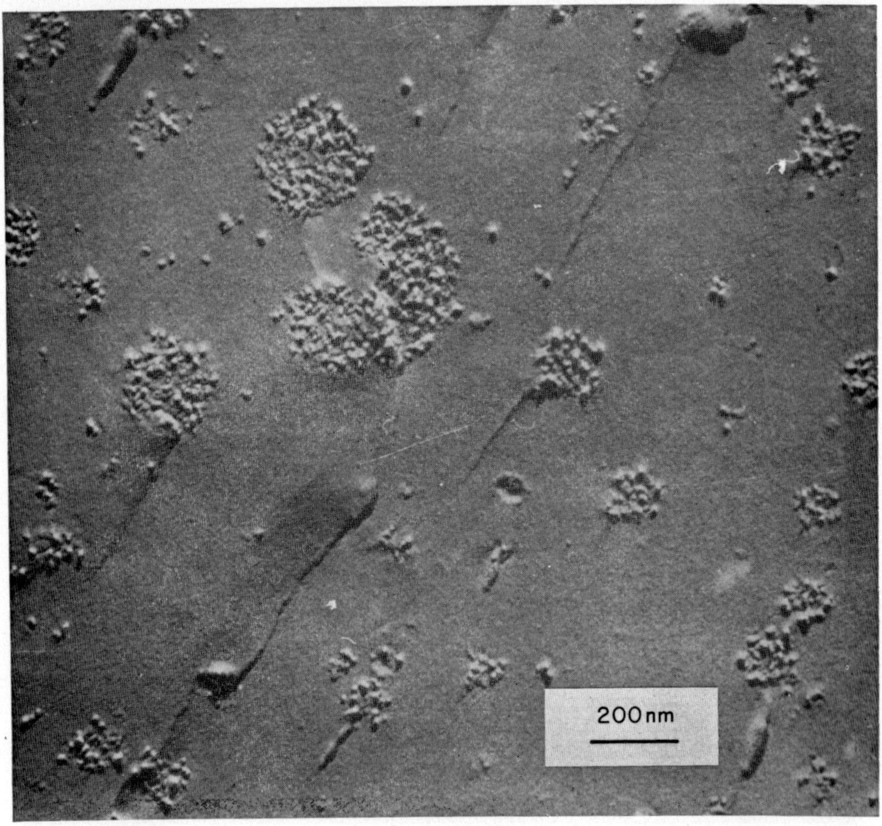

FIG. 11. Electron micrograph of replica of freeze-etched skim milk.

shaped or of irregular shape, and are distinguishable from each other and from sucrose by their refractive indices (Table 2). Van Kreveld and Michaels (1965) have shown that all the habits of α-lactose crystals found in dairy products are crystallographically equivalent to the tomahawk form, and that they arise from it by different relative growth ratios of the faces. Bushill *et al.* (1965) have prepared and characterized all the known crystalline forms of lactose by X-ray diffraction and infrared spectrometry, and have also given microscopically determined refractive index data.

Examination of many electron micrographs of frozen ice cream shows that only very occasionally can microcrystals of sucrose be found, and these are very small, ranging from 40 nm to 160 nm in size (Fig. 14).

G. The Fat–Mix and Air–Mix Interfaces

The surface area presented to the aqueous phase by the fat globule in one gram of a typical ice cream mix is 1·0–1·2 m^2 (Berger *et al.*, 1972; Nielsen, 1973) depending on the mean fat globule diameter. On homogenization, some surface-active materials in the mix are quickly adsorbed onto the naked fat globule surfaces and form a protective membrane.

Sherman (1961) assumed the membrane to be a uniform shell around the fat globule, and using rheological measurements calculated the membrane thickness to be 300 nm. Bird and Stainsby (1973) refined the method, using electron microscopical estimates of the globule diameter, and obtained an estimate of 60–90 nm for the membrane thickness. They also pointed out that

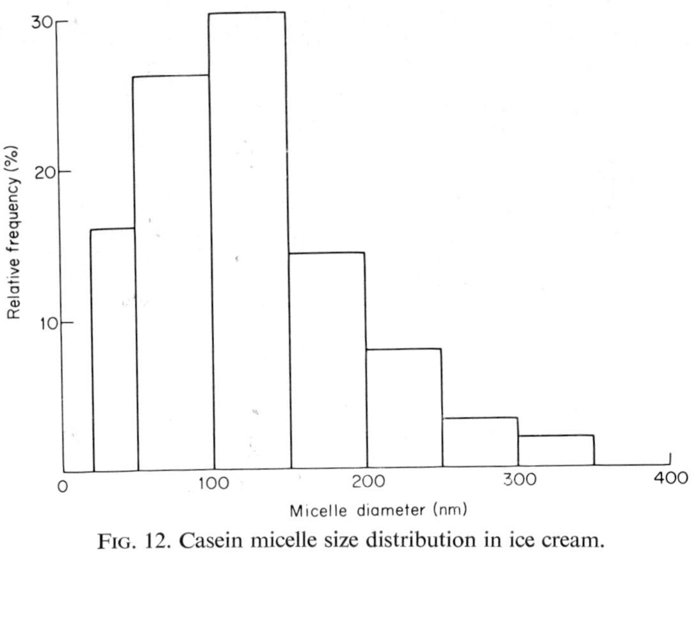

FIG. 12. Casein micelle size distribution in ice cream.

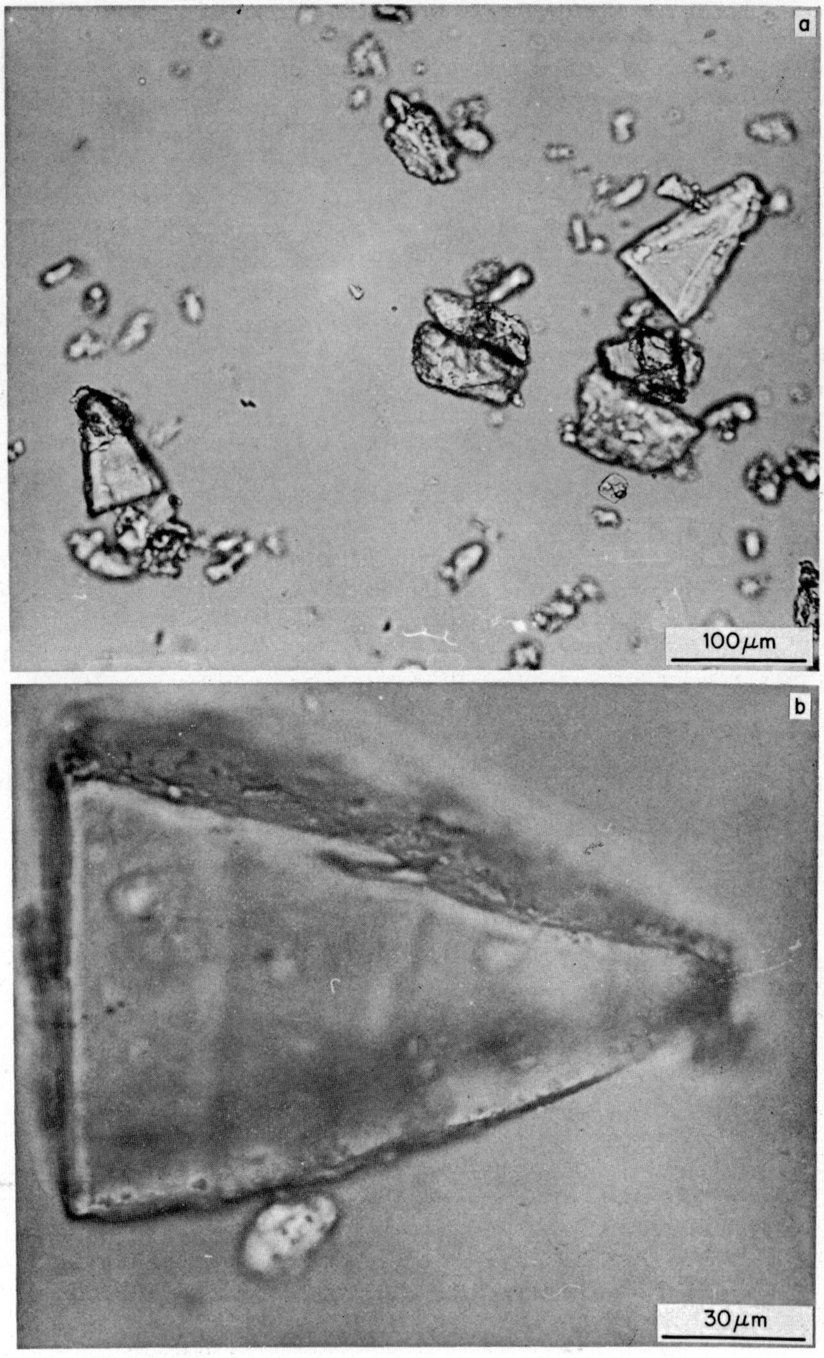
a
100μm
b
30μm

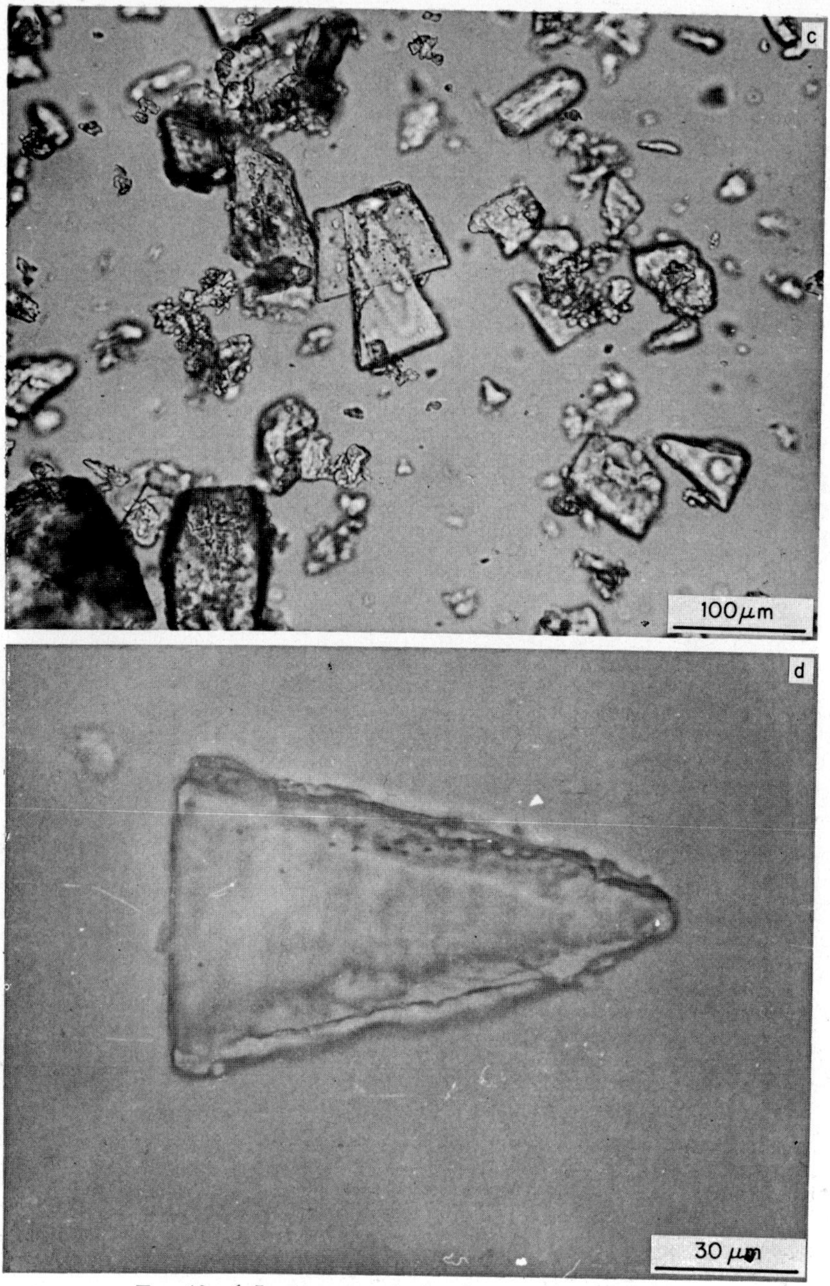

Figs 13a–d. Lactose crystals showing tomahawk shape.

these estimates tend to be high because, as electron microscopical evidence shows, some whole casein micelles become attached to fat globules thereby greatly increasing their apparent hydrodynamic thickness (Fig. 15).

Some evidence on the composition and structure of the fat globule mem-

FIG. 14. Sucrose crystals in ice cream.

brane (FGM) in ice cream has come from electron microscopical studies (Berger *et al.*, 1972), but much has also been learned from other techniques including interfacial tension methods, micro-electrophoresis, surface balance methods, bulk and surface rheological studies, and chemical separation and analytical techniques. The evidence emerging from these multi-disciplinary approaches has been summarized by Berger and White (1976).

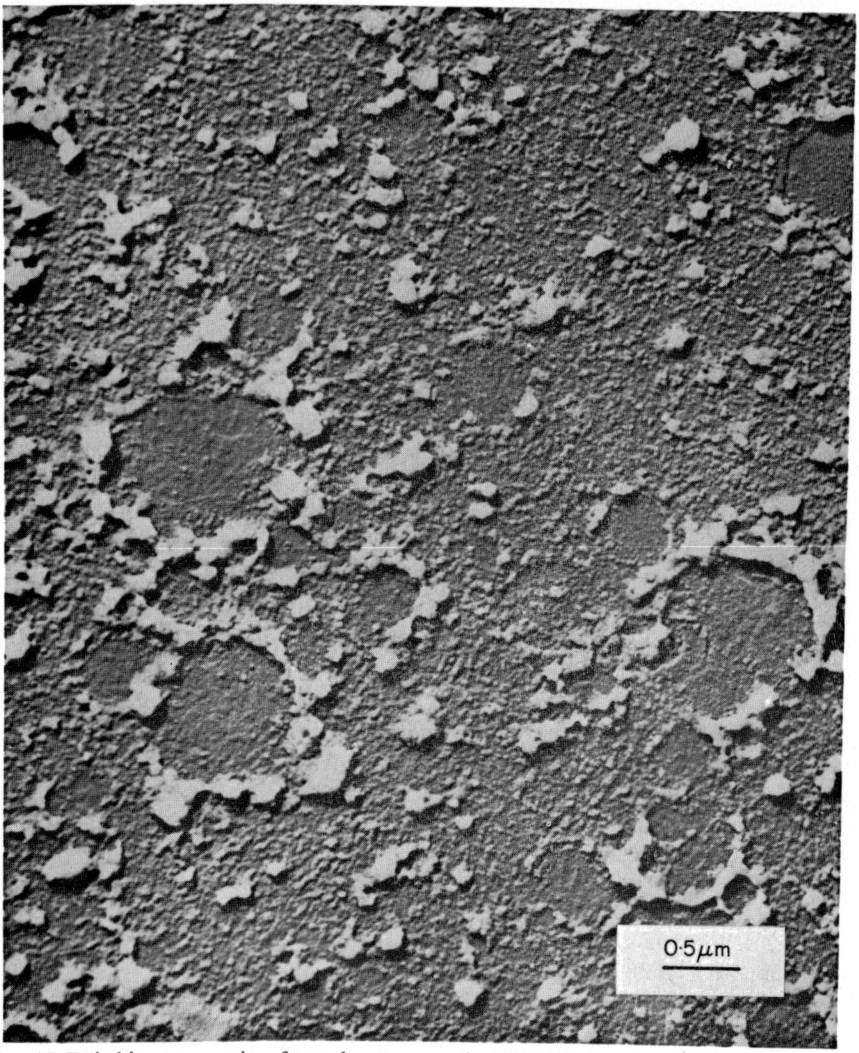

FIG. 15. Dried ice cream mix, after solvent removal of fat globules, showing casein micelles attached to fat globule 'ghosts'.

Fat globules are created in 10^{-6}–10^{-5} s in the homogenizer, and the immediate membranes formed on their surfaces are a chance affair that depends on the local micro-environment around each globule. In these circumstances, an intensely competitive situation exists at the fat–serum interface and it seems likely that the membrane has a mosaic or patchwork structure, with the various surface-active materials attached in different regions.

Further electron microscopical studies are needed to explore the non-uniformity of the FGM in ice cream.

Where the air–mix interface is concerned, little electron microscopical work has been reported other than our own. Our previous studies (Berger *et al.*, 1972) showed that the interior surface of air cells in ice cream had a smooth appearance, typical of fat surfaces, and that the air cells had many fat globules protruding into them (Fig. 3). Microscopy alone cannot fully interpret what is happening here, and other methods must be used in further studies on the subject.

ACKNOWLEDGEMENTS

The authors would like to express their thanks to Mr I. D. Boyes and Dr R. Reed of the Procter Dept. of Food and Leather Science, University of Leeds, for the preparation of Figs 3, 7, 9, 10 and 14, to Dr G. G. Jewell and Mr D. A. Cruickshank of the Food Research Association, Leatherhead, for Figs 11 and 15, to Mr G. DuBarry of the Electrical Research Association, Leatherhead, for Fig. 4, and to Miss A. V. Bell for considerable assistance with microscopical measurements.

REFERENCES

Alsafar, T. A. and Wood, F. W. (1968). *Alex. J. Agric. Res.* **16**, 135.
Arbuckle, W. S. (1940). *Missouri Agric. Exp. Sta. Res. Bull.* No. 320.
Arbuckle, W. S. (1949). *Ice Cream Tr. J.* **45** (10), 86.
Arbuckle, W. S. (1950). *Ice Cream Tr. J.* **46** (10), 106.
Arbuckle, W. S. (1960). *Ice Cream Tr. J.* **56** (10), 62.
Arbuckle, W. S. and Cremers, L. F. (1954). *Ice Cream Fld.* **64** (4), 98.
Berger, K. G. and White, G. W. (1971). *J. Fd Technol.* **6**, 285.
Berger, K. G. and White, G. W. (1976). *Dairy Ind. Int.* **41**, 199, 236.
Berger, K. G., Bullimore, B. K., White, G. W. and Wright, W. B. (1972). *Dairy Inds*, **37**, 419, 493.
Bird, A. E. and Stainsby, G. (1973). *J. Sci. Fd Agric.* **24**, 1339.
Buchheim, W. (1969). *Milchwissenschaft*, **24**, 6.
Bushill, J. H., Wright, W. B., Fuller, C. H. F. and Bell, A. V. (1965). *J. Sci. Fd, Agric.* **16**, 622.
Carroll, R. J., Thompson, M. P. and Melnychyn, P. (1971). *J. Dairy Sci.* **54**, 1245.

Cole, W. C. and Boulware, J. H. (1940). *J. Dairy Sci.* **23**, 149.
Farrall, A. W., Walts, C. C. and Hanson, R. L. (1941). *J. Dairy Sci.* **24**, 539.
Fil'chakova, N. N. (1972). *Kholodil'naya Tekhnika*, No. 9, 34.
Goulden, J. D. S. and Phipps, L. W. (1962). *Proc. 1st Int. Congr. Fd Sci. Technol.* **1**, 765.
Goulden, J. D. S. and Phipps, L. W. (1964). *J. Dairy Res.* **31**, 195.
Graham, D. E. and Phillips, M. C. (1974). 'Theory and Practice of Emulsion Technology', S.C.I. Symposium, Brunel University, U.K.
King, N. (1950). *Dairy Inds* **15**, 1052.
King, N. (1962). *Proc. 16th Int. Dairy Congr.* **C**, 48.
King, N. (1966). *Proc. 17th Int. Dairy Congr.* **C**, 289.
Knoop, E. (1972). *Milchwissenschaft*, **27**, 364.
Kreveld, A., van and Michaels, A. S. (1965). *J. Dairy Sci.* **48**, 259.
Loo, C. C. and Carleton, W. M. (1953). *J. Dairy Sci.* **36**, 64.
Nickerson, T. A. (1954). *J. Dairy Sci.* **37**, 1099.
Nickerson, T. A. (1956). *J. Dairy Sci.* **39**, 1342.
Nickerson, T. A. (1957). *J. Dairy Sci.* **40**, 309.
Nielsen, B. J. (1973). *Mod. Dairy*, **52**, (5), 10.
Parkinson, C. J. (1976). *J. Sci. Fd Agric.* **27**, 784.
Parkinson, C. J. and Sherman, P. (1972). *J. Colloid Interface Sci.* **41**, 328.
Parry, R. M. (1974). *In* 'Fundamentals of Dairy Chemistry' (B. H. Webb, A. H. Johnson and J. A. Alford, eds), p. 603. Avi. Publ. Co., Westport, Connecticut
Phipps, L. W. (1971). *Nature, Lond.* **233**, 617.
Phipps, L. W. (1974a). *J. Dairy Res.* **41**, 1.
Phipps, L. W. (1974b). *J. Dairy Res.* **41**, 339.
Phipps, L. W. (1975). *J. Phys. D. Appl. Phys.* **8**, 448.
Reid, W. H. E. (1934). *Missouri Agric. Exp. Sta. Res. Bull.* No. 340.
Reid, W. H. E. (1938). *Ice Cream Tr. J.* **34** (5), 20.
Reid, W. H. E. and Decker, C. W. (1942). *J. Dairy Sci.* **25**, 726.
Reid, W. H. E. and Decker, C. W. (1943). *Missouri Agric. Exp. Sta. Res. Bull. No. 370.*
Reid, W. H. E. and Hales, M. W. (1934). *Missouri Agric. Exp. Sta. Res. Bull.* No. 215.
Reid, W. H. E. and Hensley, R. E. (1959). *Proc. 15th Int. Dairy Congr.* **2**, 1179.
Reid, W. H. E. and Minert, K. R. (1942). *Missouri Agric. Exp. Sta. Res. Bull.* No. 339.
Reid, W. H. E., Drew, R. J. and Arbuckle, W. S. (1939). *Missouri Agric. Exp. Sta. Res. Bull.* No. 330.
Reid, W. H. E., Cooley, R. J. and Arbuckle, W. S. (1940). *Missouri Agric. Exp. Sta. Res. Bull.* No. 323.
Sawyer, W. H. (1969). *J. Dairy Sci.* **52**, 1347.
Schmidt, D. G. (1968). *Neth. Milk Dairy J.* **22**, 40.
Shama, F. and Sherman, P. (1966). *J. Fd Sci.* **31**, 699.
Sherman, P. (1961). *Fd Technol., Champaign*, **15**, 394.
Sherman, P. (1962). *Proc. 1st Int. Congr. Fd Sci. Technol.* **1**, 785.

Sherman, P. (1965). *J. Fd Sci.* **30**, 201.
Sherman, P. (1969). *J. Text. Stud.* **1**, 43.
Shimmin, P. D. and Hill, R. D. (1964). *J. Dairy Res.* **31**, 121.
Shimmin, P. D. and Hill, R. D. (1965). *Aust. J. Dairy Technol.* **20**, 119.
Stistrup, K. and Andreasen, J. (1966). *Proc. 17th Int. Dairy Congr.* **EF**, 375.
Valaer, E. P. and Arbuckle, W. S. (1961). *Ice Cream Fld.* **77** (1), 10.
Wherry, E. T. (1928). *J. Wash. Acad. Sci.* **18**, 302.
Winchell, A. N. (1943). 'The Optical Properties of Organic Compounds.' University of Wisconsin Press, Madison.
Young, F. E. and Jones, F. T. (1949). *J. Phys. Colloid Chem.* **53**, 1334.

14. Novel Protein Foods

F. O. FLINT

Procter Department of Food Science,
University of Leeds, U.K.

I. INTRODUCTION

The British Food Standards Committee (1974) in their 'Report on Novel Protein Food' group new protein-rich foods into three main categories. The first and largest group consists of conventional food sources, e.g. legumes, oil seeds and fish, which have been modified by processing and now appear in a novel form. The other groups comprise the leaf protein concentrates derived from higher plants and the protein-rich material obtained by the growth of selected micro-organisms.

This chapter is concerned chiefly with the first group of novel proteins. This includes the 'engineered' foods on which the food industry has expended most effort in the production of meat-like foods from soya and other readily available plant proteins (Gutcho, 1973).

The starting point for many of these products is the protein-rich soybean. This is the source of both the conventional food 'full fat soya flour' and also the novel foods, defatted soya meal and flour. The latter are regarded as 'novel' because they are the products of solvent extraction which constitutes a modification of a conventional food by processing (British Food Standards Committee, 1974). The soybean makes a good starting point in the study of the microscopy of novel proteins because morphological features present in the bean are readily seen in defatted meals and flours and in many of the soya-based protein products designed to extend and replace meat.

As a source of oil and animal feed, soybeans have been of commercial interest for a long period and this is reflected in the early literature. Wallis (1913) describes hand sections of the bean's seed coat (testa) and the embryo (cotyledons), and illustrates these in elegant detail. His sketch of the 'powder of soybean' shows all the cellular elements present in the soybean. This early paper includes an examination of commercial specimens of soybean cake and meal being made at a time when adulteration was not uncommon. Two of the samples he examined contained foreign particles, and Wallis describes the sand, cotton seed, rice starch, barley starch and pinewood sawdust he saw with the microscope.

Wallis made further reference to the bean, the microstructure of which is now well documented (Winton and Winton, 1932; Wallis 1955, 1965; Vaughan, 1970).

Though superficially similar to other leguminous seeds, the soybean is distinct in two respects: it contains an aleurone layer in its seed coat and its cotyledons almost always lack starch granules. In common with other legumes, a cross-section of the bean shows two main structures, the seed coat or hull (testa) and the two large cotyledons (embryo) which form the bulk of the bean. Figure 1 shows these features in detail.

The outside of the seed coat is made up of closely packed, similarly sized, rectangular palisade cells. Varietal differences probably explain the different dimensions reported in the literature for palisade cells. These range from a radial diameter of 45·5–50 μm (Wallis, 1913; Vaughan, 1970) up to 60 μm (Winton and Winton, 1932), with the corresponding cell breadth varying from 10–14 μm up to 20 μm.

The palisade cells form a double layer at the seed scar (hilum), this being the only area in the seed coat lacking the characteristic 'hourglass' cells which generally underlie the palisade cells. Hourglass cells are thick-walled, lignified cells in contact with one another only at the cell top and base (Fig. 1). They vary in height from approximately 40 μm to 120 μm, with a diameter of about 25–35 μm, the largest cells being near the hilum.

Palisade and hourglass cells are important diagnostic aids in the detection of soya products. They are chemically resistant, and when they appear as a residue in the 'crude fibre' determination this is indicative of soya material in the sample (Wallis, 1965). They are also resistant to heat and pressure, so

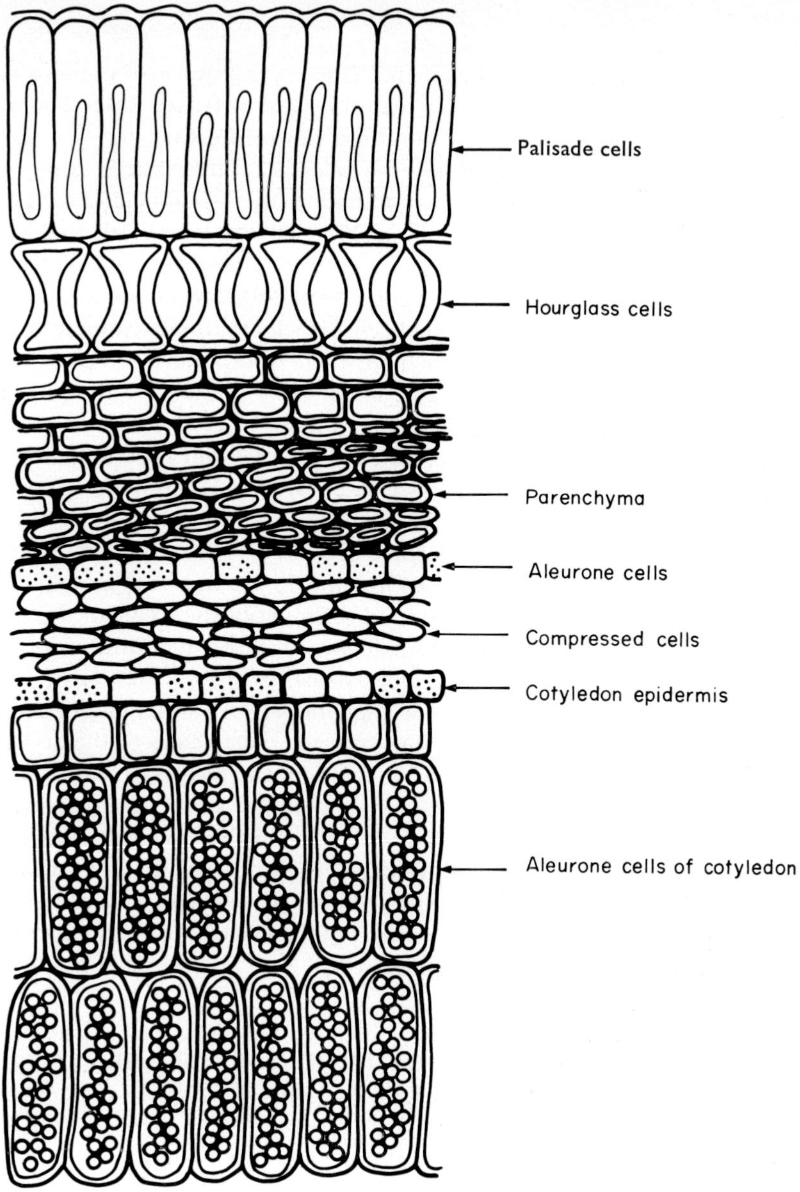

FIG. 1. Portion of soybean, cross-section.

that textured vegetable proteins prepared from soya flour also contain intact palisade and hourglass cells which can be demonstrated microscopically (Coomaraswamy and Flint, 1973; Flint and Lewin, 1976).

Below the hourglass cells are several layers of thin-walled, compressed parenchyma cells, the lower rows containing a distinct single layer of rectangular aleurone cells. These cells are about 20 μm by 10 μm in size and contain small, rounded aleurone grains (protein bodies) and some oil droplets.

The remaining 90% of the soybean consists of cotyledon tissue. The surface of the cotyledon has its own epidermal cells which are almost square in cross-section, 10–15 μm wide and 15 μm high, their outer walls being somewhat thicker than the side and inner walls. On the 'flat' side of each cotyledon, immediately below the epidermis, are three lines of cells which are radially elongated and some 50–100 μm long with a width of about 15 μm. The remaining cotyledon cells are simple parenchyma cells, rounded–polygonal in shape and ranging in size from 30 μm to 50 μm long and 15 μm to 25 μm wide. All cotyledon cells, including the epidermis, are filled with closely packed aleurone grains (protein bodies) between 4 μm and 18 μm in size (Vaughan, 1970) and numerous tiny oil droplets. Scattered throughout the tissue of the cotyledons are small prismatic crystals of calcium oxalate which are arranged in pairs end to end. The crystals are clearly visible in polarized light. Wallis (1913) described them as being 24 μm long and 4–5 μm wide, and he mapped their distribution in the cotyledon. The characteristic shape of the crystals is unusual in legumes and may be used as a diagnostic aid for soya material.

Occasional starch granules have been noted by some observers. Wallis (1955) attributes these to seed immaturity, though they may be explained as a varietal difference (Winton and Winton, 1932). Since this early work, no new features of the soybean have been shown by the light microscope. The ultrastructure of the soybean as revealed by the electron microscopy of Saio and Watanabe (1968) and Wolf and Baker (1972, 1975) is discussed in Chapter 2.

II. MICROSTRUCTURE OF PROCESSED SOYBEANS: SOYA FLAKES AND FLOURS

Raw soybeans contain a trypsin inhibitor, and inactivation of this by heat treating the beans is an invariable part of processing. In the production of full fat flours the beans are steamed, dried and then cracked to loosen the hull, much of which can then be removed by air blast. Flaking and grinding the dehulled material gives an edible full fat flour containing 21% fat and 41% protein (Horan, 1967).

Dehulling is never complete and the light microscope shows that full fat flakes and flours contain all the cellular elements present in the soybean. Protein bodies can still be seen in the aleurone and cotyledon cells, and the abundant lipids may also be demonstrated with fat-soluble dyes such as oil red O.

Full fat flours can now be made by extrusion cooking of the bean particles (Mustakas *et al.*, 1970) instead of flaking, and this is claimed to give a nutritionally improved product. Such a product would have microscopical features more in common with extruded textured soya products, e.g. protein bodies and oil bodies would lose their identity, and many cotyledon cell walls would be damaged, but seed coat cells would survive.

In the production of defatted flakes and flours the cracked beans are rolled to give thin (0·25 mm) flakes which facilitate oil extraction. Hexane is used to extract the soya oil, and this leaves the flakes with only about 1% fat, but now containing residual solvent which is removed by a combination of heat and vacuum. A minimum exposure to heat is important where protein denaturation is undesirable, e.g. in flakes to be used for protein isolation. Where flakes are used as animal feed, steam treatment of the solvent-extracted flakes is important as a means of improving nutritional quality by the destruction of the antinutritional factors present.

This cooking or 'toasting' of defatted flakes, whilst it usually improves the nutritional value, denatures the protein, and excessive denaturation can itself reduce the nutritional quality. The dispersibility of the protein is measured by the nitrogen solubility index (NSI) which reflects the degree of protein denaturation. If little or no moist heat is used after hexane extraction, the NSI figure is high, usually 50–90% depending on the process. Such a product is called a high-solubility meal or a white flake. A moderate amount of moist heat yields a medium-solubility cooked flake with an NSI from 25–50%, whereas a greater amount of moist heat results in a low-solubility toasted product in which the NSI is below 25%, usually in the range 12–25%.

Using light microscopy, Sipos and Witte (1961) and Smith (1976) have shown that the cell walls in an uncooked defatted flake are largely intact. Figures 2 and 3, which are of a medium-solubility cooked meal, confirm this and also give an indication that protein bodies within the cells are still discernible even after some heat treatment. Soya grits, obtained by the coarse grinding of soya flakes and soya flour which has a smaller particle size, are widely used in the food industry as such, and also in the form of protein concentrates. These are made by the extraction of low molecular weight components from the flakes and flours which leaves them with an increased protein content of about 70% (Wolf and Cowan, 1971). Concentrates are used by the meat industry to increase water retention and reduce shrinkage on cooking, as well as to increase the protein content of comminuted products.

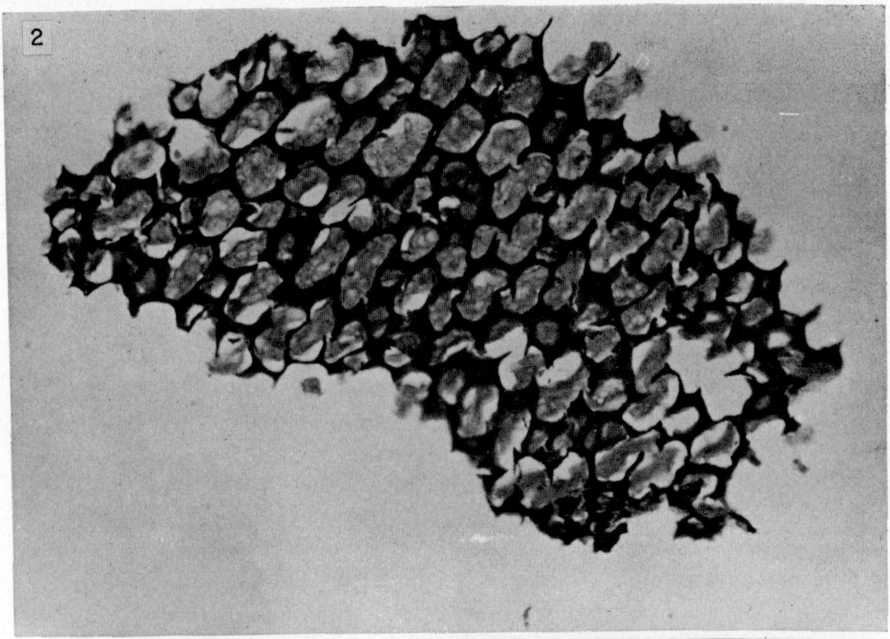

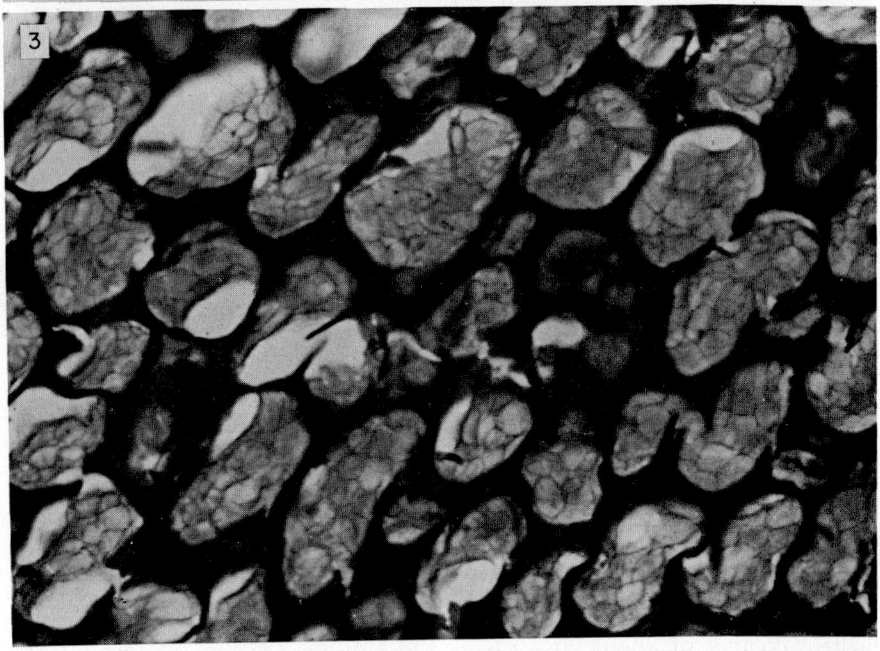

FIGS 2 and 3. Medium-solubility soya flake showing intact cell walls containing partially modified protein bodies, ×90 and ×226 respectively.

III. MICROSTRUCTURE OF TEXTURED SOYA PRODUCTS

Soya isolates, concentrates and flours are used to produce the wide range of textured proteins now available. These are used to extend and simulate meat. Here, one of the main objectives in texturizing soya material is to reproduce the chewiness and mouth-feel associated with natural meat products. But meat texture is itself variable (see Chapter 5) and the method of cooking it adds further variety. Thus a textured soya product simulating steak would be expected to have a different microstructure to one designed to imitate, say, crisply cooked bacon. The starting point for these textured products is material of plant origin the microstructure of which is totally different from meat, and yet to a greater or lesser extent a meat-like structure is engineered by one of the several methods described below.

A. Spun Proteins

The earliest successful imitation meats, or meat analogues as they are termed, were spun from soya proteins isolated from undenatured defatted white flakes. Boyer (1954) isolated soya protein by dispersing the flakes in alkali, centrifuging off insoluble material and treating the solution with acid to precipitate the protein at its isoelectric point (about pH 4·5). The precipitated protein of 95–98% purity was redispersed in aqueous alkali and used as a 'spinning dope' in a spinning technique which closely followed that used for textile fibres such as rayon. The 'spinning dope' could be coloured and flavoured at this stage, and it was also possible to add fat to it in the form of an emulsion. The dope was then forced through the porous surface of a spinneret (a die containing some 5000–15 000 tiny holes each about 0·003 inches (76 μm) in diameter) into a fixing bath of salt and acetic acid where the coagulated filaments were formed. These were led away from the spinneret at sufficient speed to keep them in a straight condition. After being led from the precipitating bath, the filaments were further drawn out, resulting in a greater degree of orientation of the protein molecules and a corresponding increase in filament strength.

Conversion of the protein filaments into a meat analogue requires that the bundles of fibres are bound together, and this can be done in one of several ways. The method chosen depends on the meat being simulated (Gutcho, 1973), e.g. in simulating the white meat of poultry all the filament groups from a battery of spinnerets are first treated with an edible binder and combined before passing through a bath of fat, whereas in simulating a pork chop the product of each spinneret is treated with fat before the combined filament groups pass through the binder. Flavouring and colouring may be added at the binder stage, but better penetration is achieved if these are incorporated in the spinning dope.

Many of these differences in processing detail can be seen in the microstructure of the spun product. The most obvious difference is that of fibre diameter, as shown in Figs 4 and 5. The fibres seen in Fig. 4 (a ham analogue) have a diameter of about 40 μm, i.e. of the same order as the muscle fibres in pork, whereas the fibres in Fig. 5 (a beef analogue) are much coarser (about 130 μm in diameter) than the meat fibres simulated. The lipid content and its location also varies. Here, the beef analogue has a high fat content (about 35%, based on dry weight) and the fat globules shown in Fig. 5 form an integral part of the fibre structure. Compare this with the fat globules in Fig. 4 which, though closely associated with the fibres, are external to them. The presence of fat inside the fibre plays an important part in whether or not the product achieves a mouth-feel similar to meat, especially when the appropriate animal fat is used.

Different spun products vary in their carbohydrate content from almost nil (Coomaraswamy and Flint, 1973) to a few per cent, and this can usually

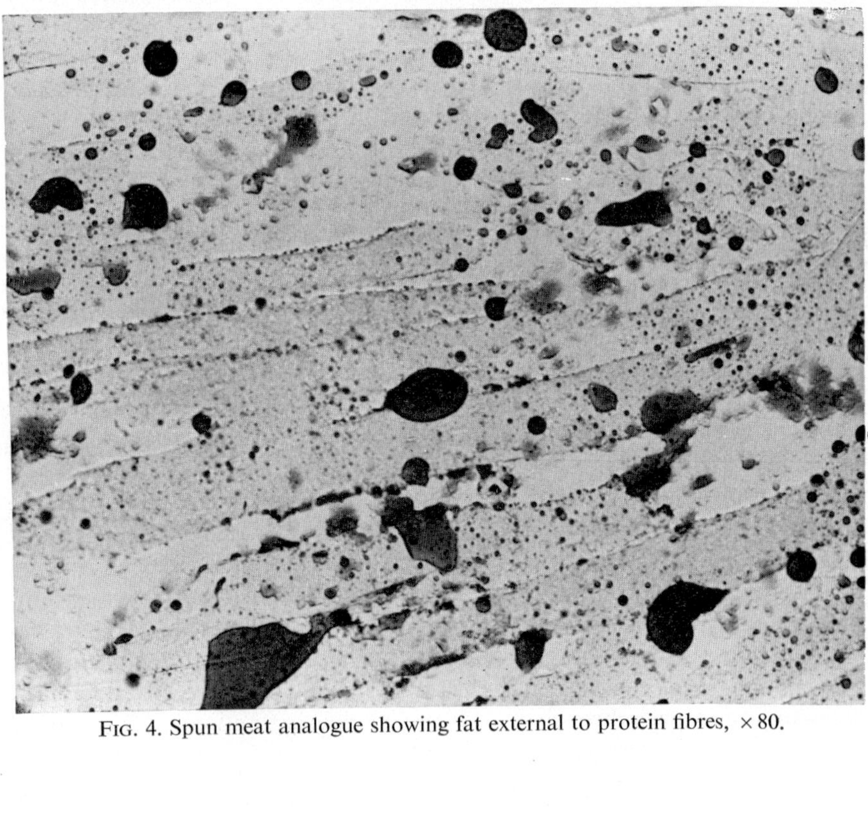

FIG. 4. Spun meat analogue showing fat external to protein fibres, ×80.

be attributed to the binder used. An interesting development is the combination of a spun soya protein with unstructured gluten (wheat protein) as shown in Fig. 6. This greatly increases the carbohydrate content of the product because of the starch which the gluten invariably contains.

Isolate spinning is capable of manufacturing a wide range of products, but it is expensive. More recently it has been found possible to convert defatted soya flakes into fibres by extruding a finely divided slurry of them through a small orifice (Wolf, 1976). These fibres would be expected to contain some of the more resistant plant cells present in the defatted flakes. This is in common with the structured soya proteins produced by extrusion cooking which are next discussed.

B. Extruded Soya

'Extrusion cooking may be defined as the process by which moistened, starchy and/or proteinaceous materials are plasticized and cooked in a tube by a combination of pressure, heat and mechanical shear.' (Smith, 1976.)

The effect of extrusion cooking on defatted soya flour is to free most of the protein from the cotyledon cells. The empty cells collapse and the protein,

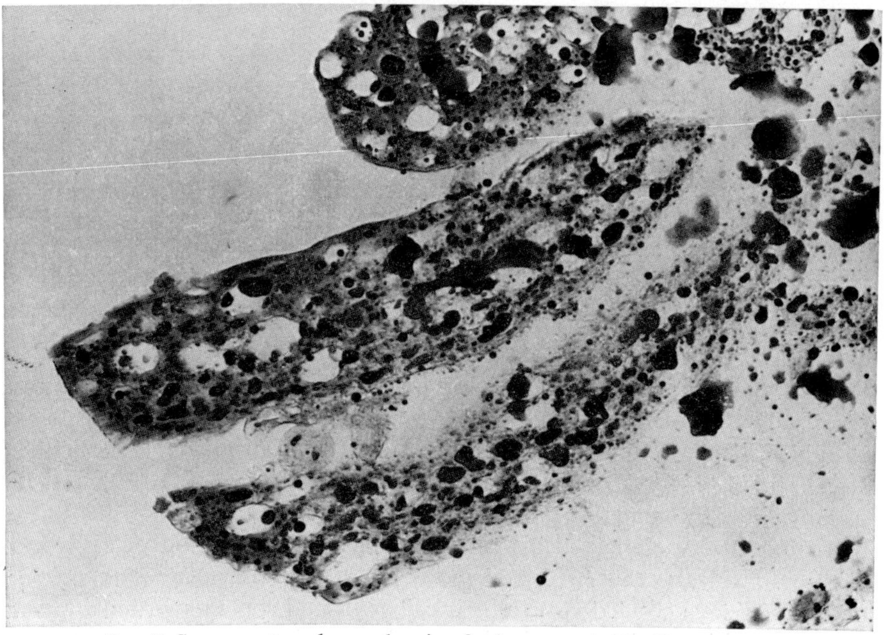

FIG. 5. Spun meat analogue showing fat incorporated in fibres, ×90.

now in a plastic state, forms a matrix in which these collapsed cells, any remaining intact cotyledon cells and all the resistant seed coat cells are embedded. This cooked material, still under pressure, is extruded through a die into the atmosphere; the resulting pressure drop causes the product to expand and gives it a spongy texture. Size and shape of the extrudate is

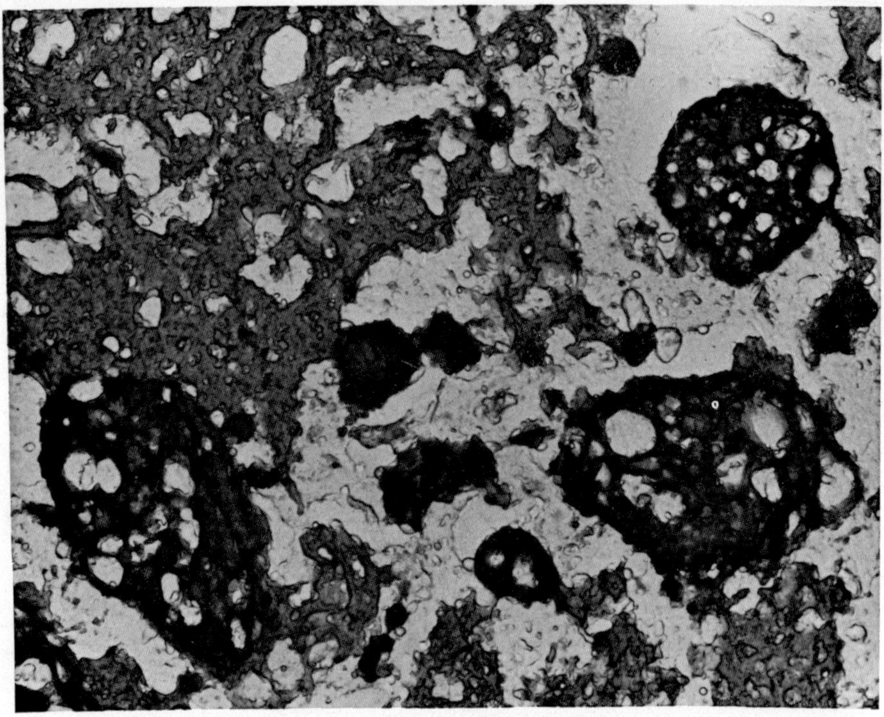

FIG. 6. Section of spun soya meat analogue with gluten binder, ×90. (Soya stained dark with paler-staining gluten.)

controlled by the pressure drop, the arrangement of the dies and the speed of the cutting knife which slices the still-moist material into manageable pieces (Wolf and Cowan, 1971).

Published photomicrographs of extruded soya products (Link, 1969; Jewell, 1973, 1974; Coomaraswamy and Flint, 1973; Smith, 1975, 1976) usually show the material to have an almost fibrous appearance due to the presence of very elongated air cells running in the direction of extrusion. Figure 7 illustrates this effect and shows how the strands of protein envelop particles of soya cotyledon with their cell walls and contents virtually intact.

The conditions of thermoplastic extrusion yield a range of products with differing densities, and this is reflected in the variety of microstructures obtainable from a single starting material like defatted soya flour. Figure 8 shows a highly expanded product. The light, open structure is associated with the crispness appropriate for these imitation bacon bits, and the presence of quite large fragments of seed coat is accommodated by the protein network without apparent detriment to the product's texture. Textured soya

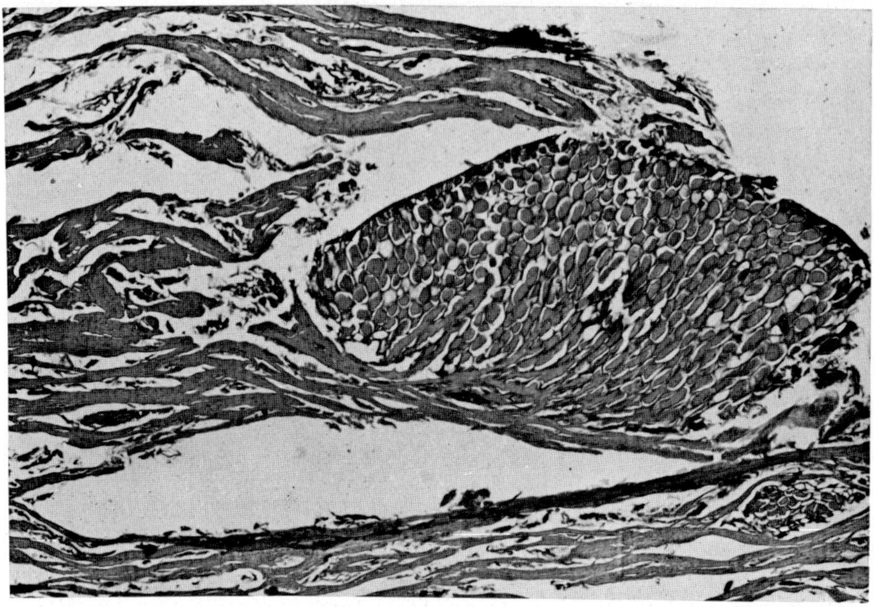

FIG. 7. Extruded soya flour showing large and small groups of cotyledon cells, ×30.

granules designed to extend the meat in hamburgers are required to be tougher and chewier and correspondingly denser, as Fig. 9 shows. In this product a high proportion of the cotyledon cell walls are now in an amorphous state, but small groups of cells still remain as seen in Fig. 12.

The flexibility of extrusion cooking enables it to deal with a wide range of raw materials which can be processed singly or in admixture. Figure 10 is an extrusion-cooked brown mustard (*Brassica juncea*) and soya, and shows intact mustard cells and palisade cells derived from soya.

One ingredient which is being used increasingly as a constituent of textured vegetable proteins is wheat protein (gluten). This merits separate consideration because it is widely available, its amino acid profile complements that of soya and it appears to have useful functional properties.

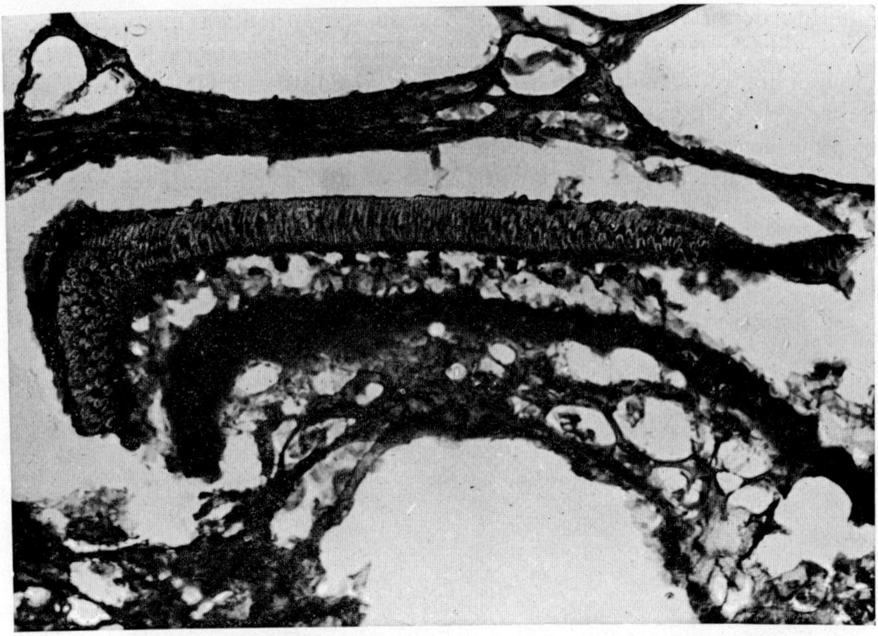

FIG. 8. Imitation bacon bits, showing a group of palisade cells, × 80.

IV. GLUTEN AS A NOVEL PROTEIN CONSTITUENT

In one sense the plant protein industry started in 1866 when Dr John Harvey Kellogg prepared 'vegetarian steaks' from fresh gluten (Robinson, 1972). The protein material derived from wheat flour by making a flour dough, allowing it to stand and washing out the starch, is termed 'vital gluten'; it still contains some starch (15–20%) and about 5% fats and lipoproteins. The direct descendant of Kellogg's 'steaks' can be found in the canned 'meatless steaks' still marketed today. This material is a dense, rubbery product, entirely lacking in any fibrous structure and with its starch fraction in a gelatinized state. Figure 11 shows the typical microstructure of these products; much of the starch is dispersed throughout the protein, but here and there individual swollen granules can be seen. The gas-retaining properties of gluten, so important in breadmaking, are not utilized in these canned glutens which have a uniformly dense structure with only the occasional gas cell present.

Gluten, both freshly isolated (vital gluten) and in a dried form, finds widespread use as a supplement to bread flours where it gives better loaf volume

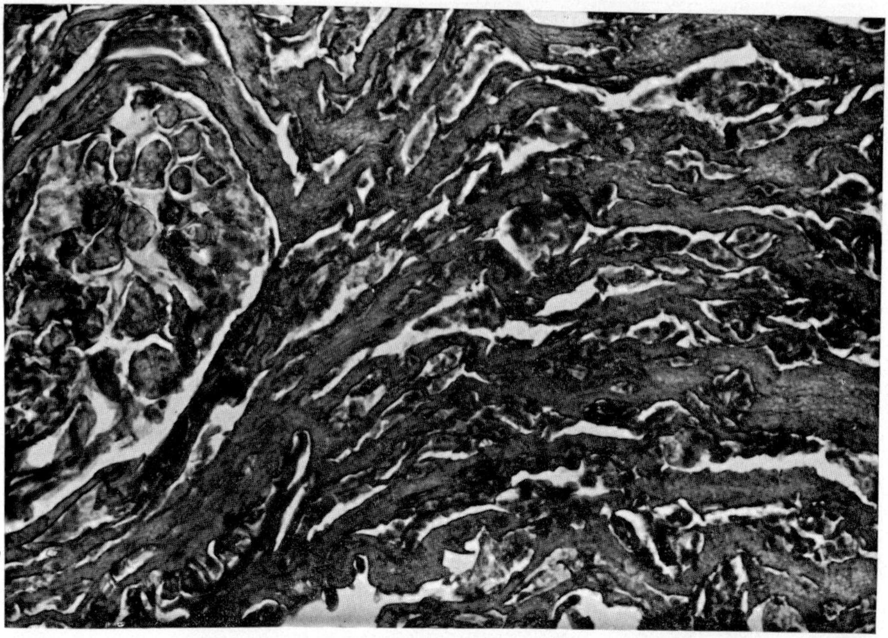

FIG. 9. Hamburger granules, ×80. Note cotyledon cells on the left.

and enables starch-reduced diet breads to be made. With its 80% protein, dried gluten can be regarded as a plant protein concentrate and is used as such in vegetarian (and non-vegetarian!) sausages. Its microscopical appearance in these products is similar to Fig. 11.

The use of gluten as a second ingredient of textured vegetable proteins has developed for both nutritional and functional reasons. Nutritionally, soya and wheat proteins are complementary; soya protein has a high lysine content which wheat lacks, wheat protein is rich in sulphur-containing amino acids which soya protein lacks. It is, however, the functional properties of water absorption and retention shown by processed gluten that have most influenced its use as an ingredient in textured vegetable proteins, particularly those designed for incorporation in comminuted meats.

Extrusion-cooked soya proteins absorb 200–300% of their weight of water. Products containing gluten will hold up to 400% water. Figures 12 and 13 show that this additional water adsorption is not due to the formation of an open spongy structure (which would allow absorbed water to be squeezed out easily in mechanical processing), and the phenomenon can be better explained in terms of the water-holding properties of the extrusion-cooked gluten, especially the gelled starch present.

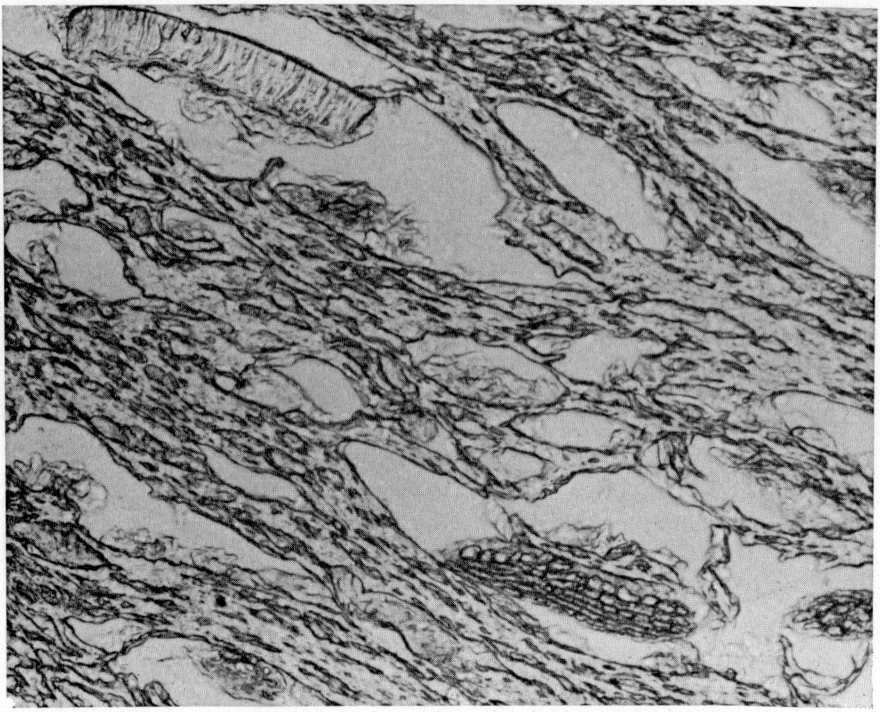

FIG. 10. Extruded defatted brown mustard and soya flour, ×90.

The sample shown in Fig. 12 was stained to show the gelatinized starch granules present, but obviously the water-holding properties will depend more on the starch fraction which has been cooked beyond the stage of retaining its morphological identity. Though unstained, the swollen starch granules can just be seen in Fig. 13. This photomicrograph shows a fragment of soya cotyledon with soya protein trapped in the darker-stained cotyledon cells. The darker areas of the protein matrix are due to soya protein and the lighter areas to gluten, an indication that in the combined products the wheat and soya protein may be incompletely mixed.

Another interesting characteristic of these and many of the soya–gluten products is that the air cells present lack the directional orientation typical of air cells present in soya extrudates. This may be due to the cohesive strength contributed by the gluten, but differences in processing technique can also play an important part. Figure 14 shows an experimental extruded blue lupin (*Lupinus angustifolius*) and gluten product which has the fibrous texture typical of many commercial soya extrudates. The palisade cells seen in Fig. 14 are morphologically distinct from those of most legumes, only the

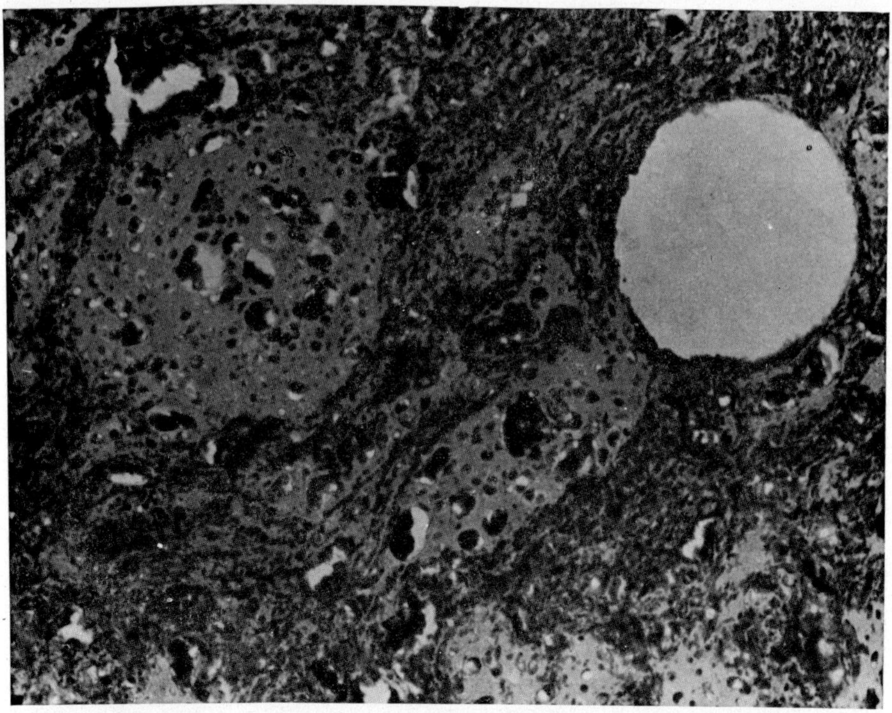

FIG. 11. Canned 'gluten steaks', ×90. The darker-staining particles are gelatinized starch granules.

lupin genus having a curved seed coat cell. Histological differences between soya and other leguminous seeds are shown by Vaughan (1970), and the main histological features of leguminous seeds are summarized in Winton and Winton (1935). These histological characteristics would serve as aids in characterization of the source of plant protein used in the manufacture of extruded products.

V. MICROSCOPY OF SOYA AND OTHER PLANT PROTEIN ISOLATES

The isolated proteins of plants present a more difficult task in microscopical identification. The dehydrated materials reflect the method of drying rather than the source of the protein; even at the ultrastructural level Wolf and Baker (1975) likened the SEM of sodium soya proteinate to that of spray-dried calcium caseinate. The soya protein isolates are virtually free of carbohydrate material, and when added to food products the protein disperses readily. Jewell (1974) demonstrated that with toluidine blue staining meat products containing a soya protein isolate showed an overall increase in staining; however this difference is too subtle to be used for the positive identification of the plant protein.

VI. NOVEL PROTEIN FOODS FROM MICRO-ORGANISMS

Of the very large number of different micro-organisms known only a tiny proportion have been studied as potential sources of protein (Hulse, 1974). These, however, span a wide range of morphologically distinct types, and unless they receive destructive processing the broad class of origin can be recognized microscopically. Examples of yeasts, moulds, bacteria (Stokes, 1958) and algae (Black, 1958) as possible sources of food protein have received attention over a relatively long period and interest continues.

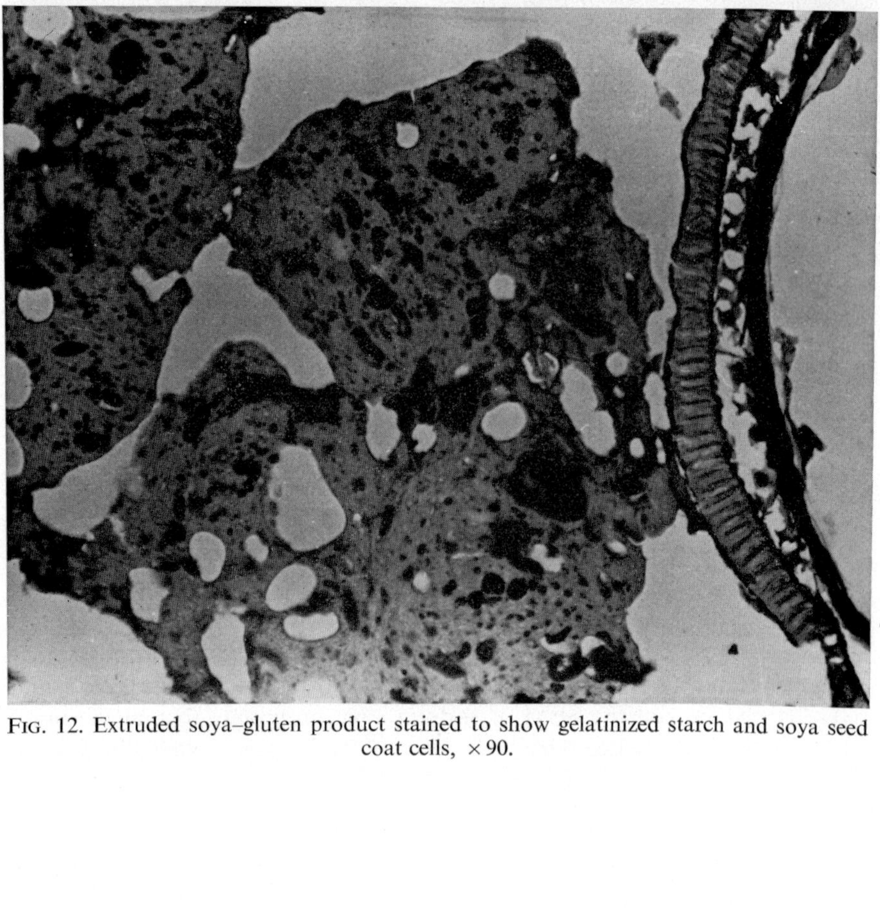

FIG. 12. Extruded soya–gluten product stained to show gelatinized starch and soya seed coat cells, ×90.

FIG. 13. Same product as Fig. 12 but stained to show soya protein, ×90. Note central group of soya cotyledon cells.

Modern developments discussed by the British Food Standards Committee (1974) include the growth of strains of *Fusarium graminearum* on starch waste and of *Aspergillus niger* on carob husks. Sources of single cell protein (SCP) mentioned range from a modern industrial process using bacteria capable of growth on methanol to the use of algae, e.g. *Chlorella spp*. and *Spirulina maxima* which are able to photosynthesize their own carbohydrates and can be grown in warm water containing sources of nitrogen and minerals.

VII. THE MICROSCOPY OF FUTURE NOVEL PROTEIN FOODS

Because of their rate of growth, micro-organisms may well become important sources of protein in the future. The unicellular types seem destined to be used as animal feed or as a human food supplement, e.g. in the protein enrichment of bread (Hulse, 1976), but the filamentous forms clearly have potential for the production of varied protein foods with a natural micro-fibrous texture.

Meanwhile, the early textured vegetable proteins are being supplemented and replaced by what are described as 'second-generation' texturizing methods (Horan and Wolff, 1976).

When they were first produced spun proteins were regarded as high-protein foods designed to replace meat. The extruded products were seen only as

FIG. 14. Extruded defatted blue lupin flour and vital wheat gluten product, ×75.

meat extenders. This situation is already changing; spun products are appearing in combination with meat as extenders and new methods of extrusion are yielding meat analogues. These new methods include a double-extrusion technique which enables denser products with a more meat-like texture to be made, not only from soya flour but also from defatted soya flakes, defatted peanut flour, vital wheat gluten and rape seed concentrates (Smith, 1975, 1976). With many protein-rich oil seed residues available it is obvious that the extruded products of tomorrow will show even greater variety of structure and content than those currently available.

Variations of spinning technology, which include the use of mixtures of proteins and polysaccharides, and innovations in the spinning method are discussed by Horan and Wolff (1976) and described in the patent literature

(Gutcho, 1973). These differences will be seen in the microstructure of the spun products of the future.

Even without further technological developments the range of novel protein products with distinct microscopical features will continue to grow. If we add to these developments further probable technological innovations the range of novel products with distinct microscopical features will be great indeed.

ACKNOWLEDGEMENTS

The extruded products (mustard and soya (Fig. 10), lupin and gluten (Fig. 14)) were provided by Dr J. F. Kefford of CSIRO Australia.

The work was made possible by the financial help given by the Sainsbury Centenary Trust.

REFERENCES

Altschul, A. M., ed. (1958). 'Processed Plant Protein Foodstuffs.' Academic Press, New York and London.

Bergeron, M. and Durand, P. (1977). *Annls Nutr. Aliment.* **31**, 261.

Black, W. A. P. (1958). *In* 'Processed Plant Protein Foodstuffs' (A. M. Altschul, ed.), p. 805. Academic Press, New York and London.

Boyer, R. A. (1954). U.S. Patent 2 682 466, 29 June 1954.

British Food Standards Committee (1974). 'Report on Novel Protein Food', F.S.C./REP/62. H.M.S.O., London.

Coomaraswamy, M. and Flint, F. O. (1973). *Analyst*, **98**, 542.

Flint, F. O. and Lewin, Y. A. (1976). *J. Fd Technol.* **11**, 137.

Gutcho, M. (1973). 'Textured Foods and Allied Products.' Noyes Data Corporation, New Jersey.

Horan, F. E. (1967). 'Proc. Int. Conf. on Soybean Protein Foods', ARS-71-35, U.S. Department of Agriculture, Washington D.C. (Cited in Wolf and Cowan, 1971.)

Horan, F. E. and Wolff, H. (1976). *In* 'New Protein Foods' (A. M. Altschul, ed.), Vol. 2, Part B, p. 261, 278. Academic Press, New York and London.

Hulse, J. H. (1974). *In* 'New Protein Foods' (A. M. Altschul, ed.), Vol. 1, Part A, p. 188. Academic Press, New York and London.

Jewell, G. G. (1973). British Food Manufacturing Industries Research Association Tech. Circ. No. 528. (Confidential to members.)

Jewell, G. G. (1974). British Food Manufacturing Industries Research Association Tech. Circ. No. 571. (Confidential to members.)

Link, H. (1969). *Fleischwirtschaft*, **49** (4), 470.

McLennan, D. G., Gow, J. G. and Stringer, D. A. (1973). *Process Biochem.* **8**, 22.

Mustakas, G. C., Albrecht, W. J., Bookwalter, G. N., McGhee, J. E., Kwolek, W. F. and Griffin, E. L. (1970). *Fd Technol., Champaign*, **24**, 1290.

Robinson, R. F. (1972). *Fd Technol., Chicago*, **26**, 59.
Saio, K. and Watanabe, T. (1968). *J. Fd Sci. Technol., Tokyo*, **15**, 290.
Sipos, E. and Witte, N. H. (1961). *J. Am. Oil Chem. Soc.* **38**, 17.
Smith, O. B. (1975). *Fd Engng*, **47**, 10, Engineered Foods Suppl. EF, 26.
Smith, O. B. (1976). *In* 'New Protein Foods' (A. M. Altschul, ed.), Vol. 2, Part B, p. 87, 104, 107. Academic Press, New York and London.
Stokes, J. L. (1958). *In* 'Processed Plant Protein Foodstuffs' (A. M. Altschul, ed.), p. 789. Academic Press, New York and London.
Vaughan, J. G. (1970). 'The Structure and Utilization of Oil Seeds.' Chapman and Hall, London.
Wallis, T. E. (1913). *Pharm. J.* **91**, 120.
Wallis, T. E. (1955). 'Textbook of Pharmacognosy', 3rd Edn. J. and A. Churchill, London.
Wallis, T. E. (1965). 'Analytical Microscopy', 3rd Edn. J. and A. Churchill, London.
Winton, A. L. and Winton, K. B. (1932). 'The Structure and Composition of Foods', Vol. 1. John Wiley and Sons, New York.
Winton, A. L. and Winton, K. B. (1935). 'The Structure and Composition of Foods', Vol. 2. John Wiley and Sons, New York.
Wolf, W. J. (1976). *In* 'Advances in Cereal Science and Technology' (Y. Pomeranz, ed.), Vol. 1, p. 354. A.A.C.C., St Paul, Minnesota.
Wolf, W. J. and Baker, F. L. (1972). *Cereal Sci. Today*, **17**, 124.
Wolf, W. J. and Baker, F. L. (1975). *Cereal Chem.* **52**, 387.
Wolf, W. J. and Cowan, J. C. (1971). 'Soybeans as a Food Source.' Butterworth, London.

15. Sugar

H. E. C. POWERS
(formerly Chief Chemist of Tate and Lyle Ltd.)

34 Wordsworth Avenue, South Woodford, London, U.K.

I. INTRODUCTION

There are a number of organic substances classified by science as sugars, e.g. sucrose, glucose (dextrose), laevulose, lactose, maltose, raffinose, etc., each differing in molecular structure. By common usage, however, the term sugar generally implies sucrose, and it will be used in this sense here with only passing reference to other sugars.

The world's annual production of sugar is now around 80 million tons, more than double that produced in 1914. Moreover, in recent years the amount has increased even more dramatically. It is pertinent here to draw attention to the fact that sugar is virtually 100% solids, whereas most other foods contain a large proportion of water. Sugar and salt are both pure, crystalline chemicals and both are included in the diet of practically the whole world population.

Research thereon has been carried on by generations of keen men, and is still developing rapidly. Moreover to the credit of the industry this has been a most co-operative effort within the period of my personal activity. As with so many industries, the advent of very high-resolution instruments, and also ingenious new techniques, has been most helpful.

With the increasing price of oil, the potential use of sugar as a base for the synthesis of organic compounds is becoming an interesting possibility. This becomes more promising in view of the fact that sugar is produced from the present energy of the sun in almost unlimited quantities, whilst the available supplies of fossilized fuels will be running down in the years ahead. Sugar is of supreme value as a guinea-pig substance for research upon crystallization mechanisms, aided as it is by the justification of benefit to the future of the world-wide industry itself. This is already showing in many fields of study, to mention but one, in the light which is being thrown upon the molecular activities in aqueous solution. Thus, as sugar has both H and OH bonds, it enters into loose combination with water molecules when allowed to stand undisturbed. Several hydrates of sucrose are known and can be crystallized within certain low temperature ranges.

II. SUGAR PRODUCTION

A. Raw Materials and Initial Processes

Sugar is a natural product produced in the sap of many plants, the most important of these being sugar cane which grows in many parts of the world which have a near-tropical climate. Sugar cane probably originated in the islands of south-east Asia, and records indicate that it was a luxury food for the advanced civilizations of the world as long as 2000 years ago. Next in importance is sugar beet, which originated as a crop less than 200 years ago. The Napoleonic wars cut off all supplies of cane sugar from France, and rewards were offered for finding an alternative sugar source from plants which could be grown locally. A scraggy root which grew on the shores of the Mediterranean was chosen, and by long and persistent research it gradually became so improved as now to be second only to cane. Small, local supplies of sugar are also derived from the sap of certain varieties of maple and also palm. In the cases of both cane and beet the sap contains from 15–20% of sugar. With the tough, fibrous cane the sap is extracted by crushing between heavy rollers, then washing to extract as much sugar as possible. With beet the roots are washed, then sliced, and the sugar extracted by diffusion. In both cases the rather dilute juice contains very many unwanted impurities, both solid and in solution. The cane juice has rather a pleasant odour, but beet is the reverse. In both cases the dilute juice is very vulnerable to microbiological development which can cause both serious losses and increasing problems of extraction and crystallization.

It will be evident that microscopy has played, and still is playing, a major part in both plant breeding and the later research upon microbic and yeast enemies. The juices are most vulnerable in their dilute state, since after evaporation the osmotic pressure is too high for most organisms to survive, except certain types which are thus termed osmophyls. So, if the cane is processed as quickly as possible after cutting, the loss will be minimal, but it has become a practice in some areas to burn the cane leaves before cutting. Then, since the cells in the cane tip are ruptured by the heat, the cane needs to be treated rapidly or this microbiological degradation takes place with its inevitable penalties. Rupture by frost can lead to similar damage, this being true with both cane and beet if left exposed to freezing conditions.

The late Pamela Scarr was an outstanding researcher in microbiological problems, and I recommend her 1968 paper in which she used phase contrast microscopy to test whether raw sugar deliveries were likely to deteriorate on long storage. Vast quantities of bagasse (cane fibre) are still burned wastefully, awaiting some economically worthwhile use. Three important papers on this topic are available. Suarez and Moser (1973) describe a study of the cell wall with electron microscopy; Withy (1964) uses fluorescent microscopy with acridine orange as a dye, the lignified cell walls being rendered yellow–green and cellulose deep-red; and Delavier and Shokrani (1974) reproduce ten transmission electron and scanning electron micrographs to show the structure of cell walls and pores.

B. Purification and Crystallization

The purification of the juices from both cane and beet is a multi-stage process. The initial step is essentially a clarification process, in which all suspended solid matter is removed by one of several filtration processes. In the cane industry certain coagulation and decantation treatments are helpful. Bennett and Fentiman (1969) developed the Talofloc process, now widely used. Chen and Picou (1972–73) report on a polymer flocculant in mud filtration, giving eight electron micrographs. In both cane and beet industries the most common and most successful process is known as carbonatation. In this, milk of lime is added to the juices (in the case of beet) or to the higher density 'melter' (in the case of cane). Carbon dioxide, generally in the form of washed flue gas, is passed through until the pH has been brought down to the region of 8·2–8·5, at which the maximum precipitation of lime has taken place. Filtration then removes not only the precipitated calcium carbonate and other calcium salts but also many other absorbed substances and suspended solids. This yields a bright, clear, but still coloured solution. To remove this colour various processes are again used; occasionally sulphur dioxide (sulphitation) is employed, but more widely the use of some form of an adsorbent such as animal charcoal, decolorizing carbons, resins, etc. is involved. Thus a water-white sugar solution is obtained.

Sugar production is a most effective form of food dehydration, in which the last stage is crystallization. In this, nature comes to the aid of the manufacturer in that the component sugar molecules are brought together in crystals—a most perfect example of mini-packaging. This takes place in large vacuum pans, in which the expert pansman in charge can continuously vary the temperature, concentration and other variables in order to produce the type of crystals desired in that particular boiling. When he has achieved this objective the very viscous 'massecuit' is allowed to flow into centrifugal machines which, rotating at speed, throw off most of the syrup, leaving the crystals. These are given a brief wash to remove still more effectively the remaining syrup. The crystals are then sent to rotating, near-horizontal 'granulators' through which warm, dry air is passed counter-current to the sugar. This yields the familiar dry crystals.

III. THE CRYSTALLINE STATE AND INDUSTRIAL PROBLEMS

Crystallinity may be considered as a particularly highly ordered form of matter, in that the component molecules (or atoms) are built in an orderly manner, reminiscent of bricks in a wall. The molecular structure of crystals in general has been the subject of intensive study for more than 50 years, particularly by such high-resolution techniques as X-ray and neutron diffraction, as well as by light and electron microscopy. Cumulatively these studies have had almost unimaginable influence upon our lives, and we can foresee the advent of far greater developments as we learn more and more about molecular architecture. Certain substances, when crystallized under very careful control, are endowed with remarkable physical properties of controlling light, time measurement, electricity, etc. With due humility we can continue to learn more and more about how to so influence the growth of crystals as to extend these abilities.

During the last 20 years a world-wide society has been built up with this objective. The British section of the society is the British Association for Crystal Growth, the interest of whom is associated with crystallization as distinct from '*post mortem*' crystallography.

A. Different Forms of Crystals Grown from Aqueous Solution

When crystals are grown whilst completely immersed in solution they assume the familiar flat-sided, many-angled, form known as polygonal (Fig. 1). If they grow whilst suspended by surface tension, or whilst attached to the sides or bottom of a container, and are unable to receive molecules on the protected side, then they grow as 'hopper' crystals, visibly distorted on the protected face (Fig. 2). In a well-grown crystal of any one substance the angles between adjacent sides will always be identical for all crystals of that substance. Measurement of these angles by means of an instrument known as a goniometer forms one method of identifying the substance.

Nevertheless, the apparent shape of a crystal of any substance may superficially vary very greatly to the human eye. Both sugar and salt can be grown in a compact form or elongated along one or other axis. The limiting forms may be described as 'acicular', 'needle' or even 'whisker' (Fig. 3). Though so different to the eye, the molecular structure of each and all of these is still the same as in the normal form. This great difference in external shape may be due to the presence of certain impurities, which may also affect the growth rate (Smythe, 1967). However, I have shown that needle forms of sucrose can be grown in pure sucrose solution under certain conditions, e.g. at higher supersaturation. When a pure sucrose solution is nucleated at 1·1 supersaturation (SS), crystals of normal axis ratio develop but these vary in growth rate (Fig. 4). At 1·3–1·6 SS normal axis ratio crystals form and grow uniformly (Fig. 5), but after long delay a crop of 'late-developer' crystals appears, some perhaps acicular others normal (Fig. 6). If the experiment is carried out in a long, vertical glass tube, sedimentation rates make this effect striking. The phenomenon merits further investigation for it may throw light upon nucleation.

Sugar 'icicles' may easily be grown by allowing a sugar solution to trickle down or over a suspended crystal in a drying atmosphere, just as with stalactites. Even more remarkable and unfamiliar crystal forms are produced when thin films of solution are allowed to nucleate and grow under controlled vapour pressure conditions (see Section III. I). No more fascinating subjects can be observed, particularly when seen in colour under polarized light.

Having devoted 20 years of my life to the study of crystallization I am in agreement with Tutton (1911), who commented that it was remarkable that no definition of life had been formulated which did not equally well apply to crystals.

FIG. 1. A particularly compact commercially grown granulated sugar.

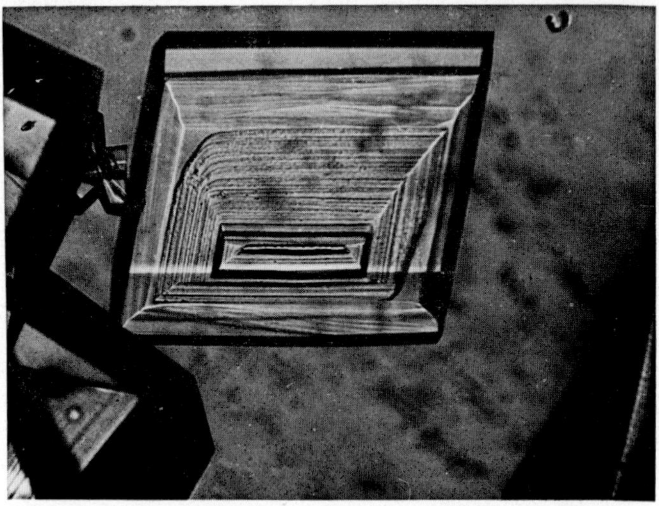

FIG. 2. A 'hopper' crystal showing change of shape from needle to normal.

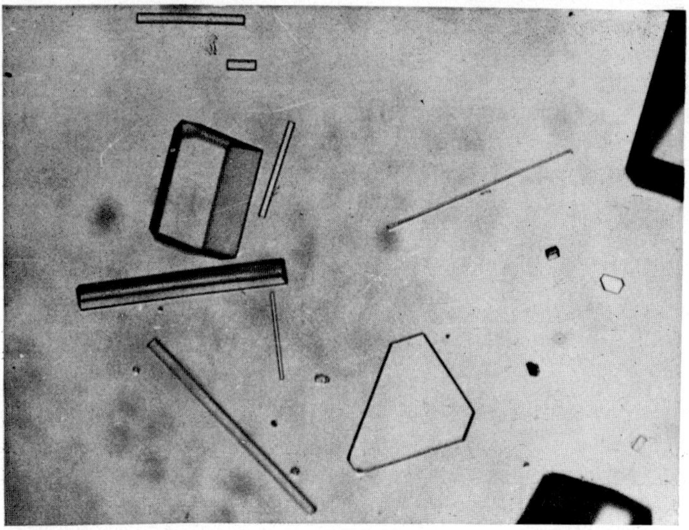

FIG. 3. Showing needle, triangular and rectangular crystal forms grown from pure sugar solution.

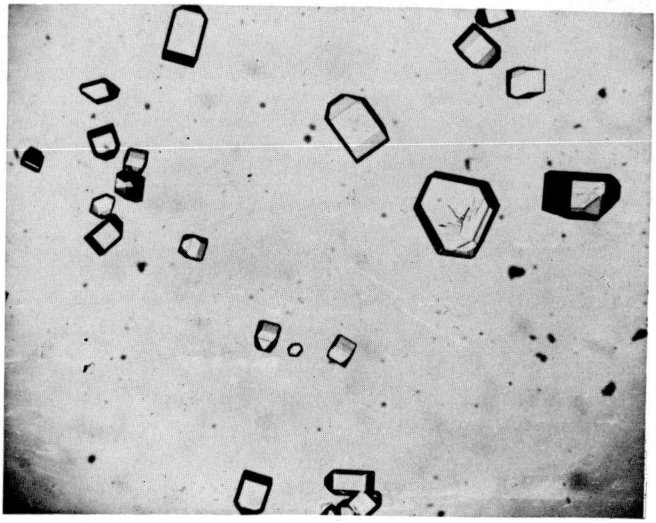

FIG. 4. Showing variation in growth rate when nucleated at 1·1 SS.

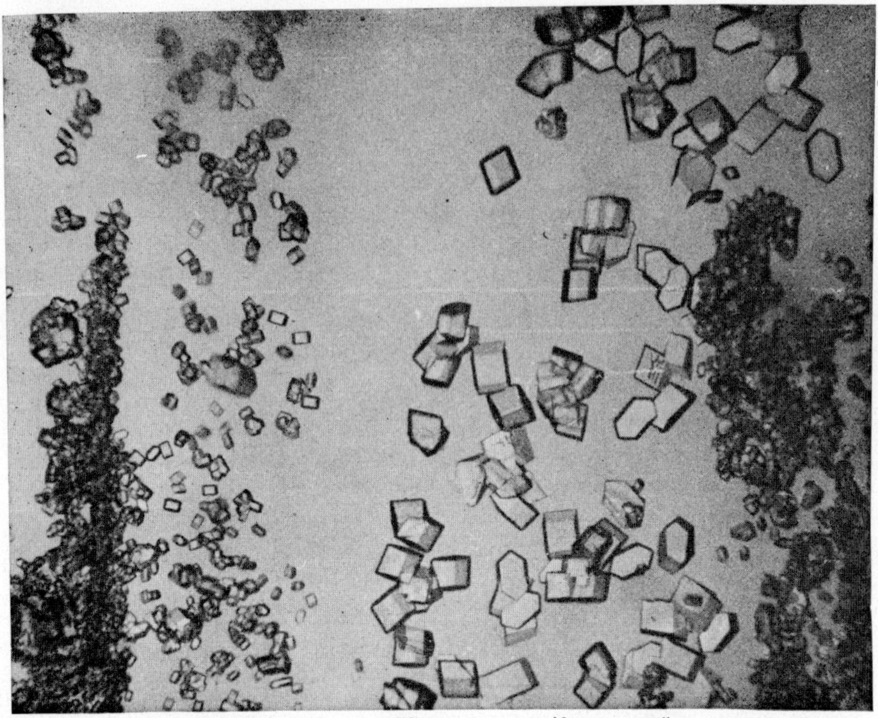

FIG. 5. Sucrose crystals nucleated at 1·4 SS grow at a uniform rate (between two glasses). After the crystals are well grown, the upper glass is moved and a second uniform crystal crop grows. (Nucleation brought about by the movement.)

B. How Crystals Grow

Though it is perhaps natural to look upon crystals as unflawed structures, in fact it is not possible to grow truly perfect-structured crystals. N. N. Sheftal* (personal communication) during 20 years of his early life tried to grow sucrose crystals suitable for piezo-electric purposes, but found them too liable to imperfections. Ammonium dihydrogen phosphate (ADP) proved more amenable and is now grown under extremely careful control for that purpose.

In a supersaturated solution, molecules or molecular groupings tend to be drawn to the surface of a growing crystal by short-range forces. Once contact has been established, they are attracted in one direction only, being free to glide in any direction on the surface. On meeting any projection (a layer front) they will only be able to glide along this until meeting an angle, at which point they will be held until orientated to the appropriate position

* Now Principal of the Institute of Crystallography, Moscow.

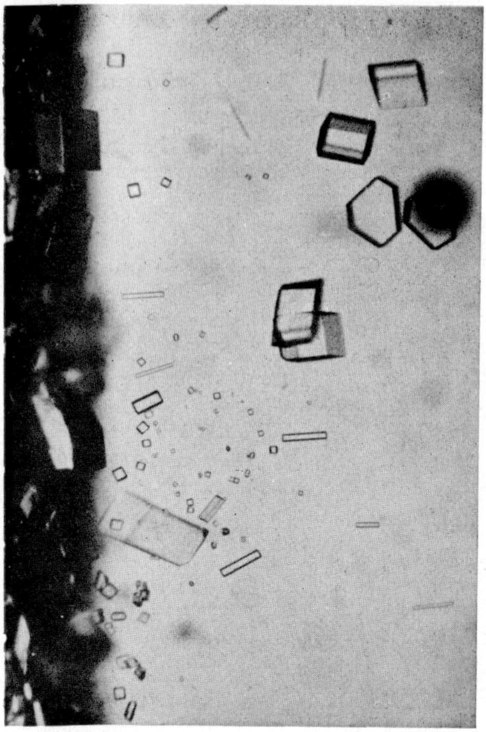

FIG. 6. Nucleated at 1·4 SS, normal uniform crystals grow first and are followed by 'late developers'.

and become part of the solid crystal. This being repeated by the constant bombardment and migration and attachment of fresh molecules from the solution, the cumulative effect is rendered visible by the advance of the layer fronts (Figs 7 and 8). Dissolution occurs by layer recession from corners and edges of the crystal, and also from 'etch pits' from which the layers may be seen spiralling up from the bottom (Powers, 1960). This advance and retreat of layers makes an excellent subject for cine photomicrography.

Since growth layers normally originate from various points on the crystal face (Fig. 9), one can see in these cine films how layers advancing from various directions tend to overlay one another, and such contacts can build in pockets of solution, solid particles (Fig. 10), molecules, and even micro-organisms. These irregularities lead to corresponding faults in the later layer overlays, and so *ad infinitum*.

If one considers how attracted molecules are free to glide until joining the actively growing layer front, it will be realized that any perfect face would not be able to bring them to anchor. There would be no three-dimensional attracting bondages to initiate the next perfect layer. To attempt to explain this, Frank (1949) put forward the possibility of self-perpetuating spiral layer fronts, and these have been demonstrated by cine and still photomicrography using sucrose and with other substances. Personally I believe

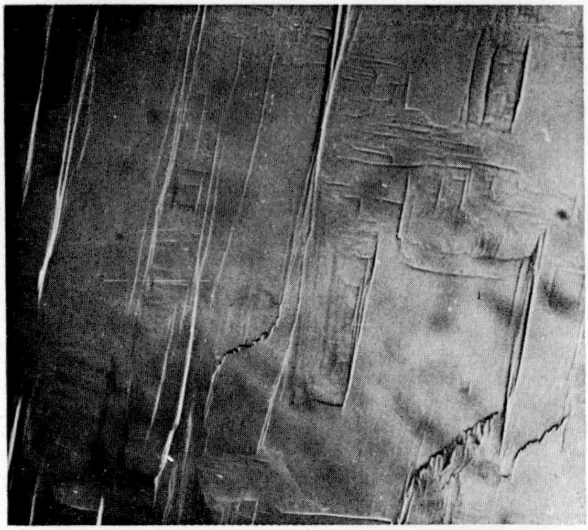

FIG. 7. Typical sucrose layer growth as seen by transmitted light.

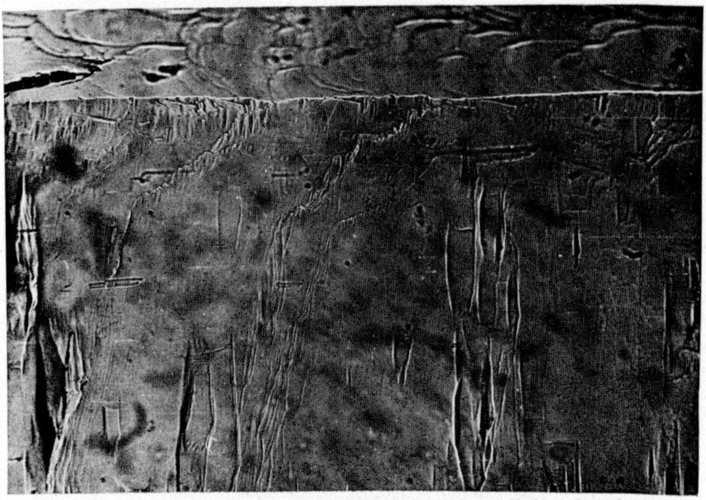

FIG. 8. Typical sucrose layer growth as seen by transmitted light, showing no overlap of layers at edges. Side face resembles spiral growth form.

that both types of growth mechanism may operate for layer growth is perpetuated by inclusions and faults, augmented by the overlying layers. In place of the usual transmitted light, growth spirals can only be seen by reflected light (different contour?) (Figs 11 and 12).

Bunn (1933) pointed out that although molecules of the substance being

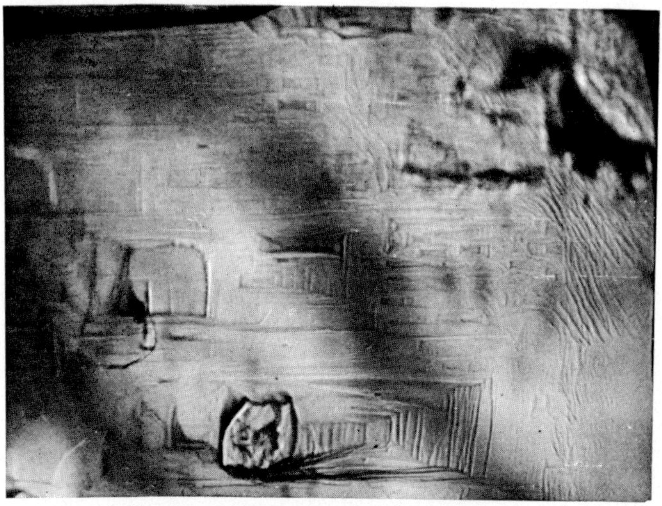

FIG. 9. Sucrose crystal dissolving. Layer recession from corners and edges, and also up from the etch pits, can be seen.

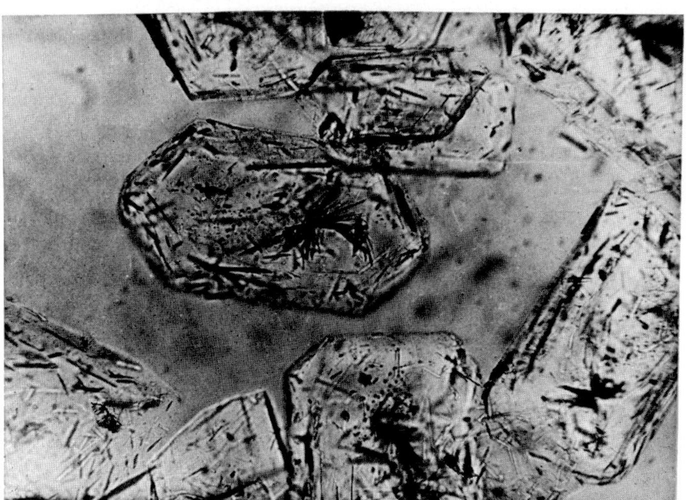

FIG. 10. Sucrose crystal with a calcium sulphate inclusion.

FIG. 11. Two perfect spiral growths as seen by reflected light.

FIG. 12. More usual view of overlapping spiral growths, seen by reflected light.

crystallized would be most strongly attracted to the growing face, some of the impurity molecules present could also be so attracted, though with less intensity. They could thus be 'overgrown' and built in. Cumulatively this also would lead to the perpetuation of faults on the later overlays.

C. Internally-held Water, Bonded Water, and Inclusions

In my studies of sugar crystals I was puzzled that when apparently dry crystals were crushed to make pulverized and icing sugars the product frequently looked slightly damp. This was particularly noticeable when the sugar being crushed consisted of crystals of a larger than normal size. Logically, this seemed to indicate that the crystals must contain some solution, even though our standard analysis only indicated a mere trace. I had already found microscopic evidence of syrup being built-in in the inch-long, coloured Kandi crystals which were, and still are, being produced as coffee sugar in Europe. I had found as much as 1% of syrup in some of these crystals, and it was reasonable to expect some to be present even in smaller crystals such as those of 'granulated' sugar. The surface of such crystals is invariably very irregular, primarily owing to the conditions of boiling in the big commercial pan. It is also considerably abraded, and fine particles of sugar adhere to the surface as the result of the crystals passing through the rotating granulator and falling from belt to hopper, etc. Thus, when looking for inclusions under the microscope, visibility is usually very poor. To overcome this difficulty I developed a simple technique. Several crystals are immersed in a small glass cell containing saturated sugar solution, and this is placed under the microscope. The solution is then warmed several degrees until one can see evidence that the crystals are just gently dissolving. This is apparent from the rounding of sharp corners and edges, and the formation of etch pits. With larger etch pits the spiral of layer recession winding up from the bottom of the pit may be observed (Fig. 9), but for this resurfacing dissolution is not carried further than clearing the surface. As soon as this has been accomplished the solution is cooled to produce supersaturation. One can then watch the formation of perfectly clear 'windows' which permit observation and photomicrography, and the recording of the interior features (Fig 13).

Numerous crystals from several refineries were thus treated, and visible fluid was present in all, varying considerably in amount and disposition. Some fairly large inclusions were randomly situated, and usually a number of smaller ones were arranged in an hourglass or a Maltese cross pattern. Notably, the centre of the cross was often nearer to one end of the crystal, indicating that growth had not been uniform. In addition, there were invariably drifts of very minute droplets forming what looked like a wisp of gossamer flicked through the crystal with no regard for axis. In due course I was able to record the origin of these on cine film and this clearly showed the sudden formation of a crack due to internal growth strain. Then one could see the very thin film of solution, which had been drawn into the crack, change gradually from a continuum to a drift of fine droplets as solution and recrystallization led to the lower free-energy conditions. Thus, with the aid of a microscope, the presence of mother syrup in the interior of crystals was clearly recorded. Further studies showed that whilst there was a considerable variation in amount of syrup, as might be anticipated, the larger the crystals were the higher was the proportion of internal water as syrup. This ranged from 0·4% in inch-long Kandi sugar down to 0·04% in granulated sugar.

At the time this was a new and unexpected finding and it clearly had marked implications for sugar analysis around the world, for the international standard methods of sugar analysis were not then capable of estimating sucrose by polarization within less than 0·1 %, and water was estimated by drying, which naturally failed to reveal the built-in solution.

FIG. 13. A cleavage 100 face of a crystal has been resurfaced; typical growth layers and also internal solution inclusions can be seen.

Fortunately I had been carrying out a good deal of research upon the effects of atmospheric vapour pressure upon sugar. I had recorded that the equilibrium relative humidity (e.r.h.) of a saturated solution of pure sugar was 86 %. Increasing impurity decreased the e.r.h. until at the lowest purity, i.e. molasses or treacle, the corresponding figure would be down to 55–50 % e.r.h. Thus it was evident that stored granulated sugar itself would have an e.r.h. corresponding to the atmosphere in the store in which it had been kept (probably around 55–65 %), and that on crushing the crystals the value would rise to over 80 %, so demonstrating that (i) the thin layer of syrup on the surface of the crystals would be of a quality near to that of molasses, and (ii) that inclusions would be nearly pure syrup, it not having been evaporated during drying (Powers, 1951).

The specific gravity of sugar crystals as quoted by various authorities had varied considerably, more than could be accounted for by reasonable experimental error. Closer study of these inclusions under the microscope provided the explanation. Not only syrup, but also air was present in the inclusions. Using the large Kandi crystals, I watched as one was slowly dissolved; when the first inclusion was 'breached' a thin stream of the coloured syrup drained downwards, and a fine bubble of air collected, then broke away and rose up to the surface of the solvent. Thus variations in the different samples of sugar used for previous estimations of specific gravity would quite probably have been the cause of the differences. Naturally, in granulated sugars, which are boiled under vacuum, one would not so easily see the air bubble release. Additionally, it is possible that since all crystals are known to have built-in faults, variation of these may affect their specific gravity. Van Hook and Beall (1976) have demonstrated that raffinose is chemisorbed into the sucrose crystal, not built into the lattice.

D. Internal Crystal Structure

Whilst still dealing with the interior of the sugar crystal, there are several features of interest regarding the structure. First, it has one cleavage plane, which lies parallel to the 100 face. A knife edge gradually pressed down on either of the end-faces will cause a split, as with a block of wood or mica. This is due to the breaking of the weakest links between sucrose molecules which give way under the stress. By contrast, the cubic salt crystal has three cleavage planes, so that a knife edge bearing down on any face will cause cleavage. One must not confuse the layers exposed by cleavage with growth layers, they are not the same.

Several of my colleagues have tried to reveal evidence of structure within the crystal. Van Hook used alcohol-etching of the surface followed by the production of electron micrographs of the etched area; these showed a mosaic structure, i.e. many rectangular blocks resembling bricks. I felt that if sufficient mosaic could be induced it might be promising in producing icing sugar. In collaboration with the Associated Electrical Industries (AEI) Research Laboratories, a number of micrographs of cleavage faces of Kandi crystals were taken. Of these only one showed definite mosaic, another showing 'moon craters' which I felt might be spherical fluid inclusions; in a third clearly marked 'ripples' in the cleavage layers were evident, which covered the whole field. Others merely showed indeterminate layer fractures. One, however, proved most exciting. Bradley (personal communication) of

AEI was in the process of increasing the resolution of their electron microscopes, and this last micrograph showed a 'splayed hand' of cards, eleven in all. Knowing the magnification, Bradley was able to calculate the height of each 'card' by dividing the height of the bunch by eleven. The result was within 10% of the size of the sucrose molecule, thus this was the first micrograph to show single molecule layers.

Later, Kamoda and Yamane (1975) submitted the three differently treated types of sugar crystal, described below, to cleavage, then took electron micrographs of the features (shadow cast with chromium, then carbon).

(i) 'Non-deformed' crystal grown slowly at 30°C in 1·05–1·15 SS solution, for one week. The micrograph revealed near-rectilinear, typical layer growth.

(ii) Crystal Kandi sugar, a commercial grade grown from selected seed, about 1·5 g each in weight, grown on in 1·05–1·15 SS at 50°C for one week. Product 10–26 mm. Electron micrographs at ×15 000 showed nipple-like projections, from 0·2 μm to 2·5 μm diameter, fairly closely packed. A typical fine, 'growth-layer-like' surface covered the whole field (compare with 'ripples' above).

(iii) Coarse crystal Superior, a commercial grade, crystals 1·0–2·4 mm, made by three successive boilings. Electron micrographs were similar to (ii), but growth layers were not visible on nipples, though elsewhere fewer but deeper layers than in (ii) were apparent.

Now Kandi crystals are grown in static tanks of liquor slowly cooled for ten days, not under vacuum, i.e. in a similar way to (i). Since the AEI were carrying out the investigation for me, I had not requested them to repeat the experiments on commercial gran. It would appear that the nipple-like developments characterize the commercial rapid boilings under vacuum. It is evident that here is another good field for research into the structure of the crystal.

Kamoda and Yamane (1975) also gave micrographs of a cleavage face on any sugar crystal after exposure to an atmosphere of 85–90% relative humidity. This causes the condensation of moisture which builds up along the layer fronts in nipples resembling that above. Since I had previously reported the presence of moisture in all sucrose crystals, the authors considered that this may be the explanation.

Twenty years later, Stansell and Jewell (1975) reproduced a micrograph of the 'fractured surface' of a crystal of granulated sugar showing long, parallel lines of orderly 'rectangular crystallites', graded in size from 0·3 μm to 1·0 μm, fading into lines which the authors record as 'growth layer lines'. On either side of these the surface showed etch markings free from crystallites. This crystal had been cleaved and lightly etched with an ethanol–water mix (G. G. Jewell, personal communication). This was Van Hook's method, which opened this sequence of investigations, and it may be a characteristic of this etch treatment, not necessarily evidence of crystallite construction. In view of the differential solubility of micrometre-sized crystals (Section III. H. 2), might one expect these to crystallize on the parent crystal in time?

I have occupied much space with a discussion of the complexities of interior structure since sucrose is so suitable a subject with which to explore possible crystal structure in general. I hope others may be attracted to devote effort to further illuminate this area of research.

E. Pulverized Sugar

This is another subject worthy of investigation and considerable information is provided by Kelly and Mak (1976) who have summarized their studies upon pulverized sugar, and also upon nucleation by dispersion of the sugar in organic fluids. They found recognizable differences between the product from vibrating mill crushing and that from rolling ball crushing. Sonic vibration was used to effect complete dispersion. The product from the hammer blow of the vibrating ball mill crushing results in a high proportion of dipolar particles. That from homogeneous nucleation was largely non-polar, possibly due to an absorbed sheath. Microscopists cannot fail to be excited by the electron micrographs, over 50 in number, reproduced by Kelly and Mak. They show the surface features of the various pulverized, precipitated and homogeneously precipitated particles, and these features are described, most appropriately in my view, in geological and similar well-known terms. Thus we have 'larval flow' and 'recrystallization belts', and 'craters which could have been formed by the impact of minute sucrose meteorites'. Two precipitated particles have an 'amorphous appearance'. Those interested in the subject should certainly study the book. Farag *et al.* (1971) have examined physical characteristics affecting dust formation.

F. Conglomeration

Within the sugar industry one of the most undesirable crystal disorders is that of conglomeration, i.e. the formation of small clumps of crystals joined together (Fig. 14). It is evident that this effect results in increased proportions of syrup being held between the component crystals. A little thought will reveal that not only will impurity be retained, but that moisture will be built in by a surface glaze during granulation.

A great change in the manner of handling both raw and refined sugar took place just after the Second World War, a change from sacks to bulk. Very soon sugar folk all over the world were finding sugar 'concreting', i.e. the formation of hard lumps due to moisture migration. At the bottom of a raw sugar delivery from Queensland the sugar was set so hard that a roadbreaker had to be used to get it out. Refined sugar silos 'set' sufficiently to prevent sugar running out freely. This led to me giving top priority to the problem of e.r.h. and moisture migration, and to the detection of the source of the moisture in apparently dry sugar. A visit to Queensland and Fiji solved the raw problem; the sugar was being loaded too moist and too hot, and this was soon remedied. (See also Section III. D for a discussion of moisture held internally in fluid inclusions.) Internal moisture can diffuse out, especially under temperature gradient conditions, but this was not the full explanation. Studying the evaporation and crystallization of syrup in thin films on the crystals as going to the granulator was not easy under the microscope, so the problem was broken down—drying on glass, on a cleaved sucrose crystal face, at different temperatures and relative humidities, as is discussed in

FIG. 14. Showing both single crystals and conglomerates.

Section III. I. Through this detailed work the cause of conglomeration became obvious.

The crystallization of sugar is very much slower than, say, that of inorganic salts. Thus, adequate time must be allowed for molecular entry into the crystal lattice. The 'engineer approach' to drying the sugar efficiently and rapidly was too vicious. Some of the syrup layer was forced through the 'crystallization belt' of conditions, and some of the syrup became built in by a sugar glaze which was extremely slow, almost unable, to crystallize even though hot and extremely supersaturated. Only very slowly, in storage, was moisture picked up from the atmosphere, to a condition in which the delayed crystallization could proceed. This caused the e.r.h. to rise from maybe 10% to 60% or more, and some of the built-in water was freed and became able to migrate through the bulk causing the concreting. Conglomerated grain has the greatest chance of holding syrup under the glaze, but the surface of commercial boiled sugar is so rough that it adds to this built-in syrup.

Conglomerates can be formed during boiling the sugar in the pan, or by separate grains sticking together after centrifuging. In both cases there will be a layer of syrup between the adhering crystals. To demonstrate this, two Kandi crystals grown into one another can be broken apart and a dry finger wiped across the freshly separated surfaces; on touching a small piece of paper the stickiness of the syrup will allow the paper to be picked up. Also, the two surfaces show rhythmic growth bands as are apparent in the hopper crystal.

There are two other serious consequences of conglomeration. Bulk density is much less with conglomerated sugar; this naturally reduces the weight capacity of plant and packaging, and hence raises costs. Also, as reported from Sweden, conglomerated sugar increases the dust due to breaking of conglomerates during handling. Hill (1965) describes a simple way of assessing 'crystal regularity index' and this is now widely used.

G. The Crystal Surface

On examining the exterior of commercially produced granulated sugar the first surprise is to find how rugged and variable is the surface (Fig. 15). This brings to mind the 'nipples', described in Section III. D, which may result from such surfaces. Certainly any attempt to calculate the surface area mathematically would be totally wrong. Bennett and Fentiman (1969) have contributed an elegant verification of this, using light and electron microscopy. It is quite clear that the cycle of growth and dissolution which the pansman uses can produce shattering irregularities on the surface. Added to this there is usually a variable amount of sugar dust and also seed crystals. Some vacuum pans produce a recognizably different surface pattern.

It was quite evident that this surface would be highly vulnerable when exposed to atmospheres varying in r.h. Indeed, this had become a feature of concern to the whole sugar world when the post-war change from bag to bulk handling and storage took place everywhere. Crystals nucleated and grown under the microscope were totally different, being remarkably clear,

FIG. 15. Showing the surface of a commercially produced crystal.

and even the growth layers could be seen clearly by carefully arranged lighting. Using transmitted light one can see one type of layer; using reflected light a totally different layer pattern is evident. It is my belief that this strange differentiation is caused by the cross-section contour of the two types. This easily accounted for the fact that layer growth had not been recorded with sucrose until I was able to publish, with micrographs, and even more revealingly with cine photomicroscopy. Also that I did not see spiral growth layers until Albon and Dunning (1957) showed that they would only be visible in reflected light. I would like to impress upon any researcher on this very fascinating subject the great efficiencies of these two techniques; verbal descriptions cannot do them justice, either with crystal growth or dissolution.

Sugar, or indeed any crystals from aqueous solution, even when apparently dry, still have a fluid layer adsorbed on the surface, although with pure crystals the layer may only be a few molecules thick. This phenomenon is nevertheless of major significance, and the researcher must fully understand how it comes about. Crystallization of sugar in commercial pans concentrates about half of the sucrose in the pan into relatively pure crystals, leaving the residual mother syrup much lower in purity. This is the primary action in preparing pure crystals. The spinning, washing and drying removes most of the impurities, but such as are left, under the influence of the 'vicious drying' will not have time to permit all the sugar to crystallize until only uncrystallizable molasses is left. Part will form a glaze too viscous to permit further crystallization. This glaze is really a 'sugar glass'. Naturally, the syrup held between adjacent crystals in conglomerates will be most affected, the fluid being built in by this glaze. Being below 10% e.r.h. on leaving the granulator, when removed to storage it takes up moisture from the surrounding atmosphere. This reduces the viscosity and allows crystallization to be resumed, but this gradual change takes days, even weeks, to equilibrate to the atmospheric relative humidity (r.h.). Moreover, any change of temperature or r.h. instigates further change. It is not surprising that the world industry as a whole suffered considerably until the phenomenon was at last understood, and the design of plant changed to minimize the vulnerability.

It will be realized that under the above circumstances the surfaces of sugar crystals can never be truly static, but are rather in a state of continual molecular change, even in apparently dry crystals. There is also evidence that certain molecules are adsorbed tenaciously by the crystal surface. This surface layer can be quite complex, providing quite a challenge to microscopist and investigator. When immersed in syrup and warmed, the layer fronts may be seen to recede from the edges and corners of the crystal, and etch pits form over the planar face, growing larger as dissolution proceeds. As these increase in size, the receding layers may be seen spiralling up from the bottom of the pits. On cooling the syrup to create supersaturation, the layers will reverse in direction and proceed to recreate the near-perfect planar face, and sharp edges and corners once again. Stansell and Jewell (1975) express the opinion that layer growth ceases to operate at higher supersaturation. It will be most interesting to see what growth form then operates. Broken or damaged areas on a crystal are 'repaired' more rapidly than sound areas, presumably by presenting increased and more active layer fronts.

I have used the term 'planar face' above, but under the microscope it can be seen that even a well-regrown face is not truly flat. On a spiral one may measure the height by focussing onto the peak, then down to the base.

H. Nucleation

One of the most fascinating aspects of crystallization is that of nucleation, the birth of the crystal, and elucidation of the mechanisms involved and of how the initial nucleus develops into the superb molecular architecture of a crystal has been a considerable challenge. Much has been learned, but more still can only be revealed in time with the aid of improved technology.

For over 100 years those working in the sugar industry were familiar with the ability of crystals of sugar to bring about copious nucleation when immersed in moving supersaturated syrup. The resulting cloud of fine crystals was given the name of 'false grain'. Perhaps the best example is that of the syrup thrown from the crystals during centrifugation which, under the microscope, is seen as well-formed and remarkably uniform crystals. In recent years this phenomenon, with other substances, has been rechristened 'breeding' and also 'secondary nucleation', but I still prefer my own earlier term, 'dynamic nucleation', which also includes the shock aspect and emphasizes the actuation by movement of fluid relative to sugar crystals, and also non-sugar surfaces.

The early experiments of Young (1911) and the Earl of Berkeley (1912) showed that even when no existing crystals were present in a supersaturated sugar solution, induced movement or shock would bring about nucleation. These were stimulating and potentially valuable observations but little was added to them until after the Second World War when, in an American paper, an author wrote of friction between suspended crystals causing fragments of the colliding crystals to be broken off, thus providing the further crop of seed crystals. I was interested in whether or not crystals suspended in so viscous a medium would meet in head-on collision and decided to commence my own attempt to learn something of the mechanics of nucleation (Powers, 1958, 1963a).

My first set-up was to show clearly that a crystal moving relative to a nuclei-free solution could initiate nucleation. The solution was made 1·2 SS at 20°C, warmed until unsaturated and kept thus for a while to ensure that any nuclei present would be dissolved ('curing' the solution). A large crystal, fixed on a short horizontal stem attached at right-angles to a vertical shaft, was inserted gently into the solution in the early stages of the curing process. The whole system was then allowed to cool to 20°C and the shaft rotated. In a short time a cloud of very fine crystals became visible in the circular orbit of the rotating crystal. Discard of nuclei from a moving crystal having been demonstrated, the next step was to construct a 'fluid wind tunnel' and carry out a series of trial runs, varying the degree of supersaturation and the speed of fluid past the fixed crystal. This sequence was carried out by two operators, the second much later and having no information of the initial results. A count assessment was made of the number of nuclei per 100 ml produced in each run. The results gave reasonable agreement.

Some years later, similar experiments were carried out by Mullin and Raven (1961) using citric acid, and by Strickland-Constable and Mason (1963) with magnesium sulphate, and gave much the same results. Thus it would appear that it is a general phenomenon.

1. Mechanism of nucleation

These initial investigations only confirmed the phenomenon of the formation and shedding of nuclei, not the mechanism. My early efforts were aimed at seeking to find out whether dendritic or similar fragile projections could grow, and hence be broken off by movement. None could be seen, even when the crystal was grown in a gelled, supersaturated solution which might be expected to protect any such projections until large enough to be visible under the microscope. However, the experiment was not without results, for the gel caused multiple 'tesselations' to grow instead of the single plane 100 face.

Dedek (1953) put forward what was an attractive explanation of the likely prelude to the actual placement of the sucrose molecules onto the crystal face. Several hydrates of sucrose are known and can be crystallized, each within a certain low temperature range (below 0°C). He suggested that in solution each sucrose molecule was attached to eight water molecules (the true number was later found to be seven by Brown and Levy (1963)) and that as the supersaturation of the solution rose the sucrose molecules displaced the water molecules and formed sucrose aggregates, this increasing with further rises in supersaturation. Dedek proposed that as these aggregates came within the attraction zone of the crystal they would be held by the surface, but would still be free to glide over the surface until individual molecules found their place and became a part of the lattice proper. On the basis of this I built up a tentative theory that during the growth there was a 'fluidized surface' of sucrose molecules free to glide over, but not to leave, the surface until disturbed by dynamic forces. Then, a group of these might be detached to form a nucleus.

2. Cine photomicrography

Continuing to develop efforts to throw light upon the mechanism of sucrose nucleation and growth, time-lapse cine photomicrographic methods were introduced in order to record dynamically the changes taking place. A number of techniques were developed enabling crystal activities to be controlled under the microscope, equipped to take both still and cine cameras. One cine sequence was carried out innumerable times, with variations. A supersaturated solution was spread out on a slide, and nucleation induced by:

(i) scratching the film with a clean needle, which produced a few, well-dispersed nuclei;
(ii) drawing a clean sugar crystal through the film, which produced a multitude of nuclei;
(iii) dipping a needle into powdered sugar, then drawing it through the film, which produced poorly dispersed, crushed sugar particles in the film, and provided evidence of the more rapid growth due to the greatly increased surface layers—until mended.

By using solutions of different SS I found that at 1·1 the nuclei varied in growth rate; the smallest visible ones appeared not to change in size although some perhaps developed after a while. Even when some crystals had grown to considerable size, some of the barely visible ones could still be seen to be unchanged. Having taken great care to exclude any undesirable matter, I was reasonably sure that they were all sugar crystals. There was a logical reason for this, in that crystals of less than ~20 μm in diameter increase in solubility with diminishing size. (It is not sufficiently widely realized that published tables of solubility of various substances should really only be interpreted as correct data for crystals of larger than ~20 μm in diameter (Powers, 1962).)

Using time-lapse techniques enables the individualism of crystal growth in solutions around 1·1 SS to be clearly shown; moreover, some crystals develop a burst of growth, sometimes in one direction only. It will be recalled (Section III. C) that the Maltese cross inclusions frequently indicate that growth has been lopsided, and the hopper illustration (Fig. 2) shows a change from needle to normal contour. I was most fortunate to be able to record two nuclei developing alongside one another; one grew in needle form, the other normally, the two developing length-wise at the same rate. When they were quite large, the needle form, for no known reason, developed lateral growth which continued until the two types were alike in habit and size.

Except when confronted by unequivocal filmed evidence, it is difficult to believe that such variations in efficiency and pattern exist in crystal growth. The phenomena described above present an exciting challenge for researchers, as do many aspects of contemporary work on sugar, and here, particularly, the scanning electron microscope will be of great use (Adl, 1976).

3. General study of nucleation

For many years workers in the field of nucleation have faced an unsolved problem: what is the size of the initial nucleus? Van Hook (1961) has calculated that it comprises between 80 and 100 sucrose molecules, but in view of the increasing solubility of sucrose crystals with decreasing size I have never been content with this, since so small a nucleus could only be possible at high SS and nucleation is easily produced even below 1·05 SS.

The true picture has been gradually and painstakingly revealed. The early concept of Dedek (1953) has been built up by Pidoux, Tikhomirof, Schliephake, Sapronov and others. At lower concentrations the sucrose exists in hydrated form, in solution. As the concentration is increased this gradually gives way to sucrose-only aggregates.

At 75% concentration the seven hydrogen bonding sites are equated with the water present as hydrated groupings. At higher concentrations, sucrose molecules with free hydrogen bonding sites are able to form groupings for sucrose only. The above researchers were able to demonstrate that in undisturbed solutions these large molecular groupings show sedimentation even under the influence of gravity. Private correspondence with others indicates that a similar concept is being favoured for other crystalline substances. That the subject is still open is indicated by a paper by Allen *et al.* (1972) who found that pH is an important consideration, the solute water aggregates being stabilised by H^+ and OH^-.

Stansell and Jewell (1975) have now provided what may be the logical answer to the problem. Using the freeze-etch technique with 74% solution, structural units of the order of 25–30 nm were seen, which in turn appeared to contain subunits of only 2·5–4·0 nm. From these units the 'building blocks' aggregating to form the true crystal are 0·2 μm in size, hence being too small to be visible under the light microscope. With this instrument Stansell and Jewell noted that a sudden appearance of nuclei of 0·5 μm was usually the first indication of true crystal nuclei being present. Note that 2·5–4·0 nm approximates to Van Hook's critical nucleus. So, in fact, these are molecular aggregates which are not of normal crystal construction; this may solve the above problem.

It is worthy of note that Garton and Head (1963, 1964) observed that a flash of crystallo-luminescence accompanied the nucleation of strongly supersaturated sodium chloride. They calculated the 'click-over' time to be less than 8×10^{-8} seconds, and remarked that at such a speed it is not surprising that imperfections occur in the nucleus. Invariably, with most if not all crystals, the larger the crystal, the more imperfect is its structure.

As Tammann (1925) recorded, very many organic substances can be converted into glasses before crystallization sets in, and I was able to demonstrate that this was true with sucrose solutions in thin layers, exposed to atmosphere of very low r.h. Great care has to be taken to exclude any potential nucleating dust, but this is relatively easy if the humidator technique described in Section VI is used.

In commercial production 'seeding' the boiling pan has for many years been carried out with powdered sugar dispersed in a higher alcohol to prevent adhesion of particles. In this it is important that the seed should be as near ideal as practicable, i.e. 'pedigree seed' should be used, and this is exemplified in the extreme care taken in producing seed for ADP upon which crystals are grown for piezo-electric purposes.

I. Thin-film Crystallization

Several lines of thought arose from my desire to study nuclear development in the thin film of syrup remaining on the crystal surface at different stages of the process, notably in drying. I also wished to examine the mechanism of dendritic crystallization, and how it arose. The possible experimental conditions fell into three groups:

(i) a syrup film between two transparent, inert surfaces, i.e. between two glass or perspex slides;
(ii) a syrup film between one inert surface and a variable atmosphere, i.e. between one glass slide and an atmosphere of controlled r.h.;
(iii) a syrup film between one cleaved sucrose surface, i.e. active participation, and an atmosphere of controlled r.h.

This work proved to be both exciting and productive. In group (i), using a sucrose solution of chosen SS which was nucleated by scratching with a needle, a few nuclei formed and grew as the usual polygonal crystals, and the process was recorded by time-lapse photography. With solution of 1·1 SS the crystals grew at different speeds, finally producing a range of sizes. Moreover, many individuals exhibited bursts of growth in one direction. This individuality of crystal growth is of great significance in that it shows the grain size variation seen in commercial pans to be due, at least in part, to this characteristic. In a sequence of papers on the topic, Nicol and Parker (1971, 1973) and Nicol and Farmer (1972), using sophisticated modern methods (radioactive seed, etc.), elegantly confirmed my evidence by cine photomicrography.

By repeating the above procedure in a solution of 1·3 SS or above, the nuclei grew uniformly, although in pan boiling secondary nucleation was greatly increased. At 1·4 SS, after the initial nuclei had grown well, some 'late-developer' nuclei appeared, sometimes all acicular, sometimes mixed with uniform 'square' crystals. Ultimately I developed a logical demonstration of this by using a six-foot-long, vertical perspex tube filled with 1·4 SS solution. After curing this, seeding was carried out by means of a captive crystal placed an inch below the surface. Long after the first crop of normal crystals had subsided down the tube, a cloud of acicular plus some square crystals could be seen falling slowly from the top. It is my hope that some young researcher may be able to follow this up. Some years later, a Russian reported a similar dual crop whilst using an inorganic material, so it is probably a widespread phenomenon.

One further striking example of deviational crystal growth should be mentioned. Using nickel sulphate of a certain concentration in a film, crystals like chisels with oblique growing tips were grown. Then, suddenly, for no known reason, a fine needle growth shot forth from the tip so rapidly that the eye could barely follow it. The oblique tip of the parent crystal continued sedately forward along the pathway of the needle, the latter being reminiscent of bomber pathfinder activity. I have a feeling that by studying exceptions to normality such as this one may learn more than by only studying the standard, well-known phenomena.

Previous to this study the sugar industry had been troubled with needle grain, and several research groups in both cane and beet put in a great deal of effort to discover the cause. They found ultimately that it was due to certain organic substances produced by early-stage microbiological degradation (see Section I. A). To this I have been able to demonstrate that changes in the physical conditions (concentration, etc.) can bring about similar changes in form. Sheftal, quite early on, showed that very small changes in, for example, temperature are quite capable of eliminating, or bringing on, a crystal face.

With powdered sugar, copious nucleation is produced, but this is too close-set for clear observation. As growth proceeds the more efficient will emerge from the mêlée. Crushed faces are more rapidly 'repaired'. Deliberate seeding with crushed spherulite (Fig. 16) nucleated normal type nuclei which outgrew the initial spherulitic fragments. These are thought provoking results.

Naturally, the best form of crystal for industry is the most compact, having least syrup-holding surface area and the highest bulk gravity. The latter characteristic gives the best yield from commercial plant and has the least storage space requirements. Worst of all is conglomerated grain which has low bulk density and the highest level of impurity and moisture-holding capacity. All of these features are best seen under low magnification.

To return to group (ii), i.e. crystallization between inert glass slide and air. This experiment opened a Pandora's box of the most fantastic growth forms. Consider the case of one seed crystal, immersed in a thin film of saturated sucrose solution on a microscope slide, inserted in a Perspex container (humidator) containing crystals of a substance of which a saturated solution has the required r.h., hermetically sealed with a touch of Vaseline, and examined under the microscope (Fig. 17). Molecules of water will migrate from the solution to the control crystals, striving to attain and maintain the desired r.h. and hence degree of supersaturation of the syrup film. Thus if common salt is used, a saturated solution of this will have an r.h. of 76% and the syrup film will evaporate to this level. Crystallization will proceed at this concentration and, as long as any solid salt remains, an r.h. of 76% will be maintained in the humidator. The initial seed crystal will continue to grow normally until its topmost surface is raised above the meniscus of the film of solution. Then, since only part of its surface remains in contact with solution, growth will continue within the contact area (Fig. 18). Thereafter the growth pattern becomes more complex, as may be seen in the illustrations. Most people are familiar with frost patterns on a window pane in winter, and they are of the same type of thin-film crystallization.

The infinite number of variations of pattern which nature can produce under these conditions is truly exciting, especially when seen in colour. The patterns may be examined by using a polarizing microscope and viewing the slide under polarized light, and even better by also inserting a half-wave quartz filter. When using polarized light only, some parts of the field will not

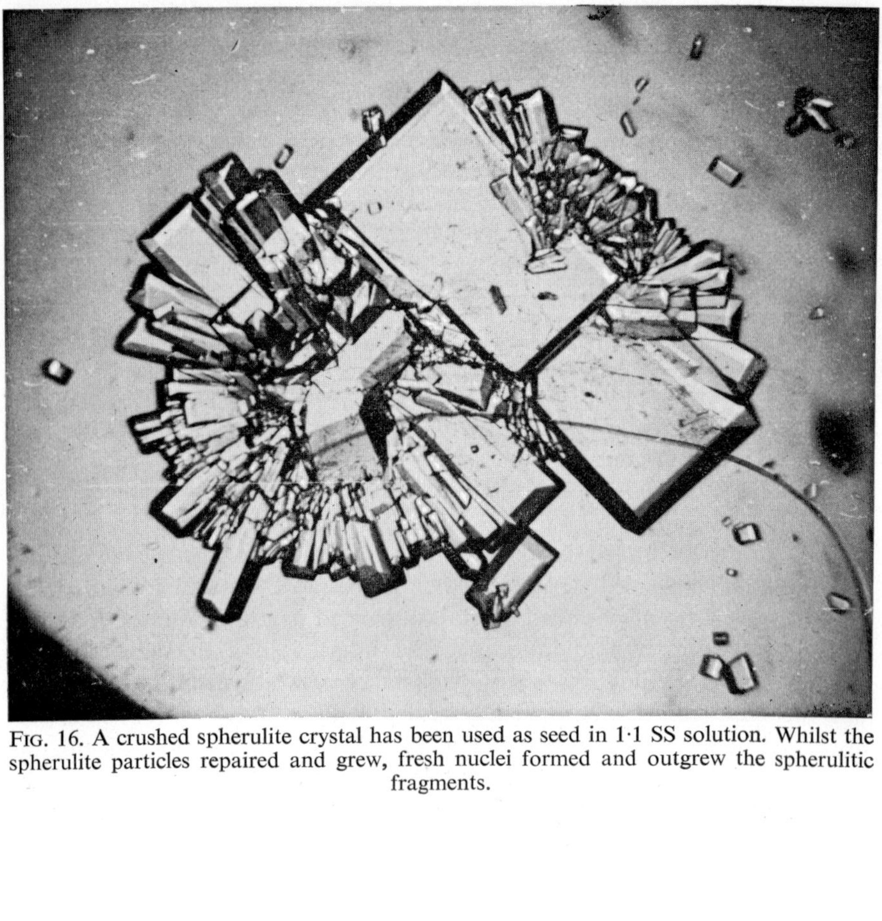

FIG. 16. A crushed spherulite crystal has been used as seed in 1·1 SS solution. Whilst the spherulite particles repaired and grew, fresh nuclei formed and outgrew the spherulitic fragments.

be visible, but with this filter every feature of the complex field will be seen in colour and may be recorded on film. Even apart from the sheer beauty of such growths, a great deal of pertinent information may be derived from the manner of growth. This can enlighten the observer even on features of growth of

FIG. 17. A home-made 'humidator'.

FIG. 18. An initial polygonal sucrose crystal develops dendritically as the film of syrup gets thinner.

polygonal crystals. For example, although the leading crystal edges may be in contact with solution on a comparatively broad front, growth bursts out in perhaps only a local area. Branching arms of dendritic growth (Figs 19 and 20) often retain the tendency to form the angles characteristic of the polygonal form, but they also show occasional curvature due to continued growth strain in one direction only. This is probably due to the film in contact with the growing face becoming exhausted and attenuated. Thus growth stops until fresh SS solution diffuses in (Fig. 21). Then growth recommences at the most efficient part of the crystal front, gathering speed as it meets the incoming SS film, when the maximal growth exhausts the contact film again. Then the cycle is repeated over and over again until the film of solution is exhausted.

I am certain that this thin-film crystallization (Figs 19–24) deserves much study. In a way it is comparable to cutting and polishing a very thin section of bone, etc. for microscopic research. Certain clearly marked features in this vast field of research have been named. Thus a much-branched pattern is termed 'dendritic', developments in spherical form are termed 'spherulitic', or, if two-dimensional, as 'radialitic' (Fig. 25). Both types may later develop rhythmically (Powers, 1963b).

These sucrose forms, though unknown to most people, are familiar to researchers in plastics, glass and other viscous materials, for such workers find very similar crystals growing in their media. Inevitably, when watching such forms developing under the microscope, one feels vaguely that primordial living forms may have developed under similar thin-film conditions. Surface tension, viscosity, adsorption, etc. are then able to play a major part in evolution. Even the simplest of living organisms is still vastly more complex than the above sugar forms, in that blending in of different molecules is entailed, but, as mentioned earlier, mankind has already learned how to grow certain crystals endowed with truly remarkable abilities.

The third of the systems explored (iii) was the commonplace situation of a film of solution between a crystal surface and air of controlled r.h., as is used in drying by granulation. In the industry the 'engineer approach' was that of driving off water as quickly and cheaply as possible. Insufficient time was given for what should have been seen as another crystallization operation—one which needed doing slowly enough to permit the sucrose molecules to attain their position as part of the lattice of the parent crystal.

I found that if a film on a cleavage face was dried slowly, in an atmosphere just below 86% r.h., much of it was able to dry successfully, some as part of the main face, some as very small crystals of normal form attached to the main face. If dried at a much lower r.h., a number of spherulitic crystals formed on the surface which would make the formation of 'sugar dust' on meeting the frictions of bulk handling much more likely.

FIG. 19. A classic example of sucrose dendritic growth in which the angles of the polygonal form are repeated in rhythmic pattern, primarily originating from angle tips. In molecular structure this would still be one crystal, as formed.

FIG. 20. An example of very fine dendritic growth. Though so dispersed, this is all one crystal, as grown, but it will gradually form a sequence of separate polygonal crystals.

FIG. 21. An excellent example of rhythmic sucrose, carbon bloomed.

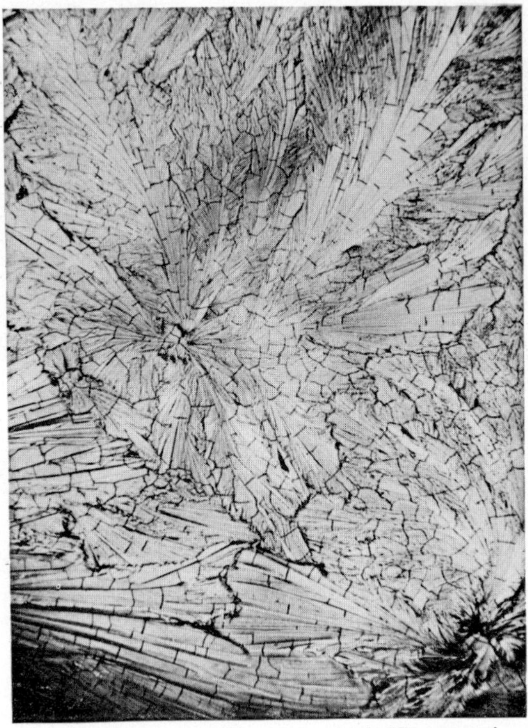

FIG. 22. Sucrose crystallized in thin film at very high SS as seen in ordinary light.

FIG. 23. As Fig. 22 but using polarized light.

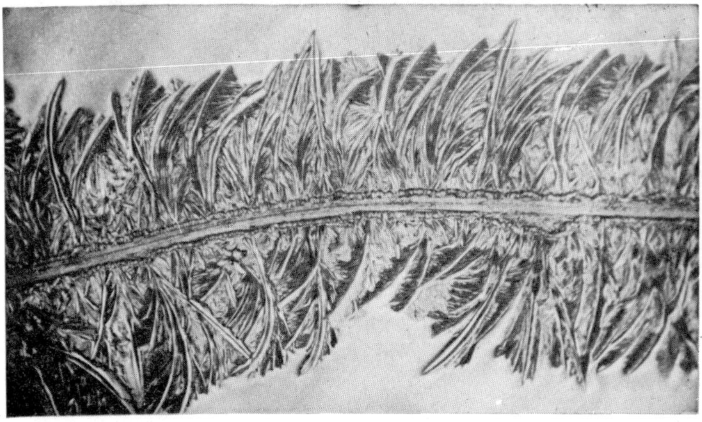

FIG. 24. A sucrose solution in a humidator at 44% r.h., nucleated by single needle scratch, shows that myriads of nuclei are formed but only the most efficient grow on.

FIG. 25. Sucrose crystals showing fantastic form. Each, as grown, forms a single crystal.

IV. TRIBOLUMINESCENCE AND OTHER REFINERY PROBLEMS

Those responsible for handling sugar in commercial quantities are highly conscious of the danger of explosion and fire, and if static sparks are formed in atmosphere laden with finely divided sugar dust the hazard becomes acute. A great deal of research has taken place into this problem, and is still continuing. If sugar crystals are crushed between two glass plates in a darkened room, a glow of violet light can be seen. This property is known by the fearsome name of triboluminescence. It is due to electric charges being released by the breaking of molecular bonds. Some years ago I received a letter from the director of a Moroccan refinery in which I had recently lectured on crystallization. He wrote that they had just started making cube sugar (which entailed cutting slabs) and that their operators were worried because they could see a violet glow in the cutting plant. I advised him as to the cause, and that it would only be dangerous if they failed to take adequate precautions against electric discharge in the plant and to prevent the accumulation of deposits of sugar dust on ledges, etc. This danger of explosion is even more acute during the crushing process in which icing sugar is made, and if an electric charge detector is used both positive and negative charges may be indicated in different parts of the plant (Kelly and Mak, 1976).

The problem of conglomeration has already been considered, especially with regard to the migrant water. Here, with particular attention to explosion, it should be remembered that after one such accident in Sweden the rigorous investigation which followed revealed that the dust was much more severe with conglomerated sugar than with clean grain. On breaking a conglomerate, one finds that minute splinters of crystal fly at the break. Examination of a sample of sugar under a low-power microscope will immediately reveal the degree of conglomeration (Hill, 1965) and a micrograph will present irrefutable evidence. Hill's term for this degree is ‘crystal regularity index’. Essentially, the sample of sugar to be graded is brushed over a sieve with suitable size apertures. This rests on a sheet of adhesive paper in such a manner that one crystal rests in each hole. The mesh is then removed and the regular rows of crystals assessed; good being given two marks, moderate one, and bad none. Low-power microscopy is adequate. Very strict care is vital with granular material to ensure that a truly representative sample is obtained, since segregation will invalidate any result.

The objective of the sugar refiner is to eliminate impurities and to produce, as efficiently as possible, an attractive sugar in those crystal sizes and forms desired by the purchaser. Crystal size is one of the important variables. This is now well and simply expressed by the terms mean aperture and coefficient of variation (MA and CV). For some time it had been evident that the old method of expressing grain sizes in terms of percentages passing through a nest of sieves was unwieldy, and it was suggested that the Rosin Ramler technique be tried. After testing this system on a number of our own sugars and finding it excellent, I was able to recommend it for trial at other factories and refineries world-wide. Within several years it was being used as the official method in most laboratories and refineries, involving two mesh sievings only. The technique was originated by Hazen, in 1914, using a special grid based upon the equation of the normal or Gaussian frequency function for plotting size frequency data. Cumulative sieved sizes are plotted on arithmetic probability paper. If the sizes between 15·87 and 84·13% fall on a straight line, then the system is applicable. For pulverized sugars, as distinct from whole boiled grain, geometric probability paper appears to be a better fit; then, the MA is the aperture corresponding to 50% on this line and the CV is

$$\frac{\text{Aperture at } 15{\cdot}87 - \text{Aperture at } 84{\cdot}13}{2\text{MA}} \times 100.$$

Several modifications to the technique have been put forward (Butler, 1974), and micrographs have since been used successfully as a means of calculating MA and CV, not only for sugars but also for precipitated $CaCO_3$, etc. (Kelly and Mak, 1973). Correct sampling is vital, as is the accuracy of the sieves used. One full-nest sieving should be carried out with new sieves, and plotting the sizes on the probability paper will detect any faulty sieve.

A. Further Refinery Operations

The control of the ultimate grain size of the product is mostly in the hands of experienced pansmen. Their first concern is to produce the appropriate 'density' or concentration of seed grain to permit the final grain size, when the pan is full, to be of the desired MA and CV. Recall the vision of the stars in the sky on a clear night, for this is how he may first see the seed crystals. He is the skilled operator still, though several systems of controlling boiling by mechanical aids are increasingly being used. The older pansmen do remarkably well by removing a sample from the vacuum pan from time to time, using a 'proof stick' to do so. He spreads the sample on his viewing glass, then holds this in front of an electric light which provides a simple but effective back-lighting. Now that the penalties of conglomerated grain are more widely realized, the built-in microscope should be frequently used to minimize that evil. Many progressive firms are actively studying continuous crystallizers and centrifuges and other forms of automation, and, in short, a more scientific approach has been, and is, increasingly apparent throughout the industry.

It may be desirable to give a brief summary of the wide range of sugars produced together with their *raison d'être*. The importance of grain size is largely due to its effect on the rapidity of dissolution. Thus icing is a finely pulverized sugar designed for confectionery requiring a coating free from grittiness. Castor, the finest boiled sugar, is used for sprinkling over fruit dishes, being smooth enough to be pleasant to the palate when presented in that manner. Icing tends to clot when used in the same way, and is also too sweet. Granulated is the general purpose sugar produced in by far the greatest quantity, and hence at basic cost. In the baking industry the right choice of grain sizes facilitates a regulated rate of dissolution and thus of incorporation into the confection.

Selection of an attractive crystal form is typified by cube or tablet sugars, in which the crystals are bonded together, forming cubic, tabular or even fancy moulded shapes. These are mostly used to grace the table for tea or coffee, etc., being uniform tablets of pleasing appearance. These 'porous' sugars derive from the very early cone-shaped blocks of sugar which were used in Britain before the centrifugal machine made obsolete the slow draining off of the old 'loaves' of sugar (hence the term 'loaf sugar') by gravity. These loaves of sugar are still used in certain countries, the blocks being well suited to carrying on camel back. When these loaves were used in Britain the shop-keepers' unfortunate assistants had the formidable task of breaking them up for the sale of manageable pieces!

Since sugar is purified to a level above 99·9% it is evident that with equal weights of different sugars the degree of sweetening power will be the same. Apparent differences may be due, for example, to the rapidity of solution reaction on the palate (try icing sugar), or visually to differences in bulk density. Conglomeration can make 10% difference. Some may think 'raw' (brown) sugar sweeter than white, this is due to the relatively thick layer of low-purity syrup which still envelopes every crystal, and which rapidly produces the sweet reaction on the palate. A great deal of study has been devoted to rendering cube sugar able to break down and dissolve rapidly, or the reverse, whichever is desired.

A very different type of sugar is a fondant in which the crystal is enveloped in syrup; the whole product can be blended to a near-solid or to a fluid according to the proportion of syrup enveloping the grain. A low-powered microscope is enlightening with any of the above, and is an indispensable adjunct in even the smallest sugar laboratory.

One interesting practical use is to decide whether a stirrer full of hot masse is under- or over-saturated. A small sample should be placed on a slide under the microscope, in a cell which permits the temperature to be raised or lowered. Using this an expert can soon tell at what temperature the syrup is just saturated. Molasses being so viscous, it is desirable for centirifugation to be carried out at that temperature; lower the temperature, the product would be more viscous, raise the temperature, and the yield would be smaller.

V. OTHER SUGARS

Throughout the preceding pages the term sugar has been applied to sucrose, in agreement with popular usage. In fact there are quite a number of carbohydrates which scientifically are also sugars, having related molecular structures and being sweet to the taste in various degrees. Several are marketed for human consumption, but they are not considered in detail here. Most of these other sugars are more difficult to crystallize than sucrose. Dextrose, perhaps the most widely used, is common in the production of false grain, and its role in the curing of syrups is very important. The temperature coefficient of the hydrate growth is small. Laevulose is even more difficult to crystallize, and I recall my excitement when finding a small bottle of this in syrup form, in a cupboard where it had probably been undisturbed for 20 years or more, in which half-a-dozen near-perfect crystals of several millimetres length could be seen. Such conditions are excellent for growing good crystals, but hardly commercial!

VI. APPARATUS AND TECHNIQUES FOR LIGHT MICROSCOPY, AND CINE AND STILL PHOTOMICROGRAPHY

In the hope that some readers may wish to join in this fascinating study I give this brief description of my apparatus and techniques. The former is an all-purpose design facilitating quick and easy changing from cine and still photomicrography to direct photography.

The stand is constructed of 1″ teak, base 8″ × 20″, upright 30″, made rigid. This stands on a foam rubber sheet which itself lies on another rubber sheet. The whole is best placed on a very steady desk at which one can sit comfortably, knees under the desk, eyes on a level with the horizontal eyepiece of the adapter between microscope and camera. Comfort is important and, particularly when working near moving machinery, etc., damping out of vibration will be vital. Illumination for transmitted light is passed from the rear of the base, through a 3″ hole in the upright, and so onto the microscope mirror. For reflected light the lamp is placed at one side and the beam adjusted as required, the angle of slope of the subject being adjusted similarly.

A polarizing microscope, stood on the teak base, may be connected with an adaptor to a Leica still or Bolex (16 mm) cine camera. Their weight can be supported by an adjustable clamp held in a vertical 10″ brass slit in the upright back. The adapter should have an eyepiece which will allow the subject to be viewed even when, for example, the time-lapse cine is continuously in operation. Before commencing, one should make sure that the true field is seen in the eyepiece, even a small displacement can be annoying at times. Another minor feature of importance is that whilst watching crystal growth one's breath can impinge upon the very vulnerable film with disastrous results. A sheet of polythene with a hole for the viewing eyepiece can be left permanently in position to deflect breath and is a perfect remedy.

Humidators are easily made to whatever size one desires, from clear Perspex, sawn to the correct sizes and sealed together with an appropriate adhesive. A suitable size would be $3\frac{1}{2}'' \times 1\frac{1}{2}'' \times \frac{3}{4}''$. The main box should be divided into three compartments by means of divisions (about half the height of the outer structure) upon which the microscope slides will rest. The two outer compartments are to contain the appropriate salt to maintain the inner atmosphere at the desired humidity (Table 1). The top rim may have a touch of vaseline to ensure hermetic sealing when the lid is in place.

With a relatively slow crystallizing substance like sugar, time intervals may be 30 seconds to five or more minutes, and an automatic control (e.g. as made by Prior of Bishop's Stortford) will facilitate continual operation for many hours. Good controls will provide for a range of intervals and exposure times. With mineral salts, growth is so much faster that self-operation is feasible. The brass slit may be used to hold a camera for the occasional copying or direct photography without the microscope, and this is a very real asset.

Always remember that the atmosphere is loaded with potentially catalytic dust, so the utmost care must be taken when operating with an exposed thin, fluid film. It is relatively easy to grow a dendritic film so thin as to be barely visible under a light microscope. To avoid wastage of time and material, one

TABLE 1. Suitable salts to maintain constant relative humidity

r.h. (%)	Substance
86	(Sucrose) $KHSO_4$
76	$NaCl$
66	$NaNO_2$
52	$Mg(NO_3)_2.6H_2O$
44	$K_2CO_3.2H_2O$
20	$KC_2H_3O_2$

must be able to assess light values, and a Gossen's Lunasix meter applied to the horizontal eyepiece gave me excellent results, even whilst filming. I advise sacrificing the first film to assess meter readings with the subject and film chosen.

For the interest of crystallographers, the fullest exposition on the sucrose crystal is given by Vavrinesz (1956).

ACKNOWLEDGEMENTS

I would like to express my appreciation of my various 'Crystal Cronies', and particularly those at the Philip Lyle Memorial Research Laboratory (Reading, U.K.) for sending me pertinent reprints. The expert assistance of Mrs A. Alford (Philip Lyle Memorial Research Laboratory) in checking the references is also acknowledged.

REFERENCES

Adl, J. (1976). 'Dust Production During Conditioning of Granulated Sugar.' M.Sc. Thesis, London University.

Albon, N. and Dunning, W. J. (1957). *Nature, Lond.* **180**, 1348.

Allen, A. T., McDonald, M. P., Nicol, W. M. and Wood, R. M. (1972). *Nature Phys. Sci.* **235**, 36.

Bennett, M. C. and Fentiman, Y. L. (1969). Proc. Symp. Inst. Chem. Eng. Industrial Crystallisation, p. 217.

Berkeley, Earl of (1912). *Phil. Mag.* **24**, 254.

Brown, G. M. and Levy, H. O. (1963). 6th Cong. Int. Union of Crystallography. *Science*, 3584.

Bunn, C. W. (1933). *Proc. R. Soc.* **141**, 567.

Butler, R. A. (1974). International Commission For Uniform Methods Of Sugar Analysis, 16th session.

Chen, J. O. P. and Picou, R. W. (1972–73). *J. Am. Soc. Sug. Beet Technol.* **2**, 100.

Dedek, J. (1953). Proc. 7th Assembly Commission Internationale Technique de Sucrerie Betterave, 85.

Delavier, H. J. and Shokrani, R. (1974). 15th Cong. Int. Soc. Sugar Cane Technol., 1475.

Farag, S. A., Schmalz, C. L. and Norman, L. W. (1971). *J. Am. Soc. Sug. Beet Technol.* **16**, 448.

Frank, F. C. (1949). *Discuss. Faraday Soc.* **5**, 48.

Garton, V. A. and Head, R. B. (1963). *Phil. Mag.* **8**, 1793.

Garton, V. A. and Head, R. B. (1964). *Phil. Mag.* **14**, 1243.

Hill, S. (1965). *Int. Sug. J.* **67**, 201.

Kamoda, M. and Yamane, T. (1975). *Proc. Res. Soc. Japan Sug. Refin. Technol.* **6**, 72.

Kelly, F. H. C. and Mak, F. K. (1973). *Int. Sug. J.* **75**, 131.
Kelly, F. H. C. and Mak, F. K. (1976). 'The Sucrose Crystal.' Singapore University Press, Singapore.
Mullin, J. W. and Raven, K. D. (1961). *Nature, Lond.* **190**, 251.
Nicol, W. M. and Farmer, C. L. (1972). *Sucr. Belge*, **91**, 55.
Nicol, W. M. and Parker, J. S. (1971). *Int. Sug. J.* **73**, 237, 355.
Nicol, W. M. and Parker, J. S. (1973). *Sucr. Belge*, **92**, 299.
Powers, H. E. C. (1951). *Int. Sug. J.* **633**, 254.
Powers, H. E. C. (1958). *Nature, Lond.* **182**, 715.
Powers, H. E. C. (1960). *Nature, Lond.* **184**, 289.
Powers, H. E. C. (1962). *Nature, Lond.* **196**, 58.
Powers, H. E. C. (1963a). *Ind. Chemist*, July, 351.
Powers, H. E. C. (1963b). *J. R. Microsc. Soc.* **82**, 23.
Scarr, P. M. (1968). *J. Appl. Bact.* **31**, 525.
Smythe, B. M. (1967). *Aust. J. Chem.* **20**, 1087.
Stansell, D. I. and Jewell, G. G. (1975). Food Research Association Leatherhead, Res. Rep. 236.
Strickland-Constable, R. F. and Mason, R. E. A. (1963). *Nature, Lond.* **197**, 897.
Suarez, G. O. and Moser, A. C. (1973). *Cuba Azucar*, (Jan.–March), 8.
Tammann, G. (1925). 'States of Aggregation.' Van Nostrand, New York.
Tutton, A. E. H. (1911). 'Crystals.' Kegan Paul, Trench and Trubner, London.
Van Hook, A. (1961). 'Crystallization, Theory and Practice.' A.C.S. Monogr. No. 152, American Chemical Society, Washington D.C.
Van Hook, A. and Beall, H. (1976). *Proc. Res. Soc. Japan Sug. Refin. Technol.* **26**, 74.
Vavrinesz, G. (1956). 'Atlas of Sugar Crystals.' Bartens, Berlin.
Withy, D. (1964). *Int. Sug. J.* **66**, 385.
Young, S. W. (1911). *J. Am. Chem. Soc.* **33**, 1375.

16. Computer-aided Identification of Food Materials

GEORGINA H. JOLLIFFE and G. O. JOLLIFFE

Department of Pharmacy, Chelsea College, London, U.K.

I. INTRODUCTION

Few marriages would appear to have dimmer prospects of success than the one between the rigorous and highly mathematical applications of computers and the descriptive, empirical and often intuitive approach of the food analyst. Nevertheless, the computer has certain features which permit it to assist in identification problems: it can have an extremely large, permanent and reliable memory; it calculates quickly and the information in its memory can be available rapidly when required. The computer is not influenced by recent experience and personal bias; it can be up-dated easily, but cannot do the physical examination. At the present time, for example, it is unable to carry out the examination of the microscopy of a powdered sample or convert narrative or pictorial information for its use. The program required to completely imitate an expert's thinking would be impractically large. However, major advances in pattern recognition (Ledley, 1964) will undoubtedly contribute to this end. There is scarcely an area of study or specialization that the computer has not touched in one way or another, although there are vast differences in the amount of involvement that exists between various fields.

It is, therefore, surprising to find a singular lack of application of the computer in food microscopy. Computer-aided identification of powdered vegetable drugs (Jolliffe and Jolliffe, 1976) introduced the concept in a closely related field.

Computers are almost automatically associated with the solution of sets of differential equations and large information matrices or other complex mathematical operations. The approach developed here, however, is based on the computer's ability to scan, describe, display and classify information from a large matrix data bank.

Food microscopy presents many problems for an analyst, primarily because it demands knowledge covering a wide field and identification of samples in the whole, fragmented or powdered condition requires a detailed knowledge acquired over many years of experience. Today fewer people are available with this expertise and it would seem desirable to use the computer to aid in identification problems.

The aim of this chapter is to present a computer program which has been designed to assist in the identification of a miscellany of fragmented or powdered food materials, and their adulterants, including the herbs and spices, and to suggest its adaptation to associated problems by suitable selection of characters in the analytical data.

Identification of powdered samples normally involves a detailed microscopic examination of the sample followed by reference to suitable keys (Kraemer, 1920; Claus, 1956), tables (Schneider, 1921; Anon, 1963; Wallis, 1967; Trease and Evans, 1978; Brain and Turner, 1975), powder atlases (Tschirch and Œsterle, 1900; Greenish and Collin, 1904; Flück *et al.*, 1935; Deryng, 1961; Eschrich, 1966; Jackson and Snowdon, 1974), punched card systems (Clarke, 1938; Phillips, 1948; Brazier and Franklin, 1961; Fish and Nelson, 1969; Nelson, 1972) or by direct reference to relevant textbooks (see later).

The use of textbooks and reference atlases, while a valuable aid in confirming identity, does not initiate a systematic approach to the problem of identification of powdered food materials. Tables of histological features are sometimes difficult for the inexperienced analyst to interpret as the description, particularly of 'inflorescences and floral members' and 'seeds and fruits', may apply to the powdered material or, sometimes, it is more appropriate to the sectional appearance of the sample. Before an atlas or table can be used the sample must first be referred to its morphological group; error at this stage is time consuming and frequently results in incorrect identification. The punched card system devised by Nelson (1972) for the rapid identification of powdered crude drugs and spices requires the assessment of over 100 characters, some of which are difficult for the inexperienced analyst to interpret, and entails many micro-scale measurements. Hence, it would seem pertinent to devise a scheme whereby an analyst relatively inexperienced in microscopy may rapidly obtain the correct identity of a powdered sample from as comprehensive a list as possible, not necessarily limited to those in current commercial demand.

CHARACTERS - Enter + if present, - if absent		No.	+/-	Input Data
CALCIUM OXALATE ...	Prisms	1		
	Needles	2		
	Sphenoids/sandy crystals	3		
	Clusters/rosettes	4		
	Crystal layer/sheath[a]	5		
SECRETION CELLS, CANALS or DUCTS		6		
ALEURONE[b] ...	$\ngtr$ 18 μm	7		
	> 18 μm	8		
STARCH GRAINS		9		
MUCILAGE		10		
CORK		11		
FIBRES		12		
TRICHOMES	Covering	13		
	Unicellular[c]	14		
	Multicellular	15		
	Glandular	16		
	Stalk Unicellular ...	17		
	Stalk Multicellular ...	18		
	Head Unicellular ...	19		
	Head Multicellular ...	20		
VESSELS/TRACHEIDS ...	Lignified	21		
	Non-lignified	22		
STOMATA	Anomocytic	23		
	Anisocytic	24		
	Paracytic	25		
	Diacytic	26		
	Actinocytic	27		
	Graminaceous	28		
STONE CELLS		29		
POLLEN		30		

Reference :

Enter 2 if +
1 if -

a crystal sheath or calcium oxalate localised in the region of the main veins

b for the purpose of this observation aleurone includes protein as grains or amorphous protein

c includes papillae

FIG. 1. Analytical data sheet.

TABLE 1. List of food materials and relevant data (pp. 596–601)

No.	Name of food in program	Latin name	Stored data for characters 1–30 (see Fig. 1)[a]						Morphological group[b]	Discrimination index
1	Absinth	*Artemisia absinthium* L.	11121	11121	12222	22212	21211	11122	5	100
2	Alfalfa	*Medicago sativa* L.	21112	12111	12222	21212	21211	11112	5	100
3	Almond	*Prunus amygdalus* Batsch.	21121	12111	12221	11111	21111	11121	4	100
4	Angelica root	*Angelica archangelica* L.	11111	21121	22111	11111	21111	11111	8	77
5	Angustura	*Galipea officinalis* Hancock	22111	21121	22111	11111	11111	11121	1	99
6	Anise	*Pimpinella anisum* L.	11121	22111	12221	11111	21211	11121	3	89
7	Areca	*Areca catechu* L.	11111	11211	12111	11111	21111	11121	4	87
8	Arrowhead	*Sagittaria* spp.	11111	11121	12111	11111	21111	11111	8	93
9	Artichoke	*Cynara scolymus* L.	11111	21111	12111	11111	21121	11111	2	99
10	Asparagus	*Asparagus officinalis* L.	12111	11111	12111	11111	21111	11221	5	99
11	Bamboo shoots (decort.)	*Phyllostachys mitis* Riv. *P. quilioi* Riv. *Sasa paniculata* Shib. and Mak.	11111	11111	12111	11111	21111	11111	5	93
12	Banana	*Musa sapientum* L.	12111	21121	12111	11111	21111	11111	3	96
13	Bayberry	*Laurus noblis* L.	11111	22121	11111	11111	21211	11121	3	100
14	Bay leaf	*Laurus noblis* L.	11111	21111	12221	11111	21211	11121	6	98
15	Beechnut (meat)	*Fagus sylvatica* L.	11121	12121	11111	11111	21111	11111	10	100
16	Beet greens	*Beta vulgaris* L. var. *crassa* Alef.	11211	11121	12212	11111	21211	11111	5	100
17	Beetroot	*Beta vulgaris* L. var. *crassa* Alef.	11211	11111	22111	11111	21111	11111	8	99
18	Brake	*Pteridium aquilinium* Kuhn.	11111	11121	12212	11111	21111	11111	9	100
19	Brazil nut (meat)	*Bertholletia excelsa* H.B.K.	11111	11211	11111	11111	31111	11111	10	88
20	Brazil nut (shell)	*Bertholletia excelsa* H.B.K.	11111	11111	13111	11111	21111	11121	10	58
21	Buchu	*Barosma* spp.	11121	21112	12221	11111	21211	11111	6	100
22	Buckwheat	*Fagopyrum esculentum* Moench	11111	12121	12111	11111	21111	11121	3	77
23	Cacao	*Theobroma cacao* L.	11121	12122	12212	11111	21111	11121	4	100
24	Canella	*Canella alba* Murray	11121	21121	33111	11111	11111	11121	1	100

25	Capsicum	*Capsicum minimum* Roxb. *C. frutescens* L.	11211	12111	12212	22112	21121	11121	3	100
26	Caraway	*Carum carvi* L.	11121	22111	12111	11111	21211	11121	3	87
27	Cardamom fruit	*Elettaraia cardamomum* Maton var. *minuscula* Burkill	22111	22121	12111	11111	21111	11121	3	97
28	Cardamom seed	*Elettaria cardamomum* Maton var. *minuscula* Burkill	21131	22121	11111	11111	21111	11121	4	97
29	Cardoon	*Cynara cardunculus* L.	11111	11121	12222	21212	21211	11111	6	100
30	Carrot	*Daucus carota* L.	11111	22123	22111	11111	21111	11121	8	90
31	Cashew nut	*Anacardium occidentalis* L.	11111	22121	11111	11111	21111	11111	10	100
32	Cassava	*Manihot* spp.	21111	21121	22111	11111	21111	11121	8	99
33	Cassia bark	*Cinnamomum cassia* Blume *C. loureirii* Nees *C. burmanni* Blume	12111	21122	22111	11111	11111	11121	1	89
34	Cassia buds	*Cinnamomum cassia* Blume *C. lourerii* Nees	22121	21122	12221	11111	21311	11121	3	100
35	Cauliflower	*Brassica oleracea* L. var. *botrytis* DC.	11111	11111	12111	11111	21121	11111	5	97
36	Celery Fruit	*Apium graveolens* L.	11121	22111	12221	11111	21211	11121	3	89
37	Celery leaf	*Apium graveolens* L.	11111	21111	12111	11111	21211	11111	6	87
38	Chamomile	*Anthemis noblis* L.	11121	11111	11222	21212	21211	11122	2	98
39	Chervil	*Anthriscus cerefolium* Hoffm.	11111	21111	12221	11111	21211	11111	6	88
40	Chicory leaf	*Cichorium intybus* L.	11111	21111	12212	11111	21211	11111	6	89
41	Chicory root	*Cichorium intybus* L.	11111	21111	22111	11111	21111	11111	8	94
42	Chives	*Allium schaenoprasum* L.	11111	21111	12111	11111	21111	11211	6	98
43	Chufa	*Cyperus esculentus* L.	11111	11121	12111	11111	21111	11121	8	74
44	Cinnamon	*Cinnamonum zeylanicum* Nees	12111	21122	22111	11111	11111	11121	1	89
45	Clove fruit	*Syzgium aromaticum* (L.) Merril and L. M. Perry	11121	22121	12111	11111	21211	11121	3	98
46	Clove flower	*Syzgium aromaticum* (L.) Merril and L. M. Perry	11121	21111	12111	11111	21211	11122	2	100
47	Clove stem	*Syzgium aromaticum* (L.) Merril and L. M. Perry	21121	21121	22111	11111	21111	11121	9	98

TABLE 1 (*continued*)

No.	Name of food in program	Latin name	Stored data for characters 1–30 (see Fig. 1)[a]	Morphological group[b]	Discrimination index
48	Coconut (endosperm)	*Cocos nucifera* L.	11111 11211 11111 11111 11111 11111	10	90
49	Coconut (shell)	*Cocos nucifera* L.	11111 11111 12111 11111 21111 11121	10	71
50	Coffee	*Coffea arabica* L.	11111 12122 11111 11111 21111 11121	4	99
51	Coriander	*Coriandrum sativum* L.	21121 22111 12111 11111 31211 11121	3	98
52	Cubebs	*Piper cubeba* L.	21111 22121 12111 11111 21311 11121	3	78
53	Cucumber	*Cucumis sativus* L.	11111 12111 12212 21212 21111 11121	3	100
54	Cumin	*Cuminium cyminum* L.	11121 22111 12212 11111 21211 11121	3	100
55	Dandelion root	*Taraxacum officinale* Wiggers	11111 21111 21212 11111 21111 11111	8	100
56	Date stone	*Phoenix dactylifera* L.	11111 11111 11111 11111 31111 11121	4	66
57	Dill	*Anethum graveolens* L.	11121 22111 12111 11111 21311 11121	3	87
58	Endive	*Cichorium endiva* L.	11111 21111 12212 11111 21211 11111	6	89
59	Fennel	*Foeniculum vulgare* Miller	11121 22111 12111 11111 21311 11111	3	98
60	Fenugreek	*Trigonella foenum-graecum* L.	11111 12112 11111 11111 21111 11111	4	100
61	Fig	*Ficus carica* L.	11121 22112 12223 11111 21311 11121	3	98
62	Galangal	*Alpinia officinarum* Hance	11111 21121 12111 11111 21311 11111	8	71
63	Garden sorrel	*Rumex acetosa* L.	11121 11111 12221 21212 21211 11111	6	99
64	Garlic	*Allium sativum* L.	21121 21111 12111 11111 21111 11221	7	100
65	Gentian	*Gentiana lutea* L.	12111 11121 21111 11111 21111 11111	8	100
66	Ginger (scraped)	*Zingiber officinale* Roscoe	11111 21121 12111 11111 12111 11111	8	99
67	Ginger (unscraped)	*Zingiber officinale* Roscoe	11111 21121 22111 11111 12111 11111	8	99
68	Grape seed	*Vitis vinifera* L.	12121 11211 11111 11111 31111 11121	4	100
69	Guarana	*Paullinia sorbilis* Mart.	11111 13121 12111 11111 21111 11121	4	55
70	Hazelnut (meat)	*Corylus* spp.	11121 11231 11111 11111 21111 11111	10	99
71	Hazelnut (shell)	*Corylus* spp.	11111 11111 11221 11111 21111 11121	10	98
72	Hemp (hemp seed)	*Cannabis sativa* L.	11111 12111 11111 11111 21111 11121	3	97
73	Honewort	*Cryptotaenia canadensis* DC. var. *japonica* Mak.	11111 21121 12221 11111 21211 11111	6	98
74	Hops	*Humulus lupulus* L.	11121 23111 11221 21212 21211 11123	3	99
75	Horseradish	*Cochlearia armoracia* L.	11111 21121 21111 11111 21111 11121	8	100
76	Hyssop	*Hyssopus officinalis* L.	11111 11121 12222 22222 21111 21112	5	85

77	Jerusalem artichoke	*Helianthus tuberosus* L.	11111	21111	21111	11111	21111	11111	8	88
78	Japanese potato	*Stachys sieboldii* Miq.	11111	11131	12111	11111	21211	11111	8	92
79	Juniper berries	*Juniperus communis* L. var. *depressa* Pursh	21111	22131	12111	11111	21111	11121	3	78
80	Kohlrabi	*Brassica oleracea* L. var. *caulorapa* DC.	11111	11111	12111	11111	21121	11121	9	96
81	Leek	*Allium porrum* L.	21111	21111	12111	11111	21111	11211	6	89
82	Lemon peel	*Citrus limonia* Osbeck	21112	21111	12111	11111	21211	11121	3	89
83	Lettuce	*Lactuca sativa* L.	11111	21111	12212	21212	21211	11111	6	100
84	Linseed	*Linum usitatissimum* L.	11111	11212	11111	11111	21111	11121	4	90
85	Liquorice	*Glycyrrhiza glabra* L. and other spp. *Glabra*	21112	11121	22111	11111	21111	11111	8	100
86	Lotus	*Nelumbo nucifera* Gaertn.	11111	11121	12111	11111	21111	11121	8	74
87	Mace	*Myristica fragrans* Houtt.	11111	21111	11111	11111	21111	11111	4	98
88	Marigold	*Calendula officinalis* L.	21121	11111	11222	21212	21211	11122	2	99
89	Marjoram	*Majorana hortensis* M.	11111	11121	11222	22222	21111	21112	5	98
90	Maté	*Ilex paraguensis* St Hilaire	21121	11111	12221	11111	21211	11121	6	100
91	Matricaria	*Matricaria chamomilla* L.	11121	11111	11221	21212	21211	11122	2	97
92	Mulberry	*Morus nigra* L.	11122	11111	11222	22112	21211	11111	6	100
93	Mullein	*Verbascum thapsus* L.	11111	11111	11212	21222	11211	11111	6	100
94	Mustard (Black/ White)	Black: *Brassica nigra* (L.) Koch White: *Sinapis alba* L.	11111	11212	11111	11111	21111	11121	4	90
95	Nutmeg	*Myristica fragrans* Houtt.	11111	21221	12111	11111	21111	11111	4	98
96	Olive stone	*Olea europaea* L.	11111	11211	12111	11111	21111	11121	4	87
97	Onion	*Allium cepa* L.	21111	21111	12111	11111	21111	11211	7	89
98	Orange leaf	*Citrus aurantium* L.	21112	21111	12111	11111	21211	11111	6	98
99	Orange peel	*Citrus aurantium* L.	21112	21111	12111	11111	21211	11121	3	89
100	Origanum	*Origanum vulgare* L.	11111	11121	12212	22222	21111	21122	5	98
101	Palm shoots	*Inodes palmetto* Cook *Oreodoxa oleracea* Mart.	12111	11111	12111	11111	21211	11111	6	99
102	Paprika	*Capsicum annuum* L.	11211	12111	12221	22232	21121	11121	3	100
103	Parsley	*Petroselinium sativum* Hoffm.	11111	21111	12331	11111	21211	11111	6	73
104	Parsnip	*Pastinaca sativa* L.	11111	22123	22111	11111	21111	11121	8	90
105	Peanut (meat)	*Arachis hypogaea* L.	11111	12121	11111	11111	21211	11111	10	89
106	Peanut (shell)	*Arachis hypogaea* L.	11111	11111	12331	11111	21111	11121	10	69

TABLE 1 (*continued*)

No.	Name of food in program	Latin name	Stored data for characters 1–30 (see Fig. 1)[a]	Morphological group[b]	Discrimination index
107	Pecan (meat)	*Carya illinoensis* Koch	11111 12121 11111 11111 21211 11111	10	89
108	Pecan (shell)	*Carya illinoensis* Koch	11111 11111 11111 11111 11111 11121	10	79
109	Pepper (Black/ White)	*Piper nigrum* L.	21111 22121 12111 11111 21311 11121	3	78
110	Peppergrass	*Lepidium sativum* L.	11111 11111 12221 11111 21121 11111	6	90
111	Peppermint	*Mentha piperita* L.	11111 11121 12222 22222 21111 21122	5	87
112	Pimento	*Pimenta officinalis* Lindley	21121 22121 12221 11111 21211 11121	3	100
113	Pineapple	*Ananas sativus* Schult.	12111 11111 12111 11111 21111 11121	3	96
114	Pine nut	*Pinus* spp.	11111 12121 11111 11111 11111 11121	10	99
115	Pistachio nut	*Pistacia vera* L.	11111 12121 12111 11111 21111 11121	10	77
116	Poppy seed	*Papaver somniferum* L.	11221 12111 12111 11111 21111 11111	4	100
117	Potato	*Solanum tuberosum* L.	11111 12121 21111 11111 21111 11131	8	99
118	Pumpkin pulp	*Cucurbita pepo* L.	11111 21121 12111 11111 21211 11111	3	86
119	Pumpkin seed	*Cucurbita pepo* L.	11111 12121 11111 11111 21111 11121	4	93
120	Purslane	*Portulaca oleracea* L.	11121 11121 12111 11111 21211 11111	5	99
121	Radish	*Raphanus sativus* L.	11111 11111 22111 11111 21111 11111	8	97
122	Red clover	*Trifolium pratense* L.	21112 12121 12221 21212 21211 11122	5	100
123	Red-rose petal	*Rosa gallica* L.	11121 11111 13221 21212 21311 11112	2	98
124	Rhubarb (petiole)	*Rheum rhaponticum* L.	11121 11121 11221 21212 12211 11111	6	100
125	Rose hips	*Rosa canina* L. and other spp. of *R.*	11121 12111 12221 11111 11211 11121	3	100
126	Rosemary	*Rosmarinus officinalis* L.	21111 11111 12212 22222 21111 12111	6	100
127	Rue	*Ruta graveolens* L.	11121 21111 11111 11111 21211 11111	6	100
128	Safflower	*Carthamus tinctorius* L.	21111 21111 12222 11111 21111 11112	2	100
129	Saffron	*Crocous sativus* L.	11111 11111 13221 11111 21111 11112	2	100
130	Sage herb	*Salvia officinalis* L.	11111 11121 12222 22222 21111 21111	5	98
131	Salsify	*Tragopogon porrifolius* L.	11111 21111 21111 11111 21111 11111	8	88
132	Sassafras	*Sassafras variifolium* (Salisbury) D. Kuntz	12111 21121 22111 11111 11111 11121	1	97
133	Savory	*Satureia hortensis* L.	31111 11121 12222 22122 21111 21112	5	98

134	Sesame	*Sesamum indicum* L.	21121	12111	11111	11111	21111	11111	4	100
135	Slippery elm bark	*Ulmus fulva* Michaux	21112	11122	22111	11111	11111	11111	1	100
136	Soybean	*Glycine max* Merr.	21111	11211	11111	11111	21111	11111	4	99
137	Spearmint	*Mentha spicata* L.	11111	11121	12222	22222	21111	21122	5	87
138	Spinach	*Spinacia oleracea* L.	11211	11111	12221	21221	21211	11111	5	100
139	Star anise	*Illicium verum* Hook. f.	21111	21211	12111	11111	21211	11121	3	100
140	Sunflower	*Helianthus annuus* L.	11111	12111	12212	11111	21111	11121	4	100
141	Swamp cabbage	*Ipomoea reptans* Poir. and Cog.	11121	11111	11111	21112	21211	11111	6	100
142	Sweet basil	*Ocimum basilicum* L.	31121	11121	12222	22112	21111	21112	5	100
143	Sweet potato	*Ipomoea batatas* Lam.	11121	21121	21111	11111	21111	11111	8	100
144	Taro	*Arum esculentum* L.	13111	21121	22111	11111	21111	11111	8	76
145	Tarragon	*Artemisia dracunulus* L.	11111	21121	12111	22112	21211	11111	5	100
146	Tawny lily	*Hemerocallis fulva* L.	11111	21121	11221	11111	21111	11212	2	100
147	Tea	*Thea sinensis* L.	11121	11111	12221	11111	21111	12121	6	100
148	Thyme	*Thymus vulgaris* L.	11111	11121	12222	22222	21111	21112	5	85
149	Tomato fruit	*Lycopersicon esculentum* Mill.	11111	12111	12221	11111	21111	11121	3	98
150	Tomato seed	*Lycopersicon esculentum* Mill.	11111	12111	11221	11111	21111	11121	4	98
151	Tree of heaven	*Ailanthus glandulosa* Desf.	11122	11111	13223	11111	21211	11111	6	100
152	Turmeric	*Curcuma domestica* Valeton	11111	21121	21221	11111	21211	11111	8	99
153	Turnip	*Brassica rapa* L. var. *rapifera* Metzg.	11111	11111	21111	11111	21111	11121	8	98
154	Vanilla	*Vanilla planifolia* Andrews	22111	12113	12221	11111	21211	11111	3	100
155	Walnut (meat)	*Juglans regia* L.	11111	12111	11111	11111	31211	11111	10	98
156	Walnut (shell)	*Juglans regia* L.	11111	11111	11111	11111	11111	11121	10	79
157	Water chestnut	*Eleocharis tuberosa* Schult.	11111	21121	12111	11111	21111	11111	8	83
158	Water cress	*Roripa nasturtium* Rusby	11111	11131	12221	11111	21121	11111	5	90
159	Yam	*Dioscorea* spp.	21121	11121	22111	11111	21111	11121	8	99
160	Yautia	*Xanthosoma* spp.	13111	21121	22111	11111	21111	11111	8	76
161	Zedoary	*Curcuma zedoaria* Roscoe	11111	21121	21221	11111	21111	11111	8	99

[a] 1 = Absent; 2 = present; 3 = not always present or difficult to identify.
[b] 1 = Barks; 2 = flowers; 3 = fruits; 4 = seeds; 5 = herbs; 6 = leaves; 7 = underground leaves; 8 = underground roots and stems; 9 = stems; 10 = nuts.

FIG. 2. Computer printout from FOODID.

```
RUN

 76/12/15. 10.14.40.
PROGRAM    FOODID

FOODID - FOOD  MATERIAL   AND   SPICE
         IDENTIFICATION BY MICROSCOPY   - EDITION 15/12/76


THIS PROGRAM CONTAINS 161 POWDERED FOOD MATERIALS


DO YOU WISH TO INSERT DATA? Y
REFERENCE:? TEST DATA 1
DATA  ? 21111,11111,12212,22222,21111,12111

**************************************************************
REFERENCE - TEST DATA 1
DATA CHECK -  21111  11111  12212  22222  21111  12111
POINTS HAVE BEEN ALLOCATED TO A MAXIMUM OF 30

INPUT DATA AGREES ENTIRELY WITH : ROSEMARY - LEAF

 25 SAGE HERB                    - HERB
    *** CHECK REF. NOS. - 1  9  14  26  27

 24 CUCUMBER                     - FRUIT
    *** CHECK REF. NOS. - 1  7  17  19  27  ETC.

 24 HYSSOP                       - HERB
    *** CHECK REF. NOS. - 1  9  14  26  27  ETC.

 24 LETTUCE                      - LEAF
    *** CHECK REF. NOS. - 1  6  17  19  23  ETC.

 24 MULLEIN                      - LEAF
    *** CHECK REF. NOS. - 1  12  17  21  23  ETC.

 24 ORIGANUM                     - HERB
    *** CHECK REF. NOS. - 1  9  26  27  29  ETC.

 24 SAVORY                       - HERB
    *** CHECK REF. NOS. - 9  14  18  26  27  ETC.

 24 THYME                        - HERB
    *** CHECK REF. NOS. - 1  9  14  26  27  ETC.


**************************************************************
*** CONFIRM  BY  COMPARISON  WITH  AUTHENTIC  SAMPLE ***
**************************************************************
```

```
MORE DATA ? Y
REFERENCE:? TEST DATA 2 (ONE DELIBERATE ERROR)
DATA  ? 11111,21112,12221,11111,21211,11111

*****************************************************
REFERENCE - TEST DATA 2 (ONE DELIBERATE ERROR)
DATA CHECK -  11111  21112  12221  11111  21211  11111
POINTS HAVE BEEN ALLOCATED TO A MAXIMUM OF 30

 29 BUCHU                        - LEAF
    *** CHECK REF. NOS. - 4

 29 CHERVIL                      - LEAF
    *** CHECK REF. NOS. - 10

 29 PARSLEY                      - LEAF
    *** CHECK REF. NOS. - 10


*****************************************************
*** CONFIRM  BY  COMPARISON  WITH  AUTHENTIC  SAMPLE ***
*****************************************************


MORE DATA ? Y
REFERENCE:? TEST DATA 3
DATA  ? 11111,11121,12222,22222,21111,21122

*****************************************************
REFERENCE - TEST DATA 3
DATA CHECK -  11111  11121  12222  22222  21111  21122
POINTS HAVE BEEN ALLOCATED TO A MAXIMUM OF 30

INPUT DATA AGREES ENTIRELY WITH : PEPPERMINT - HERB

                                :SPEARMINT - HERB

 29 HYSSOP                       - HERB
    *** CHECK REF. NOS. - 29

 29 ORIGANUM                     - HERB
    *** CHECK REF. NOS. - 14

 29 THYME                        - HERB
    *** CHECK REF. NOS. - 29


*****************************************************
*** CONFIRM  BY  COMPARISON  WITH  AUTHENTIC  SAMPLE ***
*****************************************************


MORE DATA ? N

 CP       5.055 SECS.

RUN COMPLETE.
```

A computer program is presented based upon twelve main characters (Fig. 1). In the program is a data bank of coded information corresponding to the characteristics of each food material listed (Table 1). Information in the data bank was compiled after microscopical examination of authentic powdered materials and by consulting published literature (Tschirch and Œsterle, 1900; Greenish and Collin, 1904; Winton, 1916; Kraemer, 1920; Schneider, 1921; Greenish, 1923; Mansfield, 1929; Thoms, 1931; Wallis, 1948, 1965, 1967; Winton and Winton, 1949; Claus, 1956; Deryng, 1961; Eschrich, 1966; Jackson and Snowdon, 1974; Parry, 1969; Vaughan, 1970). One point is allocated for each agreement, giving a maximum score of 30 points and the three highest scores are printed out (more if identical points have been scored by a number of foods). The program also includes characteristics of items that are present but may be difficult to identify; in these circumstances a score of one is given, regardless of the input data. This procedure makes some allowance for lack of observation by the inexperienced microscopist and for differences due to natural variation.

The computer printout (Fig. 2) indicates the food(s) that agrees with the input data (i.e. 30 points scored), and when the input data does not entirely agree with any food (i.e. the maximum of 30 points has not been scored) the computer printout indicates for the highest scores the number(s) of the character(s) that should be checked. Final confirmation of the identity of the powder should, in common with other aids to identification, be obtained wherever possible by comparison with an authentic reference standard of the named material.

II. COMPUTER PROGRAM

In the program (flow diagram Fig. 3; program Fig. 4) written in BASIC, the 12 main (30 specific) characters for 161 powdered food materials and some adulterants are coded in DATA statements in such a form that Ø represents a character definitely absent, 1 a character definitely present and 2 a character that is present but sometimes difficult for the less-skilled microscopist to confirm. For convenience, the 30 specific characters are arranged in groups of five digits with the leading zeroes suppressed. The DATA statement also includes coded information on the morphological group of the food. The foods are, in general, referred to by their common names but the corresponding botanical names are given in the list of foods in Table 1. The term 'food' is used in its widest sense and includes herbs and spices, nuts and shells, and vegetables.

The program calls for a reference to be submitted in the first instance, followed by suitable input data. The input data are required in a slightly different form from the DATA statements in the program; the numeral 1 represents an absent character and 2 an observed character. The numerals are input at the terminal in six blocks of five characters, and the program is designed to check that only the numerals 1 and 2 are typed in at the terminal, the computer itself checking that six sets of digits only have been transmitted. The necessity for the difference in data presentation is to accommodate the check in the program to obviate accidental false information being accepted for processing. As an analyst may find it difficult to interpret the data as stored in the program (i.e. with the leading zeroes suppressed) the information is presented in Table 1 in the same format as the input data, with the addition that the numeral 3 (not a valid number for input information) represents a character not always present or difficult to identify.

The output data information includes the reference and a copy of the input data for immediate checking, and also a list of the more likely identities based upon a scoring system with a maximum of 30 points. The method of scoring is as follows:

(i) the input blocks of data are added numerically to the corresponding data blocks in the program for each food;
(ii) each food is given an initial score of 30 points;
(iii) each block of the added data is examined for a numeral 2 and one point is deducted for each numeral 2 present (see Table 2), which corresponds with a disagreement between the input and output data;
(iv) the printout comprises the foods and their corresponding morphological group, together with the highest scores, subject to a minimum of three foods and a maximum of three different scores;
(v) when less than 30 points are obtained, the printout indicates which character number(s) (subject to a maximum of five) is at variance;
(vi) when more than one food receives 30 points, complete identity can be achieved by reference to Table 3;
(vii) identity is finally confirmed by comparison with an authentic reference sample.

This program will also print the complete list of foods in alphabetical order or, if preferred, the foods within a requested morphological group.

III. DISCRIMINATION INDEX

The limited input data make it impossible to guarantee a unique identification and, depending on the sample examined, the degree of confidence in the output information will vary. For this reason the concept of a discrimination

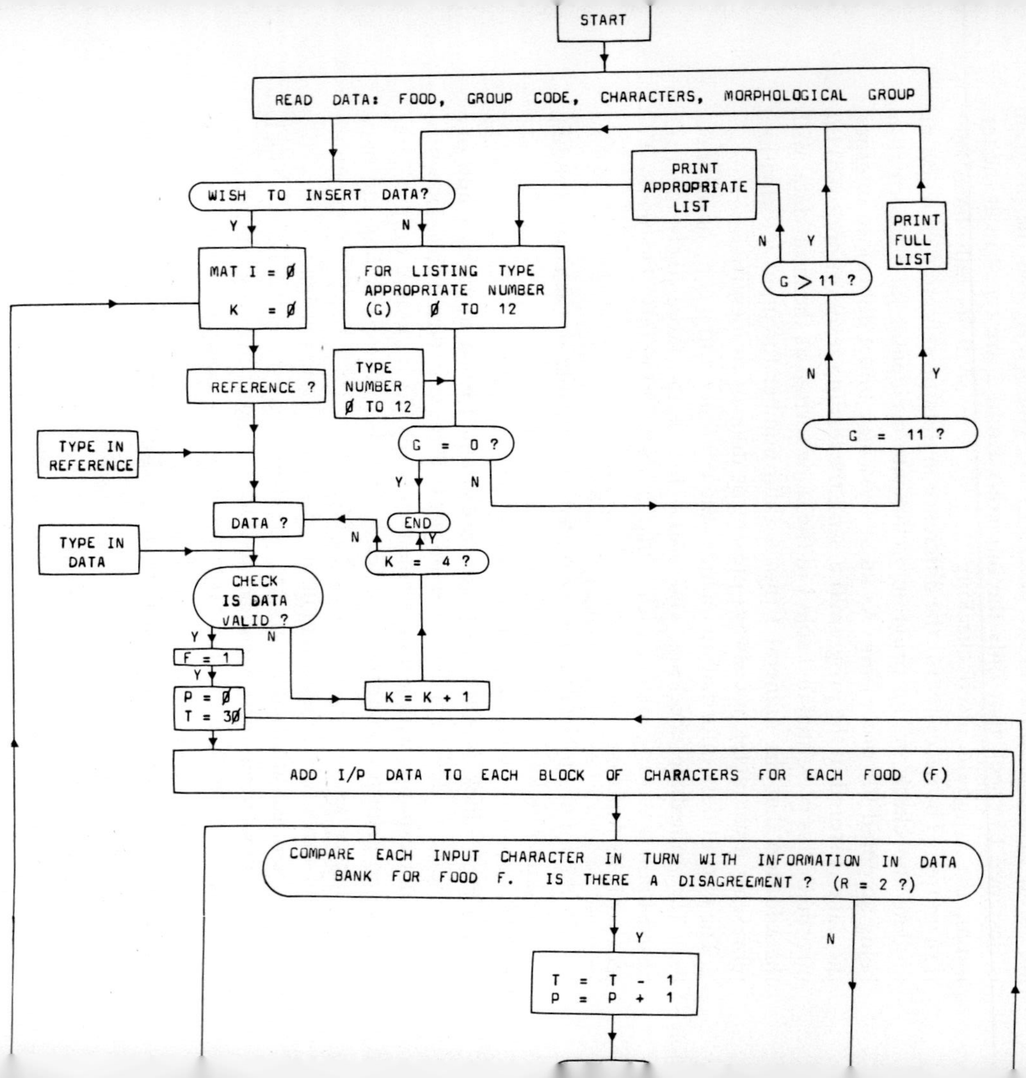
START
READ DATA: FOOD, GROUP CODE, CHARACTERS, MORPHOLOGICAL GROUP
WISH TO INSERT DATA?
Y
N
MAT I = Ø
K = Ø
FOR LISTING TYPE APPROPRIATE NUMBER (G) Ø TO 12
PRINT APPROPRIATE LIST
G > 11 ?
N
Y
PRINT FULL LIST
REFERENCE ?
TYPE NUMBER Ø TO 12
TYPE IN REFERENCE
G = 0 ?
G = 11 ?
Y
N
DATA ?
END
TYPE IN DATA
K = 4 ?
CHECK IS DATA VALID ?
F = 1
K = K + 1
P = Ø
T = 3Ø
ADD I/P DATA TO EACH BLOCK OF CHARACTERS FOR EACH FOOD (F)
COMPARE EACH INPUT CHARACTER IN TURN WITH INFORMATION IN DATA BANK FOR FOOD F. IS THERE A DISAGREEMENT ? (R = 2 ?)
T = T - 1
P = P + 1

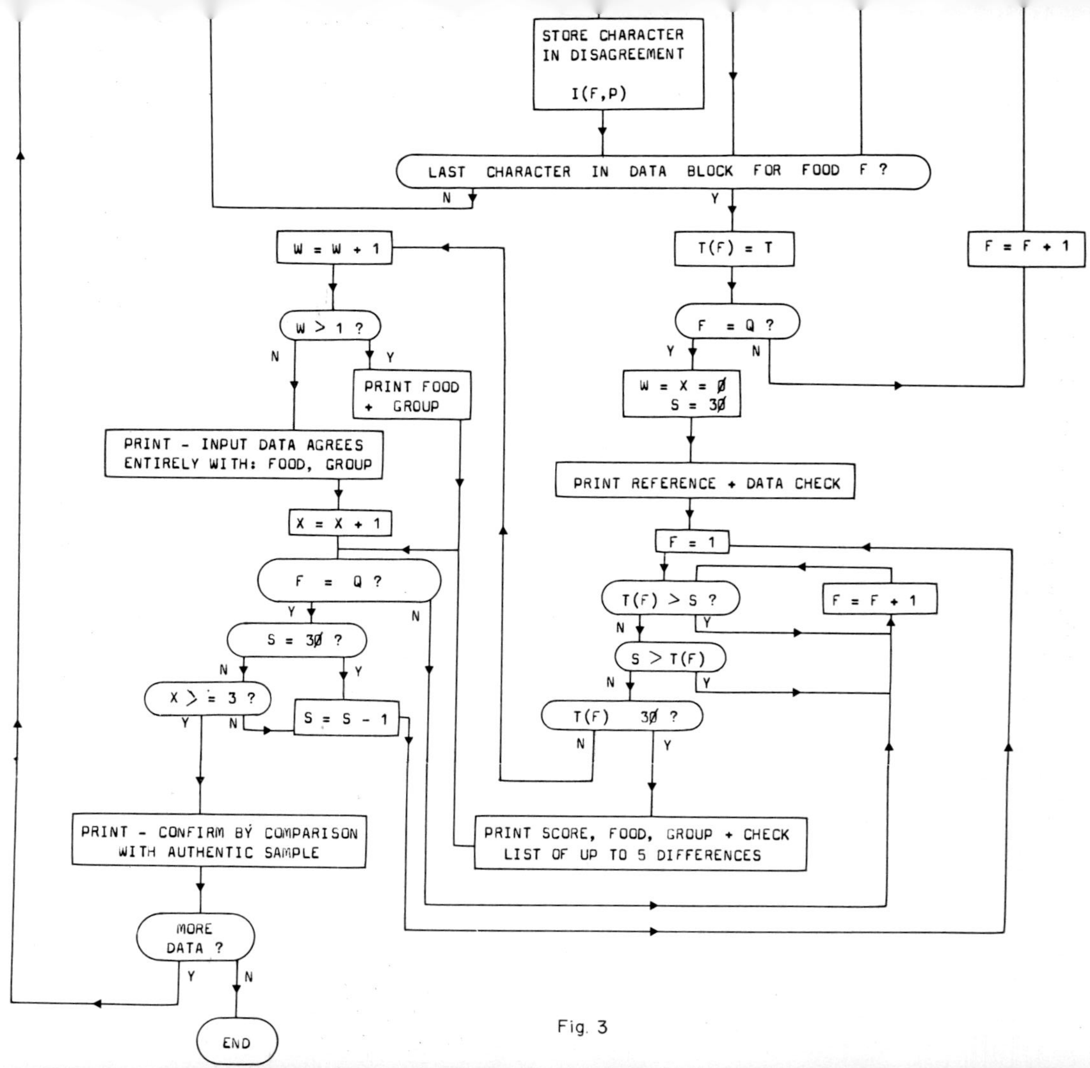

STORE CHARACTER IN DISAGREEMENT
I(F,P)
LAST CHARACTER IN DATA BLOCK FOR FOOD F ?
N
Y
W = W + 1
T(F) = T
F = F + 1
W > 1 ?
N
Y
PRINT FOOD + GROUP
F = Q ?
Y
N
W = X = Ø
S = 3Ø
PRINT - INPUT DATA AGREES ENTIRELY WITH: FOOD, GROUP
PRINT REFERENCE + DATA CHECK
X = X + 1
F = 1
F = Q ?
Y
N
T(F) > S ?
F = F + 1
S = 3Ø ?
S > T(F)
X >= 3 ?
S = S - 1
T(F) 3Ø ?
PRINT - CONFIRM BY COMPARISON WITH AUTHENTIC SAMPLE
PRINT SCORE, FOOD, GROUP + CHECK LIST OF UP TO 5 DIFFERENCES
MORE DATA ?
END

Fig. 3

FIG. 4. Computer program (FOODID) (pp. 608–613).

```
00100 PRINT 'FOODID - FOOD MATERIAL   AND   SPICE'
00110 PRINT '            IDENTIFICATION BY MICROSCOPY   - EDITION 19/1/77'
00120 REM Q = NUMBER OF FOOD MATERIALS IN DATA BANK
00130 LET Q = 161
00140 DIM V(1,6),F$(161),M(161),T(161),C(161,6),I(161,5)
00150 PRINT
00160 PRINT
00170 REM     Z  =  NUMBER OF DATA BLOCKS
00180 REM V(1,Z)=  SET OF EXPERIMENTAL VALUES
00190 REM   M(F) =  MORPHOLOGICAL GROUP
00200 REM F$(F) =  NAME OF FOOD MATERIAL
00210 REM     Q  =  NUMBER OF FOOD MATERIAL IN DATA
00220 REM   T(F) =  POINTS TOTALISER
00230 PRINT 'THIS PROGRAM CONTAINS';Q;'POWDERED FOOD MATERIALS'
00240 PRINT
00250 FOR F = 1 TO Q
00260 REM READ FOOD MATERIAL, MORPHOLOGICAL CODE AND BASIC DATA
00270 READ F$(F), M(F)
00280 REM READ IN DIAGNOSTIC CHARACTERS FOR EACH FOOD MATERIAL
00290 REM C(F,Z) = 6 BLOCKS OF 5 CHARACTERS
00300 REM 0 = CHARACTER DEFINITELY  ABSENT (1 IN INPUT DATA)
00310 REM 1 = CHARACTER DEFINITELY PRESENT (2 IN INPUT DATA)
00320 REM 2 = CHARACTER PRESENT MAY BE DIFFICULT TO OBSERVE
00330 FOR Z = 1 TO 6
00340 READ C(F,Z)
00350 NEXT Z
00360 NEXT F
00370 REM READ MORPHLOGICAL GROUP
00380 FOR G = 1 TO 10
00390 READ G$(G)
00400 NEXT G
00410 PRINT
00420 PRINT 'DO YOU WISH TO INSERT DATA';
00430 INPUT I$
00440 IF SUBSTR(I$,1,1) = 'Y' THEN 00470
00450 IF SUBSTR(I$,1,1) = 'N' THEN 03230
00460 GOTO 00420
00470 MATI = ZER
00480 LET K = 0
00490 PRINT 'REFERENCE:';
00500 INPUT N$
00510 PRINT 'DATA ';
00520 MAT INPUT V
00530 REM TEST FOR VALID INPUT DATA
00540 FOR Z = 1 TO 6
00550 LET X = V(1,Z)
00560 LET Y = X + C(F,Z)
00570 FOR B = 1 TO 5
00580 LET N = INT(X/10↑(5-B))
00590 LET X = X - (N*10↑(5-B))
00600 IF N > 2 THEN 00900
00610 IF N = 0 THEN 00900
00620 NEXT B
00630 NEXT Z
00640 REM ADD INPUT DATA TO DATA FOR EACH FOOD -
00650 REM DEDUCT 1 FROM T FOR EVERY DIFFERENCE
00660 REM FOR EACH FOOD TO OBTAIN SCORE (MAX = 30)
```

FIG. 4.—*continued*

```
00670 FOR F = 1 TO Q
00680 REM T = INITIAL SCORE OF 30
00690 LET T = 30
00700 REM P LIMITS OUTPUT OF CHARACTERS TO BE CHECKED
00710 LET P = 0
00720 FOR Z = 1 TO 6
00730 LET X = V(1,Z)
00740 LET Y = X + C(F,Z)
00750 FOR B = 1 TO 5
00760 LET R = INT(Y/10↑(5-B))
00770 LET Y = Y - (R*10↑(5-B))
00780 REM R = 2 FOR EACH ITEM OF DISAGREEMENT
00790 IF R = 2 THEN 00810
00800 GOTO 00850
00810 LET T = T - 1
00820 LET P = P + 1
00830 IF P > 5 THEN 00850
00840 LET I(F,P) = (Z-1)*5 + B.
00850 NEXT B
00860 NEXT Z
00870 LET T(F) = T
00880 NEXT F
00890 GOTO 00980
00900 PRINT
00910 LET K = K+1
00920 IF K = 4 THEN 03650
00930 PRINT 'EACH BLOCK SHOULD CONTAIN FIVE FIGURE ONES OR TWOS ONLY
00940 PRINT '6 BLOCKS OF DATA ARE REQUIRED'
00950 GOTO 00510
00960 REM EACH FOOD HAS BEEN ALLOCATED A POINT FOR EACH AGREEMENT
00970 REM AND THE TOTAL NUMBER OF POINTS SCORED IS IN LOCATION T(F)
00980 PRINT
00990 REM X IS USED AS A COUNTER TO LIMIT OUTPUT DATA
01000 LET W = X = 0
01010 REM S = MAXIMUM POSSIBLE NO. OF POINTS
01020 LET S = 30
01030 GOSUB 01620
01040 PRINT 'REFERENCE - '; N$
01050 PRINT 'DATA CHECK - ';
01060 FOR Z = 1 TO 6
01070 PRINT V(1,Z);
01080 NEXT Z
01090 PRINT
01100 PRINT 'POINTS HAVE BEEN ALLOCATED TO A MAXIMUM OF'; S
01110 PRINT
01120 FOR F = 1 TO Q
01130 IF T(F) > S THEN 01380
01140 IF S > T(F) THEN 01380
01150 IF T(F) < 30 THEN 01240
01160 LET W = W + 1
01170 IF W > 1 THEN 01200
01180 PRINT 'INPUT DATA AGREES ENTIRELY WITH : '; F$(F); ' ';
01190 GOTO 01210
01200 PRINT TAB(31);' :';F$(F); ' ';
01210 GOSUB 01580
01220 PRINT
01230 GOTO 01370
01240 PRINT S; F$(F);
01250 GOSUB 01580
01260 PRINT '     *** CHECK REF. NOS. -';
01270 FOR P = 1 TO 5
```

FIG. 4.—*continued*

```
01280 IF I(F,P) = 0 THEN 01300
01290 PRINT I(F,P);
01300 NEXT P
01310 REM LIMIT TO FIVE ONLY CHECK REF. NUMBERS
01320 REM PRINT ETC.IF MORE THAN FIVE POINTS OF DIFFERENCE ARE OBSERVED
01330 IF S > 24 THEN 01350
01340 PRINT ' ETC.';
01350 PRINT
01360 PRINT
01370 LET X = X + 1
01380 NEXT F
01390 REM LIMIT OUTPUT TO MAXIMUM OF THREE SCORES
01400 IF S = 30 THEN 01420
01410 IF X >= 3 THEN 01440
01420 LET S = S - 1
01430 GOTO 01120
01440 PRINT
01450 GOSUB 01620
01460 PRINT '*** CONFIRM  BY  COMPARISON  WITH  AUTHENTIC  SAMPLE ***'
01470 GOSUB 01620
01480 PRINT
01490 PRINT
01500 PRINT 'MORE DATA';
01510 INPUT Z$
01520 IF SUBSTR(Z$,1,1) = 'Y' THEN 00470
01530 IF SUBSTR(Z$,1,1) = 'N' THEN 03650
01540 LET K1 = K1 + 1
01550 IF K1 = 3 THEN 03650
01560 PRINT 'HAVE YOU ';
01570 GOTO 01500
01580 LET G = M(F)
01590 PRINT TAB(30);'- '; G$(G)
01600 RETURN
01610 STOP
01620 FOR L8 = 1 TO 56
01630 PRINT '*';
01640 NEXT L8
01650 PRINT
01660 RETURN
01670 DATA ABSINTH,5,10,10,1111,11101,10100,11
01680 DATA ALFALFA,5,10001,1000,1111,10101,10100,1
01690 DATA ALMOND,4,10010,1000,1110,0,10000,10
01700 DATA ANGELICA ROOT,8,0,10010,11000,0,10000,0
01710 DATA ANGUSTRURA,1,11000,10010,11000,0,0,10
01720 DATA ANISE,3,10,11000,1110,0,10100,10
01730 DATA ARECA,4,0,100,1000,0,10000,10
01740 DATA ARROWHEAD,8,0,10,1000,0,10000,0
01750 DATA ARTICHOKE,2,0,10000,1000,0,10010,0
01760 DATA ASPARAGUS,5,1000,0,1000,0,10000,110
01770 DATA BAMBOO SHOOTS (DECORT.),5,0,0,1000,0,10000,0
01780 DATA BANANA,3,1000,10010,1000,0,10000,0
01790 DATA BAYBERRY,3,0,11010,0,0,10100,10
01800 DATA BAY LEAF,6,0,10000,1110,0,10100,10
01810 DATA BEECHNUT (MEAT),10,10,1010,0,0,10000,0
01820 DATA BEET GREENS,5,100,10,1101,0,10100,0
01830 DATA BEETROOT,8,100,0,11000,0,10000,0
01840 DATA BRAKE,9,0,10,1101,0,10000,0
01850 DATA BRAZIL NUT (MEAT),10,0,100,0,0,20000,0
01860 DATA BRAZIL NUT (SHELL),10,0,0,2000,0,10000,10
01870 DATA BUCHU,6,10,10001,1110,0,10100,0
01880 DATA BUCKWHEAT,3,0,1010,1000,0,10000,10
```

```
01890 DATA CACAO, 4, 10, 1011, 1101, 0, 10000, 10
01900 DATA CANELLA, 1, 10, 10010, 22000, 0, 0, 10
01910 DATA CAPSICUM, 3, 100, 1000, 1101, 11001, 10010, 10
01920 DATA CARAWAY, 3, 10, 11000, 1000, 0, 10100, 10
01930 DATA CARDAMOM FRUIT, 3, 11000, 11010, 1000, 0, 10000, 10
01940 DATA CARDAMOM SEED, 4, 10020, 11010, 0, 0, 10000, 10
01950 DATA CARDOON, 6, 0, 10, 1111, 10101, 10100, 0
01960 DATA CARROT, 8, 0, 11012, 11000, 0, 10000, 10
01970 DATA CASHEW NUT, 4, 0, 11010, 0, 0, 10000, 0
01980 DATA CASSAVA, 8, 10000, 10010, 11000, 0, 10000, 10
01990 DATA CASSIA BARK, 1, 1000, 10011, 11000, 0, 0, 10
02000 DATA CASSIA BUDS, 3, 11010, 10011, 1110, 0, 10200, 10
02010 DATA CAULIFLOWER, 5, 0, 0, 1000, 0, 10010, 0
02020 DATA CELERY FRUIT, 3, 10, 11000, 1110, 0, 10100, 10
02030 DATA CELERY LEAF, 6, 0, 10000, 1000, 0, 10100, 0
02040 DATA CHAMOMILE, 2, 10, 0, 111, 10101, 10100, 11
02050 DATA CHERVIL, 6, 0, 10000, 1110, 0, 10100, 0
02060 DATA CHICORY LEAF, 6, 0, 10000, 1101, 0, 10100, 0
02070 DATA CHICORY ROOT, 8, 0, 10000, 11000, 0, 10000, 0
02080 DATA CHIVES, 6, 0, 10000, 1000, 0, 10000, 100
02090 DATA CHUFA, 8, 0, 10, 1000, 0, 10000, 10
02100 DATA CINNAMON, 1, 1000, 10011, 11000, 0, 0, 10
02110 DATA CLOVE FRUIT, 3, 10, 11010, 1000, 0, 10100, 10
02120 DATA CLOVE FLOWER, 2, 10, 10000, 1000, 0, 10100, 11
02130 DATA CLOVE STEM, 9, 10010, 10010, 11000, 0, 10000, 10
02140 DATA COCONUT (ENDOSPERM), 10, 0, 100, 0, 0, 0, 0
02150 DATA COCONUT (SHELL), 10, 0, 0, 1000, 0, 10000, 10
02160 DATA COFFEE, 4, 0, 1011, 0, 0, 10000, 10
02170 DATA CORIANDER, 3, 10010, 11000, 1000, 0, 20100, 10
02180 DATA CUBEBS, 3, 10000, 11010, 1000, 0, 10200, 10
02190 DATA CUCUMBER, 3, 0, 1000, 1101, 10101, 10000, 10
02200 DATA CUMIN, 3, 10, 11000, 1101, 0, 10100, 10
02210 DATA DANDELION ROOT, 8, 0, 10000, 10101, 0, 10000, 0
02220 DATA DATE STONE, 4, 0, 0, 0, 0, 20000, 10
02230 DATA DILL, 3, 10, 11000, 1000, 0, 10200, 10
02240 DATA ENDIVE, 6, 0, 10000, 1101, 0, 10100, 0
02250 DATA FENNEL, 3, 10, 11000, 1000, 0, 10200, 0
02260 DATA FENUGREEK, 4, 0, 1001, 0, 0, 10000, 0
02270 DATA FIG, 3, 10, 11001, 1112, 0, 10200, 10
02280 DATA GALANGAL, 8, 0, 10010, 1000, 0, 10200, 0
02290 DATA GARDEN SORREL, 6, 10, 0, 1110, 10101, 10100, 0
02300 DATA GARLIC, 7, 10010, 10000, 1000, 0, 10000, 110
02310 DATA GENTIAN, 8, 1000, 10, 10000, 0, 10000, 0
02320 DATA GINGER (SCRAPED), 8, 0, 10010, 1000, 0, 1000, 0
02330 DATA GINGER (UNSCRAPED), 8, 0, 10010, 11000, 0, 1000, 0
02340 DATA GRAPE SEED, 4, 1010, 100, 0, 0, 20000, 10
02350 DATA GUARANA, 4, 0, 2010, 1000, 0, 10000, 10
02360 DATA HAZELNUT (MEAT), 10, 10, 120, 0, 0, 10000, 0
02370 DATA HAZELNUT (SHELL), 10, 0, 0, 110, 0, 10000, 10
02380 DATA HEMP (HEMP SEED), 4, 0, 1000, 0, 0, 10000, 10
02390 DATA HONEWORT, 6, 0, 10010, 1110, 0, 10100, 0
02400 DATA HOPS, 3, 10, 12000, 110, 10101, 10100, 12
02410 DATA HORSERADISH, 8, 0, 10010, 10000, 0, 10000, 10
02420 DATA HYSSOP, 5, 0, 10, 1111, 11111, 10000, 10001
02430 DATA JERUSALEM ARTICHOKE, 8, 0, 10000, 10000, 0, 10000, 0
02440 DATA JAPANESE POTATO, 8, 0, 20, 1000, 0, 10100, 0
02450 DATA JUNIPER BERRIES, 3, 10000, 11020, 1000, 0, 10000, 10
02460 DATA KOHLRABI, 9, 0, 0, 1000, 0, 10010, 10
02470 DATA LEEK, 6, 10000, 10000, 1000, 0, 10000, 100
02480 DATA LEMON PEEL, 3, 10001, 10000, 1000, 0, 10100, 10
02490 DATA LETTUCE, 6, 0, 10000, 1101, 10101, 10100, 0
```

FIG. 4.—*continued*

```
02500 DATA LINSEED, 4, 0, 101, 0, 0, 10000, 10
02510 DATA LIQUORICE, 8, 10001, 10, 11000, 0, 10000, 0
02520 DATA LOTUS, 8, 0, 10, 1000, 0, 10000, 10
02530 DATA MACE, 4, 0, 10000, 0, 0, 10000, 0
02540 DATA MARIGOLD, 2, 10010, 0, 111, 10101, 10100, 11
02550 DATA MARJORAM, 5, 0, 10, 111, 11111, 10000, 10001
02560 DATA MATE, 6, 10010, 0, 1110, 0, 10100, 10
02570 DATA MATRICARIA, 2, 10, 0, 110, 10101, 10100, 11
02580 DATA MULBERRY, 6, 11, 0, 111, 11001, 10100, 0
02590 DATA MULLEIN, 6, 0, 0, 101, 10111, 100, 0
02600 DATA MUSTARD (BLACK/WHITE), 4, 0, 101, 0, 0, 10000, 10
02610 DATA NUTMEG, 4, 0, 10110, 1000, 0, 10000, 0
02620 DATA OLIVE STONE, 4, 0, 100, 1000, 0, 10000, 10
02630 DATA ONION, 7, 10000, 10000, 1000, 0, 10000, 100
02640 DATA ORANGE LEAF, 6, 10001, 10000, 1000, 0, 10100, 0
02650 DATA ORANGE PEEL, 3, 10001, 10000, 1000, 0, 10100, 10
02660 DATA ORIGANUM, 5, 0, 10, 1101, 11111, 10000, 10011
02670 DATA PALM SHOOTS, 6, 1000, 0, 1000, 0, 10100, 0
02680 DATA PAPRIKA, 3, 100, 1000, 1110, 11121, 10010, 10
02690 DATA PARSLEY, 6, 0, 10000, 1220, 0, 10100, 0
02700 DATA PARSNIP, 8, 0, 11012, 11000, 0, 10000, 10
02710 DATA PEANUT (MEAT), 10, 0, 1010, 0, 0, 10100, 0
02720 DATA PEANUT (SHELL), 10, 0, 0, 1220, 0, 10000, 10
02730 DATA PECAN (MEAT), 10, 0, 1010, 0, 0, 10100, 0
02740 DATA PECAN (SHELL), 10, 0, 0, 0, 0, 0, 10
02750 DATA PEPPER (BLACK/WHITE), 3, 10000, 11010, 1000, 0, 10200, 10
02760 DATA PEPPERGRASS, 6, 0, 0, 1110, 0, 10010, 0
02770 DATA PEPPERMINT, 5, 0, 10, 1111, 11111, 10000, 10011
02780 DATA PIMENTO, 3, 10010, 11010, 1110, 0, 10100, 10
02790 DATA PINEAPPLE, 3, 1000, 0, 1000, 0, 10000, 10
02800 DATA PINE NUT, 4, 0, 1010, 0, 0, 0, 10
02810 DATA PISTACHIO NUT, 4, 0, 1010, 1000, 0, 10000, 10
02820 DATA POPPY SEED, 4, 110, 1000, 1000, 0, 10000, 0
02830 DATA POTATO, 8, 0, 1010, 10000, 0, 10000, 20
02840 DATA PUMPKIN PULP, 3, 0, 10010, 1000, 0, 10100, 0
02850 DATA PUMPKIN SEED, 4, 0, 1010, 0, 0, 10000, 10
02860 DATA PURSLANE, 5, 10, 10, 1000, 0, 10100, 0
02870 DATA RADISH, 8, 0, 0, 11000, 0, 10000, 0
02880 DATA RED CLOVER, 5, 10001, 1010, 1110, 10101, 10100, 11
02890 DATA RED-ROSE PETAL, 2, 10, 0, 2110, 10101, 10200, 1
02900 DATA RHUBARB (PETIOLE), 6, 10, 10, 110, 10101, 1100, 0
02910 DATA ROSE HIPS, 3, 10, 1000, 1110, 0, 100, 10
02920 DATA ROSEMARY, 6, 10000, 0, 1101, 11111, 10000, 1000
02930 DATA RUE, 6, 10, 10000, 0, 0, 10100, 0
02940 DATA SAFFLOWER, 2, 10000, 10000, 1111, 0, 10000, 1
02950 DATA SAFFRON, 2, 0, 0, 2110, 0, 10000, 1
02960 DATA SAGE HERB, 5, 0, 10, 1111, 11111, 10000, 10000
02970 DATA SALSIFY, 8, 0, 10000, 10000, 0, 10000, 0
02980 DATA SASSAFRAS, 1, 1000, 10010, 11000, 0, 0, 10
02990 DATA SAVORY, 5, 20000, 10, 1111, 11011, 10000, 10001
03000 DATA SESAME, 4, 10010, 1000, 0, 0, 10000, 0
03010 DATA SLIPPERY ELM BARK, 1, 10001, 11, 11000, 0, 0, 0
03020 DATA SOYBEAN, 4, 10000, 100, 0, 0, 10000, 0
03030 DATA SPEARMINT, 5, 0, 10, 1111, 11111, 10000, 10011
03040 DATA SPINACH, 5, 100, 0, 1110, 10110, 10100, 0
03050 DATA STAR ANISE, 3, 10000, 10100, 1000, 0, 10100, 10
03060 DATA SUNFLOWER, 4, 0, 1000, 1101, 0, 10000, 10
03070 DATA SWAMP CABBAGE, 6, 10, 0, 0, 10001, 10100, 0
03080 DATA SWEET BASIL, 5, 20010, 10, 1111, 11001, 10000, 10001
03090 DATA SWEET POTATO, 8, 10, 10010, 10000, 0, 10000, 0
03100 DATA TARO, 8, 2000, 10010, 11000, 0, 10000, 0
```

FIG. 4.—*continued*

```
03110 DATA TARRAGON, 5, 0, 10010, 1000, 11001, 10100, 0
03120 DATA TAWNY LILY, 2, 0, 10010, 110, 0, 10000, 101
03130 DATA TEA, 6, 10, 0, 1110, 0, 10000, 1010
03140 DATA THYME, 5, 0, 10, 1111, 11111, 10000, 10001
03150 DATA TOMATO FRUIT, 3, 0, 1000, 1110, 0, 10000, 10
03160 DATA TOMATO SEED, 4, 0, 1000, 110, 0, 10000, 10
03170 DATA TREE OF HEAVEN, 6, 11, 0, 2112, 0, 10100, 0
03180 DATA TURMERIC, 8, 0, 10010, 10110, 0, 10100, 0
03190 DATA TURNIP, 8, 0, 0, 10000, 0, 10000, 10
03200 DATA VANILLA, 3, 11000, 1002, 1110, 0, 10100, 0
03210 DATA WALNUT (MEAT), 10, 0, 1000, 0, 0, 20100, 0
03220 DATA WALNUT (SHELL), 10, 0, 0, 0, 0, 0, 10
03230 DATA WATER CHESTNUT, 8, 0, 10010, 1000, 0, 10000, 0
03240 DATA WATER CRESS, 5, 0, 20, 1110, 0, 10010, 0
03250 DATA YAM, 8, 10010, 10, 11000, 0, 10000, 10
03260 DATA YAUTIA, 8, 2000, 10010, 11000, 0, 10000, 0
03270 DATA ZEDOARY, 8, 0, 10010, 10110, 0, 10000, 0
03280 PRINT 'TO OBTAIN FOOD MATERIAL AND SPICE LIST TYPE NUMBER -'
03290 PRINT '1 = BARKS     4 = SEEDS     7 = UNDERGROUND LEAVES'
03300 PRINT '2 = FLOWERS   5 = HERBS     8 = UNDERGROUND ROOTS/STEMS'
03310 PRINT '3 = FRUITS    6 = LEAVES    9 = STEMS    10 = NUTS'
03320 PRINT 'TYPE 11 FOR FULL LIST OR 12 TO RETURN OR 0 TO TERMINATE'
03330 INPUT G
03340 IF G = 0 THEN 03650
03350 IF G = 11 THEN 03480
03360 IF G > 11 THEN 00420
03370 PRINT
03380 PRINT 'LIST OF FOOD MATERIALS IN MORPHOLOGICAL GROUP ';
03390 GOSUB 01590
03400 PRINT '=============================================='
03410 FOR F = 1 TO Q
03420 IF M(F) <> G THEN 03440
03430 GOSUB 03550
03440 NEXT F
03450 PRINT
03460 PRINT 'TYPE ANOTHER NUMBER 0 TO 12'
03470 GOTO 03290
03480 PRINT TAB(20); 'LIST OF FOOD MATERIALS'
03490 PRINT TAB(20); '======================'
03500 PRINT
03510 FOR F = 1 TO Q
03520 GOSUB 03550
03530 NEXT F
03540 GOTO 00420
03550 REM SUBROUTINE FOR BASIC DATA PRINTOUT
03560 PRINT F$(F);
03570 FOR Z = 1 TO 6
03580 PRINT TAB(25); C(F, Z)+11111;
03590 NEXT Z
03600 PRINT
03610 RETURN
03620 STOP
03630 DATA BARK, FLOWER, FRUIT, SEED, HERB, LEAF
03640 DATA UNDERGROUND LEAF, UNDERGROUND ROOT OR STEM, STEM, NUT
03650 END
```

TABLE 2. Truth table

Characters present or absent						Corresponding coded information						
−	−	+	+	+/−	+/−	Program:	0	0	1	1	2	2
−	+	−	+	−	+	Input:	1	2	1	2	1	2
✓	×	×	✓	✓	✓	Sum =	1	2	2	3	3	4

Key: + = character present;
− = character absent;
✓ = agreement, corresponding to numbers 1, 3, 4;
× = disagreement, corresponding to number 2 only.

TABLE 3. Features to distinguish drug combinations scoring 30 points for correct input data (pp. 614–616)

Name of food	Feature
Angelica root	Latex tubes absent.
Taro	Latex tubes present; starch grains up to 11 μm.
Yautia	Latex tubes present: starch grains up to 20 μm.
Anise	Covering trichomes conical; slightly curved, warty walls.
Celery fruit	Covering trichomes as small protuberances on fragments of epicarp in lateral view.
Areca	Some fibres with associated parenchymatous cells each containing a nodule of silica.
Olive stone	Silica nodules absent.
Brazil nut (shell)	Fibres present (may be difficult to see in the powdered material); colourless palisade cells up to 1 mm long.
Coconut (shell)	Fibres present; stegmata each containing a warty silica body up to 12 μm in diameter; stone cells with yellow walls and dark brown contents.
Peanut (shell)	Fibres present showing great variety of form, including T-shaped and saw-toothed, and criss-cross arrangement; presence of thick-walled and pitted cells from the inner hypodermal region.
Date stone	Fibres absent; presence of large thin-walled cells with conspicuous reddish contents.
Pecan (shell)	Fibres and vessels absent; presence of diamond-shaped crystals in the cavity of some stone cells.
Walnut (shell)	Fibres and vessels absent; absence of crystals from the stone cells.
Juniper berries	Prisms of calcium oxalate embedded in stone cells.
Cubebs	Stone cells devoid of calcium oxalate; starch grains up to 12 μm in diameter.
Pepper (Black/White)	Stone cells devoid of calcium oxalate; starch grain up to 6 μm in diameter. (Black and white pepper may be distinguished by the absence of the strongly thickened, yellow stone cells from the hypodermal layer of white pepper.)

TABLE 3 (*continued*)

Name of food	Feature
Brazil nut (meat)	Parenchyma isodiametric, up to 100 μm in diameter.
Coconut (endosperm)	Parenchyma isodiametric, up to 50 μm in diameter and radially elongated, often attaining a length of 300 μm (average length 160 μm).
Buckwheat	Starch mainly polygonal, 2–15 μm, with distinct, central, cleft hilum.
Chufa	Starch pear-, kidney- or spindle-shaped, up to 18 μm; hilum not distinct.
Guarana	Starch grains rounded, up to 12 μm, many distorted as a result of heat used in drying.
Lotus	Starch grains elongated, up to 65 μm, one end of grain rounded and contains excentric hilum.
Pistachio	Occasional, small (2–8 μm) rounded or ovoid starch grains.
Caraway	Reticulate, lignified parenchyma absent.
Dill	Reticulate, lignified parenchyma present.
Celery leaf	Covering trichomes absent.
Chervil	Stiff, unicellular, warty, pointed, covering trichomes present.
Parsley	The cells of the upper epidermis over the main veins are frequently ballooned resulting in a papillose appearance.
Chicory leaf	No short emergence-like hairs on the margin of the leaf.
Endive	Short, emergence-like hairs, often as broad as long, present on the margin of the leaf.
Carrot	Crystalline chromatophores present in starch-containing cells.
Parsnip	Crystalline chromatophores absent from starch-containing cells.
Galangal	Starch grains usually club-shaped with hilum in broader end; some irregular forms, including hammer-shaped, also present; mainly 20–35 μm long but sometimes exceed 80 μm.
Pumpkin pulp	Starch grains up to 10 μm in diameter.
Water chestnut	Starch grains of curious forms including triangular, quadrilateral, pentagonal, spindle-shaped; small central hilum; up to 27 μm long.
Hyssop and thyme	The histological structure of hyssop is similar to that of thyme. The chief differences, and these are probably more morphological than histological, are (i) lower epidermal cells of hyssop are more sharply sinuous than those of thyme, (ii) the multicellular covering trichomes are longer in thyme (some >300 μm in length) than in hyssop, (iii) unicellular covering trichomes more numerous in thyme. Sensory character differences exist, e.g. taste which is aromatic, warm and pungent in thyme but aromatic, warm and bitter in hyssop.

TABLE 3 (*continued*)

Name of food	Feature
Jerusalem artichoke	Typical latex tubes absent, oleoresin ducts with bright yellow contents.
Salsify	Branched latex tubes forming a conspicuous network.
Leek	Lower epidermal cells with distinct striations radiating from the guard cells of the stomata; calcium oxalate prisms up to 35 μm in length.
Onion	Striations absent; calcium oxalate prisms up to 50 μm in length.
Linseed	Sclerenchyma layer, cells long and narrow with irregularly thickened, pitted walls.
Mustard (Black/White)	Sclerenchyma layer, cells with radial and inner tangential walls thickened forming 'beaker cells'. Black mustard, seed coat fragments brown. White mustard, seed coat fragments yellow.
Lemol peel	Chromatophores yellow.
Orange peel	Chromatophores orange.
Peanut (meat)	Aleurone grains 5–12 μm.
Pecan (meat)	Aleurone grains up to 5 μm.
Peppergrass	No bladder cells.
Water cress	Many lower epidermal cells of the petiole and leaf blade are modified into bladder cells.
Peppermint	With polarized light a crystal group is visible in many of the multicellular heads of the glandular trichomes.
Spearmint	Absence of crystals from the oil glands.

index is introduced in an attempt to show empirically the amount of reliance which can be placed on any given result. The discrimination index, I, is defined by the equation $I = 100 - 10(n_1 - 1) - n_2$, where n_1 is the number of foods listed with 30 points and n_2 is the number of foods with 29 points when data for a named food are input. Thus, the discrimination index would be 100 if only one food achieved 30 points with none scoring 29 points, and zero if there were 10 foods with 30 points and 10 with 29 points. A modified computer program was used to examine the input data for each of the foods in order to check the validity of the program and so calculate the corresponding values for the discrimination index (Table 1).

The success of the program is indicated by the fact that the discrimination index is 99–100 for 71 food materials (about 44%), 90–100 for 115 foods (about 71 %) and only 4 foods (about 2·5 %) have a discrimination index less than 70, the lowest score being 55 for guarana. However, the program does separate them successfully from the other powders in the list and reference to Table 3 shows that each may be differentiated by a simple, more detailed observation.

When $n_1 \geqq 2$ (see example given in Fig. 2), simple distinguishing features are shown in Table 3. However, in all instances final confirmation by comparison with an authentic sample is advisable.

IV. CHEMOMICROSCOPICAL REAGENTS

Chloral hydrate solution. Chloral hydrate (50 g) dissolved in distilled water (20 ml). Used for calcium oxalate, cork, stomata, trichomes, fibres and pollen. The powder should be 'boiled' in the presence of the mountant in order to remove unwanted cell contents.

Iodine water. Mix one volume of a weak solution of iodine (British Pharmacopoeia, 1973) with nine volumes of distilled water. Used for starch granules, which yield a characteristic blue–black colour.

Picric acid solution. A saturated solution of picric acid in water made by dissolving 1 g of picric acid in 95 ml of distilled water. Used for protein in the form of aleurone grains or amorphous protein. As many seeds and fruits contain abundant fixed oil, it is advisable to de-fat the powder by immersion in light petroleum (boiling range 40–60°C) for a few minutes before staining. In addition, the background colour can be removed, after staining, by irrigating the mount with water, when the yellow colour given by protein will be clearly visible.

Phloroglucinol solution. Phloroglucinol (1 g) plus ethanol (95 %) to 100 ml. This solution is used, in the presence of concentrated hydrochloric acid, for lignified vessels/tracheids and stone cells. If lignified, these structures will acquire a red coloration.

Ruthenium red solution. Ruthenium red (0·008 g) dissolved in 10 % solution of lead acetate (10 ml). This reagent must be freshly prepared. Most mucilages give a red colour. Since some mucilages do not stain with this reagent it is advisable to confirm a negative reaction using corallin soda solution.

Corallin soda solution. Solution (a): corallin (5 g) dissolved in 90 % ethanol (100 ml). Solution (b): sodium carbonate (25 g) dissolved in water (100 ml). Add 1 ml solution (a) to 20 ml solution (b). The mixed reagent will deteriorate after a few weeks. Most mucilages give a pink colour. (Note: lignin also stains with corallin soda solution and ruthenium red solution, therefore care must be exercised in interpreting these mounts.)

Sudan red III solution. Sudan red III (0·01 g) dissolved in 90% ethanol (5·0 ml) and glycerin (5·0 ml).

V. PROCEDURE

The powder is examined microscopically in the above reagents and the appropriate entry made on the analytical data sheet (Fig. 1). The usual format of +/− indicating present/absent is used to record the assessed characters on the data sheet. This information is converted into numerals (− = 1; + = 2) to convert into suitable input data for analysis by a computer with a suitable BASIC compiler.

The program (Fig. 4) was tested using a suitable teleprinter or visual display unit linked via an acoustic coupler and Post Office telephone line to the Kronos CDC 6600 computer situated at Imperial College, University of London.

VI. CONCLUSION

The advantage of the method described in this chapter is that a relatively inexperienced microscopist (albeit a good food analyst) can, with confidence, arrive at the correct identity of any one of the 161 fragmented food materials listed. It is quite easy to add additional foods to the program at any time if this is thought desirable, since the program is in a form that is readily understood by anyone possessing a minimal amount of knowledge of computer programming. Furthermore, the program is readily adaptable to other areas where a large matrix of data requires scanning. The important point to watch in any new application is that the parameters chosen are such as to give a high discrimination index for the majority of the items listed; in the application described in this chapter 94·4% of the foods have a discrimination index above 75.

REFERENCES

Anon (1963). 'Tables for the Identification of Powdered Drugs of Organised Structure.' Pharmaceutical Society of Great Britain, London.

Brain, K. R. and Turner, T. D. (1975). 'The Practical Evaluation of Phytopharmaceuticals.' Wright-Scientechnica, Bristol.

Brazier, J. D. and Franklin, G. L. (1961). 'Identification of Hardwoods: A Microscope Key.' Forest Products Research Bulletin No. 46, H.M.S.O., London.

British Pharmacopoeia (1973). H.M.S.O., London, p. 251.

Clarke, S. H. (1938). *New Phytol.* **37**, 369.

Claus, E. P. (1956). 'Gathercoal and Wirth Pharmacognosy', 3rd Edn. Henry Kimpton, London.

Deryng, J. (1961). 'Atlas Sproszkowanych Roślinnych Surowców Leczniczych.' Państwowy Zaklad Wydawnictw Lekarskich, Warsaw.

Eschrich, W. (1966). 'Pulver-Atlas der Drogen des Deutschen Arzneibuches.' Gustav Fischer Verlag, Stuttgart.

Fish, F. and Nelson, P. F. (1969). *J. Pharm. Pharmac.*, **21**, Suppl. 98S.

Flück, H., Schlumpf, E. and Siegfried, K. (1935). 'Pharmakognosticher Atlas zur Pharmacopoea Helvetica.' Kommissionsverlag Wepf, Basle.

Greenish, H. G. (1923). 'The Microscopical Examination of Foods and Drugs', 3rd Edn. J. and A. Churchill, London.

Greenish, H. G. and Collin, E. (1904). 'An Anatomical Atlas of Vegetable Powders.' J. and A. Churchill, London.

Jackson, B. P. and Snowdon, D. W. (1974). 'Powdered Vegetable Drugs.' Stanley Thornes, London.

Jolliffe, Georgina H. and Jolliffe, G. O. (1976). *Analyst*, **101**, 622.

Kraemer, H. (1920). 'Scientific and Applied Pharmacognosy', 2nd Edn. Chapman and Hall, London.

Ledley, R. S. (1964). *Science*, **146**, 216.

Mansfield, W. (1929). 'Microscopic Pharmacognosy.' Chapman and Hall, London.

Nelson, P. F. (1972). 'Analytical Microscopy of Vegetable Materials.' 114 Corsebar Drive, Paisley, Scotland. (Published privately.)

Parry, J. W. (1969). 'Spices', Vol. 2. Chemical Publishing Co., New York.

Phillips, E. W. J. (1948). 'Identification of Softwoods by their Microscopic Structure', Forest Products Research Bulletin No. 22, H.M.S.O., London.

Schneider, A. (1921). 'The Microanalysis of Powdered Vegetable Drugs', 2nd Edn. Pitman Press, London.

Thoms, H. (1931). 'Handbuch der Praktischen und Wissenschaftlichen Pharmazie.' Urban and Schwarzenberg, Berlin.

Trease, G. E. and Evans, W. C. (1978). 'Pharmacognosy', 11th Edn. Baillière Tindall, London.

Tschirch, A. and Œsterle, O. (1900). 'Anatomischer Atlas der Pharmakognosie.' C. H. Tauchnitz, Leipzig.

Vaughan, J. G. (1970). 'The Structure and Utilization of Oil Seeds.' Chapman and Hall, London.

Wallis, T. E. (1948). 'Practical Pharmacognosy', 5th Edn. J. and A. Churchill, London.

Wallis, T. E. (1965). 'Analytical Microscopy: Its Aims and Methods in Relation to Foods, Water, Spices and Drugs', 3rd Edn. J. and A. Churchill, London.

Wallis, T. E. (1967). 'Textbook of Pharmacognosy', 5th Edn. J. and A. Churchill, London.

Winton, A. L. (1916). 'The Microscopy of Vegetable Foods', 2nd Edn. John Wiley and Sons, New York.

Winton, A. L. and Winton, K. B. (1949). 'The Structure and Composition of Foods', Vol. 2. John Wiley and Sons, New York.

Author Index

Numbers in italics indicate the page on which the references are listed.

N

O

P

T

U

V

W

Subject Index

N

Q

R